Kirk-Othmer

ENCYCLOPEDIA
OF CHEMICAL
TECHNOLOGY

Second Edition

INDEX TO
VOLUMES 1–22
AND
SUPPLEMENT

Interscience Publishers
a division of John Wiley & Sons, Inc.
New York · London · Sydney · Toronto

Kirk-Othmer

ENCYCLOPEDIA

OF CHEMICAL

TECHNOLOGY

Second completely revised edition

INDEX TO

VOLUMES 1–22

AND

SUPPLEMENT

CONTENTS OF THE ENCYCLOPEDIA

VOLUME 1

Abherents, 1
Ablation, 11
Abrasives, 22
Absorption, 44
Acetaldehyde, 77
Acetal Resins, 95
Acetals and Ketals, 107
Acetate and Triacetate Fibers, 109
Acetic Acid Derivatives, 138
Acetoacetic Acid and Ester, 153
Acetone, 159
Acetophenone, 167
Acetylene, 171
Acid Anhydrides, 211
Acid-Base Systems, 213
Acid Halides, 222
Acids, Carboxylic, 224
Acids, Dicarboxylic, 240
Acrolein and Derivatives, 255
Acrylamide, 274
Acrylic Acid and Derivatives, 285
Acrylic and Modacrylic Fibers, 313
Acrylonitrile, 338

Actinides, 351
Adhesives, 371
Adipic Acid, 405
Adsorption, 421
Aerosols, 470
Air Conditioning, 481
Alcoholic Beverages, Distilled, 501
Alcohols, 531
Alcohols, Higher, Fatty, 541
Alcohols, Higher, Synthetic, 560
Alcohols, Polyhydric, 569
Alcohols, Unsaturated, 598
Aldehydes, 639
Algal Cultures, 649
Alkali and Chlorine Industries, 668
Alkaloids, 758
Alkanolamines, 809
Alkoxides, Metal, 632
Alkyd Resins, 851
Alkylation, 882
Alkylphenols, 901
Allyl Compounds, 916
Aluminum and Aluminum Alloys, 929

VOLUME 2

Aluminum Compounds, 1
Amides, Acid, 66
Amides, Fatty Acid, 72
Amination by Reduction, 76

Amines, 99
Amino Acids, 156
Aminophenols, 213
Amino Resins and Plastics, 225

v

Vol. 1 *(continued)*

Ammonia, 258

Ammonium Compounds, 313

Ammonolysis, 332

Amyl Alcohols, 374

Analgesics and Antipyretics, 379

Anesthetics, 393

Aniline and Its Derivatives, 411

Antacids, Gastric, 427

Anthraquinone, 431

Anthraquinone Derivatives, 438

Anthraquinone and Related Quinonoid
    Dyes, 501

Antibiotics, 533

Antifreezes and Deicing Fluids, 540

Antimony and Antimony Alloys, 562

Antimony Compounds, 570

Antioxidants, 588

Antiseptics and Disinfectants, 604

Antistatic Agents, 649

Aquametry, 673

Aromaticity, 701

Arsenic, 711

Arsenic Compounds, 718

Asbestos, 734

Ascorbic Acid, 747

Asphalt, 762

Atoms and Atomic Structure. 806

Automobile Exhaust Control, 814

Azeotropy and Azeotropic Distillation,
    839

Azine Dyes, 859

Azines, 868

Azo Dyes, 868

VOLUME 3

Bacterial, Rickettsial, and Mycotis
    Infections, Chemotherapy, 1

Bagasse, 36

Bakery Processes and Leavening
    Agents, 41

Barbituric Acid and Barbiturates, 60

Barium, 77

Barium Compounds, 80

Batteries and Electric Cells, Primary, 99

Batteries and Electric Cells, Secondary,
    161

Bearing Materials, 271

Beer and Brewing, 297

Bentonite, 339

Benzaldehyde, 360

Benzene, 367

Benzenesulfonic Acids, 401

Benzidine and Related
    Diaminobiphenyls, 408

Benzoic Acid, 420

Benzophenone, 439

Benzyl Alcohol and β-Phenylethyl
    Alcohols, 442

Beryllium and Beryllium Alloys, 450

Beryllium Compounds, 474

Bile Constituents, 480

Bioassay, 489

Biotin, 518

Bismuth and Bismuth Alloys, 527

Bismuth Compounds, 535

Bleaching Agents, 550

Blood, Animal, 567

Blood Fractionation, 576

Boron and Boron Alloys, 602

Boron Compounds, 608

Brighteners, Optical, 737

Bromine, 750

Bromine Compounds, 766

Butadiene, 784

Butanes, 815

Butyl Alcohols, 822

Butylenes, 830

Butyraldehyde, 865

Vol. 3 (*continued*)

Butyric Acid and Butyric Anhydride, 878
Cadmium and Cadmium Alloys, 884

Cadmium Compounds, 899
Caffeine, 911
Calcium and Calcium Alloys, 917

VOLUME 4

Calcium Compounds, 1
Calking and Sealing Compositions, 28
Calorimetry, 35
Camphor, 54
Candles, 58
Caramel Color, 63
Carbamic Acid, 69
Carbides, 70
Carbohydrates, 132
Carbon, 149
Carbonated Beverages, 335
Carbon Dioxide, 353
Carbon Disulfide, 370
Carbonic Esters and Chloroformic Esters, 386
Carbonium Ion and Carbanion, 394
Carbonization, 400
Carbon Monoxide, 424

Carbon Monoxide–Hydrogen Reactions, 446
Carbonyls, 490
Cardiovascular Agents, 510
Castor Oil, 524
Catalysis, 534
Cathartics, 586
Cellulose, 593
Cellulose Derivatives, 616
Cellulose Derivatives - Plastics, 653
Cement, 684
Centrifugal Separation, 710
Ceramics, 759
Ceramics, Chemical Ware, 832
Cerium and Cerium Compounds, 840
Cesium and Cesium Compounds, 855
Chemical Warfare, 869
Chloramines and Chloroamines, 908
Chloramphenicol, 928

VOLUME 5

Chlorine, 1
Chlorine Oxygen Acids and Salts, 7
Chlorocarbons and Chlorohydrocarbons, 85
Chlorohydrins, 304
Chlorophenols, 325
Chlorophyll, 339
Chlorosulfonic Acid, 357
Chocolate and Cocoa, 363
Choline, 403

Chromatography, 413
Chromium and Chromium Alloys, 451
Chromium Compounds, 473
Cinnamic Acid, Cinnamaldehyde, and Cinnamyl Alcohol, 517
Citric Acid, 524
Clays, 541
Coagulants and Anticoagulants, 586
Coal, 606
Coated Fabrics, 679

Vol. 5 (*continued*)

Coatings, Industrial, 690
Cobalt and Cobalt Alloys, 716
Cobalt Compounds, 737
Coffee, 748
Color and Constitution of Organic
    Dyes, 763

Colorimetry, 788
Color Measurement, 801
Color Photography, 812
Colors for Ceramics, 845
Colors for Foods, Drugs, and Cosmetics,
    857

VOLUME 6

Complexing Agents, 1
Computers, 25
Confectionery, 34
Contraceptive Drugs, 60
Conveying, 92
Coordination Compounds, 122
Copper, 131
Copper Alloys, 181
Copper Compounds, 265
Cork, 281
Corrosion, 289
Corrosion Inhibitors, 317
Cosmetics, 346
Cotton, 376
Cottonseed, 412
Coumarin, 425
Cresols, 434
Crotonaldehyde, 445
Crotonic Acid, 464
Cryogenics, 471
Crystallization, 482

Crystals, 516
Cumene, 543
Curare and Curare-Like Drugs, 547
Cyanamides, 553
Cyanides, 574
Cyanine Dyes, 605
Cyanocarbons, 625
Cyanoethylation, 634
Cyanohydrins, 668
Cyclohexane, 675
Cyclohexanol and Cyclohexanone, 683
Cyclopentadiene and Dicyclopenta-
    diene, 688
Data—Interpretation and Correlation,
    705
Density and Specific Gravity, 755
Dental Materials, 777
Dentifrices, 848
Detergency, 853
Deuterium and Tritium, 895
Dextrose and Starch Syrups, 919

VOLUME 7

Dialysis, 1
Diamines and Higher Amines
    Aliphatic, 22
Diarylamines, 40
Diatomite, 53

Diene Polymers, 64
Diffusion, 79
Diffusion Separation Methods, 91
Dimensional Analysis, 176
Diphenyl and Terphenyls, 191

Vol. 7 (*continued*)

Distillation, 204

Diuretics, 248

Driers, and Metallic Soaps, 272

Drilling Fluids, 287

Drycleaning, 307

Drying, 326

Drying Agents, 378

Drying Oils, 398

Dust, 429

Dyes and Dye Intermediates, 462

Dyes—Application and Evaluation, 505

Dyes, Natural, 614

Dyes, Reactive, 630

Economic Evaluation, 642

Eggs, 661

Elastomers, Synthetic, 676

Electrical Testing, 716

Electroanalytical Methods, 726

Electrochemistry, 784

Electrodecantation, 841

Electrodialysis, 846

Electrolytic Machining Methods, 866

Electron Spin Resonance, 874

VOLUME 8

Electron Tube Materials, 1

Electrophoretic Deposition, 23

Electroplating, 36

Electrostatic Precipitation, 75

Elements—Periodic System, 93

Embalming Fluids, 100

Embedding, 102

Emulsions, 117

Enamels, Porcelain or Vitreous, 155

Enzymes, Industrial, 173

Epinephrine, 231

Epoxidation, 238

Epoxides, 263

Epoxy Resins, 294

Esterification, 313

Ester Interchange, 356

Esters, Organic, 365

Ethane, 383

Ethanoic Acid, 386

Ethanol, 422

Ethers, 470

Ethylene, 499

Ethylene Oxide, 523

Evaporation, 559

Explosives, 581

Extraction, 719

Fats and Fatty Oils, 776

Fatty Acids, 811

Feeds, Animal, 857

Fermentation, 871

Ferrites, 881

VOLUME 9

Ferroelectrics, 1

Fertilizers, 25

Fibers, Man-Made, 151

Fibers, Vegetable, 171

Film Deposition Techniques, 186

Film Materials, 220

Filters, Optical, 244

Filtration, 264

Vol. 9 (*continued*)

Fire Prevention and Extinction, 286
Fire-Resistant Textiles, 300
Fish and Shellfish, 316
Flavor Characterization, 336
Flavors and Spices, 347
Flotation, 380
Fluidization, 398

Fluid Mechanics, 445
Fluorescent Pigments (Daylight), 483
Fluorine, 506
Fluorine Compounds, Inorganic, 527
Fluorine Compounds, Organic, 686
Foamed Plastics, 847
Foams, 884

VOLUME 10

Food Additives, 1
Food and Food Processing, 23
Foods, Diet, 62
Food Standards, 69
Formaldehyde, 77
Formic Acid and Derivatives, 99
Free Radicals, 114
Friction Material, 124
Friedel-Crafts Reactions, 135
Fruit Juices, 167
Fuels, 179
Fungicides, 220
Furfural and Other Furan Compounds, 237
Furnaces, Electric, 252
Furnaces, Fuel-Fired, 279
Furs and Fur Processing, 294
Gallium and Gallium Compounds, 311
Gas Cleaning, 329
Gas, Manufactured, 353
Gas, Natural, 443
Gasoline and Other Motor Fuels, 463
Gelatin, 499

Gems, Synthetic, 509
Germanium and Germanium Compounds, 519
Gilsonite, 527
Glass, 533
Glue, Animal and Fish, 604
Glutaric Acid, 618
Glycerol, 619
Glycolic Acid, 632
Glycols, 638
Gold and Gold Compounds, 681
Gravity Concentration, 695
Grignard Reaction, 721
Guanidine and Guanidine Salts, 734
Gums, Natural, 741
Hafnium and Hafnium Compounds, 754
Hair Preparations, 768
Hardness, 808
Heat Transfer, 819
Heat Transfer Media Other Than Water, 846
Helium Group Gases, 862
Heterocyclic Compounds, 895

VOLUME 11

Hexanes, 1
High-Temperature Alloys, 6
Histamine and Antihistamine Agents, 35
Hormones, 45
Hydantoin, 141
Hydrazine and Its Derivatives, 164
Hydrazoic Acid and Azides, 197
Hydrides, 200
Hydrocarbon Oxidation, 224
Hydrocarbon Resins, 242
Hydrocarbons, 262
Hydrochloric Acid, 307
Hydrogen, 338
Hydrogen-Ion Concentration, 380
Hydrogen Peroxide, 391
Hydroprocesses, 418
Hydroquinone, Resorcinol, and Pyrocatechol, 462
Hydroxylamine and Hydroxylamine Salts, 493

Hypnotics, Sedatives, Anticonvulsants, 508
Imines, Cyclic, 526
Indicators, 548
Indigoid Dyes, 562
Indium and Indium Compounds, 581
Indole, 585
Industrial Toxicology, 595
Inks, 611
Inorganic High Polymers, 632
Inorganic Refractory Fibers, 651
Inositol, 673
Insecticides, 677
Instrumentation and Control, 739
Insulation, Electric, 774
Insulation, Thermal, 823
Insulin, 838
Iodine and Iodine Compounds, 847
Ion Exchange, 871

VOLUME 12

Iron, 1
Iron Compounds, 22
Isocyanates, Organic, 45
Isoprene, 64
Itaconic Acid, 83
Ketenes and Related Substances, 87
Ketones, 101
Lactic Acid, 170
Laminated and Reinforced Plastics, 188
Laundering, 197
Lead, 207
Lead Alloys, Lead Utilization, 247
Lead Compounds, 266
Leather, 303
Lecithin, 343
Lignin, 361

Lignite and Brown Coal, 381
Lime and Limestone, 414
Linen, 461
Liquefied Petroleum Gas, 470
Liquid–Level Measurement, 481
Literature of Chemistry and Chemical Technology, 500
Lithium and Lithium Compounds, 529
Lubrication and Lubricants, 557
Luminescent Materials, 616
Microlide Antibiotics, 632
Magnesium and Magnesium Alloys, 661
Magnesium Compounds, 708
Magnetic Materials, 737
Magnetic Properties, 772
Magnetic Separation, 782

Vol. 12 (*continued*)

Magnetic Tape, 801
Maleic Anhydride, Maleic Acid, and
    Fumaric Acid, 819
Malic Acid, 837

Malonic Acid and Derivatives, 849
Malts and Malting, 861
Manganese and Manganese Alloys, 887

VOLUME 13

Manganese Compounds, 1
Margarine, 56
Marketing and Marketing Research, 66
Mass Spectrometry, 87
Mass Transfer, 99
Matches, 160
Meat and Meat Products, 167
Mechanical Testing, 184
Melting and Freezing Temperatures,
    198
Mercury, 218
Mercury Compounds, 235
Metallic Coatings, 249
Metal Surface Treatments, 284
Metal Treatments, 315
Methacrylic Compounds, 331
Methane, 364
Methanol, 370
Micas, Natural and Synthetic, 398
Microchemistry, 424
Microencapsulation, 436
Microorganisms, 457
Microscopy, Chemical, 491

Milk and Milk Products, 506
Mixing and Blending, 577
Molasses, 613
Molybdenum and Molybdenum
    Alloys, 634
Molybdenum Compounds, 645
Morpholine, 659
Naphthalene, 670
Naphthalenecarboxylic Acids, 690
Naphthalene Derivatives, 697
Naphthenic Acids, 727
Nickel and Nickel Alloys, 735
Nickel Compounds, 753
Niobium and Niobium Compounds, 766
Nitration, 784
Nitric Acid, 796
Nitrides, 814
Nitro Alcohols, 826
Nitrobenzene and Nitrotoluenes, 834
Nitrofurans, 853
Nitrogen, 857
Nitroparaffins, 864
Nitrophenols, 888

VOLUME 14

Nomenclature, 1
Nomographs, 15
Nuclear Magnetic Resonance, 40
Nuclear Reactors, 74
Nuts, 122

Ocean Raw Materials, 150
Odor Control, 170
Oils, Essential, 178
Olefin Polymers, 217
Olefins, Higher, 313

Vol. 14 (*continued*)

Organoleptic Testing, 336
Osmosis, Osmotic Pressure, and
    Reverse Osmosis, 345
Oxalic Acid, 356
Oxo Process, 373
Oxygen, 390
Ozone, 410
Packaging and Packages, 432
Packing Materials, 443
Paint, 462

Paint and Varnish Removers, 485
Paper, 494
Parasitic Infections, Chemotherapy, 532
Patents, 552
Pectic Substances, 636
Penicillins, 652
Pentanes, 707
Perfumes, 717
Peroxides and Peroxy Compounds, 746
Petroleum, 835

VOLUME 15

Petroleum (Refinery Processes), 1
Petroleum (Products), 77
Petroleum Waxes, 92
Pharmaceuticals, 112
Phase Rule, 133
Phenol, 147
Phenolic Aldehydes, 160
Phenolic Ethers, 165
Phenolic Resins, 176
Phenolsulfonic Acids, 208
Phenylacetic Acid, 213
Phenylenediamines and Toluenedi-
    amines, 216
Phonograph Record Compositions, 225
Phosphoric Acids and Phosphates, 232
Phosphorus and the Phosphides, 276
Phosphorus Compounds, 295
Photochemical Technology, 331
Photography, 355
Phototubes and Photocells, 396

Phthalic Acids and Other Benzene-
    Polycarboxylic Acids, 444
Phthalocyanine Compounds, 488
Pigments, 495
Pilot Plants, 605
Piperazine, 638
Piping Systems, 646
Plant Growth Substances, 675
Plant Layout, 689
Plant Location, 700
Plasticizers, 720
Plastics Technology, 790
Plastics Testing, 811
Platinum Group Metals, 832
Platinum Group Metals, Compounds,
    861
Plutonium, 879
Plywood, 896
Poisons, Economic, 908

xiv   CONTENTS OF THE ENCYCLOPEDIA

VOLUME 16

Polyamides, 1
Polycarbonates, 106
Polyelectrolyte Complexes, 117
Polyene Antibiotics, 133
Polyester Fibers, 143
Polyesters, 159
(Polyhydroxy)Benzenes, 190
Polymerization Mechanisms and
    Processes, 219
Polymers, 242
Polymers Containing Sulfur, 253
Polymethine Dyes, 282
(Polymethyl)benzenes, 300
Polypeptide Antibiotics, 306
Poromeric Materials, 345
Potassium, 361
Potassium Compounds, 369
Powder Metallurgy, 401
Power Generation, 436
Pressure Measurement, 470
Pressure Vessels, 482
Printing Processes, 494

Propane, 546
Propionaldehyde, 548
Propionic Acid, 554
Propyl Alcohols, 559
Propylene, 579
Propylene Oxide, 595
Proteins, 610
Psychopharmacological Agents, 640
Pulp, 680
Pumps and Compressors, 728
Pyrazoles, Pyrazolines, Pyrazolones,
    762
Pyridine and Pyridine Derivatives, 780
Pyridoxine, Pyridoxal, and
    Pyridoxamine, 806
Pyrotechnics, 824
Pyrrole and Pyrrole Derivatives, 841
Quaternary Ammonium Compounds,
    859
Quinoline and Isoquinoline, 865
Quinoline Dyes, 886
Quinones, 899

VOLUME 17

Radioactive Drugs and Tracers, 1
Radioactive Elements, Natural, 9
Radiochemical Analysis and Tracer
    Applications, 35
Radiochemical Technology, 53
Radioisotopes, 64
Radiopaques, 130
Rare Earth Elements, 143
Rayon, 168
Refraction, 210
Refractories, 227
Refractory Coatings, 267
Refractory Fibers, 285
Refrigeration, 295
Reprography, 328

Resins, Natural, 379
Resins, Water–Soluble, 391
Rhenium, 411
Rheology, 423
Riboflavine, 445
Roofing Materials, 459
Rosin and Rosin Derivatives, 475
Rubber Chemicals, 509
Rubber Compounding, 543
Rubber Derivatives 646
Rubber, Natural, 660
Rubidium and Rubidium Compounds,
    684
Safety, 694

Vol. 17 (*continued*)

Salicylic Acid and Related Compounds, 720
Sampling, 744
Seaweed Colloids, 763

Sedimentation, 785
Selenium and Selenium Compounds, 809
Semiconductors, 834

VOLUME 18

Shale Oil, 1
Shellac, 21
Shortenings and Other Food Fats, 33
Silica, 46
Silicon and Silicides, 111
Silicon Compounds, 133
Silk, 269
Silver and Silver Alloys, 279
Silver Compounds, 295
Size Measurement of Particles, 310
Size Reduction, 324
Size Separation, 366
Smokes, Fumes, and Smog, 400
Soap, 415
Sodium, 432

Sodium Compounds, 458
Soil Chemistry of Pesticides, 515
Solders and Brazing Alloys, 541
Solvent Recovery, 549
Solvents, Industrial, 564
Sorbic Acid, 589
Soybeans, 599
Spandex, 614
Sprays, 634
Stains, Industrial, 654
Starch, 672
Steam, 692
Steel, 715
Sterilization, 805
Steroids, 830

VOLUME 19

Stilbene Derivatives, 1
Stoichiometry, Industrial, 17
Streptomycin and Related Antibiotics, 33
Strontium, 48
Strontium Compounds, 50
Styrene, 55
Styrene Plastics, 85
Succinic Acid, 134
Sugar, 151
Sulfamic Acid and Sulfamates, 242
Sulfolane, 250
Sulfonamides, 255

Sulfonation and Sulfation, 279
Sulfonic Acids, 311
Sulfoxides, 320
Sulfur, 337
Sulfur Compounds, 367
Sulfur Dyes, 424
Sulfuric Acid and Sulfur Trioxide, 441
Sulfuric and Sulfurous Esters, 483
Sulfurization and Sulfur-Chlorination, 498
Surfactants, 507
Sweeteners, Nonnutritive, 593
Talc, 608

Vol. 19 (*continued*)

Tall Oil, 614

Tantalum and Tantalum Compounds, 630

Tar and Pitch, 653

Tar Sands, 682

Tartaric Acid, 733

Tea, 743

Tellurium and Tellurium Compounds, 756

Temperature Measurement, 774

Terpenes and Terpenoids, 803

VOLUME 20

Tetracyclines, 1

Textile Testing, 33

Thallium and Thallium Compounds, 63

Therapeutic Agents, Protozoal Infections, 70

Thermal Analysis, 99

Thermodynamics, 118

Thermoelectric Energy Conversion, 147

Thiamine, 173

Thiazole Dyes, 185

Thioglycolic Acid, 198

Thiols, 205

Thiophene, 219

Thiosulfates, 227

Thorium and Thorium Compounds, 248

Thyroid and Antithyroid Preparations, 260

Tin and Tin Alloys, 273

Tin Compounds, 304

Tire Cords, 328

Titanium and Titanium Alloys, 347

Titanium Compounds, 380

Tobacco, 504

Toluene, 527

Tool Materials for Machining, 566

Trademarks, 578

Transportation, 596

Transport Processes, 610

Triazinetriol, 662

Triphenylmethane and Related Dyes, 672

Turbidimetry and Nephelometry, 738

Turpentine, 748

Type Metal, 756

Ultrasonics, 773

Unsaturated Polyesters, 791

VOLUME 21

Uranium and Uranium Compounds, 1

Urea and Urea Derivatives, 37

Urethan Polymers, 56

Uric Acid, 107

UV Absorbers, 115

Vacuum Technology, 123

Vanadium and Vanadium Alloys, 157

Vanadium Compounds, 167

Vanillin, 180

Vapor-Liquid Equilibria, 196

Veterinary Drugs, 241

Vinegar, 254

Vinylidene Polymers, 269

Vinyl Polymers, 304

Vinyon and Related Fibers, 441

Viral Infections, Chemotherapy, 452

Vol. 21 *(continued)*

Viscometry, 460
Vitamins, 484
Wallboard, 601

Wastes, Industrial, 625
Water, 651

VOLUME 22

Water (Desalination), 1
Waterproofing and Water Repellency,
    135
Waxes, 156
Weed Killers, 174
Weighing and Proportioning, 220
Welding, 241
Wheat and Other Cereal Grains, 253
Wine, 307
Wolfram and Wolfram Alloys, 334
Wolfram Compounds, 346
Wood, 358

Wool, 387
Xanthates, 419
Xanthene Dyes, 430
X-Ray Analysis, 438
Xylenes and Ethylbenzene, 467
Yeasts, 507
Zinc, 555
Zinc Compounds, 604
Zirconium and Zirconium Compounds,
    614
Zone Refining, 680

SUPPLEMENT

Adamantane, 1
Adhesion, 16
Alkylphenols, 27
Amine Oxides, 32
Automobile Exhaust Control, 50
Beryllides, 73
Cancer Chemotherapy, 81
Carbon Black, 91
Carbon Fibers, 109
Cells, High Temperature, 120
Ceramic Composite Armor, 138
Chemical Cleaning, 157
Chlorine, 167
Coal, 177
Copper, 249
Cresylic Acids, Synthetic, 271
Diamond, 274

Electromigration, 278
Enzyme Detergents, 294
Ethylene, 310
Eutrophication, 315
Fertilizers, 338
Fire Extinguishing Agents, 364
Fuel Cells, 383
Gas, Natural, 421
Great Salt Lake Chemicals, 438
Halogenated Fire Retardants, 467
Heat Pipe, 488
Hydrides, 499
Information Retrieval Services and
    Methods, 510
Iron By Direct Reduction, 535
Mercury, Recovery by Electrooxida-
    tion, 542

Supplement (*continued*)

CONTRIBUTORS TO THE ENCYCLOPEDIA

Aarons, Ralph, *E. I. du Pont de Nemours & Co., Inc.,* Sulfamic acid and sulfamates (19:242)

Abbott, Norman D., *Fabric Research Laboratories,* Textile testing (20:33)

Abramo, S. V., *E. I. du Pont de Nemours & Co., Inc.,* Adipic acid (1:405)

Abramovitch, R. A., *University of Alabama,* Pyridine and pyridine derivatives (16:780)

Ackerman, James, *Sterling-Winthrop Research Institute,* Radiopaques (17:130)

Adamec, J. B., *The International Nickel Company, Inc.,* Nickel and nickel alloys (13:735)

Akerlof, G. C., *Aerospace Department, Princeton University,* Electrochemistry (7:784)

Albert, A., *Mayo Clinic,* Hormones (Anterior-pituitary-like hormones) (11:64)

Albright, Lyle F., *Purdue University,* Nitration (13:784)

Allen, C. F. H., *Rochester Institute of Technology,* Heterocyclic compounds (10:895)

Allen, E. M., *Pittsburgh Plate Glass Company,* Calcium chloride (Calcium compounds) (4:11)

Amerine, M. A., *University of California, Davis, California,* Wine (22:307)

Anderson, A. R., *Anderson Chemical Division, Stauffer Chemical Company,* Aluminum, organometallic compounds (2:25); Silicon compounds (halides) (18:166), Silicon ethers and esters (18:216)

Anderson, F. N., *Mallinckrodt Chemical Works,* Sodium compounds (Sodium, Iodide) (18:485)

Anderson, Ralph F., *International Minerals & Chemical Corporation, Bioferm Division,* Fermentation (8:871)

Antonsen, D. H., *The International Nickel Company, Inc.,* Nickel compounds (13:753)

Aplan, Frank F., *Union Carbide Corporation, Mining and Metals Division,* Flotation (9:380)

Apple, E. F., *General Electric Company,* Luminescent materials (12:616)

Arkin, D., *Chemical Projects Associates, Inc.,* Olefins (S:632)

Armstrong, Robert F., *Diamond Alkali Company,* Calcium carbonate (Calcium compounds) (4:7)

Aston, John G., *The Pennsylvania State University,* Calorimetry (4:35)

Atchison, G. J., *The Dow Chemical Company,* Bromine compounds (3:766)

Atlas, S. M., *Bronx Community College,* Fibers, man-made (Research possibilities in the fully synthetic fibers) (9:164)

Ault, W. C., *U.S. Department of Agriculture,* Fatty acids (Manufacture) (8:825); (Analysis: standards) (830); (From tall oil) (845); (Polybasic) (847); Branched-chain acids, (849)

Aunan, W. J., *American Meat Institute Foundation,* Meat and meat products (13:167)

Autenrieth, J. S., *Hercules Incorporated,* Terpene and terpenoids, (Terpene resins) (19:803)

Axe, W. Nelson, *Phillips Petroleum Company,* Hydrocarbons (Survey) (11:262)

Axtell, Oliver, *Celanese Chemical Company,* Economic evaluation (7:642)

Ayers, George W., *The Pure Oil Company,* Benzene (3:367)

Babcock, Chester I., *Natiohal Fire Protection Association,* Fire prevention and extinction (9:286)

Baber, R. L., *Nitrogen Division, Allied Chemical Corporation,* Ammonia (2:258)

Bach, R. O., *Lithium Corporation of America, Inc.,* Lithium and lithium compounds (12:529)

Bacon, J. C., *Climax Molybdenum Co.,* Molybdenum compounds (13:645)

Bacon, F. E., *Union Carbide Metals Company,* Boron alloys (Boron and boron alloys) (3:602); Chromium and chromium alloys (5:451); Manganese and manganese alloys (12:887)

Bacon, L. E., *Union Carbide Corporation, Carbon Products Division,* Baked and graphitized products, uses (Refractory applications) (Carbon) (4:238)

Badollet, M. S., *Consultant,* Asbestos (2:734)

Baker, A. J., *U.S. Department of Agriculture,* Wood (22:358)

Baker, Jr., Philip J., *Commercial Solvents Corporation,* Hydroxylamine and hydroxylamine salts, (hydroxylamine) (11:493); Nitroparaffins (13:864)

Bakker, W. T., *General Refractories Company,* Refractories (17:227)

Baldwin, R. H., *Amoco Chemicals Corporation,* Phthalic acids and other benzenepolycarboxylic acids (15:444)

Ballman, A. A., *Bell Telephone Laboratories, Inc.,* Silica (Synthetic quartz crystals) (18:105)

Banks, C. K., *M & T Chemicals Inc., a subsidiary of American Can Company,* Tin and tin alloys (Ditinning) (20:294); Tin compounds (20:304)

Bannister, D. W., *Toms River Chemical Corporation,* Dyes and dye intermediates (7:462); Dyes, reactive (630), Phenosulfonic acids (15:208)

Barbour, A. K., *Imperial Smelting Corporation, Ltd.,* Fluorinated compounds, organic (Fluorinated aromatic compounds) (9:775)

Barnes, Marion D., *Covenant College,* Sulfur (Special uses) (19:364)

Barney, A. L., *E. I. du Pont de Nemours & Co., Inc.,* Vinylidene polymers (Fluoride) (21:269)

Barnhart, Robert R., *Uniroyal Chemical division of Uniroyal, Inc.,* Rubber compounding (17:543)

Barnhart, W. S., *Pennsalt Chemicals Corporation,* Fluorine compounds, organic (Polyvinylidine fluoride) (9:840)

Barsness, D. A., *General Electric Company,* Embedding, (8:102)

Batchelor, Clyde S., *The Raybestos Division, Raybestos-Manhattan, Inc.,* Friction material (10:124)

Bates, Roger G., *National Bureau of Standards,* Hydrogen-ion concentration (11:380)

Bauchwitz, P. S., *E. I. du Pont de Nemours & Co., Inc.,* Chlorocarbons and chlorohydrocarbons (Chloroprene) (5:215)

Beach, Leland K., *Enjay Chemical Laboratory,* Propylene (16:579)

Beal, Jr., James B., *Ozark-Mahoning Company,* Fluorine compounds, inorganic (Phosphorous fluorides) (9:527)

Beamer, William H., *The Dow Chemical Company,* Radiochemical technology (17:53)

Beams, J. W., *University of Virginia,* Gas centrifugal separation (Centrifugal separation) (4:710)

Bean, D. C., *Shawinigan Chemicals Limited,* Carbon black (Acetylene black) (4:243)

Bebb, R. L., *Firestone Tire & Rubber Company,* Styrene-butadiene solution copolymers (S:910)

Beck, Karl M., *Abbott Laboratories,* Sweeteners, nonnutritive (19:593)

Beecher, B. K., *Wyandotte Chemicals Corporation,* Drying agents (7:378)

Beerbower, Alan, *Esso Research and Engineering Co.,* Radioisotopes (17:64); Solubility parameters (S:889)

Behrndt, Klaus H., *Bell Telephone Laboratories,* Film deposition techniques, (9:186)

Beller, H., *General Aniline & Film Corporation,* Acetylene, (1:171)

Benenati, R. F., *Polytechnic Institute of Brooklyn,* Stoichiometry, industrial (19:17)

Benesovsky, Friedrich, *Metallwerk Plansee A.-G., Reuttle/Tyrol,* Survey; Industrial heavy-metal carbides; Cemented carbides (Carbides) (4:92); Nitrides (13:814)

Benson, Frederic R., *Atlas Chemical Industries, Inc.,* Alcohols polyhydric (Sugar alcohols) (1:569)

Bent, Richard L., *Eastman Kodak Co.,* Alcohols (1:531); Amines, survey (Amines) (2:99)

Berenbaum, M. B., *Thiokol Chemical Corporation,* Polymers containing sulfur (Polysulfides) (16:253)

Berlow, Evelyn, *Heyden Newport Chemical Corp.,* Other polyhydric alcohols (Alcohols, polyhydric) (1:588)

Berry, G. W., *Johns-Manville Corporation,* Roofing materials (17.459)

Beyer, George L., *Eastman Kodak Co.,* Turbidimentry and nephelometry (20:738)

Biberfeld, Henry, *A. Hollander & Sons, Ltd.,* Furs and fur processing (10:294)

Bicking, Charles A., *The Carborundum Company,* Sampling (17:744)

Biederman, N., *Institute of Gas Technology,* Gas, natural (S:421)

Bikales, Norbert M., *Chemical Consultant, Executive Editor of Encyclopedia of Polymer Science and Technology,* Acrylamide (1:274); Allyl compounds (1:916); Cyanoethylation (6:634)

Bixler, Harris J., *Amicon Corporation,* Polyelectrolyte complexes (16:117)

Blackwood, Robert K., *Chas. Pfizer & Co., Inc.,* Tetracyclines (20:1)

Blanding, Warren, *Marketing Publications Inc.,* Transportation (20:596)

Block, B. P., *Pennsalt Chemicals Coproration,* Inorganic high polymers (11:632)

Blomeke, J. O., *Oak Ridge National Laboratory,* Nuclear reactors (Waste management) (14:102)

Blomquist, R. F., *Forest Products Laboratory, Forest Service, U.S. Department of Agriculture,* Adhesives (1:371)

Blum, M., *Atomergic Chemetals Company,* Potassium (16:361)

Blumenthal, Warren, B., *National Lead Company,* Zirconium and zirconium compounds, zirconium metal (22:614)

Bobalek, Edward G., *University of Maine,* Coatings, industrial (5:690)

Bodanszky, Miklos, *The Squibb Institute for Medical Research,* Hormones posterior-pituitary hormones (11:68)

Boeglin, Albert F., *International Minerals & Chemical Corporation,* Magnesium compounds (12:708)

Bogue, Robert H., *Consultant to the Cement Industry,* Cement (4:684)

Boltz, David F., *Wayne State University,* Colorimetry and fluorometry (5:788)

Bomberger, H. B., *Reactive Metals, Inc.,* Titanium and titanium alloys (20:347)

Bondley, R. J., *General Electric Company,* Electron tube materials (8:1)

Boone, James L., *U.S. Borax Research Corporation,* Refractory boron compounds (Boron compounds) (3:673)

Booser, E. R., *General Electric Company,* Bearing materials (3:271); Lubrication and lubricants (12:557)

Booth, Albert B., *Hercules Incorporated,* Terpenes and terpenoids (19:803); Survey (834)

Booth, R. B., *American Cyanamid Company,* Cyanides (Alkaline earth metal cyanides) (6:601)

Bosqui, F. L., *Consulting Engineer,* Gravity concentration (10:695); Sedimentation (17:785); Size separation (18:366)

Boutros, R. D., *Mixing Equipment Co., Inc.,* Mixing and blending (13:577)

Bovey, F. A., *Bell Telephone Laboratories, Inc.,* Nuclear magnetic resonance (14:40)

Bower, J. G., *U.S. Borax Research Corporation,* Elemental boron (Boron and boron alloys) (3:602)

Boyer, Raymond F., *The Dow Chemical Co.,* Styrene plastics (19:85)

Boynton, R. S., *National Lime Association,* Lime and limestone (12:414)

Bradley, William B., *American Institute of Baking,* Yeast-raised products (Bakery processes and leavening agents) (3:41)

Bratt, John, *Wood Ridge Chemical Corp.,* Mercury compounds (13:235)

Briggs, G. S., *Radio Corporation of America,* Phototubes and photocells (15:396)

Briggs, H. M., *South Dakota State University,* Feeds, animal (8:857)

Briggs, J. Z., *Climax Molybdenum Co.,* Molybdenum and molybdenum alloys (13:634)

Bringer, Robert, *Minnesota Mining and Manufacturing Co.,* Fluorine compounds, organic (Polychlorofluoroethylene) (9:832)

Bronkala, W. J., *Indiana General Corporation,* Magnetic separation (12:782)

Brooker, L. G. S., *Eastman Kodak Company,* Color and constitution of organic dyes (5:763)

Brown, James A., *Allied Chemical Corporation,* Fluorine compounds inorganic (Sulfur fluorides, in part) (9:664)

Brown, W. E., *The Squibb Institute for Medical Research,* Antibiotics (2:533)

Brownell, L. E., *The University of Michigan,* Filtration (9:264)

Browning, B. L., *The Institute of Paper Chemistry,* Paper (14:494)

Bruce, John N., *Consultant,* Flame (Chemical warfare) (4:888)

Brumbaugh, C. C., *Diamond Alkali Company,* Alkali and chlorine industries (1:668)

Bruno, Michael H., *International Paper Company,* Printing processes (16:494)

Brusie, J. P., *General Aniline & Film Corporation,* Propargyl alcohol, 2-butyne-1,4-diol, and 2-butene-1,4-diol, manufacture (Alcohols, unsaturated) (1:598)

Buckley, D. J., *Esso Research and Engineering Co.,* Olefin polymers (Polymers of higher olefins) (14:309)

Budenholzer, R. A., *American Power Conference, Illinois Institute of Technology,* Power generation (16:436)

Bulloff, Jack J., *State University of New York at Albany,* Reprography (17:328)

Bunn, C. W., *The Royal Institution, London,* X-Ray analysis (22:438)

Burachinsky, Bohdan V., *Interchemical Corporation,* Inks (11:611)

Burg, Richard W., *Merck & Co., Inc.,* Pyridoxine, pyridoxal and pyridoxamine (16:806)

Burkett, M. N., *Union Carbide Corporation, Carbon Products Division,* Baked and graphitized products, uses (Nuclear reactor applications) (Carbon) (4:234)

Burns, J. J., *Burroughs Wellcome & Co. (U.S.A.), Inc.,* Ascorbic acid (2:747)

Burr, Francis K., *Fabric Research Laboratories, Inc.,* Textile technology (Textile waste treatment), (S:979)

Butler, Kenneth, *Chas. Pfizer & Co., Inc.,* Penicillins (14:652)

Butters, P. A., *Murex Limited,* Niobium and niobium compounds (13:766)

Butterworth, George A. M., *Fabric Research Laboratories,* Textile testing (20:33)

Buxton, M. W., *Imperial Smelting Corporation, Ltd.,* Fluorine compounds, organic (Fluorinated aromatic compounds) (9:775)

Byrne, J. G., *University of Utah,* Metal treatments (13:315)

Byrns, Alva C., *Kaiser Aluminum & Chemical Corp.,* Fluorine compounds, inorganic (Silicon fluorides) (9:650)

Cahn, Robert, *Esso Research and Engineering Company,* Butadiene (3:784)

Cairns, E. J., *Argonne National Laboratory,* Cells, high temperature (S:120)

Cairns, T. L., *E. I. du Pont de Nemours & Company, Inc.,* Cyanocarbons (6:625)

Caldwell, D. L., *The Lummus Company,* Ethylene, (8:499) Ethylene (S:310)

Calvert, Robert, *Consulting Patent Attorney,* Patents (Practice and management) (14:552)

Camp, Frederick W., *Sun Oil Company,* Tar sands (19:682)

Campbell, Jr., G. W., *U.S. Borax Research Corporation,* Boron hydrides (Boron compounds) (3:684)

Cannell, Douglas, *The Sherwin-Williams Company,* Paint (14:462)

Cantow, Manfred J. R., *Air Reduction Co., Inc.,* Vinyl polymers (Poly (vinyl chloride)) (21:369)

Cantrill, James E., *General Electric Company,* Phenolic ethers (15:165)

Carapella, Jr., S. C., *American Smelting and Refining Company,* Antimony and antimony alloys (2:562); Arsenic (711)

Carbone, Walter E., *Wilputte Coke Oven Division, Allied Chemical Corporation,* By-Product Ammonia (Ammonia) (2:299)

Carey, Carroll L., *The Institute of Paper Chemistry,* Paper (14:494)

Carlson, O. N., *Ames Laboratory, United States Atomic Energy Commission,* Calcium and calcium alloys (3:917); Vanadium and vanadium alloys (21:157)

Carr, E. L., *Firestone Tire & Rubber Company,* Styrene-butadiene solution copolymers (S:910)

Carruthers, J. R., *Bell Telephone Laboratories, Inc.,* Semiconductors, (Manufacture of devices) (17:862); Nonlinear optical materials (S:623)

Casey, Robert S., *Consultant,* Iron compounds (12:22)

Castor, C. Robert, *Electronics Division, Union Carbide Corp.,* Gems, synthetic (10:509)

Cathey, Henry M., *Agricultural Research Service, U.S. Department of Agriculture, Beltsville, Maryland,* Plant growth substances (15:675)

Caunt, A. D., *Imperial Chemical Industries, Ltd.,* Polymers of higher olefins (S:773)

Cavallito, C. J., *Neisler Laboratories, Inc.,* Curare and curare-like drugs (6:547)

Cesare, F. C., *Uniroyal, Inc.,* Polypropylene fiber (S:808)

Cesark, Frank F., *American Cyanamid Co.,* Xanthene dyes (22:430)

Chadwick, A. F., *E. I. du Pont de Nemours & Company, Inc.,* Hydrogen peroxide (11:391)

Chadwick, D. H., *Mobay Chemical Company,* Isocyanates, organic (12:45)

Chadwick, John L., *Purvin & Gertz, Inc., Consulting Engineer,* Pentanes (14:707)

Chamberlain, J. R., *York Division, Borg-Warner Corporation,* Refrigeration (17:295)

Chandrasekaran, S., *Polytechnic Institute of Brooklyn,* Diene polymers (7:64)

Chaney, David W., *Chemstrand Research Center, Inc.,* Acrylic and modacrylic fibers (1:313)

Chang, Stephen S., *Rutgers, The State University,* Flavor characterization (9:336)

Chapman, Douglas G., *Food and Drug Directorate, Department of National Health and Welfare, Canada,* Food standards (10:69)

Chatfield, C. H., *Handy & Harman,* Solders and brazing alloys (18:541)

Chen, Hung Tsung, *Newark College of Engineering,* Thermodynamics (20:118)

Chen, W. K. W., *Celanese Plastics Company,* Electrodialysis (7:846)

Chichester, D. F., *Chas. Pfizer & Co., Inc.,* Tartaric acid (19:733)

Chin, D., *Tenneco Chemicals, Inc.,* Salicylic acid and related compounds (17:720)

Choh, Hao Li, *The Hormone Research Laboratory, University of California,* Anterior-pituitary hormones (Hormones) (11:52)

Cier, Harry E., *Esso Research and Engineering Co.,* Toluene (20:527); Xylenes and ethylbenzene (22:467)

Clapper, T. W., *American Potash & Chemical Corp.,* Chloric acid and chlorated (Chlorine oxygen acids and salts) (5:50)

Clarke, James, *University of South Carolina,* Iodine and iodine compounds, (organic compounds) (11:847)

Clauss, Francis J., *Lockheed Missiles & Space Co.,* High-temperature alloys (11:6)

Clegg, P. L., *Imperial Chemical Industries, Ltd.,* Olefin polymers (14:217)

Cody, William F., *Attorney,* Margarine (Legal aspects) (13:56)

Cofrancesco, A. J., *General Aniline & Film Corporation,* Anthraquinone (2:431); Anthraquinone derivatives (438); Anthraquinone and related quinonoid dyes (501); Dyes, natural (7:614); Polymethine dyes (Cyanine, hemicyanine and styryl dyes) (16:292)

Cohn, J. G., *Engelhard Industries, a division of Engelhard Minerals & Chemicals Corp.,* Platinum group metals, compounds (15:861)

Coler, Myron, A., *Markite Corporation,* Electrophoretic deposition (8:23)

Colton, F. B., *Division of Chemical Research, GD Searle & Co.,* Contraceptive drugs (6:60)

Colwell, C. E., *Union Carbide Corporation,* Sorbic acid (18:589)

Comiskey, P. T., *Freeport Sulphur Company,* Sulfur (19:337)

Comley, Eugene A., *Girdler Corporation,* Carbon dioxide (4:353)

Connick, Jr., William J., *U.S. Dept. of Agriculture,* Waterproofing and water repellency (22:135)

Conti, James J., *Polytechnic Institute of Brooklyn,* Transport processes (20:610)

Cooke, Giles B., *Essex Community College,* Cork (6:281)

Cooper, C. M., *Michigan State University,* Solvent recovery (18:549)

Copson, Raymond L., *Allied Chemical Corporation, Solvay Process Division,* Chromium compounds (5:473)

Corbett, L. W., *Esso Research & Engineering Co.,* Asphalt (Manufacture test procedures) (2:762)

Cottis, S. G., *The Carborundum Company,* Poly(hydroxybenzoic acid) (S:741)

Coulter, K. E., *The Dow Chemical Company,* Styrene (19:55)

Cowan, J. C., *U.S. Department of Agriculture, Agricultural Research Service,* Soybeans (18:599)

Coxe, C. D., *Handy & Harman,* Silver and silver alloys (18:279)

Coyle, Thomas D., *National Bureau of Standards,* Silica (Introduction) (18:46)

Crawford, Joseph, *Ingersoll Milling Machine Company,* Electrolytic machining methods (Electrolytic machining) (7:866)

Cremers, L. F., *National Dairy Products Corporation,* Milk and milk products (13:506)

Crooks, Jr., H. M., *Parke, Davis & Company,* Epinephrine, (8:231)

Cross, N. O., *Esso Research and Engineering Co.,* Radioisotopes (17:64)

Crouthamel, Carl E., *Argonne National Laboratory,* Radiochemical analysis and tracer applications (17:35)

Cupas, Chris A., *Western Reserve University,* Friedel-Crafts reactions (10:135)

Curci, Ruggero, *University of Bari, Italy,* Peroxides and peroxy compounds (Reaction mechanism) (14:820)

Curran, J. H., *Union Carbide Corporation, Carbon Products Division,* Baked and graphitized products, uses (mechanical applications) (Carbon) (4:228)

Cusick, Charles F., *Honeywell, Inc.,* Pressure measurement (16:470)

Cuthbertson, G. R., *Uniroyal, Inc.,* Polypropylene fiber (S:808)

Daby, E., *Ford Motor Co.,* Automobile exhaust control, (Photochemical smog) (S:50)

Dalter, Raymond S., *Carlisle Chemical Works, Inc.,* Sulfurization and sulfur-chlorination (19:498)

Dalton, P. B., *General Aniline & Film Corporation,* Alcohols, unsaturated (propargyl alcohol, 2-butyne-1,4-diol, and 2-butene-1,4-diol, properties and uses) (1:598)

Daniel, Arthur F., *United States Army Signal Research and Development Laboratory,* Primary cells (Military types) (Batteries and electric cells, primary) (3:111)

Dante, Mark F., *Shell Chemical Company,* Solvents, industrial (18:564)

Danziger, W. J., *The M. W. Kellogg Company, Division of Pullman Incorporated,* Heat-transfer media other than water (Heat-transfer media) (10:846)

Darby, J. R., *Monsanto Company,* Plasticizers (15:720)

Daul, G. C., *ITT Rayonier, Inc.,* Rayon (17:168)

Davis, Robert E., *American Potash & Chemical Corporation,* Cesium and cesium compounds (4:855); Rubidium and rubidium compounds (17:684)

Davis, William H., *Texas National Bank of Commerce,* Propylene (16:579)

Dawson, Jr., P. P., *The University of Texas,* Nomographs (14:15)

Day, John D., *Bechtel Corporation,* Microwaves (S:563)

Dayton, B. B., *The Bendix Corporation,* Vacuum technology (21:123)

Dean, P. J., *Bell Telephone Laboratories, Inc.,* Semiconductors (Theory and properties of devices) (17:834)

de Becze, George I., *St. Thomas Institute for Advanced Studies,* Enzymes, industrial (8:173); Yeasts (22:507)

DeGarmo, Oliver, *Monsanto Company,* Coumarin (6:425)

Deinet, A. J., *Heyden Newport Chemical Corporation,* Benzaldehyde (3:360)

de la Breteque, Pierre, *Soc. Francaise pour l'Industrie, de l'Aluminium, Div. of Swiss Aluminium, Ltd.,* Gallium and gallium compounds (10:311)

Del Rio, Carlos, *Universidad Nacional de Mexico,* Vitamins (Nicotinic acid) (21:509)

DePalma, J. J., *Eastman Kodak Company,* Filters, optical (9:244)

De Ritter, Elmer, *Hoffmann-La Roche, Inc.,* Riboflavine (17:445)

Dermer, Otis, C., *Oklahoma State University,* Imines, cyclic (11:526)

deStevens, George, *Ciba Pharmaceutical Company,* Analgesics and antipyretics (2:379); Diuretics (7:248)

Deutsch, Zola G., *Consulting Engineer*, Alkali and chlorine industries (1:668)

Devlin, T. J., *Esso Production Research Co.*, Literature (documentation) (12:511)

Dewey, Robert H., *Commercial Solvents Corporation*, Alkanolamines (alkanoamines from nitro alcohols) (1:824); Nitro alcohols (13:826)

DiBella, E. P., *Heyden Newport Chemical Corporation*, Benzaldehyde (3:360)

Dickson, Allan D., *U.S. Dept. of Agriculture*, Malts and malting (12:861)

Diddams, Donald G., *Sterling Drug Inc.*, Vanillin (21:180)

Dmuchovsky, B., *Monsanto Company*, Maleic anhydride, maleic acid, and fumaric acid (Maleic anhydride) (12:819)

Doak, G. O., *North Carolina State of The University of North Carolina at Raleigh*, Antimony compounds (2:570); Arsenic compounds (2:718); Bismuth compounds (3:535)

Doedens, J. D., *The Dow Chemical Company*, Chlorophenols, (5:325)

Doelp, L. C., *Houdry process and Chemical Co.*, Hydroprocesses (11:446)

Domingues, G. S., *Geigy Chemical Corporation, Ardsley, N.Y.*, Stilbene derivatives (Economic aspects under Stilbene dyes) (19:1)

Donovan, J. R., *Monsanto Company*, Chlorosulfonic acid (5:357)

Dorfman, Ralph I., *Syntex Research, Division of Syntex*, Hormones (Sex hormones; Nonsteroidal estrogens) (11:127)

Dorsky, J., *The Givaudan Corporation*, Acetophenone (1:16)

Dorwart, R. A., *York Division, Borg-Warner Corporation*, Refrigeration (17:295)

Doub, Leonard, *Parke, Davis & Co.*, Sulfonamides (19:261)

Dougherty, Harry W., *Merck, Sharp & Dohme*, Vitamins (Nicotinic acid; Biological aspects of nicotinic acid) (21:509)

Downing, R. C., *E. I. du Pont de Nemours & Co., Inc.*, Fluorine compounds, organic (History of the organic fluorine industry, Fluorinated hydrocarbons) (9:704)

Downing, R. S., *The Sherwin-Williams Company*, Paint and varnish removers (14:485)

Doying, E. G., *Union Carbide Corporation, Carbon Products Division*, Activated carbon (carbon) (4:149)

Doyle, John E., *Castle Company*, Sterilization (18:805)

Doyle, John R., *State University of Iowa*, Iron compounds (12:22)

Draganov, S. M., *U.S. Borax Research Corporation*, Boron compounds (boron halides) (3:680)

Drake, Jr., George L., *Cotton Finishes Laboratory, U.S. Department of Agriculture*, Fire-resistant textiles (9:300); Textile technology (Fire resistant textiles) (S:944)

Dressler, H., *Koppers Company, Inc.*, Benzenes (Polyhydroxy) (16:190)

Dressler, R. G., *Trinity University*, Water (Sources) (21:651)

Drury, J. S., *Oak Ridge National Laboratory*, Nuclear reactors (Isotope separation) (14:85)

Dryden, I. G. C., *British Coal Utilisation Research Association*, Coal (5:606)

Dudey, Norman D., *Argonne National Laboratory*, Radiochemical analysis and tracer applications (17:35)

Dugan, Jr., L. R., *Michigan State University*, Antioxidants (2:588)

Dumbaugh, William H., *Corning Glass Works*, Silica (Vitreous) (18:73)

Duncan, J. J., *General Aniline & Film Corporation*, Preparation for dyeing (Dyes-application and evaluation) (7:505)

Duncker, Charles, *Monsanto Chemical Company*, Benzoic acid (3:420)

Dunlop, A. P., *The Quaker Oats Company,* Furfural and other furan compounds (Furfural) (10:237)

Dunn, Hugh, *Interchemical Corporation,* Inks (11:611)

Dutcher, James D., *The Squibb Institute for Medical Research,* Polyene antibiotics (16:133)

Dux, J. P., *American Viscose Division, FMC Corporation,* Vinyon and related fibers (21:441)

Eastman, G. Yale, *Radio Corporation of America,* Heat pipe (S:488)

Eberlin, Walter R., *The Children's Hospital of Philadelphia,* Hormones (Adrenal-cortical hormones) (11:77)

Eckey, E. W., *E. W. Eckey Research Laboratory,* Ester interchange (8:356)

Economos, George, *Allen-Bradley Company,* Ferrites (8:881)

Economy, James, *The Carborundum Company,* Inorganic refractory fibers (11:651); Phenolic fibers (S:667); Poly(Hydroxybenzoic)acid (S:741)

Eddinger, R. T., *FMC Corporation,* Coal (to synthetic crude oil) (S:177)

Eddleman, George H., *Imperial Metal & Chemical Co.,* Type metal (20:756)

Edwards, F. G., *The Dow Chemical Company,* Vinylidene polymers (Chloride) (21:275)

Edwards, John O., *Brown University,* Peroxides and peroxy compounds (Inorganic; Reaction mechanisms) (14:746)

Ehrich, Felix Frederick, *E. I. du Pont de Nemours & Co., Inc.,* Pigments (Organic) (15:555)

Ehrlich, John, *Parke, Davis & Company,* Chloramphenicol (4:928)

Eibeck, Richard E., *Allied Chemical Corporation,* Fluorine compounds, inorganic (Sulfur fluorides, in part); Fluorine compounds, organic (Perfluoroalkylsulfur fluorides) (9:676)

Eichel, F. G., *The Givaudan Corporation,* Acetophenone (1:167)

Eickner, H. W., *U.S. Department of Agriculture,* Wood (22:358)

Elam, Edward U., *Tennessee Eastman Company,* Esters, Organic (8:365)

Elkin, E. M., *Canadian Copper Refiners, Ltd.,* Selenium and selenium compounds (17:809); Tellurium and tellurium compounds (19:756)

Ellern, Herbert, *UMC Industries, Inc.,* Matches (13:160); Pyrotechnics (16:824)

Ellestad, R. B., *Lithium Corporation of America, Inc.,* Lithium and lithium compounds (12:529)

Ellickson, R. T., *University of Oregon,* Atoms and atomic structure (2:806)

Elliott, J., *Toms River Chemical Corporation,* Dyes, reactive (7:630); Phenosulfonic acids (15:208)

Elliott, Martin A., *Illinois Institute of Technology,* Gas manufactured (10:353)

Ellis, David W., *University of New Hampshire,* Fluorescent pigments (daylight) (Theory) (9:483)

Ellis, J. R., *General Aniline & Film Corporation,* Dyes-application and evaluation (Theories of dyeing; dyeing machinery, acid dyes, direct dyes; Insoluble azo dyes; Miscellaneous types of dyes; Pigments) (7:505)

Ellison, Jr., Samuel P., *University of Texas,* Petroleum (Origin) (14:835)

Elsea, R., *Carrier Air Conditioning Company, a division of Carrier Corporation,* Air conditioning (1:481)

Elslager, Edward F., *Parke, Davis & Company,* Therapeutic agents, protozoal infections (20:70)

Ely, James K., *Interchemical Corporation,* Inks (11:611)

Elzinga, E. R., *Esso Research and Engineering Co.,* Proteins from petroleum (S:836)

Engdahl, Richard B., *Battelle Memorial Institute,* Smokes, fumes, and smog (18:400)

Engel, George T., *U.S. Department of the Interior, Bureau of Mines,* Mercury (13:218)

Enos, Jr., Herman I., *Hercules Incorporated,* Rosin and rosin derivatives (17:475)

Entrikin, John B., *Centenary College of Louisiana,* Microchemistry (13:424)

Ericson, K. R., *The Procter & Gamble Company,* Alcohols, higher, fatty (1:542)

Ervin, H. O., *American Gilsonite Co.,* Gilsonite (10:527)

Eslyn, W. E., *U.S. Department of Agriculture,* Wood (22:358)

Ettling, B. V., *Washington State University,* Hydroquinone, resorcinol, pyrocatechol (11:462)

Etzel, Gastão, *Camphor & Allied Products Limited,* Camphor (4:54)

Famulener, Keith, *Ampex Corporation,* Magnetic tape (12:801)

Farber, M., *Uniroyal, Inc.,* Polypropylene fiber (S:808)

Farrow, G., *Fiber Industries, Inc.,* Polyester fibers (16:143)

Fassett, David W., *Eastman Kodak Company,* Industrial toxicology (11:595)

Faust, J. P., *Chemical Division, Olin Corporation,* Water (Treatment of swimming pools) (22:124)

Fenter, John R., *Dept. of the Air Force,* Ceramic composite armor (Ceramic and adhesives) (S:138)

Fenton, Edward A., *American Welding Society,* Welding (22:241)

Ferguson, D. E., *Oak Ridge National Laboratory,* Nuclear reactors (14:74)

Ferstandig, L. L., *Halocarbon Products Corporation,* Fluorine compounds, organic (Fluoroethanols; Polybromotrifluoroethylene) (9:751)

Fetzer, W. R., *Consultant,* Caramel color (4:63)

Feuge, R. O., *Southern Utilization Research and Development Division, Agricultural Research Service, U.S. Department of Agriculture,* Cottonseed (6:412)

Fiedelman, Howard W., *Morton International, Inc.,* Sodium compounds (Sodium chloride) (18:468)

Filachione, E. M., *Agricultural Research Service, U.S. Department of Agriculture, Wyndmoor, Pa.,* Lactic acid (12:170)

Finch, Roland, *U.S. Department of the Interior Fish and Wildlife Service, Bureau of Commercial Fisheries,* Fish and shellfish (9:316)

Fiorino, Nicholas A., *Honeywell, Inc.,* Instrumentation and automation (11:739)

Fischer, E., *Radio Corporation of America,* Phototubes and photocells (15:396)

Fischer, Roland B., *The Dow Chemical Company,* Tool materials for machining (20:566)

Fisher, Edward W., *Garlock Inc.,* Packing materials (14:443)

Fisher, J. G., *Tennessee Eastman Company, a division of Eastman Kodak Company,* Thiazole dyes (Disperse and cationic dyes containing the thiazole nucleus) (20:191)

Fissore, Oscar F., *Wilputte Coke Oven Division, Allied Chemical Corporation,* By-product ammonia (Ammonia) (2:299)

Fleckenstein, Lee J., *Eastman Kodak Co.,* Aldehydes (1:639); Amides, acid (2:66)

Fleming, R. W., *The Wm. S. Merrell Company, Division of Richardson-Merrell, Inc.,* Histamine and antihistamine agents (11:35)

Flint, Gerhard, *Great Salt Lake Minerals & Chemicals Corporation,* Great Salt Lake chemicals (S:438)

Florio, Patrick A., *Pfister Chemical, Inc.,* Oxalic acid (14:356)

Folkins, Hillis O., *The Pure Oil Company,* Carbon disulfide (4:370)

Fowler, Don G., *Lead Industries Association, Inc.,* Lead compounds (Control of industrial hazards) (12:301)

Fowles, G. W. A., *University of Reading, Reading, England,* Vanadium compounds (21:167)

Fox, W. B., *Allied Chemical Corporation,* Fluorine compounds, inorganic (Oxygen fluorides) (9:631)

Franz, J. E., *Monsanto Company,* Maleic anhydride, maleic acid, and fumaric acid (12:819)

Frear, G. L., *Nitrogen Division, Allied Chemical Corporation,* Ammonia (2:258)

Freas, J. G., *General Aniline & Film Corporation,* Sulfur dyes; Vat dyes; Leuco esters of vat dyes (Dyes-application and evaluation) (7:505)

Freedman, Leon D., *North Carolina State of The University of North Carolina at Raleigh,* Antimony compounds (2:570); Arsenic compounds (2:718); Bismuth compounds (3:535)

Freeman, Frank H., *Kerr Manufacturing Company,* Dental materials (6:777)

Frey, F. W., *Ethyl Corporation,* Lead compounds, organic, (organolead compounds) (12:282)

Fried, Josef, *University of Chicago,* Hormones (Survey) (11:45)

Friedberg, A. L., *University of Illinois,* Enamels, porcelain or vitreous (8:155)

Frielfeld, M., *General Aniline & Film Corporation,* Glycols (1,4-Butylene glycol and butyrolactone) (10:667)

Frishman, Daniel, *Technical Advisor to Reid-Meredith, Inc.,* Furs and fur processing (10:294)

Frysinger, Galen R., *ESB Incorporated,* Fuel cells (S:383)

Fryth, Peter W., *Hoffmann-LaRoche, Inc.,* Vitamins (survey) (21:484)

Fuchs, Julius, *E. I. du Pont de Nemours & Co., Inc.,* Sulfuric and sulfurous esters (19:483)

Fugate, W. O., *American Cyanamid Company,* Acrylonitrile (1:338)

Fuller, G., *Imperial Smelting Corporation, Ltd.* Fluorine compounds, organic (Fluorinated aromatic compounds) (9:775)

Funderburk, Lance, *Toms River Chemical Corp.,* Quinoline dyes (16:886)

Gage, J. C., *Imperial Chemical Industries, Ltd.,* Toxicity (Chlorocarbons and chlorohydrocarbons) (5:92)

Gale, W. A., *American Potash & Chemical Corporation,* Chloric acid and chlorates (Chlorine oxygen acids and salts) (5:50)

Gall, John F., *Pennsalt Chemicals Corporation,* Fluorine compounds, inorganic (Aluminum fluoride; Calcium fluoride; Hydrogen fluoride) (9:529)

Gantz, G. M., *General Aniline & Film Corporation,* Evaluation of dyes-instrumental methods (Dyes-application and evaluation) (7:505)

Garfield, Eugene, *Institute for Scientific Information,* Information retrieval services and methods (S:510)

Garfinkel, M., *Chemical Projects Associates, Inc.,* Olefins (S:632)

Gaylord, N. G., *Gaylord Associates, Inc.,* Allyl compounds (1:916)

Gaylord, W. M., *Union Carbide Corporation, Carbon Products Division,* Baked and graphitized products, uses (Textile and fibrous applications) (under Carbon) (4:241)

Gearhart, William M., *Eastman Chemical Products, Inc.,* Crotonaldehyde, crotonic acid (6:445)

Geigy AG, J. R., *Geigy Chemical Corporation,* Ardsley, New York, Stilbene derivatives (Stilbene dyes) (19:1)

Gerarde, H. W., *Esso Research and Engineering Company,* Hydrocarbons (Toxicity) (11:293)

Gergel, M. G., *Columbia Organic Chemicals Company,* Fluorine compounds, organic (Heptafluorobutyric acid) (9:773); Iodine and iodine compounds, (organic compounds) (11:847)

Gershon, S. D., *Lever Brothers Company,* Thioglycolic acid (20:198)

Gerstner, H. G., *Colonial Sugars Company,* Sugar (Cane sugar) (19:166)

Gibbs, George B., Hydrocarbon processing, petroleum (products) (15:77)

Gilbert, Everett E., *Allied Chemical Corp.,* Sulfonation and sulfation, (19:279); Sulfonic acids (311)

Gillette, Leslie A., *Pennsalt Chemicals Corporation,* Amyl alcohols (2:374)

Gilmont, Roger, *Roger Gilmont Instruments, Inc., and Polytechnic Institute of Brooklyn,* Vapor-liquid equilibria, (21:196)

Gittinger, Jr., L. B., *Freeport Sulphur Company,* Sulfur (19:337)

Glatz, Alfred C., *National Aeronautics and Space Administration,* Thermoelectric energy conversion (20:147)

Glavis, F. J., *Rohm & Haas Company,* Acrylic acid and derivatives (1:285); Methacrylic compounds (13:331)

Glennie, D. W., *Crown Zellerbach Corporation,* Lignin (12:361)

Glicksman, Martin, *General Foods Corporation,* Gums, natural (10:741)

Gloger, W. A., *National Lead Company,* Pigments (Inorganic) (15:495)

Gloor, W. E., *Hercules Incorporated,* Olefin polymers (Polyethylenes made by the Ziegler process) (14:217)

Goerz, Jr., David J., *Bechtel Corporation,* Microwaves (S:563)

Goetz, Milton M., *Chemical Construction Corp.,* Plant layout (15:689)

Goetzinger, N. J., *American Potash & Chemical Corporation,* Rare earth elements (17:143)

Goheen, D. W., *Crown Zellerbach Corporation,* Lignin (12:361)

Good, L. F., *Freeport Sulphur Company,* Sulfur (19:337)

Goodman, I., *Imperial Chemical Industries, Ltd., and University of Manchester Institute of Science and Technology,* Polyesters (16:159)

Gordon, David A., *Geigy Industrial Chemicals Division of Geigy Chemical Corporation,* UV Absorbers (21:115)

Gordon, Maxwell, *Smith, Kline & French Laboratories,* Psychopharmacological agents (16:640)

Gorman, Marvin, *Lilly Research Laboratories, Eli Lilly and Company,* Macrolide antibiotics (12:632)

Gottesman, R. T., *Tenneco Chemicals, Inc.,* Salicylic acid and related compounds (17:720)

Gottlieb, Robert, *Stauffer Chemical Company,* Aluminum halides (under Aluminum compounds) (2:17)

Gould, T. R., *Johns-Manville Corporation,* Refractory fibers (17:285)

Gower, A. H., *Chemical Division, Olin Corporation,* Water (Treatment of Swimming pools) (22:124)

Granito, Charles E., *Institute for Scientific Information,* Information retrieval services and methods (S:510)

Gray, G. R., *Baroid Division, National Lead Company*, Drilling fluids (7:287)

Gray, William T., *Leeds & Northrup Company*, Temperature measurement (19:774)

Green, Ralph V., *E. I. du Pont de Nemours & Co., Inc.*, Carbon monoxide (4:424)

Greenbaum, Sheldon B., *Diamond Shamrock Chemical Co.*, Vitamins (Vitamin B_{12}; Vitamin D) (21:542)

Greene, Gerald U., *New Mexico Institute of Mining and Technology*, Cadmium compounds (3:899)

Greene, L. Wilson, *Consultant*, Incendiaries (Chemical warfare) (4:895)

Greene, Robert L., *Purvin & Gertz, Inc., Consulting Engineer*, Pentanes (14:707)

Gregg, Robert A., *Uniroyal, Inc.*, Spandex (18:614)

Gregor, John R., *Peninsular Grinding Wheel Co.*, Abrasives (1:22)

Gresky, A. T., *Oak Ridge National Laboratory*, Nuclear reactors (Chemical reprocessing) (14:91)

Greubel, Paul W., *Interchemical Corporation*, Inks, (11:611)

Griemsmann, J. W. E., *Polytechnic Institute of Brooklyn*, Pipeline heating (S:694)

Griffin, W. C., *Atlas Chemical Industries, Inc.*, Emulsions (8:117)

Grim, Ralph E., *University of Illinois*, Clays, uses (5:560)

Grimmel, Harry W., *American Hoechst Corporation*, Indigoid dyes (11:562)

Grisar, J. M., *The Wm. S. Merrell Company, Division of Richardson-Merrell, Inc.*, Histamine and antihistamine agents (11·35)

Griswold, J., *Allied Chemical Corporation*, Fluorine compounds, inorganic (Sodium fluorides) (9:662)

Gross, Donald J., *University of North Carolina*, Electroanalytical methods (7:726)

Gross, W. H., *The Dow Chemical Company*, Magnesium and magnesium alloys (12:661)

Grote, I. W., *University of Chattanooga*, Cathartics (4:586)

Grove, Jr., C. S., *Syracuse University*, Laminated and reinforced plastics (12:188)

Grummitt, O., *Western Reserve University*, Drying oils (7:398)

Gruntfest, I. J., *General Electric Company*, Ablation (1:11)

Guenther, Ernest, *Fritzsche Brothers, Inc.*, Flavors and spices (9:347)

Guest, H. R., *Union Carbide Chemicals Company*, Acrolein and derivatives (1:255)

Guiseley, Kenneth B., *Marine Colloids, Inc.*, Seaweed colloids (17:763)

Gustafson, R. E., *Cameron and Jones, Inc.*, Shale oil (18:1)

Guth, J. H., *National Dairy Products Corporation*, Milk and milk products (13:506)

Guthrie, C. E., *Oak Ridge National Laboratory*, Nuclear reactors (Safety in nuclear facilities) (14:108)

Hack, C. H., *National Lead Company*, Lead alloys, lead utilization (12:247)

Hackler, W. C., *North Carolina State of The University of North Carolina*, Ceramics (Thermal treatment of ceramics) (4:783)

Hadley, C. P., *Radio Corporation of America*, Phototubes and photocells (15:396)

Haefling, J. A., *Ames Laboratory, United States Atomic Energy Commission*, Calcium and calcium alloys (3:917)

Hahnel, E. C., *General Aniline & Film Corporation*, (Propargyl alcohols, 2-butyne-1,4-diol, and 2-butene-1,4-diol, properties and uses (Alcohols, unsaturated) (1:598)

Hainsworth, Ernest, *Tea Research Institute of East Africa*, Tea (19:743)

Hajny, G. J., *U.S. Department of Agriculture*, Wood (22:358)

Halbedel, H. S., *The Harshaw Chemical Company*, Fluorine compounds, inorganic (Boron fluorides, in part; Halogen fluorides) (9:562)

Hall, J. W., *Union Carbide Corporation, Linde Division,* Nitrogen (13:857)

Hall, N. T., *Pennsalt Chemicals Corporation,* Fluorine compounds, organic (Poly-(vinylidene fluoride) (9:840)

Hall, R. C., *General Aniline & Film Corporation,* Textile printing (Dyes· application and evaluation) (7:505)

Hals, L. J., *Minnesota Mining & Manufacturing Co.,* Fluorine compounds, organic (Fluoroethers and amines) (9:753)

Hamann, Edmund H., *Fritzsche Brothers, Inc.,* Flavors and spices (9:347)

Hamil, H. F., *Celanese Chemical Company,* Hydrocarbon oxidation (11:224)

Hamilton, J. Kelvin, *Rayonier, Inc.,* Cellulose (4:593)

Hammack, James M., *National Fire Protection Association,* Fire extinguishing agents (S:364)

Hamme, John V., *North Carolina State of The University of North Carolina,* Ceramic raw materials (Ceramics) (4:762)

Hammond, George S., *California Institute of Technology,* Free radicals (10:114)

Hammond, W. A., *W. A. Hammond Drierite Company,* Calcium sulfate (Calcium compounds) (4:14)

Handman, S. E., *The M. W. Kellogg Company,* Piping systems (15:646)

Hankin, Jerome W., *Bechtel Corporation,* Microwaves (S:563)

Hann, R. A., *U.S. Department of Agriculture,* Wood (22:358)

Hansen, Charles, *PPG Industries, Inc.,* Solubility parameters (S:889)

Happel, J., *New York University,* Methylacetylene (S:547)

Hardie, D. W. F., *Imperial Chemical Industries, Ltd.,* Chlorocarbons and chlorohydrocarbons (Survey) (5:85); Methyl chloride (100); Methylene chloride (111); Chloroform (119); Carbon tetrachloride (128); Ethyl chloride (140); Other chloroethanes (148); Dichloroethylenes (178); Trichloroethylene (183); Tetrachloroethylene (195); Chloroacetylenes (203); Chlorinated paraffins (231); Chlorinated benzenes (253); Benzene hexachloride (267); Chlorinated naphthalenes (297) Survey; Methyl chloride; Methylene chloride; Chloroform; Carbon tetrachloride; Ethyl chloride; Other chloroethanes; Vinyl chloride; Dichloroethylenes; Trichloroethylene; Tetrachloroethylene; Chloroacetylenes; Chlorinated benzenes; Benzene hexachloride; Chlorinated paraffins, Chlorinated naphthalenes (Chlorocarbons and chlorohydrocarbons) (5:85)

Harding, William B., *The Bendix Corporation,* Metallic coatings (13:249)

Hardy, Edgar E., *Monsanto Research Corporation,* Fluorine compounds, organic (Fluorinated carboxylic acids, in part) (9:767); Isocyanates, organic (12:45); Phosgene (S:674)

Hargreaves, II, C. A., *E. I. du Pont de Nemours & Company, Inc.,* Neoprene (Elastomers, synthetic) (7:705)

Harner, H. R., *The Eagle-Picher Company,* Germanium (General) (10:519)

Harrer, T. S., *Allied Chemical Corporation,* Sulfuric acid and sulfur trioxide (19:441)

Harrington, R. V., *Corning Glass Works,* Glass (10:533)

Harris, Elbert E., *Merck & Co.,* Pyriodoxine, pyridoxal, and pyridoxamine (16:806)

Harris, George C., *Hercules Incorporated,* Rosin and rosin derivatives (17:475)

Harris, G. H., *The Dow Chemical Company,* Xanthates (22:419)

Harris, Stanton, A., *Merck & Co., Inc.,* Pyridoxine (16:806)

Harris, T. L., *Cadbury Brothers, Limited,* Chocolate and cocoa (5:363)

Hart, Andrew W., *The Dow Chemical Company,* Alkanolamines from olefin oxides and ammonia (Alkanolamines) (1:809); Diamines and higher amines aliphatic (7:22); Imines, cyclic (11:526); Iodine and iodine compounds (11:847)

Hart, Harold, *Michigan State University,* Carbonium ion and carbanion (4:394)

Hartford, Winslow, H., *Allied Chemical Corporation, Solvay Process Division,* Chromium compounds (5:473)

Harwood, H. J., *Durkee Famous Foods,* Amides, fatty acid; Amines, fatty (Amines) (2:72); Amines, lower aliphatic (2:116)

Hasek, Robert H., *Tennessee Eastman Company,* Ketenes and related substances, (Ketenes) (12:87)

Hass, Henry B., *Chemical consultant,* Chlorine (S:167)

Häusermann, H., *J. R. Geigy S. A.,* Brighteners, optical (3:737)

Hausner, Henry H., *Consulting Engineer,* Powder metallurgy (16:401)

Hay, D. Robert, *Drexel University of Technology,* Zone refining (22:680)

Hay, John O., *The Harshaw Chemical Co., Division of Kewanee Oil Co.,* Manganese compounds (13:1)

Hay, R. G., *Gulf Research & Development Co.,* Olefins, higher (14:313)

Hayes, A. C., *School of Textiles, North Carolina State University,* Silk (18:269)

Hayes, E. R., *Shawinigan Chemicals Limited,* Acetaldehyde (1:77)

Hayes, Kenyon J., *Norwich Pharmacal Company,* Nitrofurans (13:853)

Healy, Robert M., *American Potash & Chemical Corporation,* Cerium and cerium compounds (4:840)

Hechenbleikner, Ingenuin, Carlisle Chemical Works, Inc., Sulfurization and sulfur chlorination (19:498)

Hector, Annemarie, *Carl-Engler- und Hans-Bunte-Institut für Mineralol- and Kohleforschung der Technischen Hochschule Karlsruhe,* Carbon monoxide-hydrogen reactions (4:446)

Hedrick, Glen W., *Naval Stores Laboratory, U.S. Department of Agriculture,* Rosin and rosin derivatives (17:475)

Heichelheim, H. R., *Texas Technological College,* Methane (13:364)

Helling, C. S., *U.S. Department of Agriculture,* Soil chemistry of pesticides (18:515)

Helvy, F. A., *Radio Corporation of America,* Phototubes and photocells (15:396)

Hemming, Charles B., *U.S. Plywood-Champion Papers, Inc.,* Plywood (15:896)

Hemstreet, R. A., *Central Research Laboratory, Air Reduction Company, Inc.,* Helium-group gases (10:862)

Hendrick, G. W., *Naval Stores Laboratory, U.S. Department of Agriculture,* Rosin and rosin derivatives (17:475)

Hendricks, James F., *Worthington Corporation,* Pumps and compressors (Compressors) (16:741)

Henrie, T. A., *Bureau of Mines, U.S. Dept. of the Interior,* Mercury, recovery by electrooxidation (Mercury) (S:542)

Henry, Edward C., *General Electric Company,* Ferroelectrics (9:1)

Hensel, Walter G., *Patent Attorney,* Trademarks (20:578)

Hensley, Lee C., *GAF Corporation,* Polymethine dyes (16:282)

Herbig, James A., *The National Cash Register Co.,* Microencapsulation (13:436)

Herman, H., *University of Pennsylvania,* Metal treatments (13:315)

Herting, David C., *Tennessee Eastman Co.,* Vitamins (Vitamin E) (21:574)

Heydinger, E. R., *Marathon Oil Company,* Petroleum (Resources) (14:856)

Hickson, J. L., *International Sugar, Research Foundation, Inc.,* Sugar (Derivatives) (19:221); Economics (233); Special sugars (237)

Higgins, D. G., *Waldron-Hartig Division, Midland-Ross Corporation,* Coated fabrics (5:679)

Hiki, N., *Ford Motor Co.,* Automobile exhaust control (Photochemical smog) (S:50)

Hileman, Jack C., *El Camino College,* Carbonyls (4:489)

Hilker, L. D., *National Dairy Products Corporation,* Milk and milk products (13:506)

Hill, E. S., *Fiber Industries, Inc.,* Polyester fibers (16:143)

Himmelblau, D. M., *The University of Texas,* Fuels (combustion calculations) (10:191)

Himmelfarb, David, *Boston Naval Shipyard,* Fibers; vegetable (9:171)

Hinckley, Alfred A., *Metal Hydrides Division, Ventron Corporation,* Hydrides (11:200)

Hindersinn, R. R., *Hooker Chemical Corp.,* Halogenated fire retardants (S:467)

Hiscock, B. F., *The Dow Chemical Company,* Styrene (19:55)

Hodgins, T., *Purdue University,* Fluorine compounds, organic (9:686)

Hoehn, Willard M., *G. D. Searle & Company,* Bile constituents (3:480)

Hoffman, Harold L., *Hydrocarbon processing,* Petroleum (Products) (15:77)

Hoffmann, Conrad E., *E. I. du Pont de Nemours & Co., Inc.,* Viral infections; chemotherapy (21:452)

Hogan, John E., *The Okonite Company,* Insulation, electric (Cable coverings) (11:802)

Hogan, J. P., *Phillips Petroleum Company,* Olefin polymers (High-density (linear) polyethylene) (14:242)

Hoglund, R. L., *Union Carbide Corporation,* Diffusion (7:79); Diffusion separation methods (91)

Hoh, G. L. K., *E. I. du Pont de Nemours & Company, Inc.,* Hydrogen peroxide (11:391)

Hoiberg, Arnold J., *The Flintkote Company,* Asphalt (2:762)

Holcomb, Dysart E., *United Gas Corporation,* Gas, natural (10:443)

Holden, Geoffrey, *Shell Chemical Company,* Isoprene (12:64)

Holderried, John A., *Michigan Chemical Corporation,* Amines, dialkylchloroalkyl (Amines) (2:139)

Holland, C. D., *Texas A & M University,* Distillation (7:204)

Hollowell, J. L., *E. I. du Pont de Nemours & Co., Inc.,* Poromeric material (16:345)

Holmes, W. C., *Consulting Chemist,* Ammonium compounds (2:313)

Hopfan, Seymour, *Memorial Hospital for Cancer and Allied Diseases,* Radioactive drugs and tracers (17:1)

Horne, Ralph A., *Woods Hole Oceanographic Institute,* Water (Properties) (21:668)

Horner, L. I., *Celanese Fibers Company, Division of Celanese Corporation of America,* Acetate and triacetate fibers (1:109)

Horowitz, Emanuel, *National Bureau of Standards,* Plastics testing (15:811)

Horsley, Lee H., *The Dow Chemical Company,* Propylene oxide (16:595)

Hort, E. V., *General Aniline & Film Corporation,* Propargyl Alcohol; 2-Butyne-1,4-diol; 2-Butene-1,4-diol; properties and uses (Alcohols, unsaturated) (1:598); Glycols (1,4-Butylene glycol and γ-butyrolactone) (10:667); Pyrrole and pyrrole derivatives (16:841)

Hoskins, W. M., *University of California, Berkeley,* Bioassay (3:489)

Howe, H. E., *American Smelting and Refining Company,* Bismuth and bismuth alloys (3:527); Cadmium and cadmium alloys (884); Thallium and thallium compounds (63)

Howells, T. A., *The Institute of Paper Chemistry,* Paper (14:494)

Høyrup, H. E., *Bryggeriforeningen, Copenhagen,* Beer and brewing (3:297)

Hoyt, C. H., *Crown Zellerbach Corporation,* Lignin (12:361)

Hubbard, H. L., *Monsanto Company,* Chlorinates biphenyl and related compounds (Chlorocarbons and chlorohydrocarbons) (5:289); Terphenyls, hydrogenated derivatives (both under Diphenyl and terphenyls) (7:191)

Huber, Jr., John, *Stepan Chemical Company, Maywood Chemical Division,* Caffeine (3:911)

Hubinger, D. C., *E. I. du Pont de Nemours & Co., Inc.,* Dimethylacetamide (under Acetic acid derivatives) (1:145)

Huebotter, E. E., *Baroid Division, National Lead Company,* Drilling fluids (7:287)

Hughes, F. R., *Radio Corporation of America,* Phototubes and photocells (15:396)

Humfeld, G. P., *Radio Corporation of America,* Phonograph record compositions (15:225)

Hunt, R. H., *Shell Oil Company,* Petroleum (Composition) (14:845)

Huntington, H. B., *Rensselaer Polytechnic Institute,* Electromigration (S:278)

Hurd, Charles D., *Northwestern University,* Carbohydrates (4:132)

Hussong, R. V., *National Dairy Products Corporation,* Milk and milk products (13:506)

Hutchins, III, J. R., *Corning Glass Works,* Glass (10:533); Glycols (676)

Hutchins, J. E., *Eastman Chemical Products, Inc.,* Other glycols (Glycols) (10:676)

Hutzler, G. J., *Spencer Kellogg Division of Textron, Inc.,* Castor oil (4:524)

Hyman, Herbert H., *Argonne National Laboratory,* Helium-group gases (Compounds) (10:888)

Hynes, John B., *Hynes Chemical Research Corporation,* Fluorine compounds, organic (Fluorinated carboxylic acids; Miscellaneous organic fluorine compounds) (9:767)

Illy, Hugo, *Toms River Chemical Corp.,* Quinoline dyes (16:886)

Ingle, George W., *Monsanto Company,* Color measurement (5:801)

Ingwalson, R. W., *Velsicol Chemical Corp.,* Nitriles (S:590)

Innes, W. B., *American Cyanamid Company,* Automobile exhaust control (2:814)

Irwin, William E., *Miles Chemical Company,* Citric acid (5:524); Malic acid (12:837)

Isacoff, Harry, *International Flavors & Fragrances (U.S.),* Cosmetics (6:346)

Isenberg, I. H., *The Institute of Paper Chemistry,* Paper (Fiber composition) (14:494)

Iverson, M. L., *Atomics International,* Boron halides (Boron compounds) (3:680)

Izard, E. F., *E. I. du Pont de Nemours & Co., Inc.,* Ester interchange (8:356)

Jackson, F. L., *Freeport Sulphur Company,* Sulfur (19:337)

Jackson, H. W., *National Dairy Products Corporation,* Milk and milk products (13:506)

Jackson, R. B., *Allied Chemical Corporation,* Fluorine compounds, inorganic (Oxygen fluorides) (9:631)

Jacobs, Joseph J., *Jacobs Engineering Co.,* Potassium compounds (16:369); Sodium sulfates (Sodium sulfates) (18:502)

Jakobi, W. W., *Gould-National Batteries, Inc.,* Introduction, secondary cells (3:161); Secondary cells, alkaline (172)

Jaquiss, Donald B. G., *General Electric Company,* Phenolic aldehydes (15:160)

Jarvis, N. L., *U.S. Naval Research Laboratory,* Fluorine compounds, organic (Sulfur chemistry of fluorochemicals) (9:707)

Johnson, N. J., *Union Carbide Corporation, Carbon Products Division,* Carbon (Baked and graphitized, uses) anode applications (4:205)

Johnson, R. F., *Swiss Federal Institute of Technology*, Azo dyes (2:868)

Johnson, R. N., *Union Carbide Corporation*, Polymers containing sulfur (Polysulfone resins) (16:272)

Johnston, Lee G., *American Institute of Laundering*, Laundering (12:197)

Jolly, S. E., *Sun Oil Company*, Naphthenic acids (13:727)

Jones, Haydn, *Hizone Laboratories, Wilmette, Illinois*, Embalming fluids (8:100)

Jones, Martha B., *Royal Crown Cola Company*, Carbonated beverages (4:335)

Jones, Reuben, G., *Eli Lilly and Company*, Thyroid and antithyroid preparations (Antithyroid substances) (20:268)

Jordan, Donald G., *Consultant*, Pilot plants (15:605)

Joslyn, Maynard A., *University of California*, Vinegar (21:254)

Jukes, T. H., *University of California, Berkeley*, Choline (5:403)

Julian, James, *J. M. Worthington Corporation*, Pumps and compressors (Compressors) (16:741)

Kallgren, R. W., *The Dow Chemical Company*, Antifreezes and deicing fluids (2:540)

Kaltenecker, L. V., *Chemical Construction Corp.*, Plant location (15:700)

Kamienski, C. W., *Lithium Corporation of America, Inc.*, Lithium and lithium compounds (12:529)

Kaswell, Ernest R., *Fabric Research Laboratories*, Textile testing (20:33)

Kawasaki, Edwin P., *Republic Steel Corporation*, Pickling of steel (S:684)

Kearney, P. C., *U.S. Department of Agriculture*, Soil chemistry of pesticides (18:515)

Kehde, H., *The Dow Chemical Company*, Styrene (19:55)

Kehe, Henry J., *The B. F. Goodrich Company*, Diarylamines (7:40)

Keilin, Bertram, *Amicon Corporation*, Osmosis, osmotic pressure, and reverse osmosis (Osmosis) (14:345)

Keith, Jr., F. W., *The Sharples Company*, Centrifugal separation (4:710)

Keller, Arthur G., *Louisiana State University*, Bagasse (3:36); Molasses (13:613)

Keller, E. H., *E. I. du Pont de Nemours & Co., Inc.*, Glycolic acid (10:632)

Keller, W. D., *University of Missouri, Columbia*, Clays survey (5:541)

Kemp, Gordon, *American Cyanamid Company*, Bacterial, rickettsial, and mycotic infections; Chemotherapy (3:1)

Kendall, Donald B., *Toledo Scale Company, Division of Reliance Electric Company* Weighing and proportioning (22:220)

Kern, Joyce C., *Glycerine Producers' Association*, Glycerol (10:619)

Kertesz, Z. I., *Food and Agriculture Organization of the United Nations*, Food additives (in part), food and food processing (10:23)

Keskkula, Henno, *The Dow Chemical Co.*, Styrene plastics (19:85)

Keutgen, W. A., *Union Carbide Corporation*, Phenolic resins (15:176)

Kharasch, Norman, *Department of Chemistry, University of Southern California*, Sulfur compounds (19:367)

Kieffer, Richard, *University of Vienna*, Survey; Industrial heavy-metal carbides; Cemented carbides (Carbides) (4:69); Nitrides (13:814)

Kihlgren, T. E., *The International Nickel Company, Inc.*, Nickel and nickel alloys (13:735)

Kirk, B. S., *Central Research Laboratory, Air Reduction Company, Inc.*, Helium-group gases (10:862)

Kirk, James C., *Continental Oil Company*, Cyclohexane (6:675)

Kirshenbaum, Isidor, *Esso Research and Engineering Company,* Butadiene, (3:784); Density and specific gravity (6:755)

Klarmann, Emil G., *Lehn & Fink, Inc.,* Antiseptics and disinfectants (2:604)

Kleckner, W. R., *Diamond Alkali Company,* Hydrochloric acid (11:307)

Klemantaski, S., *British Steel Corporation,* Slagceram (S:876)

Klimstra, P. D., *G. D. Division of Chemical Research, G. D. Searle & Co.,* Contraceptive drugs (6:60)

Klug, E. D., *Hercules Powder Company,* Cellulose derivatives (4:616)

Kneen, Eric, *Kurth Malting Company,* Malts and malting (12:861)

Knepper, W. A., *United States Steel Corporation,* Iron (12:1)

Knuth, Charles J., *Chas. Pfizer & Co., Inc.,* Itaconic acid (12:83)

Kobberger, Robert, *Worthington Corporation,* Pumps and compressors (16:728)

Koeppen, R. C., *U.S. Department of Agriculture,* Wood (22:358)

Kolari, O. E., *American Meat Institute Foundation,* Meat and meat products (13:167)

Kolodny, Edwin R., *American Cyanamid Company,* Acrylamide (1:274)

Kooi, E. R., *Corn Products Company,* Dextrose and starch syrups (6:919)

Kopelman, Bernard, *Sylvania Electric Products, Inc.,* Wolfram and wolfram alloys (22:334)

Kouris, C. S., *National Aniline Division, Allied Chemical Corporation,* Aniline and its derivatives (2:111); Diarylamines, used as dye intermediate (Table 1 under Diarylamines) (7:40)

Kovach, George P., *Foster Grant Co., Inc.,* Abherents (1:1)

Kralovec, R. D., *E. I. du Pont de Nemours & Co., Inc.,* Cyclohexanol and cyclohexanone (6:683); Glycolic acid (10:632)

Krantz, Jr., John C., *University of Maryland,* Anesthetics (2:393); Antacids, gastric (427)

Krett, O. J., *National Dairy Products Corporation,* Milk and milk products (13:506)

Kriegel, W. W., *North Carolina State of The University of North Carolina,* Scope of ceramics (Ceramics) (4:759)

Krum, Jack K., *The R. T. French Co.,* Vanillin (21:180)

Kuhajek, Eugene, *Morton International, Inc.,* Sodium compounds; Sodium chloride (18:468)

Kulka, Marshall, *Uniroyal (1966) Ltd.,* Quinoline and isoquinoline (16:865)

Kyle, H. E., *Union Carbide Corporation,* Oxo process (14:373)

Landberge, M. J., *General Aniline & Film Corporation,* Paper coloring (Dyes-application and evaluation) (7:505)

Lane, John C., *Ethyl Corporation,* Gasoline and other motor fuels (10:463)

Langwill, Katheryn E., *Refined Syrups & Sugars, Inc.,* Confectionery (6:34)

Lanier, H., *Kennecott Copper Corp.,* Copper (6:131); (S:249)

Lannefeld, Theodore E., *Fabric Research Laboratories, Inc.,* Textile technology (Adhesives for carpet backing and apparel) (S:932)

Lanngguth, Robert P., *Monsanto Company, Inorganic Chemicals Division,* Enzyme detergents (S:294)

Lanson, H. J., *Lanson Chemicals Corporation,* Drying oils (7:398)

Lapkin, M., *Olin Mathieson Chemical Corporation,* Epoxides (8:263)

Larson, H. L., *Union Carbide Corporation, Carbon Products Division,* Metallurgical applications, Carbon (baked and graphitized, uses) (4:230)

Laskin, A. I., *Esso Research and Engineering Co.*, Proteins from petroleum (S:836)

Lassen, Harry G., *Ford Motor Co.*, Automobile exhaust control (S:50)

Laubengayer, A. W., *Cornell University*, Germanium, (10:519)

Laudise, R. A., *Bell Telephone Laboratories, Inc.*, Silica (Synthetic quartz crystals) (18:105)

Lavanchy, A. C., *The Sharples Company*, Centrifugal separation (4:710)

Lawley, Alan, *Drexel University*, Zone refining (22:680)

Leary, J. A., *Los Alamos Scientific Laboratory, University of California*, Plutonium (15:879)

Lederman, S. J., *Hooker Chemical Corporation*, Phenol (15:147)

Lee, G. Fred, *University of Wisconsin*, Eutrophication (S:315)

Lee, Jr., R. E., *General Electric Company*, Bearing materials, (3:271); Lubrication and lubricants (12:557)

Leeds, M. W., *Air Reduction Company*, Secondary and tertiary acetylenic and ethylenic alcohols and glycols (Alcohols, unsaturated) (1:618); Vinyl polymers (Poly-(vinyl alcohol)) (21:353)

Leekely, R. M., *The Institute of Paper Chemistry*, Paper (Pigment coating) (14:494)

Leidigh, W. Jared, *Bechtel Corporation*, Microwaves (S:563)

Lemke, Charles H., *E. I. du Pont de Nemours & Co., Inc.*, Sodium (18:432)

Le Monnier, Ernest, *Celanese Corporation of America*, Ethanoic acid (Acetic acid, acetic anhydride) (8:386)

Lenz, Robert W., *University of Massachusetts*, Polymerization mechanisms and processes (16:219)

Leonard, E. F., *Columbia University*, Dialysis (7:1)

Leonhardt, G. E., *National Institute of Drycleaning*, Drycleaning (7:307)

Lescisin, G. A., *Union Carbide Corporation, Chemicals Division*, Ethanol (8:422)

Lesh, James B., *Armour Pharmaceutical Company*, Thyroid and antithyroid preparations (Survey; Thyroid preparations; Thyrocalcitonin) (20:260)

Leston, Gerd, *Koppers Company*, Alkylphenols (S:27); Cresylic acids, synthetic (S:271)

Lewin, S. Z., *New York University*, Refraction (17:210)

Lewis, R. B., *Shell Oil Company*, Asphalt (Economic aspects) (2:762)

Leyes, Charles E., *Newark College of Engineering*, Esterification (8:313)

Lichtenstein, I., *The Lummus Company*, Ethylene (8:499)

Lichtenwalter, G. D., *Shell Chemical Company*, Chlorohydrins (5:304)

Liggett, L. M., *Speer Carbon Company*, Carbon (Baked and graphitized products, manufacture) (4:158)

Linden, Henry R., *Institute of Gas Technology*, Gas manufactured (10:353); Coal (Gasification) (S:198)

Lindner, Victor, *U.S. Army Picatinny Arsenal*, Explosives (Propellants) (8:659)

Lindsay, J. D., *Texas A & M University*, Distillation, (7:204)

Lindstrom, R. E., *Bureau of Mines, U.S. Dept. of the Interior*, Mercury, recovery by electrooxidation (Mercury) (S:542)

Lipson, M., *Division of Textile Industry, CSIRO, Geelong, Victoria, Australia*, Wool (22:387)

Liss, Raymond L., *Monsanto Company*, Enzyme detergents (S:294)

Liu, Benjamin Y. H., *University of Minnesota*, Dust (Engineering) (7:429)

Lockwood, Lewis B., *Miles Chemical Company,* Citric acid, (5:524); Malic acid (12: 837)

Loening, Kurt L., *Chemical Abstracts Service,* Hydrocarbons (Nomenclature) (11:288); (14:1)

Logsdon, R. F., *American Air Filter Co., Inc.,* Gas cleaning (10:329)

Loibl, Fred, *National Institute of Drycleaning,* Drycleaning (7:307)

Long, G. Gilbert, *North Carolina State of The University of North Carolina at Raleigh,* Antimony compounds (2:570); Arsenic compounds (718); Bismuth compounds (3:535)

Lord, E., *The Cotton Silk & Man-Made Fibres Research Association, Shirley Institute, Manchester,* Cotton (6:376)

Lotts, A. L., *Oak Ridge National Laboratory,* Nuclear reactors (Fuel element fabrication) (14:87)

Louderback, H. B., *E. I. du Pont de Nemours & Co., Inc.,* Cyclohexanol and cyclohexanone (6:683); Formic acid and derivatives (Formamide and dimethylformamide) (10:103)

Lowenheim, F. A., *M & T Chemicals, Inc.,* Electroplating (8:36)

Lurie, Arnold, P., *Eastman Kodak Company,* Acetic acid derivatives (1:138); (Acetamide) (142); (Acetanilide) (148); (Acetoacetic acid) (153); Benzidine and related diaminobiphenyls (3:408); Butyraldehyde (865); Butyric acid and butyric anhydride (878); Cyanohydrins (6:668); Ethane, (Ethanethiol (Ethyl mercaptan) (8:383); Halogenated derivatives (Ethanoic acid (Halogenated) (415); Ethers (470); Ketones (12:101)

Lushbough, Channing, H., *Mead Johnson & Company,* Foods, diet (10:62)

Luthy, Max, *The Givaudan Corporation,* Acetophenone (1:167)

Lyon, R. N., *Oak Ridge National Laboratory,* Nuclear reactors (14:74)

Lyons, John W., *Monsanto Company,* Rheology (17:423)

Macaluso, P., *Stauffer Chemical Co.,* Sulfur compounds (19:371)

MacGregor, W. S., *Crown Zellerbach Corporation,* Sulfoxides (19:320)

MacInnis, M., *Sylvania Electric Products, Inc.,* Wolfram compounds (22:346)

MacIntosh, R. M., *Tin Research Institute, Inc.,* Tin and tin alloys (20:273)

MacKenzie, G. D., *General Refractories Company,* Refractories (17:227)

MacMasters, M. M., *Kansas State University,* Wheat and other cereal grains (22:253)

MacWilliam, E. A., *Eastman Kodak Company,* Photography (15:355)

Madaus, John H., *MSA Research Corporation,* Potassium (16:361)

Magaha, Jr., Ernest P., *Biological Laboratories, United States Army, Fort Detrick,* Toxic chemicals (Introduction and defense); Summary (Chemical warfare) (4:869)

Magee, Richard J., *American Cyanamid Co.,* Veterinary drugs (21:241)

Mageli, Orville L., *Lucidol Division, Wallace & Tiernan, Inc.,* Peroxides and peroxy compounds, organic (Peroxides and peroxy compounds) (14:766)

Maher, P. K., *Davison Chemical Division, W. R. Grace & Co.,* Silica (Amorphous) (18:61)

Manley, T. C., *The Welsbach Corporation,* Ozone (14:410)

Mann, Roger H., *Shell Chemical Company,* Isoprene (12:64)

Margrave, J. L., *Rice University,* Selenium and selenium compounds (Properties, compounds) (17:809)

Marion, W. W., *Iowa State University,* Eggs (7:661)

Mark, Herman F., *Polytechnic Institute of Brooklyn,* Diene polymers (7:64); Research possibilities in the fully synthetic fibers (Fibers man-made) (9:164); Polymers (16:242)

Markland, W. R., *Chesebrough-Pond's Inc.,* Hair preparations (10:768)

Marks, Ernest M., *Kind & Knox Gelatin Co.,* Gelatin (10:499)

Marshall, Jr., W. R., *University of Wisconsin,* Drying (7:326)

Martell, Arthur E., *Illinois Institute of Technology,* Complexing agents (6:1)

Marth, E. H., *University of Wisconsin,* Milk and milk products (13:506)

Martin, A. R., *National Institute of Drycleaning,* Drycleaning (7:307)

Martin, Donald R., *The Harshaw Chemical Company,* Fluorine compounds, inorganic, boron trifluoride (9:554)

Martin, James W., *William Zinsser & Co.,* Shellac (18:21)

Martin, Jerome L., *Commercial Solvents Corporation,* Nitroparaffins (13:864)

Mast, W. C., *The Goodyear Tire & Rubber Company,* Rubber derivatives (17:646)

Masur, Charles, J., M.D., *Lederle Laboratories,* Cancer chemotherapy (S:81)

Matsuguma, Harold J., *Picatinny Arsenal, Department of the Army,* Nitrobenzene and nitrotoluenes (13:834); Nitrophenols (13:888)

Matsuyama, George, *Beckman Instruments, Inc.,* Indicators (11:548)

Mattson, V. L., *Kerr-McGee Corporation,* Uranium and uranium compounds (21:1)

Mausteller, J. W., *MSA Research Corporation,* Oxygen generation systems (S:658)

Mavrovic, Ivo, *Consulting Engineer,* Urea and urea derivatives (21:37)

Mawhinney, Matthew H., *Consulting Engineer,* Furnaces, fuel-fired (10:279)

McAdam, J. R., *Cyanamid of Canada, Ltd.,* Cyanamides (6:553)

McBee, E. T., *Purdue University,* Fluorine compounds, organic (9:686)

McBride, David L., *United States Steel Corporation,* Iron by direct reduction (S:535)

McCoy, J. B., *American Potash & Chemical Corporation,* Thorium and thorium compounds (20:248)

McDonald, Emma J., *U.S. Department of Agriculture,* Sugar (Sugar analysis) (19:155)

McGannon, Harold E., *United States Steel Corporation,* Steel (18:715)

McGinn, Charles E., *Allied Chemical Corporation,* Textile technology (Solvent dyeing) (S:973)

McGinnis, R. A., *Spreckels Sugar Co.,* Sugar (Properties of sucrose) (19:151); (Beet sugar) (19:203)

McGlashan, M. L., *The University, Exeter, U.K.,* Units (S:984)

McIlhenny, W. F., *The Dow Chemical Company,* Ocean raw materials, (14:150)

McKay, J. E., *The Bunker Hill Co.,* Lead (12:207)

McKee, R. C., *The Institute of Paper Chemistry,* Paper (Paper and paperboard containers) (14:494)

McKeon, J. T., *The M. W. Kellogg Company,* Piping systems (15:646)

McKetta, John J., *The University of Texas,* Dimensional analysis (7:176); Nomographs (14:15)

McKusick, B. C., *E. I. du Pont de Nemours & Co., Inc.,* Cyanocarbons (6:625)

McLoud, E. S., *Consultant,* Waxes (22:156)

McMahon, Jr., Charles J., *University of Pennsylvania,* Hardness (10:808)

McNeil, Donald, *Coal Tar Research Association, Gomersal, Leeds, England,* Cresols (6:434); Tar and pitch (19:653)

Meals, Robert, *General Electric Co.,* Silicon compounds (Silicones) (18:221)

Mecca, S. B., *Schuylkill Chemical Co.,* Uric acid (21:107)

Mehaffey, J., *Sherwin-Williams Laboratory, Western Reserve University,* Drying oils (7:398)

Mesiah, Raymond N., *FMC Corporation,* Triazinetriol (20:662)

Metcalf, Robert L., *University of California,* Insecticides (11:677); Poisons, economic (15:908)

Meth-Cohn, Otto, *University of Salford, Salford, England,* Thiophene (20:219)

Meyer, D. H., *Amoco Chemicals Corporation,* Phthalic acids and other benzenepoly-carboxylic acids (15:444)

Meyer, L. H., *E. I. du Pont de Nemours & Co., Inc.,* Tritium (Deuterium and tritium) (6:910)

Meyer, Robert, *Societe des Usines Chimiques Rhone-Poulenc,* Barbituric acid and barbiturates (3:60)

Miccioli, Bruno R., *The Carborundum Company,* Refractory coatings (17:267)

Michaels, Alan S., *Amicon Corporation,* Polyelectrolyte complexes (16:117)

Miller, F. M., *University of Maryland,* Indole (11:585)

Miller, H. C., *Super-Cut, Inc.,* Diamond, natural (Carbon) (4:283)

Miller, Phil H., *Union Carbide Corporation,* Glycols (Ethylene glycol, propylene glycol, and their derivatives) (10:638)

Miller, R. J., *California Research Corporation,* Acetone (1:159)

Miller, R. W., *Eastman Chemical Products, Inc.,* Alcohols, higher, synthetic (1:560)

Millet, M. A., *U.S. Department of Agriculture,* Wood (22:358)

Mills, G. Alexander, *Houdry Process and Chemical Company,* Catalysis (4:534)

Miltner, A. J., *E. I. du Pont de Nemours & Co., Inc.,* Glycolic acid (10:632)

Mitchell, Jr., John, *E. I. du Pont de Nemours & Co., Inc.,* Aquametry (2:673)

Mitchell, R. L., *Rayonier, Inc.,* Cellulose (4:593); Rayon (17:168)

Mjos, K., *Jefferson Chemical Company, Inc.,* Piperazine (15:638)

Modell, Walter, *Cornell University Medical College,* Cardiovascular agents (4:510)

Modic, F. J., *General Electric Company,* Embedding (8:102)

Moline, S. W., *Union Carbide Corporation,* Sorbic acid (18:589)

Montesinos, Mary Jane, *Tennessee Corporation,* Copper compounds (6:265)

Montgomery, P. D., *Monsanto Company,* Cyanides (Hydrogen cyanide) (6:574)

Mooney, R. B., *Imperial Chemical Industries, Ltd.,* Cyanides (Alkali metal cyanides) (6:585)

Moore, Robert E., *Sun Oil Company,* Adamantane (S:1)

Moore, W. E., *U.S. Department of Agriculture,* Wood (22:358)

Moores, R. G., *General Foods Corporation,* Coffee (5:748)

Morin, Robert, *Lilly Research Laboratories, Eli Lilly and Company,* Macrolide antibiotics (12:632)

Morley, S. G., *University of Nottingham, Wolfson Institute,* Carbon fibers (S:109)

Morral, F. R., *Cobalt Information Center, Battelle Memorial Institute,* Cobalt and cobalt alloys (5:716); Cobalt compounds (5:737)

Morrell, C. E., *Enjay Chemical Intermediates Laboratory,* Butanes (3:815); Butylenes (3:830); Propane (16:546)

Morrison, George O., *Technical Consultant,* Vinyl polymers (Poly(vinyl acetals)) (21:304)

Morrison, H. R., *Linde Company, Division of Union Carbide Corporation,* Cryogenics (6:471)

Morrow, G. S., *Shell Chemical Company,* Sulfolane (19:250)

Morse, Shirley K., *General Aniline & Film Corporation,* Aminophenols (2:213)

Morton, C. J., *Purdue University,* Fluorine compounds, organic (9:686)

Moser, Frank H., *Holland-Suco Color Company, Division of Chemetron Corp.,* Phthalocyanine compounds (15:488)

Moss, J., *General Aniline & Film Corporation,* Disperse dyes; Fiber blends (Dyes-application and evaluation) (7:505)

Mraz, Richard G., *Hercules Powder Company,* Alkyd resins (1:851)

Mrozik, Helmut, *Merck & Co., Inc.,* Parasitic infections, chemotherapy (14:532)

Muder, Richard E., *Koppers Company, Inc.,* Benzene from coal (Benzene) (3:367)

Muir, William M., *Harris Research Laboratories, Inc.,* Acids, dicarboxylic (1:240)

Mullin, C. R., *The Dow Chemical Company,* Photochemical technology (15:331)

Mullin, J. W., *University of London,* Crystallization (6:482)

Mullins, L. J., *Los Alamos Scientific Laboratory, University of California,* Plutonium (15:879)

Mulryan, H. T., *United Sierra Division Cyprus Mines Corporation,* Talc (19:608)

Murray, Royce W., *University of North Carolina,* Electroanalytical methods (7:726)

Muzyczko, T. M., *The Richardson Company,* Chemical cleaning (S:157)

Myerholtz, R. W., *Amoco Chemicals Corporation,* Olefin polymers (High-density (linear) polyethylene, in part) (14:242)

Myers, Jack, *University of Texas,* Algal cultures (1:649)

Naphtali, Leonard M., *Polytechnic Institute of Brooklyn,* Computers (6:25)

Nathan, Charles C., Corrosion inhibitors (6:317)

Neisel, R. H., *Johns-Manville Research & Engineering Center,* Insulation, thermal (11:823)

Nelson, Jr., L. A., *Freeport Sulphur Company,* Sulfur (19:364)

Nelson, W. L., *University of Tulsa,* Petroleum (refinery processes) (15:1)

Nesbitt, E. A., *Bell Telephone Laboratories,* Magnetic materials (12:737)

Neu, E. L., *Great Lakes Carbon Corporation,* Diatomite (7:53)

Neukom, Hans, *Swiss Federal Institute of Technology,* Pectic substances (14:636)

Neumark, H. R., *Allied Chemical Corporation,* Fluorine (9:506)

Neville, G. H. J., *British Titan Products Co., Ltd.,* Titanium compounds (inorganic) (20:380)

Neville, Roy G., *Bechtel Corporation,* Microwaves (S:563)

Newman, D. J., *Chemical Construction Corporation,* Nitric acid (13:796)

Newsom, H. C., *U.S. Borax Research Corporation,* Boron compounds (Boric acid esters, organic boron compounds) (3:652)

Nicholson, D. G., *East Tennessee State University,* Chlorine (5:1)

Niegowski, S. J., *The Welsbach Corporation,* Ozone (14:410)

Nielsen, N. Christian, *Central Methodist College,* Coordination compounds (6:122)

Nielsen, Ralph H., *Wah Chang Corporation,* Hafnium and hafnium compounds (10:754)

Nieneker, Darle L., *Jefferson Chemical Co., Inc.,* Morpholine (13:659)

Nies, Nelson, P., *U.S. Borax Research Corporation,* Boron oxides, boric acid, and borates (Boron compounds) (3:608)

Nieuwenhuis, H. K., *Chemical Projects Associates, Inc.,* Olefins (S:632)

Norris, F. A., *Swift & Company,* Fats and fatty oils (8:776)

Northcott, J., *National Aniline Division, Allied Chemical Corporation,* Aniline and its derivatives (2:411)

Nowak, R. M., *The Dow Chemical Company,* Unsaturated polyesters (20:791)

Nursten, Harry E., *Massachusetts Institute of Technology,* Azine dyes (2:859)

Oeda, Haruomi, *Ajinomoto Co., Inc.,* Monosodium glutamate (Amino acids) (2:198)

O'Flaherty, F., *University of Cincinnati,* Leather (12:303)

Olah, George A., *Western Reserve University,* Friedel-Crafts reactions (10:135)

Olin, A. D., *Toms River Chemical Corporation,* Dyes and dye intermediates (7:462); Dyes, reactive (630)

O'Neal, M. J., *Shell Oil Company,* Petroleum (Composition) (14:845)

O'Neil, F., *General Aniline & Film Corporation,* Leather dyeing (Dyes-application and evaluation) (7:505)

Oroshnik, William, *Chemo Dynamics,* Vitamins (Vitamin A) (21:490)

Orr, Jr., Clyde, *Georgia Institute of Technology, School of Chemical Engineering,* Size reduction (18:324)

Orsino, Joseph A., *National Lead Company,* Secondary cells (lead-acid) (Batteries and electric cells, secondary) (3:249)

Orton, Denis G., *Southern Dyestuff Company, a division of Martin Marietta Corporation,* Sulfur dyes (19:424)

Othmer, Donald F., *Polytechnic Institute of Brooklyn,* Azeotropy and azeotropic distillation (2:839); Data-interpretation and correlation (6:705); Thermodynamics (20:118); Water (Water supply and desalination) (22:1); Pipeline heating (S:694)

Pacifico, Carl, *Management Supplements, Inc.,* Marketing and marketing research (13:66)

Packowski, George W., *Joseph E. Seagram & Sons, Inc.,* Alcoholic beverages (distilled) (1:501)

Palmour III, Hayne, *North Carolina State of The University of North Carolina,* Ceramics (Properties of ceramics) (4:793)

Pan, L. C., *Chemical Construction Corporation,* Sodium compounds (Sodium nitrate; Sodium nitrite) (18:486)

Papée, D., *Cie de Produits Chimiques et Electrometallurgiques, Péchiney,* Aluminum compounds (Aluminum oxide (alumina)) (2:41)

Papin, Roger, *Carbonisation et Charbons Actifs,* Bentonite (3:339)

Papkoff, Harold, *The Hormone Research Laboratory, University of California,* Hormones, (Anterior-Pituitary hormones) (11:52)

Parché, M. Constance, *The Carborundum Company,* Carbides (Silicon carbide) (4:114)

Pardey, W. N., *General Aniline & Film Corporation,* Dyes-application and evaluation (Evaluation of dyes-colorfastness tests) (7:505)

Parker, Albert, *Consultant Fuel Technologist,* Fuels (Survey) (10:179)

Parmerter, Stanley M., *Corn Products Company,* Starch (18:672)

Parrish, W., *International Business Machines Corporation,* X-Ray analysis (Fluorescence spectrography) (22:456)

Paschkis, Victor, *Columbia University,* Furnaces, electric (10:252)

Patai, Saul, *The Hebrew University of Jerusalem,* Aromaticity (2:701)

Patel, Ghanshyam R., *Pfister Chemical, Inc.,* Oxalic acid (14:356)

Pattison, V. A., *Hooker Chemical Corp.,* Halogenated fire retardants (S:467)

Peacock, William H., *American Cyanamid Company,* Stains, industrial (18:654)

Pearlson, W. H., *Minneosta Mining & Manufacturing Co.,* Fluorine compounds, organic (Fluoro ethers, and amines) (9:753); Fluorine compounds, organic (Other perfluorocarboxylic acids) (773)

xliv CONTRIBUTORS TO THE ENCYCLOPEDIA

Peckham, Jr., G. T., *Clinton Corn Processing Co.,* Lactic acid (12:170)

Pentz, Jr., C. A., *Union Carbide Corporation, Chemicals Division,* Ethanol (8:422)

Perlman, D., *University of Wisconsin,* Streptomycin and related antibiotics (19:33)

Perrin, John, *The University of Wisconsin,* Pharmaceuticals (15:112)

Petrarca, Anthony E., *The Ohio State University,* Information retrieval services and methods, (S:510)

Pfafflin, James R., *University of Windsor, Ontario, Canada,* Water (Sewage; Water reuse) (22:104)

Pichler, Helmut, *Carl-Engler-und Hans-Bunte-Institut fur Mineralol-und Kohleforschung der Technischen Hochschule Karlsruhe,* Carbon monoxide-hydrogen reactions (4:466)

Pidgeon, L. M., *University of Toronto,* Barium (3:77); Strontium (19:48)

Pigott, K. A., *Mobay Chemical Company,* Urethan polymers (21:56)

Pilorz, B. H., *Shell Chemical Company,* Allyl chloride (Chlorocarbons and chlorohydrocarbons) (5:205)

Platt, Alan E., *The Dow Chemical Co.,* Styrene plastics (19:85)

Plimmer, J. R., *U.S. Department of Agriculture,* Soil chemistry of pesticides (18:515); Weed killers (22:174)

Plueddemann, Edwin P., *Dow Corning Corporation,* Silicon compounds (Silylating agents) (18:260)

Poffenberger, Noland, *The Dow Chemical Company,* Diphenyl and terphenyls (Diphenyl) (7:191); Phenol (Hydrolysis of monochlorobenzene toluene oxidation) (15:147)

Pollak, Peter I., *Merck & Co., Inc.,* Thiamine (20:173)

Pollock, Lyle, *Phillips Petroleum Company,* Liquefied petroleum gas (12:470)

Pomeranz, Y., *U.S. Dept. of Agriculture,* Wheat and other cereal grains (22:253)

Port, John H., *Cleveland Refractory Metals Division, Chase Brass & Copper Co., Inc.,* Rhenium (17:411)

Porter, C. A., *American Mineral Spirits Company, A Division of Union Oil Company of California,* Hexanes (11:1)

Poster, Arnold, R., *Metals Sintering Corp.,* Powder metallurgy (16:401)

Posteraro, Anthony F., *College of Dentistry, New York University,* Dentrifrices (6:848)

Powell, B. D., *Cadbury Brothers, Limited,* Chocolate and cocoa (5:363)

Powell, R. S., *Allied Chemical Corporation,* Thiazole dyes (Direct and vat dyes containing the thiazole nucleus) (20:185)

Powers, Paul O., *Amoco Chemicals Corporation,* Hydrocarbon resins (11:242)

Powers, R. A., *Union Carbide Corporation, Consumer Products Division,* Corrosion (6:289)

Pregel, Boris, *Canadian Radium & Uranium Division Canrad Precision Industries, Inc.,* Radioactive elements, natural (17:9)

Preisman, L., *Pittsburgh Plate Glass Company,* Barium compounds (3:80); Strontium compounds (19:50)

Preston, J., *Chemstrand Research Center, Inc., Monsanto Company,* Polyimides (S:746)

Preston, R. W. G., *Imperial Chemical Industries Limited,* Alkylphenols (1:901)

Proctor, J. F., *E. I. du Pont de Nemours & Co., Inc.,* Deuterium and tritium (6:895)

Pryde, E. H., *U.S. Department of Agriculture,* Fatty acids (Industrially important reactions; Economics aspects) (8:811)

Pupp, C., *Ozark-Mahoning Company,* Fluorine compounds, inorganic (Phosphorus fluorides) (9:635)

Quin, J. P., *Imperial Chemical Industries, Ltd.* Alkali metal cyanides (Cyanides) (6: 585)

Quinn, John F., *Monsanto Chemical Company,* Benzene-sulfonic acid (3:401)

Raff, R., *Washington State University,* Hydroquinone, resorcinol, pyrocatechol (11:462)

Rairdon, Charles T., *Johns-Manville Products Corporation,* Calking and Sealing compositions (4:28)

Raizman, Paula, *Monsanto Company,* Courmarin (6:425)

Ramadanoff, D., *Union Carbide Corporation, Products Division,* Baked and graphitized products, uses (Electrical applications) (Carbon) (4:222)

Raphaelian, L. A., *Olin Mathieson Chemical Corporation,* Hydrazine and its derivatives (Hydrazine) (11:164)

Rau, Eric, *FMC Corporation,* Sodium compounds (Carbonates) (18:458)

Reck, Richard A., *Armour Industrial Chemical Company,* Quaternary ammonium compounds (16:859)

Redfern, J. P., *Stanton Instruments Limited,* Thermal analysis (20:99)

Reed, H. W. B., *Imperial Chemical Industries Limited,* Alkylphenols (1:901)

Reed, Robert M., *Girdler Corporation, Subsidiary of The Chemical & Industrial Corporation,* Carbon Dioxide (4:353); Hydrogen (11:338)

Reeves, H. E., *National Institute of Drycleaning,* Drycleaning (7:307)

Reichle, Walter T., *Polymer Research and Development Union Carbide Corporation,* Hydrazoic acid and azides (11:197)

Reid, J. David, *U.S. Dept. of Agriculture,* Waterproofing and water repellency (22:135)

Reiger, M. M., *Warner Lambert Company,* Thioglycolic acid (20:198)

Remde, H. F., *Johns-Manville Research & Engineering Center,* Insulation, thermal (11:823)

Repka, Jr., Benjamin C., *Hercules Incorporated,* Olefin polymers (Polypropylene) (14:282)

Retallick, W. B., *Houdry Process and Chemical Co.,* Hydroprocesses (11:418)

Revilock, J. F., *Union Carbide Corporation, Carbon Products Division,* Baked and graphitized products, uses (Chemical applications) (Carbon) (4:216)

Reyerson, L. H., *The American Institute of Chemists,* Linen (12:461)

Rheineck, A. E., *North Dakota State University of Agriculture and Applied Science,* Drying oils (7:398)

Rhum, David, *Air Reduction Co.,* Vinyl polymers (Poly (vinyl acetate)) (21:317)

Ricci, J. E., *New York University,* Phase rule (15:133)

Rice, J. K., *Cyrus Wm. Rice and Company, NUS Corp.,* Steam (18:692)

Richardson, R. J., *Allied Chemical Corporation,* Polyamides (Fibers) (16:46)

Ricksecker, R. E., *Chase Brass & Copper Company,* Copper alloys (6:181)

Ries, P. D., *Union Carbide Corporation, Carbon Products Division,* Baked and graphitized products, uses (Arc carbon applications (Carbon) (4:206)

Riesser, G. H., *Shell Chemical Company,* Chlorohydrins (5:304)

Riley, E. C., *Eastman Kodak Company,* Dust (hygiene) (7:453)

Ringk, William F., *Benzol Products Division, Cowles Chemical Company,* Benzyl alcohol (Benzyl alcohol and β-phenylethyl alcohol) (3:442); Cinnamic acid, cinnamaldehyde and alcohol (5:517); Phenylacetic acid (15:213)

Rinkenbach, Wm., H., *Consulting Chemist,* Explosives (8:581)

Roach, John D., *National Lead Company,* Zirconium (Zirconium hydrides) (22:625)

Roberts, J. R., *United States Gypsum Co.,* Wallboard (21:601)

Roberts, L. M., *Research-Cottrell, Inc.,* Electrostatic precipitation (8:75)

Robson, H. L., *Olin Mathieson Chemical Corporation,* Chlorine monoxide, hypochlorous acid, hypochlorites; (Chlorine oxygen, acids and salts) chlorous acid, chlorites, and chlorine dioxide (Chlorine oxygen acid and salts) (5:7)

Robson, H. L., *Olin Mathieson Chemical Corporation,* Bleaching agents (3:550); Chloramines and chloroamines (4:908); Chlorous acid, chlorites, and chlorine dioxide (Chlorine oxygen acids and salts) (5:27)

Rochow, E. G., *Harvard University,* Elements-periodic system (8:93)

Rochow, T. G., *Central Research Division, American Cyanamid Company,* Microscopy, chemical (13:491)

Rockwell, F. H., *Diamond Alkali Company,* Alkali and chlorine industries (1:668)

Roebuck, A. H., *Kaiser Aluminum & Chemical Corporation,* Corrosion (6:289)

Rogers, T. H., *The Goodyear Tire & Rubber Company,* Rubber, natural (17:660)

Rollet, Michel, *Société des Usines Chimiques Rhone-Poulenc,* Barbituric acid and barbiturates, (3:60)

Rollinson, C. L., *University of Maryland,* Aluminum compounds (2:1); Calcium compounds (4:1)

Rosato, D. V., *Consultant, Plastics World,* Laminated and reinforced plastics (12:188)

Rose, J. B., *Imperial Chemical Industries, Ltd.,* Polymers of higher olefins (S:773)

Rosenwald, R. H., *Universal Oil Products Company,* Alkylation (1:882)

Ross, R. D., *Thermal Research & Engineering Corp.,* Wastes-industrial (21:625)

Ross, Sydney, *Rensselaer Polytechnic Institute,* Adsorption, theoretical (1:421)

Rothemund, P., *Consulting Chemical Engineer,* Chlorophyll (5:339)

Rouse, Jr., Ben P., *Tennessee Eastman Company,* Cellulose derivatives-plastics (4:653)

Rubens, L. C., *The Dow Chemical Company,* Unsaturated polyesters (20:791)

Rugger, George, *Dept. of the Army,* Ceramic composite armor (Fabrication and plastics) (S:138)

Runyan, Walter R., *Texas Instruments, Inc.,* Silicon and silicides (Pure silicon) (18:111)

Rushton, J. H., *Purdue University,* Mixing and blending (13:577)

Russell, Charles C., *Koppers Company, Inc.,* Carbonization (4:400)

Russell, Jr., G. A., *General Refractories Company,* Refractories (17:227)

Ryer, F. V., *Lever Brothers Co., Inc.,* Soap (18:415)

Ryland, Ada L., *E. I. du Pont de Nemours & Co., Inc.,* Mass spectrometry (13:87)

Ryland, Charles, *Coors Porcelain Company,* Ceramics, chemical ware (4:832)

Ryznar, John W., *Nalco Chemical Company,* Aluminates (Aluminum compounds) (2:6)

Saeman, Walter C., *Olin Mathieson Chemical Corp.,* Aluminum sulfate and alums (Aluminum compounds) (2:58)

Sage, Maurice S., *Sage Laboratories Inc.,* Aerosols (1:470)

Saltman, William M., *The Goodyear Tire & Rubber Company,* Elastomers, synthetic (7:676)

Santmyers, Donald, *E. I. du Pont de Nemours & Co., Inc.,* Sulfamic acid and sulfamates (19:242)

Sartoretto, Paul, *W. A. Cleary Corporation,* Lecithin (12:343)

Saunders, J. H., *Mobay Chemical Company,* Fluorine compounds, organic (Fluorinated carboxylic acid, in part) (9:767)

Schaefer, F. C., *American Cyanamid Company,* Cyanamides (6:553)

Scheibel, E. G., *Cooper Union School of Engineering and Science,* Extraction (8:719)

Scheibel, E. G., *The Cooper Union for the Advancement of Science and Art,* Extraction (8:761)

Schiffman, Louis F., *Amchem Products, Inc.,* Metal surface treatments (Chemical and electrochemical conversion treatments) (13:292)

Schildknecht, C. E., *Gettysburg College,* Vinyl polymers (Vinyl ethers, monomers and polymers) (21:412)

Schlechten, A. W., *Colorado School of Mines,* Zinc and zinc alloys (22:555)

Schnell, H., *Farbenfabriken Bayer AG,* Polycarbonates (16:106)

Schoengold, Morris D., *Esso Research and Engineering Co.,* Literature (sources) (12:500)

Schonhorn, Harold, *Bell Telephone Laboratories,* Adhesion (S:16)

Schooley, Arthur T., *The B. F. Goodrich Co.,* Microplants (S:557)

Schroeder, William F., *Archer Daniels Midland Company,* Shortenings and other food fats (Shortenings) (18:33)

Schule, Elmer C., *Allied Chemical Coporation,* Polyamides (Plastics) (16:88)

Schultz, M. L., *Radio Corporation of America,* Phototubes and photocells (15:396)

Schultz, Peter C., *Corning Glass Works,* Silica (Vitreous) (18:73)

Schultze, Henry C., *Union Carbide Corporation,* Ethylene oxide (8:523)

Schumacher, Joseph C., *American Potash & Chemical Corporation,* Perchloric acid and perchlorates (Chlorine oxygen acids and salts) (5:61)

Schwander, H. R., *J. R. Geigy AG Basel,* Stilbene derivatives (Stilbene dyes) (19:1)

Schwartz, A. M., *Harris Research Laboratories, Inc.,* Detergency (6:853)

Schwarzkop, Alex J., *National Aeronautics and Space Administration,* Metal surface treatments, case hardening (13:304)

Schweitzer, Carl E., *E. I. du Pont de Nemours & Co., Inc.,* Acetal Resins (1:95)

Schwenzfeier, Jr., C. W., *The Brush Beryllium Company,* Beryllium and beryllium alloys (3:450); Beryllium compounds (3:474)

Scobie, A. G., *Shawinigan Chemicals Limited,* Calcium carbide (Carbides) (4:100)

Sconce, J. S., *Hooker Chemical Corporation,* Sodium compounds, (Sodium sulfides) (18:510)

Scott, A. H., *National Bureau of Standards,* Electrical testing (7:716)

Scott, Kenneth A., *Sun Oil Company,* Naphthalenecarboxylic acids (13:690)

Seaborg, Glenn T., *U.S. Atomic Energy Commission,* Actinides (1:351)

Seamster, A. H., *The Dow Chemical Company,* Ion exchange (11:871)

Sears, J. K., *Monsanto Company,* Plasticizers (15:720)

Seeley, Sherwood B., *The Joseph Dixon Crucible Company,* Natural graphite (Carbon) (4:304)

Seglin, L., *FMC Corporation,* Coal (To Synthetic crude oil) (S:177)

Selwood, P. W., *University of California, Santa Barbara,* Magnetic properties (12:772)

Seymour, Raymond B., *University of Houston,* Plastics technology (15:790)

Shacter, J., *Union Carbide Corporation,* Diffusion (7:79); Diffusion separation methods, (7:91)

Shanley, Edward S., *Arthur D. Little, Inc.,* Peroxides and peroxy compounds (Inorganic) (14:746)

Shapiro, H., *Ethyl Corporation,* Lead compounds, organic (Organolead compounds) (12:282)

Shapiro, S. H., *Armour Industrial Chemical Co.,* Amine oxides (S:32)

Shaver, Forrest W., *The B. F. Goodrich Company,* Rubber chemicals (17:509)

Shea, J. W., *Union Carbide Corporation Carbon Products Division,* Baked and graphitized products, uses (Electrode applications) (Carbon) (4:227)

Shedlovsky, Leo, *Colgate-Palmolive Company,* Foams (9:884)

Sheers, E. H., *American Cyanamid Company*, Guanidine and guanidine salts (10:734)

Shell, G., *Envirotech Corp.*, Pollution (S:711)

Shell, H. R., *U.S. Dept. of the Interior, Bureau of Mines*, Micas, natural and synthetics (13:398)

Sheppard, Chester S., *Lucidol Division, Wallace & Tiernan, Inc.*, Peroxides and peroxy compounds (Organic) (14:766)

Sherratt, S., *Imperial Chemical Industries, Ltd.*, Fluorine compounds, organic, Polytetrafluoroethylene (9:805)

Shiftan, Ernest, *International Flavors & Fragrances, Inc.*, Perfumes (14:717)

Shimotake, H., *Argonne National Laboratory*, Cells, high temperature (S:120)

Shor, A. Louis, *American Cyanamid Co.*, Veterinary drugs (21:241)

Shreve, R. Norris, *Purdue University*, Amination by reduction (2:76)

Sidi, H., *Heyden Newport Chemical Corporation*, Benzyl chloride, benzal chloride, and benzotrichloride (Chlorocarbons and chlorohydrocarbons) (5:281)

Siegmund, J. M., *Allied Chemical Corporation*, Fluorine (9:506)

Sievenpiper, Frederic L., *Allied Chemical Corporation*, Textile technology (Solvent dyeing) (S:973)

Silberberg, I. H., *Texas Petroleum Research Committee*, Dimensional analysis (7:176)

Silver, Raymond P., *Hercules Powder Company*, Alkyd resins (1:851)

Silvernail, Walter L., *American Potash & Chemical Corporation*, Cerium and cerium compounds (4:840); Rare earth elements (17:143); Thorium and thorium compounds (20:248)

Simeral, J. E., *Union Carbide Corporation*, Sorbic acid (18:589)

Simms, E. G., *Ohio Ferro-Alloys Corporation*, Silicon and Silicides (Metallurgical silicon and silicides) (18:125)

Simon II, D. E., *NUS Corporation div., Cyrus Wm. Rice and Company*, Water (industrial) (22:65)

Singer, J. J., *Holland-Suco Color Company, Division of Chemetron Corporation*, Pigments (Dispersed pigment concentrates) (15:589)

Singley, J. E., *University of Florida*, Copper compounds (6:265); Water (Munciple water treatment) (22:82)

Sjöstrom, Loren B., *Arthur D. Little, Inc.*, Organoleptic testing (14:336)

Skaperdas, George T., *The M. W. Kellogg Company, Division of Pullman Incorporated*, Heat transfer (10:819)

Skau, Evald L., *U.S. Dept. of Agriculture*, Melting and freezing temperatures (13:198)

Skelland, A. H. P., *University of Notre Dame*, Mass transfer (13:99)

Skochdopole, R. E., *The Dow Chemical Company*, Foamed plastics (9:847)

Skolnik, Leonard, *The B. F. Goodrich Company*, Tire cords (20:328)

Skougstad, Marvin W., *U. S. Geological Survey*, Water (Analysis) (21:688)

Slack, A. V., *Tennessee Valley Authority*, Fertilizers (9:25; S:338)

Smith, Jr., A. A., *American Smelting and Refining Co.*, Indium and indium compounds (11:581)

Smith, Eric, *Olin Mathieson Chemical Corporation*, Hydantoin (11:141)

Smith, James S., *Sylvania Electric Products, Inc.*, Wolfram and wolfram alloys (22:334)

Smith, Richard F., *GAF Corporation*, Pyrrole and pyrrole derivatives (16:841)

Smith, Samuel, *Minnesota Mining and Manufacturing Company*, Soil release finishes (Textile technology) (S:964)

Smith, W. R., *Cabot Corporation*, Carbon black (Carbon) (4:243)

Snider, O. E., *Allied Chemical Corporation*, Polyamides (Fibers) (16:46)

Solotorovsky, Morris, *Rutgers University*, Bacterial, rickettsial, and mycotic infections, chemotherapy (3:1)

Souders, Mott, *Chemical Engineer*, Fluid mechanics (9:445)

Specht, E. H., *Rohm & Haas Company*, Acrylic Acid and derivatives (1:285)

Spring, S., *Oxford Chemical Division, Consolidated Foods Corp.*, Metal surface treatments (Cleaning, pickling, and related processes) (13:284)

Springer, D. B., *The International Nickel Company, Inc.*, Nickel compounds (13:753)

Stamberger, P., *Technical Consultant*, Electrodecantation, (7:841)

Standiford, Ferris C., *W. L. Badger Associates, Inc.*, Evaporation (8:559)

Standish, W. L., *E. I. du Pont de Nemours & Co., Inc.*, Adipic acid (1:405)

Stanley, R. H., *Titanium Intermediates Ltd.*, Titanium compounds (Organic) (20:424); Nitrogen fixation (S:604)

Stansbury, Jr., H. A., *Union Carbide Chemicals Company*, Acrolein and derivatives (1:255)

Stearns, Richard S., *Sun Oil Company*, Viscometry (21:460)

Stefanucci, A., *General Foods Corporation*, Coffee (5:748)

Steigman, Joseph, *Polytechnic Institute of Brooklyn*, Acid Anhydrides (Acid-base systems) (1:211)

Stenger, V. A., *The Dow Chemical Company*, Bromine (3:750); Bromine compounds (3:766); Sodium compounds (Sodium bromides) (18:484)

Steunenberg, R. K., *Argonne National Laboratory*, Cells, high temperature (S:120)

Stevens, Chapin E., *Management Research Consultants*, Surfactants (19:507)

Stevens, F. R., *Ames Laboratory of the U.S. Atomic Energy Commission*, Vanadium and vanadium alloys (21:157)

Stewart, Jr., A. Theodore, *Bechtel Corporation*, Microwaves (S:563)

Stewart, Albert E., *Mallinckrodt Chemical Works*, Aluminum acetate (Aluminum compounds) (2:11)

Stewart, George H., *Gonzaga University*, Chromatography (5:413)

Stewart, G. T., *The Dow Chemical Company*, Packaging and packages (14:432)

Stewart, Robert D., *American Potash & Chemical Corporation*, Perchloric acid and perchlorates (Chlorine oxygen acids and salts) (5:61)

Stimpson, E. G., *National Dairy Products Corporation*, Milk and milk products (13:506)

Stokes, C. A., *Consultant*, Carbon black (S:91); Research management (855)

Stoll, Max, *Firmenich & Cie.*, Oils, essential (14:178)

Stonecipher, W. D., *Hercules Incorporated*, Turpentine (20:748)

Stonehouse, A. James, *The Brush Beryllium Company*, Beryllides (S:73)

Stoops, Robert F., *North Carolina State of The University of North Carolina*, Ceramic forming processes (Ceramics) (4:776)

Straschil, H. K., *Engelhard Industries, a division of Engelhard Minerals & Chemicals Corp.*, Platinum group metals, compounds (15:861)

Strelzoff, Samuel, *Chemical Construction Corporation*, Nitric acid (13:796); Pressure vessels (16:482)

Strong, Laurence E., *CBA Project, Earlham College*, Blood fractionation (3:576)

Stuart, David M., *Neisler Laboratories, Inc.,* Coagulants and anticoagulants (5:586)

Stubbings, R. L., *Institute of Leather Technology,* Leather (12:303)

Sullivan, Edward A., *Ventron Corporation,* Hydrides (S:499)

Sullivan, R. A., *National Dairy Products Corporation,* Milk and milk products (13:506)

Sumpter, W. C., *Western Kentucky State College,* Indole (11:585)

Surprenant, Donald T., *Union Carbide Corporation,* Film materials (9:220)

Sutter, R. C., *Diamond Alkali Company,* Hydrochloric acid (11:307)

Sutton, G. P., *Envirotech Corp.,* Pollution (S:711)

Svoboda, Gordon H., *Eli Lilly and Company,* Alkaloids, history, preparation, and use (1:778)

Sweeny, W., *E. I. du Pont de Nemours & Co., Inc.,* Polyamides (General) (16:1)

Swiss, Jack, *Westinghouse Electric Corporation,* Insulation, electric (11:774)

Sylvan, S., *American Air Filter Co., Inc.,* Gas cleaning (10:329)

Tarkow, Harold, *U.S. Department of Agriculture,* Wood (22:358)

Tate, Dan C., *U.S. Plywood Champion Papers, Inc.,* Tall oil (19:614)

Tate, R. W., *Delavan Manufacturing Company,* Sprays (18:634)

Taub, David, *Merck & Co.,* Steroids (18:830)

Taylor, Alfred H., *Airco Industrial Gases Division, Air Reduction Company, Inc.,* Oxygen (14:390)

Taylor, Donald F., *Fansteel Inc.,* Tantalum and tantalum compounds (19:630)

Taylor, W. I., *Ciba Pharmaceutical Company,* Alkaloids, survey (1:758)

Tertian, R., *Cie de Produits Chimiques et Electrometallurgiques, Pechiney,* Aluminum oxide (Alumina) (Aluminum compounds) (2:41)

Terwilliger, Jr., R. C., *Carrier Air Conditioning Company, a division of Carrier Corporation,* Air conditioning (1:481)

Tesi, A. F., *Celanese Fibers Company, Division of Celanese Corporation of America,* Acetate and triacetate fibers (1:109)

Tesoro, Giuliana, C., *J. P. Stevens & Company,* Antistatic agents (2:649)

Theimer, E. T., *International Flavors & Fragrances, Inc.,* β-Phenylethyl alcohol (Benzyl alcohol and β-phenylethyl alcohol) (3:442)

Thiessen, Gilbert, *Koppers Company, Inc.,* Naphthalene (13:670)

Thirtle, J. R., *Eastman Kodak Company,* Color photography (5:812); Phenylenediamines and toluenediamines (15:216); Quinones (16:899)

Thomas, I. M., *Anderson Chemical Division, Stauffer Chemical Company,* Alkoxides, metal (1:832)

Thomas, Jr., A. D., *The University of Texas,* Mechanical testing (13:184)

Thompson, A. Paul, *Eagle-Picher Industries, Inc.,* Lead compounds (Inorganic compounds) (12:266); Health and safety factors (Lead compounds) (Toxicity) (12:299); Zinc and zinc alloys (22:555); Zinc compounds (22:604)

Thompson, Arthur C., *Nalco Chemical Company,* Aluminum compounds (Aluminates) (2:6)

Thompson, D. C., *E. I. du Pont de Nemours & Company, Inc.,* Elastomers, synthetic (Neoprene) (7:705); Vinylidene polymers (Fluoride) (21:269)

Tischer, Thomas N., *Eastman Kodak Company,* Silver compounds (18:295)

Tomezsko, Edward S. J., *The Pennsylvania State University,* Calorimetry (4:35)

Tomlinson, II, G. H., *Domtar Limited,* Pulp (16:680)

Torgeson, D. C., *Boyce Thompson Institute for Plant Research, Inc.,* Fungicides (Agricultural fungicides) (10:220)

Toubes, Benjamin, *Victor Chemical Works, a division of Stauffer Chemical Company,* Aluminum compounds (Aluminum formate) (2:14)

Towers, Richard B., *Link-Belt Company,* Conveying (6:92)

Towle, P. H., *Amoco Chemicals Corporation,* Phthalic acids and other benzenepoly-carboxylic acids (15:444)

Townsend, H. N., *Union Carbide Corporation, Carbon Products Division,* Baked and graphitized products, uses (Aerospace applications) (Carbon) (4:202)

Tracey, Michael V., *CSIRO (Commonwealth Scientific and Industrial Research Organization) Division of Food Preservation, Australia,* Proteins (16:610)

Trefener, W. S., *General Refractories Company,* Refractories (17:227)

Treibl, Hans G., *Standard Naphthalene Products, Co., Inc.,* Naphthalene derivatives (13:697)

Tso, T. C., *Agricultural Research Service, Beltsville, Maryland, U.S. Department of Agriculture,* Tobacco (20:503)

Tsu, K., *American Cyanamid Company,* Automobile exhaust control (2:814)

Tucker, James W., *Victor Chemical Works, a division of Stauffer Chemical Company,* Chemical leavening (Bakery processes and leavening agents) (3:41)

Tuddenham, W. M., *Kennecott Copper Corp.,* Copper (S:249)

Tuemmler, William B., *FMC Corporation,* Carbonic esters and chloroformic esters, (4:386)

Tumerman, L., *National Dairy Products Corporation,* Milk and milk products (13:506)

Turi, Paul, *Sandoz, Inc.,* Insulin (11:838); Hypnotics, sedatives, anticonvulsants (11:508); Succinic acid (19:134)

Turk, Amos, *The City College of the City University of New York,* Odor control (14:170)

Turk, Stanley D., *Phillips Petroleum Co.,* Thiols (20:205)

Turner, N. Joe, *Boyce Thompson Institute for Plant Research, Inc.,* Fungicides (Industrial fungicides) (10:228)

Turner, Errett S., *Bell Telephone Laboratories, Inc.,* Patent (Literature) (14:583)

Tuttle, John B., *Standard Oil Company (New Jersey)* Candles (4:58); Petroleum waxes (15:92)

Twiggs, H. C., *Eastmont Chemical Products, Inc.,* Glycols (other glycols) (10:676)

Ullyot, Glenn E., *Smith, Kline & French Laboratories,* Psychopharmacological agents (16:640)

Unger, W. E., *Oak Ridge National Laboratory,* Nuclear reactors (Special engineering for radiochemical plants) (14:115)

Vachet, P., *Cie de Produits Chimiques et Electrometallurgiques, Pechiney,* Aluminum and aluminum alloys (1:929)

Vahlteich, H. W., *Consultant,* Margarine (13:56)

Vahrman, Mark, *The City University, London,* Lignite and brown coal (12:381)

Valko, Emery I., *Lowell Technological Institute,* Antistatic agents (2:649)

van den Dool, H., *International Flavors & Fragrances (Nederland), N. V.,* Benzophenone (3:439)

Van Dyke, Charles H., *Carnegie-Mellon University,* Silicon compounds (Introduction) (18:133); Silanes (18:172)

Van Horn, W. M., *The Institute of Paper Chemistry,* Paper (Biological and industrial waste problems) (14:494)

Van Lare, E. J., *Eastman Kodak Company,* Color and constitution of organic dyes (5:763); Cyanine dyes (6:605)

Van Wagenen, H. D., *The Procter & Gamble Company,* Alcohols, higher, fatty (1:542)

Van Wazer, John R., *Monsanto Company,* Phosphoric acids and phosphates; Phosphorous and the phosphides; Phosphorous compounds (15:232)

Veldhuis, M. K., *U.S. Fruit and Vegetable Products Laboratory, U.S. Department of Agriculture, Winter Haven, Florida,* Fruit juices (10:167)

Vermeulen, Theodore, *University of California,* Adsorption, industrial (1:459)

Vigneron, Maurice, *Société de Chimie Organique et Biologique A.E.C. (formerly l'Alimentation Equilibrée Commentry)* Amino acids, survey (2:156)

Voedisch, Robert W., *Lawter Chemicals, Inc.,* Fluorescent pigments (Daylight) (9:483)

Vogel, T., *General Aniline & Film Corporation,* Basic dyes, fiber-reactive dyes (Dyes-application and evaluation) (7:505)

von Fischer, William, *Consultant,* Coatings, industrial (5:690)

Von Halle, E., *Union Carbide Corporation,* Diffusion (7:79); Diffusion separation methods (7:91)

von Rosenberg, Annelle E., *Esso Research and Engineering Co.,* Radioisotopes (17:64)

Wagner, Arthur F., *Merck Sharp & Dohme Research Laboratories, Merck & Co., Inc.,* Vitamins (Vitamin K) (21:585)

Wagner, Frank A., *B. F. Goodrich Chemical Company,* Resins, water-soluble (17:391)

Wagner, Jr., Frank S., *Celanese Chemical Company,* Glycols (1,3-Butylene glycol) (10:660)

Walker, A. B., *Research-Cottrell, Inc.,* Electrostatic precipitation (8:75)

Walker, J., Frederic, *Consultant,* Formaldehyde (10:77); Formic acid and derivatives (Formic acid) (10:99)

Wallace, John G., *E. I. du Pont de Nemours & Co., Inc.,* Epoxidation (8:238)

Walsh, D., *New York University,* Methylacetylene (S:547)

Walton, B. C., *Chemical Construction Corporation,* Pressure vessels (16:482)

Wamser, Christian A., *Allied Chemical Corp.,* Thiosulfates (20:227)

Ward, D. J., *Universal Oil Products Company,* Cumene (6:543)

Washburn, T. N., *Oak Ridge National Laboratory,* Nuclear reactors, (Fuel element fabrication) (14:87)

Waugh, T. D., *Arapahoe Chemicals, Division of Syntex Corp.,* Grignard reaction (10:721)

Way, Stuart, *Westinghouse Electric Corp.,* Coal (Power from coal by gasification and magnetohydrodynamics) (S:217)

Weidlen, Jr., E. R., *Union Carbide Corporation,* Inositol (11:673)

Weil, B. H., *Essor Research and Engineering Co.,* Literature (Documentation) (12:511)

Weinstock, B., *Ford Motor Co.,* Automobile exhaust control (Photochemical smog) (S:50)

Weissler, Alfred, *American University,* Ultrasonics (20:773)

Wells, A. F., *Imperial Chemical Industries, Limited,* Crystals (6:516)

Wenner, Wilhelm, *Hoffmann-La Roche, Inc.,* Malonic acid and derivatives (Malonic acid) (12:849)

Wentorf, Jr., Robert H., *General Electric Research Laboratory,* Diamond, synthetic (Carbon) (4:294); (S:274)

Wertz, John E., *University of Minnesota,* Electron spin resonance (7:874)

Weschler, Joseph R., *Ciba Products Company,* Epoxy resins (8:294)

Wessling, R., *The Dow Chemical Company,* Vinyldiene polymers (Chloride) (21:275)

Weyl, W. A., *The Pennsylvania State University,* Colors for ceramics (5:845)

Whaley, Thomas P., *International Minerals & Chemical Corporation,* Magnesium compounds (12:708)

Wheaton, R. M., *The Dow Chemical Company,* Ion exchange (11:871)

Whetstone, R. R., *Shell Development Company,* Chlorinated derivatives of cyclopentadiene (Chlorocarbons and chlorohydrocarbons) (5:240)

Whitaker, G. C., *Harshaw Chemical Company,* Driers and metallic soaps (7:272); Fluorine compounds, inorganic (Ammonium fluoride, lithium fluoride, magnesium fluoride, potassium fluorides) (9:548)

Whitby, Kenneth T., *University of Minnesota,* Dust (engineering) (7:429)

White, Wayne E., *The Ozark-Mahoning Company,* Ferrous fluoride, Ferric fluoride (Fluorine compounds, inorganic) (9:625); Inorganic (Antimony fluorides (9:549); Barium fluoride (9:551); Beryllium fluorides (9:552); Bismuth fluorides (9:553); Cadmium fluoride (9:572); Cobalt fluorides (9:582, 583); Copper fluorides (9:583); Germanium fluorides (9:584); Iron fluorides (9:625); Lead fluorides (9:626); Mercury fluorides (9:628); Nickel fluorides (9:628); Nitrogen fluorides (9:629); Phosphorus fluorides (9:635); Silver fluorides (9:661); Tantalum fluorides (9:635); Silver fluorides (9:661); Tantalum fluorides (9:681); Tin fluorides (9:682); Titanium fluorides (9:683); Zinc fluorides (9:684); Zirconium fluorides (9:685)

Whitehead, Jr., Robert C., *Honeywell, Inc.,* Liquid-level measurement (12:481)

Whitney, Roy P., *The Institute of Paper Chemistry,* Paper (14:494)

Wickson, E. J., *Enjay Laboratories,* Fatty acids (Trialkylacetic) (8:851); Propyl-alcohols (Iso) (16:564)

Wiese, H. K., *Esso Research and Engineering Company,* Cyclopentadiene and dicyclopentadiene (6:688)

Wiewiorowski, T. K., *Freeport Sulphur Company,* Sulfur (19:337)

Wilcock, D. F., *General Electric Company,* Bearing materials (3:271)

Wilder, O. H. M., *American Meat Institute Foundation,* Blood, animal (3:567)

Wiley, Paul F., *The Upjohn Company,* Pyrazoles, pyrazolines, pyrazolones (16:762)

Wilkinson, S., *The Wellcome Research Laboratories, Beckenham, Kent, England,* Polypeptide antibiotics (16:306)

Wilkinson, Jr., J. M., *General Aniline & Film Corporation,* Acetylene (1:171)

Williams, Jr., Roger, *Roger Williams Technical & Economic Services, Inc.,* Marketing and marketing research (13:66)

Williams, William A., *United States Army, Edgewood Arsenal,* Tonic chemicals (Chemical warfare); (4:869)

Wills, John H., *Philadelphia Quartz Company,* Silicon compounds (Synthetic inorganic silicates) (18:134)

Windheuser, John J., *The University of Wisconsin,* Pharmaceuticals (15:112)

Windholz, Thomas B., *Merck & Co.,* Steroids (18:830)

Wink, W. A., *The Institute of Paper Chemistry,* Paper (Physical properties) (14:494)

Winter, E. A., *Tennessee Corporation,* Copper compounds (6:265)

Wise, Edmund M., *Consultant,* Gold and gold compounds (10:681); Platinum group metals (15:832)

Witten, Benjamin, *United States Army, Edgewood Arsenal,* Tonic chemicals (Chemical warfare), (4:869)

Witterholt, Vincent G., *E. I. du Pont de Nemours & Co., Inc.,* Triphenylmethane and related dyes (20:672)

Wocasek, Joseph J., *Celanese Chemical Company,* Propionaldehyde (16:548); Propionic acid (16:554); Propyl alcohols (normal) (16:559)

Wohnsiedler, H. P., *American Cyanamid Company,* Amino resins and plastics (2:225)

Wohrer, Luis C., *The Carborundum Company,* Phenolic fibers (S:667)

Wolinski, L. E., *E. I. du Pont de Nemours & Company, Inc.,* Fluorine compounds, organic (Poly(vinyl fluoride)) (9:835)

Wood, A. S., *GAF Corporation,* Vinyl polymers (Poly(pyrrolidine)) (21:427)

Wood, W. S., *Sun Oil Company,* Safety (17:694)

Woodman, J. F., *Rohm and Haas Company,* Methacrylic compounds (13:331)

Woodroof, J. G., *University of Georgia, College of Agriculture,* Nuts (14:122)

Woodruff, H. Boyd, *Merck & Co., Inc.,* Microorganisms (13:457)

Woods, W. G., *U.S. Borax Research Corporation,* Organic boron compounds, in part (Boron compounds) (3:707)

Woodward, Jr., H. F., *Gulf Oil Corporation,* Methanol, (13:370)

Woodward, John D., *University of Reading, England,* Biotin (3:518)

Woolf, Cyril, *Allied Chemical Corporation,* Perchlorofluoroacetones (Fluorine compounds, organic) (9:754); Trifluoroacetic acid (Fluorine compounds, organic) (9:771)

Wooster, Charles B., *State University of New York, College at New Paltz,* Ammonolysis (2:332)

Work, Robert W., *North Carolina State University,* Fibers (man-made) Survey (9:151)

Wyart, John W., *Celanese Chemical Company,* Solvents, industrial (18:564)

Yeager, Ernest B., *Western Reserve University,* Introduction; Primary cells, fuel cells (Batteries and electric cells primary) (3:111)

Yeager, John F., *Union Carbide Corporation,* Introduction; Primary cells (Batteries and electric cells, primary) (3:111)

Young, Clyde, T., *University of Georgia, College of Agriculture,* Nuts (14:122)

Young, Harland H., *Swift & Company,* Glue, animal and fish, (10:604)

Zenhäusern, A., *Swiss Federal Institute of Technology,* Azo dyes (2:868)

Zenz, Frederick A., *Squires International, Inc.,* Absorption (1:44); Fluidization (9:398)

Zienty, M. F., *Miles Laboratories, Inc., Chemicals Division,* Malic acid (12:837)

Zisman, W. A., *U.S. Naval Research Laboratory,* Fluorine compounds, organic (Surface chemistry of fluorochemicals) (9:707)

Zollinger, H., *Swiss Federal Institute of Technology,* Azo dyes (2:868)

Zuckerman, Samuel, H., *Kohnstamm & Co., Inc.,* Colors for foods, drugs, and cosmetics (5:857)

Zweidler, Reinhard, *J. R. Geigy AG, Basel,* Brightners, optical (3:737); Stilbene derivatives (Optical brighteners of the stilbene series) (19:13)

Zwick, D. M., *Eastman Kodak Company,* Color photography (5:812)

WARNING

NITROBENZENE and NITROTOLUENES

Care should be taken in any manufacturing process using mixtures of nitric acid, nitrobenzene, and water as these mixtures are explosive over a wide range of conditions. An explosion of such a mixture destroyed a commercial nitration plant in 1960; see C. M. Mason, R. W. Van Dolah, J. Ribovich, *J. Chem. Eng. Data* **10,** 173 (1965).

Please insert a copy of this sheet opposite page 837 of *Encyclopedia of Chemical Technology,* 2nd ed., Vol. 13 (1967).

GENERAL INDEX

GENERAL INDEX

Entries are indicated by volume and page numbers: italic numbers indicate volumes; numbers that follow colons indicate pages.

Titles of articles are indicated by a bold face number: Acrolein and derivatives, *1*:**255**.

An entry such as "Polyfluorocarbon resins. See *Fluorocarbon*." means that all entries for polyfluorocarbon resins will be found under Fluorocarbon.

For an entry such as "Aluminum bromide, *2*:24. (See also *Aluminum tribromide*.)" specific material will be found in the subentries and related material under the cross reference.

The following abbreviations are used in the Index:

addn	addition	lab	laboratory
anal	analysis, analytical	manuf	manufacture
bp	boiling point	mech	mechanical
chem	chemical	mol	molar, molecular
compn	composition	mp	melting point
compd	compound	phys	physical
conc	concentrated, concentration	prepn	preparation
decompn	decomposition	prod	product, production
detn	determination	prop	property
deriv	derivative	soln	solution
dil	dilute	temp	temperature
elec	electric(al)	vol	volume
ind	industrial	wt	weight

A

ACTH. (See also *Hormones.*)
general prop, 56-57
ACTH, *7*:267. (See also *Adrenel-cortical hormones.*)
AEPD. See *2-Amino-2-ethyl-1,3-propanediol.*
AMP. See *2-Amino-2-methyl-T-propanol.*
AMPD. See *2-Amino-2-methyl-1,3,-propanediol.*
A-Norcortisone, *18*:883
A-Nortestosterone, *18*:884
APHA color scale, *5*:809
API scale of specific gravity, *6*:758, 759β-Apo-2-carotenal
Abaca, *9*:171
analysis, *9*:173
characteristics, *9*:174, 175
Abbe number, *17*:217
Abherents, *1*:**1**
in food, *10*:16
Abies sibirica, *14*:192
Abietane,
derivatives of, *17*:482
Abietic acid, *16*:690; *17*:482
phys prop, *1*:236
surfactants, *19*:545
tall oil, *19*:626
Abitilguanide, *21*:457
Ablation, *1*:**11**
Abrasives, *1*:**22**
fused alumina as, *2*:56
silicon carbide as, *4*:128
Abrasive wheels
use of furfural in, *10*:244
Abscisic acid, *15*:685
Absinthe. See *Wormwood.*
Absorbed water, *21*:682
Absorption, *1*:**44**
aquametry by, *2*:682
equivalent theoretical plate, *7*:95, 97
Absorption spectrum, *5*:788
Acacia
pill formulation, *15*:124
tablet binder, *15*:121
tablet coater, *15*:122
Acaricides, *11*:713
veterinary, *21*:250
rubber from butyraldehyde, *3*:873

Accelerators. (See also *Catalysis.*)
Accelerators
electron, *S*:581
Accroides, *17*:382
Acenaphthalene, *13*:678
from coal tar, *19*:657, 669
diacylation, *19*:139
Acenaphthenes
in naphthalene feedstocks, *11*:454
Acenocoumarin, *5*:600, 601
Acenocoumarol, *6*:432
1,4-Acepleidanedione, *19*:139
Acetaldehyde, *1*:**77**; *2*:413, 600; *6*:466, 468, 670, 693; *7*:412, 414; *8*:241, 429, 588; *10*:663
butadiene from, *3*:803
by-product of ethylene oxide manuf, *8*:531
content in distilled alcoholic beverage, *1*:510
as denaturant, *8*:450, 453
hydroammonolysis of, *2*:347
hydrogenation to prepare ethanol, *8*:437
from lactic acid, *12*:174
manuf, *1*:84, 203
manuf of acetic anhydride, *8*:410
manuf from ethyl alcohol, *8*:447
NMR spectra, *14*:53
occurence, *1*:77
oxidation to form peroxyacetic acids, *8*:253
oxidation to produce acetic acid, *8*:393
phenolic resins, *15*:176
phys prop, *1*:78
polymers of, *1*:99
in prepn of acetyl chloride, *1*:140
in prepn of acrolein, *1*:268
prepn from ether, *8*:479
prepn from ethylene oxide, *8*:530
in prepn of glyoxal, *10*:646
in prepn of lactonitrile, *1*:347
in prepn of pentaerythritol, *1*:590
production from ethylene, *8*:520
reaction to form butadiene, *6*:453
reaction with crotonaldehyde, *6*:452
reactions, *1*:79
synthesis of crotonaldehyde, *6*:446
in vinegar, *21*:256
and vinyl alcohol, *21*:357
vinyl acetate manuf, *21*:327

5

Acetaldehyde-aniline, *17*:518
Acetaldehyde cyanohydrin, *6*:668, 672; *12*:178, 184
  prepn of, *1*:83
Acetaldehyde diethyl acetal. See *Acetals.*
  poly(vinyl acetal) resins from, *21*:310
Acetaldehyde dimethyl acetal, *1*:107
Acetaldehydedisulfonic acid, *19*:287
Acetaldehyde hemiperoxyacetate, *8*:241
Acetaldehyde monoperoxyacetate, *8*:241
Acetaldehyde oxime,
  prepn of, *1*:82
Acetaldol, *4*:82; *6*:446
  as denaturant, *8*:450, 453
  prepn of 1,3-butylene glycol, *10*:660
Acetal resins. See *Polyformaldehydes.*
Acetal resins, *1*:**95;** *12*:191
  in phonograph records, *15*:230
  prepn from ethyl alcohol, *8*:428
Acetals, *1*:**107;** *3*:874
N,N-Acetals, *12*:98
*O,N*Acetals, *12*:98
Acetamide, *6*:650
  hydrolysis of, *2*:69
  phys prop, *2*:68
  prepn, *1*:144; *2*:858
  in prepn of acetanilide, *1*:150
  prop, *1*:143
  saponification, *1*:143
  specifications, *1*:144
  uses, *1*:144
Acetamidine
  thiamine synthesis, *20*:178
4-Acetamido-6-amino-1-phenol-2-1-sulfonic acid, *2*:222
p-Acetamidobenzenestibonic acid, *2*:585
p-Acetamidobenzenesulfonamide, *7*:258
*p*Acetamidobenzenesulfonyl chloride. See *N-Acetylsulfanilyl chloride.*
2-Acetamido-4-chlorophenol, *2*:215, 222
3-Acetamido-4-hydroxybenzenearsonic acid. See *Acetarsone.*
2-Acetamido-5-nitrothiazole, *20*:85
3-Acetamidoperchlorylbenzene, *9*:604
o-Acetamidophenol
  in prepn of 2-amino-4-nitrophenol, *2*:217
  in prepn of 2-amino-5-nitrophenol, *2*:217
p-Acetamidophenol
  in prepn of 4-acetamido-6-amino-1-phenol-2-sulfonic acid, *2*:222
5-Acetamido-1,3,4-thiadiazole-2-sulfonamide, *19*:259
3-Acetamido-2,4,6-triiodobenzoic acid, *17*:132

α-Acetamido-*p*-toluenesulfonyl chloride, *19*:270
2-Acetamido-*p-toluenesulfonyl chloride, 19*:270
  form, *2*:67
5-Acetamido-2,4,6-triiodo-N-methylisophthalamic acid, *17*:133
3-Acetamido-2,4,6-triiodo-5-(*N*-methylacetamido)-benzoicacid, *17*:133
5-Acetamido-2,4,6-triiodo-5-(*N*-methylisophthala-mic acid, *17*:133
Acetanilide, *1*:148; *2*:414; *7*:478; *8*:407; *10*:799
  chemical prop, *1*:150
  form, *2*:68
  grades, *1*:152
  phys prop, *1*:148; *2*:68
  prepn, *1*:151
  prepn from acetyl chloride, *1*:139
  sulfonation, *19*:299
  uses, *1*:152
Acetanisole. See *p-Methoxyacetophenone.*
Acetanisole, *1*:170. ((See also *p-Methoxyaceto-phenone.*)
Acetarsone, *2*:731
  amebicide, *20*:93
Acetate
  spectral transmittance curve, *9*:253
Acetate and triacetate fibers, *1*:**109;** *9*:155
  Chromspun, *1*:109
  flame resistance, *9*:312
  optical brighteners for, *3*:747
Acetate-butyrate films
  photography, *15*:364
Acetate fiber
  drycleaning of, *7*:307
  dyeing on blends, *7*:566
  prepn for dyeing, *7*:518
Acetate fibers
  anthraquinone dyes for, *2*:503
Acetate flake
  chemical prop, *1*:115
  manuf, *1*:116
  phys prop, *1*:115
Acetates
  fatty amine solubilities, *2*:132, 133
Acetazolamide, *19*:259
  as diuretic, *7*:259, 260
  as veterinary drug, *21*:252
Acetenyl chloride. See *Monochloroacetylene.*
Acetenyl chloride, *5*:204
Acethydrazide, *13*:856
Acetic acid, *2*:13, 174; *8*:386, 812. ((See also *Ethanoic acid.*)

acetophenone from, 3:374
carbon monoxide in prepn of, 4:435
composition of vapor over, 6:733
from crotonic acid, 6:465, 466
dehydration of, 2:850
in detergents, 6:858
in dyeing, 7:538
in dyeing of wool blends, 7:566
in dyeing of wool with acid dyes, 7:523, 525,
  528
in esterification, 8:346
in ethane oxidation, 11:226
as inhibitor in pickling, 6:323, 325, 328
in manuf of Tetryl, 8:623
from methanol, 13:372
from oxidation of acetaldehyde, 1:80
phenolic aldehydes, 15:162
phys prop, 1:212
prepn from acetaldehyde, 1:80
in prepn of acetamide, 1:140
in prepn of 4-acetamido-6-amino-1-phenol-
  2-sulfonic acid, 2:222
in prepn of acetanilide, 1:151
in prepn of acetylacetone, 12:157
in prepn of acetyl chloride, 1:140
in prepn of acrylic anhydride, 1:290
in prepn of aluminum triacetate, 2:12
prepn from ethylene, 8:520
in prepn of ethylidene diacetate, 1:83
prepn from ketene, 12:91
in prepn of lead dinitrosoresorcinate, 8:591
reaction with succinic acid, 19:144
reaction with titanium ethoxide, 1:842
as starch modifier, 10:17
sulfoxides, 19:328
in vinegar, 21:256
vinyl acetate manuf, 21:321
from wood hydrolysis, 22:383 405
Acetic acid amide. See Acetamide.
Acetic acid derivatives, 1:**138**
Acetic acid-dimethyl pyrone
  cooling curve, 13:203
Acetic acid esters, 8:378
Acetic anhydride, 2:213, 215; 6:610, 621;
  7:43; 8:251, 405
  in cellulose triacetate prod, 1:114
  in prepn of dimethyl acetamide, 1:146
  vinyl acetate manuf, 21:327
Acetimidine, 12:856
Acetoacetamide
  prepn from acetamide, 1:144
Acetoacetanilide, 2:414, 423

azo coupling, 2:871
phys prop, 1:235
in urine of diabetics, 1:154
Acetoacetic acid, 1:153
  oxidation by liver of crotonic acid to form,
    6:465
Acetoacetic acid and ester, 1:153
  from methylene chloride, 5:114
Acetoacetic ester, 1:**153**; 6:670; 9:605
  sodium in prod, 18:451
Acetobutyraldol
  prepn, prop, 3:874
α-Acetobutyrolactone, 1:156, 159
Acetoglycerol, 10:629
Acetoguanamine, 10:930
Acetohexamide, 19:259
Acetoin, 12:105, 118, 145; 13:515
  in vinegar, 21:256
Acetol, 12:105, 143
  prepn from propargyl alcohol, 1:600
Acetomeroctol, 2:620
1-Acetonaphthone, 12:107
2-Acetonaphthone, 12:107 7:41, 43, 44, 414;
  8:450, 539; 10:640, 663; 12:104
  chloroform from, 5:124
  chloroform reaction with, 5:122
  fluorination of, 9:509
  manuf of acetic anhydride, 8:410, 427
  manuf from ethyl alcohol, 8:455
  in manuf of isoprene, 12:77
  mass spectrometry, 13:92
  from milk, 13:572
  in nail enamel, 6:368
  in prepn of diacetone alcohol, 12:138
  prepn of isopropyl ether, 8:489
  in prepn of mesityl oxide, 12:136
  reaction with, 9:652
  reaction with ethylene oxide, 8:529
  reaction with isoprene, 12:69
  in refrigeration, 17:310
  from shale oil, 18:18
  stain solvent, 18:661
  from sucrose, 19:224
  thermochemical additive constants, 8:681
Acetone, 1:**159**; 2:413, 415, 600; 6:640, 670,
  672, 693, 696: 7:41, 43, 44, 414; 8:450, 539;
  10:640, 663; 12:104
  from isopropyl alcohol, 16:575
Acetone-chloroform. See Chlorobutanol.
Acetone cyanohydrin, 6:455, 668, 671, 672;
  12:120; 13:333
Acetonedicarboxylic acid, 5:527; 6:630

Acetone monocarbocylic acid. See *Acetoacetic acid.*
Acetone peroxide
  flour improver, *22*:293
Acetonitrile, *S*:590, 594
  from acetamide, *1*:143
  in butadiene separation, *3*:806
  perchlorylation reaction, *9*:604, 693
  in prepn of acetanilide, *1*:150
  in prepn of acrylonitrile, *1*:347
Acetonylacetone. See *2,5-Hexanedione.*
Acetonylacetone, *10*:901; *12*:158, 161
δ-Acetonyllevulinic acid, *12*:160
Acetophenetidin(e)
  as antipyretic, *2*:390
  phys prop, *2*:68
Acetophenone, *1*:**167;** *6*:670, 693; *8*:408; *12*:107, 110, 113, 131
  from acetyl chloride, *3*:374
  in manuf of benzoic acid, *3*:425
  photoinitiator, *16*:519
  prepn, *10*:147
  styrene by-product, *19*:62
  as synthetic flavor, *9*:354
Acetopropanol, *12*:105
Acetosal. See *Tetrachloroethane.*
Acetovanillone, *21*:193
  from lignin, *21*:189
2-Acetoxyacrylonitrile, *12*:89
3β-Acetoxy-19-hydroxycholest-5-ene, *18*:854
1-Acetoxybutadiene, *20*:23
2-Acetoxy-1,3-butadiene, *12*:111
1-Aceto-3-butyrolactone
  prepn, *8*:530
5-Acetoxy-3-chloro-2-pentanone, *20*:180
18β-Acetoxy-11,17α-dimethoxy-3-oxo-16β-methoxycarbonyl-2-3-*se co*-20α-yohimbane
  in prepn of reserpine, *1*:802
4-Acetoxy-2-fluorobiphenyl, *9*:797
2-Acetoxyisoprene, *12*:111
5-Acetoxymercurivanillin, *21*:192
Acetoxymercuri-2-ethylhexylphenolsulfonate, *2*:620
2-Acetoxymercuri-4-(1,1,3,3-tetramethyl-butyl)phenol, *2*:620
2-Acetoxypiperylene, *12*:111
17-Acetoxypregnenolone, *18*:853
3β-Acetoxypregn-16-en-20-ones, *18*:842
3β-Acetoxypregn-5,16-dien-20-one, *18*:840
3β-Acetoxypregn-9(11),16-dien-20-one, *18*:840
17α-Acetoxyprogesterone, *6*:78
8-Acetoxyquinoline, *2*:631

Acetoxysilane
  silicone rubber cure, *18*:230
α-Acetoxystyrene, *12*:111
Acetpyrogallol, *16*:198
Acetrizoic acid, *17*:132
N-Acetyl-N-(3-amino-2,4,6-triiodophenyl)-2-methyl-β-alanine, *17*:137
O-Acetyl-β-D-xylose *16*:688
1-Acetylacetamidoanthraquinone, *2*:450
Acetylacetanilide. See *Acetoacetanilide.*
Acetylacetic acid. See *Acetoacetic acid.*
Acetylacetone
  titanium chelates, *20*:466, 470
  titanium (III) chelates, *20*:470
Acetylacetone, *6*:614; *12*:148, 149, 150, 155, 156
Acetyl acetone, *6*:614; *9*:605; *12*:148; *19*:328
  in dental materials, *6*:826
  prepn, *1*:141
  reaction with boron trifluoride, *9*:556
Acetyl,-β-methylcholine chloride, *5*:410
  as cardiovascular agent, *4*:520
p-Acetylaminobenzenesulfonyl chloride. See *N-Acetylsulfanilyl chloride.*
N-Acetylaniline. See *Acetanilide.*
1-Acetylanthrapyridone, *2*:450
Acetylbenzene, *1*:167; *12*:107. (See also *Acetophenone.*)
N-Acetylbenzenesulfonamide, *4*:924
Acetylbenzoyl, *12*:140, 144
Acetylbenzoylaconine
  as cardiovascular agent, *4*:523
Acetylbenzoylmethane, *12*:148
Acetyl bromide
  phys prop, *1*:224
  prepn from acetyl chloride, *1*:139
Acetyl cellulose. See *Cellulose acetate.*
Acetyl chloride, *1*:138-142; *6*:613; *10*:663; *12*:155
  acetophenone from, *3*:374
  from methyl chloride, *5*:105
  reaction with ethylene, *8*:527
  reaction with ethylene oxide, *8*:530
  sulfuryl chloride, *19*:402
Acetylcholine, *5*:409; *6*:549
Acetylcholine chloride, *2*:409
2-Acetylcoumarone, *10*:908
Acetyl cyanide, *1*:139
4-Acetyl-4-(2-cyanolthyl) heptanedinitrile, *6*:640 *2*:450
Acetyldigitoxin
  as cardiovascular agent, *4*:515

1-Acetyl-7 *H*-dibenz [*f,i,j*]-isoquinolin-2,7(3*H*.)-dione, *2*:450
1-Acetyl-2,2-dimethylethylene oxide, *12*:137
N-Acetyl-4,4'-dinitrodiphenylamine, *15*:218
  acetaldehyde reaction with, *5*:228
  acetone reaction with, *5*:228
  in aquametry, *2*:673, 690
  butadiene from, *3*:803
  from calcium carbide, *4*:100
  chlorine reactions with, *5*:161, 182
  from chloroform, *5*:122
1-*O*-Acetyldihydrovitamin K$_{1(20)}$, *21*:589
  chloroprene from, *5*:220, 224
  comparison with ethylene as a raw material, *8*:520
  1,1-dichloroethane from, *5*:q49
  1,2-dichloroethylene from, *5*:182
  dried by alumina, *2*:52
  esterification, *8*:344
  hexachloroethane reaction with, *5*:200
  hydrogenation, *14*:326
  hydrogen chloride reactions with, *5*:149, 173
  in manuf of isoprene, *12*:77
  in manuf of vinyl chloride, *5*:173
  oxalic acid from, *14*:359
  in prepn of ketones, *12*:126
  in prepn of vinyl fluoride, *9*:835
  in prepn of vinylidene fluoride, *9*:841
  reactions with formaldehyde, *10*:86, 667, 668, 670
  removal during manuf of ethylene oxide, *8*:539
  succinic acid from, *19*:143
  1,1,2,2-tetrachloroethane from, *5*:161
  tetrachloroethylene from, *5*:200
  1,1,2-trichloroethane from, *5*:158
  toxicity of, *11*:297
Acetylene, *1*:**171**; *2*:415, 583; *6*:574, 694
  1,1-dichloroethane from, *5*:149
  gas welding, *22*:245
  industrial uses, *S*:634
  oxygen in synthesis, *14*:405
  polymerization, *20*:447
  2-pyrrolidone reaction, *21*:432
  vinyl acetate manuf, *21*:321
Acetylene black, *4*:244, 278
  in electric cells, *3*:112
  furnace black technology, *S*:95
  process for, *S*:98
Acetylenedicarbonitrile, *6*:695
Acetylenedicarboxylic acid
  phys prop, *1*:232
  prepn from butynediol, *1*:606

Acetylenedicarboxylic acid esters, *6*:695
Acetylene dichloride. See *1,2-Dichloroethylene.*
Acetylene diureine-formaldehyde, *2*:248
Acetylenes
  nomenclature, *11*:288
Acetylene tetrabromide. See *1,1,2,2-Tetrabromoethane.*
Acetylene tetrabromide, *3*:776
  in polystyrene foams, *19*:120
Acetylene tetrachloride. (See also *1,1,2,2-Tetrachloroethane.*)
  in decontaminating agents, *4*:886
Acetylenic acids
  definition, *8*:817
Acetylenic alcohols
  secondary, *1*:619
  tertiary, *1*:621
Acetylenic glycols
  tertiary, *1*:619
Acetylenics
  phys prop, *1*:811
  removal from butadiene, *3*:804
  uses, *1*:823
N-Acetylethanolamine
Acetyl fluoride
  phys prop, *1*:224
  prep from acetyl chloride, *1*:139
  reaction with boron trifluoride, *9*:558
N-Acetylglucosamine, *14*:671
Acetylides, *4*:72
  prepn, *1*:173
Acetyl iodide, *8*:473
  phys prop, *1*:224
  prepn from acetyl chloride, *1*:139
Acetylisobutyryl, *12*:140
Acetylisovaleryl, *12*:140
4-Acetyl-4-methylheptanedinitrile, *6*:638
1-Acetyl-2-methylnaphthohydroquinone, *21*:595
3-Acetyl-6-methyl-1,2-pyran-2,4(3*H*)-dione. See *Dehydroacetic acid.*
3-Acetyl-6-methyl-2*H*-pyran-2,4,(3*H*)-dione-, *1*:157
3-Acetyl-6-methyl-1,2-pyran-2,4(3*H*)-dione, *10*:923
1-*O*-Acetyl-2-methyl-1,4-naphthohydroquinone, *21*:589
2-Acetylnaphthalene, *12*:107
Acetyl nitrate, *8*:407
Acetyl perchlorate
  polyester reactions, *16*:173
Acetyl peroxide, *8*:407; *19*:143
N-Acetylphthalimide, *1*:144
Acetylpropionyl, *12*:140
β-Acetylpyridine, *21*:526

3-Acetylpyridine, *21*:526
Acetylpyridine adenine dinucleotide, *21*:527
4-Acetylpyrogallol, *16*:198
Acetylsalicylic acid, *17*:727, 731
  derivatives, *17*:733
  o-hydroxybenzoic acid from, *S*:741
  manuf, *17*:732
  specifications, *17*:733
  uses, *17*:733
β-Acetylstyrene, *12*:107
N'-Acetylsulfamethoxypyridazine, *19*:269, 277
Acetylsulfanilyl chloride, *19*:267
N-Acetylsulfanilyl chloride
  phys prop, *1*:152
  prepn from acetanilide, *1*:150, 152
Acetylsulfathiazole, *1*:152
Acetyl sulfide
  prepn from acetyl chloride, *1*:139
Acetylsulfuric acid, *1*:117
Acetyl tetrafluoroborate, *9*:558
N-Acetyl-*p*-toluenesulfonamide, *4*:924
Acetyl triallyl citrate, *5*:540
Acetyl-tri-*n*-butyl citrate, *5*:539
Acetyl tributyl citrate
  as cellulosic plasticizer, *4*:657
Acetyl triethyl citrate, *5*:539
  as cellulosic plasticizer, *4*:657
Acetyl tryptophan, *2*:167
Acetyl ureas
  as hypnotics, *11*:510
2-Acetyl-1,3,3,5-tetramethylcyclohex-5-en-1-ol,
  12:137
Acheson furnace, *4*:185
γ Acid, *13*:729
ε Acid, *13*:722
1,2,4-Acid, *13*:725
Acid anhydrides, *1*:**211**
  aquametry by, *2*:678
  esterification of, *8*:335
Acid Anthracene Red G, *2*:895
Acid-base reactions
  indicators for, *11*:548
Acid-base systems, *1*:**213**
Acid Black 1, *2*:895; *5*:867; *7*:496, 534
Acid Black 1 (CI), *5*:867, *7*:525, 535
Acid Black 2, *2*:862
Acid Black 48
  prepn, *2*:515
Acid Blue dyes, *20*:724
Acid Blue 1 (CI), *7*:525
Acid Blue 1 (CI), *7*:525
Acid Blue 102, *2*:863, 7:526
Acid Blue 113, *2*:896

Acid Blue 15 (CI), *7*:526
Acid Blue 158A, *2*:887
Acid Blue 25
  prepn, *2*:513
Acid Blue 40, *7*:491
Acid Blue 43
  prepn, *2*:514
Acid Blue 45
  prepn, *2*:514
Acid Blue 59, *2*:863; *7*:522, 526
Acid Blue 7 (CI), *7*:522, 526
Acid Blue 78
  prepn, *2*:513
Acid Blue 9 (CI), *7*:526
Acid Blue 90 (CI), *7*:526
Acid Blue 98, *2*:863
Acid chlorides
  aquametry by, *2*:677
  esterification, *8*:337
Acid conversion syrups, *6*:927
Acid disazo dyes, *2*:895
Acid dyes, *7*:521
  in paper coloring, *7*:583
  in textile printing, *7*:575
  used in coloring hair, *10*:792
  uses of, *7*:476
Acid Fuchsine, *20*:715
Acid Fushsin D, *5*:871
Acid Green 1, *5*:872; *7*:522
Acid Green 25 (CI), *5*:868, 873
  prepn, *2*:513
Acid Green 3, *7*:525
Acid Green 3 (CI), *7*:525
Acid halides, *1*:**222**
Acidic chalcides
  as catalysts, *10*:161
Acidity
  of water, *21*:692
Acid Magenta, *20*:715
Acid monoazo dyes, *2*:880
Acid nitriles, *S*:590
Acidolysis, *8*:361
Acid Orange XX, *7*:525
Acid Orange 20, *2*:880, *5*:872
Acid Orange 24, *2*:877, 895; *5*:868
Acid Orange 7, *2*:874; *5*:869; *7*:496, 525
Acid Orange 72, *7*:522
Acid pickling
  inhibitors in, *6*:321
Acid polyazo dyes, *2*:895
Acid potassium tartrate, *19*:736
Acid Red (CI), *7*:525
Acid Red 1, *7*:522

Acid Red 14, 2:886
Acid Red 163, 2:896
Acid Red 214, 2:888
Acid Red 26, 5:869
Acid Red 33, 5:871
Acid Red 52, 7:526
Acid Red 53, 2:905
Acid Red 73, 7:525
Acid Red 73(CI), 7:525
Acid Red 87, 5:870, 873
Acid Red 88, 2:880; 5:872
Acid Red 89, 7:522
Acid Red 92, 5:870
Acid Red 97, 2:895
Acid retardation, 11:890
Acids
    Arrhenius definition of, 1:214
    Bronsted definition of, 1:215
    carboxylic. See Acids, carboxylic.
    conjugate, 1:215
    content in distilled alcohol beverages, 1:510
    dicarboxylic. See Acids, dicarboxylic.
    in ester-acid interchange, 8:361
    as hazardous chemicals, 9:292
    from hydrolysis of esters, 8:365
    Lewis definition, 1:218
    monocarboxylic. See Acids, carboxylic.
Acids, carboxylic, 1:**224**
Acids, dicarboxylic, 1:**240**
Acid value, 8:794
Acidum isonicotinicum, 21:527
Acidum nicotinicum, 21:509
Acid Violet 14, 2:882
Acid Violet 17, 7:525, 526
Acid Violet 19, 7:507
Acid Violet 43 (CI), 5:872
    prepn, 2:514
Acid Violet 6B, 5:867
Acid Violet 6B, 20:717
Acid Violet 6B, 10:793
Acid Violet 7, 2:881
Acid Violet 78, 2:888
Acid Violet 9, 7:522
Acid Yellow 1, 5:872, 874; 7:522
Acid Yellow 11, 5:872
Acid Yellow 17 (CI), 7:525
Acid Yellow 23, 2:881
Acid Yellow 3, 5:871; 7:522
Acid Yellow 36, 5:872; 7:525
Acid Yellow 36 (CI), 5:872
Acid Yellow 40, 2:881
Acid Yellow 5, 7:526

Acid Yellow 7, 7:526
Acid Yellow 73, 5:871
Acid Yellow 99, 2:888
cis-Aconitase, 15:918
Aconite
    as cardiovascular agent, 4:523
Aconitine
    as cardiovascular agent, 4:523
Aconitric acid, 1:232
    from citric acid, 5:526
    in prepn of itaconic acid, 12:83
Acoustic drying, 20:787
Acridine, 10:899, 920
Acridine dyes, 2:638; 7:471, 473
Acridine Orange
    structural formulas, 5:769
Acridine Orange N, 7:533
Acridinic acid, 1:240
    phys prop, 1:240
Acriflavine
    as antiseptic, 2:638
Acrilan, 1:313
Acrolein, 6:694; 7:413; 19.142
    addition to methyl vinyl ether, 21:416
    photochemical smog, S:52
    epichlorohydrin from, 5:320
    in manuf of epoxy resins, 8:299
    in prepn of methylthiopropionic aldehyde,
        2:165
    reaction with butadiene, 3:785
    as starch modifier, 10:17
    in synthesis of glycerol, 10:624, 663
Acrolein and derivatives, 1:**255**
Acrolein cyanohydrin, 1:266
Acrolein dimethyl acetal
    prepn from acrolein, 1:262
Acrolein-pentaerythritol resins, 19:495
Acrylamide, 1:**274**
    as raw material in amino resin manuf, 2:227
Acrylamide polymers
    prepn, 1:306
    prop, 1:308
Acrylamide-sodium acrylate copolymer
    as drilling fluid flocculant, 7:305
Acrylamide sulfate
    prepn from acrylonitrile, 1:340
Acrylases, 2:176
Acrylate elastomers, 7:699
Acrylate esters, 1:206
Acrylate polymers
    in synthetic waxes, 22:170
Acrylates, 8:313

UV absorbers, *21*:117

Acrylic acid, *1*:**285**; *6*:694; *19*:142
  chloroprene reaction with, *5*:218
  in drilling fluids, *7*:301
  grafts, *16*:7
  succinic acid from, *19*:144
  in synthetic oil products, *7*:415
  vinyl acetate copolymer, *21*:331

Acrylic acid esters, *6*:694
  phys prop, *1*:287
  prepn, *1*:291, 296
  prepn from acetylene, *1*:206
  in prepn of acrylic acid, *1*:294

Acrylic acid esters, polymers
  phys prop, *1*:309
  prepn, *1*:306

Acrylic acid polymers, *1*:303
  prepn, *1*:305
  prop, *1*:308

Acrylic and modacrylic fibers, *1*:**313**

Acrylic anhydride
  phys prop, *1*:285
  prepn, *1*:295
  prepn from acrylic acid, *1*:290
  reactions, *1*:290

Acrylic coatings, *13*:359

Acrylic-divinyl benzene linkage, *11*:875

Acrylic ester interpolymers, *16*:346

Acrylic-ethylene dimethacrylate linking, *11*:875

Acrylic fibers, *9*:157
  dyeing with basic dyes, *7*:533
  dyeing with direct dyes, *7*:546
  dyeing of fiber blends, *7*:570
  prepn for dyeing, *7*:518
  quaternized anthraquinone dyes for, *2*:508

Acrylic lacquers, *5*:704

Acrylic plastics, *13*:331
  in dentistry, *6*:827

Acrylic resins, *12*:191
  in latex paints, *22*:153

Acrylics
  in packing, *14*:452
  in papermaking, *14*:519
  plasticizers for, *15*:726

Acrylics, thermosetting
  in white baking finishes, *5*:698

Acryloid B66
  application in fluid bed technology, *9*:400
  butadiene-styrene polymers, *19*:95
  butadiene-styrene terpolymer, *19*:73
  cellulose reaction with, *4*:642

conversion to glycidamide, *8*:244
copolymer bisulfite adducts, *19*:288
in cyanoethylation of cellulose, *6*:652
in fiber-forming materials, *9*:169
increased dye affinity, *11*:501
Michael addn hydrogen cyanide, *6*:574, 575, 579, 583
in nail enamels, *6*:370 647, 652-662, 673, 695; *12*:114
prepn from ethylene, *8*:520
in prepn of monosodium glutamate, *2*:206
prod from ethylene oxide, *8*:553
reaction with ethylene amine, *7*:29
reaction with gelatin, *10*:501, 670
steroid synthesis, *18*:858
styrene copolymer, *19*:73, 90
sulfolane solvent, *19*:253
sulfoxides, *19*:334

Acrylonitrile, *1*:**338**; *2*:413; *3*:786; *6*:634-639, 643,647, 652-662, 673, 695; *12*:114
  alkyl vinyl ether copolymers, *21*:423
  poly(vinyl alcohol) copolymer, *21*:358
  unsaturated polyester diluent, *20*:820

Acrylonitrile-butadiene
  in packing, *14*:451

Acrylonitrile-butadiene copolymers (NBR), *7*:682

Acrylonitrile-butadiene elastomer
  prop of flexible polymer, *9*:863

Acrylonitrile-butadiene interpolymer, *16*:346

Acrylonitrile-chloroprene copolymers
  plasticizers for, *13*:832

Acrylophenone, *12*:107, 161

Acryloyl chloride, *7*:635
  phys prop, *1*:285
  prepn of, *1*:290, 295
  in prepn of acrylic esters, *1*:300
  reactions, *1*:290

Acryloyl isocynate
  in prepn of acrylamide, *1*:281

Acrysol GS
  neutralized polyacrylic acids, *S*:941
  polyurethans, *21*:61

Actinides, *1*:**351**
  peroxy compounds of, *14*:754

Actinium, *17*:27. ((See also *Actinides.*)
  radioactive elements, natural, *17*:9

Actinium compounds
  crystal structure, *1*:366
  phys prop, *1*:366

Actinocin, *16*:331

Actinolite. See *Asbestos.*

13

Adhesives

in ceramics, 4:772
Actinomotry, 15:334
Actinomycetes
  in lakes, S:320
Actinomycin D, 12:648
Actinomycin D-dactinomycin, S:86
Actinomycins, 16:330
Activated alumina, 2:51
  as adsorbent, 1:460
  capacity as a dyeing agent, 7:385, 391
Activated bauxite
  as adsorbent, 1:460
Activated carbon, 4:149
  bottling water purified with, 4:341
  carbon dioxide purified with, 4:364
  as catalyst in prepn of acetyl chloride, 1:140
  to decolorize MSG solution, 2:209
  in fluid bed technology, 9:400
  in glycerol manuf, 10:623
Activated charcoal. See Activated carbon.
Activity
  thermodynamic defn, 20:138
Activity coefficient
  for binary systems, 21:208
  equations for, 6:729; 7:786-882
  temperature and pressure dependence, 21:212
  thermodynamic defn, 20:139
Acumycin, 12:633, 634
N-Acyl-N-alkyltaurates, 19:521
Acylals, 8:366
Acylaminoanthraquinone vat dyes, 2:520
2-Acylamino-4-alkylurete
  prepd from guanidine, 10:736
Acylation
  aliphatic amines, 2:104
  amides, 2:70
N-Acylaziridines
  reduction with LiAlH₄, 11:535
1-Acylaziridines, 11:528
Acyl chlorides
  phys prop, 1:223
  reactions of, 1:223
N¹-Acyl-N-N-dimethyl-1,3-diaminopropane
  oxide, S:43
Acylguanidine
  prepd from guanidine, 10:736
Acyloins, 8:370
Acyloxyboranes, 3:662
5-Acylsalicylic acid, 17:723
N-Acylsarcosinates, 19:514

Acylureas
  prepn from isocynates, 12:49
Adamantanamine
  as influenza preventative, 13:460
Adamantane, S:1
1-Adamantanecarboxylic acid, S:7
1,3-Adamantanecarboxylic acid, S:8
1,3-Adamantanediols, S:6
Adamantane hydrochloride
  antiviral agent, 21:455
  from oxo process, 14:373
  polyhydric, 1:569
  reaction with acid chlorides, 8:337
  reactions with organic acids, 8:315
  in soaps, 18:429
  sulfated, 19:523, 528
  unsaturated, 1:598
Adamantanes
  antiviral agents, 21:455
1-Adamantanol, S:7
Adamantylacetic acid, S:9
Adamantylchloroacetic acid, S:9
Addition
  bisulfite to olefins, 14:329
  electrophilic in olefins, 14:318
  free-radical in olefins, 14:318
Additives
  color, for foods, drugs and cosmetics, 5:858, 860, 862, 879
  for food, 4:337
Adenochrome, 5:591
Adenochrome monosemicarbazone sodium
  salicylate. See Carbazochrome salicylate.
Adenosine monophosphate, 2:70, 179
Adenosine triphosphate, 2:170, 179; 9:321
  metabolism of urea herbicides, 22:191
  nitrogen fixation, S:609
Adenosylhomocystiene, 2:190
S-Adenosyl-L-methionine, 21:522
Adenosylmethionine, 2:190
Adenylic acid
  nicotinic acid linkage, 21:512
Adherography, 17:339
Adhesion, 1: 376; S:16; SV:1
Adhesive
  in ceramic composite armor, S:138
  joint strength, S:16
Adhesive binder
  use of lignosulfates as, 12:371
Adhesives, 1:371
  alcohol used in, 8:447
  amino resins used in, 2:245

aziridines as, *11*:541

for carpet backing and apparel, *S*:932

castor oil in prepn of, *4*:533

for ceramic armor, *S*:153

clays used in, *5*:578

delayed-tack, *15*:764

emulsion, *15*:764

epoxy resins used as, *8*:311

hot-melt, *15*:764

microencapsulation of, *13*:451

polyimide, *S*:769

poly(vinyl acetate), *21*:341

poly(vinyl alcohol), *21*:362

poly(vinyl butyral), *21*:314

pressure-sensitive, *15*:765

silicate, *18*:163

solution, *15*:765

Adipamide, *1*:409

Adipic acid, *1*:**405**; *12*:129

Adipic acid, *1*:405

from cyclohexane, *19*:143

hexamethylenediamine salt, *16*:31

polyesters, *21*:62

polymerization, *16*:89

prepn from cyclohexane, *6*:663, 675, 685, 686

prepn of urethane polyesters, *7*:694

as stabilizer, *2*:15

thermoplastics, *15*:794

Adipic anhydride, *1*:408

as starch modifier, *10*:17

Adiponitrile, *6*:582

prepn from acrylonitrile, *1*:341

Adipose tissue

in meats, *13*:168

Adipoyl bis(dibenzyl phosphate), *1*:409

Adipoyl chloride, *1*:408

3,3'-(Adipoyldiimino)bis(2,4,6-triiodo-benzoic acid, *17*:138

5,5'-(Adipoyldiimino)bis(2,4,6-triiodo-*N*-methyl-isophthalamic acid, *17*:135

assay, *11*:57

Admex 710

as cellulosic plasticizer, *4*:657

Adrenal-cortical hormones, *11*:77-92; *18*:830

Adrenalin(e), *2*:188, 190. ((See also *Epine-phrine.*)

Adrenalone, *8*:234

Adrenochrome, *8*:237; *16*:641

Adrenochrome monosemicarbazone sodium salicylate, *5*:591

Adrenocorticotropic hormone. See *ACTH*.

Adrenocorticotropin. See *ACTH*.

general prop, *11*:56

method of isolation, *11*:57

Adrenolutin, *16*:641

Adrenosterone, *11*:93

Adsorbents, *1*:460

Adsorption, *1*:**421**

in detergency, *6*:880

diffusion in, *13*:143

of liquid wastes, *21*:629

mass transfer in, *13*:100

in odor control, *14*:172

theory of adhesion, *S*:18

Adsorption chromatography, *1*:468

aluminas used in, *2*:53

Adsorption indicators

chloride ion determined by, *5*:6

Aeration

effects of on yeast cultures, *22*:539

means of odor and taste control in water, *22*:102

sewage treatment, *22*:110

Aerosol OT. See *Dioctyl sodium.*

Aerosol OT. See *Dioctyl sodium sulfosuccinate.*

prepn from fatty acids, *8*:822

in waxes, *22*:156

Aerosolization, *20*:786

Aerosol particles

prop, *7*:436

sampling, *7*:444

size, *7*:430

transport, *7*:449

Aerosols, *1*:**470**. ((See also *Dust.*)

fluorinated hydrocarbons in, *15*:125

Aerosporin

amebicides, *20*:82

Aerozine-50

prop, *8*:688

Aflatoxin B₁

and tobacco, *20*:523

Aflatoxins

in nuts, *14*:139

Aftershave lotions, *6*:354

Agar, *17*:391, 763-768

as cathartic, *4*:591

in confectioneries, *6*:52

in dental impression materials, *6*:799

Agarase

microorganism producing, *8*:210

Agar jellies, *6*:52

Agarose, *17*:765, 768

Agathic acid, *17*:386

Age hardening
  of metals, *13*:325
Aggregative fluidization, *9*:411, 418
Aglycones, *18*:838
Agriculture
  water for, *22*:8
Agstone, *12*:453
Air
  as aerosol propellent, *1*:475
  in bleaching fats, *13*:180
  buoyant effect of and weighing, *22*:228
  dried by alumina, *2*:52
  as lubricant, *12*:606
  as nuclear reactor coolant, *14*:81
  separation of helium-group gases, *10*:874
Air bearings, *3*:290
Air cleaners, *8*:92; *10*:349
Air conditioners
  cost, *1*:500
  design conditions, *1*:484
  mass transfer in, *13*:100
  principles, *1*:481
  systems of, *1*:490
  terms, *1*:482
  use of ferroelectric ceramics in, *9*:1
  uses in industry, *1*:496
Air conditioning, *1*:**481**
Aircraft deicing, *2*:559
Aircraft industry
  use of adhesives in, *1*:397
Aircraft lubrication
  cyclohexane used, *6*:687
  use of triethylene glycol, *10*:648
Air dehumidification systems
Air humidification
  mass transfer in, *13*:100
Air oxidation
  in odor control, *14*:173
Air pollution, *18*:400; *S*:730
  and city planning, *18*:414
  community problems, *14*:171
  control measures, *18*:407
  effects of contaminants, *18*:404
  and solvents, *18*:581
Air refreshers
  use of aerosol products in, *1*:478
Air sanitation systems
  use of triethylene glycol, *10*:648
Akuammicine, *1*:761
Alabaster, *4*:16
Alamine-borane, *3*:729
Alanine
  from acetaldehyde, *1*:83

  in silk, *18*:271
  in tobacco, *20*:508
  in wine, *22*:316
β-Alanine, *2*:157; *6*:673
Alba Red, *5*:871
Albert effect, *15*:388
Albies alba, *14*:192
Albite, *4*:769
Albomycin, *16*:329
Albumin, *2*:216
  in eggs, *7*:663
  iodinated, *20*:260
  in lithography, *17*:330
Albumin, blood
  animal, *3*:571
  human. See *Blood fractionation, 3*:576.
Albuminoid chloramines
  hydrolysis constant of, *4*:913
Albumins
  in cereal grains, *22*:268
Al-chlorite, *5*:551
Alchols, *1*:**531**
  unsaturated, *1*:**598**
Alcohol
  unoxidized in vinegar, *21*:256
Alcoholic beverages
  distilled, *22*:518
Alcoholic beverages, distilled, *1*:**501**
Alcoholometry, *8*:459
Alcohols
  alkylation of ammonia, *2*:106
  autooxidation of secondary, *11*:401
  dehydration, to prep ethers, *8*:474
  dehydrogenation to synthesize esters, *8*:347
  in ester interchange, *8*:356
  ethoxylated, *19*:528
  higher, fatty, *1*:**542**
  higher synthetic, *1*:**560**
  from hydrolysis of esters, *8*:365
  as hypnotics, *11*:510
  from oxo process, *14*:373
  polyhydric, *1*:**569**
  reaction with acid chlorides, *8*:337
  reactions with organic acids, *8*:315
  in soaps, *18*:429
  sulfated, *19*:523, 528
  solubility parameters, *S*:894, 895
Alcohols, acetylenic
  secondary, *1*:619
  tertiary, *1*:621
Alcohols, higher, fatty
Alcohols, polyhydric
  used in alkyd resins, *1*:861

Alcohol sulfates, *19*:304
Alcoholysis, *8*:356
Aldactone
   corticosteroid analog, *11*:91
Aldehyde C$_{14}$
   flavoring for carbonated beverages, *4*:344
Aldehyde C$_{16}$, *4*:344
   as flavor in carbonated beverages, *4*:344
Aldehyde G.m.b.H process
   for liquid-phase oxidation of ethylene, *1*:86
Aldehyde-amines, *17*:511
Aldehyde collidine, *16*:803
Aldehyde oil, *7*:414
Aldehydes, *1*:**639**
   esterification, *8*:346
   as hypnotics, *11*:512
   from oxo process, *14*:373
   in photochemical smog, *S*:50
   reaction with ethyleneamines, *7*:25
   solubility parameters, *S*:893
   synthesis from *N*-acylaziridines, *11*:535
Aldehydine, *21*:514
Aldgamycin, *12*:634
Aldohexuronic acids, *19*:224
Aldol, *2*:600
   1-butanol from, *3*:826
   in synthesis of 1,3-butylene glycol, *10*:664
Aldolase
   in milk, *13*:515
Aldol condensation
   def, *1*:642
Aldol-α-napthylamine, *17*:518
Aldoses
   iodine analysis, *19*:162
Aldosterone, *18*:861
   biosynthesis of, *11*:79
   and edema, *7*:267
   phys prop, *11*:79
Aldrin, *5*:244; *8*:249; *18*:525
   adsorption, *18*:522
   peracetic acid reaction with, *5*:246
   in water, *21*:701
Ale, *3*:299
Alen acid, *13*:714
Aleuritic acid, *18*:24
Alevaine, *3*:30
Alexandrite glass, *5*:849
Alfalfa
   as animal feed, *8*:864
Alfin
   catalyst, *S*:928
Alfin polybutadiene, *S*:929
Alfin polymers, *S*:929

Alfol process
   for synthetisizing higher alcohols, *1*:560
Algae. (See also *Algal cultures.*)
   blue-green, *13*:472; *17*:763
   blue-green and nitrogen fixation, *S*:609
   brown, *13*:471; *17*:763
   green, *13*:471; *17*:763
   red, *13*:471; *17*:763
   slime molds, *13*:472
   synthesis of protein in, *2*:159
   toxicity of blue-green, *S*:322
   yellow, *13*:471
Algal cultures, *1*:**649**
Algarrobella, *12*:322
Algicides
   and eutrophication, *S*:333
   for swimming pools, *22*:130
Algin, *17*:763, 768-774
   in dentifrices, *6*:849
Alginase
   microorganism producing, *8*:210
Alginates, *17*:391
Alginic acid, *6*:802
   disintegrant, *15*:121
Algol Scarlet G, *7*:486
Alicyclic amines, *2*:109
Aliphatic acids
   chlorinated as herbicides, *22*:199
   metabolism of herbicidal chlorinated, *22*:200
Aliphatic alcohols
   ethoxylated, *19*:536
Aliphatic aldehydes
   prepn, *2*:106
   reactions with aziridines, *11*:530
Aliphatic amines, *2*:103
Alite, *4*:688
Alizarin, *7*:464, 486, 620.((See also *Hydroxy-anthraquinone.*)
   manuf, *2*:440, 474
   as product of fusion of 2-aminoanthra-quinone, *2*:454
   pigments, *15*:577
   synthetic prod, *19*:442
   zirconium complexes, *22*:655
Alizarin Irisol R, *5*:872
Alizarin Saphirol, *7*:465
Alizarin Yellow C, *16*:198
Alizarin Yellow R, *11*:552
Alizarine Blue SAP, *7*:486
Alizarine Cyanine Green, *7*:465
Alizarine Cyanine Green F, *5*:868, 873

Alizarine Cyanine Green G
  prepn, 2:513
Alizarine Direct Blue A, 7:486
Alizarine lakes, 15:577
Alizarine Light Gray BBL, BBLW
  prepn, 2:515
Alizarine Pure Blue B
  prepn, 2:513
Alizarine Red WA, 7:527
Alizarine Saphirol A
  prepn, 2:513
Alizarine Saphirol B, 7:522
  prepn, 2:474, 514
Alizarine Saphirol 5E
  prepn, 2:513
Alizarine Sky Blue, 7:465
Alizarine Violet 1R
  prepn, 2:488
Alizarine Yellow GG, 2:883
Alizurol Purple, 5:872
Alizurol Purple SS, 5:871. See *Solvent Violet 13.*
Alkadienes
  nomenclature of, 11:288
Alkali Blue, 15:588, 592
Alkali Blue, 20:674
Alkali and chlorine industries, 1:**668**
Alkaligens
  hydrocarbon utilization, S:839
Alkali lignins, 12:372
Alkali metal-chalcogen cells, S:123
Alkali metal cyanides, 6:585
Alkali metal halides
  zone refining of, 22:698
Alkali metal-halogen cells, S:123
Alkali metals
  reactions with aziridines, 11:531
  in water, 21:698
Alkaline earth metal cyanides, 2:601
Alkaline-earth metals
  in water, 21:698
Alkaline hyposulfite, 2:433
Alkaline silver cyanide
  metalplating bath, 6:813
Alkalinity
  of swimming pool water, 22:131
Alkaloids, 1:**758**
  berberine, 20:79
  holarrhena, 20:78
  ipecac, 20:74
  in nuts, 14:132
  pregnane, 18:841

  salamandra, 18:844
  solanum, 18:842
  veratrum, 18:844
Alkanes
  nomenclature, 11:288
Alkanolamides, 1:812, 819
  sulfated, 19:526
  surfactants, 19:513
Alkanolamines, 1:**809;** 13:831
  manuf, 2:82
  polystyrene stabilizers, 19:90
  in soaps, 18:431
Alkanolamine titanates, 20:466
Alkatrienes
  nomenclature of, 11:288
Alkatriynes
  nomenclature, 11:288
Alkenes
  nomenclature of, 11:288
  sulfated, 19:302
Alkenylsuccinic anhydride, 12:826
Alkenynes
  nomenclature, 11:288
Alkomide, 21:247
Alkoxides, metal, 1:**832**
2-Alkoxy-4,6-bis(alkylamino)-s-triazines, 22:185
5-Alkoxy-7-chloro-5a,6-anhydro tetracycline, 20:13
2-Alkoxy-2-mercaptothiazolidine
  from xanthates, 22:424
2-Alkoxy-2-methyl-3-chlorotetrahydrofuran, 20:180
Alkoxysilanes, 18:189, 191
Alkyd resins, 1:**851;** 12:191
  benzoic acid in manuf of, 3:430
  chain-growth, 16:229
  isophthalic acid, 15:461
  phthalic anhydride, 15:455
  pigments, 15:522
  tall oil, 19:625
  use of glycerin, 10:627
  use of neopentylene glycol, 10:677
Alkyds
  medium oil, 1:877
  short-oil, 1:876
N-Alkylacrylamides, 19:140
Alkylaluminum chlorides
  manuf, 2:36
Alkylaluminum hydrides
  manuf, 2:35
Alkylaluminums. ((See also *Aluminum alkyls.)*
Alkylamines, 2:99-138

tert-Alkylamines, *2*:125
    props, uses, *2*:125
    reactions, *2*:126
Alkyl-2-aminopropionates, *19*:566
N-Alkylanilines
    in hydrorefining, *11*:437
Alkylarenesulfonates. ((See also *Surfactants.*)
    with isobutane, *3*:818
    of olefinic gases, *15*:2
    with olefins, *14*:328
    in paint removers, *14*:491
    synthesis of aliphatic amines, *2*:104, 106
    synthesis of aromatic amines, *2*:115
Alkylating agents
    anticancer agents, *S*:81
Alkylation, *1*:**882**; *15*:41
    exhaustive in amine oxide prepn, *S*:40
Alkylbenzenesulfonates
    biodegradable, *19*:582
    surfactants, *19*:513, 515
Alkylbenzyldimethylammonium chlorides,
    19:563
Alkylbis(2-hydroxyethyl)amine oxide, *S*:34, 37
2-Alkyl-1-carboxymethyl-1-(2-hydroxyethyl)-2-
    imidazolinium salts, *19*:566
Alkylchlorosilanes
    water-repellent agents, *22*:144
Alkyldimethylbenzylammonium chlorides
    in prepn of synthetic higher alcohols, *1*:560
    specifications, *2*:37
    swimming pool algicide, *22*:130
Alkylenimines, *10*:906, See *Imines, cyclic.*
Alkylenimines, *10*:906
Alkyl halosulfurates, *19*:485
Alkylhydrazines, *11*:167-196
    hydrolysis, *19*:486
Alkyl hydrogen sulfates, *19*:483
2-Alkyl-1-(hydroxyethyl)-2-imidazolines, *19*:560
Alkyl hypochlorites, *5*:24
Alkylidene acetates, *8*:407
4,4'-Alkylidenedimorpholines, *13*:665
Alkylidene mono- and bis-acetoacetic esters,
    1:158
2-Alkyl-2-imidazolines, *19*:556
Alkyl-3-iminodipropionates, *19*:566
Alkylisoquinolines, *16*:881
Alkyllithium
    as catalyst, *S*:911
    methylacetylene isomer, *S*:547
Alkylmagnesium halides, *10*:721
Alkylmercuric compounds, *13*:236, 244
N-Alkylmorpholine oxides, *S*:43
Alkylnaphthalenes, *13*:678, 691, 692

    isomerization of, *13*:691
    oxidation, *13*:693, 694, 695
    oxidation with dichromate, *13*:693
Alkylnaphtalenesulfonates, *13*:678. ((See also
    Surfactants.)
Alkyl nitrates
    as catalyst in manuf of amines, *8*:455
    as catalyst in production of ethylene, *8*:503
    hot-dipped coatings, *13*:261
    in prepn of ethers, *8*:474
    C-N rotation, *14*:72
Alkyl orthophosphates
    surfactants, *19*:529
Alkylphenols, *S*:**27**
    hydrocardanol, *1*:915
Alkylphenols, *1*:**901**
    ethoxylated, *19*:527, 533
    sulfated, *19*:527
Alkylphenoxypoly(ethyleneoxy)ethanols, *19*:533
Alkyl phosphates
    in polystyrene, *19*:90
Alkylpoly(ethyleneoxy)ethanols, *19*:536
Alkyl polyphosphates
    surfactants, *19*:529
N-Alkyl-1,3-propanediamines, *19*:560
4-Alkylpyridines
    oxidation of, *21*:530
Alkylpyridinium halides
    surfactants, *19*:564
Alkylquinolines, *16*:873
Alkylsuccinimides, *19*:140
    alkylating agents, *19*:487
    surfactants, *19*:523
Alkyl sulfates, *19*:483
Alkyl sulfites, *19*:484
2-Alkylthio-4,6-bis(alkylamino)-*s*-triazines, *22*:185
dl-Alkylthiosuccinic anhydride, *12*:823
Alkyltrimethylammonium salts
    surfactants, *19*:564
N-Alkyltrimethylenediamines, *19*:555
Alkyl vinyl ethers
    interpolymers, *21*:422
Alkyl vinyl ethers, *1*:205, *21*:415
Alkynes
    nomenclature, *11*:288
Allantoin, *10*:78; *11*:141; *21*:112
    in beet sugar, *19*:211
    from uric acid, *21*:108
Allene, *S*:547
Allergen, *4*:528
Allergens
    in nuts, *14*:133
Allethrin, *12*:162

Allithiamine, *20*:183

Allitol, *1*:572

Allobarbital, *3*:61, 70

DL-Allobiotin, *3*:520

Allocinnamic acid, *5*:517

Allo-ocimene, *1*:633

Allophanates
  prepn from isocyanates, *12*:49

Allophane, *5*:552
  in ceramic products, *5*:561

Allose, *4*:136

Alloxan, *21*:113
  from uric acid, *21*:107
  from uric acid, *21*:109

Alloxantin, *21*:113
  from uric acid, *21*:111

Alloys
  brazing, *18*:541
  cast alloys for machining, *20*:571
  electromigration of, *S*:289
  electroplating of, *8*:66
  iron wolfram, *22*:344
  type metal, *20*:756
  welding of, *22*:245

Alloys, high-temperature. See *High-temperature alloys.*

Alloys, low-melting, *3*:532

Allspice, *9*:352, 365, 373. ((See also *Pimenta
  berry.)*

Allspice, *9*:352, 365
  essential oils as flavor, *9*:373

All-*trans*-β-carotene *21*:503

All-*trans*-3-dehydroretinol, *21*:501

All-*trans*-3,7-dimethyl-9-2',6',6'-trimethyl-1'-cycloh-
  exen-1'-yl)2,4,6,8-nonatetrae ne-1-ol, *21*:490

All-*trans*-retinal
  from vitamin A, *21*:506

Allylacetone, *12*:105

Allyl acrylate
  phys prop, *1*:288

Allyl alcohol, *6*:695.   (See also *Allyl compounds.)*
  mass spectrometry, *13*:92
  in prepn of allyl ether, *8*:494
  prod from propylene oxide, *8*:277
  in synthesis of glycerol, *10*:624

Allyl alcohol oxide, *8*:247

Allylamine, *6*:695
  prepn from acrylonitrile, *1*:341
  in prepn of acrylonitrile, *1*:347
  prepn of allyl ether, *8*:494

Allylamines
  reactions, uses, *2*:123

Allyl amyl xanthate
  in ore flotation, *22*:427

α-Allylbenzoin
  flavor and perfume materials, *9*:358
  photoinitiator, *16*:518
  prod and sales of, *8*:376
  as synthetic flavor, *9*:355

Allyl bromide, *3*:776

5-Allyl-5-*sec*-butylbarbituric acid, *3*:61, 70

Allyl caproate

Allyl chloride, *6*:562; *5*:205
  dichlorohydrins from, *5*:316
  epichlorohydrin from, *5*:320
  glycerol from, *16*:254
  manuf, *1*:922
  in prepn of acetone, *1*:161
  in prepn of allyl alcohol, *1*:922
  in prepn of allyl ethers, *8*:475
  in synthesis of glycerol, *10*:624
  toxicity of, *5*:213

Allyl chloroformate
  prop, *4*:386

Allyl cinnamate, *5*:520

Allyl compounds, *1*:**916**

Allyl crotonate, *6*:469

2-Allyl cyclohexanone, *8*:244

Allyl 2,3-dichloropropyl ether, *8*:494

Allyl ether, *8*:493
  phys prop, *8*:471
  prepn, *8*:475

N-Allylethylenediamine, *7*:28

Allyl ethyl ether
  phys prop, *8*:471
  prepn from allyl chloride, *1*:918

Allyl glycidyl ether
  phys prop, *8*:471
  prepn, *8*:494

Allyl hexanoate. See *Allyl caproate.*

Allylidene diacetate, *1*:264

Allyl iodide
  in prepn of allyl ether, *8*:494

5-Allyl-5-isobutylbarbituric acid, *3*:61

5-Allyl-5-isopropylbarbituric acid, *3*:61, 70

Allyl isothiocyanate
  as denaturant, *8*:450
  prepn from allyl bromide, *1*:918
  as synthetic flavor, *9*:354

Allyl mercaptan
  occurrence, *20*:206

Allylmercaptomethylpenicillin, *14*:653

4-Allyl-2-methoxyphenol, *15*:170

5-Allyl-5-(1-methylbutyl)-barbituric   acid,
  3:61, 70, 72

5-Allyl-5-(1-methylbutyl)-2-thiobarbituric acid, *3*:61, 70
1-Allyl-3,4-methylene dioxybenzene, *15*:170
Allyl methyl maleate, *12*:824
5-Allyl-1-methyl-5-(1-methyl-2-pentynyl)barbituric acid, *3*:61, 70
Allyl nitrile, *6*:695
Allyl reactions
  in olefins, *14*:318
Allyl sodium, *8*:494
  in alfin catalyst, *S*:928
Allylthiourea
  photography, *15*:361
Allylurea
  in chlormerodrin synthesis, *13*:242
4-Allylveratrole, *15*:170
  flavor and perfume materials, *9*:358
Allyl vinyl ether
  copolymers of, *21*:421
Allyl vinyl sulfoxide, *19*:331
Almond, bitter, *14*:193
Almond oil, *8*:450, 778, 784
Almonds, *14*:145
Alnico, *12*:761
Aloe, *9*:178; *17*:387
  as a cathartic, *4*:589
Aloe-emodin, *2*:466
Aloin, *4*:589
Alpha particles, *17*:14, 54, 66, 69
Alphasone acetophenide
  as veterinary drug, *21*:251
Alphazurine FG, *5*:868
Alstyrine, *16*:784
Altaite, *12*:280; *19*:759
Altrose, *4*:136
Alum, *7*:525; *2*:63.   (See also *Aluminum compounds.*)
  coagulant in water treatment, *22*:67
  in corundum manuf, *10*:513
  in glycerol manuf, *10*:621
Alumina, *2*:46, 50; *2*:46, 48, 50; *8*:540; *17*:227; *2*:41-58.   (See also *Aluminum compounds; Aluminum oxide; Bauxite; Corundum.*)
  alkalized and sulfur oxide recovery, *S*:359
  in ceramic armor, *S*:146
  isopropylation of phenol, *S*:30
  methylation of phenol, *S*:272
  prepn of activated, *2*:51
  in sodium-sulfur cells, *S*:124
Alumina hydrate
  manuf of aluminum trifluoride, *9*:531, 542
  pigments, *15*:577, 562

α-Alumina monohydrate. See *Boehmite.*
β-Alumina monohydrate, *2*:42. See *Diaspore.*
  phys prop, *2*:46
Alumina-silica fibers, *17*:285
  phys prop, *4*:74
Alumina, sintered
Alumina trihydrate
  prepn from sodium aluminate, *2*:8
β-Alumina trihydrate, *2*:8, 42
α-Alumina trihydrate, *2*:42, 43, 44, 45
  commercial product, *2*:50
  phys prop, *2*:46
  in prepn of adsorbent alumina, *2*:52
  in prepn of calcined alumina, *2*:54
β-Alumina trihydrate, *2*:42, 45, 47
  phys prop, *2*:46
Aluminates, *2*:6-10
Aluminides
  coating of refractory metals, *11*:33
Aluminized steel, *13*:261
Aluminohydrides, *11*:217
Aluminon, *17*:721
  as adsorbent, *1*:460
Aluminosilicate 11sieves"
Aluminous cement, *4*:696
Aluminum, *1*:**929;** *2*:1
  analysis for in water, *21*:697
  anodized, dyeing of, *2*:217
  in bearing materials, *3*:271
  in ceramic-faced armor, *S*:139
  chromate treatment, *13*:298
  in chromium metal prod, *5*:454
  cladding with, *13*:276
  coatings, *13*:252
  copper alloys with, *6*:207, 250
  corrosion resistance to water, *6*:302
  deoxidizer for steel, *18*:724
  effects on brasses, *6*:230, 241
  in egg pasteurization, *7*:668
  electromigration of, *S*:286
  in electron tubes, *8*:2
  electroplating, *8*:43
  electropolishing, *9*:564
  etching, *13*:290
  ferro alloy, *18*:751
  hardness of, *S*:74
  as hazardous chemical, *9*:290
  heating of, *10*:289
  heat of combustion of, *4*:898
  impact tests, *13*:189
  in incendiary agents, *4*:897
  inhibitors for, *6*:342

manganese compounds, *13*:3
in manuf of aluminum chloride, *2*:18
manuf from clay, *2*:62
methyl chloride, reactions with, *5*:104
in modified stainless steels, *11*:21
in prepn of organoaluminum compds, *2*:26
reactions, *2*:1
reaction with trichloroethylene, *5*:188
from seawater, *14*:156
in superalloys, *11*:21, 27
in type metal, *20*:757
use of cryolite in manuf, *9*:540, 544
welding of, *22*:247, 250
in zinc analysis, *22*:561
in zinc alloys, *22*:600
Aluminum acetate, *2*:11
in sodium aluminate prepn, *2*:6
in water-repellent cloth, *22*:139
Aluminum acetylsalicylate, *17*:733
Aluminum-alizarin lake, *2*:5
Aluminum alkoxide
as catalyst, *8*:357
Aluminum alkyls, *2*:26-41. (See also *Alkylaluminums.*)
Aluminum alloys, *1*:**955**
welding, *22*:247, 250
Aluminum and aluminum alloys, *1*:**929**
Aluminum anilide, *2*:413
Aluminum aspirin, *17*:733
Aluminum benzoate, *15*:490
Aluminum-bismuth alloy, *1*:959
Aluminum borohydride, *3*:702
composition of, *6*:227
Aluminum brass, inhibited
Aluminum bromide, *2*:24. ((See also *Alumi-
num tribromide.)*
carbon tetrachloride reaction with, *5*:131
chloroform reaction with, *5*:122
in heat transfer, *10*:857
H₂S reaction, *19*:377
tetrachloroethylene reaction with, *5*:198
Aluminum bronzes, *6*:193, 207, 250
casting of, *6*:254
prop of, *6*:255, 256, 259
Aluminum butoxide
as catalyst, *1*:82
Aluminum butylate; *7*:414, *8*:346
Aluminum chlorhydroxyallantoinate, *21*:113
Aluminum chloride, *2*:17-24; *6*:678; *12*:131,
155
as catalyst in manuf of amines, *8*:455
as catalyst in production of ethylene, *8*:503
hot-dipped coatings, *13*:261

in prepn of ethers, *8*:474
Aluminum chloride hexahydrate
in prepn of aluminum diacetate, *2*:12
Aluminum chloride-pyridine
as catalyst, *15*:162
Aluminum-chlorine cells, *S*:133
Aluminum chromate, *5*:499
Aluminum compounds, *2*:**1**
Aluminum diacetate, *2*:11
prepn, *2*:12
Aluminum difluoride, *9*:531
Aluminum diformate, *2*:14
Aluminum dihydroxyallantoinate, *21*:113
Aluminum dihydroxyaminoacetate
as antacid, *2*:430
Aluminum diphenylamide, *7*:43, 44
Aluminum distearate, *7*:276
Aluminum-2-ethylhexoate, *7*:278
Aluminum ethoxide
as catalyst, *1*:82
Aluminum fluoride
prepn, *9*:655
prepn, *2*:4
Aluminum fluosilicate, *6*:802; *9*:658
Aluminum foil, *2*:2
as thermal insulator, *11*:832
Aluminum formate, *2*:14
in prepn of aluminum triformate, *2*:14
Aluminum formoacetate, *2*:14, 16
Aluminum hexacyanoferrate(II), *12*:26
Aluminum hexaurea sulfate triiodide, *2*:619
Aluminum hydrate
pigments, *15*:516
Aluminum hydride, *11*:205; *2*:572
as diuretic, *7*:253
prepn of aluminum diacetate, *2*:12
prepn of aluminum triformate, *2*:14
in prepn of organoaluminum compounds,
2:27
in propellants, *8*:660
Aluminum hydroxide, *2*:2, 3
Aluminum hydroxide gel
as antacid, *2*:429
Aluminum industry
carbon refractories in, *4*:239
Aluminum iodide, *2*:24
Aluminum isobutylate, *8*:346
Aluminum isopropoxide, *1*:640; *2*:12; *12*:110;
16:567; *1*:844
alcoholysis catalyst, *20*:488
Aluminum isopropylate. See *Aluminum isopropoxide.*

Aluminum lactate, *12*:183, 184
Aluminum laurate, *9*:893
Aluminum-lead alloy, *1*:958
Aluminum-lithium alloys, *12*:537, 538
Aluminum-magnesium alloy, *1*:958
  in Ziegler synthesis of aluminum alkyl, *2*:34
Aluminum monoacetate, *2*:11
  prepn, *2*:12
Aluminum monofluoride, *9*:531
Aluminum myristate, *9*:893
Aluminum naphthenate, *7*:273
Aluminum nitrate, *2*:25
Aluminum nitride, *13*:823; *4*:74; *12*:628;
  13:823
Aluminum oleate, *7*:273
  source of gallium, *10*:311
Aluminum ore. (See also *Bauxite.*)
Aluminum organometallic compounds, *2*:26-
41
Aluminum oxide, *2*:41-58. (See also *Alumina;*
  Bauxite; Corundum.)
  as abrasive, *1*:25
  composition, *2*:56
  in conversion of aniline to diphenylamine,
    7:14
  flame-retardant polyesters, *20*:831
  in glass forming, *10*:538, 542, 544
  hydrated, in polystyrene, *19*:90
  in opaque armor, *S*:150
  phys prop, *4*:74
  prop, *10*:513, 756
  refractory, *17*:231, 232
  refractory, *17*:244
  in synthetic gem prod, *10*:510
  titanium grinding, *20*:365
  in transparent armor, *S*:152
  zero point of charge, *9*:384
Aluminum oxide hydrate
  as paper pigment, *14*:516
Aluminum palmitate, *7*:283
Aluminum perchlorate, *5*:71
Aluminum permanganate, *13*:33
Aluminum phenoxide, *12*:69
  alkylation catalyst, *S*:29
  as catalyst in Friedel-Crafts reactions,
    10:161
  phenolic aldehydes, *15*:162
Aluminum phosphate
  in dental cement, *6*:779
Aluminum phosphates, *15*:275
Aluminum phosphide, *15*:293
  phosphatized, *13*:296

pigments, *15*:546, 564
  as protective covering for cables, *11*:815
  in pyrotechnics, *16*:826
Aluminum resinate, *7*:279
Aluminum salicylate, *17*:728
Aluminum salts
  as antiseptic, *2*:622
  tanning of fur skins, *10*:301
Aluminum *tert*-butoxide, *12*:113
  as antiperspirant, *6*:355
  as catalyst, *8*:407; *15*:161
  as catalyst in commercial alkylation with
    ethylene, *1*:884
  as catalyst for isomerizing *n*-butane, *3*:817
  as catalyst in prepn of ethyl chloride, *8*:501
  catalyst in prepn of tris-2-chloroethyl phos-
    phate, *8*:530
  cumene synthesis, *15*:150
  dimethyl sulfate reaction with, *5*:107
  in dye industry, *7*:481
  ethylbenzene prod, *19*:58
  in Friedel-Crafts reactions, *10*:138
  in heat transfer, *10*:857
  as hydroisomerization catalyst, *11*:451
  reaction with ethylene oxide, *8*:528
  reaction with ethyl ether, *8*:479
  reaction with silicon tetrafluoride, *9*:652
  sulfolane reaction, *19*:251
Aluminum silicate, *18*:157
  pigments, *15*:525
  in papermaking, *14*:501
  slag-former in welding, *22*:242
Aluminum silicates, *2*:1, 413
  in ceramics, *4*:771
  pigments, *15*:511
Aluminum-silicon alloys, *1*:956; *18*:545
Aluminum silicon-bronze, *6*:220
Aluminum soaps, *7*:281
Aluminum sodium chlorohydroxylactate,
  12:183
Aluminum sodium lactate, *12*:183
Aluminum stearate, *7*:276, 285, 304; *9*:893
  as abherent, *1*:9
  in paints, *14*:475
Aluminum subacetates, *2*:11
Aluminum sulfamate, *19*:248
Aluminum sulfate, *2*:58-65; *9*:542, 892
  as antiperspirant, *6*:355
  coagulant for water treatment, *22*:94
  fire-fighting foams, *S*:368
  in food processing, *10*:10
  in manuf metallic soap, *7*:274

and nitrogen fixation, S:617
from sulfuric acid, 19:481
in tanning furs, 10:301
Aluminum tantalide, 19:648
Aluminum telluride, 19:763
Aluminum tetraisopropoxide, S:617
Aluminum thiophenoxide
alkylation catalyst, S:30
Aluminum titanate, 20:413
Aluminum-titanium alloys, 20:368
Aluminum triacetate, 2:11. (See also *Aluminum acetate.*)
prepn, 2:11
Aluminum trichloride. See *Aluminum chloride.*
Aluminum trifluoride, 9:529, 544
manuf, 9:531
prop, 9:529
Aluminum trifluoride trihydrate, 9:532
Aluminum triformate, 2:14
Aluminum triiodide, 2 4
Aluminum-zinc alloy, 1:958
Aluminum-zirconium alloys, 22:620
in reduction of nitro compounds, 2:98.
(See also *Mercury.*)
Alumoweld, 13:277
Alunite
source of potassium alum, 2:64
Amagat's law, 20:135
Amalgam alloy
in dental materials, 6:807
Amalgamation, 13:218, 232
Amalgams
in reduction of nitro compounds, 2:98. (See also *Mercury.*)
Amandin
in almonds, 14:125
Amantadine hydrochloride, S:11
Amaranth, 5:866, 873
in carbonated beverages, 4:342, 343, 344
Amaranth Lake, 5:869
Amaromycin, 12:633, 634
Amaryllidaceae alkaloids, 1:770
Amatol, 8:627
characteristics, 8:632, 633, 634
Amber, 17:382
succinic acid from, 19:134
Ambergris, 14:725
Amberlite 1R-112
in alkylation of phenol, 1:894
in ceramics, 4:772
Amblygonite, 12:530
Ambrein, 14:725
Ambrette seed, 14:194

Ambrettolite, 18.24
Amebiasis, 20:70
m-aminophenol as catalyst in prepn of, 2:213, 214
antiamebic drugs, 20:73
experimental chemotherapy, 20:71
Amebicides. (See also *Protozoal infections.*)
aziridine structure in, 11:542
American fossil resin, 17:383
Americium. (See also *Actinides.*)
as fission product, 14:106
Americium-241, 17:92
Americium-243
U.S. prodn, 21:6
Americium compounds
crystal structure, 1:366
phys prop, 1:367
Amesite, 5:547
Amiben
herbicide, 22:201
metabolism, 22:202
soil chemistry, 18:536
Amidase, 14:670
Amide hydrolase, 14:697
Amides
cyanoethylation, 6:647
esterification, 8:339
as herbicides, 22:198
as hypnotics, 11:516
metabolism of herbicide amides, 22:198
reduction of, 2:107
Amides, acid, 2:**66**
Amides, fatty acid, 2:**72**
Amidohalophosphates, 15:323
Amidol
photography, 15:373
Amidomycin, 16:336
Amidone, 2:149, 150
prepn, 2:149, 150
Amidosulfuric acid, 9:677; 19:242, 304, 306, 491. (See also *Sulfamic acid.*)
sulfating agent, 19:528
sulfonation, 19:282
swimming pool stabilizer, 22:127
Amidrazone dyes
for color photography, 5:832
Aminals, 12:98
Amination by reduction, 2:**76**
Amine-boranes, 3:728
Amine D, 17:504
as cardiovascular agents, 4:521
Amine oxidase inhibitors
Amine oxides, S:**32**

surfactants, *19*:556

Amine perchlorates, *5*:74

Amines, *2*:**99**
  alkylation, *19*:488
  carbonylation, *6*:643
  dialkylchloroalkyl, *2*:139
  fatty, *2*:127
  as inhibitors in acid pickling, *6*:321, 322,
    323, 329
  lower aliphatic, *2*:116
  surfactants, *19*:555

Amines, aromatic
  alkylation, *1*:895

Amine sulfamates, *19*:246

Aminitrozole
  amebicides, *20*:85

Amino Naphthol Red 6B, *2*:881

4-Amino-N¹⁰-methyl-peteroylglutamic acid,
  S:85

Amino R acid, *13*:715

Amino S acid, *13*:715

p-Aminoacetanilide, *15*:217

4'-Aminoacetanilide, *15*:217, 222

Aminoacetic acid. See *Glycine*.

Amino acids, *2*:**156**. ((See also *Proteins*.)
  sonochemical reactions, *20*:781

α-Aminoacrylic acid
  in wool, *22*:401

1-Aminoadamantane hydrochloride
  antiviral agent, *21*:455

1-Aminoadamatane hydrochloride, *S*:11

5-Amino-2-(*p*-aminoanilino)benzenesulfonic acid,
  7:49

7-Amino-2-(7-12*p*-(4-amino-6-sulfo-1-naphthyl-
  ạazo)-(phenylazo13-8-hydroxy-6-sulfo-2-naph-
  thylazo)-1-naphthol-3-sulfonic acid, *7*:470

5-Amino-2-anilinobenzenesulfonic acid, *7*:49

Aminoanthraquinone, *16*:50

1-Aminoanthraquinone, *7*:478; *2*:448, 449,
  450, 451, 452
  prod, *2*:457

2-Aminoanthraquinone, *7*:478; *2*:356, 357, 448
  prod of, *2*:458
  thiazole dyes, *20*:190

Aminoanthraquinones
  di-, *2*:447

1-Aminoanthraquinone-2-sulfonic acid,
  19:307

Aminoazonaphthalene, *7*:465

Aminoazotoluene, *15*:217

Aminobenzaldehydes, *3*:366

p-Aminobenzaldehyde thiosemicarbazone
  antiviral agent, *21*:454

9-Aminobenzanthrone, *2*:495

p-Aminobenzenestibonic acid, *2*:585

p-Aminobenzenesulfonamide, *19*:261

m-Aminobenzenesulfonic acid. See *Metanilic
  acid*.

o-Aminobenzenesulfonic acid. See *Orthanilic
  acid*.

p-Aminobenzenesulfonic acid. See *Sulfanilic
  acid*.

3-Aminobenzenesulfonic acid, *19*:299

4-Aminobenzenesulfonic acid, *19*:299, 311

2-Aminobenzenethiol, *6*:646

p-Aminobenzoate, *6*:356
  UV absorbers, *21*:117

m-Aminobenzoic acid, *3*:435
  phys prop, *1*:238

o-Aminobenzoic acid. *3*:434.   (See also *Anthranilic acid*.)
  dyes, *19*:433

p-Aminobenzoic acid, *19*:261, 274; *3*:34, 435
  antithyroid agent, *20*:270
  dyes, *19*:433
  phys prop, *1*:238
  yeast requirement, *22*:517

p-Aminobenzoic acid (PABA)

2-Aminobenzothiazole
  thiazole dyes, *20*:192

2-Aminobenzothiazoles, *10*:916

1-Aminobenzotriazole, *10*:918

3-Aminobenzotrifluoride, *9*:778

α-Aminobenzylpenicillin, *3*:15, 16; *14*:664, 698

Aminobiphenyls, *19*:298

Aminoboranes, *3*:732

Aminoborazines, *3*:731

2-Amino-1-bromo-3-chloroanthraquinone, *7*:478;
  2:487

1-Amino-2-bromo-4-hydroxyanthraquinone, *2*:459
  prepn, *2*:479

1-Amino-2-bromo-4-(*p*-toludino)anthraquinone,
  7:478

1-Amino-1-butanol, *3*:868, 874, See *Butyraldehyde-ammonia*.

2-Amino-1-butanol
  phys prop, *1*:8, 26

1-Amino-4-bromo-2-anthraquinone sulfonic
  acid, *2*:452; *7*:478
  prod, *2*:459
  use in dyeing, *2*:515

β-Aminobutyric acid, *6*:466

γ-Aminobutyric acid, *2*:189; *16*:645

2-Amino-4-carbamoyl butyric acid, *2*:157, 199

2-Amino-3-carbamoylpropionic acid, *2*:157

ε-Aminocaproic acid
  polyamide from, *16*:2
ω-Aminocaproic acid, *16*:30
2-Aminocarbazole, *3*:412
Aminocarboxylates
  surfactants, *19*:513
7-Aminocephalosporanic acid, *14*:699
α-Amino-*o*-cresols
2-Amino-3-chloroanthraquinone, *2*:480
1-Amino-5-(and 8)-chloroanthraquinone,
  7:478
2-Amino-7-chloroanthraquinone
  prepn, *2*:95
1-Amino-5-chloroanthraquinone, *7*:478
2-Amino-5-chlorobenzoxazole, *16*:677
2-Amino-4-carbamoylbutyric acid. See *Glutamine.*
5-Amino-1,3-dihydroxybenzene, *16*:202
  prepd from guanidine, *10*:736
2-Amino-4,6-dichlorophenol, *2*:216
3-Amino-3,6-dideoxy-D-mannose, *16*:138
4-Amino-4,6-dideoxy-D-mannose, *16*:138
2-Amino-4-chloro-5-nitrophenol, *2*:215
  antiamebic agents, *20*:88
2-Amino-4-chloro-6-nitrophenol, *2*:216
2-Amino-6-chloro-4-nitrophenol, *2*:216
2-Amino-4-chlorophenol, *16*:676; *2*:216, 222
  in prepn of 6-amino-4-chloro-1-phenol-2-
  sulfonic acid, *2*:222
6-Amino-4-chloro-1-phenol-2-sulfonic acid,
  7:478; *2*:222
2-Amino-6-chloro-1-phenol-4-sulfonic acid,
  2:223
6-Amino-4-chloro-*m*-toluenesulfonic acid,
  7:478
2-Amino-5-chloro-*p*-toluenesulfonic acid,
  7:478
2-Aminocymene
  condensation with dialkylaminoalkyl chlorides, *2*:143
Aminodichlorotriazine, *10*:931
2-Amino-4,7-dihydroxypteridine, *10*:932
2-Amino-3,5-diiodobenzoic acid, *16*:82
2-Amino-1,3-dibromoanthraquinone, *2*:451,
  480
1-Amino-2,4-dibromoanthraquinone, *2*:451,
  459, 479; *7*:478
3-Amino-2,5-dichlorobenzoic acid, *18*:536;
  22:201
1-Amino-2,4-dihydroxyanthraquinone, *2*:458,
  460
2-Amino-4,6-dihydroxpteridine, *10*:932
p-Aminodimethylaniline
  from guaicol, *21*:185

6-Amino-3,3-dimethyl-7-oxo-4-thia-1-azabicyclo-
  (3.2.0.)heptane-2-carboxylic acid, *14*:655
4-Amino-2,6-dimethylpyrimidine, *19*:268
2-Amino-4,6-dimethylpyrimidine, *12*:157
2-Amino-4,6-dinitrophenol, *2*:216. See *Picramic acid.*
Aminodiol, *13*:826
p-Aminodiphenylamine, *15*:218; *7*:42, 565
p-Aminodiphenylamine,
  in hair coloring, *10*:795
p-Aminodiphenylamine hydrochloride
  in hair coloring, *10*:795
Aminodiphenylamines, *15*:220
p-Aminodiphenylamine sulfonic acid
  in hair coloring, *10*:795
4-Aminodiphenylamine-2-sulfonic acid,
  15:218, 222
Aminoditolyls, *7*:42
Amino epsilon acid, *13*:715
2-Aminoethanesulfonic acid, *19*:311, 521.
  (See also *Taurene.*)
2-Aminoethanethiol. (See also *2-Mercapto-
  ethylamine.*)
2-Aminoethanethiol, *10*:804
2-Aminoethanethiol hydrochloride, *6*:646
Aminoethylethanolamine
  phys prop, *1*:811
2-Aminoethyl hydrogen sulfate, *19*:486, 494
N-Aminoethylpiperazine, *15*:644
2-Amino-2-ethyl-1,3-propanediol, *18*:660
  phys prop, *1*:826
  uses, *1*:830
3-Amino-α-ethyl-2,4,6-triiodohydrocinnamic acid,
  17:136
3-Amino-2-fluoropyridine, *9*:798
Aminoform. See *Hexamethylenetetra*
Amino G acid, *7*:478; *13*:715
Aminoguanidine, *10*:740, 902
Aminoguanidine bicarbonate
  synthesis of amitrole, *22*:212
  prepn of tetracene, *8*:592
1-Aminoguanidine hydrogen carbonate
1-Amino-4-guanidovaleric acid, *2*:157
Amino H acid, *13*:716
Aminohaloboranes, *3*:736
  effectiveness as antioxidant, *2*:590
7-Aminoheptanoic acid, *16*:36
6-Aminohexanoic acid, *13*:882; *16*:2
1-Aminohydantoin, *13*:856
6-Amino-6,11-dehydrodibenzo[*be*]*thiepine*, *10*:934
2-Amino-2-(hydroxymethyl)-1,3-propanediol,
  18:660
4-Amino-5-hydroxy-1-napthalenesulfonic acid, *13*:725

2'-Amino-9,10-dihydro-1,2-iminodianthraquinone, 2:454

3-Amino-4-hydroxyarsenosobenzene hydrochloride, 2:730. See *Oxophenarsine hydrochloride*.

2-Amino-3-hydroxybutyric acid. See *Threonine*.

4-Amino-3-hydroxy-1-napthalenesulfonic acid, 13:725

5-Amino-6-hydroxy-2-napthalenesulfonic acid, 13:725

6-Amino-4-hydroxy-2-napthalenesulfonic acid, 13:724

7-Amino-4-hydroxy-2-napthalenesulfonic acid, 13:724

8-Amino-4-hydroxy-2-napthalenesulfonic acid, 13:724

4-Amino-5-hydroxy-1,3-napthalenedisulfonic acid, 13:725

4-Amino-5-hydroxy-1,7-napthalenedisulfonic acid, 13:725

3-Amino-5-hydroxy-2,7-napthalenedisulfonic acid, 13:725, 726

2-Amino-6-hydroxypurine, 10:913

Aminohydroxynapthalenedisulfonic acids, 13:725

Aminohydroxynapthalenesulfonic acids, 13:724

2-Amino-3-*p*-hydroxyphenylpropionic acid. See *Tyrosine*.

2-Amino-3-hydroxypropionic acid. See *Serine*.

Aminohydroxyxanthenes, 22:435

2-Amino-3-imidazolepropionic acid. See *Histidine*.

2-Amino-1,2'-iminodianthraquinone, 2:455

1-Amino-3-imino-isoindolenines, 16:206

Aminoiminomethanesulfinic acid, 10:799

Aminoiminomethanesulfinic acid, 10:799

2-Amino-3-indolylpropionic acid. See *Tryptophan*.

2-Amino-5-iminopyrrolenines, 16:206

2-Aminoindan, 2:388

Aminoketone dyes, 7:471, 475

α-Aminolevulinic acid, 2:189

Aminolysis, 2:368

2-Amino-3-mercaptopropionic acid. See *Cysteine*.

5-Amino-2-mercaptothiazoles, 10:915

Aminomercury chloride, 2:619

3-Aminomesitol, 15:372

1-Amino-2-methoxy-5-methylbenzene. See *Cresidine*.

6-Amino-2-methylbenzothiazole, 16:296

2-Amino-3-methylbutyric acid. See *Valine*.

4-(Aminomethyl)-5-hydroxy-6-methyl-3-pyridinemethanol-, 16:806

4-Amino-3-methyl-1-naphthol hydrochloride vitamin K$_7$, 21:585

2-Amino-3-methylpentanoic acid. See *Isoleucine*.

2-Amino-2-methyl-l-propanol (AMP)
phys prop, 1:826
in soaps, 18:431
uses, 1:829

2-Amino-2-methyl-1-proponal
as diuretic, 7:253

2-Amino-2-methyl-1,3-propanediol (AMPD)
phys prop, 1:826

2-Amino-4-methylmercaptobutyronitrile
as intermediate in prepn of methionine, 1:271
uses, 1:830

2-Amino-4-methylpentanoic acid. See *Leucine*.

1-Amino-2-methyl-2-propanol, 6:648

5-(Aminomethyl)pyrimidine, 20:178

2-Amino-6-(methylsulfonyl)benzothiazole, 20:192, 193

2-Amino-4-methylthiobutyric acid. See *Methionine*.

4-Aminomorpholine
prepn from bis(2-chloroethyl)ether, 8:487

5-Amino-1-naphthalenesulfonic acid, 13:710

3-Amino-1,5-naphthalenedisulfonic acid, 7:478; 13:715

8-Amino-1,6-naphthalenedisulfonic acid, 13:714

4-Amino-1,3-naphthalenedisulfonic acid, 13:714

5-Amino-1,3-napthalenedisulfonic acid, 13:715

6-Amino-1,3-napthalenedisulfonic acid, 7:478; 13:715

7-Amino-1,3-naphthalenedisulfonic acid, 7:478; 13:715

2-Amino-1,5-napthalenedisulfonic acid, 13:715

4-Amino-1,5-napthalenedisulfonic acid, 13:715

4-Amino-1,6-napthalenedisulfonic acid, 13:715

8-Amino-1,6-napthalenedisulfonic acid, 13:715

4-Amino-1,7-napthalenedisulfonic acid, 13:715

4-Amino-2,6-naphthalenedisulfonic acid, 13:714

1-Amino-2,7-naphthalenedisulfonic acid, *13*:714

3-Amino-2,7-napthalenedisulfonic acid, *13*:715

2-Amino-1-naphthalenesulfonic acid, *13*:713; *7*:478

4-Amino-1-naphthalenesulfonic acid, *13*:709; *19*:3

6-Amino-1-naphthalenesulfonic acid, *13*:713

7-Amino-1-naphthalenesulfonic acid, *13*:714

8-Amino-1-naphthalenesulfonic acid, *7*:478; *13*:713

6-Amino-2-naphthalenesulfonic acid, *13*:709, 714

5-Amino-2-naphthalenesulfonic acid, *13*:712

7-Amino-2-naphthalenesulfonic acid, *13*:714

8-Amino-2-naphthalenesulfonic acid, *13*:712

8-Amino-1,3,5,6-napthalenetetrasulfonic acid, *13*:717

4-Amino-1,3,5-napthalenetrisulfonic acid, *13*:716

6-Amino-1,3,5-napthalenetrisulfonic acid, *13*:716

8-Amino-1,3,5-napthalenetrisulfonic acid, *13*:716

4-Amino-1,3,6-napthalenetrisulfonic acid, *13*:716

7-Amino-1,3,6-napthalenetrisulfonic acid, *13*:717

8-Amino-1,3,6-napthalenetrisulfonic acid, *13*:716

Aminonaphthalenesulfonic acids, *13*:709

Aminonapthalenetetrasulfonic acids, *13*:716

4-Aminonaphthalimide
derivatives as optical brighteners, *3*:745

4-Amino-1-naphthol

4-Amino-1,8-naphthol,-2,'4'-dimethylphenylimide dye, *9*:496
menadione derivatives, *21*:597

5-Amino-1-naphthol, *13*:724

8-Amino-2-naphthol, *13*:724

8-Amino-1-naphthol-3,5,disulfonic acid, *13*:725

8-Amino-1-naphthol-3,6-disulfonic acid, *2*:871; *13*:725

8-Amino-1-naphthol-5,7-disulfonic acid, *7*:469; *13*:725. See *Chicago acid.*

1-Amino-8-naphthol-2,4-disulfonic acid, *7*:469. See *Chicago acid.*

7-Amino-8-naphthol-3,6-disulfonic acid, *13*:726

8-Amino-1-naphthol-3,6-disulfonic acid monosodium salt, *7*:478

Aminonaphthols, *13*:724

7-Amino-1-naphthol-3-sulfonic acid
sodium salt, *7*:478

5-Amino-1-naphthol-3-sulfonic acid, *13*:724

6-Amino-1-naphthol-3-sulfonic acid, *13*:724
dyes from, *2*:897, 899
sodium salt, *7*:478

7-Amino-1-naphthol-3-sulfonic acid, *13*:724
azo dyes, *2*:882

8-Amino-1-naphthol-5-sulfonic acid, *13*:725

1-Amino-8-naphthol-4-sulfonic acid, *13*:725

1-Amino-2-naphthol-4-sulfonic acid, *13*:725

1-Amino-2-naphthol-4-sulfonic acid, (1,2,4-acid), *7*:478

1-Amino-2-naphthol-6-sulfonic acid, *13*:725

2-Aminonitriles, *6*:669

p'-Amino-*p*-nitrobenzanilide, *15*:222

4-Amino-4'-nitrodiphenyl sulfide, *19*:270

2-Amino-4-(5-nitro-2-furyl)thiazole hydrobromide, *13*:856

4-Amino-2-nitrophenol, *2*:218
in hair coloring, *10*:795

2-Amino-4-nitrophenol, *19*:432; *2*:217
in hair coloring, *10*:795
prepn, *2*:95

2-Amino-5-nitrophenol, *2*:215, 217

6-Amino-4-nitro-1-phenol-2-sulfonic acid, *2*:222

4-Amino-6-nitro-1-phenol-2-sulfonic acid, *2*:222

2-Amino-6-nitro-1-phenol-4-sulfonic acid, *2*:223

Aminonitrostilbenedisulfonic acid, *19*:7

4-Amino-4'-nitro-2,2'-stilbenedisulfonic acid, *19*:2, 7

2-Amino-5-nitrothiazole, *21*:248
thiazole dyes, *20*:194

9-Aminononanoic acid
polyamide from, *16*:3

6-Amino-2-octylundecanedioic acid, *1*:253

13-Amino-18-norestradiol-3-methyl ether, *18*:873

3-Amino-2-oxazolidinone, *13*:856

2-Aminooxazoline, *6*:562

ω-Aminopelargonic acid, *14*:420; *16*:3

6-Aminopenicillanic acid, *14*:654, 697
reactions of, *14*:663

Aminopenicillins, *14*:685

2-Aminopentanedioic acid. See *Glutamic acid.*

m-Aminophenol, *2*:213, 214

o-Aminophenol, *6*:614, 649; *2*:215
developing agent, *15*:371
dye for furs, *10*:306

in hair coloring, *10*:795
from *o*-nitrophenol, *2*:97
in prepn of *o*-(2,4-dinitroanilino)phenol, *2*:218
prepn with electrolytic cell, *2*:97
3-Aminoperchlorylbenzene, *9*:604
p-Aminophenol, *2*:218
 4-amino-1-phenol-2,6-disulfonic acid, *2*:224
 dye for furs, *10*:306
 effectiveness as antioxidant, *2*:590
 in hair coloring, *10*:795
 p,p'-iminodiphenol, *2*:220
 p-methylaminophenol, *2*:221
 p-(2-Naphthylamino)phenol, *2*:221
 in prepn of *p*-(4-aminoanilino)phenol, *2*:218
 in prepn of *p*-(4-amino-1-naphthyl-amino)phenol, *2*:218
 in prepn of *p*-anilinophenol, *2*:219
 in prepn of *p*-(4-dimethylaminoanilino) phenol, *2*:220
 in prepn of *p*-(2,4-dinitroanilino)phenol, *2*:220
 in prepn of *n*-(*p*-hydroxyphenyl)glycine, *2*:220
6-Amino-1-phenol-2-sulfonic acid, *2*:224
6-Amino-1-phenol-2,4-disulfonic acid, *2*:224
4-Amino-1-phenol-2,6-disulfonic acid, *2*:224
p-Aminophenol hydrochloride, *2*:219, 413
 in hair coloring, *10*:795
-Aminophenols, *2*:218, 221
Aminophenols, *2*:**213**
 diaminophenols, *2*:221
 *meta*aminophenols, *2*:213
 *ortho*aminophenols, *2*:215
p-Aminophenols, *2*:218, 221. (See also *Aminophenols.*)
4-Aminophenol-2-sulfonic acid
5-Aminophenol-2-sulfonic acid
 in hair coloring, *10*:795
2-Aminophenol-4-sulfonic acid
 in hair coloring, *10*:795
4-Amino-1-phenol-2-sulfonic acid
 in prepn of 4-amino-6-nitro-1-phenol-2-sulfonic acid, *2*:222
2-Amino-1-phenol-4-sulfonic acid, *2*:223
4-Amino-*m*-phenylene dithiocyanate, *20*:193
p-Aminophenylnaphthotriazolesulfonic acids, *19*:4
3-Amino-1-phenyl-2-pyrazoline, *6*:650
1-Amino-3-phenylpropane
 in hydro refining, *11*:437
2-Amino-3-phenylpropionic acid. See *Phenylalanine.*

4-Aminophthalic anhydride
 polyimides from, *S*:750
Aminophylline
 as cardiovascular agent, *4*:521
 as diuretic, *7*:253
Aminopolycarboxylic acids
 as chelating agents, *6*:7
20α-Aminopregn-5-en-3β-ol-3-acetate, *18*:876
 prepn, *8*:280
1-Amino-2-propanol
2-Aminopropionic acid, *2*:157
3-Aminopropionic acid, *2*:157
3-Aminopropionitrile, *6*:648
3-Aminopropylcellulose, *6*:660
 from propylenimine, *11*:534
2-Aminopropyl *tert*-butyl peroxide
Aminopterin
 cancer-inhibiting effects, *S*:81
 folic acid antagonist, *21*:489
3-Amino-5-pyrazolones, *10*:911
2-Aminopyridine, *2*:391; *6*:615; *9*:798; *16*:802
 condensation with dialkylchloroalkylamines, *2*:143
 manuf, *2*:367
 prepn from propargyl alcohol, *1*:601
2-Aminopyrimidine
3-Aminopyridine, *16*:802
4-Aminopyridine, *9*:798
Aminopyrine, *1*:159
3-Aminopyrine, *9*:798
4-Aminopyrogallol hydrochloride, *16*:209
Aminopyrroles, *16*:850
Aminoquinoline, *16*:893
Aminoquinolines, *16*:875
Amino resin polymerization
 kinetics of, *2*:235
Amino resins
 commercial type, *2*:246
 identification, *2*:244
 monomers, *2*:229, 230
 polymerization, *2*:229, 230
 uses, *2*:244
Amino resins and plastics, *2*:**225**
5-Aminosalicylic acid
 sulfonation, *19*:298
p-Aminosalicylic acid, *3*:30; *17*:734
 m-aminophenol as catalyst in prepn of, *2*:213, 214
 tuberculostatic drug, *19*:40
Aminosidine, *19*:42
1-Aminosorbitol, *19*:225
4-Aminostilbene, *19*:14

4-Aminostilbene-2-sulfonic acid
  Aminostilbenenaphthotriazoles, *19*:2
  in prepn of monoazole optical brighteners,
    3:742
Aminosuccinic acid. See *Aspartic acid.*
2-Aminothiazole, *10*:915
5-Aminotetrazole, *10*:918
4-Amino-2,3,5,6-tetrafluorobenzoic acid,
    9:795
    reaction with *N*-acetylsulfanilyl chloride,
      1:152
    thiazole dyes, *20*:192
5-Amino-2-thiazolecarbonamide, *6*:588
5-Amino-2-thiazolethiols, *10*:915
2-Amino-3,4,6-trichlorophenol, *2*:217
2-Amino-4,6,7-trihydroxypteridine, *10*:932
4-Amino-3,5,6-trichloropicolinic acid, *22*:210
4-Amino-2,3,5-trichloropyridine, *22*:210
2-Amino-6-thiocyanatobenzothiazole, *20*:193
o-Aminothiophenols, *10*:916
5-Aminothio-2-thiazolecarboxamide, *6*:588
o-Aminotoluene. See *o-Toluidine.*
2-Amino-*s*-triazine-4,6-dithiol, *6*:567
2-Amino-*s*-triazine-4,6-diol
  prepn, *6*:568
3-Amino-1,2,4-triazole, *10*:917
  herbicide, *22*:212
4-Amino-1,2,4-triazole, *10*:917
2-Amino-4,6,7-trihydroxypteridine, *10*:932
ω-Aminoundecanoic acid, *15*:794
11-Aminoundecanoic acid
  polymerization, *16*:92 n
Aminovinyl ketones, *12*:90
Amisometradine
  as diuretic, *7*:253, 254
  soil chemistry, *18*:536
Amitrole
  in hair-coloring prepn, *10*:797
  in hair-treatment prepns, *10*:802
  in hair-waving lotion, *10*:788
  herbicide, *22*:212
  hydrogen use in manuf of, *11*:370
  in manuf of ethyleneamines, *7*:31
  in manuf ammonium nitrate, *2*:324
  metabolism of, *22*:212
  in prepn of ammonium bicarbonate, *2*:315
  yeast prod, *22*:534
Ammelide, *22*:187
  cyanuric acid from, *20*:667
  prepn, *6*:568
  from urea, *21*:38
Ammeline, *10*:931

cyanuric acid from, *20*:667
Amminozirconium halogenides, *22*:624
Ammonal, *8*:631-633
Ammonia, *6*:331
Ammonia, *2*:**258**
  alkylation of, *2*:106
  in ammoniation of superphosphate, *9*:117
  ammonolytic reactions, *2*:332
  aquatic plant nutrients, *S*:325
  arylation of, *2*:112
  carbon dioxide reaction with, *4*:356
  carbon disulfide reactions with, *4*:372
  as flame retardant, *9*:304
  catalytic agent for automobile exhaust, *S*:66
  chlorine reaction with, *4*:916
  coal-tar distilling, *19*:670
  converter, *13*:805
  cryogenic properties, *6*:472
  diacyl derivatives of, *2*:70
  in diarylamines, *7*:40
  as fertilizer, *S*:338
  free ammonia in urea, *21*:54
  In fertilizer industry, *9*:53
  fumes, *18*:403
  graphite compounds with, *4*:311
  high pressure oxidation, *13*:804
  hypochlorite reactions with, *4*:916
  hypochlorous acid reaction with, *4*:916
  liquid as plasticizer, *15*:773
  in manuf of acrylonitrile, *1*:346
  in manuf of alkanolamines, *1*:814
  in manuf of ammonium nitrate, *8*:596
  in manuf of guanidine, *10*:739
  reactions with acrylic esters, *1*:292
  in manuf of ammonium dithiocarbamate,
    4:372
  dried by alumina, *2*:52
  in manuf hydrogen cyanide, *6*:576, 577
  manuf melamine, *6*:583
  mercury compounds, *13*:248
  oxidation compression system, *13*:805
  oxidation to nitric acid, *13*:800
  oxidation. See *Nitric acid.*
  prepn of difluoroacetic acid, *9*:770
  produced in kidneys, *7*:252
  prop, *8*:688
  properties compared to water, *21*:675
  reaction on anthraquinonesulfonic acids,
    2:456
  reactions with boranes, *3*:693
  reaction with chlorine trifluoride, *9*:589
  reaction with dialkylchloroalkylamines,
    2:142

reaction with ethylene oxide, 8:528
reactions with formaldehyde, 10:85
reaction with perchloryl fluoride, 9:603
scrubbing and sulfur oxide recovery, S:359
in Solvay process, 2:318
synthesis of and nitrogen fixation, S:605
in water, 21:700
Ammonia alum, 2:26, 64
Ammonia, aqua, 9:58
Ammonia battery, 3:135
Ammonia cyanoethylation, 6:648
  in Castner process, 6:591
  in deuterium and tritium, 6:903
  prod potassium cyanide, 6:598
Ammonia dynamites, 8:640
Ammonia liquor, 2:299
Ammonia-soda process, 1:712. (See also
  Solvay process; Sodium carbonate.)
Ammonia synthesis gas
  from carbon monoxide, 4:438
Ammoniated mercury ointment, 13:240
Ammoniated superphosphate, 9:32
Ammonium acetate, 2:313, 314
  cyanoethylation of, 6:648
  in dye application, 7:525, 528
  in manuf of acetamide, 1:144
Ammonium acid fluoride. See Ammonium bi-
  fluoride.
Ammonium acid oxalate, 14:369
Ammonium acid purpurate
  test for uric acid, 21:107
Ammonium acid urate, 21:112
Ammonium adipate, 1:415
Ammonium alkyl sulfates, 19:244
Ammonium aluminum sulfate. See Ammonia
  alum.
Ammonium aluminum sulfate, 2:64
  in food processing, 10:10
  in manuf of corundum, 10:513
  in tanning fur skins, 10:301
Ammonium benzoate, 3:433
Ammonium biborate, 3:647
Ammonium bicarbonate, 2:315
Ammonium bichromate
  sensitizer, 16:514
Ammonium bifluoride, 9:548, 589, 620
Ammonium bisulfate. See Ammonium hydro-
  gen sulfate.
Ammonium bromide, 2:262, 263, 317, 319
  prepn of diphenylamine, 7:40
Ammonium carbamate, 4:69, 356
  ammonia-carbon dioxide reaction, 21:40

equilibrium conversion, 21:41
properties of, 21:41
in urea fertilizer technology, S:344
urea from, 21:37
Ammonium carbonate, 12:117; 2:316
  in amino acid synthesis, 2:165
  in manuf ammonium sulfate, 2:329
  in manuf glycine, 2:350
Ammonium chlorate
  thermochemical constants as propellant,
  8:681
Ammonium chloride, 2:316-319
  analysis for, 2:313
  by-product soda ash manuf, 1:734
  catalyst in prepn m-aminophenol, : 2:213
  coal-tar distillation, 19:661
  as diuretic, 7:251
  in electric cells, 3:111
  as fertilizer, 9:77
  flux, 18:542
  in prepn of glutamic acid, 2:203
Ammonium chlorite, 5:31
Ammonium chloroiridate, 15:847, 856
Ammonium chloropalladate, 15:868
Ammonium chloroplatinate, 15:834, 848, 864;
  19:467
Ammonium chlororhodate, 15:857
Ammonium chlororuthenate, 15:847
Ammonium chlorozincate, 22:608
Ammonium chromate, 5:497
Ammonium chrome alum, 5:494
Ammonium chromium rhodanilate, 5:494
Ammonium compounds, 2:**313**. (See also
  Quaternary ammonium compounds.)
Ammonium cryolite, 9:542
Ammonium cyanate
  from urea, 21:39
Ammonium cyanide, 6:599; 2:207
Ammonium cyclamate, 19:601
Ammonium dicarbonatozirconate, 22:640
Ammonium dichromate
  manuf of, 5:484
  phys prop of, 5:481
Ammonium difluorophosphate, 9:643
Ammonium dihydrogen phosphate
  in dental materials, 6:796
  nonlinear optical materials, S:624
Ammonium dithiocarbamate, 10:915
  prepn, 4:372
Ammonium diuranate
  fuel pellet manuf, 21:27
  thermal decomposition, 21:32

uranium refining, *21*:25

Ammonium ferrous ferrocyanide, *2*:310

Ammonium fluoberyllate, *3*:461, 475

Ammonium fluoborate, *9*:564, 565

Ammonium "fluoborax", *9*:558

Ammonium fluoride, *2*:413; *7*:41; *9*:548, 589.
  (See also *Ammonium bifluoride.*)
  in prepn of boron trifluoride, *9*:558, 620
  in prepn of nitrogen trifluoride, *9*:629

Ammonium fluoroantimonite, *9*:549

Ammonium fluorohafnate, *10*:764

Ammonium fluorosulfate, *9*:679

Ammonium fluosilicate, *9*:620, 655
  phys prop, *9*:658, 659

Ammonium formate, *10*:105; *12*:116

Ammonium fructoheptonate, *19*:230

Ammonium glutamate, *2*:205

Ammonium hexachloroosmate(IV), *15*:874

Ammonium hexachloropalladate(IV), *15*:868

Ammonium hexachloroplatinate(IV), *15*:864, 867

Ammonium hexachloroplumbate(IV), *12*:271

Ammonium hexachlororuthenate(IV), *15*:859

Ammonium hexachlorotitanate, *20*:388

Ammonium hexacyanoferrate(II), *12*:26

Ammonium hexacyanoferrate(III), *12*:31

Ammonium hexafluorogermanate, *9*:585

Ammonium hexafluorotitanate, *9*:684

Ammonium hexofluogallate(III), *10*:322

Ammonium hydrogen fluoride. See *Ammonium bifluoride.*

Ammonium hydrogen sulfate, *2*:329; *19*:243

Ammonium hydrogen sulfate, *2*:329; *19*:243
  in pulping of wood, *2*:295. ((See also *Pulp.*)
  in pulpmaking, *12*:369

Ammonium hydrosulfide, *2*:330

Ammonium hydroxide
  in paints, *14*:469

Ammonium lactate, *12*:176

Ammonium magnesium sulfate
  in manuf of corundum, *10*:513

Ammonium metaphosphate, *15*:249

Ammonium metavanadate, *21*:160

Ammonium molybdate, *13*:656; *15*:869

Ammonium molybdate
  thioglycolic acid analysis, *20*:200
  zirconium separation, *22*:658

Ammonium monofluorophosphate, *9*:642

Ammonium naphtherate, *7*:284
  in Amatol, *8*:627
  blasting explosives, *8*:642
  as diuretic, *7*:251
  as explosive, *8*:581, 593, 595

explosive characteristics, *8*:633
  in prepn of Cyclonite, *8*:623
  sensitivity characteristics, *8*:632
  stability characteristics, *8*:634

Ammonium nitrate, *2*:320-329; *6*:513
  in Ammonal, *8*:631
  thermochemical additive constant, *8*:681
  prod, *9*:61
  as fertilizer, *9*:59-65
  melt granulation, *S*:341
  yeast prod, *22*:534

Ammonium nitrate carbonate, *9*:65

Ammonium nitrate sulfate, *9*:65

Ammonium nitrite, *2*:329
  prod from waste pickle liquor, *2*:309, 311

Ammonium oleate soap
  in dye base, *10*:796

Ammonium orthophosphates, *15*:239, 249

Ammonium oxalate, *14*:368

Ammonium oxalatoferrate, *14*:368

Ammonium ozonide, *14*:415

Ammonium parawolframate, *22*:352

Ammonium parawolframate, *22*:356
  from scheelite, *22*:338

Ammonium pentaborate, *3*:648

Ammonium pentafluorostannite, *9*:682

Ammonium perchlorate, *5*:67, 69; *8*:627
  decomposition as propellant, *8*:669, 671
  explosive characteristics, *8*:633
  in explosives, *8*:644
  in rocket propellants, *8*:660
  sensitivity characteristics, *8*:632
  stability characteristics, *8*:634

Ammonium peroxydisulfate, *14*:756
  in prepn of polyacrylonitrile, *1*:319

Ammonium peroxysulfate, *10*:5
  polyvinylpyrrolidone gel, *21*:431
  as starch modifier, *10*:17

Ammonium perrhenate, *17*:414

Ammonium persulfate, *14:756*

Ammonium phosphate, *9*:32, 78, 79
  in dentrifices, *6*:848
  fermentation of molasses, *22*:523
  as fertilizer, *S*:338
  as fertilizer, *9*:125
  fire retardant on wood, *22*:376
  in matches, *13*:162
  to phosphorylate cellulose, *9*:306
  prepn for use in liquid fertilizer, *9*:136
  yeast prod, *22*:534

Ammonium phosphate-ammonium chloride
  as fertilizer, *9*:130

Ammonium phosphate nitrate
  as fertilizer, 9:129; 15:290
Ammonium phosphates, 15:274.   (See also
  Ammonium orthophosphates.)
Ammonium phosphate sulfate
  as fertilizer, 9:129
Ammonium phosphate-urea
  as fertilizer, 9:129
Ammonium picrate, 8:618
  explosive characteristics, 8:633
  in Picratol, 8:628
  sensitivity characteristics, 8:632
  stability characteristics, 8:634
Ammonium polyphosphate, 15:274, 290;
  9:130
  as fertilizer, S:342
Ammonium polyphosphates, 15:239, 248
  crystalline, 15:249
Ammonium polysulfide
  p,p'-iminodiphenol, 2:220
  p-nitroaniline reduction, 15:217
  polysiloxanes, 18:225
  prepn, 8:474
  sulfoxides, 19:331
Ammonium polysulfides, 2:331
Ammonium pyrrolidine dithiocarbamate,
  21:697
Ammonium radical, 2:264
Ammonium reineckate. See Reinecke salt.
Ammonium saccharin, 19:595
Ammonium salicylate, 17:729
Ammonium sulfamate, 19:243. ((See also
  Ammonium amino sulfate.)
  as flame retardant for textiles, 9:304
  as an herbicide, 22:214
Ammonium sulfate, 2:299, 302, 312, 329-331;
  6:512, 513; 7:528, 563; 19:243
  as ferroelectric, 9:5
  as fertilizer, 9:65
  fire retardant on wood, 22:376
  manuf, 9:66
  in manuf ammonium chloride, 2:318
  in prepn of ammonia alum, 2:64
  in prepn of glutamic acid, 2:203
  prod from waste pickle liquor, 2:309, 311
  prod, 9:66
  prop, 9:67, 550
  in premetalized dyeing, 7:577
  recovery from coke oven gas, 2:302
  from sulfuric acid, 19:481
  in textile printing, 7:576
Ammonium sulfatozirconate, 22:651
Ammonium sulfide, 2:330,

tetrasulfur tetranitride, 19:406
Ammonium sulfite
  action with thiodiglycol, 10:648
  ammonium thiosulfate from, 20:239
  glycerol prod, 22:529
  in manuf ammonium chloride, 2:318
Ammonium syngenite, 9:119
Ammonium tartrate
  fermentation, 19:134
Ammonium tetrachloroiodide, 2:263
Ammonium tetrafluoroberyllate, 9:552, 553
  as ferroelectric, 9:5
Ammonium tetraoxalate, 14:369
Ammonium tetraphenylborate, 16:384
  choline assay with, 5:407, 408
Ammonium tetrasulfatozirconate, 22:662
Ammonium tetrathiocanato diamminechrom-
  ate(III), 5:494
Ammonium tetrathiowolframate, 22:354
Ammonium thallium arsenate
  as ferroelectric, 9:5
Ammonium thiocyanate, 2:227; 6:455; 7:37;
  9:312; 10:616, 758
  manuf of, 4:372
  photography, 15:378
  in textile printing, 7:576
  zinc analysis, 22:561
Ammonium thioglycolate, 10:786
Ammonium thiosulfate, 20:239
  analysis, 20:240
  developers, 16:502
  manuf, 20:239
  photography, 15:382
  uses, 20:240
Ammonium thiosulfate
  specifications, 20:240
Ammonium tricarbonatozirconate, 22:639,
  656
Ammonium trifluorostannite, 9:682
Ammonium trifluorozincate, 9:684
Ammonium trinitrophenolate. See Ammonium
  picrate.
Ammonium trisulfatocerate, 4:851
Ammonium urate, 21:107
Ammonium vanadate, 21:171
Ammonium wolframate, 22:351
Ammonium zirconyl carbonate, 22:633, 639
Ammonolysis, 2:332
  of anthraquinonesulfonic acids, 2:456
Ammoxidation, 22:485
Amobarbital, 3:61, 70, 72
Amodiaquine
  amebicides, 20:84

amebicide, *20*:87

Amosite. See *Asbestos.*

Ampere
  definition of, *S*:990
  and nitrogen fixation, *S*:610
  synergism of, *S*:474

Amphetamine
  phenylacetic acid, *15*:214

d-Amphetamine, *16*:672

Amphibole. See *Asbestos.*

Amphotercin B, *3*:8, 35; *16*:133, 138, 141
  veterinary fungicide, *21*:249

Ampicillin, *14*:664, 687
  amebicides, *20*:82
  veterinary uses, *21*:244

Amprolium, *10*:19
  as anticoccidial, *21*:247

Amygdalin, *6*:672; *3*:360
  hydrolysis in nuts, *14*:138

Amyl acetate, *2*:379
  as drycleaner and solvent, *7*:318
  prod and sales, *8*:376
  specifications, *8*:373

sec-Amyl acetate
  prepn, *8*:342

tert-Amyl alcohol, *2*:375

Amyl alcohols, *2*:**374**

Amylase
  action and uses, *8*:182, 187
  in milk, *13*:514

α-Amylase, *15*:685

α-Amylase, *S*:297, 301
  action and uses, *8*:182, 185, 188, 190, 192
  in beer brewing, *3*:308
  microorganism producing, *8*:210

β-Amylase, *8*:190, 193; *12*:863
  in beer brewing, *3*:308
  starch cleavage, *18*:676

Amylases, *16*:622
  fungal in grain mashes, *22*:522
  from yeast, *22*:518

α-Amylases
  in food products, *10*:9
  starch cleavage, *18*:676
  uses, *8*:191

Amyl butyrate. See *Isopentyl butyrate.*

Amyl chloride
  prepn of *n*-amyl ethers, *8*:493

n-Amyl chloroformate
  prop, *4*:386

α-Amylcinnamaldehyde. (See also *α-Pentyl-cinnamaldehyde.*)
  flavor and perfume materials, *9*:358

Amylenes. See *Pentylenes.*

n-Amyl ether, *8*:493
  phys prop, *8*:471

sec-Amyl ether, *8*:493
  phys prop, *8*:471

t-Amyl ethyl ether
  prepn from ethyl alcohol and 2-methyl-2-butene, *1*:898

Amyl lactate
  prepn, *8*:353

Amyl nitrate, *12*:119
  as cardiovascular agent, *4*:521
  veterinary uses, *21*:252

Amyloglucosidase, *8*:183; See *Glucoamylase.*
  action and uses, *8*:190
  in grain mashes, *22*:522

Amyloglucosidase-*β*-amplase
  action & uses, *8*:183

Amylo-1,6-*d*-glucosidase
  uses, *8*:190, 194

Amyloids
  in nuts, *14*:129

Amylopectin, *18*:672, 679

Amylose, *18*:672, 679
  as indicator, *11*:557

p-tert-Amylphenol
  stabilizer for poly(vinyl formal), *21*:307

Amylphenols, *2*:626

Amyl salicylate See *Isopentyl butyrate.*

α-Amyl-3-methoxy-4-hydroxycinnamaldehyde
  from vanillin, *21*:181

Anabasine, *1*:765
  synthesis of nicotinic acid, *21*:513

Anacardic acid
  in cashew nutshells, *14*:133

Analcite
  in ion exchange, *11*:871

Analgesics, *11*:508

Analgesics and antipyretics, *2*:**379**

Analog computer, *6*:25, 32

Anaphoresis, *8*:23

Anaphylaxis
  relation of histamine to, *11*:35
  veterinary, *21*:252

Anatase, *20*:392
  in ceramics, *4*:773

Anauxite, *5*:545

Andalusite
  in ceramics, *4*:771
  refractory, *17*:231, 233

Androestone, *18*:879

Androgens, *11*:93

Androstane, *18*:840

Androstano-[3,2-c]pyrazoles, *18*:866

Androsta-1,4,6-triene-3,17-dione, *18*:851

Androstrane
  structure of, *11*:79

Androst-4-ene-3,16-dione, *18*:854

Androst-4-ene-3,17-dione, *11*:93

Andrusson process
  manuf hydrogen cyanide, *6*:577
  manuf sodium cyanide, *6*:592

Anergen, *2*:151

Anesthetics, *2*:**393**; *11*:508

Anethole, *15*:162, 163, 170; *10*:729; *19*:819, 821; *9*:355
  as denaturant, *8*:450
  as synthetic flavor and perfume materials, *9*:358

α-Angelica lactone, *19*:225

Angelica root, *14*:194
  essential oil as flavor, *9*:352

Angiography, *17*:132

Anglesite, *12*:208
  zero point of charge, *9*:384

Angolamycin, *12*:633, 634

Angolide, *16*:336

Angostura, *1*:771

Anhydrite, *4*:14, 17, 24; *9*:88, 94; *19*:343, 356

Anhydroaglycone, *12*:644

N',N-Anhydrobis(β-hydroxyethyl)biguanide, *21*:457

Anhydrocarboxyglycine
  shrinkproofing wool, *22*:409

5a,6-Anhydro-7-chlortetracycline, *20*:14

Anhydroenneaheptitol, *1*:596; *10*:922

Anhydroerythromycin, *12*:642

Anhydroformaldehyde aniline, *2*:234

Anhydrone
  as dyeing agent, *7*:387

Anhydrone magnesium perchlorate. See *Anhydrone*

Anhydro-oleandomycin, *12*:643

Anhydrosorbitol esters, *19*:545
  ethoxylated, *19*:546

Anhydrotetracycline, *20*:22

4a,12a-Anhydrotetracycline, *20*:20

5a,6-Anhydrotetracycline, *20*:14

Anhydrous β-D-glucose, *6*:919

Anhydrous borax, *3*:642. (See also *Sodium tetraborate.*)

Anhydrous cupric sulfate
  prepn, *6*:277

Anhydrous α-D-glucose, *6*:919

Anhydrous perchloric acid, *5*:62

Anhydrous sodium sulfate, *7*:525

as assistant in dyeing wool with acid dyes, *7*:523

Anhydrovitamin A, *21*:495

Anil, *1*:644

Anileridine, *2*:384, 385; ,

Anilides
  from aniline, *2*:414

Aniline, *6*:649; *2*:**411**; *7*:478
  as antioxidant, *2*:600

Aniline Black, *2*:864
  as dye for furs, *10*:307
  prod on cotton, *7*:565
  in textile printing, *7*:576

Aniline dyes
  from shale, *18*:18

Aniline hydrochloride, *2*:214, 412, 419
  in dyeing of furs, *10*:307
  in prod of Aniline Black, *7*:565

Aniline oil, *7*:478
  in prod of Aniline Black, *7*:565

Aniline phosphate, *2*:412

Aniline Pink, *7*:465

Aniline point
  of solvents, *18*:574

Aniline sulfate, *2*:419, 424

Aniline trichloracetate, *8*:418

Aniline-2-sulfonic acid. See *Orthanilic acid.*

Aniline-3-sulfonic acid. See *Metanilic acid.*

Aniline-4-sulfonic acid. See *Sulfanilic acid.*

β-Anilinoacrolein anil hydrochloride, *6*:615

2-Anilinoethanol, *18*:660

8-Anilino-1-naphtholenesulfonic acid, *7*:49

6-Anilino-1-naphthol-3-sulfonic acid, *7*:49

p-Anilinophenol, *2*:214, 219, 413; *7*:40

3-Anilinoprop-2-enanil, *16*:288

3-Anilinopropionitrile, *6*:649

γ-Anilinopyrotartaric acid lactam, *12*:84

C-Anilino-N-phenylsuccinimide, *12*:842

Anils
  from aniline, *2*:414

Animal diseases
  use of drugs, *21*:242

Animal feeds. See *Feeds, animal.*

Animal glue. See *Glue.*

Animal hides. See *Leather.*

Anisaldehyde, *1*:645; *15*:160, 162, 163

p-Anisaldehyde
  flavor and perfume materials, *9*:358

Anise, *14*:194
  essential oil as flavors, *9*:352, 363
  star, *14*:194

Anise oil
  as denaturant, *8*:450

Anise seed, 9:365, 368
Anisic acid
  phys prop, 1:238
p-Anisic acid, 15:163
Anisidine, 5:600, 601
o-Anisidine, 7:478
  in hair coloring, 10:795
p-Anisidine, 6:649
  dye synthesis, 20:726
p-Anisidine hydrochloride, 6:636
Anisidines, 2:422; 19:298
Anisil, 12:140
Anisole, 15:148, 165, 171; 19:487
  from methylation of phenol, S:272
  perchlorylation reaction, 9:604
Anisomycin
  amebicide, 20:80
Anisotropy
  diamagnetic, 14:47
Anisoyl chloride, 15:163
p-Anisyl chloroacetate, 15:163
Annealing
  in alloys, 11:12
  of brasses, 6:221
  of metals, 13:323
  stress-relief, of copper alloys, 6:239
Annofoline, 1:763
Annotinine, 1:759, 763
Anobial, 2:617
Anodes
  Anodic coating, 13:250
  in electroplating, 4:53
  graphite, 4:190, 205; 5:54
  titanium, in chlorate manuf, 5:54
Anodic control
  of electrochemical corrosion, 6:290, 318
Anodic inhibitors, 6:318
Anodic oxidation, 13:300
"Anomalous water", 21:682
Anopheles mosquitos
  controls, 11:678
Anorthite, 4:2, 4, 769
  in ceramics, 4:769
Anorthoclase, 4:769
Anovlar, 6:68, 87
Anserine, 2:190
  in fish, 9:319
Antacids, gastric, 2:427
Antazoline.HCL
  as antihistamine, 11:36
Antergan, 2:152
Anthacene Chromate Brown EBA, 7:527
Anthanthrone Orange, 15:587

Anthelmintics, 6:605
  veterinary, 21:249
Antheridiol, 18:845
Anthocyanin
  in wines, 22:312, 316
Anthocyanose, 8:202
Anthophyllite. ((See also Asbestos.)
Anthophyllite, 19:608
  in ceramics, 4:772
Anthracene, 2;432, 433, 434
  from coal tar, 19:657
  in coal tar, 19:668
  in dye intermediates, 7:469
  photodimerization, 15:334
Anthracene cathartics, 4:588
Anthracene oil
  carbon black from, S:95
  coal-tar distillate, 19:668
Anthracite. ((See also Coal.)
  characteristics as furnace fuel, 10:280
  filters in water treatment systems, 22:70
  from coal tar, 19:657
  in fluorination of organic compounds, 9:694, 698
  gasification of, 10:361
  in manuf of calcium chloride, 4:104
  manuf of perchloryl fluoride, 9:606
  recarburizer for steel, 18:724
  sulfur dioxide, 19:419
Anthradipyrazoles
  antiamebics, 20:88
1,2,9,10-Anthradiquinone, 2:470
Anthraflavin, 2:466
Anthragallol, 2:466
Anthrahydroquinone, 2:433
Anthramine, 7:49
Anthramycin methyl ether
  amebicide, 20:82
Anthranilic acid, 21:522
  alkaloids from, 1:771
  phys prop, 1:238
  saccharin synthesis, 19:598
Anthranol, 2:433
Anthrapurpurin, 2:466
Anthra[1,9]-pyrazol-6(2H)-one, 2:449. See 1,9-Pyrazolanthrone.
Anthrapyrazolone dyes, 2:527
Anthrapyrimidine dyes, 2:527
Anthrapyrimidine Yellow, 15:587
Anthraquinhydrone, 16:556
Anthraquinoid dyes, 20:188
Anthraquinone, 2:431; 7:478
  autooxidation of, 11:397

Anthraquinone acid dyes, *2*:511
Anthraquinoneacridone dyes, *2*:527
Anthraquinonecarbazole dyes, *2*:463, 523
Anthraquinone derivatives, *2*:**438**
Anthraquinonedisulfonic acid, *19*:315
1,4,-Anthraquinonedisulfonic acid
　prepn, *2*:479
1,5,-Anthraquinone disulfonic acid, *2*:440, 443
　manuf, *2*:445
1,8-Anthraquinone disulfonic acid, *2*:440, 443
　manuf, *2*:445
Anthraquinonedisulfonic acids. See *Anthra-
　quinonesulfonic acids*
Anthraquinone dyes, *2*:**501;** *7*:471, 475;
　19:424
1-Anthraquinonesulfonic acid, *7*:478
　polysiloxanes, *18*:225
2-Anthraquinonesulfonic acid, *2*:357, 440, 441,
　442, 443
　manuf, *2*:446
　sodium salt, *7*:469
Anthraquinonesulfonic acids, *2*:433, 439, 481,
　482; *19*:298
2-Anthraquinonesulfonic acid, sodium salt,
　7:469
Anthraquinonethiazole dyes, *2*:527
Anthraquinone wool dyes, *2*:511
Anthraquinon-1-ylmercury sulfate, *2*:444
Anthrarufin, *2*:466
Anthrasol Printing Yellow I 3G, *2*:894
Anthraxylon. See *Vitrain.*
Anthrimides, *2*:462
Anthrone, *2*:433
　sugar analysis, *19*:164
Antianemia preparations. See *Iron compounds;
　Pharmaceuticals.*
Antibac, *2*:617
Antibacterial macrolides, *12*:632
Antibacterials
　as veterinary drugs, *21*:243
Antibiotics, *2*:**533**. (See also *Pharmaceuticals.*)
　as amebicides, *20*:80
　as anticoccidials, *21*:246
　in cancer therapy, *S*:81, 86
　β-lactam, *14*:652
　macrolide, *16*:332
　neomycins, *19*:41
　paromomycins, *19*:41
　from streptogramin, *16*:331
　from *Streptomyces* spp, *16*:326, 328
　streptomycins, *19*:33
　"synergistic", *16*:331
　tetracyclines, *20*:1

　as veterinary drugs, *21*:243
Antiblocking agents. See *Abherents.*
Anticaking agents
　in food products, *10*:4
Anticancer agents, *S*:81
Anticoagulant
　bishydroxy coumarin as, *6*:432
Anticoagulants, *5*:599
　antagonists of, *5*:602
Anticoccidials, *21*:246
Anticonvulsants, *11*:**508**
Antidepressants, *16*:659
Antiencephalitis drugs
　aziridine structure in, *11*:542
Antiepileptics, *11*:508
Antiferromagnetism, *12*:737
Antifertility agents. See *Contraceptive drugs.*
Antifoaming agents. ((See also *Foam.*)
　in food products, *10*:4
Antifolic acid antagonists
　cancer-inhibiting effects, *S*:81
Antifreeze
　ethylene glycol as, *10*:642
Antifreezes, *2*:**540**
Antifriction bearings, *3*:295. ((See also *Bear-
　ing materials.*)
Antigenic drift
　in influenza viruses, *21*:452
Antigens
　in viruses, *21*:452
　ultrasonic extraction, *20*:781
Antihistamine agents. ((See also *Histamines.*)
　commercially available, *11*:36-41
Antihistamine agents, *11*:**35**
Antihistamines, *2*:151; *16*:674
　use of lithium amide in, *12*:541
Antihistomonal agents, *21*:246
Anti-icing agents
　in gasoline, *10*:477
Anti-inflammatory agents
　veterinary, *21*:250
Antiknock agent
　methylcyclopentadrenyl manganese tricar-
　bonyl as, *6*:699
Antiknock agents
　economic aspects, *12*:293
Antiknock gasoline
　ethylene bromide in, *3*:751
Antimalarial, *2*:151
Antimalarials
　aziridine structure in, *11*:542
Antimetabolites
　anticancer agents, *S*:81

in cancer therapy, *S*:84
Antimicrobial food preservatives, *10*:12
Antimonial lead, *12*:249, 255
  prop, *12*:257
Antimonic(V) acid, *2*:574, 585
Antimony, *2*:**562**; *17*:124, 125
  antiamebic agents, *20*:94
  effects on brasses, *6*:227, 241
  effects on copper, *6*:195, 210
  electrodeposition of zinc, *22*:587
  in germanium photodiode, *15*:421
  manganese compounds, *13*:4
  photoemitters, *15*:403
  removal from lead ore, *12*:226
  silane compounds, *18*:200
  in tin-lead alloys, *18*:543
  in tin oxide glass coatings, *10*:592
  in type metal, *16*:497
  white opaque pigments, *15*:506
  in type metal, *20*:756
Antimony(III) bromide. See *Antimony tribromide.*
Antimony(III) chloride. See *Antimony trichloride.*
Antimony(III) iodide, *2*:576
Antimony(III) nitrate, *2*:578
Antimony(III, V) oxide. See *Antimony dioxide.*
Antimony(III) perchlorate trihydrate, *2*:578
  schistosomiasis treatment, *14*:545
Antimony(III) sulfate, *2*:578
Antimony(III) sulfide. See *Antimony trisulfide*
Antimony(V) chloride, *2*:576. (See also *Antimony pentachloride.*)
Antimony(V)chlorofluoride, *9*:799
Antimony(V) fluoride. See *Antimony pentafluoride.*
Antimony(V) oxide. See *Antimony pentoxide.*
Antimony(V) sulfide, *2*:577
Antimony bromides, *2*:576, 577
Antimony chlorides, *2*:575, 576
Antimony compounds, *2*:**570**. (See also *Stib-* entries.)
Antimony dioxide, *2*:574
Antimony fluoride
  as catalyst, *8*:361
Antimony fluorides, *2*:576
Antimony fluorosulfate, *9*:679
Antimony-gold alloy, *10*:692
Antimony hydride, *2*:563, 570, 572. (See also *Stibine.*)
Antimony isopropoxide, *1*:848
Antimony lactate, *12*:184
Antimonylcatechol, *2*:579, 580

Antimonyl chloride, *2*:575. (See also *Antimony oxychloride.*)
Antimony lithium thiomalate, *2*:585; *14*:546
Antimonyl sulfate, *2*:578
Antimony oxide
  as catalyst, *8*:361
  as electrode, *7*:739
  as ester-exchange catalyst, *16*:167
  and fire-retardancy, *S*:476
  flame retardant, *15*:762
  as flame retardant, *S*:953
  flame-retardant polyesters, *20*:831
  as opacifier in enameling, *8*:167
  pigment, *15*:496
  in polystyrene, *19*:90
  in producing flame-retardant textiles, *9*:305
  tartar emetic, *19*:740
Antimony oxides, *2*:573
Antimony oxychloride
  in producing flame-retardant textiles, *9*:305
Antimony pentachloride, *2*:576; *19*:392
Antimony pentafluoride, *2*:576
  reaction with wolfram hexachloride, *22*:347
Antimony pentasulfide, *2*:577
  in fluorination of organic compounds, *9*:694, 698
  manuf of perchloryl fluoride, *9*:606
Antimony pentoxide, *2*:574, 580; *9*:550
  as ester-exchange catalyst, *16*:175
  as glass former, *10*:538
Antimony potassium tartrate, *19*:736
Antimony sesquisulfide, *2*:577
Antimony silicooxide
  pigments, *15*:507
Antimony sodium gluconate, *2*:585; *14*:546
Antimony sodium tartrate
  as drug, *2*:584
  schistosomiasis treatment, *14*:545
Antimony sodium thioglycollate, *2*:585
Antimony sulfide, *13*:223
  in explosive priming compositions, *8*:653
  in matches, *13*:160
Antimony sulfides, *2*:577
Antimony tetrafluoride monofluorosulfate, *9*:680
Antimony tetroxide, *2*:563, 574. (See also *Antimony oxide.*)
Antimony thioglycollamide, *2*:585
Antimony tribromide, *2*:576
Antimony trichloride, *2*:575. ((See also *Antimony chlorides.*)
  extractive distillation agent, *22*:473
  reactions, *2*:579, 580

thionyl chloride, *19*:399
Antimony trifluoride, *2*:575, 576; *9*:549
  in fluorination of organic compounds, *9*:694
Antimony triiodide, *2*:576
Antimony trioxide, *2*:564, 573, 581
  fire-retardant paint, *22*:376
  flame retardant, *15*:801
  as glass former, *10*:538
  pigments, *15*:506
  in prepn of tartar emetic, *2*:578
  reactions, *2*:579, 580
Antimony trisulfide, *2*:564, 570, 577
Antimony vermilion, *15*:525
Antimony yellow, *15*:535
Antimycin A, *12*:648
Antiobiotics
  biosynthesis of β-lactam, *14*:668
Antioxidants, *2*:**588**
  alkylphenols as, *S*:27
  in cosmetics, *6*:371
  in fats, *8*:781; *13*:181
  impurity in copper, *S*:265
  use in food preservation, *10*:14
Antiozonants, *17*:526
Antipellagra vitamin, *21*:509
Antiperspirant preparations
  aluminum diacetate, *2*:13
  aluminum formate, *2*:16
  aluminum formoacetate, *2*:16
Antiperspirants, *6*:354
Antiplasticizers, *15*:775
Antipyretics. See *Analgesics and antipyretics.*
Antipyrine, *1*:159; *2*:391; *18*:432, 451
  from sodium, *18*:432
Antirachitic activity
  and Vitamin D, *21*:549
Antirust agents
  head cyanamide as, *6*:559
  in gasoline, *10*:477
Antirust paints
  lead cyanamide in, *12*:269
Antiseptics, *2*:**604;** *18*:806
Antiseptics and disinfectants, *2*:**604**
Antispasmodic drugs, *2*:151
Antistatic agents, *2*:**649**
Antisyphilitics, *13*:242
Antithyroid preparations, *20*:260, 268; *10*:195
  extrathyroid effects, *20*:271
  inorganic ions, *20*:269
  iodide, *20*:268
  methods of testing, *20*:271
  1-methyl-2-imidazolethiol, *20*:268
  naturally occurring, *20*:269

organic, *20*:270
  prepd from guanidine, *10*:737
  thiouracil, *20*:268
Antitrypsins
  in soybeans, *18*:602
Antitumor agents
  acrylamine derivatives, *1*:783
Antivitamins, *21*:489
Antlerite, *6*:135, 170
Apatite, *9*:83; *12*:620
  as abrasive, *1*:25
  in ceramics, *4*:772
Apatites
  phosphatic, *15*:280
Apional, *16*:209
Apocyrine, *21*:193
Apoenzymes, *17*:445
Apoglucinic acid, *19*:211
Apomorphine
  veterinary emetic, *21*:252
α,β-Apooxytetracyclines, *20*:14
Apples
  malic acid in, *12*:837
Apple wax, *22*:161
Apricot kernel oil, *8*:778, 784
Aprobarbital, *3*:61, 70
Apron conveyors, *6*:100
Aquadrol, *20*:472
Aquametry, *2*:**673**
  psychrometry, *1*:481
Aqua regia
  in gold refining, *10*:684
Aquatic plant
  source of nutrients, *S*:325
Aqua Vitae, *1*:502
Arabinogalactan, *10*:751. See *Larch gum.*
Arabinose, *4*:134
D-Arabinose, *19*:228
L-Arabinose
  in prepn of mannitol, *1*:574
Arabitol. See *Alcohols, polyhydric.*
D-Araboascorbic acid, *2*:751
Arabogalactan
  from wood, *22*:383
D-Arabonic acid, *19*:228, 230
Arabonic lactone
  structure, *4*:138
Arachic acid. See *Arachidic acid.*
Arachic acid, *8*:813
  phys prop, *1*:231
Arachidonic acid, *8*:815
  in eggs, *7*:664
Arachidyl alcohol. See *1-Duodecanol.*

Arachidyl alcohol, *1*:543
Arachin
  in peanuts, *14*:125
Aragonite, *4*:2,8; *12*:415
Aralkylsuccinic anhydrides, *12*:823
Arboviruses, *21*:452
L-Arcanose, *12*:639
Arc furnaces, *10*:257
Arcs, carbon, *4*:206
Arc welding
  use of helium and argon, *10*:880, 884
Arecoline, *1*:765; *14*:533
  from areca nut, *14*:141
  veterinary drug, *21*:252
Arenes
  nomenclature, *11*:288
Arenesulfonic acids. ((See also *Sulfonic acids.*)
  in paint removers, *14*:487
Argentic compounds. See *Silver compounds.*
Argentite, *18*:284
Argentometric titration
  in acetaldehyde determination, *1*:90
Argentous fluoride. See *Silver fluoride.*
Arginine, *2*:157
  in wine, *22*:316
L(+)-Arginine salt, *3*:522
Argols, *19*:735
Argon, *10*:862
  cryogenic prop of, *6*:472, 474
  dried by alumina, *2*:52
  in electron tubes, *8*:2
  as lubricant, *12*:606
  in makeup gas in ammonia synthesis, *2*:273
  in phototubes, *15*:398
  phys prop, *10*:868
  plasma arcs, *15*:345
  separation from air, *10*:875
  separation of isotopes, *7*:138, 166
  shielding gas in welding, *22*:244
  in tobacco smoke, *20*:516
Argyrodite, *10*:519
Ariboflavinosis, *17*:448
Aristol, *2*:619
Aristolochic acid, *1*:759, 769
Armagnac, *1*:516
Arnel, *1*:109. (See also *Acetate and triacetate fibers.*)
Arnica, *10*:802
Aroclors, *5*:291, 292
Aromatic compounds, *2*:701
  classes, *2*:702
Aromatic hydrocarbons
  carbanion alkylation of, *1*:893

  carbonium ion alkylation of, *1*:888
  free-radical alkylation of, *1*:893
Aromaticity, *2*:**701**
Aromatic oils
  as lubricants, *12*:564
Aromatics
  in commercial hexane, *11*:2
β-Aroylacrylic acids, *12*:822
Arrhenius
  theory of acids and bases, *1*:214
Arsanilic acid, *10*:19
  poultry feed additive, *2*:715
Arsenates, *2*:722
  as corrosion inhibitor, *6*:324
  effects on brasses, *6*:228, 241
  effects on copper, *6*:195, 210
  manganese compounds, *13*:4
  removal from lead ore, *12*:226
  restriction of use in embalming fluids, *8*:100
  silane compounds, *18*:200
Arsenic, *2*:**711**
  analysis for in water, *21*:698
  antiamebic agents, *20*:93
Arsenicals, *11*:680
Arsenic-76, *17*:123
  electrodeposition of zinc, *22*:587
  veterinary uses, *21*:252
Arsenic(III) acetate, *2*:724
Arsenic(III) bromide, *2*:721
Arsenic(III) chloride, *2*:721
Arsenic(III) fluoride, *2*:720
Arsenic(III) iodide, *2*:721
Arsenic(III) oxide. See *Arsenic trioxide.*
Arsenic(III) sulfate, *2*:724
Arsenic(III) sulfide, *2*:723
Arsenic(V) fluoride, *2*:721
Arsenic(V) oxide, *2*:722
Arsenic(V) sulfide, *2*:724
Arsenic acid, *2*:722
  as cotton defoliant, *2*:715
  dye oxidant, *20*:693
  as a herbicide, *22*:214
Arsenic acids
  Béchamp reaction in prepn of, *2*:725
  Bart reaction in prepn of, *2*:725
Arsenic compounds, *2*:**718**
  medical uses, *2*:729
Arsenic dioxide, *2*:723,
Arsenic fluorides, *2*:720, 721
Arsenic fluorosulfate, *9*:679
Arsenic-gold alloy, *10*:690
Arsenic iodides, *2*:721

Arsenic monosulfide, 2:724. (See also *Arsenic sulfides.*)
  pigments, 15:536
Arsenic oxides, 2:565, 721, 722
Arsenic pentafluoride, 2:721; 9:587. (See also *Arsenic fluorides.*)
  p-anilinophenol as antioxidant for hydrocarbons used in, 2:220
  antioxidants in, 2:598
  control of automobile exhausts, 2:814
Arsenic pentasulfide, 2:724
Arsenic pentoxide, 2:722
  as glass former, 10:538
  as insecticide, 11:680,
  soil chemistry, 18:537
Arsenic red, 2:723
Arsenic sesquioxide, 2:721
Arsenic sesquisulfide, 2:723
Arsenic sulfide
  pigments, 15:536
Arsenic sulfides, 2:723, 724
Arsenic tetroxide, 2:723
Arsenic tribromide, 2:729
Arsenic trichloride, 2:721
  chamber process, 19:459
Arsenic trifluoride, 2:720; 9:635
  reaction with wolfram hexachloride, 22:347
Arsenic trihydride. See *Arsine.*
Arsenic triiodide, 2:721
Arsenic trioxide, 2:280
  as glass former, 10:538
  as insecticide, 11:680
  medical use, 2:730
  metallurgy, 2:713
  properties, 2:721
  rodenticide, 15:916
  soil chemistry, 18:537
  structure, S:1
  uses, 2:715
Arsenic trisulfide
  in shellac, 18:30
Arsenious oxide. See *Arsenic trioxide.*
Arsenolite, 2:721
Arsenopyrite, 2:712, 720; 19:350
Arsenoso compounds, 2:726
Arsenous acid. See *Arsenic trioxide.*
Arsenous arsenate, 2:723
Arsenous oxide. See *Arsenic trioxide.*
Arsenous sulfide. See *Arsenic monosulfide.*
Arsenyl bromide, 2:720
Arsenyl chloride, 2:720
Arsenyl sulfate, 2:724
Arsine, 2:716, 718; 13:4; 19:379

  in chemical warfare, 4:872
  as corrosion inhibitor, 6:324
  manganese complexes, 13:7
  zirconium compound adducts, 22:658
Arsinic acids, 2:725
Arsino phosphotungstic acid
  uric acid test, 21:111
Arsonic acids, 2:724, 726
Arsphenamine, 2:729
Arsthinol, 2:730
  amebicide, 20:94
1-Arterenol. See *Norepinephrine.*
Artificial sweetening agents. See *Sweetening agents.*
Arubren, 5:233
Arylamides
  antioxidants, 17:523
  pigments, 15:569
1-Aryl-3-amino-5-pyrazolone
  structural formula, 5:824
Arylation
  of ammonia, 2:112
  of amines, 2:112
N-Aryl-α-chlorobenzylidenimine, 7:42
Aryl fluorosulfates, 9:677
Aryl halides
  prepn of amines from, 2:112
3-Arylhydantoins, 10:912
Arylhydrazines
  prepn of, 11:170
  prop of, 11:170
1-Aryl-3-hydroxy-5-pyrazolone imine
  structural formula, 5:824
Arylide maroons, 15:576
Arylide reds, 15:576
Arylides
  pigments, 15:569
Arylimino ether
  prepn, 7:42
Aryloxyalkonoic acids
  chlorinated as herbicides, 22:179
N-Aryl-*p*-phenylenediamines, 15:223
Aryl phosphates
  in polystyrene, 19:90
Arylsuccinimides, 19:140
2-Aryltetrazoles, 10:919
Arylthioureas, 10:916
Arysulfonylguanidine
Asbestos, 2:**734**
  amosite, 11:834
  in ceramics, 4:772
  chrysotile, 5:547; 10:129; 14:449; 19:467
  contact process, 19:467

felted, *11*:833
fibers, *20*:34
filler, *15*:802
as gasket material, *14*:449
manuf of, *4*:704
in packing, *14*:449
prop as embedding filler, *8*:110
in resins, *15*:793
Asbestos blankets, *11*:833, 834
Asbestos-cement board, *21*:607
Asbestos-cement products. (See also *Cement.*)
Asbestos fiber
  used in reinforced plastics, *12*:189
Asbestos millboard, *11*:833
Asbestos paper, *11*:832, 833
Asbestos textiles, *2*:745
Ascaridole, *15*:338
Ascarite, *7*:396
Ascharite, *3*:628, 648
Ascophyllan, *17*:783
Ascorbic acid, *2*:**747**
  analysis for phosphates in water, *21*:695
  developing agent, *15*:372
  as food preservative, *10*:14
  growth inhibitor, *15*:681
  in nuts, *14*:131, 138
  use in carbonated beverages, *4*:342
Ascorbyl palmitate
  as synergist, *2*:592
Ascospore
  formation in yeast, *22*:513
1-Asparaginase
  in cancer therapy, *S*:89
Asparagine
  in beet sugar, *19*:211
Aspartate, *21*:521
Aspartic acid, *2*:157, 189. (See also *Amino acids.*)
  biotin and, *3*:524
  in tobacco, *20*:508
  in wine, *22*:316
Aspartyl-phenylalanine methyl ester
Aspergillus
  hydrocarbon utilization, *S*:839
Asphalt, *2*:**762;** *15*:9, 790; *19*:653. (See also *Bitumen.*)
  cracked, *2*:766, 777
  ethyleneamines used as wetting and emulsifying agent, *7*:35
  manuf, *15*:2
  rock, *2*:764
  water-resistant prop, *22*:149
Asphaltenes, *2*:784

Asphalts
  carbon fiber precursors, *S*:110
Asphalts, airblown *2*:766, 772, 795
Asphaltum
  in stains, *18*:657, 661
Aspidium oleoresin, *14*:533, 543
Aspidospermine, *1*:773
Aspirin, *2*:390, 430; *17*:726, 731. (See also *Analgesics and antipyretic.*)
Asplund process
  for hardboard, *21*:606
Astrakanite
  from Great Salt Lake, *S*:455
Astrazon Blue B, *20*:702
  I.G. Farben methods, *16*:298
  manuf, *20*:704
Astrazon dyes, *16*:296
Astrophloxine 5G, *7*:533
Astroquartz, *17*:290
Atabrine, *2*:149, 151; *14*:543
  manuf from α-acetobutyrolactone, *1*:159
Atacamite, *6*:135
Atomic coordinates
  x-ray determination, *22*:448
Atomization
  mechanics of, *18*:634
  pneumatic, *18*:636
Atomizers, *18*:634-654
Atoms and atomic structure, *2*:**806**
Atoxyl, *2*:729; *3*:2; *19*:141
Atranorin, *15*:160
Atrazine, *22*:187
Atropine, *15*:323, *1*:766, 778, 804
  as cardiovascular agent, *4*:520
  as nerve-agent antidote, *4*:885
  prepn by Grignard reaction, *10*:728
  veterinary blocking agent, *21*:252
Attapulgite, *5*:552
  in ceramic products, *5*:561
  in drilling fluids, *5*:575; *7*:294
  as pigment, *14*:515
Attapulgus clays
  as thixotropic binders, *4*:34
Attar of roses, *14*:729
Attractants
  as insecticides, *11*:729
  for insecticides, *11*:724
Attritus, *5*:608, 609
Aubepine, *15*:163
Audio tape, *12*:807
Auer metal, *4*:848
Auger effect
  x-ray fluorescence analysis, *22*:460

Auramine, 7:497, 533; 20:75
Auremetine, 20:75
Aureolin, 5:745; 15:536
Aurine, 20:693, 720
Austenite, 18:766, 770
Austenitic stainless steel
Austenitic steels, 5:471
  corrosion resistance to water, 6:302
Autoclaves, 18:811
Autolyzates
  yeast extracts, 22:531
Automobile cooling
  corrosion inhibitors in, 6:343
Automobile exhaust control, 2:**814**; S:**50**
Automobiles
  fuel cells for electric, S:411
  and nitrogen fixation, S:608
Automotive antistall additive
  use of ethylene glycol ethers in, 10:643
Automotive coatings, 5:698
Automotive industry
  p-anilinophenol as antioxidant for hydro-
    carbons used in, 2:220
  antioxidants in, 2:598
  control of automobile exhausts, 2:814
  use of acetal resins in, 1:105
  use of adhesives in, 1:397
  use of aluminum in, 1:986
  use of embedment in, 8:103
Autoradiography, 17:43
Autunite, 21:8
  synthetic, 15:678
  use of magnesium in, 12:706
Auxin and Auxins, 9:30; 15:676
Auxochrome, 5:766
Auxochromes
  in indigo, 11:564
  pigments, 15:568
Avenin, 22:268
Avertin, 15:922
Avicennite, 20:65
Avidin, 3:519, 523
  in eggs, 7:664
Avitaminosis, 21:488
3H,2a-Azaacenaphthylene, 10:936
Azaanthracenes, 10:933
7H,6a-Azabenz[a]anthracene, 10:936
3-Azabenzanthrone, 10:899
3H,4H,3a-Azabenzonaphthalene, 10:936
12H,11a-Azabenzo[4,5,]pentalo-[2,1-a]phenan-
  threne, 10:937
1-Azabicyclo[2.2.2]octane, 10:937
8-Azabicyclo[3.2.1]octane, 10:899

Azacycloheptane, 10:934
Azacyclonol, 16:673
Azacyclopropane, 10:898. See Ethylenimine.
6-Azaequilenin, 18:872
6-Azaestrone, 18:871
9-Azafluorene. See Carbazole.
3H,9-Azafluorene, 10:897; 11:593
3H,9-Azafluorene, 10:897
9-Azafluorene, 10:909
1-Azaindene, 11:585
3a-Azaindene, 10:897, 936
Azalomycin F
  amebicide, 20:82
8aH,4a-Azanaphthalene, 10:899, 936
8aH4a-Azanaphthalene, 10:899
Azanaphthalenes, 10:932
4H,4a-Azanopholene, 10:936
3H,3a-Azapentalene, 10:936
3H,4H,3a-Azaphenalene, 10:936
10bH,4a-Azaphenanthrene, 10:936
2-Azaphenanthrene, 10:933
5-Azaphenanthrene, 10:899, 934
Azaphenanthrenes, 10:933
Azaserine, 2:188
  amebicide, 20:80
  cancer therapy, S:85
  oxygen in synthesis of, 14:405
Azelaic acid, 1:232, 247, 250
  in dental material, 6:827
  polyesters, 20:490
  prepn, 8:821
Azelaic acid esters
  prod and sales of, 8:377
Azeotrope
  binary systems, 21:214
  definition, 21:208
Azeotropic distillation, 2:**839**
Azeotropic mixtures
  composition and boiling points, 8:324
Azetidine, 11:526
2-Azetidinones, 10:907
2-Azidohexafluoro-2-propanol, 9:760
3-Azidopropionitrile, 6:650
Azimidobenzene, 10:917
Azine, 7:522, 533
Azine dyes, 2:**859**; 7:471, 474; 19:432
  for color photography, 5:830
Azine fluoride, 9:630
Azines, 2:**868**
Azino group, 2:868
Azinphosmethyl, 15:528, 910
Aziridine. See Ethylenimine.
Aziridine, 10:898

copolymerization, *16*:174
series of cyclic imines, *11*:526
1-Aziridineproprionitride, *6*:648
Aziridines
  analytical tests for, *11*:538
  flammability and stability, *11*:537
  isomerization, *11*:535
  polymerization, *11*:535
  ring opening reactions, *11*:533
  storage and handling, *11*:537
Aziridinylbenzoquinones
  in tumor therapy, *11*:542
Azlactones, *10*:914
Azobenzene, *2*:415; *3*:412; *9*:591
α,α,-Azobis-(α,γ-dimethylvaleronitrile), *12*:121
2,2'-Azobisisobutyramidine hydrochloride, *9*:836
Azobisisobutyronitrile, *15*:796
  initiator of free-radical reactions, *10*:115
  polyester initiator, *20*:828
2,2'-Azobisisobutyronitrile, *9*:836, 857
Azo coupling, *2*:870
Azodicarbonamide, *9*:833, 837
  flour improver, *22*:293
1,1'-Azodicyclohexane-carbonitrile
  photoinitiator, *16*:518
Azodiphenyl-4,4'-dicarboxylic acid
  aromatic amines in manuf of, *2*:111
  for color photography, *5*:832

used in hair coloring, *10*:792
from phenol, *15*:148
in prepn of azo dyes, *2*:894
Azo dyes, *2*:**868**; *7*:471, 472, 533
Azo Eosine G, *7*:484
9-Azofluorene, *10*:909
Azoic dyes, *2*:889; *7*:471, 472, 547
Azoisobutane, *9*:591
Azoisobutyronitrile
  addition reactions, *20*:209
  as maturing agent, *10*:6
  microencapsulation of, *13*:454
Azole. See *Pyrrole.*
Azole, *10*:898
Azoles
  as optical brighteners, *3*:741
Azomethine dyes, *15*:220
Azo pigments, *2*:907
Azo Phloxine GA Extra Conc, *7*:525
Azopyrroles, *16*:850
Azoxybenzene, *2*:415, *3*:412
  in prepn of *p*-aminophenol, *2*:219
  prepn, *2*:77
Azulenes, *10*:927
Azulmic acid, *6*:576, 602
Azure blue, *15*:541
Azurite. See *Copper compounds.*
Azurite, *15*:541

B

Babassu oil, *8*:778, 784, 790
Babbitt metals, *3*:275
Bacillus
  detergent enzymes from, *S*:297
  hydrocarbon utilization, *S*:839
Bacitracin, *14*:675; *16*:308
  amebicides, *20*:80
  in bacterial infections, *3*:8
  as veterinary drug, *21*:244
Bacitracins, *16*:315-317
Backmann process for bleaching powder, *5*:19
Backus process
  carbon dioxide recovered by, *4*:359
Bacteria, *13*:461-468
  analysis for in water, *21*:703
  classification of *3*:17
  utilization of hydrocarbons by, *S*:839
Bacteria subtilis, *2*:209
Bacterial culture
  uses, *8*:190
Bacterial infections
  chemotherapy, *3*:**1**
Bacterial proteasis
  uses, *8*:190
Bactericides
  aziridine structure in, *11*:542
Bacteriological acid indicator
  *N-(p*hydroxyphenyl)glycine as, *2*:220
Baddeleyite, *22*:631
  in ceramics, *4*:772
Badische acid, *13*:714
Baeyer's acid, *13*:714, 721
Baffles
  in vacuum technology, *21*:137
Bagasse, *3*:**36;** *19*:184
  in insulating board, *21*:602, 617
Baghouse fume, *12*:222
Bahr's salt, *6*:631
Bainite, *11*:15; *13*:328; *18*:776, 771
Baked products
  use of lecithin in, *12*:356
Bakelite, *15*:792; *16*:272
Baker's yeast, *22*:533
Bakery processes and leavening agents, *3*:**41**
Bakery products
  use of lactic acid in, *12*:179, 181

  use of propionates as preservative, *10*:13
Baking acids, *3*:51
Baking industry
  humidity conditions within, *1*:486
  use of abherents in, *1*:310
Baking powder, *3*:51, 53
  sodium alum used in, *2*:65
Baking soda. See *Sodium bicarbonate.*
Ballas, *4*:285
  as abrasive, *1*:25
Ball bearings, *3*:295
Ball clays
  in ceramics, *4*:764
Balling scale of specific gravity, *6*:758
Ballistics
  terminal, *S*:147
Ball mills
  in manuf of lead oxides, *3*:257
Balm of Gilead, *17*:387
Balsam
  Canada, *17*:383
  pigments, *15*:538
  tolu, *14*:194
Balsam of Peru, *17*:388
Balsam of Tolu, *17*:388; *20*:527
Balsams, *17*:387, 389
BAMP. See *4-sec-Butyl-2- (α methylbenzyl) phenol.*
Banbury mixers, *4*:662, 665; *15*:593
Bandamycin, *12*:633, 634
"Bandrowski's base," *10*:307; *15*:220
  carbamate herbicide, *22*:193
  prepn from butynediol, *1*:606
Barban
Barbane
Barbital, *3*:60, 61, 67, 70
Barbiturates. (See also *Barbituric acid.*)
  as anticonvulsants, *11*:519
Barbiturates, *3*:60
  as hypnotics, *11*:510
  sodium in synthesis, *18*:451
Barbituric acid, *19*:328
  as a hypnotic, *11*:509
  prepn, *1*:846
  urea derivatives, *21*:39

44

Barbituric acid and barbiturates, *3*:**60**

Barite. *22*:563. See *Barium sulfate; Barium compounds.*

Barium, *3*:**77**; *17*:126, 137, 140
  in Brazil nuts, *14*:133
  cesium alloys with, *4*:862
  in electron tubes, *8*:2
  organometallics in phonograph records, *15*:229
  in phototubes, *15*:398

Barium *3*:81, 92; *7*:287, 291, 292; *3*:81, 92; *15*:515; *19*:51, 351
  in ceramics, *4*:772
  from ocean, *14*:164
  as paper pigments, *14*:517
  photography, *15*:365
  pigments, *15*:499, 503, 505, 515, 523, 534, 562
  zero point of charge, *9*:384

Barium acetate, *3*:82

Barium aluminate
  in cathodes, *8*:2

Barium borate, *3*:651

Barium bromate, *3*:770

Barium bromide, *3*:87, 770
  phys prop, *1*:238

Barium carbonate, *3*:82
  in cathodes, *8*:2
  in drilling fluids, *7*:301, 302
  pigments, *15*:516
  prepn from acetylene, *1*:173
  rodenticide, *15*:916
  sodium sulfide by-product, *18*:512

Barium chlorate, *5*:60

Barium chloride, *3*:86; *6*:512
  phosphor, *15*:352

Barium chlorite, *5*:31

Barium chromate, *5*:481, 498
  pigment, *15*:535

Barium compounds, *3*:**80**

Barium cyanamide, *6*:559

Barium cyanide, *6*:599, 604

Barium cyanurate, *20*:663, 670

Barium dihydrogen cyanurate, *20*:663

Barium feldspar
  in glass-ceramics, *10*:548

Barium ferrate(IV), *12*:40

Barium ferrites, *3*:85. (See also *Iron compounds.*)

Barium fluoride, *9*:551

Barium fluorozirconate, *22*:633

Barium fluosilicate, *9*:655
  as insecticide, *11*:682

phys prop, *9*:658, 660

Barium formate, *3*:86

Barium hexacyanoferrate(II) *12*:27

Barium hexacyanoferate(III), *12*:31

Barium hexafluorogermanate, *9*:585

Barium hexafluorostannate(II), *9*:682

Barium hydride, *11*:204

Barium hydroferrate(II), *12*:40

Barium hydrogen cyanurate, *20*:663, 670

Barium hydrosulfide, *3*:95, 97

Barium hydroxide, *2*:579; *3*:90; *10*:636; *12*:113
  cesium alum reaction with, *4*:864
  cesium sulfate reaction with, *4*:863
  in drilling fluids, *7*:301
  sugar recovery, *19*:216

Barium hypochlorite, *5*:15

Barium iodide, *3*:88

Barium manganate, *13*:12

Barium manganate(V), *13*:11

Barium minerals
  in ceramics, *4*:772

Barium monofluorophosphate, *9*:642, 643

Barium nitrate, *3*:88
  in explosive priming compositions, *8*:653
  pigments, *15*:541
  thermochemical constant, *8*:681

Barium oxide, *2*:200; *3*:89
  as drying agent, *7*:391

Barium perchlorate, *5*:69
  as drying agent, *7*:395

Barium perferrate, *12*:40

Barium peroxide, *14*:747, 751

Barium perxenate, *10*:893

Barium potassium chromate, *5*:499, 510

Barium resinate
  pigments, *15*:562

Barium saccharate, *19*:217

Barium sodium niobate
  nonlinear optical materials, *S*:625, 629

Barium stearate, *7*:280
  in fire extinguishers, *7*:285

Barium strontium niobate
  nonlinear optical materials, *S*:630

Barium sulfate, *3*:91; *6*:826

Barium sulfhydrate, *3*:95

Barium sulfide, *3*:81, 95
  pigments, *15*505, 515, 523

Barium sulfite, *3*:97

Barium tetracyanoplatinate(II), *15*:863

Barium tetrafluorozincate, *9*:684

Barium thiosulfate, *20*:241, 242

Barium titanate, *3*:85, 97; *8*:30; *20*:412

dielectric constant vs temperature, *4*:824
dielectric props of, *4*:823, 824
domain structures, *9*:2
electronic uses, *4*:775
ferroelectric prop, *9*:9, 13
in glass-ceramics, *10*:548
manuf, *9*:15
Barium titanate ceramics
prop, *9*:20
Barium yellow, *15*:535
Barium zirconate, *22*:656
in glass-ceramics, *10*:548
Barium zirconium niobate
ferroelectric prop, *9*:12
Barley. (See also *Wheat and other cereal grains.)*
classes of, *22*:263
livestock food, *22*:257
origin of, *22*:253
prod of malt, *12*:861
technology, *22*:269
U.S. prod, *22*:259
Barley malts
in beer brewing, *3*:300
Barometer
mercury, *13*:218
Barrel, of petroleum
defn, *14*:835
Barrels, *14*:438
Barton process
manuf of lead oxides, *3*:257
Bart reaction, *2*:725
Barylite, *3*:453
Barytes. See *Barium sulfate; Barium compounds.)*
Basalt
as abrasive, *1*:25
BASF chlorine cell, *1*:689
BASF cracking process
in acetylene manuf, *1*:186
Basic Blue 10, *2*:865
Basic Blue 12, *2*:865
Basic Blue 12 (CI), *7*:533
Basic Blue 24, *2*:867
Basic Blue 25, *2*:867
Basic Blue 6, *2*:864
Basic Blue 9, *2*:866
Basic Blue 9 (CI). See *Methylene Blue.*
Basic carbonate white lead, *15*:500
Basic colors
Basic copper carbonate. See *Copper compounds.*
Basic copper carbonate. See *Malachite.*

Basic dyes, *7*:53
in paper coloring, *7*:582
used in coloring hair, *10*:792
uses of, *7*:476
Basic Green 5, *2*:867
Basic lead sulfate, *15*:541
Basic lead sulfates, *12*:279
Basic Orange 2, *2*:874
Basic Red 1 (CI), *9*:497
Basic Red 2, *2*:861
Basic Red 2 (CI). See *Safranine.*
Basic Orange 10 (ci), *7*:533
Basic Orange 14 (CI), *7*:533
Basic silicate white lead, *15*:503
Basic slag
as fertilizer, *9*:109
Basic sulfate white lead, *15*:502
Basic tellurium nitrate, *19*:767
Basic tellurium selenate, *19*:767
Basic tellurium sulfate, *19*:767
Basic tellurium tellurate, *19*:767
Basic Violet 1, *7*:496
Basic Violet 2
phenylation, *20*:699
Basic Violet 3 (CI), *7*:533
Basic Violet 10, *5*:870, 873
Basic Violet 10 (CI). See *Rhodamine B.*
in pigment manuf, *9*:497
Basic Violet 14 (CI), *7*:507
Basic Yellow 1, *20*:194
Basic Yellow 2, *7*:497
Basic Yellow 2 (ci), *7*:533
Basic Yellow 12, *20*:194
Basic zinc chromate, *15*:534
Basic zinc sulfoxylate, *19*:422
Basil
essential oil as flavor, *9*:352
sweet, *14*:194
Basolan Chrome Brilliant Red 3BM, *7*:527
Basramite
commercial slagceram, *S*:884
Basse taille
enameling technique, *8*:155
Bassorin, *10*:746
Bast fibers, *9*:171
Bastnasite, *17*:146
cerium from, *4*:843, 845
Bathochromic shift, *15*:568
Bathophenanthroline
α-tocopherol assay, *21*:581
Batik printing, *7*:580
Bating
in leather manuf, *12*:312

Batrachotoxın, *18*:832
Batteries, *3*:**99**
  carbon black used in, *4*:277, 280
  fuel cells, *3*:139
  graphite used in, *4*:332
Batteries and electric cells, primary, *3*:**99**
  secondary, *3*:**161**
Battery chargers
  fuel cells as, *S*:404
Battery oxides, *12*:274
Baumé scale, *6*:758
Bauxite, *1*:931; *2*:1, 8
  calcined, *17*:227
  in ceramics, *4*:763, 773
  desulfurization, *15*:46
  desulfurization catalyst, *11*:434
  in manuf of aluminum chloride, *2*:18
  in manuf of aluminum sulfate, *2*:60
  products derived from, *2*:50
  refractory, *17*:231, 232
  wax decolorizer, *15*:97
Bauxite cement, *4*:696
Bauxite-residue red, *15*:519
Bay, *14*:195
Bayberry wax, *22*:164
Bayer hydrate. See *α-Alumina trihydrate.*
Bayerite. See *β-Alumina trihydrate.*
Bayer process
  in prod of alumina, *1*:27, 937
Bayer α-trihydrate. See *α-trihydrate.*
Bay leaves, *9*:369
Bayluscide, *15*:911
Bay oil
  as denaturant, *8*:450
Bay rum
  use of alchol in, *8*:447
1-B-D-Arabinofuranosylcytosine, *S*:86
Bearing materials, *3*:**271,** 896
  carbon-graphite, *4*:228
  copper alloys as, *6*:263
Bearings
  high-temperature, *3*:286
Bearings, gas, *3*:290
Béchamp process
Béchamp reaction, *2*:725
Becquerelite, *21*:8
Beef tallow, *8*:790
Beer
  chillproofing, *8*:196
  yeast fermentation, *22*:526
Beer and brewing, *3*:**297**
Beerbower equations, *S*:891
Beer-Lambert Law, *15*:332; *17*:69

Beer's law of absorption, *5*:788
Beeswax, *22*:158, 166
  in candles, *4*:58; *15*:105
  in dental material, *6*:840, 841
  in eye shadows, *6*:368
  in food processing, *10*:10
  in lipsticks, *6*:364
Beet molasses
  in prepn of monosodium glutamate, *2*:200
Beet pulp
  as animal feed, *8*:867
Beet residue
  manuf cyanide, *6*:592
Beet sugar, *19*:203-220
Behenic acid, *1*:231; *8*:813
Behenolic acid, *8*:816
Beidellite, *5*:548
Beilstein's rule, *3*:372
Belgian retort process
  for zinc metal, *22*:572
Belladine, *1*:771
Belt conveyors, *6*:95, 96, 101
Bemegride, *21*:535
Benactyzine, *2*:149, 151
Benactyzine hydrochloride, *2*:153. See *β-Di-ethylaminoethyl benzilate.*
Benadryl, *2*:151
Benares hemp, *9*:184
Bendroflumethiazide, *7*:264
  reaction with TNT, *8*:612
m-Benezenedisulfonic acid, *3*:402, 406; *15*:211
Benhenium emboate, *14*:534
Benite, *8*:651, 695
Benoxinate hydrochloride, *2*:405, 408
Bensulide, *18*:536
Bentone, *3*:355
Bentonite, *3*:**339**
  in ceramics, *4*:766
  as dental abrasive, *6*:849
  in detergents, *6*:858
  in drilling fluids, *7*:293, 299; *5*:575
  fining of wines, *22*:331
  in hydrocracking, *11*:422
  pigments, *15*:512
  as thixotropic binder, *4*:33
  zero point of charge, *9*:384
Benzactyzine, *16*:676
Benzalacetone, *12*:107
Benzalacetophenone, *12*:107; *19*:327
Benzal chloride, *5*:281
  cinnamic acid from, *5*:519
  photochemical, *15*:338
Benzaldehyde, *8*:244

Benzaldehyde, 2:369; 3:**360**; 6:670, 672, 693; 7:478; 12:143
  in carbonated beverages, 4:344
  carbon monoxide in prepn of, 4:433
  chlorination, 19:402
  cinnamic acid from, 5:518
  as denaturant, 8:450
  in detn of pentalrythritol content, 1:591
  natural occurrence, 1:645
  phenylacetic acid, 15:214
  photochemical, 15:338
  photoinitiator, 16:519
  prepn from benzyl ether, 8:495
  in prepn of l-ephedrine, 1:789
  reaction with acetaldehyde, 1:82
  reaction with acetic anhydride, 8:407
  removal from nuts, 14:138
  as synthetic flavor, 9:354, 355
  from toluene, 20:562
Benzaldehyde cyanohydrin, 6:668, 672
Benzaldehyde phenylhydrazone, 10:918
Benzaldehyde-2,4-disulfonic acid, 19:300; 20:700
  Patent Blue VF dyes, 20:708
  Sulfo Rhodamine B synthesis, 22:434
Benzaldehyde-2-sulfonic acid, 19:300
Benzaldoxime, 10:152
Benzalethylamine, 12:51
Benzalkonium chloride, 2:633, 635; 15:128; 19:563
Benzalocetophenone, 12:107, 113, 114
Benzamide
  form, 2:68
  phys prop, 2:68
1-Benzamidohexafluorodimethylacrylic acid, 9:763
4-Benzamidopyridine
  form, 2:67
5-Benzamido-1-naphthol, 19:305
8-Benzamido-1-naphthol, 19:305
Benzanilide, 9:591
  form, 2:68
  phys prop, 2:68
Benzanthrone. See 7H-Benz[de] anthracen-7-one.
9-Benzanthronesulfonic acid, 2:496
Benzathine penicillin G, 3:23
Benzaurin
  structural formulas, 5:769
1-Benzazole, 11:585
Benz[b][1,8]phenanthrolines
7H-Benz (de)anthracen-7-one, 7:478, 492
7H-Benz[de] anthracen-7-one, 2:494
  manuf of, 2:498
  reactions of, 2:494
Benzedrine
  phenylacetic acid, 15:214
Benzene, 2:144; 3:**367**; 7:481
  alkylation, 10:137; 15:150
  alkylation of, 2:22
  alkylation with butylbenzenes, 1:890
  alkylation with dodecylbenzene, 1:890
  alkylation with ethylbenzene, 1:889
  alkylation with isopropyl ether, 8:488
  alkylation with propylene, 1:889
  allyl chloride reactions with, 5:208
  azeotrope with n-hexane, 11:2
  benzyl chloride reaction with, 5:283
  as by-product in ethylene prod, 8:510
  carbon monoxide reaction with, 4:433
  carbon tetrachloride reaction with, 5:132
  in catalytic reformate, 15:31
  chloroform reaction with, 5:122
  chlorosulfonation, 19:299
  in coal tar, 19:657, 669
  from coke-oven gas, 4:415, 419
  in commercial hexanes, 11:1
  complex with LCNE, 6:626
  dehydrogenated from cyclohexane, 6:677
  as denaturant, 8:450, 452
  1,2-dichloroethane reaction with, 5:151
  in dye industry, 7:480, 481
  dye intermediates from, 7:482
  in dye intermediates, 7:469
  ethyl chloride reactions with, 5:143, 146
  fluorination, using cobaltic fluoride, 9:698
  by hydrodealkylation, 11:453
  hydrogenation to prep cyclohexane, 6:679
  industrial sulfonation, 19:308
  in manuf of cumene, 6:544
  manuf of ethylbenzene, 22:500
  in manuf of maleic anhydride, 12:828
  methyl chloride reactions with, 5:104
  methylene chloride reaction with, 5:114
  monosulfonation, 19:293
  nitration of, 13:836
  nitronium perchlorate reaction with, 5:74
  oxychlorination, 15:150
  oxyhydrochlorination of, 5:169
  perchlorylation reaction, 9:603
  from phenol, 15:148
  phenol from, 15:149
  in prepn anthraquinone, 2:435
  in prepn 2-benzoyl-5-sulfobenzoic acid, 2:442
  in prepn of diphenyl, 7:191

reaction with antimony trichloride, 2:575
reaction with cobalt trifluoride, 9:785, 786, 790
reaction with phthalic anhydride, 10:162
from shale oil, 18:18
as solvent, 2:143
stain solvent, 18:661
structure, 2:704
from toluene, 20:543, 556
toxicity of, 11:298
tubular reactor nitration, 13:837
vapor-phase nitration, 13:837
vinyl chloride reaction with, 5:172
Benzeneboronic acid, 3:716
Benzenecarboxamide. See Benzamide.
Benzenecarboxylic acid. See Benzoic acid.
1,2-Benzenediacetic acid. See o-Phenylenedi-acetic acid.
Benzenediazonium-2-carboxylate, 19:329
Benzenediazonium chloride, 6:642, 693; 17:723
Benzenediazonium perchlorate, 5:74
1,3-Benzenedicarboxylic acid. See Isophthalic acid.
1,2-Benzenediol, 11:462-472
  alkylation of, S:30
  hydrous zirconia and, 22:654
  surfactants, 19:582
  zirconium chelates, 22:655
1,3-Benzenediol, 19:164, 11:462, 472-483; 16:202. (See also Resorcinol.)
  resins, 15:176
  zirconium compounds, 22:655
1,4-Benzenediol, 11:462, 483-492
Benzenediols, 15:148. (See also Hydroquinone; Pyrocatechol; Resorcinol.)
o-Benzenedisulfonamide
  reaction with dialhylchloroalkyl amines, 2:143
p-Benzenedisulfonic acid, 3:406
o-Benzenedithiol
  tetrachloroethylene reaction with, 5:199
p-Benzenedithiol
  melting point, 20:207
Benzene hexabromide, 3:374
Benzenehexacarboxylic acid. See Mellitic acid.
Benzene hexachloride, 3:374; 5:267; 15:337
  as an insecticide, 11:694
  toxicity of, 5:278
  x-ray analysis, 22:453
Benzenehexol, 16:190, 212, 213
Benzenepentol, 16:190, 212
Benzenepolycarboxylic acids, 15:444

trans-Benzenepropenoic acid. See Cinnamic acid.
Benzenes
  chlorinated, 5:253
Benzenesulfinic acid, 6:645
Benzenesulfonamide, 19:328
  in pigment manuf, 9:497
  reaction with oxygen difluoride, 9:632
Benzenesulfonic acid, 3:401; 6:678; 8:326; 9:677; 19:312
  in manuf of ethyl ether, 8:481
Benzenesulfonic anhydride, 3:402, 406; 10:154
Benzenesulfonyl chloride 1:223; 3:402, 406
  phys prop, 1:224
Benzenesulfonyl fluoride, 9:677
1,2,3,4,5-Benzenetetracarboxylic acid, S:961
1,2,4,5-Benzenetetracarboxylic acid, 15:480, 796; 18:232
Benzenetetrols, 16:190, 209-211
Benzenethiol, 6:646; 19:393
  prop, 20:207
Benzene-toluene
  solvent refining, 15:95
Benzenetricarboxylic acids, 15:473
1,2,3-Benzenetriol
  zirconium complexes, 22:655
1,2,4-Benzenetriol, 22:181
Benzenetriols, 15:148. (See also Poly(hy-droxy)benzenes; Pyrogallol;
Benzethonium chloride, 2:633; 10:780
Benz[g]indolo [2,3-a]quinolizine, 10:936
Benzhydrol, 10:730; 20:678
Benzhydryl chloride, 16:674
2-(Benzhydryloxy)-N, N-dimethylethyl-amine, 2:151
Benzidine, 3:408; 6:508; 7:193; 10:909
  diazotization of, 2:870
  dyes from, 2:897
  dyes, 19:434
  rearrangement, 7:491, 535; 3:40
Benzo Fast Copper Red GGL, 2:904
Benzidine Orange, 15:573
Benzidine dihydrochloride, 3:412
Benzidine-2,2'-disulfonic acid, 3:418
Benzidine hydrochloride, 7:478
Benzidine monofluorophosphate, 9:642
Benzidine (p,p'-Diaminodiphenyl), 7:535
Benzidine sulfate, 7:478
Benzidine sulfone, 3:418
Benzidine sulfone-3,3'-disulfonic acid, 3:418
Benzidine-3-sulfonic acid, 3:417
Benzidine yellows, 15:571
Benzidine-3,3'-disulfonic acid, 3:418

Benzil, *10*:929; *12*:67, 141, 143
Benzil, *12*:140, 146
  ester-exchange catalyst, *16*:175
Benzilic acid, *2*:149; *16*:676; *19*:330
Benzimidazole, *16*:51
Benzidine-3,3'-dicarboxylic acid, *3*:419
  as nucleus in cyanine dye, *6*:607
Benzimidazoles, *10*:899, 912; *15*:221
2-Benzimidazolethiol, *2*:600
Benzin, *3*:368
  in prepn of benzil, *12*:147
1,2-Benzisothiazol-3(2*H*)-one, *19*:595. (See al-
  so *Saccharin.*)
Benzo[*b*] [*1,10*] phenanthrolines
  antiamebics, *20*:8810Benzo[*lmm*] [3,8] phen-
    anthrolines
  antiamebics, *20*88
Benzocyclopentanones, *12*:822
Benzodiazepines, *16*:655, 671
1,3-Benzodiazines, *10*:912, 928
1,4-Benzodiazines, *10*:929
1,2,3-Benzodithiazole chloride, *10*:916
Benzo[*e*]permidines
  antiamebics, *20*:88
Benzo Fast Blue B, *7*:535
Benzo Fast Blue FR, *7*:485
Benzo Fast Blue R, *7*:535
Benzo Fast Pink 2BL, *7*:535
Benzo Fast W, *2*:882
Benzo Fast Yellow 4GL EX, *7*:483
Benzo Fast Yellow 5GL, *7*:535
Benzofuran, *10*:899, 907, 908
3(*H*)-Benzofuranone, *10*:908
2-Benzofuryl methyl ketone, *10*:908
Benzoguanamine, *6*:569
  in amino resin manuf, *2*:27
  stabilizer, *21*:390
p-Benzohydroquinone, *7*:739
Benzoic acid, *3*:**420;** *7*:478, 543
  carbon monoxide in prepn of, *4*:433
  esterification, *8*:334
  freezing-point diagram, *13*:200
  as food preservative, *10*:12
  as fungicide, *10*:235
  in margarine, *13*:58
  as modifier in alkyd resins, *1*:866
  oxidation of toluene, *15*:155
  from phenylacetic acid, *15*:214
  photochemical, *15*:338
  from phthalic acid, *15*:456
  prepn from benzotrifluoride, *9*:779
  prepn from benzyl ether, *8*:495
  in resins, *17*:382

  sonochemical reactions, *20*:778
  as synthetic flavor, *9*:354
  terephthalic acid from, *15*:466
  from toluene, *20*:531
  from toluene, *20*:560
Benzoic acids
  chlorinated as herbicides, *22*:200
  metabolism of chlorinated herbicidal,
    22:201
Benzoic anhydride, *2*:213, 215; *3*:431
  phys prop, *1*:212
Benzoic sulfimide, *19*:595
Benzoin, *3*:365; *4*:139; *12*:107, 146, 147;
    14:722; *17*:388
  photoinitiator, *16*:518
Benzoin methyl ether
  photoinitiator, *16*:518
Benzoinum, *17*:388
  from coal tar, *19*:669
Benzo[*lmm*] [3,8]phenanthrolines
Benzole, (See also *Benzine.*)
Benzonitrile, *8*:244; *9*:591; *19*:403; *S*:595
  from chlorobenzene, *5*:256
  manuf from toluene by ammonolysis, *2*:344
  perchlorylation reaction, *9*:604
  polyimides from, *S*:749
Benzonitrile oxide, *10*:919
Benzonyl, *7*:469
9,10-Benzophenanthrene, *7*:191
Benzophenone, *3*:**439**
  from benzoyl chloride, *3*:374
  ester-exchange catalyst, *16*:175
  flavor and perfume materials, *9*:358
  in Grignard reactions, *10*:730
  photochemical polymerization, *15*:339
  photochemical reduction, *15*:335
  photoinitiator, *16*:519
  in soaps, *14*:736
Benzophenone-*o*-carboxylic acid, *3*:436
Benzophenone oxime, *9*:591
3,4,3',4'-Benzophenonetetracarboxylic dianhydride
Benzopinacol, *10*:730; *12*:109; *15*:335
Benzo-*p*-toluidide
4 *H*, 1-Benzopyran, *10*:898
  form, *2*:67
1,2'-Benzopyrone. See *Coumarin.*
1 *H*-Benzo 12*b13* pyrrole, *11*:585
p-Benzoquinine, *7*:739
Benzo12*c13*quinoline, *10*:934
Benzo12*h13*quino124,3-*b13*quinolines
  antiamebics, *20*:88
Benzoquinolines, *16*:876

p-Benzoquinone, 2:415; 6:695
  polyester inhibitor, 20:823
  reaction with aziridines, 11:530
Benzoquinones, 16:902; 18:859
Benzoquinonium, 6:549
Benzoquinonium chloride, 6:550
Benzoselenazole
  as nucleus in cyanine dye, 6:607
o-Benzosulfimide, 19:595
Benzothiadiazine
  diuretics, 4:522
Benzothiazole
  as nucleus in cyanine dye, 6:607
Benzothiazole dyes, 19:433
  absorption curves, 6:608
  structure, 6:609 ·
Benzothiazoles, 10:899, 916
Benzothiazole-2-sulfonamide
  as diuretic, 7:259
Benzothiazole sulfonamides
  as accelerators, 17:510
9-(2-Benzothiazolinylidene)-2-[2benzothiazolyth-
  ethiodide methylene]-3-ethyl-4-thiazolidin-
  one, 6:620
Benzothionyl disulfide
  microencapsulation of, 13:454
Benzothiophene, 10:899
1-Benzothiopyran, 19:324
Benzotriazole, 15:219
Benzotriazoles, 10:917
  polypropylene light stabilizers, S:813
Benzotrichloride, 5:281
  in manuf of benzoic acid, 3:424
  photochemical, 15:338
  in prepn of acetyl chloride, 1:140
Benzotrifluoride, 9:568, 591, 775, 778, 780
Benzoxathiol-2-thione derivatives
  in hair tonics, 10:802
Benzoxazole
  as nucleus in cyanine dye, 6:607
  optical brightener, 16:51
2-Benzoxazolinone-5-sulfonic acid, 2:223
Benzoyl, 12:140
Benzoylacetanilide
  developer in color photography, 5:819
Benzoylacetic acid
  prepn lobelanine, 1:795
Benzoylacetone, 8:408; 12:149, 152
  reaction with boron trifluoride, 9:556
Benzoylacetylmethane, 12:155
N-Benzoylbenzenesulfonamide, 4:924
o-Benzoylbenzoic acid, 15:446; 3:436; 7:478;
  15:446

in manuf anthraquinone, 2:431, 432, 435
in manuf of benzophenone, 3:440
1-Benzoyl-3-carbethoxy-4-piperidone, 2:144
Benzoyl chloride, 3:430; 8:244, 346; 19:402
  benzophenone from, 3:374
  from benzotrichloride, 5:283
  photochemical, 15:338
  phys prop, 1:224
  in prepn of acetyl chloride, 1:140
Benzoyl dibenzanthronyl, 7:469, 478
N-Benzoyl-S,S-dimethylsulfilimine,   19:323,
  330
Benzoylglycine, 21:536
Benzoyl isocyanate, 19:330
1-Benzoylnaphthalene, 2:495
Benzoyl peroxide, 2:623; 6:783 7:698; 10:6;
  14:767; 15:796
  as catalyst, 8:494
  curing agent, 18:228
  in dental material, 6:832
  flour bleaching agent, 22:293
  as hazardous chemical, 9:291, 794
  polyester initiator, 20:826
  in polymerization of ethylene, 8:501
  polymerization of vinyl fluoride, 9:836
  poly(vinyl acetate) initiator, 21:332
  vinylidene fluoride elastomers curing agent,
    21:270
Benzoylpropionylmethane, 12:148
N-Benzoylsulfamic acid, 19:244
O-Benzoylthiamine disulfide, 20:182
Benzoylthiamine monophosphate, 20:182
2-Benzoyl-5-sulfobenzoic acid, 2:442
Benzthiazide
  as diuretic, 7:263, 264
N-Benzyl-N,N'-dimethyl-N-phenylethylenedi-
  amine, 2:152
N-Benzyl-β-chloropropionamide
  as anticonvulsant, 11:523
N-Benzylacetamide, 19:270
Benzyl acetate, 8:377, 380
  in soaps, 14:736
Benzylacetone, 6:670
Benzyladenine, 15:681
Benzyl alcohol, 2:413; 3:**442**
  flavor and perfume materials, 9:358
  photochemical, 15:338
  in soaps, 14:736
  as synthetic flavor, 9:354
  vat dyeing wool, 22:404
Benzylamine, 9:591
  polyester scission, 16:178

p-Benzylaminophenol
  as stabilizer, *8*:489
Benzylaniline
Benzyl benzoate, *3*:432
  parenteral solvent, *15*:128
  photochemical, *15*:338
  prepn from benzyl ether, *8*:495
  specifications, *8*:373
  as synthetic flavor, *9*:354
Benzylbenzyl alcohol, *3*:443
Benzyl bromide
  in chemical warfare, *4*:877
  in manuf of benzoic acid, *3*:425
Benzyl butyrate
  prod and sales of, *8*:377
Benzylcellulose, *4*:638
Benzyl chloride, *2*:166, 420; *5*:281; *6*:562;
  8:495
  benzaldehyde manuf from, *3*:362
  from benzene, *3*:374
  p-chlorophenol reaction with, *5*:331
  in manuf benzyl alcohol, *3*:443
  photochemical, *15*:338
  from toluene, *19*:402
  from toluene, *20*:562
Benzyl chloroformate, *2*:171
Benzyl cinnamate, *5*:520
  flavor and perfume materials, *9*:358
  prod and sales of, *8*:377
Benzyl cyanide, *16*:673. (See also *Phenylaceto-*
  nitrile.)
  hydrolysis, *15*:214
Benzyldichlorosilane, *12*:69
Benzyldimethylamine-borane, *3*:728
Benzyldimethylphenylammonium nitrophen-
  oxide
Benzyldimethyl(2-phenoxyethyl)ammonium 3-
  hydroxy-2-naphthoate, *14*:534
Benzyl ether, *3*:443; *8*:471, 495
  flavor and perfume materials, *9*:358
  phys prop, *8*:471
N-Benzylhexamethylenetetrammonium chloride, *10*:97
Benzylideneacetone, *12*:107
α-Benzylideneacetophenone, *12*:107
Benzylidene chloride. See *Benzal chloride.*
2-Benzyl-2-imidazoline, *4*:519
Benzyl iodide
  from benzyl chloride, *5*:283
Benzyl isobutyrate, *14*:736
Benzyl isocyanate
  use of dimethylacetamide in prepn, *1*:147
S-Benzylisothiourea, *19*:318

Benzyl lactate
  prepn, *8*:353
Benzyl magnesium chloride, *10*:731
Benzyl mercaptan, *5*:283
Benzyl methyl ether
  phys prop, *8*:471
Benzyl *p*-nitrobenzoate, *3*:443
2-Benzyloxyethyl acrylate
  phys prop, *1*:288
p-Benzyloxyphenol, *15*:171
Benzylpenicillin, *14*:653, 666, 669, 700; *15*:213
  in bacterial infections, *3*:1
Benzyl perchlorate, *5*:73
Benzyl phenyl ether
  from benzyl chloride, *5*:283
Benzyl phenyl ketone, *12*:107, 146
Benzyl phenyl sulfone, *6*:642
1-Benzylpiperidine, *2*:369
Benzyl propionate, *8*:377; *9*:358; *14*736
Benzyl salicylate, *17*:730
  from benzyl chloride, *5*:283
  flavor and perfume materials, *9*:358
Benzyltrimethylammonium hydroxide, *12*:114
Benzyl xanthic acid
  esters of, *22*:422
Benzyne, *10*:918
Bephenium hydroxynaphthoate, *14*:534
Berberine, *1*:769; *20*:79
Berberine sulfate, *20*:79
Bergamot, *14*:184, 195
  essential oil as flavor, *9*:352
Bergamot oil
  as denaturant, *8*:450
Bergius process
  synthetic crude oil from coal, *S*:180
Beriberi
  and thiamine, *20*:173
Berkelium. See *Actinides.*
Berlin green, *12*:31
Berlin white, *15*:539
Berl saddles, *1*:59, 65
  in controlled flash evaporation, *22*:34
Bermudez asphalt, *2*:765
Bernoulli's theorem, *9*:467
Beryl, *2*:1; *3*:450, 453, 454
Beryl flotation process, *9*:388
Beryllia. See *Beryllium oxide.*
Beryllides, *3*:475, 478; *S*:**73**
Beryllium
  in copper alloys, *6*:209, 212, 250
  in election tubes, *8*:2
  heat of combustion of, *4*:898
  manganese compounds, *13*:3

in neutron prod, *17*:39
powder metallurgy, *16*:432
as propellant, *8*:661, 693
in research reactors, *14*:84
from seawater, *14*:156
sweeteners, *19*:595
zone refining, *22*:697
Beryllium-9
in nuclear reactors, *14*:76
Beryllium acetates, *3*:474
Beryllium and beryllium alloys, *3*:**450**
Beryllium bromide, *3*:475
Beryllium carbide, *3*:452, 468, 469, 474
phys prop, *4*:74
Beryllium carbonates, *3*:474
Beryllium chloride, *3*:459, 475
reaction with ethyl ether, *8*:479
Beryllium compounds, *3*:**474**
Beryllium fluoride, *3*:460, 461, 475; *9*:552
and nuclear reactor fuels, *14*:82
in prepn of glass, *10*:538
Beryllium hydride, *3*.475; *11*:204
Beryllium hydroxide, *3*:452, 454
Beryllium iodide, *3*:475
Beryllium nitrate, *3*:476
Beryllium nitride, *3*:452, 476
Beryllium oxide, *3*:476. (See also *Beryllium and beryllium alloys.*)
Beryllium oxide, *4*:803, 805, 818
in ceramic armor, *S*:151
hardness of, *S*:74
insulator in lithium-chlorine cells, *S*:128
in prepn of beryllium fluoride, *9*:552
refractory, *17*:244
Beryllium oxyfluoride, *9*:552
Beryllium sulfate, *3*:477
Berylometer, *3*:452
Bessemer process, *18*:720
BET theory, *1*:431
Betaine, *6*:592; *5*:405
Beta particle, *17*:14, 66
absorption, *17*:69
Betts process
of electrolytic lead refining, *12*:236
Beverages
carbonated, *4*:335
uses of industrial enzymes, *8*:191
BHA
antioxidant, *2*:593, 597
BHC. See *Benzene hexachloride.*
γ-BHC
adsorption, *18*:522

BHT
antioxidant, *2*:593, 597, 599, 601
Biacetyl, *12*:140, 144
ester-exchange catalyst, *16*:175
as flavoring, *13*:56
Biacylguanidine
prepn from guanidine, *10*:736
Bialamicol, *20*:88
amebicide, *20*:88
p, p-Bianiline. See *Benzidine.*
o,p'-Bianiline, *3*:412
4,4'-Bi-*o*-anisidine, *3*:417
Biazzi process
manuf of explosives, *8*:595
o,o'-Bibenzoic acid. See *Diphenic acid.*
4,4'-Bibenzoic acid, *16*:182
4,4'-Bi-7*H*-benz12*de*13anthracene-7,7'-dion- e,
7:469, 478
Bibenzoyl. See *Benzil*
Bibenzyl. See *Dibenzyl.*
Bibutyryl, *12*:140
hexachlorocyclopentadiene reaction with,
5:244
Bichlorendo, *S*:485
Bicyclo (2,2,1)heptadiene-2,5
Bicyclo (2, 2, 1)-5-heptene-2,3-dicarboxylic
acid anhydride, *5*:692
Bicyclohexyl, *7*:198, 200
Bieberite, *5*:742
Bifluoroboric acid
catalyst, *15*:167
Bi-2-furoyl, *10*:241; *12*:140
Biglycidyl, *5*:319
Biguanide
antiviral agent, *21*:457
Biguanide sulfate, *6*:567
Biguanidine nitrate, *10*:738
Biisobutyryl, *12*:140
Bile acids, *18*:830
Bile constituents, *3*:**480**
Bile extract
in food processing, *10*:8
Bile pigments, *16*:855
Bilineurine. See *Choline.*
Bilinewrine. See *Choline.*
Bimethallyl
prepn by Grignard reaction, *10*:728
Bimolecular radical reactions, *14*:823
Binary alloys, *6*:536
Binary mixtures
distillation separation, *7*:95, 217
Binary systems
correlation of, *21*:220

types of, *21*:213

vapor-liquid equilibria of, *21*:205

Binders, thixotropic

metallic soaps as, *4*:33

Bingham body, *21*:462

Bingham plastics, *20*:646

Bioassay, *3*:**489**

of proteolytic enzymes, *S*:303

Biocells, electrochemical, *3*:154

Biocytin, *3*:519

Biodegradation, *19*:582

Biological assay. See *Bioassay.*

Biological degradation

of waste liquids, *21*:629

Biological nitrogen fixation, *S*:608

Biologicals

pharmaceuticals, *15*:117

Biosorption

sewage treatment, *22*:110

Biot

definition of, *S*:1005

Biotex, *S*:295

Biotin, *2*:160; *3*:**518**; *8*:175; *21*:485, 489. (See also *Enzymes, industrial.*)

sulfur in, *19*:369

yeast requirement, *22*:517

D-Biotin

in prepn of glutamic acid, *2*:204

Biotinidase, *3*:519

Biotin methyl ester, *3*:522

D-Biotinol, *3*:523

Biotin *S*-oxide, *3*:522

Biotin sulfoxide, *3*:522

ε-*N*-Biotinyl-L-LYSINE, *3*:519

ε-*N*-Biotinyl-L-lysine. See *Biocytin.*

Biotite, *13*:399, 736

rubidium in, *17*:685

Biphenamine hydrochloride, *10*:775

p,p'-Biphenol, *2*:600

Biphenyl, *3*:373, 398. See *Diphenyl.*

Biphenyl benzene, *7*:570

4,4'-Biphenyldiamine. See *Benzidine.*

2,4'-Biphenyldiamine, *3*:412

2,2'-Biphenyldicarboxylic acid. See *Diphenic acid.*

Biphenylene ketone, *12*:107, 132

Biphenylhexol, *16*:191

Biphthalyl, *15*:448

Bipivaloyl, *12*:140

Bipropionyl, *12*:140

Bipyridine

synthesis from pyridine, *22*:207

2,2'-Bipyridine, *22*:207

2,2'-Bipyridine(α,α'-dipyridyl), *21*:580

4,4'-Bipyridine, *22*:207

Biquinolines, *16*:876

Biradicals, *10*:114

Birch

sweet, *14*:195

Birch black, *15*:544

Birnessite, *13*:21

1,3-Bis-(acetoxyimino)-2-propanone, *6*:630

Bisacodyl, *4*:588

2*H*-4,6-Bis(alkylamino)-*s*-triazines, *22*:186

N,*N'*-Bisalkylcarbamylpiperazine, *15*:639

Bis(amino)boranes, *3*:733

Bis(2-aminoethyl)sulfide, *11*:533

Bis(4-aminophenyl) ether, *S*:756

polyimide enamels, *S*:769

polypyromellitimide of, *S*:756, 760

Bis(*p*-aminophenyl)methane, *2*:234

Bis(4-aminophenyl) methane

polypromellitimide of, *S*:762, 766

2,2-Bis(4-aminophenyl)propane

Bis(4-aminophenyl) sulfide

Bis(*p*-aminophenyl)sulfone, *19*:267, 270

Bis(4-aminophenyl) sulfone

Bisazocyano acid, *17*:658

Bisazoles

as optical brighteners, *3*:743

Bis(2-azolyl)stilbenedisulfonic acids, *19*:15

2,2'-Bis-1,3-benzdithiolene

from tetrachloroethylene, *5*:199

4,4'-Bis(2-benzoxazolyl)stilbenes, *19*:14

Bisbenzylisoquinoline alkaloids, *1*:770

Bis(biphenyl)chromium(I) iodide, *5*:480

1,4-Bis(bromoacetoxyl)-2-butene

as fungicide, *10*:235

Bis(2-bromoethyl)ether

phys prop, *8*:471

2,2-Bisbromomethyl-1,3-propanediol, *20*:831

as fire retardant, *S*:467

as fire retardant, *S*:482

Bis(2-butoxyethyl)adipate, *15*:764

Bis(2-butoxyethyl)ether, *15*:883

Bis(Tri-*n*-butylphosphine) nickel dibromide, *13*:759

Bis(β-carboxypropionyl) peroxide, *9*: 812, 813

1,5-Bis(chlorendo)cyclooctane, *S*:485

2,2-Bis(chlorodifluoromethyl)-1,3-oxazolidine, *9*:760

2,2-Bis(chlorodifluoromethyl)-3-trifluoromethyl-trifluorooxetane, *9*:764

Bis-4-chlorobutyl ether, *16*:254

phys prop, *8*:471

prepn, *8*:492

Bis-4-chlorobutyl formal, *16*:254, 256
1-Bis(2-chloroethyl)-amino-1-oxo-2-aza-5-oxa-
    phosphoridin, *S*:83
Bis(1-chloroethyl) ether
Bis(2-chloroethyl) ether, *8*:486
Bis(2-chloroethyl) formal, *7*:697; *16*:253, 256
Bis(2-chloroethyl) formal polymer, *16*:260
Bischloroethyl nitrosourea
    alkylating agent, *S*:84
Bis(chloroisopropyl) ether, *8*:281
1,3-Bis(2-chloroethyl)-1-nitrosourea, *S*:84
Bis(2-chloroethyl) sulfide, *4*:872
Bis(2-chloroethyl) sulfide (mustard)
Bis(2-chloroethyl) sulfone
1,2-Bis(2-chloroethylthio)ethane
Bis(2-chloroethylthioethyl) ether
1,6-Bis(4-chlorophenyl)diguanidinohexane,
    2:637
Bis(4-chlorophenyl) sulfone, *19*:270
Bis(5-chloro-2-hydroxyphenyl)methane, *2*:631.
    See *Dichlorophene.*
Bis(2-chloroisopropyl) ether, *8*:491
Bis(chloromethyl) ether, *10*:85
    prepn of three-block polymers, *S*:917
3,3-Bis(chloromethyl)-1-oxacyclobutane, *15*:795
3,3-Bis(chloromethyl)oxethane, *1*:592
Bis(*p*-chlorophenyl)iodonium chloride, *2*:619
Bis(*p*-chlorophenyl) sulfone, *15*:795
Bis(2-chloropropyl) ether
Bis(3-chloropropyl) ether
Bis(β-chlorovinyl)stibine trichloride, *2*:583
Bischofite, *3*:758
Bis(cinnamylidenehexanediamine), *7*:696
1,4-Bis(2-cyanoethoxy)butane, *10*:670
N,N-Bis(2-cyanoethyl)acetamide, *6*:650
N,N-Bis(2-cyanoethyl glycine, *6*:650
Bis(2-cyanoethyl)phenylphosphine, *6*:651
Bis(2-cyanoethyl)phosphine, *6*:651
1,4-Bis(2-cyanoethyl)piperazine, *15*:639
Bis(2-ethylhexyl) adipate, *1*:244
    plasticizer, *15*:728, 751, 756, 763, 769, 774
N,N-Bis(2-cyanoethyl)-(*o,p*)-toluenesulfon-
    amide,- *15*:759
N,N-Bis(2 cyanoethyl)-*p*-toluenesulfonamide,
    6:650
    prod and sales of, *8*:377
Biscyclopentadienylcobalt. See *Coballcene.*
Biscyclopentadienyldimethyltitanium, *20*:440,
    444
Bis(cyclopentadienyl) di(phenylselenide),
    22:660
Biscyclopentadienyldiphenyltitanium, *20*:428
Biscyclopentadienyldiphenyltitanium, *20*:443

Biscyclopentadienyliron, *6*:696. See *Ferrocene.*
Biscyclopentadienylmanganese, *13*:42
Bis(cyclopentadienyl)nickel, *13*:757
Biscyclopentadienyltitanium, *20*:443, 447
Biscyclopentadienyltitanium(II), *20*:435
Biscyclopentadienyltitanium(III)   chloride,
    20:442
Biscyclopentadienyltitanium allyl complexes,
    20:442
Bis(cyclopentadienyl)titanium carbonyl, *4*:490
Biscyclopentadienyltitanium   dibromide,
    20:426
Biscyclopentadienyltitanium   dichloride,
    6:699; *20*:428, 440
Bis(cyclopentadienyl)titanium diphenyl, *7*:70
Biscyclopentadienyltitanium diphenyl, *20*:426,
    435
Biscyclopentadienyltitanium diphenyl, *20*:435
Biscyclopentadienyltitanium propyl chloride,
    20:440
Bis(cyclopentadienyl) zirconium dichloride,
    22:644, 649
Biscyclopentadienylmagnesium
    titanium halide reactions, *20*:434
Bisdehydrodoisynolic acid, *18*:864
Bisdehydrothio-*p*-toluidine, *20*:186
Bis(4,6-diamine-2-*s*-triazinyl)amine, *6*:571
Bis(2,3-dibromo-2-butene-1,4-diol), *1*:605
4,4'-Bis(2,6-di-*tert*-butylphenol, *1*:913
2,2-Bis(3,4-dicarboxyphenyl)propane
    cross-linked polyimides from, *S*:753
    polyimide from, *S*:767
    prepn of monomers, *S*:753
N,N'-Bis(dichloroacetyl)-1,8-octanediamine), *6*:85
Bis(2,4-dichlorobenzoyl)peroxide, *18*:229
Bis(2,3-dichloropropyl)ether, *8*:494
Bis(3,5-dichloro-2-hydroxyphenyl) sulfide. *8*
    Bithionol.
1,4-Bis(dicyanomethylene)cyclohexane, *6*:628
4,4'-Bis(diethylamino)benzophenone, *20*:700,
    712, 721
Bis(diethylboryl)sulfate, *3*:722
Bis(diisopropylphenyl)methane
    in heat transfer, *10*:857
p,p'-Bis(dimethylamino)benzophenone, *12*:132
4,4'-Bis(dimethylamino)benzophenone, *7*:478;
    12:107
Bis(dimethylamino)borane, *3*:733, 734
1,1-Bis(dimethylamino)disilane, *18*:196
Bis(dimethylthiocarbamoyl) disulfide, *2*:601;
    19:771
Bisdisilanyl ether, *18*:189, 194
Bis(2,3-epoxypropyl) suberate, *1*:250

Bis123-ethylbenzothiazole(2)13carbocyanine i-
odide, *16*:283
Bis123-ethylbenzothiazole(2)13trimethine cyanine iodide, *16*:283
4-Bis(3-ethyl-2-benzothiazolinylidenemethyl)
methylene-2-3-dimethyl-5-0-40-1-phenyl-2-
pyrazolinium iodide, *6*:619
2,5-Bis(3-ethyl-2-benzothiazolinylidenemethyl)-
3,4,4−trimethyl-4H-imidazolium iodide,
6:618
Bis(2-ethylhexyl) phthalate, *15*:721
plasticizer, *15*:751, 769
prod and sales of, *8*:376
Bis(2-ethylhexyl) sebacate, *1*:244; *8*374; *9*:714;
15:762, 769, 774; *20*:490
prod and sales of, *8*:377
2,2-Bis(ethylsulfonyl)propane, *20*:215
1,2-Bis(ethylthio)propane
thiol addition, *20*:209
2,17-Bisethynyl-A-norandrostanediol, *18*:885
(Bis-4-flavylo) trimethinecyanine perchlorate,
16:284
Bis glycine hydriodide-iodine
as disinfectant, *2*:619
Bis(Heptafluoropropyl) ether, *9*:689
Bis(1*H*1*H*-heptafluorobutyl) 3-methylglutar-
ate, *9*:732
Bis(2-(-*n*-hexaxyethoxy) ethyl succinate,
15:736
Bishydroxycoumarin, *5*:600, 601; *15*:919
Bis(2-hydroxyethyl)alkylamine oxides
toxicity of, *S*:43
Bis(2-hydroxyethyl)alkylamines, *S*:40
Bis(2-hydroxyethyl)dodecylamine oxide, *S*:46
N,N-Bis(2-hydroxyethylhydrazine)
N,N-Bis(2-hydroxyethyl)lauramide, *S*:37
Bis(1-hydroxyethyl) peroxide, *2*:396; *8*:479.
See *Ether peroxide.*
1,4-Bis(2-hydroxyethyl)piperazine, *15*:639
Bis(2-hydroxyethyl) sulfide, *8*:529
prepn, *8*:529
Bis-2-hydroxyethyl sulfoxide, *19*:331
Bis(22-Hydroxy-3-α-methylcyclohexyl)-meth-
yl-phenyl 13 methane. See *2,2' Methylnebis
[6-(methylcyclohexyl)-**cresol**].*
2,2-Bis(hydroxymethyl)-1,3-propanediol. See
Pentaerythritol.
Bis(hydroxyphenyl)alkanes
in soaps, *18*:429
1,1-Bis(4-hydroxyphenyl)cyclohexane polycarbonate, *16*:114
1,3-Bis(4-hydroxyphenyl)-5,7- dimethylada-
mantane, *S*:14

1,1-Bis(*p*-hydroxyphenyl)ethane
from acetaldehyde, *1*:83
3,3-Bis(*p*-hydroxyphenyl)phthalide. See *Phenolphthalein.*
as cathartic, *4*:588
2,2-Bis(*p*-hydroxyphenyl)propane, *1*:912. See
Bisphenol A.
1,4-Bis(2-hydroxy-3,4,6-trimethylphenyl)-1,4-
butanedione, *19*:139
Bisindenyltitanium chloride, *20*:443
2,2'-Bisindole indigo, *11*:565
Bis(2-iodoethyl) ether, *8*:487
Bis(2-isocyanatoethyl) carbonate, *21*:103
Bis(2-isocyanatoethyl) fumarate, *21*:103
Bis(isopropylphenyl)methane
in heat transfer, *10*:857
Bismaleimide
Bismanol, *3*:534
Bismark brown
in hair coloring, *10*:793
1,4-Bis(methanesulfonoxy)butane, *S*:83
Bis(2-methoxyethyl) phthalate
Bis(methoxymethyl) ether, *8*:477
Bis(2-methoxymethoxyethyl) adipate
phys prop, *1*:409
Bis(methoxymethyl)urea, *2*:233
N,N'-Bis(methoxymethyl)uron, *2*:231, 232
3,5-Bis(methoxymethyl)uron, *2*:232
Bis(3-methylbutyl) azelate, *1*:244
Bis(methylene dioxide)
steroid synthesis, *18*:876
1,1-Bis(3-methyl-4-hydroxphenyl)ethane. See
4,4'-Ethylidenedi-o-cresol
Bis(3-methyl-1,4-naphthoquinone-2-yl)-
methane, *21*:593
Bis(methylphenyl)methane
in heat transfer, *10*:857
Bis(methylthio)methane, *19*:321
Bis(1-methyltrimethylene) pyroborate, *10*:661
Bismite, *3*:529
Bismuth
effects on brasses, *6*:227
effects on copper, *6*:195
impurity in copper, *S*:265
in manganese compounds, *13*:4
in tin-lead alloys, *18*:543
in type metal, *20*:757
Bismuth(III) bromide. See *Bismuth tribromide.*
Bismuth(III) chloride. See *Bismuth trichloride.*
Bismuth(III) fluoride. See *Bismuth trifluoride.*
Bismuth(III) iodide. See *Bismuth triiodide.*
Bismuth(III) nitrate pentahydrate, *3*:542
Bismuth(III) perchlorate, *3*:542

Bismuth(III) sulfate, *3*:542
Bismuth(III) sulfide, *3*:542
Bismuth(V) fluoride. See *Bismuth pentafluoride.*
Bismuth-212, *17*:9
Bismuthates, *3*:537
Bismuth and bismuth alloys, *3*:**527**
Bismuth carbonate (basic), *3*:546
Bismuth chromate, *5*:499
Bismuth compounds, *3*:**535**
Bismuth glance. See *Bismuth(III) sulfide.*
Bismuth-gold alloy, *10*:690
Bismuth *N*-glycolyl-*p*-arsanilate, *3*:547
Bismuth halides, *3*:538
Bismuthides, *3*:536
Bismuthine, *3*:536
Bismuthinite, *3*:527; *22*:338
Bismuth oxides, higher, *3*:537
Bismuth oxybromide, *3*:542
Bismuth oxychloride, *3*:530, 541
  formation of, *5*:8
Bismuth oxyfluoride, *3*:541
Bismuth oxyiodide, *3*:541, 542
Bismuthoxy *p*-*N*-glycoloylarsanilate, *20*:93
Bismuth pentafluoride, *3*:541; *9*:553
Bismuth perchlorate, *5*:71
Bismuth phosphate
  plutonium extraction, *15*:883
Bismuth phosphomolybdate
  catalyst, *16*:586
Bismuth potassium tartrate, *3*:547
Bismuth salicylate (basic), *3*:547
Bismuth salts, *3*:542
Bismuth sesquioxide. See *Bismuth trioxide.*
Bismuth sesquisulfide. See *Bismuth(III) sulfide.*
Bismuth sodium triglycollamate, *3*:546
Bismuth stannate, *20*:318
Bismuth subcarbonate, *3*:546
Bismuth subnitrate, *3*:542, 547
  veterinary uses, *21*:252
Bismuth subsalicylate, *3*:547
Bismuth tribromide, *3*:540
Bismuth trichloride, *3*:540
Bismuth trifluoride, *3*:539; *9*:553
Bismuth trihydride, *3*:536
Bismuth triiodide, *3*:540
Bismuth trioxide, *3*:537; *12*:143
  amebicides, *20*:93
Bismuthyl nitrate, *3*:542
Bismuthyl sulfate, *3*:542
Bismutite, *3*:529
Bis(*o*-nitrophenyl)butadiyne, *11*:563
Bis(*o*-nitrophenyl) disulfide, *19*:299

Bisnorbornyleneimides
"Bis" oils, *7*:409
Bis(pentachloroethyl) tetrachlorosuccinate, *19*:137
Bis(1*H*, 1 *H*-pentadecafluorooctoxy)-bis(*t*-butoxy)silane, *9*:729 , 730, 731
Bis(1*H*, 1 *H*-pentadecafluorooctyl)dodecenyl succinate, *9*:729, 730, 731
Bis(1 *H*, 1 *H*-pentadecafluorooctyl)3-methylglutarate, *9*:729
Bis(1*H*, 1*H*-pentadecafluorooctyl)tetrachlorophthalate, *9*:729, 730, 731
Bis(pentafluorophenyl) ether, *9*:793
Bis(pentafluorophenyl)mercury, *9*:795, 796
Bis(pentafluorophenyl)methanol, *9*:795
Bis(pentafluorophenyl) sulfide, *9*:792
Bis(phenethyl) ether, *8*:477
Bisphenol A, *1*:912; *8*:294, 361; *12*:195; *15*:148, 173, 187, 747, 793; *16*:106, 272, 517
  phenolic resins, *15*:187, 206
Bisphenol A polycarbonate, *16*:106
Bisphenols
  polysulfides, *19*:493
Bis(2-phenoxyethyl) ether, *8*:487
N,*N*'-Bis(phenylamino)methane, *2*:234
3,5-Bis(phenylazo)salicylic acid, *17*:723
Bis(2-phenylbenzopyran)trimethinepyranine perchlorate, *16*:284
Bis(*N*-phenyl-2-naphthylamine) sulfide. See *1,1'-Thiobis12-phenyl-2-naphythylamine13.*
Bis(thiocarbamoyl)sulfides, *17*:514
Bis(triazinylamino)stilbenedisulfonic acids, *19*:15
Bistriazinyl brighteners, *3*:739
Bistributyltin oxide, *10*:235
Bis(tributyltin) oxide, *20*:321
Bis(tributyltin) sulfide, *20*:321
Bis(3,5,6-trichloro-2-hydroxphenyl)methane. See *Hexachlorophene.*
1,4-Bis(trichloromethyl)benzene, *14*:549
Bis(trichloromethyl) disulfide, *19*:374
Bis(trichloromethyl) sulfone, *10*:235
Bistriethyllead, *12*:290
O,*N*-Bis(trifluoroacetyl)hydroxylamine; *9*:795
1,1-Bis(trifluoromethyl)-2-nitroethanol, *9*:757, 762
2,2-Bis(trifluoromethyl)-2-propiolactone, *9*:764
2,2-Bis(trifluoromethyl)-5-aryl-4-oxazolidones, *9*:757
Bistrimate, *3*:546
Bis(trimethylsilyl)acetamide, *18*:262
Bis(trimethylsilyl)trifluoroacetamide, *18*:262

Bis(triphenylphosphine) nickel nitrosylbromide, *13*:759

Bistyryl. See *1,4-Diphenyl-1,3-butadiene.*

Bistyryl, *5*:521

Bis(1*H*, 1*H*-undecafluorohexyl 3-methylglutarate, *9*:729

4,4'-Bis(6-triazinyl)stilbene, *19*:14

Bithionol, *2*:631

Bitter almond
  essential oil as flavor, *9*:352

Bitumen, *19*:653. (See also *Asphalt, Tar sands.*)
  coking, *19*:712
  in roofing materials, *17*:462

Bituminous coal. See *Coal, coals.*

Biurea, *10*:917; *22*:293

Biuret, *2*:226; *10*:917
  determination of in urea, *21*:53
  in urea, *9*:69; *21*:51
  from urea, *21*:38

Biurets
  as hypnotics, *11*:510, 516
  prepn from isocyanates, *12*:49

5,5'-Bivanillin
  from vanillin, *21*:182

Bixbyite, *13*:18

Bixin, *7*:615

Blanc fixe, *3*:94; *15*:515

Black 83 (CI), *7*:535

Black ash. See *Barium sulfide.*

Black blasting powder, *8*:651, 695

Blackbody
  carbon blacks as, *4*:276

Blackbody radiation, *10*:834

Black cyanide. See *Calcium cyanide.*

Black lead. See *Graphite.*

Black oxide of iron, *12*:37

Black oxide of lead, *12*:273

Black powder, *8*:647, 582
  manuf, *8*:648
  in seismic explosives, *8*:645

Blast furnace gas
  cleaning by electrostatic precipitator, *8*:90
  composition of, *4*:438

Blast-furnace process
  for manuf of iron, *12*:11

Blast furnaces
  carbon linings in, *4*:239

Blasting explosives, *8*:635
  manuf, *8*:636

Blasting gelatin, *8*:640
  manuf, *8*:636

Bleach
  semitropical, *5*:20

Bleaches. See *Bleaching agents.*

Bleaching
  of fats, *13*:180
  of hair, *10*:790, 798

Bleaching agents, *3*:**550**, (See also *Chloramines.*)
  activators for, *5*:30, 34
  in food products, *10*:4
  for pulp, *5*:40
  hydrogen peroxide as, *11*:408
  liquid, *5*:12. (See also *Sodium hypochlorite solutions.*)
  optical brighteners, *3*:737
  powder, *5*:16, 18. (See also *Semitropical bleach.*)
  repressors for, *5*:30
  for tallow, *5*:44
  tropical, *5*:20

Bleaching of furs, *10*:304

Bleaching powder, *3*:556; *5*:16, 18. (See also *Semitropical bleach*)
  in manuf of chloroform, *5*:124

Bleaching sequences, *5*:34, 35

Bleach liquor, *3*:556; *5*:14

Blende. See *Zinc blende.*

Blended whiskey, *1*:514

Blending, *13*:**577**

Blister agents. See *Vesicant agents.*

Blister copper, *6*:144

Blood
  animal, *3*:**567**
  cryogenic preservation of, *6*:477
  human, *3*:576
  water content, *21*:652

Blood albumin
  animal, *3*:571
  human, *3*:576

Blood dilution studies
  tracer used in, *6*:605

Blood fractionation, *3*:**576**

Blood, human
  clotting mechanism, *5*:586
  sodium citrate in processing of, *5*:539

Blood meal
  as animal feed, *8*:866

Blood plasma
  animal, *3*:573
  human, *3*:576

Blood proteins, *3*:576

Blood sugar. See *Dextrose.*

Blowing agents, *17*:532. (See also *Foamed plastics.*)

Blue 45 (CI), *7*:522

Blue basic lead sulfate, *15*:541
Blue bice, *15*:541
Blue black, *15*:544
Blue fining
  vinegar manuf, *21*:266
Blue frit, *15*:541
Blue gas process, *10*:363, 425
Blue malachite, *15*:541
Blueprinting, *17*:356
Blue verditer, *15*:541
Blutene chloride. See *Tolonium chloride.*
Board feet, *22*:359
Boart, *4*:285
Boea, *17*:386
Boehmite, *1*:937; *5*:553; *13*:18
Boehmite, *2*:8, 42, 43
Boehmite, *16*:517
  in alumina gels, *2*:43
Bohn-Schmidt reaction, *2*:470
  in manuf of purpurin, *2*:473
Bohr magneton
  as a "unit", *S*:997
Bohr radius
  as a "unit", *S*:997
Boilers. (See also *Steam.*
  steam, *18*:696
Bois de rose
  Brazilian, *14*:196
Bois de rose oil
  essential oils as flavor, *9*:352
Bolekic acid, *8*:787
Boltzmann spin temperature, *14*:45
Bomb calorimeter, *4*:45
Bombs
  fire, *4*:892
  incendiary, *4*:899
  magnesium, *4*:899
  oil, *4*:899
Bond length
  by x-ray analysis, *22*:453
Bond order
Bond shift
  and NMR, *14*:73
Bone ash
  phosphorus, *15*:280
Bone black. See *Bone char.*
Bone char, *4*:149, 152; *15*:545
  as adsorbent, *1*:460
  sugar refining, *19*:188
  wax decolorizer, *15*:97
Bone glue. See *Glue.*
Bone meal
  as animal feed, *8*:866

Bone oil
  as denaturant, *8*:450
Bone phosphate of lime, *9*:40, 394
Bones, calcined
  in ceramics, *4*:772
BON Maroons, *15*:575
BON Reds, *15*:575
Bonofan. See *Tetrachloroethane.*
Bookbinding
  use of adhesives in, *1*:395
Boracetic anhydride
  reactions with anthraquinone derivatives,
    2:449, 468
Boracic acid. See *Boric acid.*
Boracite, *3*:628
Borane
  and hydroboration, *S*:505
Boranes, *3*:684
Borates, *3*:609, 617, 662
  analysis for in water, *21*:696
Boratozirconium chloride, *22*:639
Borax, *12*:588. (See also *Boron compounds.*)
Borax, *3*:618, 623
  anhydrous, *3*:642
  in dental stones, *6*:788
  in dentifrices, *6*:848
  in enamel frits, *8*:158
  fire-retardant on wood, *22*:376
  as flame retardant for textiles, *9*:304
  as fungicide, *10*:230
  as a herbicide, *22*:214
  photography, *15*:373
  in prepn of boron trifluoride, *9*:558
  sodium tetraborate decahydrate, *3*:618, 623
  sodium tetraborate pentahydrate, *3*:639
  source of micronutrient fertilizer, *9*:146
  tellurium recovery, *19*:759
Borazines, *3*:729
Borazoles, *3*:729
Borazon, *3*:679
  as abrasive, *1*:28
Bordeaux mixture, *15*:583
Bordeaux mixture, *10*:222; *11*:725
  as fungicide, *6*:266
  soil chemistry, *18*:538
Boric acid, *2*:467; *3*:612; *7*:414; *9*:559. (See
  also *Boron oxide.*)
Boric acid, *8*:450
  corrosion of zirconium, *22*:619
  dehydration catalyst in prepn of ethers,
    8:474
  Duff reaction, *15*:162
  in enamel frits, *8*:158

esters of, *3*:652
fire-retardant on wood, *22*:376
as fungicide, *10*:230
as flame retardant for textiles, *9*:304
in matches, *13*:161
in nickel electroplating bath, *8*:63
pigments, *15*:538
in powders, *6*:360
in prepn of boron trifluoride, *9*:558
prepn of fluobaric acid, *9*:655
reaction with glycols, *10*:639
from seawater, *14*:156
to stabilize aluminum monoacetate, *2*:11, 13
Boric anhydride. See *Boric oxide.*
Boric oxide, *3*:609. (See also *Boric acid.*)
    as catalyst in manuf of 1-butanol, *8*:456
    esterification of, *3*:654, 659
    in prepn of trifluoroboroxine, *9*:556
    reaction with 1,3-butylene glycol, *10*:661
Borides, metal, *3*:673
Borinates, *3*:716
Borinic acids, *3*:716
Borneol, *19*:819, 822
    prepn, *4*:56, 57
Borneo tallow, *5*:387; *8*:778, 784
Bornite, *6*:134, 135; *9*:393
Bornyl chloride, *4*:55, 57; *19*:831
Bornyl cinnamates, *5*:520
Boroglycerides, *10*:629
Borohydrides, *3*:701; *11*:210
    availability of, *11*:217
Boron
    analysis for in water, *21*:696
Boron and boron alloys, *3*:**602**
    content in soils, *9*:35
    deficiency in nuts, *14*:131
    ferroalloy, *18*:751
    fiber-reinforced composites, *11*:651
    heat of combustion of, *4*:898
    in heterochain inorganic polymers, *11*:643
    ignition mixtures, *16*:826
    inorganic homochain polymers, *11*:635
    manganese compounds, *13*:3
    in modified stainless steels, *11*:21
    in nickel-base superalloys, *11*:28
    as plant nutrient, *9*:29
    as propellant, *8*:661, 693
    whiskers, *15*:802
Boron-10
    separation, *14*:86
    use of fluoborates with, *9*:566
Boron acetate, *3*:662
Boron-aluminum composites, *22*:345

"Boron anomaly," *3*:620
"Boron anomaly," *3*:620
Boronates, *3*:716
Boron bases, *3*:704
Boron carbide, *3*:605, 676; *4*:74
    in ceramic armor, *S*:138
    strength value of, *4*:818
Boron compounds, *3*:**608**
Boron compounds
    as herbicides, *22*:214
    as insecticides, *11*:682
Boron-epoxy composites, *22*:345
Boron esters, *3*:652
Boron fibers
    prop of composites, *11*:665
    reinforced composites, *11*:664
    thermal instability, *11*:663
    by vapor deposition, *11*:661
Boron filaments
    wolfram alloys in manuf of, *22*:345
Boron fluoride, *7*:417. (See also *Boron trifluoride.*)
    as catalyst for acidolysis, *8*:362
    as catalyst in alkylation of phenols, *1*:895
    as catalyst for isomerizing *n*-butane, *3*:817
    reaction with butyrolactone, *10*:674
Boron fluoride dihydrate, *9*:563
Boron halides, *3*:655, 680-683
Boron hydrides, *3*:684, 706
Boronic acids, *3*:716
"Boronium" salts, *3*:721
Boron monoxide, *3*:609
Boron nitride, *3*:678, 679; *4*:74; *12*:588; *13*:822
    from cyanuric acid, *20*:666
    refractory, *17*:245
    strength value of, *4*:818
Boron nitride fibers
    high-temperature treatment, *11*:670
    in metal matrices, *11*:668
    prop, *11*:667
Boron oxide. (See also *Boric acid.*)
    as glass formers, *10*:538, 541, 542
Boron phosphate, *3*:650; *15*:275
Boron silicides, *18*:129
Boron tribromide, *3*:682
Boron trichloride, *3*:659, 665, 680
    sulfur monochloride, *19*:392
Boron trifluoride, *7*:41; *9*:554, 699, 700 (See also *Boron fluoride.*) *12*:21, 112, 128, 154, 293
    and Boron-10 separation, *14*:86
    as catalyst, *8*:408, 429, 475, 476, 529

as catalyst in Friedel-Crafts reactions, *10*:159

neutron diffraction analysis, *22*:453

prepn from fluorosulfonic acid, *9*:677

as process gas, *7*:128

reaction with ethyl ether, *8*:479

silica reaction, *18*:57

Boron trifluoride-*S,S,S*-tri-*n*-butyl trithiophosphate, *9*:561

Boron trifluoride etherate, *8*:479; *9*:569, 760; *15*:167

synthesis of vitamin $K_{1(20)}$, *21*:589

Boron trifluoride ethyl stearate

catalyst for *dl*-α-tocopheryl acetate prod, *21*:579

Boron trifluoride monoethylamine

Boron triiodide, *3*:682

Boron trioxide. See *Boric oxide and* Boron oxide.

Bort

as abrasive, *1*:25

Botrytis

paraffin degradation, *S*:839

"Bottled" gas, *8*:385

Bouguer-Lambert law

application to optical filters, *9*:246

Bouguer's (Lambert's) law of absorption, *5*:788

Bouillon

yeast, *22*:531

Boules, *10*:512

Bourbonal. See *Ethylvanillin.*

Bourbon whisky, *1*:513

Bournonite, *6*:135

Bovine mastitis, *21*:243

BPL. See *Bone phosphate of lime.*

Brabender Plastograph, *21*:390

Bradykinin

tissue hormone, *11*:50

Bragg equation

x-ray diffraction, *22*:438

Braggite, *15*:842

Bragg's law

x-ray fluorescence analysis, *22*:460

Brake fluids

use of ethylene glycol in, *10*:643

Brake linings

use of furfural in, *10*:128, 243

Branch fibers

in meats, *13*:168

Brandy, *1*:515

Brannerite, *21*:7

Brass, *6*:208, 213, 249. (See also *Copper alloys.*)

from copper and calamine, *22*:555

furnaces used for heating, *10*:288

Brassidic acid, *8*:814

Brass products. See *Copper alloys.*

Brass, red, *6*:214, 224

Brassylic acid, *1*:253

Bratton-Marshall reagent, *4*:933

Braunite, *13*:8

Bravoite, *21*:159

Brayton cycle

unconventional powerplants, *S*:68

Brazilein, *10*:307

Brazilin, *10*:307

Brazil nuts, *14*:145

Brazilwood, *7*:618

Brazing. See *Solders; Brazing alloys.*

Bread. See *Bakery processes.*

yeast, *22*:527

Bread making

action of yeast in, *22*:528

Brein, *16*:313

Bremsstrahlung, *17*:40, 89

Bretylium tosylate, *4*:519

Breuolin, *16*:313

Brevibacterium, *2*:204

Brewers' dried yeast

as animal feed, *8*:861

Brewers' grains

as animal feed, *8*:861

Brewers' malt, *8*:192, 208; *12*:861, 877

Brewers' yeast, *3*:303. (See also *Beer.*)

Brewery. (See also *Beer.*)

Brewing

of beer, *3*:297

Brick. (See also *Ceramics.*)

clays used in, *5*:566

dolomite, *17*:227

fireclay, *17*:227

forsterite, *17*:228

high-alumina, *17*:227

insulating in pipe linings, *15*:647

magnesite, *17*:227

paving, *5*:566

semisilica, *17*:227

silica, *17*:227

silica in steel furnaces, *18*:728

zirconia, *17*:228

Brick clays, *4*:768

Brighteners, optical, *3*:737

Brightstocks, *15*:59

Brilliant Blue FCF, *5*:866, 873

used in hair coloring, *10*:793

Brilliant Blue lake, *5*:867

Brilliant Croceine 3BA, 7:525
Brilliant Green, 20:702
  as bactericide, 20:731
  manuf, 20:703
Brilliant Indocyanine G, 20:717
Brilliant Indocyanine GA, 7:526
Brilliant Indocyanine 6B, 20:717
Brilliant Lake Red R, 5:870
Brilliant Melting Blue GA, 7:526
Brilliant Melting Red RRA, 7:522
Brilliant Milling Blue BA, 7:526
Brilliant Milling Green B
  manuf, 20:705
Brilliant Milling Red RRA, 7:522
Brilliant Sulpho Flavine FFA, 7:526
Brilliant Wool Blue FFR, 20:717
Brilliant Wool Blue G Extra, 20:724
Brilliant Yellow, 7:465
Brilliant Yellow 6G Base
Brilliant sulphur Flavine FFA, 7:526
Brimstone, 19:338. (See also *Sulfur.*)
Brine
  chemicals recovered from, 3:631
  manuf of salt, 8:559, 578
Brine, in ocean water
  recovery by desalination, 7:860
Brinell hardness test, 10:810
Brines
  analysis of, 21:706
Brisance
  tests for, 8:655
Britannia metal
  composition, 2:569
British anti-lewisite. See *2,3-Dimercapto-1-propanol.*
Brittania Violet, 7:464
Brix degrees, 19:158
Brobenzoxaldine
  amebicides, 20:84
Brochantite, 6:135, 169
Brodie's reagent
  in prepn of graphite oxide, 4:312
Broenner's acid, 13:714
Bromacil, 22:213
Bromal, 3:780; 8:420
  from acetaldehyde, 1:83
Bromamide, 3:766
Bromamine acid, 2:452; 7:478. See *1-Amino-4-bromo-2-anthraquinones ulfonic acid.*
Bromelain. See *Bromelin.*
Bromelin, 5:594, 595, 598; 10:9
  source of enzymes, 8:205
  uses, 8:190

Bromethol, 3:780
Brometone, 3:780
Bromic acid, 3:769
Bromides. (See also *Bromine compounds.*)
  as hypnotics, 11:518
Bromimide, 3:766
Bromine, 2:320; 3:**750**
  carbon monoxide reaction with, 4:432
  chloroprene reactions with, 5:217
  in dye industry, 7:480
  ethyl chloride reactions with, 5:142
  as fission product, 14:83
  graphite compound with, 4:312
  methyl chloride reactions with, 5:104
  methylene chloride reaction with, 5:114
  in prepn of bromoacetic acid, 8:419
  reaction with acetaldehyde, 1:83
  recovery from Great Salt Lake, S:465
  from seawater, 12:721
  in seawater, 14:155
  swimming pool water disinfectant, 22:127
  tetrachloroethylene reaction with, 5:198
  thiophene reaction, 20:222
  trichloroethylene reactions with, 5:188
Bromine-82, 17:124
Bromine chloride, 3:768
Bromine compounds, 3:**766.** (See also *Bromides.*)
Bromine dioxide, 3:769
Bromine iodide
  methylene chloride reaction with, 5:114
Bromine monoxide, 3:769
Bromine pentafluoride, 9:586
  prop, 9:593
  reactions, 9:588, 590
Bromine trifluoride, 9:586
  phys prop, 9:592
  reactions, 9:587, 588, 589
Bromisovalum, 11:517
p-Bromoacetanilide
  phys prop, 1:152
  prepn from acetanilide, 1:150
Bromoacetic acid, 3:776; 8:418
  phys prop, 1:235
Bromoacetone, 12:106
  in chemical warfare, 4:877
α-Bromoacetophenone, 1:170
Bromoacetyl bromide
  prepn from acetyl chloride, 1:139
2-Bromoacetyl-5-nitrofuran, 13:856
Bromo acids, 6:364
1-Bromoadamantane, S:5
p-Bromoaniline, 2:415; 3:775

2-Bromoanthraquinone, 5:229
Bromoanthraquinones, 2:478
  chemical reactions, 2:479
  manuf, 2:482
  prop, 2:478
  uses, 2:487
5-Bromoaspirin, 3:776
Bromobenzene, 3:776
m-Bromobenzenesulfonic acid, 3:402
Bromobenzine, 2:413
3-Bromobenzotrifluoride, 9:778
Bromobenzyl cyanide
  in chemical warfare, 4:877
1-Bromobicyclo(2,2,2)octane, S:7
2-Bromobiphenyl. See Bromodiphenyl.
4-Bromobiphenyl. See 4-Bromodiphenyl.
1-Bromobutane, 3:777
2-Bromobutanoic acid, 3:881,882. See α-Bromobutyric acid.
2-Bromo-4-tert-butylphenol, 3:776
5-Bromo-3-sec-butyl-6-(hydrocymethyl)uracil, 22:213
5-Bromo-3-sec-butyl-6-methyluracil, 22:213
α-Bromobutyraldehyde
  prepn, 3:875
α-Bromobutyric acid, 3:881, 882
  phys prop, 1:235
β-Bromobutyric acid, 6:466
4-Bromobutyric acid
  prepn, 3:879
α-Bromobutyrolactone, 10:674
3-Bromocamphor, 3:776
p-Bromochlorobenzene, 3:776
  from chlorobenzene, 5:255
  manuf by ammonolysis, 2:356
1-Bromo-3-chloro-2-butene
  from chloroprene, 5:217
Bromochlorodifluoromethane, 5:126; 9:749
  fire extinguishing agent, S:375
Bromochloroethane
  vinylidene chloride from, 21:276
1-Bromo-2-chloroethane, 3:777
Bromochloromethane, 3:774
  fire extinguishing agent, S:375
  from methylene chloride, 5:114
1-Bromo-3-chloropropane, 2:146; 3:774
2-Bromo-2-chloro-1,1,1-trifluoroethane, 2:401; 3:780; 9:750
7-Bromocholesterol
  synthesis of Vitamin D_3, 21:564
Bromocinnamic acids, 5:518
3-Bromocoumarin, 6:427

Bromocresol green, 11:552
Bromocresol purple, 11:552
4-Bromocumene, 3:776
2-Bromo-p-cymene, 3:776
Bromodichloromethane, 3:776
2-Bromo-2,2-diethylacetamide
  as hypnotic, 11:516
Bromodiethylacetylurea. See Carbromal.
Bromodiethylbenzene, 3:776
Bromodifluoroacetic acid, 9:770
4-Bromo-1,2-difluorobenzene, 9:783
4-Bromo-1,3-difluorobenzene, 9:784
1-Bromo-2,5-difluorobenzene, 9:784
Bromo-3,4-difluorobenzene
  in prepn of 1,2,3-trifluorobenzene, 9:784
1-Bromo-3,5-dimethyladamantane, S:9
5-Bromo-2,2-dimethyl-1,3-dioxa-4,6-cyclohexanedione, 19:327
Bromodimethylstibine, 2:581
6-Bromo-2,4-dinitroaniline, 3:775
Bromodiphenhydramine HCl
  antihistamine, 11:36
2-Bromodiphenyl, 3:776
4-Bromodiphenyl, 3:776
1-Bromododecane, 3:778
Bromoethane, 3:772. See Ethyl bromide.
Bromoethanoic acid. See Bromoacetic acid.
2-Bromoethyl acrylate
  phys prop, 1:289
2-Bromoethylamine, 9:760
Bromoethylbenzene, 3:776
Bromoethylene. See Vinyl bromide.
o-Bromofluorobenzene, 9:781
Bromofluorohydrocarbons
  prop, 9:748
  manuf, 9:749
Bromoform, 3:776, 780; 8:420
  from chloroform, 5:122
  in prepn of flame-retardant finish, 9:308
n-Bromoformamide, 10:107
Bromohydrins, 5:305
1-Bromoisopropyl acrylate
  phys prop, 1:289
p-Bromoisopropylbenzene, 3:776
5-Bromo-3-isopropyl-6-methyluracil, 22:213
Bromoisovalerylurea
  as hypnotic, 11:517
Bromoisovalum, 3:780
Bromomethane. See Methyl bromide.
1-Bromo-2-methylanthraquinone, 2:486
2-Bromo-3-methylanthraquinone, 2:486
3-Bromo-2-methyl-1-butene, 12:68

Bromomethyldimethylchlorosilane, *18*:262
4-Bromo-3-methylisothiazole-5-carboxaldehyde
  thiosemicarbazone,
  antiviral agent, *21*:455
8-Bromomethylquinoline, *2*:153
α-Bromonaphthalene, *3*:776
1-Bromonorbornane, *S*:7
2-Bromooctane
  reaction with allyl ether, *8*:494
Bromopentafluorobenzene
  prepn of pentafluorobenzonitrile, *9*:787, 791,
  795
4-Bromoperchlorylbenzene, *9*:604
p-Bromophenacyl bromide, *3*:775
p-Bromophenol, *3*:777
2-Bromophenol
  hydrolytic desulfonation, *19*:315
Bromophenol blue, *11*:552
Bromophenols, *2*:625
α-Bromophenylacetic acid, *15*:214
2-Bromo-4-phenylphenol, *3*:777
p-Bromophenylurea
  in soil, *18*:534
Bromopicrin, *13*:864
Bromoplatinic acid, *15*:865
Bromoplatinous acid, *15*:864
Bromoprene, *5*:229
1-Bromopropane, *3*:778
2-Bromopropane, *3*:778
1-Bromo-2-propanone, *12*:106
3-Bromopropene. See *Allyl bromide.*
α-Bromopropionic acid, *3*:777
3-Bromopropyl acrylate
  phys prop, *1*:289
3-Bromopropyne, *3*:778; *15*:913
  nematocide, *15*:913
2-Bromopyridine, *16*:803
3-Bromopyridine, *3*:777; *16*:803
4-Bromopyrogallol, *16*:192
5-Bromosalicylic acid, *3*:777
Bromosilane, *18*:186
β-Bromostyrene, *5*:518
N-Bromosuccinimide, *3*:756, 777; *19*:140, 144
  7-dehydrocholesterol synthesis, *21*:561
  in prepn of allyl compds, *1*:923
  thiophene reaction, *20*:222
7-Bromotetracycline, *20*:20
Bromotetrafluoroaniline, *19*:791
2-Bromo-1,1,2-tetrafluoroethane, *3*:780
Bromotetrafluroanisole, *9*:791
Bromotetrafluorophenol, *9*:791
4-Bromotetrahydro-3-furanone, *1*:605
Bromotheophyllinate

  as diuretic, *7*:253
2-Bromothiophene, *3*:777
3-Bromothiophene, *20*:222
Bromothymol blue, *9*:897; *11*:552
Bromotoluenes, *3*:777
Bromotrichloromethane, *12*:68
2-Bromo-1,1,3-triethoxybutane, *6*:447
2-Bromo-1,1,2-trifluoroethyl methyl ether,
  3:780
Bromotrifluoromethane, *9*:296, 749
  fire extinguishing agent, *S*:375
Bromoxynil
  herbicide, *22*:203
4-Bromo-2,3,5,6-tetrafluorophenyl, *9*:791
1-Bromo-4-chlorobenzene
Brompheniramine maleate
  antihistamine, *11*:36
Bromyrite, *3*:758
Bronchography, *17*:140
Brönsted acids, *10*:161
Brønsted system of acids and bases, *1*:215
Bronze, *20*:289. (See also *Copper alloys.*)
  corrosion resistance to water, *6*:302
  electroplating, *8*:67
  phosphorus alloys, *15*:294
  pigments, *15*:548
444Bronze, *6*:194
Bronze blue, *15*:539
Bronze, silnic
  magnetic prop, *6*:220
Bronzes, olympic, *6*:193
Bronzes, tungsten. See *Wolfram compounds.*
Brookite, *20*:390, *20*:392
  in ceramics, *4*:773
Broomroot, *9*:180
Brown coal, *12*:381
Brownprinting, *17*:358
Broxiquinolina
  amebicide, *20*:83
Brucella bacteria
  chemotherapy against, *3*:27
Brucine, *1*:795; *2*:166
Brucine alkaloid
  as denaturant, *8*:450
Brucite, *5*:551; *12*:733
Brunswick greens, *15*:536
Brushes, electrical carbon, *4*:193
  materials for, *4*:222, 332
"BTX raffinate"
Bubble caps
  for distillation columns, *7*:207, 208
Bubble plates, *9*:471

Bucherer reaction, 2:165
    in prepn of amines, 2:114
Buchwheat middlings
    as animal feed, 8:867
Bucket elevators, 6:115
Buckingham Pi Theorem, 7:179
Buckthorn
    cascara sagrada from, 4:590
Buclizine.HCl
Budding
    in yeast, 22:513
Buddle, 10:715
Bufalin, 18:845
Buffering, 1:40; 11:381, 383
Bufodienolides, 18:845
Bufotenine, 16:641
Bufotoxin
    as cardiovascular agent, 4:513
Buhrstone
    as abrasive, 1:24
Building materials
    surface-burning characteristics, S:471
Built soaps, 12:203. See *Soaps*
Bulbocapnine, 1:769
Bulk factor, 6:757
Bulk modulus, 17:425
Bulk producers
    as cathartics, 4:591
Bullvalene, 14:73
Bulsulfan
    alkylating agent, S:83
Bumping concentrator tables, 10:716
Buna N. See *Acrylonitrile-butadiene copolymers.*
Buna S, 7:677
Bunamidine hydrochloride, 14:544
Bunte salts
    thiosulfates, 20:243
Buoyancy
    definition, 9:448
Buoyancy principle, 6:770
Buquinolate
    anticoccidial, 21:248
Bureau International pour la Standardisation des Fibres Artficielles
    determination of commercial weight of acetate, 1:125
Burger's vector, 13:319
Burgundy mixture
    as fungicide, 6:266
Burnt sienna, 15:518
Burnt umber, 15:525
Burrow's solution, 2:12

Burton Union fermentation system, 3:315
Butabarbital, 3:61, 70
Butacaine sulfate, 2:405, 407
Butadiene, 3:784; 7:73, 414; 9:797
    from acetaldehyde, 1:83
    acrylonitnle-styrene terpolymer, 19:73, 95
    as by-product in ethylene prod, 8:510
    chlorine reaction with, 5:222
    chloroprene from, 5:222, 224
    copolymer with 2-vinylpyridine, 8:711
    elastomer, 19:145
    manuf from ethyl alcohol, 8:454
    in manuf of epoxy resins, 8:299
    photoisomerization, 15:332
    polymerization to *cis*-1,4-polybutadiene, 7:71
    polymers of, 7:73
    prepn, 6:453
    in prepn of anthraquinone, 2:434
    in prepn erythritol, 1:571
    prepn from glycos, 10:640
    styrene copolymer, 19:73, 91
    usc of furfural in manuf, 10:244
1,3-Butadiene
1,3-Butadiene
    synthesis of crotonaldehyde, 6:446
    toxicity of, 11:296
Butadiene-acrylonitrile
    latex, 17:598
Butadiene dioxide, 8:249
Butadiene monoxide, 3:789
Butadiene-styrene
1,3-Butadiene-1-sulfonic acid, 3:793
    drying oil, 7:414, 421
Butadiene-vinylpyridine, 7:683
Butadienylacetone
    from isochloroprene, 5:228
Butadienyl mercaptan
    from isochloroprene, 5:228
Butadienyl monoacetate, 3:792
Butadiyne, 1:605
Butaldol. See *Butyraldol.*
Butanal. See *Butyraldehyde.*
Butanamide. See *Butyramide; n-Butyramide.*
Butane. (See also *n*-Butane; *i*-Butane; Butanes.)
    2-butene azeotrope, 19:252
    characteristics as fuel, 10:280
    in manuf of formaldehyde, 10:89
    prop for gas manuf, 10:389
    refrigerant, 17:309
i-Butane
    as aerosol propellent, 1:475

butadiene from, *3*:799, 802
butenes from, *3*:799
as component of natural gas, *10*:448
in hydroisomerization, *11*:446
as motor fuel, *10*:492
oxidation of, *11*:235
prepn, *3*:817
as a refrigerant, *22*:51
1,4-Butanediamine. See *Tetramethylenediamine.*
1,4-Butanedicarboxylic acid. See *Adipic acid.*
Butanediol. (See also *Butylene glycol.*)
1,3-Butanediol, *6*:450
polyester from, *16*:180
1,4-Butanediol, *12*:62
epoxy resins, *19*:147
2,3-Butanediol
acetylation, *8*:331
butadiene from, *3*:803
prepn from diacetyl, *12*:145
1,4-Butanediol acrylate
phys prop, *1*:289
Butanediol bischloroformate, *10*:669
2,3-Butanedione, *12*:140, 144
Butanephosphonic acids, *3*:819
Butanes. (See also *Butane.*)
from catalytic reforming, *15*:33
isomerization with aluminum chloride, *2*:21
toxicity of, *11*:294
Butane-1,2,3,4-tetracarboxylic acid, *5*:217
Butanethiol
from butyl alcohol, *3*:825
occurrence, *20*:206
1-Butanethiol. (See also *Thiols.*)
1-Butanethiol
uses, *20*:214
1,2,4-Butanetricarboxaldehyde, *1*:640
1,2,4-Butanetriol, *1*:595, 603
Butanoic acid. See *Butyric acid.*
n-Butanoic acid. See *Butyric acid.*
Butanoic anhydride. See *Butyric anhydride.*
Butanol. (See also *Butyl alcohol.*)
manuf from ethyl alcohol, *8*:455
from milk, *13*:572
n-propanol azeotrope, *19*:252
1-Butanol, *7*:28. (See also *n-Butyl alcohol.*)
in esterification of fatty acids, *8*:818
from oxo process, *14*:373
reaction with acetaldehyde, *1*:83
stain solvent, *18*:659
1-Butanol. See *n-Butyl alcohol.*
2-Butanol. See *sec-Butyl alcohol.*
2-Butanone. See *Methyl ethyl ketone.*

2-Butanone, *O*-(2-cyanoethyl)oxime, *6*:645
Butanoyl chloride, *3*:882. See *Butyrl chloride.*
Butarsen, *2*:730
2-Butenal. See *Crotonaldehyde.*
Butene, *3*:830. (See also *Butylene.*)
in manuf of maleic anhydride, *12*:828
1-Butene, *2*:413 (See also *Butylenes.*)
polymerization of, *7*:66
polymers, *14*:309
2-Butene. (See also *Butylenes.*)
alkylation, *15*:42
2-Butenealdehyde, *2*:415
trans-Butenedioic acid. See *Fumaric acid.*
from butadiene, *3*:792
Butenediol, *1*:611
Butenediol dinitrate, *1*:613
2,3-Butene oxide, *19*:489
2-Butene-1,4-diol, *1*:611
hexachlorocyclopentadiene reaction with, *5*:241
manuf, *1*:613
phys prop, *1*:612
in prepn of erythritol, *1*:571
reactions, *1*:612
uses, *1*:614
3-Butene-2-one, *12*:105
1-Butene-3-yne. See *Monovinylacetylene.*
3-Butenoic acid
prepn, *8*:494
2-Butenyl acrylate
phys prop, *1*:288
1-Buten-2-ol acetate, *12*:111
1-Buten-3-ol
phys prop, *1*:629
Butenes. (See also *Butylenes.*)
butadiene from, *3*:794, 795
3-Butene-1-ynylmagnesium bromide, *9*:761
Butethal, *3*:61
Butolic acid, *18*:26
2-Butoxy-*N*-(2-diethylaminoethyl)cinchoninamide
- hydrochloride. See *Dibucaine hydrochloride.*
2:405, 408
2-Butoxy-*N*-(2-diethylaminoethyl)cinchoninamide
- hydrochloride, *2*:405, 408
2-Butoxyethyl acrylate
phys prop, *1*:288
5-Butoxy-2-methyl-3-hydroxy-3-pentene, *1*:623
Butryaldehyde
in prepn of 2-ethylhexanoic acid, *8*:849
reaction with allyl ether, *8*:494
Butter, *13*:548-551; *18*:33
Butterfat, *8*:778, 783
Buttermilk, *13*:528

as animal feeds, 8:865
in packing, 14:451
Butyl(IIR), 17:545
Butyl acetate
in lacquer industry, 3:828, 829
in nail enamels, 6:368, 369
prod, 8:376
prop, 8:379
prepn, 1:82
specifications, 8:373
n-Butyl acetate, 18:660
tert-Butyl acetate, 12:114
batch esterification, 8:329
use in dehydration of acetic acid, 2:851
prepn, 8:342
prop, 8:380
specifications, 8:373
Butyl acetyl ricinoleate
plasticizer, 20:484
N-tert-Butylacrylamide
prepn from acrylonitrile, 1:340
Butyl acrylate
alcoholysis, 20:488
phys prop, 1:301
polymerization, 1:303
toxicity data, 1:302
vinyl acetate copolymer, 21:331
sec-Butyl acrylate
phys prop, 1:287
Butyl alcohol. (See also Butanol.)
n-Butyl alcohol, 3:**822**
n-butyraldehyde from, 3:871
butyric acid from, 3:880
latent solvent, 18:577
manuf from crotonaldehyde, 6:456
in prepn of methyl ethyl ketone, 12:125, 133
prepn of n-butyl ether, 8:492
solvent recovery, 18:562
sec-Butyl alcohol, 3:822
hydrolysis of sulfuric esters, 19:486
n-propanol azeotrope, 19:252
tert-Butyl alcohol, 3:822
in prepn of tert-butyl hypochlorites, 5:24
as solvent in cyanoethylation, 6:635
t-Butyl allyl ether, 8:494
Butylals. See Acetals.
Butylamine
in synthesis of 1-butylaziridine, 11:536
n-Butylamine, 2:348
prop and uses, 2:124
tert-Butylamine, 6:648
t-Butylamine, 9:591
t-Butylamine-borane, 3:729

Butylamines
from butyraldehydes, 3:868
n-Butyl p-aminobenzoate, 2:405, 407
2-Butylanthraquinone, 2:438
Butylated hydroxyanisole, 2:593, 597 15:171
as food preservative, 10:14
use in fats, 8:781
Butylated hydroxytoluene, 2:593, 597; 8:781
as food preservative, 10:14
1-Butylaziridine
synthesis of, 11:536
sec-Butylbenzene, 12:126
in prepn of methyl ethyl ketone, 12:133
Butyl benzoate, 3:432
use in commercial dyeing of triacetate, 1:135
p-tert-Butylbenzoic acid, 3:436
as modifier in alkyd resins, 1:866
Butyl benzyl phthalate
plasticizer, 15:726, 741, 751, 753, 757, 764
from toluene, 20:562
n-Butyl benzyl phthalate, 3:442
Butylbiphenyl, 19:318
n-Butyl bromide, 3:777
Butyl butyrate
prepn, 1:82
tert-Butylcatechol
as inhibitor, 16:518
polyester inhibitor, 20:823
p-tert-Butylcatechol
styrene finishing, 19:68
Butyl chloride, 2:413
prepn from butyl ether, 8:492
t-Butyl chloride
prepn of t-butyl ether, 8:492
prepn of 3,3-dimethyl-1-butyl chloride,
8:500
n-Butyl chlorofomate, 4:386
Butylchoral hydrate
n-Butyl crotonate, 6:468, 469
tert-Butyl 2-cyanoethyl peroxide, 6:645
Butyl (2-cyanoethyl)phenylphosphinate, 6:651
Butyl decyl phthalate
prod and sales of, 8:376
N-Butyl-2,3-dehydroemetine, 20:77
Butyl 3,4-dihydro-2,2-dimethyl-4-oxo-1,2H-
pyran-6-carboxylate, 10:924
Butylene
sulfation, 19:490
Butylene. Butene.
1,3-Butylene glycol, 10:660. (See also Butanediol.)
polyesters, 21:62
1,4-Butylene glycol, 10:667

diisocyanate polymers, *21*:57
polyesters, *21*:62
2,3-Butylene glycol
  from alcoholic fermentation, *21*:256
  from wood hydrolysis, *22*:383
1,3-Butylene glycol arsenite, *10*:662
1,3-Butylene glycol borate, *10*:662
1,3-Butylene glycol carbonate, *10*:662
1,3-Butylene glycol cyclic carbamate, *10*:662
1,3-Butylene glycol cyclic sulfate, *10*:662
1,3-Butylene glycol cyclic sulfite, *10*:662
1,3-Butylene glycol diacetate, *10*:662
1,3-Butylene glycol dicarbamate, *10*:662
1,3-Butylene glycol dinitrite, *10*:662
1,3-Butylene glycol esters, *10*:662
1,3-Butylene glycol monoacetate, *10*:662
1,3-Butylene glycol monocarbamate, *10*:662
Butylene oxide, *8*:288
  block coplymers, *19*:553
  in paint removers, *14*:488
Butylenes, *3*:**830**; *8*:504. (See also *Butenes.*)
  sulfation, *19*:302
  toxicity of, *11*:296
2,3-Butylene sulfite, *19*:489
Butylethanolamine
  phys prop, *1*:811
n-Butyl ether, *8*:491
  phys prop, *8*:471
sec-Butyl ether, *8*:492
  phys prop, *8*:471
t-Butyl ether, *8*:492
  phys prop, *8*:471
tert-Butyl ethinyl sulfide, *12*:97
n-Butylethylbarbituric acid, *18*:451
2-Butyl-2-ethylglutarimide
  as hypnotic, *11*:515
Butyl ethyl ketene, *12*:88
tert-Butyl ethyl malonate, *12*:850
n-Butylethylmalonic ethyl ester, *18*:451
n-tert-Butylformamide, *6*:575
n-Butyl glyoxylate
  tetracycline synthesis, *20*:23
Butyl heptadecyl ketone, *3*:879
n-Butyl heptafluorobutyrate, *9*:773
t-Butyl hydroperoxide, *14*:774; *8*:285
tert-Butyl hydroperoxide, *6*:645
Butyl *p*-hydroxybenzoate
  as food preservative, *10*:12
  from butyl alcohol, *3*:824
  polyester initiator, *20*:826
  styrene-butadiene copolymer degradation, *S*:914
tert-Butyl hypochlorite, *5*:24

Butylideneacetone, *12*:105
N-Butylideneaniline
  from butyraldehyde, *3*:868
4,4'-Butylidenebis(6-*tert*-butyl-*m*-cresol)-, *2*:600
N-Butylidene-*n*-butylamine, *6*:640
4-Butyliminomethylhexanenitrile, *6*:640
Butyl iodide
  prepn from butyl ether, *8*:492
tert-Butyl iodide
  from butyl alcohol, *3*:825
N-Butyl isocyanate, *12*:46, 47
n-Butyllactamide, *12*:186
Butyl lactate
  in artists' oils, *14*:480
  prepn, *8*:353
n-Butyl lactate, *12*:184, 185
Butyllithium, *7*:685; *12*:156
  in polymerization of acetaldehyde, *1*:99
n-Butyllithium, *9*:785, 787, 788, 792; *12*:548
  thiophene reaction, *20*:222
sec-Butyllithium, *12*:550
tert-Butyllithium, *12*:551
Butylmagnesium bromide, *8*:494, 495
Butylmagnesium chloride
  in prepn of 2-hexanone, *1*:139
tert-Butylmagnesium chloride
  vinyl chloride polymerization, *21*:377
tert-Butyl mercaptan, *20*:205
  uses, *20*:215
Butylmercaptomethylpenicillin, *14*:653
Butyl methacrylate
  alcoholysis, *20*:488
  in dental material, *6*:830
n-Butyl methacrylate
  as fluorescent coating system, *9*:499
tert-Butylnaphthalene, *9*:714
Butylnaphthalenesulfonates, *13*:679
N-sec-Butyl-*p*-nitroaniline, *15*:218
Butyl octyl phthalate
  prod and sales of, *8*:376
Butyl oleate
  as cellulosic plasticizer, *4*:657
  prod and sales of, *8*:377
  specifications, *8*:373
n-Butyl oleate, *8*:256
  prepn, *8*:359
tert-Butyl orthovanadate
  in aquametry, *2*:679
t-Butyl peroxybenzoate, *18*:229
  polyester initiator, *20*:826
t-Butyl peroxyisobutyrate
  initiator, *21*:423

t-Butyl peroxyoctoate
  polyester initiator, *20*:826
tert-Butyl peroxypivalate, *8*:854
m-tert-Butylphenol, *1*:906
  phenolic resins, *15*:185
n-Butylphenol, *2*:626
o-tert-Butylphenol, *1*:906
p-t-Butylphenol, *S*:30
p-tert-Butylphenol, *1*:906; *19*:672
  phenolic resins, *15*:184, 206
  resins, *15*:176
  thermoplastics, *15*:792
Butylphenols
N-Butyl-2-phenylacetamide, *2*:368
  from isobutylene, *3*:844
Butylphenylphenol, *19*:318
Butylphenylphosphinate, *6*:651
n-Butyl phthalate
  kinetics of formation, *8*:319
"Butyl phthalyl butyl glycolate," *15*:751
tert-Butyl propionate, *16*:556
Butyl propyl ketone, *3*:879
Butyl rubber, *15*.797, *17*.572
  as calking compound, *4*:30
  latex, *17*:592
  phenolic resins, *15*:186
  prepn, *2*:23; *7*:688
  prepn of flexible polymers, *9*:863
  roofing, *17*:461
  use of isoprene in manuf, *12*:64, 73, 80
Butyl stearate
  as cellulosic plasticizer, *4*:657
  plasticizer, *20*:484
  prod and sales of, *8*:377
  prop, *8*:382
t-Butylstyrene
  polyester reactive diluent, *20*:822
n-Butyl tallate, *8*:256
N-t-Butyltetracyclines, *20*:10
Butyl titanate
  titanium chelates, *20*:464
  titanium sandwich structures, *20*:444
N-Butyl vinyl ether, *21*:416
  copolymers of, *21*:421
tert-Butyl xanthate
  decomposition of, *22*:421
Butyl zirconate, *1*:849
Butynediol, *1*:602; *10*:667
  in acid inhibition, *6*:329
  manuf of 1,4-butanediol, *10*:670
Butyne-1,4-diol
  synthesis of barban, *22*:193
2-Butyne-1,4-diol

prepn from formaldehyde, *10*:86, 667
2-Butyne-1,4-disulfonic acid, *1*:607
2-Butynoic acid, *8*:816
3-Butyn-1-ol, *9*:758
3-Butyn-2-one, *12*:105
Butyraldehyde, *3*:**865**; *10*:663
  from oxo process, *14*:373
  in prepn of 2-ethyl-1-hexanol, *1*:99
  in prepn of trimethylolpropane, *1*:595
  reaction with acetaldehyde, *1*:82
n-Butyraldehyde, *2*:348
Butyraldehyde-ammonia, *3*:868, 874
Butyraldol
  prepn and prop, *3*:874
Butyraldoxime
  prop, *3*:875
Butyrals *3*:874 See *Acetals.*
Butyramide, *3*:881
  prop, *3*:881
n-Butyramide
  form, *2*:68
  phys prop, *2*:68
Butyranilide
  prop, *3*:882
Butyric acid, *3*:**878**; *6*:465, 466; *12*:95
  oxidation, *19*:143
  phys prop, *1*:229
  as synthetic flavor, *9*:354
  from wood hydrolysis, *22*:383
Butyric acid esters, *8*:381
Butyric amide. See *Butyramide.*
Butyric anhydride, *3*:**878**
  phys prop, *1*:212
Butyrin, *3*:882; *10*:619, See *Glycerol tributyrate.*
Butyroin, *12*:106
Butyrolactone, *15*:800
  prepn from tetrahydrofuran, *10*:249
β-Butyrolactone, *12*:95
γ-Butyrolactone, *9*:844; *10*:667, 669, 673
  chlorinated aryloxybutyric acids syntheses, *22*:180
  copolymerization, *16*:174
  oxidation, *19*:143
Butyrone, *12*:104
Butyronitrile, *9*:591
  from butyl alchol, *3*:825
  butyric acid from, *3*:880
1-Butyro-3-palmeto-2-stearin, *10*:619
Butyrophenone, *6*:670
Butyrophenones, *16*:657
3-Butyro-2-stearo-1-palmitin, *10*:619
Butyrylacetone, *12*:148, 149

C

CASING technique, *S*:20

CB-1A, *4*:528

CCDT. See *California Chassis Dynamometer Test.*

CGS system, *S*:1005
 information retrieval, *S*:513

CI Direct Black 38, *7*:536
 flame retardant for cellulose derivatives, *S*:950

CIE Trichromatic colorimetric coordinate system, *5*:802

CI 10020, *5*:872

CI 10316, *5*:867, 872, 874

CI 11855, *7*:497

CI 12075, *5*:869

CI 12085, *5*:871

*CI*12156, *5*:867

CI 13058, *5*:871

CI 13065, *5*:872

CI 14030, *7*:491

CI 14155, *7*:491

CI 14600, *5*:872

CI 14700, *5*:867, 869

CI 15050, *2*:887

CI 15510, *5*:869; *7*:496

CI 15585, *5*:869

CI 15620, *5*:872

CI 15630, *5*:869, 870

CI 15800, *5*:870

CI 15850, *5*:869

CI 15880, *5*:871

CI 15985, *5*:867, 871

CI 16150, *5*:869

CI 16155, *5*:872

CI 16185, *5*:866, 869, 873

CI 17200, *5*:871

CI 18820, *5*:872

CI 19140, *5*:867, 871, 873

CI 20170, *5*:868

CI 20470, *5*:867

CI 22590, *7*:497, 539

CI 26100, *5*:870, 873

CI 26360, *2*:896

CI 29000, *7*:491

CI 30235, *7*:497

CI 31600, *7*:470

CI 34010, *7*:492

CI 40001, *7*:497

CI 41000, *7*:497

CI 42052, *5*:868

CI 42053, *5*:866, See *Fast Green FCF.*

CI 42085, *5*:866, 868

CI 42090, *5*:867; *5*:866, 868, 873. See *Brilliant Blue FCF.*

CI 42095, *5*:866

CI 42535, *7*:496

CI 43350, *5*:871

CI 45170, *5*:870, 873. See *Rhodamine B.*

CI 45179B, *5*:871

CI 45380, *5*:870, 873

CI 45380A, *5*:870, 873

CI 45410, *5*:870

CI 45410A, *5*:870

CI 45430, *5*:866, 869

CI 47000, *5*:871, 874

CI 47005, *5*:871

CI 50315, *2*:863

CI 50320, *2*:863

CI 50335, *2*:863

CI 50420, *2*:862

CI 59040, *5*:868, 873

CI 60725, *5*:871. See *Solvent Violet 13.*

CI 60730, *5*:872

CI 61505, *7*:470, 497

CI 61565, *5*:868

CI 61570, *5*:868, 873

CI 62055, *2*:513

CI 62105, *2*:513

CI 62125, *7*:491

CI 63000, *2*:514

CI 63010, *2*:514

CI 65005, *2*:515

CI 69525, *7*:470

CI 69825, *5*:868, 873

CI 73000, *5*:868

CI 73015, *5*:866, 868, 874

CI 73360, *5*:870, 874

CI 772, *5*:869

CMHEC. See *Carbonymethylhydronyethylcellulose.*

C-Nor-D-homoestrone, *18*:885

C-Nor-D-homoprogesterone, *18*:885

COED process, *S*:180, 186
CS, *4*:877
CS acid, *13*:722
Caapi, *1*:772
Cable insulation, *11*:802-814
Cabreuva, *14*:196
C acid, *13*:715, 723
Cacodyl, *2*:726
Cacodylic acid
  soil chemistry, *18*:537
Cacodyl oxide, *2*:726
Cadaverine, *2*:179
Cade, *14*:196
Cadmium, *3*:**884**
  chromate treatment, *13*:298
  in copper alloys, *6*:207, 210
  from Daniell cell, *22*:559
  in electric cells, *3*:172, 240
  electrodeposition of zinc, *22*:587
  in electroplating, *8*:55, 70
  electromigration of, *S*:287
  organometallics in phonograph records,
    15:229
  phosphatized, *13*:296
  in photovoltaic cells, *15*:420
  prod of zinc metal, *22*:575
  red pigments, *15*:522
  separation from zinc, *22*:562
  tin coatings, *20*:289
  in type metal, *20*:758
  in vacuum devices, *8*:2
  yellow pigments, *15*:534
  in zinc alloys, *22*:600
Cadmium acetates
  prepn and prop, *3*:903
Cadmium alloys, *3*:281, **884, 896**
Cadmium bromides, *3*:904
Cadmium and cadmium alloys, *3*:**884**
Cadmium carbonate, *3*:903
Cadmium chloride, *12*:588
  prepn and prop, *3*:904
Cadmium chromate, *5*:498
Cadmium compounds, *3*:**899**
Cadmium fluoride, *9*:572
Cadmium-gold alloy, *10*:690
Cadmium hexafluorostannate, *9*:682
Cadmium hydroxide
  prepn and prop, *3*:905
Cadmium iodide, *12*:588
  prepn and prop, *3*:905
Cadmium lithopones, *3*:909; *15*:535

Cadmium-mercury sulfides
  pigments, *15*:524
Cadmium monofluorophosphate, *9*:642
Cadmium naphthalene-1-sulfonate, *3*:907
Cadmium niobate, *9*:6
  in glass-ceramics, *10*:548
Cadmium nitrate
  properties, *3*:906
Cadmium oxide
  in photovoltaic cell, *15*:420
  prepn, prop and uses, *3*:907
Cadmium pentachlorophenate, *3*:902
Cadmium peroxide, *14*:751
Cadmium phosphors, *3*:902
Cadmium plating, *3*:896
Cadmium reds, *3*:910; *15*:522
Cadmium ricinoleate, *3*:907
Cadmium salicylate, *3*:907
Cadmium selenide, *12*:626
  nonlinear optical materials, *S*:630
  photoconductor, *15*:418
Cadmiun stearate, *7*:280
Cadmium sulfamate, *19*:244
Cadmium sulfate
  pigments, *15*:523
  prepn, prop and uses, *3*:907
Cadmium sulfide
  cadmium analysis, *22*:562
  as glass colorant, *5*:850; *9*:252
  pigments, *15*:523, 534
  photoconductor, *15*:416
  prepn and prop, *3*:908
Cadmium sulfide-cadmium selenide
  photoconductive cell, *15*:410
Cadmium sulfide, selenium
  colorant for glass, *9*:252
Cadmium sulfoselenide, *17*:830
  pigments, *15*:522
Cadmiun sulfotelluride
  pigments, *19*:771
Cadmium telluride, *12*:626
  photoconductor, *15*:418
Cadmium titanate, *20*:413
Cadmium yellows, *15*:527, 534
Cadmium zirconate
  in glass-ceramics, *10*:548
Cadmous compounds, *3*:900
Cadoxene
  as cellulose solvent, *4*:601
Caffeine, *1*:764; *3*:**911**

Cafferine, *21*:535
   in carbonated beverages, *4*:336
   as cardiovascular agent, *4*:521
   as diuretic, *7*:252
   in tea, *19*:750, 752
Cajeput *14*:196
Calabar bean, *1*:772, 782
Calabash curare, *1*:773
Calamine, *22*:563
   as insecticide, *6*:383; *11*:681
Calamus, *14*:196
   essential oils as flavor, *9*:352
Calaverite, *19*:759
Calcia, *12*:418. See *Calcium oxide; Lime.*
Calciferol, *21*:549
Calcination
   of petroleum coke, *4*:169
Calcite, *12*:415, 416, 616; *13*:31; *19*:51; *22*:563
Calcite, *4*:2, 8; *7*:292; *9*:384
   sulfur associated with, *19*:343
   in talc, *19*:608
   zero point of charge, *3*:393
Calcitonin, *20*:266
Calcium, *3*:**917**
   cesium alloys with, *4*:862
   cesium chloride reaction with, *4*:860
   content in soils, *9*:35
   ferroalloy, *18*:751
   as plant nutrient, *9*:28, 34
   reaction with zirconium dioxide, *22*:616
   removal from water, *22*:98
   use of sodium in prod, *18*:450
   in water, *21*:694
   and water hardness, *22*:73
   in wood, *22*:366
Calcium alloys, *3*:923-925
Calcium aluminate
   cement, *17*:233
Calcium aluminate glass, *10*:545
Calcium aluminum silicate, *10*:4
Calcium and calcium alloys, *3*:**917**
Calcium arsenate, *12*:229, 457
   as insecticide, *2*:715
   molluscacide, *15*:910
   soil chemistry, *18*:537
Calcium arsenide
   prepn of arsine, *2*:719
Calcium benzenesulfonate, *3*:404
Calcium bisulfite
   in pulp making, *12*:369
Calcium borates, *3*:649
   fire-retardant paint, *22*:376

Calcium boride, *3*:605
Calcium carbide, *4*:100-114; *9*:400, 602; *22*:616
   in aquametry, *2*:673, 690
   manuf of acetylene from, *1*:174
   in manuf of calcium cyanamide, *6*:554, 555, 556
   smoke from furnaces, *18*:402
Calcium carbonate, *4*:2,7; *6*:802, 804; *7*:425. (See also *Lime.*)
   as abrasive, *1*:28
   as antacid, *2*:429
   in cathodes, *8*:2
   as dental abrasive, *6*:849
   in dental materials, *6*:836
   as filter aid, *9*:268
   fire-retardant paint, *22*:376
   in limestone, *12*:414
   in matches, *13*:164
   as paint extender, *14*:466
   as paper pigment, *14*:516
   in papermaking, *14*:501
   pigments, *15*:508, 562
   prop as embedding filler, *8*:110
   removal by hydroxyacetic acid, *10*:635
   scales from seawater treatment, *22*:77
   water treatment systems, *22*:73
   in welding, *22*:242
Calcium chelate cements, *6*:825
Calcium china clay, *15*:562
Calcium chlorate
   chlorine dioxide from, *5*:40
Calcium chloride, *2*:318; *3*:11-14; *4*:9,11; *6*:512, 555, 556
   in Belknap process of gravity separation, *10*:702
   as brine in refrigerant, *10*:860
   by-product in ethylene oxide manuf, *8*:531
   by-product of soda ash manuf, *1*:733
   calcium prod from, *3*:920
   as diuretic, *7*:251
   in drilling fluids, *7*:302
   as a drying agent, *7*:381, 383, 384, 385, 387, 388, 391
   in food processing, *10*:10
   in glycerol manuf, *10*:621
   in manuf metallic soaps, *7*:274
   monobasic, *5*:16, 18
   pigments from, *15*:516
   in prepn of acetyl chloride, *1*:140
   in rock salt, *18*:472
   water additive in firefighting, *S*:379
   as water correcting agent, *10*:17

Calcium chlorite, 5:31
Calcium chlorophyll, 5:348
Calcium chromite-chromate, 5:479
Calcium citrate
  as emulsifier in cheese, 10:8
  in food processing, 10:8, 10
Calcium compounds, 4:1
  from seawater, 14:156
Calcium cyanamide, 4:100, 103, 114;
  6:602,553; 9:77. ((See also *Cyanamide.*)
  manuf, 6:555
  in manuf of cyanamide, 6:563
Calcium cyanide, 6:599, 601
  manuf, 6:602
  prepn, 6:554, 555, 556, 557
  in prepn of sodium hexacyanoferrate(II),
    12:29
  prop, 6:601
Calcium cyclamate, 19:598
Calcium cyclohexylsulfamate, 10:10
Calcium di(dichloroisocyanurate), 20:664
Calcium dihydrogen phosphate
  as water-correcting agent, 10:17
Calcium disodium EDTA
  as food additive, 10:16
  in margarine, 13:59
Calcium disulfide, 9:400
Calcium 2-ethylhexoate, 7:276, 278
Calcium ferrate(VI), 12:40
Calcium fluoride, 4:4; 9:509, 512, 544, 573-
  582, 636; 12:236, 616
  in prepn calcium cyanamide, 6:555, 556, 557
Calcium fluosilicate, 9:658, 660
Calcium formate
  in manuf of formic acid, 10:101
Calcium gallate(III), 10:325
Calcium germanate glass, 10:545
Calcium germanide, 10:523
Calcium gluconate
  in food processing, 10:8
Calcium glycerophosphate
  in food processing, 10:8
Calcium halophosphate phosphor, 12:620
Calcium hexacyanoferrate(II), 12:27
Calcium hydride
  in aquametry, 2:691
  reaction with zirconium dioxide, 22:616
  use of sodium in prod, 18:450
  uses, 3:924; 8:361
Calcium hydrogen orthophosphate, 15:239
Calcium hydroxide, 2:300; 4:4; 12:415.   (See
  also *Lime.*)

  as abrasive, 1:29
  bleaching compounds from, 5:16, 18
  in dentistry, 6:825
  in manuf of propylene oxide, 8:280
  as water-correcting agent, 10:17, 636
  water softened with, 4:3
Calcium hypochlorites, 5:16, 21
  addition to municipal water supplies, 22:85
  as bleaching agent, 3:557
  cracking of wool grease, 22:414
  prepn of, 5:22
  superchlorination of swimming pools,
    22:130
  swimming pool water disinfectant, 22:126
Calcium hypophosphite
  in food processing, 10:8
Calcium iodate
  as bleaching agent in flour, 10:6
Calcium iodide
  carbon tetrachloride reaction with, 5:131
Calcium ions
  in detergency, 6:891
Calcium lactate, 8:252; 12:183
Calcium-lead alloys, 12:259
Calcium levulinate, 19:225
Calcium lignosulfonate. See *Goulac.*
  in drilling fluids, 7:296
  in electrolytic refining of copper, 6:159
Calcium linoleate, 7:276
Calcium-magnesium-silicon alloy, 18:129
Calcium malate, 12:847
  fermentation, 19:134
Calcium manganate(IV), 13:34
Calcium-manganese-silicon alloy, 12:901
Calcium metaphosphate, 9:110
Calcium metaphosphates, 15:249
Calcium metasilicate
  pigments, 15:513
Calcium molybdate, 13:650
Calcium monofluorophosphate, 9:642, 643
Calcium monosilicide, 18:176
Calcium naphthenate, 7:276, 278
Calcium neodecanoate, 7:279
Calcium nicotinate, 21:527
Calcium nitrate
  as diuretic, 7:251
  as drilling fluids, 7:302
  as fertilizer, 9:76
  in heat transfer, 10:857
  in nitric phosphate process, S:357
Calcium orthophosphates, 15:239, 263

Calcium orthoplumbate, *12*:276
  pigments, *15*:535
Calcium orthosilicate
  zirconium dioxide recovery, *22*:652
Calcium oxalate, *14*:370
Calcium oxide, *12*:132, 415. (See also *Calcia;*
    Lime.)
  accelerator, *17*:509
  in cement formation, *4*:652, 692, 695, 699,
    700, 703
  in ceramics, *4*:770
  catalyst in fatty acid prod, *8*:826
  in ceramic armor, *S*:150
  in chemical ware, *4*:835
  curing agent, *21*:262
  ethyl chloride reaction with, *5*:142
  in glass forming, *10*:538, 542
  prepn, *4*:3
  in prepn of trifluroboroxine, *9*:556
  reaction to form methylimidazoline, *7*:26
  refractory, *17*:244
  as water correcting agent, *10*:17
Calcium perchlorate, *5*:69
Calcium permanganate, *8*:687; *13*:34
Calcium peroxide, *10*:5; *14*:750
  in dentifrices, *6*:848
  prod, *18*:450
Calcium perrhenate, *17*:413
Calcium phosphate, *10*:4. (See also *Calcium*
    hydrogen phosphate; Calcium dihydrogen
    phosphate; Calcium orthophosphate.)
  as fertilizer, *9*:32
  in food processing, *10*:8
  prepn, *9*:32
  prod, *15*:263
  tablet filler, *15*:121
  yeast prod, *22*:536
Calcium phosphates, *15*:273
Calcium phosphide, *15*:293
Calcium phthalocyanine, *15*:490
Calcium pimelate, *6*:685
Calcium plumbate
  pigments, *15*:535
Calcium plumbite
  pigments, *15*:535
Calcium polyphosphates, *15*:249
Calcium potassium hexacyanoferrate(II),
  12:28
Calcium propionate
  as food preservative, *10*:13
Calcium pyrophosphate, *15*:249
  as dental abrasive, *6*:849

Calcium resinate, *7*:279; *17*:500
  pigments, *15*:562
Calcium saccharin, *19*:596
Calcium salicylate, *17*:729
Calcium silicate, *10*:4; *18*:158
  in cement formation, *4*:685, 691, 692, 695,
    700
  filler, *21*:272
  in prepn of trifluoroboroxine, *9*:556
  as thermal insulator, *11*:833
Calcium silicate insulation, *7*:62
Calcium silicide, *18*:129
  in explosive priming compositions, *8*:653
Calcium sorbate, *10*:13
Calcium stearate, *7*:276, 280; *10*:4
  use as abherent, *1*:8
  lubricant, *15*:121
  as release agent in food, *10*:16
Calcium stearyl-2-lactylate, *10*:5; *12*:182
  flour improver, *22*:294
Calcium sulfate, *4*:14-27; *6*:802, 803, 804, 826.
  (See also *Gypsum.)*
  calcium tartrate prepn, *19*:738
  as drying agent, *7*:392
  in food processing, *10*:8, 10
  as paper pigment, *14*:517
  phosphoric acid, *15*:261
  pigment, *15*:496, 509, 516, 520, 562
  for prod of phosphoric acid, *9*:88
  in prod of phosphoric acid for fertilization,
    9:87
  removal of scale by ammonium bicarbonate,
    2:315
  scales from seawater treatment, *22*:77
  in sodium chloride deposits, *18*:471
  sulfur dioxide from, *19*:449
  as filter aid, *9*:268, 899
  sulfur from, *S*:358
  tablet filler, *15*:121
  as water-correcting agent, *10*:17
  yeast prod, *22*:536
Calcium sulfate dihydrate, *6*:784, 787
  in dental stones, *6*:787, 788
β-Calcium sulfate hemihydrate. See *Plaster of*
  Paris.
Calcium sulfide
  hydrogen sulfide from, *19*:380
Calcium sulfoaluminate, *4*:690
Calcium sulfoaluminate cement
  prepn, *4*:695
Calcium superphosphate
  yeast prod, *22*:534

Calcium tallate, 7:276, 278
Calcium tartrate, 19:736, 738
Calcium tetrafluorozincate, 9:684
Calcium tetrahydrogen diorthophosphate, 15:239
Calcium thiocyanate, 6:602
Calcium thioglycolate, 20:201, 203
  in depilatories, 6:357
Calcium thiosulfate, 20:241, 242
Calcium 3-5-5-trimethylhexoate, 7:279
Calcium ultraphosphates, 15:249
Calcium urate, 21:112
Calcium wolframate, 22:337
Calcium zirconate, 22:656
Calcium zirconylsilicate, 22:652
Calcium-45, 17:121
Calf leather
  dyeing of, 7:591
Caliche, 18:487
California Chassis Dynamometer Test, 2:825
Californium-252
Californium. See *Actinides.*
  isolation of, 14:119
Calking and sealing compositions, 4:**28**
Calking compounds, 4:28
  drying oils in, 7:424
  fish oil used in, 7:402
Callier Q factor, 9:260
Calomel. (See *Mercurous chloride.*)
Calorimetry, 4:**35**
  differential scanning, 20:99
Calorizing, 13:265
Calycanthine, 1:776
2,3,-Camphanedione, 4:55
  prepn, 4:55
Camphanic acid
  prepn, 4:55
2-Camphanol. See *Isoborneol.*
  furnaces for, 10:291
2-Camphanone. See *Camphor.*
Camphene, 4:55, 57; 19:810, 812, 814, 820, 830
Camphor, 4:**54**; 10:802; 14:196; 19:819, 831; 12:106
Camphor, 12:106
  as cardiovascular agent, 4:523
  as denaturant, 8:450
  plasticizer, 15:721, 791, 801
α-Camphorene, 19:818
Camphoric acid
  phys prop, 1:236
  prepn, 4:55
Camphorquinone, 4:55
  prepn, 4:55

Camphorsulfonate, 20:75
Camphorsulfonic acid
  prepn, 4:55
Canadian International Paper process for manuf chlorine dioxide, 5:41
Canadine, 1:769
Cananga, 14:197
Cancer chemotherapy, S:**81**
Candela
  definition of, S:991
Candelilla wax, 22:158, 163
  as abherent, 1:7
Candicidin, 16:138
Candida tropicalis
  from sucrose, 19:224
Candlenut oil, 8:778, 784
Candles, 4:**58**
  use of metallic soaps in, 7:285
Candy. (See also *Confectionary.*)
Cane fibers
  in drilling fluids, 7:298
Cane sugar. See *Sugar.*
Canfieldite, 20:276, 519
Cannabinol, 10:924
Cannel coal
  characteristics, 10:280
Canning
  of food, 10:39
  of meats, 13:176
Cannizzaro reaction, crossed
  of aldehydes, 1:641
  prepn of pentaerythritol, 1:82, 590
Cantala, 9:171, 177
Capracyl, 2:889
Capraldehyde, 9:358
Capric acid, 8:812
  in dental material, 6:827
  specs, 8:836
Caprinitrile
  stain solvent, 18:661
Caproic acid, 8:812; 14:536
  in dental material, 6:827
  phys prop, 1:229
Caprolactam, 2:166
  additive, 16:92
  commercial prod, 11:501
  nylon degradation, 16:10
  polymer from, 16:29
  prod, 15:338
  prod, 16:47
ε-Caprolactam, 6:650, 685; 9:591; 13:882
  polymerization, 16:89
  as solvent, 2:238

Caprolactone
    polyesters, *21*:62
ε-Caprolactone
    polyesters from, *16*:173
    polymerization of, *16*:181
Caproleic acid, *8*:814
Caproyl-resorcinol, *10*:147
Capryl
    in papermaking, *14*:520
Capryl alcohol, *7*:304. See *2-Octanol.*
    phys prop, *1*:543
    from ricinoleic acid, *4*:531
Caprylic acid, *8*:812; *12*:646
    in dental material, *6*:827
    phys prop, *1*:229
    specs, *8*:836
Capsaicin, *9*:369; *21*:181
Capsicum, *9*:365, 369; *14*:197
    in carbonated beverages, *4*:337
    in hair prepns, *10*:802
Captan, *3*:785
    as fungicide, *10*:225, 234
Captodiamine, *16*:675
Caramel color, *4*:**63**
    in carbonated beverages, *4*:64, 342; *6*:49;
        19:224
Caranday wax, *22*:161
Carane, *19*:815
Caraway seed, *9*:370, 352, 363, 365
Carbachol. See *Carbamylcholine chloride.*
Carbamates
    as herbicides, *22*:191
    as insecticides, *11*:711
    metabolism as herbicides, *22*:193
    in soil, *18*:527
    soil chemistry, *18*:539
Carbamic acid, *4*:**69**
    amide of, *21*:37
Carbamic acid esters, *21*:39. ((See also *Ure-thans.*)
Carbamide. See *Urea.*
Carbaminoylcholine chloride. See *Carbamyl-choline chloride.*
2-Carbamoylethylamines, *19*:486
Carbamoylethylcellulose, *6*:660
2-Carbamoylethyl hydrogen sulfate, *19*:486
Carbamoylguanazole, *2*:244
Carbamyl asparate, *2*:178, 179
Carbamylcholine chloride, *5*:411
    as cardiovascular agent, *4*:520
Carbamyl glutamate, *2*:177
Carbamyl halides
    prepn from isocyanates, *12*:49

Carbamyl phosphate, *2*:177
Carbanilide, *2*:414
    form, *2*:67, 68
    halogenated, *18*:429
    phys prop, *2*:68
    prepn from isocyanates, *12*:53
Carbanions, *4*:394, 397
Carbanthrene Blue, *5*:868, 873
Carbarsone, *2*:731
    amebicide, *20*:93
Carbaryl
    corn earworm control, *11*:679
    insecticidal carbamate, *22*:194
    in soil, *18*:527
Carbazochrome salicylate, *5*:591
Carbazole, *2*:143; *6*:649; *7*:43, 44; *10*:897, 899,
        909; *11*:593
    from coal tar, *19*:669
    polymers, *15*:800
    prepn of tetranitrocarbazole, *8*:622
    sulfonation, *19*:300
Carbazole indophenol, *19*:427
9-Carbazolepropionitrile, *6*:649
Carbazole-1,3,6-trisulfonic acid, *19*:300
N,3-Carbazolylquinone imine, *10*:910
Carbenes, *2*:784
α-Carbethoxy-γ-butyrolactone, *8*:53
5-Carbethoxymethylypyrimidine, *20*:178
3-Carbethoxycoumarin, *6*:429
2-Carbethoxycyclopentanone, *12*:130
Carbides, *4*:**70-132**
    cemented, *16*:430
    and silicides, *18*:125
Carbides, metal
    as hazardous chemicals, *9*:292
Carbide tools
    furnaces for fusion, *10*:289
Carbinoxamine maleate
    antihistamine, *11*:36
Carbitol solvent
    prepn, *8*:455
Carbocyanines, *6*:607, 609, 611
Carbodiimides
    prepn from isocyanates, *12*:52
    reaction with aziridines, *11*:530
Carbohydrases, *8*:176
    classification and uses, *8*:189, 190
Carbohydrates, *4*:**132**
    added to water supplies, *22*:85
    nitric acid oxidation of, *14*:359
    in nuts, *14*:129
    sulfated, *19*:306
    in wood, *22*:365

in yeasts, *22*:516
yeast fermentation of, *22*:517
Carbohydrazide
from cyanuric acid, *20*:666
Carboids, *2*:784
Carbolan
azo dyes, *2*:882
Carbolic acid. See *Phenol.*
Carbolic oil
coal-tar distillate, *19*:664, 668
β-Carboline, *10*:910
Carbolines, *10*:910
Carbo mineralis. See *Graphite.*
Carbomycin, *311; 12*:634, 645
amebicide, *20*:80
in chemotherapy, *12*:657
Carbon, *4*:**149.** (See also *Diamond; Graphite.*)
activated as odor and taste control agent in water, *22*:102
activated and sulfur recovery, *S*:361
adsorption by activated in water pollution control, *S*:720
in Bessemer process, *18*:721
carbon dioxide reactions with, *4*:356, 440
from carbon monoxide, *4*:431
cast alloys, *20*:571
catalytic agent for automobile exhaust, *S*:66
compounds with boron, *3*:707
decolorizing in sugar refining, *19*:189
in electron tubes, *8*:2
manganese compounds, *13*:4
in manuf of aluminum chloride, *2*:18
in manuf of carbon disulfide, *4*:374, 376
organic in water, *21*:701
from petroleum, *15*:23
in pig iron, *18*:718
pigments, *15*:541, 542
porous, *4*:220
reaction with hydrogen, *11*:344
reduction of sodium carbonate, *18*:439
reduction of zinc oxide, *22*:572
refractory, *17*:232, 245
residue in diesel fuel, *10*:487
rubber pigment, *17*:553
and silicon, *18*:125
steel, *18*:751
in steel, *18*:716
in talc, *19*:608
in wood, *22*:364
zirconium metal prod, *22*:616
Carbon, activated, *4*:149-158
in sugar purification, *19*:213
in water treatment systems, *22*:121

Carbonado, *4*:285
as abrasive, *1*:25
Carbon amber glass, *5*:849
Carbon arc, *4*:206-216
Carbonate
in electroplating process, *8*:53
Carbonated beverages, *4*:**335**
Carbonatite, *13*:768
Carbon, baked, *4*:158-243
Carbonbenzoxyglycine, *2*:171
Carbon black, *2*:599, 600; *4*:167, 243-282; *S*:**91**
as antistatic agent, *2*:667
collection of particles, *7*:442
as filler in SBR, *7*:689
as hazardous chemical, *9*:291
in matches, *13*:164
in phonograph records, *15*:229
pigments, *15*:543
and plasticizers, *15*:769
plastics, *15*:817
in processing of neoprene, *7*:712
in propellants, *8*:681
rubber filler, *15*:767, 802
vinylidene fluoride copolymer filler, *21*:271
Carbon cloth, *4*:241, 242
Carbon dichlorosulfide, *19*:373
Carbon dioxide, *4*:**353**
Carbon dioxide, *2*:414; *4*:**353;** *8*:479, 534, 540
as aersol propellent, *1*:475
ammonolysis of, *2*:362
aquatic plant nutrients, *S*:326
beet sugar purification, *19*:210
in Bessemer process, *18*:723
in carbonated beverages, *4*:345
carbon reactions with, *4*:356, 440
cloud seeding with solid, *21*:661
in commercial oxygen, *14*:401
as component of natural gas, *10*:448
in drilling fluids, *7*:290
dry-ice by-product of soda ash manuf, *1*:734
fire extinguishing agent, *S*:367
in foam mat drying, *7*:359
as fumigant in mixture with ethylene oxide, *8*:552
from gasification of solid fuels, *10*:356
as heat transfer agent, *10*:856
as lubricant, *12*:606
in manuf of acetic anhydride, *8*:408
in manuf of ammonium chloride, *2*:318
in natural gas, *S*:426
as nuclear reactor coolant, *14*:81
in prepn of ammonium bicarbonate, *2*:315

reaction with trialkylaluminum, 2:32
removal in ammonia synthesis, 2:277, 278
shielding gas in welding, 22:244
from sludge gas, 22:111
as source for algal cultures, 1:651
sugar refining, 19:188
in tobacco smoke, 20:516
waste, 21:626
Carbon dioxide gas
  in fire-extinguishing systems, 9:295-298
Carbon dioxide transfer
  biotin and, 3:524
Carbon disulfide, 2:414; 4:370; 7:43; 8:271;
  10:164; 19:372
  as catalyst, 8:407, 419
  chlorine reactions with, 5:3, 133
  from coal tar, 19:670
  fluorination of, 9:698
  in fumigating nuts, 14:139
  prepn of Báhr's Salt, 6:631
  in rayon process, 17:175
  reaction with ethyl ether, 8:479
  in soil, 18:539
  soil fumigant, 15:912
  stain solvent, 18:662
  sulfur dioxide, 19:419
  sulfur solvent, 19:342, 349
  from carbon tetrachloride, 5:132
Carbon electrodes
  in arc furnaces, 10:258
Carbon fabrics, 4:24
Carbon felt, 4:242, 243
Carbon fibers, S:109
  in ceramic composite armor, S:143
  heat-and flame-resistant, S:959
  in metal matrices, 11:668
  from polyacrylonitrile, 11:666
  prop, 11:667
  prop of composites, 11:668
  from rayon, 11:666
Carbon-graphite bearing materials, 3:275, 284
Carbon-graphite products, 4:159
Carbonic acid
  diamide of, 21:37
Carbonic anhydrase, 7:258; 19:259
  inhibition of, 7:259
Carbonic esters and chloroformic esters, 4:386
Carbonitriding, 13:311
Carbonitriles, S:590
Carbonium ion and carbanion, 4:394
Carbonium ions, 4:394
  in hydrocracking, 11:431
  mechanism in hydroisomerization, 11:449

Carbonization, 4:400
Carbonizing
  of wool, 22:402
Carbon monosulfide, 19:372
Carbon monoxide, 2:414; 4:424; 6:576; 8:805
  from acetaldehyde decomposition, 1:80
  in ammonia manuf, 2:276, 280
  in automobile exhaust, S:50, 57
  in Bessemer process, 18:723
  cesium hydroxide reaction with, 4:863
  in commercial oxygen, 14:401
  fluid-bed synthesis reactor for, 9:436, 437
  fumes, 18:403, 406
  from gasification of solid fuels, 10:356
  from lactate acid, 12:174
  methyl chloride reaction with, 5:105
  in phosphorus manuf, 15:285
  in prepn of ethanol, 8:437
  prepn from methyl ether, 8:476
  in prepn of polyketones, 12:132
  in prepn of trialkylacetic acids, 8:853
  reduction of zinc oxide, 22:571
  in prod acetic acid, 8:397
  reaction with allyl compounds, 1:919
  reaction with ethers, 8:346
  reaction with hydrogen, 11:344
  in synthetic prepn of monosodium glut-
    amate, 2:207
  in tobacco smoke, 20:516
  waste, 21:626
Carbon monoxide
Carbon monoxide-hydrogen reactions, 4:446
Carbon oxychloride, S:674
Carbon oxysulfide, 8:271; 19:372
  carbon disulfide from, 4:381
  from carbon disulfide, 4:373
Carbon products
  baked and graphitized, 4:158, 202
Carbons, decolorizing, 4:149
Carbon suboxide, 12:88, 93, 156
Carbon subsulfide, 19:371
Carbon sulfides, 19:371. (See also Carbon
  disulfide.)
Carbon sulfotelluride, 19:765
Carbon tetrabromide, 3:777; 6:618
  from carbon disulfide, 4:372
  from carbon tetrachloride, 5:131
  reaction with isoprene, 12:68
Carbon tetrachloride, 5:128-141; 10:164
  from carbon disulfide, 4:372; 5:3
  carbon monoxide flammability influenced
    by, 4:430
  chain-transfer agent, 21:374

liver fluke treatment, *14*:549
methylene chloride from, *5*:115
photochemical, *15*:337
in prepn of acetyl chloride, *1*:140
in prepn of 1,1,1,3-tetrachloropropane, *8*:502
reaction with acetaldehyde, *1*:83
in roundworm chemotherapy, *14*:536
stain solvent, *18*:661
toxicity of, *5*:93, 96, 136
trichloroethylene reaction with, *5*:188
used in dry cleaning, *7*:315
Carbon tetrafluoride, *9*:508, 509
Carbon tetraiodide, *9*:700
from carbon tetrachloride, *5*:131
Carbon-12
in nuclear reactors, *14*:76
Carbon(13)
Carbon-13
chemical shifts, *14*:69
Carbon-14, *17*:92, 120
phys prop, *17*:5
synthesis, *17*:4
uses, *17*:8
Carbon-14 monoxide
carbonyl reactions with, *4*:503
prepn of, *4*:433
1,1'-Carbonylbisaziridine, *2*:248
Carbonyl bromide, *4*:432
Carbonyl chloride, *S*:674. See *Phosgine.*
Carbonyl compounds
reaction with metal alkoxides, *1*:843
Carbonyl condensation
with nitroparaffins, *13*:871
Carbonyl cyanide, *6*:630
Carbonyl difluoride
in prepn of isocyanates, *12*:54
Carbonyl fluoride, *9*:509, 690, 696
Carbonylhemoglobin, *3*:581, 582
Carbonyl nickel
in electric cells, *3*:197
Carbonyls, *4*:433, **489**
Carbonyl sulfide, *2*:414; *4*:432; *19*:372, 382
in soil, *18*:539
Carbonyl telluride, *19*:765
Carbopol
hydrophilic carboxy vinyl polymer, *S*:941
Carboranes, *3*:698
Carbowaxes, *22*:171
lubricant, *15*:121
o-Carboxy-α,α-bis(4-hydroxyphenyl)toluene, *16*:169

o-Carboxybenzenesulfonamide, *19*:597
Carboxyethylcellulose, *6*:660
Carboxyhemoglobin, *3*:581, 582
Carboxylase
in yeast protein, *22*:516
Carboxylates
surfactants, *19*:513
Carboxylation
of olefins, *14*:332
p-Carboxylbenzenesulfonamide
as diuretic, *7*:260
Carboxylic acid amides, *2*:66
reaction with ethylenediamine, *7*:26
surfactants, *19*:550
Carboxylic acids. (See also *Acids, carboxylic.*)
prepn from trialkyaluminum, *3*:32
Carboxylic esters
surfactants, *19*:537
Carboxylmethylstarch
4-Carboxyl-1-methyl pyridinium chloride, *22*:209
4-Carboxyl-1-methyl-2-pyridone, *22*:208
o-(Carboxymethoxy)benzoic acid, *17*:722
2-(Carboxymethylamino)propionic acid, *2*:592
Carboxymethylcellulose, *17*:399, 791
films from, *13*:832
as food additive, *10*:7
as gum, *10*:741
surfactants, *19*:513
Carboxymethyldextran, *17*:330
Carboxymethyl guar gum, *10*:741
Carboxymethylhydroxyethyl cellulose, *4*:645
Carboxymethyl locust bean gum, *10*:741
Carboxymethylstarch
as gum, *10*:741
Carboxymethyltrimethylammonium chloride hydrazide
chlorophyll reactions with, *5*:345
(Carboxymethythio)succinic acid, *2*:592
Carboxypeptidase, *2*:176; *14*:675
o-Carboxy-α-toluic acid. See *Homophthalic acid.*
o-Carboxy-α-toluic acid, *1*:239
Carboxyvinyl polymer
as gum, *10*:741
3-Carboxy-5-pyrazolone, *10*:911
Carbromal, *3*:776, 780; *11*:517
Carburan, *21*:8
Carburizing, *13*:304-308
Carbyl sulfate, *19*:286
Carbyl sulfate
prepn, *8*:502

Cardamom, *9*:370, 352, 365; *14*:197
Cardanol, *1*:915
  phenolic resins, *15*:184
Cardenolides, *18*:845
Cardiac glycosides, *4*:513
Cardiovascular agents, *4*:**510**
Cardol
  in cashew nutshells, *14*:133
Cardox blasting system, *4*:367
3-Carene, *19*:810, 815; *20*:753
Carinamide, *19*:258
Carisoprodol, *2*:389
Carminic acid, *7*:626, 627
Carnallite, *3*:758; *16*:370
  cesium content of, *4*:857
  from Great Salt Lake, *S*:447, 455
  in manuf of magnesium chloride, *12*:718
  processing, *16*:377
Carnallite
  rubidium from, *17*:687
Carnauba wax, *22*:158, 161
  as adherent, *1*:7
  in dental material, *6*:836, 840, 841
  in food processing, *10*:10
  in lipsticks, *6*:364
  in mascara, *6*:367
  tablet fabrication, *15*:123
Carnitine, *2*:188
Carnosine
  in fish, *9*:319
Carnot cycle, *20*:124
Carnotite, *21*:8
  uranium ore, *21*:2
  vanadium from, *21*:160
Caroa, *9*:180
  analysis, *9*:173
Caro's acid, *19*:143
Carotene, *19*:803
α-Carotene
  as a provitamin A, *21*:501
β-Carotene, *8*:781
  as a provitamin A, *21*:501
  degradation to vitamin A, *21*:491
  in determining egg yolk color, *7*:668
  in margarine, *13*:57, 60
  in nuts, *14*:131
  market for, *21*:508
γ-Carotene
  as a provitamin A, *21*:501
Carotenes
  reaction with boron trifluoride, *9*:556
Carotenoids, *7*:615
  of provitamin A, *21*:505

Carpaine, *1*:765
Carr-Price test
  for vitamin A₂, *21*:501
  for vitamin A, *21*:505
Carrageenan, *7*:299; *17*:391, 763, 774-784
Carrageeninase
  microorganism producing, *8*:210
Carrolite
  phys prop, *5*:722
Carrot seed, *14*:197
Carthartics, *4*:**586**
Cartridge brass
  composition of, *6*:214, 218
  magnetic prop, *6*:220
  stress relief for, *6*:240
Cartridge brass
  annealing temperature, *6*:224
Cartridge cases, combustible, *2*:628; *4*:55;
    8:697
Carvenone
  from camphor, *4*:55
Carvone, *12*:107; *19*:832
  as synthetic flavor, *9*:354
l-Carvone, *19*:817
d-Carvoxime-*l*-carvoxime
Caryophyllene, *15*:334; *19*:833
Cascade design
  principles, *7*:94
Cascara sagrada, *4*:590
Cascarilla, *14*:197
Case hardening, *13*:304-315
Casein, *2*:15; *9*:499
  as animal feed, *8*:865
  cyanoethylation of, *6*:663
  effect of formamide on, *10*:105
  in milk, *13*:513, 567
  in paint removers, *14*:489
  in papermaking, *14*:517
  and wool prod, *22*:400
  iodinated, *20*:260
  iodinated, *20*:262
  prepn of iodinated, *20*:264
  resin, *17*:394
  shrinkproofing wool, *22*:409
Caseinates, *13*:567
Casein glue
  as adhesive, *1*:387
Casein hydrolyzate
  in prepn of glutamic acid, *2*:204
Casein hydrolyzates, *13*:567
Caseins
  "synthetic," *19*:113

Casella F acid, *13*:714
Cashew nuts, *14*:145
Casinghead gas, *10*:448
Cassaine, *1*:759, 773
Cassella acid, *7*:478
Cassella acid, *13*:721, 723
Cassel's green, *13*:11
Cassel yellow, *15*:536
Cassia oil. See *Cinnamon oil.*
Cassiterite, *19*:632; *20*:276, 305; *22*:338
  zero point of charge, *9*:384
Castanin
  in chestnuts, *14*:125
Cast copper alloys, *6*:244
Cast iron. See *Iron.*
Castner process
  manuf of sodium cyanide, *6*:591
Castor beans, *4*:526
Castoreum, *14*:726
Castor oil, *4*:**524;** *8*:778, 784, 786, 790, 791
  in alkyd resins, *1*:864
  drying oil, *7*:398, 404, 410
  ethoxylated, *19*:548
  hydrogenated, *22*:164
  polyurethans from, *21*:63
  sulfated, *19*:526
Castor pomace, *4*:528
Castrix, *15*:918
Catalase, *9*:897; *15*:683
  action and uses, *8*:130, 190
  amitrole inhibition, *22*:212
  in milk, *13*:514
Catalases, *10*:9
Catalysis, *4*:**534;** *11*:891
  acid, *6*:20
  redox, *6*:21
  use of fluidization in, *9*:405
Catalyst. (See also *Catalysts.*)
  cobalt-molybdenum-alumina in hydrore-
    fining, *11*:438
  development for hydrorefining, *11*:441
  for esterification reactions, *8*:326
  Isomax, *11*:431
  Lindlar, *19*:826
  nickel-tungsten-alumina in hydrorefining,
    11:438
  use of boron trifluoride as, *9*:561
Catalysts
  in hydrocarbon synthesis, *4*:452
  in industrial processes, *4*:538
  metallic soaps as, *7*:283
  Natta, *15*:797

platinum-rhodium, *15*:835
  thermal analysis, *20*:113
  use in automobile exhaust control, *S*:62
  Ziegler, *14*:323; *15*:797
  Ziegler-Natta, *14*:309; *16*:241
Catalytic asphalt, *2*:775
Catalytic convertor
  for automobile exhaust control, *2*:824
Catalytic cracking, *15*:2, 4, 26
Catalytic hydrogenation
  of alkyl cyanides, *2*:107
  of amines, *2*:78
  aniline prepn by, *2*:79
Catanac
  as antistatic agent, *2*:666
Cataphoresis, *8*:23
Catechin, *7*:614
Catechol, *2*:629. (See also *Pyrocatechol.*)
  alkylation of, *S*:30
  jelling agent, *21*:355
  oxidation-reduction resins, *16*:211
  surfactants, *19*:582
Catecholamine
  and nicotine in smoking, *20*:520
Catecholamines
Catechol oxidase, *19*:746
Catechu, *9*:353
Cathepsin
  action and uses, *8*:186
Cathodic coating, *13*:250
Cathodic control
  of electrochemical corrosion, *6*:290, 318
Cathodic inhibitors, *6*:318, 320
Cathodoluminescence, *12*:617
Cathomycin
  amebicides, *20*:82
Cation exchanger, acid-regenerated
  use as catalyst in esterification, *8*:326
Cationic dyes
  in textile printing, *7*:577
Cation rearrangement
  and NMR, *14*:73
Catnip, *14*:198
Cauassu wax, *22*:161
Caustic soda. See *Sodium hydroxide.*
  in chemical cleaning, *S*:164
Cavitation, *20*:775
Cayenne, *9*:369
Cebu maguey
  analysis, *9*:173
Cedarleaf. See *Thuja.*
Cedarwood, *14*:198

Cedarwood oil, 22:383
Cedrene, 19:833
Cedrol, 19:833
Cedryl acetate
  prod and sales of, 8:377
Celcon, 1:96
  properties, 1:100, 102
Celery seed, 9:352, 363, 365, 370; 14:198
Celesite
  zero point of charge, 9:384
Celestine, 19:48
Celestite, 7:292; 19:50
Celliton Fast Red Violet RN, 7:486
Celliton Scarlet B, 2:905
Cellobiose
  sweeteners, 19:593
Cellon. See Tetrachloroethane.
Cellophane, 9:228
  manuf, 9:233
  manuf from cellulose xanthates, 22:427
  properties, 9:222-226
  use of glycerol in manuf, 10:627
  water permeability, 16:129
Cellophane flakes
  in drilling fluids, 7:298
Cellosolve solvent
  prepn, 8:455
Cellothyl. See Methylcellulose.
Cells, blood
  animal, 3:574
  human. See Blood fractionation.
Cells, electric. See Batteries.
Cells, fuel. See Batteries.
Cells, high-temperature, S:120
Cellular glass
  as thermal insulator, 11:829
Cellular materials. See Foamed plastics.
Celluloid, 15:721, 791, 801
  as dental plastic, 6:827
Cellulose, 4:146, 593
Cellulose, 4:146, 593; 6645; 16:680, 684
  action and uses, 8:182
  alkali fusion of, 14:358
  carbon fiber precursor, S:110
  cinnamic resins, 16:517
  cotton, 6:405, 406, 412, 414, 423
  cyanoethylation of, 6:634, 643, 647
  degradation, 19:224
  deoxidizer in shield metal-arc welding,
    22:242
  flame resistance, 9:302
  in algae, 17:763

  in nuts, 14:129
  in rayon process, 17:174
  in wood, 22:365
  pulp fibers, 20:34
  radiation degradation, 17:72
  reaction with propylene oxide, 16:598
  styrene graft copolymer, 19:91
  sulfation, 19:306
  tablet filler, 15:121
  thiazole dyes, 20:185
  use of fungicide on, 7:283
  x-ray analysis, 22:454
α-Cellulose
  in resins, 15:793
Cellulose acetate, 1:109; 4:603, 632
  as filter, 9:253
  cigarette filters, 20:517
  compatability with adipate esters, 1:416
  in paint removers, 14:489
2-Chloro-5-nitrobenzoic acid, 3:437
  in reverse osmosis, 22:58
  man-made fiber, 9:154
  manuf by decompression expansion, 9:855
  plastics based on, 4:653
  plasticized, 15:726, 757, 791
  plastics from, 19:147
  prod and sale of, 8:376
  properties, 9:860
  spinning, 21:443
Cellulose acetate butyrate, 3:881; 4:636
Cellulose acetate-butyrate, 15:758, 759, 791
  plasticizers for, 15:732
  plastics based on, 4:653
Cellulose acetate butyrate film, 9:230
  properties, 9:222, 226
Cellulose acetate fibers. See Acetate and tri-
    acetate fibers.
Cellulose acetate film, 9:230
  properties, 9:222, 226
  trade names, 9:242
Cellulose acetate phthalate, 4:637
Cellulose acetate-propionate, 4:636; 15:756
  plastics based on, 4:653
Cellulose acetate sulfate, 4:637
Cellulose butyrate
  aluminum formate as stabilizer in, 2:15
  as optical filter, 9:253
Cellulose, chemical, 4:593
  manuf of, 4:606, 608
Cellulose derivatives, 4:616
  plastics based on, 4:653
  as fabric coatings, 5:681
  as plastics, 15:791

Cellulose derivatives-plastics, 4:**653**
Cellulose diacetate, 12:814
  use as recording tape, 12:807
Cellulose dinitrate, 8:597
Cellulose fibers
  optical brighteners for, 3:745
Cellulose nitrate, 4:602, 603, 625; 6:405;
  8:597; 8:582, 597; 15:790, 801
  as blending agent in alkyd resins, 1:867
  as fluorescent coating system, 9:499
  as propellant, 8:696
  as propellant for guns, 8:700
  characteristics, 8:632
  compatability with adipate esters, 1:416
  diphenylamine as stabilizer of, 7:48
  ethyl ether as solvent for, 8:486
  fiber, 17:168
  in Benite, 8:695
  in film former in nail enamels, 6:368, 369,
    370
  in lacquers, 7:424
  in liquid lipsticks, 6:365
  in matches, 13:164
  in manuf of blasting gelatin, 8:636, 640
  in paint removers, 14:489
  in phonograph records, 15:225
  in propellants, 8:659
  manuf of, 4:628
  plasticizer for, 15:721, 732, 736, 754, 756
  plastics based on, 4:653
  process, 17:169
  process of burning, 8:668
  prod of rayon, 9:153
  thermochemical additive constants, 8:681
Cellulose nitrate lacquers
  solvents for, 18:576
  as industrial coating, 5:703
Cellulose phosphate, 9:306
Cellulose, regenerated
  carbon disulfide in manuf of, 4:383
Celluloses, 10:9
  microorganism-producing, 8:210
Cellulose sulfate. See Sodium cellulose sulfate.
Cellulose triacetate
  as film base, 16:502
  in photography, 15:364
Cellulose triacetate fiber. See Acetate and
  triacetate fibers.
Cellulose triacetate film, 9:230
  properties, 9:222, 226
Cellulose trinitrate. See Cellulose nitrate.

Cellulose xanthate, 6:405; 18:512.   (See also
  Rayon.)
Cellulose xanthates
  rayon manuf, 22:427
Cement.   (See also Concrete.)
  erz, 4:696
Cement, 4:**684**
  application of fluid-bed technology, 9:400
  calcium aluminate, 17:233
  portland in wallboards, 21:613
Cement
Cement
Cementation
  of copper leach solutions, S:253
  of powders, 16:411
Cementation of metals, 13:263
Cement copper, 6:141, 170
Cemented carbides, 4:92
Cementite, 4:73; 18:763
Cement, portland, 4:5, 684
  clays used in, 5:580
  composition, 2:5
  lime in manuf of, 12:431, 458
  prepn, 2:5
  white, 4:692
Cements
  silicate, 18:163
  thermal analysis, 20:113
Cementstone, 12:415
Cement stucco, 4:705
Centrifugal separation, 4:**710**
Centrifugal still
  used in evaporation, 8:565
Centrosome
  in yeast, 22:512
Cephaeline, 1:793; 20:75
Cephalin, 12:343
  in eggs, 7:664
  in nuts, 14:132
Cephalodrine, 14:684
Cephaloridine, 14:665
Cephalosporinase, 14:684
Cephalosporins. See Penicillins.
Ceragyrite, 18:284
Ceramic armor. See Ceramic composite armor.
Ceramic bonding
  of grinding wheels, 1:33, 35
Ceramic catalyst
  in prepn of acetaldehyde, 1:86
Ceramic coatings. See Enamels, porcelain or
  vitreous.
Ceramic composite armor, S:**138**

Ceramics, 2:55; 4:**759-832**
  armor, S:146
  clays used in, 5:561
  cobalt pigments for, 5:744
  colorants for, 5:845, 851
  as electric insulation, 11:800
  ferroelectric, 9:1
  for machining tools, 20:574
  furnaces for heat-treating, 10:273, 274
  kiln, 10:291
  pigments for, 5:851
Ceramics, chemical ware, 4:**832**
Ceramics industries
  use of litharge, 12:274
Cereals. See *Wheat and other cereal grains.*
Cereclors, 5:232, 234
Ceres Blue I, 20:722, 725
Ceresin, 22:169
  in dental material, 6:840, 841, 842
  in liquid lipsticks, 6:364
  in paint removers, 14:488
Ceria. See *Ceric oxide.*
  methylation of phenol, S:273
Ceric ammonium nitrate. See *Ammonium hexanitratocerate.*
Ceric ammonium sulfate. See *Ammonium trisulfatocerate.*
Ceric oxide, 17:143; 4:841, 850
  as abrasive, 1:28
  as glass colorant, 5:847
  in ceramic pigments, 5:852
Ceric oxide, titania
  colorant for glass, 9:252
Ceri mimbi wax, 22:165
Cerin
  in cork, 6:284
Cerium, 4:**840;** 14:143; 17:127, 144, 150
  electronic configuration, 1:360
  ionic radii, 1:370
Cerium alloys, 4:846-869
Cerium and cerium compounds, 4:**840**
Cerium boride, 4:847
Cerium carbide, 13:859
Cerium compounds, 4:849-854
  in carbon-arc electrodes, 4:211
Cerium hexaboride, 3:676
Cerium hydride, 4:847; 11:207
Cerium hydroxide, 4:850
Cerium nitride, 4:847
Cerium oxide
  in thorium oxide mantle, 20:248
Cerium sulfate, 4:851

Cerium sulfide
  as ceramic material, 4:774
Cermets, 16:431. (See also *Carbides; Nitrides.*)
  in bearing materials, 3:296
Ceroplastic acid, 8:813
Cerotic acid, 8:813
  tall oil, 19:621
  spec, 1:554
Cerous compounds. See *Cerium compounds.*
Certified food colors, 5:857
Cerulean blue, 5:745; 15:541
Cerussite, 12:208
Cervantite, 2:564, 573
Ceryl alcohol. See *1-Hexacosanol.*
Ceryl alcohol, 1:543
Cesium, 4:137; 17:78, 126, 134, 137
  from seawater, 14:156
  in photoelectric cells, 8:3
  in phototubes, 15:398
  photoemitter, 15:402
Cesium alloys, 4:862
Cesium alum, 2:64; 4:864, 17.607
Cesium aluminum sulfate. See *Cesium alum.*
Cesium and cesium compounds, 4:**855**
Cesium antimonide
  photocathode, 15:404
Cesium azide
  cesium metal from, 4:860
Cesium carbonate, 17:686
Cesium chloride
  phosphor, 15:352
Cesium chlorogermanate, 10:523
Cesium chromate, 5:498
Cesium compounds, 4:862-868
Cesium cyanide, 6:599
Cesium fluorogermanite, 9:585
Cesium fluosilicate, 9:658
Cesium gallate(III), 10:325
Cesium heptafluorozirconate, 9:685
Cesium hexachlorostannate(IV), 17:686
Cesium hexachlorotitanate, 20:388
Cesium hexafluorogermanate, 9:585
Cesium hexafluorotitanate, 9:684
Cesium hydride, 11:202
Cesium-ion engines, 4:865
Cesium nonachlorodithallate(III), 20:67
Cesium pentafluorozirconate, 9:685
Cesium perchlorate, 5:69
Cesium peroxide, 14:749
Cesium superoxide, 14:764
Cesium tetraphenylborate, 16:384

Cesium trifluorozincate, 9:684
Cesium-133
  definition of second, S:990
Cesium-137
  as fission product, 14:105, 107
Cesspools, 22:105
Cetane
  in diesel fuel, 10:485
Cetane number, 10:485
Cetoleic acid, 8:814
Cetrimide, 2:633
N-Cetyl-N-ethylmorpholinium ethosulfate
N-Cetyl-N-D-glucosyl-N-methylammoni- um
    iodide, 19:226
N-Cetyl-N-methylglucamine, 19:226
Cetyl alcohol, 1:543
  as opacifying agent, 6:353
  by sodium reduction, 18:451
  composition of commercial mixture, 1:553
  evaporation control, 21:665
  in hair preps, 10:774, 777
  in shampoo, 6:363
  phys prop, 1:543
  specs, 1:554
Cetyl chloride, 19:231
Cetyl-oleyl alcohol
  composition of commercial mixture, 1:553
  specs, 1:554
Cetylpyridinium chloride, 10:780; 16:804
Cetyl-stearyl alcohol
  composition of commercial mixture, 1:553
  specs, 1:554
Cetyltrimethylammonium bromide
  in inhibition of hydrogen sulfide corrosion,
    6:338
Cevitamic acid. See Ascorbic acid.
Ceylon cinnamon oil, 12:135
Ceylon graphite, 4:308, 309, 310, 314, 324
Chaksine, 1:773
Chalcanthite, 6:276
Chalcedony
  as abrasive, 1:25
Chalcocite, 6:135, 278; 9:389
  separation from pyrite, 9:393
Chalcomycin, 12:633, 634, 646
Chalcone. See Benzalacetophenone.
Chalcone, 12:107
Chalcopyrite, 6:134, 135; 9:393; 13:736;
    19:350, 758; 22:563
D-Chalcose, 12:639
Chalk, 12:415. (See also Calcium carbonate.)
  in dentifrices, 6:848

Chamet bronze
  composition of, 6:227
Chamois, 12:334
Chamomile
  essential oils as flavor, 9:352
  German, 14:198
  Roman, 14:198
Chamosite, 5:547; 12:6
Champlevé(enameling technique), 8:155
Channel blacks, 4:244.   (See also Carbon
    black.)
  manuf of, 4:256
Char
  sugar decolorizer, 19:188
Charcoal.   (See also Carbon; Coke.)
  in black powder, 8:647
  in cigarette filters, 20:519
  in manuf of carbon disulfide, 4:374
  manuf cyanide, 6:591
  in pyrotechnics, 16:826
  phosphorus manuf, 15:280
  prod of, 22:384
  prod of brass, 22:555
  propionic acid and, 16:555
  wood, 4:166
Charcoal black, 15:544
Chaulmoogric acid
  phys prop, 1:236
Chavicine, 11:470
Chebulinic acid, 12:323
Cheese, 13:531-538
Cheese rind
  as animal feed, 8:865
Chelants
  in chemical cleaning, S:164
Chelating agents, 6:1.   (See also Complexing
    agents.)
  in cosmetics, 6:371
Chelation, 11:890
Chelerythrine, 1:770
Chelidamic acid, 2:592
Chelidonic acid, 10:923
Chelidonine, 1:769
Chemical Abstracts
Chemical-Biological Activities, S:518
Chemical cellulose, 4:593
  manuf of, 4:606, 608
Chemical cleaning, S:**157**
Chemical cotton
  manuf of, 4:609
Chemical dictionaries, 12:502
Chemical documentation, 12:511

Chemical equilibrium, *20*:141
  equilibrium constants, *20*:141
  gaseous systems, *20*:142
  simultaneous reactions, *20*:143
  standard free energy, *20*:141
  heterogeneous systems, *20*:143
Chemical poisons
  in cancer chemotherapy, *S*:82
Chemical potential, *20*:137
Chemical statistics, *12*:508
Chemical ware
  ceramic, *4*:832
Chemical warfare, *4*:**869**
Chemiluminescence, *9*:485; *12*:617
Chemisorption
  theory, *1*:447
Chemopapain
  uses, *8*:190
Chemosterilants, *11*:717, 718
Chemotherapy
  of parasitic infections, *14*:532
Chenodeoxycholic acid, *3*:481, 483
Chenopodium, *14*:198
  oil of, *14*:533
Cherry bark, *9*:353
Chert
  as abrasive, *1*:25
Chestnut wood, *12*:317, 323
Chevreul's salt, *6*:279
Chicago acid, *7*:469; *13*:725
Chicago Blue 6B, *2*:903
Chichibabin hydrocarbons, *22*:482
Chicory, *5*:757
Chigger repellant
  benzil, *12*:147
Chile nitrate. See *Sodium compounds.*
Chile saltpeter. See *Sodium compounds.*
Chillies, *9*:369
China clay, *15*:511, 540.   (See also *Kaolin.*)
  in ceramics, *4*:766
  in papermaking, *14*:501
Chinawood oil, *8*:780, 785.   (See also *Tung oil.*)
Chinese blue, *15*:539
Chinese insect wax, *22*:167
Chinese vegetable tallow, *8*:778, 784
Chinese vermilion, *15*:525
Chiniofon, *2*:632
Chiniofon, *2*:632
  amebicide, *20*:83
Chinosol, *2*:631
Chiolite, *9*:534

Chloquinate
  amebicide, *20*:87
Chlorafin 42, *5*:234
Chlorahydrate, *6*:611
Chloral.   (See also *Hypnotics and sedatives.*)
  from acetaldehyde, *1*:83
  in prepn of dichloroacetic acid, *8*:418
  manuf, *8*:435
  manuf from ethyl alcohol, *8*:454
  nitroparaffin reaction, *13*:831
  prepn from bis(2-chloroethyl) ether, *8*:487
  reaction with ketones, *12*:112
  vanillin synthesis, *15*:163
  p-xylene recovery, *22*:475
Chloral
  chlorobenzene reaction with, *5*:256
  from trichloroethylene, *5*:186
Chloral betaine
  as hypnotic, *11*:513
Chloral choline
  as hypnotic, *11*:513
Chloral hydrate
  as a hypnotic, *11*:509, 512
  prepn of liquid glue, *10*:616
β-Chlorallyl alcohol, *1*:600
α-Chloralose, *15*:922
Chlorambucil
  alkylating agent, *S*:83
  as contraceptive drug, *6*:85
Chloramide. See *Chloramine.*
Chloramine.   (See also *Monochloramine.*)
  formation of, *5*:2
Chloramine-B, *4*:923
Chloramine T, *3*:557, *19*:597
  as disinfectant, *2*:616, 618
Chloramine Yellow, *20*:187
Chloramines
  as bleaching agents, *3*:557
Chloramines and chloroamines, *4*:**908**
Chloramphenicol, *4*:**928, 929**; *13*:463, 826
  amebicides, *20*:80
  as veterinary drug, *21*:245
Chloramphenicol monosuccinate, *19*:146
Chloran TM
  flame-retardant polyesters, *20*:831
Chloranil, *2*:415; *5*:334; *10*:223
  in black pigment dyes, *16*:212
  dye oxidant, *20*:693
  reaction with ethylenimine, *11*:529
  steroid synthesis, *18*:870
Chloranilic acid, *16*:912
Chlorantine Fast Green *BLL,* *2*:901
Chlorate candle, *S*:658

Chlorate cells, 5:52
Chlorates
    as hazardous chemicals, 9:292
    manuf of, 5:52
    oxygen generation systems, S:658
    tests for, 5:46, 58
Chlorathen citrate
    antihistamine, 11:37
Chlorazanil
    as diuretic, 7:254
Chlorazol Sky Blue FF, 2:903, 904
Chlorbetamide
    amebicides, 20:90
Chlorcyclizine. HCl
    antihistamine, 11:36
Chlordane, 5:248
    chlordan from, 5:248
    chlorine reactions with, 5:248, 249
    heptachlor from, 5:249
    sulfuryl chloride reaction with, 5:248
Chlordiazepoxide, 16:671
Chlorella
    methane utilization, S:838
Chlorendic acid, 5:241; 15:801
    fire retardants, S:467, 477
    polyesters, S:478
Chlorendic anhydride, 1:863, 5:241
    flame-retardant polyesters, 20:831
Chloretone. See Chlorobutanol.
Chloretyl, 12:145
Chlorfenvinphos
    ectoparasite control, 21:250
Chloric acid, 5:27, 50
    methanol reaction with, 5:41
Chloride
    analysis in water, 21:694
Chlorin, 5:342
Chlorinated benzenes, 5:253
    chlorophenols from, 5:328
Chlorinated biphenyls
    and related compounds, 5:289
Chlorinated hydrocarbons, 5:85-303
    as electric insulation, 11:789
Chlorinated lime. See Bleaching powder.
Chlorinated naphthalenes, 5:297
Chlorinated paraffins, 5:231
Chlorinated rubber
    as blending agents in alkyd resins, 1:867
Chlorinated trisodium phosphate, 3:557
Chlorination
    of municipal water, 22:85
    photosensitized, 15:336

    in sewage treatment, 22:113
    of water, 22:81
    in water treatment systems, 22:118
Chlorine, 1:671; 5:1
    addition to water supplies, 22:85
    ammonia reaction with, 4:916
    available, 5:9
    as bleaching agent, 3:555
    in bleaching fats, 13:180
    bottling water purified by, 4:340
    carbon monoxide reaction with, 4:432
    in chemical warfare, 4:870
    oxidation-reduction potentials of, 5:3
    cyanuric acid reaction with, 4:918
    as disinfectant, 2:615
    dried by alumina, 2:52
    in dye industry, 7:480
    electrodeposition of zinc, 22:589
    in ethylene oxide manuf, 8:531
    in gold refining, 10:684
    manuf, 1:671-707
    in manuf acetyl chloride, 1:140
    in manuf of aluminum chloride, 2:18
    in odor control, 14:175
    as odor and taste control agent in water,
        22:102
    photolysis, 15:335
    as plant nutrient, 9:31
    in prepn of Hypalon, 7:695
    reaction with ammonia, 2:261
    refrigerant, 17:303
    from seawater, 14:156
    shrinkproofing wool, 22:408
    stabilization, 19:248
    swimming pool water disinfectant, 22:124
    tests for, 5:5
    thiophene reaction, 20:222
    p-toluenesulfonamide reaction with, 4:923
    in water purification, 2:297
Chlorine available, 5:9
    in chloramines, 4:909
    swimming pools disinfected by, 4:926
Chlorine battery, 3:135
Chlorine dioxide, 5:35, 50
    addition to municipal waters, 22:87
    action of light on, 5:36, 37
    in flour, 22:294
    hydrate of, 5:37
    as bleaching agent, 3:560
    as bleaching agent in flour, 10:6
    hydrogen peroxide and, 5:29
    oxidant in water treatment, 22:72

prepn of, *5*:3
repression of, *5*:29
from sodium chlorate, *5*:37, 59
sodium hydroxide reaction with, *5*:37
swimming pool water disinfectant, *22*:129
tall oil recovery, *19*:616
Chlorine heptoxide, *5*:4
    formation of, *5*:4
Chlorine hexoxide, *5*:50
Chlorine monofluoride, *9*:586, 699
Chlorine monoxide, *5*:7
    prepn of, *5*:3, 8
    uses of, *5*:10
Chlorine oxygen acids and salts, *5*:7
Chlorine trifluoride, *9*:524, 585, 586
    disposal, *9*:594, 699, 700
    properties, *9*:592
    as propellant, *8*:691
    properties, *8*:688
    reactions, *9*:590, 591
Chlorine trioxide
    prepn of, *5*:4
Chlorine-36, *17*:121
Chlorinolysis, *5*:132
    of hydrocarbons, *5*:91
Chlorisondamine
    as ganglionic blockader, *4*:519
Chlorite, *5*:551
    layer structure, *5*:542
Chlorite, *19*:608
Chlorites, *5*:3, 28
    hydrogen peroxide and, *5*:29
Chlormadinone, *6*:81, 87
    veterinary drug, *21*:251
Chlormerodrin, *13*:236, 242
3-Chloro-*N,N*-dibutylpropylamine
    prop, *2*:141
2-Chloro-*N,N*-diethylethylamine, *2*:139
    prop, *2*:141
    reaction, *2*:140
3-Chloro-*N,N*-diethylpropylamine, *2*:139
    prop, *2*:141
    reaction, *2*:140
2-Chloro-*N,N*-diisopropylethylamine, *2*:139
    prop, *2*:141
2-Chloro-*N,N*-dimethylethylamine, *2*:139
    prop, *2*:141
    reaction, *2*:140
2-Chloro-*N,N*-dimethylpropylamine, *2*:139
    prop, *2*:141
    reaction, *2*:140
3-Chloro-*N,N*-dipropylpropylamine
    prop, *2*:141

p-Chloro-*N*-methylaniline, *2*:356
N-Chloro-*N*-methyl-*p*-toluenesulfonamide, *4*:923
3-Chloro-*N,N*,2-trimethylpropylamine, *2*:139
    prop, *2*:141
Chloroacetaldehyde
    from acetaldehyde, *1*:83
    from paraldehyde, *1*:83
    prepn, *1*:83
    from vinyl chloride, *5*:172
Chloroacetaldehyde diethyl acetal, *1*:107. See
    Diethyl chloroacetal.
N-Chloroacetamide
    thiophene reaction, *20*:222
Chloroacetamidomethanol, *2*:633
N-Chloroacetanilide
    hydrolysis constant of, *4*:913
Chloroacetic acid, *8*:415-417
    ammonolysis of, *2*:350
Chloroacetic acid
    dyes, *19*:433
    from ethylene chlorohydrin, *5*:307
    phys prop, *1*:235
    in prepn of *N*-(*p*-hydroxyphenyl)glycine,
        2:220
    reaction with ethylenediamine, *7*:27, 29
    reaction with pentafluorophenol, *9*:793
    sodium 2,4-dichlorophenoxide reaction
        with, *5*:325
    thioglycolic acid synthesis, *20*:199
    2,4,5-trichlorophenol reaction with, *5*:335
Chloroacetic anhydride, *8*:416
Chloroacetone, *8*:281; *12*:106
α-Chloroacetophenone, *1*:170
    in chemical warfare, *4*:872, 877
ω-Chloroacetophenone, *10*:147
ω-Chloroacetophenone, *12*:107
N-Chloro-*m*-acetotoluide
Chloroacetyl chloride
    from acetyl chloride, *1*:139
    from 1,2-dichloroethylene, *5*:182
    from methylene chloride, *5*:114
    phys prop, *1*:224
    from trichloroethylene, *5*:186
Chloroacetylene
    from *β,β*-dichloroacrolein, *21*:416
Chloroacetylenes, *5*:203
2-Chloroacrolein
    prepn from acrolein, *1*:259
β-Chloroalkylamines, *10*:907
N-Chloro-*tert*-alkyl cyanamides, *4*:924
2-Chloroallyl-*N,N*-diethyldithiocarbamate
    as herbicide, *22*:194

2-Chloroallyl acrylate
  phys prop, *1*:288
2-Chloroallylidene diacetate
  phys prop, *1*:264
2-Chloroallyl propionate, *16*:556
N-Chloroamidine, *14*:542
Chloroamines. See *Chloramines*.
2-Chloro-4-amino-6-alkylamino-s-triazines,
  18:533
2-Chloro-4-amino-6-isopropylamino-s-triazine,
  22:187
2-Chloro-4-aminotoluene-5-sulfonic acid,
  15:575
2-Chloro-5-aminotoluene-4-sulfonic acid,
  15:576
7-Chloro-5a,6-anhydrotetracycline, *20*:12
3-Chloro-2-anthraquinonecarboxylic acid,
  2:488
3-Chloro-2-anthroic acid, *2*:433
N-Chloroanilides, *4*:922
m-Chloroaniline, *2*:424
o-Chloroaniline, *6*:649; *2*:415, 423
p-Chloroaniline, *2*:424
  p-chlorophenyl isocyanate reaction, *22*:190
  manuf by ammonolysis, *2*:356
3-Chloroaniline, *7*:42
4-Chloroaniline
  pigments, *15*:572
Chloroanilines
  in acid inhibition, *6*:329
4-Chloro-o-anisidine, *2*:893
3-(O-Chloroanilino)propionitrile, *6*:649
2-Chloroanthracene-3-carboxylic acid, *2*:433
4-Chloroanthranilic acid, *15*:575
1-Chloroanthraquinone, *2*:480
  uses of, *2*:487
2-Chloroanthraquinone, *5*:218; *7*:478
  ammonolysis of, *2*:356
  manuf, *2*:483
  melting point curve, *2*:443
  in prod aminoanthraquinones, *2*:457
  uses, *2*:487
Chloroanthraquinones
  chemical reactions, *2*:479
  manuf, *2*:482
  properties, *2*:478
  uses, *2*:487
Chloroauric acid
  in oxidation of ethylene, *1*:87
1-Chloroaziridines, *11*:528
N-Chloroazodicarbonamide. See *Chloroazodin*.
Chloroazodin, *2*:616, 617; *4*:921

p-Chlorobenzaldehyde
  dyes from, *20*:717
m-Chlorobenzaldehyde, *6*:670
o-Chlorobenzaldehyde, *3*:365; *6*:670
  dyes from, *20*:704
  dyes from, *20*:705
p-Chlorobenzaldehyde, *3*:365; *6*:670
  from chlorobenzene, *5*:256
Chlorobenzene, *2*:22, 419; *5*:253; *7*:193, 478
  ammonolysis of, *2*:353, 369
  carbon monoxide reactions with, *4*:433
  chlorination of, *5*:255, 259
  chlorosulfonation, *19*:299
  phenol synthesis, *15*:150
  phenyl ether from, *15*:172
  in prepn of diphenylamine, *7*:41
  toxicity of, *5*:264
p-Chlorobenzenediazonium hexafluorophosphate, *9*:635, 645
4-Chloro-m-benzenedisulfonyl chloride,
  15:210
Chlorobenzene-p-sulfonic acid
  from chlorobenzene, *5*:255
Chlorobenzenes
  prepn of, *3*:373, 398; *5*:253-267
4-Chloro-1,3-benzenesulfonamide
  as diuretic, *7*:260
p-Chlorobenzenesulfonic acid, *3*:406; *15*:211
p-Chlorobenzenethiol
  uses, *20*:214
2-Chloro-1,3,5-benzenetrisulfonyl chloride,
  15:210
2-Chlorobenzidine, *3*:415
m-Chlorobenzoic acid, *3*:437
m-Chlorobenzoic acid
  phys prop, *1*:238
o-Chlorobenzoic acid, *3*:437
  phys prop, *1*:238
p-Chlorobenzoic acid, *3*:437; *5*:218
  phys prop, *1*:238
2-Chlorobenzoic acid, *7*:42
p-Chlorobenzonitrile, *2*:344
4-Chlorobenzophenone
  dyes from, *20*:726
Chlorobenzotrichlorides, *5*:287, 288
3-Chlorobenzotrifluoride, *9*:778
5-Chloro-2-benzoxazolinone-7-sulfonic acid,
  2:222
4'-Chloro-o-benzoylbenzoic acid, *2*:441; *7*:478
Chlorobenzyl chlorides, *5*:287, 288
o-Chlorobenzylidene malononitrile
  in chemical warfare, *4*:872

p-Chlorobenzylphenol, 2:627

Chlorobiphenyls, 5:289

2-Chloro-bis(alkylamino)-s-triazines
  analysis as herbicides, 22:186

2-Chloro-4,6-bis(alkylamino)-s-triazines, 22:183

2-Chloro-4,6-bis(amino)-s-triazines, 22:186

1-Chloro-1,3-butadiene, 5:228

4-Chloro-1,2-butadiene. See Isochloroprene.

4-Chloro-1,2-butadiene, 5:228

2-Chloro-1,3-butadiene. See Chloroprene.

2-Chloro-4,6-bis(ethylamino)-s-triazine, 22:186

2-Chloro-4,6-bis(isopropylamino)-s-triazine,
  22:188

1-Chloro-3-bromo-5,5-dimethylhydantoin, 22:127

Chlorobutanol
  antimicrobial agent, 15:128
  from chloroform, 5:122
  as hypnotic, 11:511

4-Chlorobutanol, 10:249

4-Chloro-2-butanone

4-Chloro-1-butanol, 5:310
  from isochloroprene, 5:228

1-Chloro-2-butene, 3:790, 840

Chlorobutenes

1-Chloro-2,5-difluorobenzene, 9:784
  from butadiene, 3:790

4-Chloro-2-buten-1-ol, 1:613

Chlorobutyl, 17:573

4-Chlorobutyl alcohol, 8:492

Chlorobutyl rubber, 7:688

4-Chloro-2-butynyl N-m-chlorophenylcarbam-
  ate

4-Chloro-2-butynyl N-(3-chlorophenyl)car-
  bamate, 1:606
  as herbicide, 22:193

α-Chlorobutyraldehyde, 3:870

N-Chlorobutyranilide
  hydrolysis constant of, 4:913

β-Chlorobutyric acid, 6:466

γ-Chlorobutyric acid, 10:673

Chlorobutyric acids, 3:879

8-Chlorocaffeine, 3:912

2-Chlorocamphene, 4:55, 57

Chlorocarbanilides
  as antibacterial agents, 2:637

Chlorocarbonate. See Ethyl chloroformate.

Chlorocarbonates, 4:386

Chlorocarbons, 5:85

Chlorocarbons and chlorohydrocarbons, 5:85

6-Chlorocarbonyl-2,3-dichloroquinoxaline, 7:631

3-Chloro-2'-carboxydiphenylamine, 7:42

Chlorocarvacrol, 2:628

4-Chlorocatechol, 18:532

Chloroceric acid, 4:841

Chlorocholine chloride, 5:411

3-Chlorocoumarin, 6:427

Chlorocresol
  in paint removers, 14:491

4-Chloro-m-cresol, 6:434

4-Chloro-o-cresol, 6:442

6-Chloro-o-cresol, 6:439

Chlorocyclohexane, 6:677

α-Chlorocyclohexanone, 12:122

4-Chloro-4-cyclohexene-1,2-dicarboxylic
  from chloroprene, 5:217

3-Chlorocyclopentene, 12:68

7-Chloro-5a(11a)-dehydrotetracycline, 20:12

p-Chlorodiazobenzene, 2:873

N-Chlorodibenzylamine, 4:924

2-Chloro-4-(diethylaminoethylamino)toluene,
  14:547

5-Chloro-7-[(diethylamino)methyl]-8-quinolinol,
  20:84

2-Chloro-n,n-diethylpropylamine

5-Chloro-7-tt123-(diethylamino)propyl]3amino)-
  methyl)- 8-quinolinol, 20:84

1-Chloro-2-(N,N-diethyl)propylamine
  prop, 2:141

Chlorodifluoroacetic acid, 9:770

Chlorodifluoroamine
  prepn, 9:589

1,1,1-Chlorodifluoroethane, 9:841

1-Chloro-1,1-difluoroethane
  as aerosol propellant, 1:473
  from 1,1,1-trichloroethane, 5:155

Chlorodifluoromethane, 5:126
  refrigerant, 17:307

2-Chloro-1,4-dihydroxyanthraquinone, 2:479

4-Chloro-2,5-dihydrobenzoic acid
  from chloroprene, 5:218

10-Chloro-5,10-dihydrophenarsazine, 10:929

6-Chloro-3,4-dihydro-7-sulfamyl-1,2,4-benzothia-
  diazine-1,1-dioxide, 7:262

3-Chloro-2,5-dihydrothiophene-1,1-dioxide
  from chloroprene, 5:217

Chlorodiiodomethane
  formation of, 5:114

1-Chloro-2-dimethylaminopropane, 16:670

2-Chloro-10-(3-dimethylaminopropyl)phenothi-
  azine, 9:780

2-Chloro-4-dimethylamino-6-methylpyrimidine,
  15:918

1-Chloro-5,5-dimethylhydantoin, 4:913

2-Chloro-3,3-dimethyl-3H-indole, 6:615

1-Chloro-3,5-dimethyl-2,6-octadiene, *12*:68

Chlorodimethylphenols, *5*:337

   in prepn of *o*-(2,4-dinitroanilino)phenol, *2*:218

1-Chloro-2-(*N,N*-dimethyl)propylamine

   prop, *2*:141

   in prepn of *p*-(2,4-dinitroanilino)phenol, *2*:220

2-Chloro-3,5-dinitrobenzenesulfonic acid

   in prepn of 6-amino-4-nitro-1-phenol-2-sulfonic acid, *2*:222

2-Chloro-3,5-dinitrobenzotrifluoride, *9*:779

2-Chloro-1,4-dinitro-2-butene

   from chloroprene, *5*:217

   from chloroprene, *5*:217

2-Chloro-1,3,2-dioxaborolane, *3*:668

6-Chloro-2,4-dinitrophenol, *2*:216

4-Chloro-2,6-dinitrophenol, *2*:216

m-Chlorodiphenylamine, *16*:649

3-Chlorodiphenylamine, *7*:42

Chlorodisilanes

   from silicone synthesis, *18*:233

7-Chloro-6-epi-5a(11a)-dehydrotetracycline, *20*:12

1-Chloro-2,4-dinitrobenzene, *7*:478

4-Chloro-2,3-epoxy-1-butanol

   in prepn of erythritol, *1*:571

1-Chloro-2,3-epoxypropane, *5*:304, 318

Chloroethane. (See also *Ethyl chloride*.)

   as aerosol propellant, *1*:473

   in Du Pont process of gravity separation, *10*:703

Chloroethanesulfonic acid, *19*:231

2-Chloroethanesulfonic acid, *19*:284

2-Chloroethanesulfonyl chloride, *19*:289

2-Chloroethanethiol, *9*:763

Chloroethanoic acid. See *Chloroacetic acid*.

2-Chloro-1-ethanol, *5*:304, 306

β-Chloroethyl acetate

   prepn, *8*:353

2-Chloroethyl acetate

   prepn, *8*:530

2-Chloro-4-ethylamino-6-isopropylamino-*s*-triazine, *22*:187

2-Chloroethyl acrylate

   phys prop, *1*:289

Chloroethylbenzene

   from vinyl chloride, *5*:173

2-Chloroethyl chloroformate, *4*:386

Chloroethylene. See *Vinyl chloride*.

1-Chloroethyl ethyl ether

   phys prop, *8*:471

   prepn from ethyl ether, *8*:479

2-Chloroethyl ethyl ether

   phys prop, *8*:471

β-Chloroethyl ethyl ketone, *8*:500

β-Chloroethyl formate, *8*:353

2-Chloroethyl hydrogen sulfate, *19*:491

2-Chloroethyl *p*-aminobenzoate, *2*:406

2-Chloroethylsulfonic acid

   from ethyl chloride, *5*:143

2-Chloroethylsulfuryl chloride

   from 1,2-dichloroethane, *5*:151

1-*β*-Chloroethylthio-3-methyl-3-phospholene-1-oxide, *12*:69

2-Chloroethyltrimethylammonium chloride, *5*:411; *15*:683

2-Chloroethyl vinyl ether, *21*:414, 417

β-Chloroethyl vinyl ether, *8*:487

Chloroethyne. See *Monochloroacetylene*.

Chloroethyne, *5*:204

o-Chlorofluorobenzene, *9*:781

p-Chlorofluorobenzene, *9*:781

Chlorofluorocarbene, *9*:757

Chlorofluorocyclohexanes

2-Chloro-2-methyl-3-butene, *12*:67

   prepn, *9*:776

Chlorofluorohydrocarbons, *9*:743

Chloroform, *2*:416; *5*:119-139; *6*:642; *8*:479; *12*:121

   from acetaldehyde, *1*:83

   reaction with acetaldehyde, *1*:83

   as anesthetic, *2*:397

   carbon tetrachloride from, *5*:133

   as denaturant, *8*:450

   methylene chloride from, *5*:115

   pentachloroethane reaction with, *5*:165

   prepn from ethanol, *8*:429

   in prepn of carbon monoxide, *4*:437

   sweeteners, *19*:595

   stain solvent, *18*:661

   toxicity of, *5*:93, 97, 126

Chloroformamide, *4*:69; *8*:271

1-Chloroformamidine hydrochloride, *6*:561

N-Chloroformanilide

   hydrolysis constant of, *4*:913

Chloroformic esters, *4*:386

Chloroformylacetic acid, *12*:850

Chlorogenic acid, *15*:681

Chlorogenic acids

   in coffee, *5*:751, 752

Chloroheptafluoronaphthalene, *9*:797

2-Chloroheptafluorotoluene, *9*:785

4-Chloro-2-hexene

3-Chloro-1-hexyne, *1*:620

Chlorohydrin
  decomposition-rate constant, *21*:389
  in manuf of propylene oxide, *8*:280
Chlorohydrin process
  for propylene oxide, *16*:600
Chlorohydrins, *5*:**304**
Chlorohydrocarbons, *5*:**85**
α-Chlorohydrocinnamonitrile, *6*:642
Chlorohydroquinone
  developing agent, *15*:371
6-Chloro-17-hydroxy-1,2 α-methylenepregna-
  4,6-diene-3,20-dione acetate, *6*:82
1-Chloro-4-hydroxyanthraquinone, *2*:485, 488
2-Chloro-3-hydroxybutyraldehyde, *5*:311
α-Chloro-β-hydroxybutyric acid, *6*:466
3-Chloro-5-hydroxy-2-pentanone
  thiamine synthesis, *20*:178
2-Chloro-4-hydroxyphenoxyacetic acid,
  22:182
6-Chloro-17-hydroxypregna-4,6-diene-3,20-dione
  acetate. See *Chlormadinone*.
2-Chloro-3-hydroxypropionaldehyde, *5*:311
p-Chloroiodobenzene*p*-dichlorobenzene
  binary cooling curve, *13*:204
1-Chloro-4-iodo-3-methoxy-1-butene
  from α-chloroprene, *5*:228
5-Chloro-7-iodo-8-quinolinol, *2*:632; *20*:83.
  See *Iodochlorhydroxyquin*.
α-Chloroisobutyric acid, *3*:882
β-Chloroisopropyl alcohol
  in prepn of bis(2-chloroisopropyl)ether,
  8:491
11a-Chloroisotetracycline, *20*:18
  hydrolysis constant of, *4*:913
N-Chloro-*m*-nitroacetanilide
  hydrolysis constant of, *4*:913
Chloro-*m*-nitrobenzene, *7*:478
p-Chloro-*m*-xylenol, *6*:355
Chloromaleic anhydride, *6*:694
  reaction with isoprene, *12*:66
Chloromelamine
  as disinfectant, *2*:617
o-(Chloromercuri)phenol, *10*:222; *13*:245
5-Chloromercurivanillin, *21*:192
Chloromerodrin
  as diuretic, *7*:256
p-Chlorometacresol, *10*:774
Chloromethane. See *Methyl chloride*.
Chloromethanol, *10*:85
3-Chloro-7-methoxy-9-(1-methyl-4-diethyl-
  aminobutylamino)acridine dihydrochloride,
  2:151; *14*:543

Chloromethoxypropyl mercuric acetate,
  10:232
1-(Chloromethoxy)-2-methyl-2-nitropropane,
  13:827
1-Chloro-2-methylanthraquinone, *2*:438, 486,
  488
5-Chloro-2-methylbenzoxazole, *2*:215
Chloromethyl-*o*-benzoylbenzoic acid, *2*:486
1-Chloro-3-methyl-1,2-butadiene, *1*:622
2-Chloro-3-methyl-1,3-butadiene, *1*:622; *5*:228
2-Chloromethyl-1,3-butadiene, *12*:67
1-Chloro-3-methyl-2-butene, *1*:631; *12*:67
3-Chloro-3-methyl-1-butyne, *1*:622
1-Chloro-3-methyl-4-(2-cyclopentenyl)-2-butene,
  12:68
Chloromethyldichlorostibine, *2*:582
Chloromethyldimethylborane, *3*:710
Chloromethyldimethylchlorosilane, *18*:262
4-Chloro-2-methylphenol, *18*:532
4-Chloro-2-methylphenoxyacetic acid
  as herbicide, *22*:181
3'-Chloro-2-methyl-*p*-valerotoluidide, *22*:198
1-Chloro-4-methylthiaxanthone, *14*:546
  manuf, *2*:483
  in manuf of 2-chloro-anthraquinone, *2*:485
Chloromethylation
  in ion-exchange resin synthesis, *11*:875
  of xylenes, *22*:483
11a-Chloro-6-methylenetetracycline, *20*:9
Chloromethyl ether, *20*:23
Chloromethyl ethers
  from nitro alcohols, *13*:827
α-Chloromethyl methyl ether
  reaction with isoprene, *12*:68
3-Chloromethylpent-1-yn-3-ol
  as hypnotic, *11*:512
Chloromethylphosphonic acid, *9*:307
3-Chloro-1-propyne, *1*:600
Chloromethylstyrene
4-Chloro-1-methylthiaxanthone, *14*:546
  uranium extraction, *21*:15
Chloromycetin. See *Chloramphenicol*.
Chloronaphthalenes, *5*:297
  toxicity of, *5*:302
Chloronapthalene
  solvent, *15*:722
  in prepn of alcohol, *8*:453
N-Chloro-*o*-nitroacetanilide
N-Chloro-*p*-nitroacetanilide
  hydrolysis constant of, *4*:913
p-Chloro-*o*-nitroaniline
  pigments, *15*:574

4-Chloro-2-nitroaniline, *2*:353
  diazotates derived from, *2*:893
o-Chloro-*p*-nitroaniline
  pigments, *15*:574
2-Chloro-4-nitroaniline, *2*:353
Chloronitrobenzenes, *o* and *p*-, *7*:478
Chloro-*o*-nitrobenzene, *7*:478
1-Chloro-2(and 4)-nitrobenzene, *7*:478
1-Chloro-3-nitrobenzene, *7*:478
1-Chloro-4-nitrobenzene, *7*:478; *19*:270
4-Chloro-3-nitrobenzenesulfonic acid, *2*:223;
  15:211
2-Chloro-5-nitrobenzenesulfonic acid
  synthesis of aminodiarylamines, *7*:42
2-Chloro-4-nitrobenzoic acid, *3*:437
5-Chloro-2-nitrobenzoic acid, *3*:437
2-Chloro-5-nitrobenzotrifluoride, *9*:779
4-Chloro-5-nitrobenzotrifluoride, *9*:779
4-Chloro-2-nitrophenol, *2*:216
4-Chloro-6-nitro-1-phenol-2-sulfonic acid,
  2:222
6-Chloro-2-nitro-1-phenol-4-sulfonic acid,
  2:223
7-Chlorooxytetracycline, *20*:7
Chloropalladite, *15*:868
Chloropalladous acid, *15*:868
Chloropentafluoroacetone, *9*:755
Chloropentafluorobenzene, *9*:788, 792
2-Chloro-3-pentene, *12*:68
4-Chloroperchlorylbenzene, *9*:604
m-Chloroperoxybenzoic acid, *8*:244
Chlorophacinone, *15*:920
o-Chlorophenol
  polyester fiber solvent, *16*:155
p-Chlorophenol, *6*:645
4-Chloro-α-phenol-o-cresol
  from *p*-chlorophenol, *5*:331
  in prepn of *p*-aminophenol, *2*:219
Chlorophenol red, *11*:552
Chlorophenols, *2*:625, 626; *5*:**325**
2-Chlorophenothiazine, *16*:667
Chlorophenoxamide, *20*:70
o-Chlorophenylacetic acid, *15*:214
p-Chlorophenyldiazothiourea, *15*:917
N-Chlorophenyl diguanidino compounds,
  4:924
m-Chlorophenyl isocyanate
p-Chlorophenyl isocyanate, *12*:46
  synthesis of barban, *22*:193
m-Chlorophenyl isocyanate, *12*:46, 47
o-Chlorophenyl isocyanate, *12*:46

p-Chlorophenyl isocyanate
  p-chloroaniline reaction, *22*:190
  dimethylamine reaction, *22*:188
p-Chlorophenylmagnesium bromide, *10*:725,
  726
Chlorophenylmethylisoxazolylpenicillin, *14*:698
p-Chlorophenyl-*p*-chlorobenzenesulfonate
  from *p*-chlorophenol, *5*:331
p-Chlorophenylsilane, *18*:181
Chlorophyll, *5*:**339**; *19*:751
  need for iron in formation of, *9*:29
  magnesium in, *9*:28
  as a measure of aquatic plant populations,
    S:323
  in tea, *19*:752
Chlorophyllase, *5*:344
  in tea, *19*:751
Chlorophyllide
  in tea, *19*:751
Chlorophyllides, *5*:344
Chlorophyllin salts, *5*:348
Chloropicrin, *5*:187; *13*:883
  nematocide, *15*:912
  sterilant, *18*:817
Chloroplatinic acid, *15*:864; *19*:467
  as catalyst, *12*:69
  in silicone rubber cure, *18*:229
Chloroplatinous acid, *15*:863
Chloroprene, *5*:215-231. (See also *Neoprene*.)
  plasticizers for, *13*:832
  polymerization, *7*:705
  prepn from acetyl chloride and monovinyl
    acetylene, *1*:139, 206
  from vinyl chloride, *5*:171
α-Chloroprene, *5*:228
Chloroprene polymers. See *Neoprene*.
Chloroprene rubber. See *Neoprene*.
Chloroprene sulfone
  from chloroprene, *5*:217
Chloropromazine, *2*:150; *9*:780. See *2-Chloro-
  10-(3-dimethylaminopro pyl)phenothiazine.*
Chloropropanediols, *5*:312
3-Chloro-1,2-propanediol, *10*:905
Chloropropanols, *5*:304, 309
1-Chloro-2-propanone. See *Chloroacetone.*
1-Chloropropene, *5*:206
N-Chloropropionanilide
  hydrolysis constant of, *4*:913
α-Chloropropionic acid, *12*:178
  phys prop, *1*:235
β-Chloropropionic acid
  phys prop, *1*:235
  prepn, *8*:502

β-Chloropropionitrile, *10*:501
3-Chloropropionitrile, *6*:651
β-Chloropropionyl chloride, *10*:150
3-Chloropropionyl chloride
    in prepn of acryloyl chloride, *1*:295
3-Chloropropyl acrylate
    phys prop, *1*:289
2-Chloropropylbenzene
    from allyl chloride, *5*:208
2-Chloro-1,3-propylene sulfite, *19*:484
3-Chloro-2-propyn-1-ol
    prepn from propargyl alcohol, *1*:600
2-Chloropyridine, *9*:798; *16*:673
Chloroquinaldol
    amebicides, *20*:83
    as veterinary drug, *21*:246
Chloroquine, *20*:76, 85, 87
5-Chloro-8-quinolinol, *20*:83, 84
2-Chloroquinoline, *9*:800
4-Chlororesorcinol
    in manuf of purpurin, *2*:475
Chlorosilane, *18*:186
Chlorosilanes
    conversion to siloxanes, *18*:237
    halosulfurates, *19*:494
    from silicone synthesis, *18*:233
Chlorosiloxanes, *18*:169
Chlorosmic acid, *15*:874
    analysis of herbicidal *s*-triazines, *22*:186
Chlorostyrene, *19*:83
    foams, *19*:119
    fire retardant, *S*:483
    reactive diluents, *20*:821
N-Chlorosuccinimide, *19*:140, *4*:908, 913, 921;
    12:126; *19*:140. (See also *Succinchlorimide.*)
    methacycline synthesis, *20*:8
    thiophene reaction, *20*:222
N-Chlorosulfamate
    hydrolysis constant of, *4*:913
N-Chlorosulfamic acid, *4 917; 19*:244
6-Chloro-7-sulfamyl-1,2,4-benzothiadiazine-1,1-
    dioxide, *7*:262
6-Chloro-7-sulfamoyl-2*H*-1,2,4-benzothiadiazine--
    1,1-dioxide
    as cardiovascular agent, *4*:522
Chlorosulfinic acid, *5*:39
N-Chlorosulfonamides, *4*:922
Chlorosulfonation, *19*:279, 298
Chlorosulfonic acid, *2*:452; *5*:**357**; *10*:161;
    19:375, 395, 401, 404, 491, 494; *20*:223
    cyclamate synthesis, *19*:599
    in military smoke agents, *4*:903
    in prepn of chlorine heptoxide, *5*:5

prepn of bis(chloromethyl) ether, *8*:47
    reaction with acetanilide, *1*:150
    saccharin, *19*:597
    sulfation, *19*:304
    sulfonic acids, *19*:313
    sulfonation, *19*:282, 285
    toluene reaction with, *4*:923
4-Chlorosulfonyldiphenyl, *16*:275
Chlorosulfonyl isocyanate
    olefin condensation, *16*:34
Chlorosulfurate esters, *19*:494
Chlorosulfuric acid. See *Chlorosulfonic acid.*
Chlorosulfurization
    of oils, *19*:498
Chlorosulfurous acid, *5*:39
2-Chlorotetrafluorobenzoic acid, *9*:785
2-Chloro-1,4,4a,9a-tetrahydro-9,10-anthraquinone
    from chloroprene, *5*:218
Chlorotheophyllinate, *16*:674
8-Chlorotheophylline, *16*:675
Chlorothiazide, *4*:522; *7*:262; *19*:259
2-*N*-Chloro-4-thiazolines, *4*:924
3-Chlorothiopene
    from chloroprene, *5*:217
4-Chlorothiophenol, *19*:314
Chlorotholidone
    as diuretic, *7*:261
Chlorothymol, *2*:628
    hydrolysis constant of, *4*:913
    as denaturant, *8*:450
o-Chlorotoluene, *6*:439; *12*:68
p-Chlorotoluene, *2*:344
α-Chlorotoluene, *7*:478.   (See also *Benzyl
    chloride.*)
2-Chlorotoluene
    hydrolytic desulfonation, *19*:315
Chlorotoluenes
    chlorination of toluene, *20*:532
4-Chloro-*o*-toluidine
    diazotates derived from, *2*:893
3-Chloro-*p*-toluidine
    starling control, *15*:921
2-Chloro-*s*-triazine
2-Chloro-6-(trichloromethyl)pyridine
    used in fertilization, *9*:78
Chlorotrifluoroborates, *9*:557
2-Chloro-1,1,1-trifluoroethane
    from trichloroethylene, *5*:188
Chlorotrifluoroethylene, *7*:695; *9*:590; *9*:690;
    15:764
    manuf of polymer, *9*:832, 842
    vinylidene fluoride copolymers, *21*:269

Chlorotrifluoroethylene
  in packing, *14*:453
Chlorotrifluoromethane
  ozone in, *14*:411
  refrigerant, *17*:308
5-Chlorotrifluoropyrimidine, *9*:799
Chlorotrifluorosuccinic acid, *19*:137
Chlorotrimethylsilane, *18*:237
Chlorous acid, *5*:27
  cellulose oxidized by, *4*:596
  decomposition of, *5*:28
β-Chlorovinyldichlorarsine, *2*:727
β-Chlorovinyl methyl ketone
  prepn from acetyl chloride, *1*:139
Chlorowax, *5*:232
Chlorowaxes, *5*:232, 234
Chloroxuron
  urea herbicide, *22*:190
6-Chloro-3,4-xylenyl *N*-methylcarbamate,
  15:922
4-Chloro-3,5-xylenol, *19*:672
p-Chloro-*n*-xylenol
  as fungicide, *10*:235, 780
Chlorparaffins, *5*:231-240
Chlorpheniramine maleate
  antihistamine, *11*:37
Chlorphenoxamide
  amebicides, *20*:90
Chlorpromamide, *19*:259
Chlorpromazine, *16*:640, 649
Chlortetracycline, *10*:19; *20*:2, 5, 12, 14, 28
  amebicide, *20*:80
  in bacterial infections, *3*:1
  as food preservative, *10*:15
  ¹⁴C incorporation, *20*:20
  as veterinary drug, *21*:244
Chlorthenoxazine, *2*:392
  as diuretic, *7*:261
β-Chlorvinyl ethyl ethynyl carbinol
  as hypnotic, *11*:511
Chlosudimeprimylium
Chocolate
  coatings, for confectionery, *6*:41
  use of lecithin in, *12*:355
Chocolate and cocoa, *5*:**363**
Cholanic acid, *3*:481, 483
Cholecalciferol, *21*:549; *21*:553
  synthesis of, *21*:564
Cholecystography, *17*:135, 138, 139
Choleic acids, *3*:484, 486
Cholesta-5,7-dien-3β,25-diol, *21*:553

Cholesterol, *8*:781; *18*:835, 854, 857. (See also
  Sterols.)
  from blood fractionation, *3*:517, 584, 585
  lanolin alcohols, *19*:549
  metabolism of, *3*:486, 487
  synthesis, *18*:833
  in synthesis of androgens, *11*:93
  in synthesis of corticosteroids, *11*:79
  synthesis in man, *3*:486
  in synthesis of estrogens, *11*:107
  synthesis of testosterone, *11*:98
  in nuts, *14*:132
  vitamin D activity, *21*:550
Cholesterol acetate dibromide, *18*:836
Cholesterol biosynthesis inhibitor, *2*:150
Cholesteryl acetate, *21*:562
Cholest-4-en-7β-yl-*p*-toluenesulfonate, *18*:878
Cholest-5-en-3β-ol. See *Cholesterol.*
Cholic acid, *3*:481, 483, 484
  in food processing, *10*:8
Choline, *2*:190; *5*:**403;** *15*:683
  as catalyst for cyanoethylation, *6*:634
  in lecithin, *12*:343
  in nuts, *14*:132
Choline bitartrate, *5*:408
Choline bromide, *5*:407
Choline carbonate, *5*:407
Choline chloride, *2*:369; *5*:403, 406, 407
Choline dihydrogen citrate, *5*:403, 407
Choline theophyllinate
  as diuretic, *7*:253
Chondrosamine, *4*:142
Chromandione, *10*:150
Chromanone, *10*:926
Chromanones, *12*:822
Chromate oxidation, *5*:511
Chromate treatment, *13*:298
Chromatography, (See also *Adsorption chroma-
  tography; Gas chromatography; Gas-liquid
  chromatography; Thin-layer chromatography,*
  5:**413**
  aquametry by, *2*:684
  gas-liquid partition, *21*:232
  mass transfer in, *13*:100
Chromatography radial, *5*:421. (See also
  Chromatography.)
Chromazol Violet 2BH, *7*:527
Chrome. See *Chromium; Chromium com-
  pounds.*
Chrome Brown C, *7*:527
Chrome Violet, *20*:720
Chrome alum. See *Chromic potassium sulfate.*

Chrome dyes, 7:465
Chrome greens. See *Chromium Compounds.*
Chrome ore
  slagceram nucleating agents, S:881
Chromene, 10:920
Chromeoxane Blue R
  synthesis, 20:725
Chromeoxane Brilliant Violet SBA, 20:719
Chromic. (See also *Chromium compounds.*)
Chromic acetate, 5:495; 7:566
Chromic acetylacetonate, 5:477, 480, 495
Chromic acid.   (See also *Chromic oxide.*)
  electrolytic chromium from, 5:454
  oxidation of 3-aminoperchlorylbenzene,
    9:604
  phenylacetic acid, 15:214
  reaction with TNT, 8:612
Chromic acid anhydride. See *Chromium tri-
  oxide.*
Chromic ammonium fluoride, 5:492
Chromic anhydride
  reaction with ethyl ether, 8:479
Chromic benzoate, 5:495
Chromic bisulfite, 5:493
Chromic borate, 5:493
Chromic bromide, 5:492
Chromic carbonate, 5:493
Chromic chloride, 5:480, 491, 492, 493
  hexahydrate, 5:480
Chromic chromate, 5:499
Chromic citrate, 5:495
Chromic fluoborate, 5:492
Chromic fluoride, 5:491, 492
Chromic formate, 5:495
Chromic hydroxide, 5:477
Chromic iodate, 5:493
Chromic lactate, 5:496
Chromic naphthenate, 5:496
Chromic nitrate, 5:493
Chromic oleate, 5:495
Chromic oxalate, 5:496
Chromic oxide, 5:476, 480; 10:514.
    (See also *Chromic acid.*)
  as abrasive, 1:28
  anethole oxidation, 15:162
  catalyst, 19:460
  catalyst in prepn of nitrogen fertilizer, 9:54
Chromic oxide
  chromium metal from, 5:453
  colorant for glass, 9:252
  in glass-ceramics, 10:548
  manuf of, 5:507

  modifier, 17:286
  as pigment, 14:472
  in synthetic gems
Chromic oxide catalyst, 2:54
Chromic oxide green, 5:507, 508, 538
Chromic phosphate, 5:494
Chromic potassium fluoride, 5:491
Chromic potassium sulfate, 2:64; 5:454, 480
  photography, 15:361
Chromic stearate, 5:496
Chromic sulfate, 5:493
  basic, 5:488
  in ruby manuf, 10:514
Chromic 8-hydroxyquinolate, 5:495
Chromite
  in ceramics, 4:771
Chromites, 5:496
Chromium, 5:**451;** 7:527, 12:601
  cast alloys, 20:571
  copper alloys with, 6:250
  in copper alloys, 6:210, 212
  corrosion resistance to water, 6:302
  as dental alloy, 6:810
  effect of addition in steels, 11:18
  in electron tubes, 8:3
  in electroplating, 8:38, 40, 55
  ferroalloy, 18:751
  peroxy compounds, 14:753
  in plain carbon and low-alloy steels, 11:17
  as a refractory metal, 11:31
  in rolled zinc, 22:601
Chromium(IV) chloride, 5:481
Chromium acetate
  in textile printing, 7:577
Chromium alloys, 5:463, 472
Chromium alums, 5:494
Chromium and chromium alloys, 5:**451**
Chromium boride, 5:491; 16:432
Chromium carbide, 4:75, 83, 84; 5:491
  for machining tools, 20:573
Chromium carbonyl, 4:489; 5:475, 480, 489
Chromium catalysts, 5:512
  in tanning leather, 12:331
Chromium compounds, 5:**473.** (See also *Chromic, chromous entries.*)
  bichromate, 2:435
  as ceramic pigments, 5:852, 853
  chrome greens, 5:509, 853; 14:472; 15:536
  chrome lignosulfonate, 12:372
  chrome oranges, 15:532
  in tanning, 5:510; 12:232, 331, 335
  chrome yellow, 15:530, 536

Chromium fluoride
    azo dyes aftertreated with, *2*:883
    aftertreatment of direct dyes, *7*:538
Chromium-gold alloy, *10*:691
Chromium hydrate, *15*:538
Chromium iodide (swaged)
Chromium monofluorophosphate, *9*:642
Chromium oxide. See *Chromic oxide.*
Chromium oxide green, *15*:537
Chromium pigments, *5*:507
Chromium plating, *5*:506; *9*:897
Chromium sesquioxide. See *Chromic oxide.*
Chromium silicide, *5*:491; *18*:129
Chromium sulfate
    in leather tannage, *12*:332
Chromium thiocyanate, *5*:495
Chromium trioxide, *9*:794
    manuf, *5*:488
    phys prop, *5*:481, 482
Chromium wolframate, *5*:495
Chromium-51, *17*:122
Chromizing, *13*:264
Chromocitronine R, *7*:535
Chromogen Red B, *22*:435; 430
Chromone, *10*:149
Chromophores, *5*:766
    pigments, *15*:568
Chromoselenic alum, *2*:64
Chromosomes
    yeast, *22*:512
    in yeast, *22*:514
Chromotrope FB, *2*:886
Chromotropic acid, *2*:886; *13*:723
    analysis of phenoxyalkanoic acids, *22*:181
Chromous. (See also *Chromium compounds.)*
Chromous acetate, *5*:480, 491
Chromous ammonium sulfate, *5*:480
Chromous chloride, *5*:480, 491
Chromous fluoride, *5*:491
Chromous oxide, *5*:476
Chromous sulfate, *5*:491
Chromous sulfide, *5*:476
Chromoxane Pure Blue BA, *7*:527
Chromspun. See *Acetate and triacetate fibers.*
Chromyl chloride, *5*:481, 497; *19*:404
Chronoamperometry, *7*:733
Chronopotentiometry, *7*:734
Chrysaniline
    manuf of, *20*:710
Chrysazin, *2*:466
Chrysean, *6*:588
Chrysocolla, *6*:135, 169

Chrysoidine
    polyacrylonitrile fiber dye, *20*:729
Chrysoidine methyl violet
    used in hair coloring, *10*:793
Chrysolite
    in ceramics, *4*:772
Chrysophanic acid, *2*:466
Chrysophenine, *2*:897
Chrysophenine G, *2*:899; *7*:465
Chrysophenine Y, *7*:535
Chrysotile. See *Asbestos.*
Chugaev reaction
    for provitamins D, *21*:568
    for xanthate decomposition, *22*:422
Chymopapain
    action and uses, *8*:187
Chymotrypsin, *5*:594, 595, 598
Ciba Bordeaux B, *3*:775
Cibalan, *2*:889
Cibalan Brilliant dyes, *2*:882
Cibaneutren dyes, *2*:894
Cibaphasol process
    for dyeing wool, *22*:404
Cicosanoic acid, *8*:813
Cigarette filters
    use of triacetin, *10*:630
Cilchrome color
    photographic process, *5*:836
Ciment fondu, *4*:696
Cinchona, *1*:772, 783
Cinchona alkaloids, *1*:782
Cinchona bark
    cardiovascular agents from, *4*:516
Cinchonamine, *1*:772
Cinchonic acid. See *Cinchoninic acid.*
Cinchonidine, *1*:783
Cinchonine, *1*:783; *2*:166; *16*:784
    as denaturant, *8*:450
Cinchonidine sulfate
Cinchonine hydrochloride
    analysis of wolfram, *22*:343
Cinchoninic acid
    phys prop, *1*:240
Cinchophen, *2*:391
Cineole
1,8-Cineole, *19*:831
    in oil of eucalyptus, *14*:201
Cineoles, *19*:820, 823, 831
Cinnabar, *13*:218, 222, 247; *15*:524
    mercury recovery from, *S*:542
    soil-release agent, *S*:969
Cinnamalacetic acid, *5*:521

Cinnamaldehyde, *1*:645; *5*:520-522; *8*:407; *19*:138
  as denaturant, *8*:450
  flavor and perfume material, *9*:358
  prepn, *1*:82
  as synthetic flavor, *9*:354
  as synthetic flavor, *9*:354, 355
Cinnamal diacetate, *5*:522
Cinnamene, *19*:55
Cinnamic acid, *8*:407
  in perfumes, *14*:722
  phys prop, *1*:239
  reaction with lime, *19*:143
  in resins, *17*:382
  as synthetic flavor, *9*:354
  vinylpyrrolidone copolymer, *21*:433
Cinnamic acid, cinnamaldehyde and cinnamyl alcohol, *5*:**517**
Cinnamic acid dibromide, *5*:518
Cinnamic alcohol. See *Cinnamyl alcohol.*
Cinnamic ester
  resins, *16*:517
Cinnamide, *5*:517
Cinnamon, *9*:352, 363, 365, 370; *14*:197, 199
Cinnamon oil, *8*:450
  as denaturant, *8*:450
Cinnamoyl chloride, *5*:517
Cinnamyl acrylate
  phys prop, *1*:288
Cinnamyl alcohol
  cinnamaldehyde from, *5*:521
  flavor and perfume materials, *9*:358
  in perfumes, *14*:729
Cinnamyl bromide, *5*:522
Cinnamyl chloride, *5*:522
Cinnamyl cinnamate, *5*:520, 522
Cinnamyl magnesium chloride, *10*:731
Cinnolines, *10*:899, 920, 928
Circuit breakers
  use of sulfur hexafluoride in, *9*:671
Circulin A, *16*:322
Cirramycin, *12*:635
Cirrasol PT
Cis-1,4-Polybutadiene
  graft block copolymers, *S*:917
Citraconic acid, *1*:253; *5*:527
  in synthetic oil prepn, *7*:415
Citraconic anhydride, *5*:526; *12*:83
Citral, *1*:645
Citral, *12*:132; *19*:819, 827
  in manuf of pseudoionone, *1*:846
  as synthetic flavor, *9*:354, 355

Citric acid, *5*:**524;** *7*:527; *8*:781
  as antioxidant, *2*:592, 594, 596
  in carbonated beverages, *4*:338
  as chelating agent, *6*:8
  in confectionery, *6*:46
  in cotton, *6*:405
  as corrosion inhibitor, *6*:329
  descaling ferrous metals, *S*:163
  in detergents, *6*:858
  as food preservative, *10*:14
  in margarine, *13*:59
  phys prop, *1*:234
  in prepn of itaconic anhydride, *12*:85
  removal from tart fruit juices, *7*:861
  in sparkling wines, *22*:325
  as stabilizing agent in food, *10*:16; *10*:635
  from sucrose, *19*:224
  in vinegar, *21*:256
Citric acid cycle. See *Krebs cycle.*
Citronella
  Java, *14*:199
Citronella grass
  essential oil in; *14*:181
Citronellal, *19*:819; 823; 826; 828
  as synthetic flavor, *9*:354
Citronella oil
Citronellol, *1*:645; *14*:735; *19*:819, 826, 828
d-Citronellol, *19*:813
Citronellol acetate
  in perfumes, *14*:735
Citronellol isobutyrate
  in perfumes, *14*:735
Citronellol propionate
  in perfumes, *14*:735
Citronellyl acetate
  prod and sales, *8*:377
Citrovorum factor
  of yeast, *22*:516
Citrus Red No. 2, *5*:867
Citrullus colocynthis
  cathartic from, *4*:587
Citrus juices
  deacidification, *7*:861
Citrus oils; *9*:353
Citrus pulp, dried
  as animal feed, *8*:868
Civet, *14*:726
Civetone, *14*:726; *18*:24
L-Cladinose, *12*:639
Clad metals, *13*:273
Claisen-Schmidt reaction
  aldehydes, *1*:642

Claisen solution, *1*:902
Clamoxyquin
  amebicides, *20*:84
Clamoxyquin hydrochloride, *20*:85
Clamoxyquin pamoate, *20*:70, 85
Clapeyron equation
  binary systems, *21*:206
  Clary, See *Sage clary.*
Clathrates, *21*:40
  commercial recovery of *m*-xylene, *22*:493
  butanes, *3*:817
  from trisilane, *18*:176
  of xylene isomers, *22*:475
Claudetite, *2*:721
Clausius-Clapeyron equation, *21*:198
Claus process for sulfur, *19*:353; *4*:376
Clausthalite, *12*:278
Clavacin, *10*:923
Clay
  aluminum compounds, *2*:5
  in limestone, *12*:422
  manuf of aluminum from, *2*:62
  manuf of aluminum sulfate from, *2*:60, 62
  refractory, *17*:233
  separation by flotation, *9*:395
  wax decolorizer, *15*:97
Clayden effect, *15*:388, 393
Clays, *5*:**541**. (See also *Ceramics; Kaolin; Bentonite.*)
  in ceramics, *4*:763
  colloidal in paints, *14*:468
  kaolin as pigments, *14*:515
  from ocean, *14*:166
  as paint extender, *14*:466
  in paint removers, *14*:489
  pencil, *4*:331
  as thixotropic binders, *4*:33
Cleaning. See *Detergency.*
Cleansing cream (cosmetic), *6*:348
Clefamide, *20*:90
Clemizole. HCl
  antihistamine, *11*:37
Cleveite, *21*:7
Cleve's acid, *13*:709
1,6-Cleve's acid, *13*:712
1,7-Cleve's acid, *13*:712
Cliachite, *5*:553
Climate
  and water, *21*:652
Cliradon, *2*:384
Cloisonné, *8*:155
Clomiphene
  as an antiestrogen, *11*:137

Clopidol, *21*:247
Cloran
  fire retardant, *S*:480
Clorpan, *5*:233
Clostridia
  in canned meats, *13*:176
Clostridium acetobutylicum
  in prepn of acetone, *1*:162
Clostridium botulinum
  and radiation, *13*:178
Clostridium pasteurianum
Cloud seeding, *21*:651
Clove, *9*:352
  essential oil as flavor, 9:363, 365, 371
Clove oil
  as denaturant, *8*:450
Cloxacillin, *14*:687
Clupanodonic acid, *8*:815
"Cluster", *17*:55
Clutch friction materials, *10*:128
Coacervation
  of gelation, *10*:502
Coagulants and anticoagulants, *5*:**586**
Coagulants (blood), *5*:586
Coagulation
  in water-treatment systems, *22*:90
Coal. (See also *Anthracite.*)
Coal, *5*:**606**
Coal, *S*:**177**
  Belknap chloride cleaning process, *10*:702
  benzene from, *3*:382
  carbonization of, *4*:400
  Chance cleaning process, *10*:703
  combustion calculations, *10*:214
  cyclopentadiene from, *6*:688
  gasification of, *S*:198
  hydrocracking of, *11*:419, 420
  pigments, *15*:546
  power from, *S*:217
  prod of zinc metal, *22*:573
  reserves, *10*:185
  synthetic fuels from, *19*:684
  use in ammonia manuf, *2*:299
  use in water filtration systems, *22*:119
Coal-chemical recovery, *4*:415
Coal-conversion processes, *S*:180
Coal gas. (See also *Carbonization.*)
Coalite carbonization process, *4*:421
Coal, paper, *12*:383
Coal, powdered
Coals
Coal tar, *19*:653-682
  benzene from, *3*:386

cresols manuf from, 6:436
  as denaturant, 8:450
  dyes from, 7:480
  recovery from coke-oven gas, 4:415, 419
Coal tar acids, 19:671
Coal tar disinfectants, 2:624
Coal-tar naphtha
  stain solvent, 18:661
Coal tar pitch, 4:167, 181
Coated abrasives, 1:29
Coated fabrics, 5:679
Coating formulations
  use of isocyanates in, 12:62
Coatings. (See also *Lacquers, Paint.*)
  aliphatic polyurethan, 21:102
  amino resins as, 2:253
  automotive, 5:698
  aziridines in, 11:54
  bituminous for waterproof masonry, 22:153
  carbon blacks used in, 4:276, 280
  castor oil in prepn of, 4:532
  for confectionery, 6:40
  corrosion-resistant, for metals, 6:304
  dispersed concentrates, 15:598
  epoxides used in, 8:259
  epoxy resins used as, 8:306
  epoxy for waterproof masonry, 22:153
  for fabrics, 5:679
  for home appliances, 5:693
  hydrocarbons in, 11:280
  pitch-epoxy, 19:678
  pitch-polyurethan, 19:678
  plasticizers for surface, 15:721
  polyurethan, 21:100
  poly(vinyl alcohol)resins, 21:361
  poly(vinyl butyral), 21:314
  silicate, 18:164
  silicone-based for waterproof masonry,
    22:153
  solution, 15:765
  surface, 15:765, 862
  for wood, 5:708
Coatings, chemically resistant
  ethyleneamines and formaldehyde resins as,
    7:35
Coatings, industrial, 5:**690**
Coatings industry
  alkyd resin classification, 1:853
Coatings, metallic. See *Metallic coatings.*
Cobalamin, 21:546

Cobalt, 5:**716**, 8:29, 12:600
  cast alloys, 20:571
  catalysts in oxo process, 14:375
  in catalysts for hydrocarbon synthesis,
    4:452, 461, 465
  in cemented carbides, 4:92; 20:572
  in chlorate candle, 5:660
  in copper alloys, 6:199, 210
  as dental alloy, 6:810
  electrodeposition of zinc, 22:588
  electromigration of, 5:287
  in electron tubes, 8:3
  in electroplating, 8:58
  as neutron monitor, 17:37
  from ocean floor, 14:165
  in porcelain enameling, 8:163
  peroxy compounds, 14:754
  steel, 18:751
  in superalloys, 11:31
  toluene oxidation, 15:155
  wolfram alloys, 22:344
Cobalt(II) dithiocyanates
  methanol clathrates of, 13:372
Cobalt acetate
  ester-exchange catalyst, 16:170
Cobalt acetates, 5:737, 738
Cobalt acetylacetonate, 5:738
Cobalt alloys, 5:732-736
  corrosion resistance to water, 6:302
Cobalt aluminate, 5:745
  pigments, 15:541
Cobaltamines, 5:743
Cobalt ammonium phosphate, 15:541
Cobalt and cobalt alloys, 5:**716**
Cobalt arsenite, 15:541
Cobalt blue, 15:541
Cobalt borate, 3:651
Cobalt bullet," 5:746
Cobalt carbides, 5:738
Cobalt carbonates, 5:738
Cobalt carbonyl
  in acetic acid synthesis, 13:372
  in reaction of acetaldehyde, 1:84
Cobalt carbonyls, 4:489
Cobaltcene, 6:699
Cobalt chloride, 12:588
  water color standard, 21:690
Cobalt chromate, 5:498
Cobalt chromite, 5:496, 745
Cobalt compounds, 5:**737**. (See also *Cobaltic,
  Cobaltous entries.*)

in ceramic pigments, 5:853
as driers in alkyd resins, 1:865
as glass colorants, 5:847
Cobalt cyanides, 5:743
Cobalt 2-ethylhexoate, 7:276, 278, 283
Cobalt ethylhexanoate, 8:849
Cobalt ferrate(VI), 12:40
Cobalt ferrite, 8:893
Cobalt fluosilicate, 5:739; 9:658
Cobalt-gold alloy, 10:690
Cobalt green, 15:538
Cobalt hexafluorostannate, 9:682
Cobalt hydrocarbonyl, 8:495
  catalyst, 14:375
Cobalt hydroxides, 5:740
Cobaltic carbonyls, 5:738
Cobaltic fluoride, 9:583, 589
  in fluorination of hydrocarbons, 9:698
Cobaltic hydroxide
  lignin catalyst, 21:187
Cobaltite, 2:720; 5:722
Cobalt linoleate, 7:276, 278
Cobalt metatitanate, 20:413
Cobalt molybdate
  coal-tar refining, 19:669
  in prepn of furan, 1:603
Cobalt monofluorophosphate, 9:642
Cobalt naphthenate, 7:274, 276, 278, 283
Cobalt neodecanoate, 7:279
Cobalt nitrates, 5:740
Cobalt octoate
  polyester accelerator, 20:826
Cobalt orthotitanate, 20:413
Cobaltous acetate, 10:9
Cobaltous bromide, 5:739
Cobaltous chloride, 5:739; 9:589; 10:9
Cobaltous fluoride, 5:739; 9:583
Cobaltous iodides, 5:739
Cobaltous sulfate, 10:9
Cobalt oxalate, 5:740
Cobalt oxide
  automobile exhaust control by, 2:822
  colorant for glass, 9:252
  in dental cement, 6:779
  desulfurization catalyst, 15:46
  in enameling, 8:158, 163
  pigments, 15:541
Cobalt oxides, 5:740
Cobalt phosphate
  pigments, 15:541
Cobalt phosphates, 15:275
Cobalt phthalocyanine, 15:488

Cobalt resinate, 7:279
Cobalt silicate, 5:741, 745
Cobalt soaps, 5:740, 744
Cobalt stannate, 5:745
Cobalt stearate, 7:280
Cobalt sulfamate, 19:244
Cobalt sulfates, 5:742
Cobalt sulfide
  in hydrogenation, 14:384
Cobalt sulfides, 5:742
Cobalt sulfites, 5:742
Cobalt tallates, 7:276, 278
Cobalt telluride, 19:772
Cobalt tetracarbonyl hydride
  catalyst, 14:375
Cobalt thiocyanate, 5:746
Cobalt titanates, 20:413
Cobalt trifluoride, 9:587
  reaction with benzene, 9:785, 786, 790
Cobalt 3-5-5-trimethylhexoate, 7:279
Cobalt ultramarine, 15:541
Cobalt violet, 15:541
Cobalt yellow, 5:745; 15:536.   (See also
  Aureolin.)
Cobalt-zirconium
  catalyst, 22:635
Cobalt-59, 17:3
Cobalt-60, 17:3, 58, 78, 123
  polyester cure, 20:829
  as fission product, 14:107
  primary ionization, 17:55
Coberine, 5:388
Cocaine, 1:766, 778, 785
  as anesthetic, 2:404
Cocaine hydrochloride, 2:405, 406
Cocarboxylase, 21:488
  thiamine derivative, 20:182
Coccidiosis, 21:246
Cochineal, 7:626
Cocoa, 5:363-402
Cocoa butter, 8:778, 784, 786; 18:45
Coco amine acetate
  solubility, 2:133
Coco diethanolamide, 19:551
Coco monoethanolamide, 19:552
Coconut oil, 8:778, 784, 790, 791
  in alkyd resins, 1:864
  composition, 8:831
  prod of fatty acids, 8:827
  in soap, 18:417
  specifications, 8:832
Codecarboxylase, 16:806
  in yeast, 22:516

Codehydrase I, *21*:512

Codehydrogenase II, *21*:513

Codeine, *1*:769, 778, 797
  as analgesic, *2*:380

Cod liver oil, *8*:782
  in prepn of chamois, *10*:301; *12*:334

Coefficient of friction, *10*:126

Coefficient of thermal expansion, *6*:761

Coefficients, mass-transfer, *13*:114

Coenzyme A, *2*:190; *8*:175
  in cholesterol synthesis, *18*:833
  occurrence, *20*:205
  sulfur in, *19*:369

Coenzyme I, *16*:780; *21*:512

Coenzyme II, *16*:780; *21*:513

Coenzyme Q, *16*:210

Coesite, *18*:49, 55

Coffearine, *21*:535

Coffee, *5*:**748**
  decaffeinated, *3*:913

Coffee aroma
  components in, *5*:753, 754

Coffee flavor, *9*:353

Coffee substitutes, *5*:757

Coffinite, *21*:8

Cognac, *1*:516

Cohesive energy density, *S*:889

Cohune oil, *8*:778, 784

Coir, *9*:171, 180
  for iron, *12*:11

Coke
  in ammonia manuf, *2*:272
  calcination of, *4*:169
  calcined, *15*:22
  hard from coal, *19*:654
  calcium sulfate reaction with, *4*:703
  carbon dioxide produced from, *4*:358
  characteristics as furnace fuel, *10*:280
  in dressing of lead ore, *12*:222
  gasification of, *10*:361
  direct reduction of iron, *S*:536
  petroleum, *4*:165, 169
  phosphorus manuf, *15*:280
  pitch, *4*:166
  prepn by carbonization, *4*:400
  prop of, *4*:414
  sizes of, *4*:414

Coke-oven gas, *4*:415
  carbon monoxide produced from, *4*:438
  characteristics, *10*:280
  composition of, *4*:438

Coke ovens, *4*:407

Coking
  in petroleum refining, *15*:2, 20, 22
  bitumen upgrading, *19*:712

Colace. See *Dioctyl sodium.*

Colace. See *Dioctyl sodium sulfosuccinate.*

Colchicine, *1*:759, 778, 786

Colchicum glucoside colchicoside, *1*:786

Cold creams, *6*:348

Cold-molded materials, *11*:800

Cold-working
  in alloys, *11*:12

Colemanite, *3*:615, 627, 628, 629, 630, 649

Coliform groups
  tests for in water, *21*:703

Coliforms
  fecal, *21*:705

Colistin, *16*:319, 321, 323
  amebicide, *20*:82

Colistins
  in bacterial infections, *3*:1

Collagen, *10*:499; *12*:334

Collagen fibers
  endomysial in meats, *13*:168

Collagen formation
  ascorbic acid in, *2*:758

Collagenous fibers
  in meat, *13*:167

Collidines, *16*:780, 801

Colloidal silica. See *Silica.*

Colloids
  dialysis, *7*:2
  electrodecantation of, *7*:841

Colocynth, *4*:587

Cologne
  in aerosol containers, *1*:479

Colognes
  use of alcohol in, *8*:447

Colophonium resin, *15*:476

Colophony, *17*:379

Color
  of water, *21*:690

Color additives for foods, drugs and cosmetics, *5*:858, 860, 862, 879

Color and constitution of organic dyes, *5*:**763**

Colorants.   (See also *Pigments.*)
  for ceramics, *5*:845, 851
  for glass, *5*:846
  for glass filters, *9*:252

Color conversion filters, *9*:256, 258

Colorfastness
  of textiles, *7*:593

Colorimetry, *5*:**788**
  fluorescent, *9*:491
  sugar analysis, *19*:163
  amebicide, *20*:80
Colorimetry and fluorometry, *5*:**788**
Coloring agents
  in food products, *10*:5
Color measurement, *5*:801
  tinctorial power, *4*:65
Color photography, *5*:**812**
Color(s)
  certified, for foods, *5*:857
  for confectionery, *6*:46
  D&C, *5*:857, 858, 864
  definition of, *5*:801
  Ext D&C, *5*:857, 858, 864
  FD&C, *5*:857, 858, 864
  measurement of, *5*:801
  of organic dyes, *5*:763
  primary, *5*:801
  "striking," *5*:845
Colors, food
  for carbonated beverages, *4*:342, 343
Colors for ceramics, *5*:**845**
Colors for foods, drugs, and cosmetics, *5*:**857**
Color temperature, *19*:800
Colostrum, *13*:517
Columbium. See *Niobium.*
Colza, *8*:779, 785
Colza oil, *8*:779, 785, 786, 790, 791
Comanic acid, *10*:923
Combustion calculations, *10*:191-220
Comfort chart, *1*:485
Comino seed. See *Cumin.*
Comminution. See *Grinding.*
Commutator materials, *4*:222
Comol, *12*:766
Compaction
  solid waste disposal, *S*:739
Complexing agents, *6*:**1**
Complexing agents, *6*:2; *15*:255
  in cosmetics, *6*:371
Complexometric titrations, *7*:778
Composites. (See also *Lamination.*)
  as packing materials, *14*:456
  strength of fiber reinforced, *11*:652
Composting
  solid waste disposal, *S*:739
  of waste-solids, *21*:648
Compreg, *22*:380
Compression molding of plastics, *15*:802

Compressors, *16*:**728**
  casting of silicon brasses, *6*:255
  magnetic characteristics of silicon bronzes,
    6:220
  prop of silicon brasses, *6*:255, 260
  silicon bronzes, *6*:193, 200, 201, 210, 212
  stress relief for silicon bronzes, *6*:240
Compton effect, *17*:55
Compton scattering, *22*:438
Computer cards, *12*:811
Computers, *6*:**25**; *11*:767, 768
  in literature searching, *12*:518
Computer tape, *12*:810
Conalbumin
  in eggs, *7*:664
Conarachin
  in peanuts, *14*:125
Concrete. (See also *Cement.*)
  use of hydroxyacetic acid in, *10*:635
  use of silicon tetrafluoride in making, *9*:653
Concrete
  watertight, *22*:151
Conduction furnaces, *10*:276
Conductometric titration, *7*:776-781
Conessimine
  amebicide, *20*:78
Conessine, *1*:775; *18*:841, 866
Confectionery, *6*:**34**
  hard candy, *6*:47
  humidity conditions during manuf, *1*:486
  use of lecithin in, *12*:355
  use of sorbitol in manuf, *1*:583
Congo Orange *R*, *7*:485
Congo Red, *2*:877, 895, 896, 898; *7*:465, 535,
    534; *9*:897
  from benzidine, *3*:408, 409
  gelling agent, *21*:355
Congo copal, *17*:383
Conhydrine, *1*:765
γ-Coniceine, *1*:765
Conidendrin, *11*:470
Coniferin
  vanilla from, *21*:183
Coniferyl alcohol, *17*:388
Conine, *1*:765
Conjugated acid oils, *8*:784, 786
Conjugate fibers, *9*:162
Consistency, of fluids
  definition, *9*:446
Consolidation coal process, *S*:185
Consolidation trickling, *10*:700
Constantan, *19*:787

Contact plating, 8:69
Contact poisons
    against insects; 11:677
    of plant origin, 11:683
    synthetic organic, 11:689
Container glass, 10:558, 571
Contraceptive drugs; 6:**60**
Control. See *Instrumentation and control.*
Convallaria majalis
    cardiovascular agents from, 4:513
Convallatoxin
    as cardiovascular agent, 4:513
Convection, 10:819, 821
Convection-type furnaces, 10:275
Convective transfer theory
    of mass transfer, 13:139
Converters
    for copper smelting, 6:150
Conveying, 6:**92**
Conveyor elevators, 6:117, 118
Conveyors, oscillating, 6:110
Convolvulin, 4:587
Cooling devices
    thermoelectric, 20:168
Cooling towers
    high-temperature, S:960
    hydrocarbon assimilation, S:840
    and thermal pollution, S:722
Cooperite, 15:842
Coordination catalysts, 6:**122**; 7:686
    of vanadium, 21:177
Coordination numbers, 6:3
Coordination polymers, 11:648
Co-oxidation
Copaiba, 14:200; 17:389
Copal
    in dental material, 6:836
    sulfur in, 19:370
    thioglycolic acid reaction, 20:198
    in wine, 22:316
Copolymerization, 16:232-235
    of vinyl chloride, 21:377
    of vinyl ethers, 21:421
    of vinylidene chloride, 21:280
Copper, 6:**131**, 157, 175, 182
    alloys as pigments, 15:548
    analysis in sparkling wines, 22:324
    benzyl chloride reaction with, 5:283
    cladding with, 13:275
    chloroform reaction with, 5:122
    content in soils, 9:35
    corrosion inhibitors for, 6:343

corrosion resistance to water, 6:302
    in dental amalgam, 6:807
    in dental alloy, 6:814
    in dental solder, 6:822
    dopant in wolfram, 22:341
    electrodeposition of zinc, 22:587
    electromigration of, S:286
    in electroplating, 8:58
    furnaces used for heating, 10:288
    in gold extraction and refining, 10:683
    hydrocarbon control in automobile exhaust,
        S:63
    impact tests, 13:189
    in microwave tubes, 8:3
    from ocean floor, 14:165
    phosphorus alloys, 15:294
    in photoengraving, 16:522
    as plant nutrient, 9:30
    properties, 6:265
    as protective covering for cables, 11:816
    removal from swimming pool water, 22:132
    in rolled zinc, 22:601
    separation by flotation; 9:395
    separation from lead ore, 12:212
    silver alloys, 18:292
    in superalloys, 11:21
    thermocouples, 19:787
    tin coatings, 20:289
    tin electroplating, 8:67
    in type metal, 20:757
    welding of, 22:249
    welding of, 22:250
    wolfram alloy, 22:345
    x-ray fluorescence spectrography of, 22:456
    in zinc alloys, 22:600
    zinc analysis, 22:516
    zinc purification, 22:584
    zone refining, 22:696
Copper(III) ferrate, 6:271
Copper(II) hexacyanoferrate(II) 12:27
Copper(II) hexacyanoferrate(III), 12:31
Copper(I) chloride
    in analysis of hexacyanoferrate(II), 12:30
Copper acetate; 2:165
    glucose analysis, 19:162
    sugar analysis, 19:159
Copper acetates, 6:267
Copper acetoarsenite, 15:538
Copper acetylide, 8:581; 10:86
    catalyst in manuf propargyl alcohol, 1:610
Copper alloys, 6:**181**
    cast, 6:244-265

corrosion resistance to water, 6:302
stress-corrosion cracking of, 6:238
stress-relief annealing of, 6:239
welding, 22:249, 250
wrought, 6:181-244
Copper ammines, 6:269
Copper ammonium chromate, 5:499
Copper arsenate, 6:269
Copperas. See Ferrous sulfate.
Copper benzoate
  benzoic acid oxidation, 15:155
Copper-beryllium alloy
  photoemitters, 15:408
Copper blue, 15:541
Copper bromides, 6:273
Copper carbonate
  as fungicide, 10:222
  pigments, 15:541
Copper carbonate
  Benedict's solution, 19:162
Copper carbonates, 6:269
Copper chlorides, 6:271-273
Copper chlorophyllin, 5:348
Copper chromate, basic, 5:499
Copper chromates and chromites, 6:270
Copper chromite, 5:496; 7:41; 19:824
  as catalyst in prepn of fatty alcohol, 8:822
Copper-complex
  stilbene dyes, 19:5
Copper compounds, 6:265. (See also Cupric, Cuprous entries.)
Copper cyanide, 15:538
Copper cyanides, 6:270
Copper dibasic acetate, 15:538
Copper dimethyldithiocaibamate, 15:910
Copper ferricyanide
  photography, 15:386
Copper ferricyanides and ferrocyanides, 6:271
Copper fluosilicate, 9:660
Copper formate, 6:271
Copper glance, 6:278
Copper-gold alloys, 10:688
Copper hexafluorostannate, 9:682
Copper ions
  properties, 6:265
Copper lactate, 12:184
Copper leaching, 6:168
Copper-lead alloys, 12:259
Copper-liquor solution
  in manuf of carbon monoxide, 4:438
Copper-lithium alloys, 12:538
Copper molybdinum
  separation by flotation, 9:395

Copper naphthenate, 7:275, 276, 278, 284
  as fungicide, 10:231, 235
  in waterproofing material, 22:136
Copper nitrates, 6:274
Copper nitrites, 6:274
Copper oleate
  as textile preservative, 10:232
Copper oxide
  catalyst, 19:143
  in catalyst in prepn of nitrogen fertilizer, 9:54
  colorant for glass, 9:252
  in electric cells, 3:122
  as lubricant, 12:603
  reaction with ammonia, 2:261
  reaction with hydroxyacetic acid, 10:635
  vanillin synthesis, 15:163
Copper oxides, 6:274-276
Copper oxychloride
  as fungicide, 10:222
Copper oxychloride sulfate, 6:273
Copper perchlorate, 5:71
Copper-3-phenylsalicylate
  as fungicide, 10:232
Copper pheophytin, 5:349
Copper phosphate
  as dental cement, 6:779, 781
Copper phosphates, 6:276; 15:275
Copper-phosphorus alloys, 18:547
Copper phthalocyanine, 7:635; 12:590; 15:488, 490
  sulfonated guanidine salts, 15:492
Copper phthalocyanine blue, 15:580
Copper phthalocyanine sodium 3,3'-disulfonate, 15:491
Copper phthalocyanine tetrasulfonyl chloride, 7:635
Copperplating
  citric acid in, 5:538
Copper polychlorophthalocyanine green, 15:580
Copper, polygonization, 6:184
Copper, porphyry, 6:136
Copper-pyridine
  perchlorates detected with, 5:80
Copper 8-quinolinate
  as fungicide, 10:231, 233, 235
  as leather and plastic fungicide, 10:233
Copper resinate, 7:279
  as fungicide, 10:232
  pigment, 15:538
Copper ruby glazes, 5:853

Copper silicate, *18*:158
   as fungicide, *10*:222
Copper silicon, *18*:129
Copper sodium chromate, *5*:499
Copper sulfamate, *19*:244
Copper sulfate, *9*:388, 393
   algae control in water, *22*:84
   in chemical cleaning, *S*:159
   in fungicides, *10*:222, 230
   as molluscide, *14*:544; *15*:910
   in nuclear reactors, *14*:83
   spectral transmittance curve, *9*:256
   in water-treatment systems, *22*:66
   as a weed killer, *22*:176
Coppersulfate, acid as metalplating bath,
   6:806, 813
Copper sulfates, *6*:276-278
Copper sulfide
   hydrometallurgical treatment of, *S*:260
Copper sulfide ores
   separation by flotation, *9*:393
Copper sulfides, *6*:278
Copper sulfites, *6*:279
   swimming pool algicide, *22*:130
Copper tallate
   as textile preservative, *10*:232
Copper, tough pitch, *6*:156, 175, 177, 182, 185
Copper tungstate
   as catalyst in prepn of acetyl chloride, *1*:140
Copper utensil polishes
   use of hydroxyacetic acid in *10*:635
Copper zeolite
   as fungicide, *10*:222
Copra meal
   as animal feed, *8*:864
Coprantine Blue RRL, *2*:903
Copying. See *Reprography*.
Coquina, *12*:416
Coral rubber, *12*:550
Cordials, *1*:517
Core oils, *7*:425
Corfam, *16*:346, 350
Coriander, *9*:352, 365, 371; *14*:200
Cork, *6*:**281**
   as a thermal insulator, *11*:829
Corkboard, *6*:286
Corn, *12*:863; *22*:253-304
   as animal feed, *8*:857, 859
   used in manuf of distilled alcoholic bev-
     erages, *1*:521
Corn earworm
   control, *11*:679

Corn oil, *8*:778, 784, 786, 790
   in drilling fluids, *7*:299
   in margarine, *13*:59
   parenteral solvent, *15*:128
   prod, *8*:827
   starch by-product, *18*:681
Corn starch
   disintegrant, *15*:121
Corn syrup
   in confectionery, *6*:38
   manuf, *6*:926
Coronadite, *13*:21
Corona stabilization, *9*:668C3016
"Corrodkote" test, *8*:72
Corrosion, *6*:**289**
   of copper and copper-base alloys, *6*:232
Corrosion fatigue
   of copper alloys, *6*:240
Corrosion inhibitor
   aniline as, *2*:418
   in antifreezes, *2*:551, 556
Corrosion inhibitors, *6*:**317**
   benzoic acid, *3*:429
   chromium compounds as, *5*:507, 510
   sodium benzoate, *3*:429
Corrosion resistance
   of plated metals, *8*:72
Corticosteroids
   analogs, *11*:87
   biologic action of, *11*:80
   biosynthesis of natural, *11*:79
   chemical synthesis of, *11*:81
   metabolism of natural, *11*:81
   structure of, *11*:77
Corticosterone
   biosynthesis of, *11*:79
   chemical synthesis of, *11*:81
   cortisol, *7*:267
   phys prop, *11*:79
Cortisol
   biosynthesis of, *11*:79
   chemical synthesis of, *11*:81
   phys prop, *11*:79
   synthesis of commercial, *11*:84
Cortisone
   chemical synthesis of, *11*:81
   discovery, *18*:831
   from ergosterol, *18*:836
   osmium tetroxide; *15*:835
   phys prop, *11*:79
   synthesis, *18*:859
   synthesis of commercial, *11*:84

Cortisone acetate, *18*:859
  veterinary drug, *21*:251
Corundum. (See also *Aluminum oxide.*)
  in synthetic gem manuf, *10*:510
Corylin
  in filberts, *14*:125
Corynantheine, *16*:784
Corynebacteria
  hydrocarbon utilization, *S*:839
Corynebacterium, *2*:204
Corynebacterium diphtherial
  chemotherapy against, *3*:29
Cosmetic industry
  use of metallic soaps, *7*:285
Cosmetics, *6*:**346**
  amine oxides as, *S*:45
  colors for, *5*:857
  optical brighteners for; *3*:748
  microencapsulation of, *13*:455
  use of emulsifiers, *8*:49
  use of fatty alcohols in, *1*:556
  use of glycerin, *10*:628
  use of natural gums in, *10*:741
  use of lecithin in, *12*:357
Cosmetics
Cosmetology. See *Cosmetics.*
Cosmic rays
  as background radiation, *14*:109
Costra, *18*:487
Costus, *14*:200
Cottage cheese, *13*:530
Cotton, *6*:**376**
  bleaching agents for, *3*:562
  calcium arsenate for boll weevil control, *2*:715
  cyanoethylation of, *6*:647
  dyeing with basic dyes, *7*:533
  dyeing with leuco esters of vat dyes, *7*:562
  dyeing of raw stock, *7*:556
  effect of detergency on, *6*:868, 873, 874, 891
  fibers, *4*:598
  as filter medium, *9*:283
  flame resistance, *9*:300, 302
  Gossypium arboreum, *6*:377
  Gossypium barbadense, *6*:378
  Gossypium herbaceum, *6*:377
  Gossypium hirsutum, *6*:378
  in manuf of cellulose acetate flake, *1*:116
  as packing material, *14*:455
  pima, *6*:379
  prepn for dyeing, *7*:518
  used for reinforced plastics, *12*:189
  work of rupture valve, *1*:131

Cotton, chemical
  manuf of, *4*:609
Cotton, classification, *6*:409
Cotton fabrics
  use of sulfur dyes, *7*:552
Cotton fiber
  effect of detergents on, *6*:866; 874; 880, 883
Cottonseed, *6*:**412**
  as animal feed, *8*:862
  chemical cotton from, *4*:609
Cottonseed hulls
  in drilling fluids; *7*:297
Cottonseed oil, *8*::33, 39, 779, 784, 786, 790, 791, 802
  in alkyd resins, *1*:864
  composition, *8*:831
  in margarine, *13*:59
  parenteral solvent, *15*:128
  prod of fatty acids, *8*:827
  specifications, *8*:832
Cotton wax, *22*:161
  in copper smelting, *6*:153
Cottrell electrostatic precipitator, *8*:75, 78, 80
Couepic acid, *8*:816. See *Licanic acid.*
"Couette" constant, *21*:466
Coulombic hydration, *21*:678
Coulometry, *7*:769
Coumafuryl, *15*:919
Coumalic acid, *12*:842
Coumalins, *10*:922
Coumaran, *10*:908
Coumarandione, *10*:150
Coumaranone, *10*:908
p-Coumaric acid, *15*:681; *17*:382
Coumarin, *6*:**425;** *10*:908; *14*:729; *15*:681
  derivatives as optical brighteners, *3*:744
  optical brightener, *16*:51
  separation from vanillin, *21*:191
  as synthetic flavor, *9*:354
Coumarin anticoagulants, *5*:601
  antagonists of, *5*:603
Coumarin-3,6-disulfonic acid, *6*:427
o-Coumarinic acid lactone. See *Coumarin.*
Coumarin-6-sulfonic acid, *6*:427
Coumarone, *10*:899, 907, 908
  in coal tar, *19*:657
  resins from coal tar, *19*:670, 671
Coumarone-indene
  polymers, *15*:799
Coumarone-indene resins
  in dental materials, *6*:836
  manuf, *11*:246

polymerization, *11*:245
starting materials, *11*:243
Coumarones, *10*:907
Coumatetralyl, *15*:919
Countercurrent extraction, *8*:720
Couplers
  in color photography, *5*:818, 822
Coupling
  ¹9F-¹9F, *14*:67
  nuclear and geometry effects, *14*:56
Coupling components
  azoic, *2*:890, 891, 892, 893
Coupling constants
  proton-proton spin, *14*:60
Covellite, *6*:135, 279; *9*:393; *15*:541
Coverings, thermoset,
Cozymase, *21*:512
Cozymase I, *21*:512
Cozymase II; *21*:513
Cracked asphalts. See *Asphalt.*
Cracking
  active sites for, *11*,427
  catalytic, *15*:2, 4, 26
  of coal, *S*:180
  of hydrocarbon fuels, *10*:415
  hydrocracking, *11*:418
  thermal, *15*:18
  thermal, *S*:390
Cracking units, catalytic, *4*:580
Cranberry wax, *22*:161
Craters, carbon arc, *4*:209, 210, 212
Cream of tartar. See *Potassium hydrogen tar-trate.*
Creams (cosmetic), *6*:348
  manuf, *6*:351
Crease resistance in textiles, *1*:265
Creatine, *2*:190
  in fish, *9*:319
Creep, *13*:194. See *High-temperature alloys.*
  in polyurethans, *21*:82
  primary in fibers, *20*:41
Creep rupture, *13*:189, 194
Creosote, *19*:679
  frothing agent, *9*:385
  pitch blends, *19*:679
Creosote
  as a fungicide, *10*:229
Creosote oil
  gasoline from, *11*:422
Creosotes
  effect on wood, *22*:373
  wood preservatives, *22*:379

Cresidine, *2*:423
Creslan, *1*:313
Cresol, *2*:622, 624. (See also *Cresols.*)
  antimicrobial agent, *15*:128
m-Cresol, *6*:434, 436; 508
  menthol from, *19*:824
  alkylation, *S*:27
o-Cresol, *6*:434, 436
  in manuf of coumarin, *6*:429
  phenolic resins, *15*:184
p-Cresol, *2*:423, *6*:434, 436, 508
  alkaline fusion, *19*:315
  alkylation with isobutylene, *1*:895
  alkylation of, *S*:27
  chlorination, *19*:402
  phenolic resins, *15*:206
  synthetic, *S*:271
p-Cresol benzidine, *6*:508
Cresol carboxylic acid, *20*:720
Cresol red, *11*:552
Cresols, *6*:434; *7*:478
  aldehyde polymerization, *15*:179
  in coal tar, *19*:668, 672
  cresylic acid, *6*:434
  cresylic acid as frothing agent; *9*:385
  cresylic acids from coal tar, *19*:672
  cresylic acid in solvent refining; *15*:57
  petroleum sweetening, *15*:52
  plasticizer, *15*:749
  resins, *15*:176, 188
Cresorcinol, *11*:481
o-Cresotinic acid
  formaldehyde dyes from, *20*:725
p-Cresyl acetate, *14*:729
Cresyl diphenyl phosphate
Cresylic acid. See *Cresols.*
Cresylic acid
  electrolytic zinc prod, *22*:586
Cresylic acids, synthetic, *S*:**271**
m-Cresyl-(*m*-tolyl) acetate, *10*:235
Cresylmercuric cyanide, *10*:222
o-Cresyl phosphate
  in manuf of coumarin, *6*:429
  in synthesis coumarin, *6*:442
Crimping
  of polypropylene fibers, *S*:827
Crinine, *1*:771
Cristobalite, *17*:227; *18*:9, 54, 78, 83
Crocein acid, *13*:721
Crocetin, *7*:615
Crocidolite. See *Asbestos.*
Crocoite, *5*:451, 473

Crocus. See *Iron oxide.*
Cromophtal dyes, *2*:909
Crookesite, *20*:65
Crossed Tishchenko. See *Tishchenko reaction, crossed.*
Croton, *3*:871
Crotonaldehyde, *1* 270, 271; *6*:445; *7*:414; *8*:436; *10*:663; *12*:89, 129
Crotonaldehyde, *3*:802; *6*:694
  1-butanol from, *3*:826
  from isobutylene, *3*:842
  nitroparraffin reaction, *13*:831
  physical properties, *1*:255
  prepn from acetaldehyde, *1*:82
  prepn of crotonic acid, *6*:466
  vinylpyrrolidone copolymer, *21*:433
Crotonamide, *6*:468
Crotonanilide, *6*:468
Crotonic acid, *6*:**464**
Crotonic acid, *6*:455, 694
  butyric acid from, *3*:880
  in synthetic oil prepn, *7*:415
Crotonic acid dichloride; *6*:466
Crotonic aldehyde. See *Crotonaldehyde.*
Crotonic anhydride, *6*:455, 468
Crotonophenone, *12*:107
Crotonyl peroxide, *6*:468
Crotylacetone, *12*:105
Crotyl chloride, *3*:790, 840
Crotylideneacetone, *12*:105, 129
Crotyl phenyl ether, *15*:168
Crucibles
  graphite, *4*:231, 329
Crude oil. (See also *Petroleum.*)
  benzene from, *3*:375
  coal reserves for synthetic, *S*:180
  distilling, *13*:730
  distillation for gas manuf, *10*:472
  synthetic from coal, *S*:178
Crustecdysone, *18*:849
Cryobiology, *6*:476
Cryogenic fluids, *12*:607
Cryogenic process
  for oxygen, *14*:394
Cryogenics, *6*:**471**; *17*:295
  use of helium, *10*:878
Cryolite, *1*:936; *2*:1; *9*:534, 535
  natural, *1*:936
  properties, *9*:536
  synthetic, *9*:541
  synthetic, *1*:936
  as source of fluorine, *9*:509
  uses, *9*:544, 653

Cryolithionite, *9*:534
Cryopumps, *21*:137
Cryostat assembly, *4*:40, 41
Cryotron, *6*:478
Crypotolepine, *1*:771
Cryptomelane, *13*:21, 29
Cryptopine, *1*:769
Cryptopleurine, *1*:771, 776
Cryptoxanthin
  as a provitamin A, *21*:501
Crystal Violet, *7*:533; *20*:673
  absorption spectrum, *20*:686
  acidification of, *20*:696
  alkylation, *20*:699
  basification, *20*:697
  manuf, *20*:711
  structures of, *20*:685
  substituent effects, *20*:689
  synthesis, *20*:675
  via route of Caro and Kern, *20*:713
Crystal Violet Lactone
  manuf of, *20*:712
Crystal counters
  diamonds as, *4*:289
Crystal lattices
  radiation effects, *14*:79
Crystalline
  defn, *6*:516
Crystalline materials
  hardness testing, *10*:815
Crystallization, *6*: **482, 492**
  fractional, *22*:680
  mass transfer in, *13*:100
  pilot plant, *15*:630
Crystallography, See *Crystals.*
Crystalloids
  dialysis, *7*:2
Crystals, *6*:516
  cubic, *22*:445
  ferroelectric, *9*:1
  hexagonal, *22*:445
  lattice point, *22*:443
  reciprocal lattice, *22*:445
  tetragonal, *22*:445
  twinning of, *6*:525
  unit cell, *22*:442
Crystal violet methyl orange, *9*:897
Cuam
  as cellulose solvent, *4*:601
Cubeb, *14*:200
Cuene
  as cellulose solvent, *4*:601
Cularine, *1*:769

Cumene, 6:**543**
  commercial prod, 1:892
  (isopropylbenzene) in ethylbenzene, 22:500
  (isopropylbenzene) reaction with isoprene, 12:68
  in manuf acetophinone, 1:169
  α-methylstyrene manuf, 19:81
  α-methylstyrene prod, 19:81
  phenol synthesis, 15:149
  prepn from propylene by benzene alkylation, 1:889
  toxicity of, 11:302
Cumene hydroperoxide, 14:774; 15:150
Cumene hydroperoxide process
  for acetone, 1:160
Cumin, 14:200
  essential oils as flavor, 9:352, 365, 371
Cuminal, 3:366
Cummingtonite
  zero point of charge, 9:384
p-Cumylphenol, 1:912
Cunninghamella
  hydrocarbon utilization, S:839
Cunningham slip correction, 7:436
Cupferron
  zinc analysis, 22:561
Cupolas
  carbon refractories for, 4:240
Cuprammonium carbonate. (See also Copper compounds; Cuprammonium.)
  as fungicide, 10:232
Cuprammonium fluoride
  as fungicide, 10:232
Cuprammonium hydroxide
  as fungicide, 10:232
Cuprammonium ion, 2:264
Cuprammonium process, 17:170
Cupric. (See also Copper compounds.)
Cupric acetate, 6:643
Cupric acetoarsenite, 6:269
Cupric ammine carbonate
  copper prod, S:256
Cupric chloride
  petroleum sweetening, 15:50
Cupric ferricyanide
  dye oxidant, 20:693
Cupric fluoride, 9:583
Cupric fluosilicate, 9:658
Cupric hydroxide, 6:275
  to form cuprammonium ion, 2:264
Cupric sulfide
  pigments, 15:541

Cuprite, 6:35, 274
Cuprobam, 15:910
Cupro-cupric sulfite, 6:279
Cupro-nickel alloys, 6:241
Cupro-nickels, 6:193, 199
Cuprous. (See also Copper compounds.)
Cuprous ammonium acetate, 2:281
Cuprous ammonium formate, 2:281
Cuprous ammonium sulfate
  in butadiene separation, 3:806
Cuprous chloride, 2:356, 7:28
  as catalyst in manuf of acrylonitrile, 1:345
  in electric cells, 3:133
  in manuf of ethylenediamine, 7:31
Cuprous cyanide, 9:787
  as insecticide, 11:683
  in copper sulfide treatment, S:262
Cuprous fluoride, 9:583
Cuprous fluosilicate, 9:658
Cuprous iodide, 6:273
Cuprous iodomercurate, 13:240
Cuprous nicotine cyanide, 6:271
Cuprous nicotine thiocyanate, 6:271
Cuprous oxide
  as fungicide, 10:222
  colorant for glass, 9:252
  as molluscicide, 14:545
  pigments, 15:524, 536
  reaction with ethyl ether, 8:479
Cuprous oxychlorohydrate, 6:272
Cuprous pentafluorothiophenoxide, 9:793
Cuprous pentatetrafluorothiophenoxide, 9:792
Cuprous pyrite, 19:350
Cuprous telluride, 19:772
Cuprous 2,3,5,6-tetrafluorothiophenoxide, 9:792
Cuprous thiocyanate
  in explosive priming compositions, 8:653
Cuprous xanthate, 22:423
Curare, 6:**547**
  calabash, 1:773, 787
  pot, 1:787
  tube, 1:770, 787
Curare alkaloids, 1:786
Curare and curare-like drugs, 6:**547**
Curare, tube, 1:770
Curcuma, 9:375
Curcumin, 7:614
Curie point (temperature), 9:4
Curing
  of animal skins and hides, 12:308
Curium, 1:352. (See also Actinides.)
  as fission product, 14:106

Curium compounds
  crystal structure, 1:366
  phys prop, 1:366
Curium-244
  US prod, 21:6
Current Index to Conference Papers, S:512
  durable flame retardants, S:949
Current-potential curves
  for corrosion processes, 6:294
Curry powder, 9:371
Curvularin, 12:648
Cuskohygrine, 1:764
Cusparine, 1:771
Cuticle removers, 6:371
Cyameluric triamide, 6:571 See Melem.
Cyanacethydrazide, 14:534
Cyanacobalamin, 21:542
Cyanamide, 6:554, 560
  manuf, 6:563
  in prepn of dicyandiamide, 6:568
  properties, 6:560
  reaction with diethylenetriamine, 7:29
Cyanamides, 6:**553**
  prepn of amines from, 2:109
Cyan couplers
  for color photography, 5:825
Cyanic acid, 6:574
  ammonia reaction, 21:37
  from cyanuric acid, 20:662
  from urea, 21:39
Cyanide
  analysis for in water, 21:699
Cyanide copper plating solutions, 8:59
Cyanide gas
  in fumigating nuts, 14:139
Cyanide gold-plating solutions, 8:60
Cyanide indium plating solutions, 8:61
Cyanides, 6:**574**
  hypochlorite determination with, 5:46
  organic, S:590
Cyanide silver-plating solution, 8:64
Cyanide zinc electroplating solution, 8:66
Cyaniding, 13:310
Cyanine dyes, 6:**605; 7:533
  for color photography, 5:817
Cyanine dyes
  structural formulas, 5:769
Cyanines, 16:283-291
Cyanoacetamide, 12:151, 850, 856
Cyanoacetic acid, 8:416; 12:850
  phys prop, 1:235
Cyanoacetic acid hydrazide, 14:534

Cyanoacetyl chloride, 12:850
Cyanoacetylcoumarone
  developer in color photography, 5:819
Cyanoanthraquinones, 19:316
Cyanocarbons, 6:**625**
Cyanocobalamin, 8:175.   (See also Enzymes,
  industrial.)
  prod of, 21:546
4-Cyano-2,6-diidophenol, 22:203
4-Cyano-2,2-dimethylbutyraldehyde, 6:640
2-Cyano-1,3-dinitrobenzene, 3:372
3-Cyano-4,6-diphenyl-2-pyridone, 12:151
2-Cyanoethanesulfonic acid
  prepn from acrylonitrile, 1:343
β-Cyanoethylalkylamines, 19:560
Cyanoethylated cotton, 6:659
Cyanoethylated polyacrilonitrile, 6:642
α-Cyanoethylated polyacrylonitrile, 6:642
Cyanoethylation, 1:342; 6:634
  prepn of nitriles, S:593
Cyanoethylcellulose, 4:642; 6:636, 645, 647
  manuf, 6:652
2-Cyanoethylphosphine, 6:651
Cyanoethylsucrose, 6:664
2-Cyanoethyl-1-piperidinecarbodithioate, 6:646
Cyanoform, 6:629
Cyanogen, 6:574; 12:147
Cyanogen bromide, 2:582; 3:777; 6:588
Cyanogen chloride, 2:414; 6:574, 588
  in chemical warfare, 4:872
  formation of, 5:3
  polymerization, 20:667
Cyanogenetic glucoside
  in bitter almonds, 14:133
Cyanogen iodide, 6:588
Cyanoguanidine. See Dicyandiamide.
Cyanohydrins, 6:**668**
  prepn from aldehydes, 1:643
1-Cyano-2-hydroxymethylguanidine, 6:567
Cyanol FF, 7:483
β-Cyanopropionaldehyde, 2:206; 6:639
3-Cyanopropionaldehyde
  prepn from acrylonitrile, 1:342
β-Cyanopropionic acid, 19:141
3-Cyanopyridine, S:599
  hydrolysis of, 21:526
4-Cyanopyridine, 21:528
  amine oxide from, S:43
5-Cyanopyrimidine, 20:178
4-Cyanothiazole, 14:542
2-Cyano-1,2,3,4-tetrahydro-2-methyl-1-oxophen-
  anthrene, 19:138

Cyanuric acid, 20:662-670
  analysis, 20:668
  chemical prop, 20:663
  chlorine reaction with, 4:918
  prepn, 1:147
  from urea, 21:38
Cyanuric chloride, 2:901; 6:574; 9:799; 10:931
  hydrolysis, 20:666
  optical brighteners, 19:15
  in prepn of triallyl cyanurate, 1:923
  reaction with ethylenimine, 11:528
  stilbene dyes, 19:6
  s-triazine synthesis, 22:183
Cyanuric fluoride, 9:799
Cyanuric triazide, 8:592
Cyasterone, 18:849
Cyclamates, 19:598-601, 603
  in carbonated beverages, 4:339
Cyclamic acid, 19:599
Cycle stock, 14:836
Cyclialkylation, 10:142
Cyclic gem-Diperoxides, 14:785
Cyclic ketals
  prepn from ethylene oxide and ketones,
    8:529
Cyclic triperoxides, 14:785
Cyclizine, 16:674
Cycloalkadienes
  nomenclature, 11:288
Cycloalkenes
  nomenclature, 11:288
Cycloartenol, 18:838
Cyclobarbital, 3:61, 70
1,3-Cyclobutanedione ketene dimers, 12:94,
  156
Cyclobutanes
  NMR of, 14:62
Cyclobutanone, 12:106
Cyclobutene
  photoisomerization, 15:332
Cyclobuxine D, 18:842
Cyclocoumarol, 5:600, 601
Cyclodehydrohalogenation
  in producing epoxide, 8:267
Cyclodiene insecticides, 5:240
Cyclodienes
  as insecticides, 11:696
  in soil, 18:515
Cyclodiolefins
  nomenclature, 11:288
1,5,9-Cyclododecatriene, 3:787
Cycloeucalenol, 18:842

9-Cycloheptadecen-1 one, 14:726
Cycloheptaglucanase
  microorganism producing, 8:210
Cycloheptanone, 6:670; 12:106
Cycloheptatriene
  titanium compounds, 20:433
1,3-Cyclohexadiene
Cyclohexaglucanase
  microorganism producing, 8:210
Cyclohexamylose
  dialysis; 16:125
Cyclohexane, 8:504; 6:675
  catalytic reforming, 15:32
  in commercial hexanes, 11:1
  in hydrocracking, 11:431
  inversion and NMR, 14:70
  manuf, 6:679
  oxidation, 15:156; 19:143
  photochemistry, 15:338
  in polymers, 11:4
  properties, 6:676
  reaction, 6:675
  structure of, 5:268
  toxicity of, 11:295
Cyclohexanecarboxamide
  form, 2:67
Cyclohexanecarboxylic acid
  from benzoic acid, 20:561
Cyclohexanedimethanol
  polyester from, 16:170
1,4-Cyclohexanedimethanol, 8:361; 10:680;
  16:144
1,2-Cyclohexanedione, 12:139, 140, 142, 144
1,4-Cyclohexanedione, 6:628; 19:142
1,3-Cyclohexanediones, 12:156
Cyclohexanehexols, 11:673
Cyclohexanesulfonic acid, 6:678
Cyclohexanethiol
  spectral studies, 20:207
1,2,3-Cyclohexanetriols, 16:192
Cyclohexane-1,3-dione, 12:149
Cyclohexane-1,3,5-triols, 16:203
Cyclohexanol, 6:**683**
  NMR of, 14:57
  oxidation of cyclohexane, 15:156
  from phenol, 15:148
  prepn from cyclohexane, 6:677
  in prepn of fatty alcohol, 8:822
Cyclohexanol and cyclohexanone, 6:**683**
Cyclohexanone, 2:369; 10:663; 6:683; 12:106;
  12:111; 15:148, 338; 6:640, 670
  prepn from cyclohexanes, 6:675

in prepn of adipic acid, *1*:411
sulfuric esters, *19*:487
Cyclohexanone oxime, *15*:148, 338; *6*:677;
9:591, *15*:148, 338
caprolactam from, *16*:29
Cyclohexene
epoxidation; *8*:249
reaction with iodine monofluoride, *9*:590
3-Cyclohexenecarboxaldehyde, *1*:258
Cyclohexene oxide
block copolymers, *19*:553
1,2-Cyclohexene oxide
phys prop, *8*:265
2-Cyclohexenone
photochemical addition, *15*:334
1-(1-Cyclohexenyl)pyrrolidine, *6*:641
5-(1-Cyclohexen-1-yl)-1,5-dimethylbarbituric acid,
3:61, 70
Cycloheximide, *10*:226; *20*:77
Cycloheximide (β-[2-(3,5-Dimethyl-2-oxo-
cyclohexyl)-2-hydroxyethyl]-glutarimide)
as fungicide, *10*:226
Cyclohexone
oxidation of cyclohexane, *15*:156
Cyclohexyl acetate, *12*:108; *12*:111
Cyclohexylacetic acid, *15*:214
Cyclohexyl acrylate
phys prop, *1*:289
Cyclohexylamine, *2*:82, 124, 416, 417; *6*:332,
648; *12*:116
cyclamates, *19*:599
3-Cyclohexylaminopropionitrile, *6*:648
4-Cyclohexylamino-2,3,5,6-tetrafluorobenzene-
sulfonamide, *9*:793
Cyclohexylammonium cyclamate, *19*:599
Cyclohexyl benzene, *7*:198, 200
1-Cyclohexyl-1,3-butanedione, *12*:148
Cyclohexyl butyrate
as synthetic flavor, *9*:354
4-Cyclohexylcyclohexyl acrylate
phys prop, *1*:289
Cyclohexylhydroxylamine, *19*:600
4-(1-Cyclohexyliminoethyl)hexanenitrile, *6*:640
Cyclohexyl isocyanate, *12*:46, 47; *19*:600
Cyclohexyl nitrate, *6*:77
N-Cyclohexyl-*p*-toluenesulfonamide, *15*:764
N-Cyclohexylpiperidine; *12*:116
Cyclohexyl silicate, *18*:219
Cyclohexyl stearate
plasticizer, *20*:484
Cyclohexylsulfamic acid, *10*:11; *19*:248, 306
Cyclohexyl vinyl ether, *21*:417

Cyclones, liquid, *4*:747
disc, *4*:741
Cyclonite, *8*:622; *10*:97; *15*:774
characteristics, *8*:632, 633, 634
prepn from formaldehyde, *1*:644
Cyclononanone, *12*:130
Cyclooctadiene, *12*:824
reduction, *16*:37
1,5-Cyclooctadiene, *7*:686
Cycloocta-1,5-diene, *3*:787; *15*:861
Cyclooctanone oxime, *16*:37
Cyclooctatetraene
prepn from acetylene, *1*:174
Cyclooctene, *8*:248
Cycloolefins
nomenclature, *11*:288
Cycloparaffins
nomenclature; *11*:288
oxidation, *19*:143
in petroleum, *14*:848
Cyclopentadecanone, *12*:106
Cyclopentadiene, *6*:627, **688**
chlorinated derivatives of, *5*:240
from coal tar, *19*:670
hexachlorocyclopentadiene from, *5*:243
polymers of, *7*:77, 414, 415, 417
reaction with isoprene, *12*:66
sodium hypochlorite reaction with, *5*:243
Cyclopentadiene and dicyclopentadiene, *6*:**688**
Cyclopentadiene resins, *11*:250
Cyclopentadienyl azobenzene, *6*:693
Cyclopentadienyl Grignard, *10*:727
Cyclopentadienylphenyltitanium, *20*:443
Cyclopentadienylsodium, *20*:444
Cyclopentadienyltitanium(III) dichloride,
20:442
Cyclopentadienyltitanium(IV) alkyls, *20*:440
Cyclopentadienyltitanium ethoxydichloride,
20:435
Cyclopentadienyltitanium phenyl, *20*:443
Cyclopentadienyltitanium trichloride, *20*:440
Cyclopentadienyltrichlorosilane, *6*:699
Cyclopentadienyltriphenyltitanium, *20*:443
Cyclopentane
in commercial hexanes, *11*:1
fluorination, *9*:698
Cyclopentanecarboxylic acid
phys prop, *1*:236
1,2-Cyclopentanedione, *12*:140
1,3Cyclopentanediones, *12*:156
Cyclopentanone, *6*:670, 693; *12*:106, 111, 129
prepn from adipic acid, *1*:407
from succinic acid, *19*:142

1;2-Cyclopentene oxide
  phys prop, 8:265
Cyclopenthiazide, 7:263, 264
Cyclopentyl magnesium bromide, 10:730
Cyclophosphamide
  alkylating agent, S:83
Cyclopropanecarboxylic acid
  phys prop, 1:236
Cyclopropene acids
  in cottonseed oil, 6:414
Cyclopropyl methyl ether
  phys prop, 8:471
Cycloselenopropane, 17:825
Cycloserine, 3:30
  amebicide, 20:80
Cyclosilicates, 18:47
Cyclotetramethylenetetramine, 8:623, 624
  characteristics, 8:632
Cyclotetramethylenetetranitramine
Cyclothiazide, 7:264
Cyclotol, 8:629
  characteristics, 8:632
Cyclotrimethylenetrinitramine, 13:787. (See
  also Cyclonite.)
  in propellants, 8:660
Cylinder oil, 14:836
Cylindrite, 20:276
Cymarin
  as cardiovascular agent, 4:513
Cymene
  cresols from, S:273

p-Cymene, 4:57; 10:143; 19:815
  p-cresol from, S:271
Cypress, 14:200
Cyproheptadine.HCl
  antihistamine, 11:37
Cyproquinate
  anticoccidial, 21:248
Cyrtolite, 10:756
Cystathionine, 2:190
Cysteic acid
  sulfur in, 19:369
  from wool, 22:401
Cysteine
  as antioxidant, 2:601
  occurrence, 20:205
L-Cysteine, 10:5
Cystine, 1:767; 2:157
  occurrence, 20:205
  penicillin synthesis, 14:668
  in wool, 22:399
  yellowing of wool, 22:406
Cytarabine, 21:455
  cancer therapy, S:86
Cythioate
  ectoparasite control, 21:250
Cytoflav, 17:455
Cytokinins, 15:680
Cytosine arabinoside, 21:455
Cytosine arabinoside, S:86
Czochralski technique, S:152

D

DANC
  as decontaminating agent, *4*:886; 918; 920
D-acid, *13*:722
Dacron, *10*:645
Dahlia, *7*:464
Dahl's acid II, *13*:715
Dahl's acid III, *13*:715
  chlorinated aliphatic acid herbicide, *22*:199
Dairy industry
  use of hydroxyacetic acid in, *10*:634
Dairy products. See *Milk and milk products.*
Dakin's solution, *2*:616
Dalapon
Dalton's law, *20*:135
Dambose, *11*:673
Dam-Karrer color test
  for Vitamin $K_{1(20)}$oxide, *21*:590
Dammar, *17*:384
  in nail enamels, *6*:369
Dammar gum
  in matches, *13*:165
Dammarolic acid, *17*:385
Damylanhydrolide, *12*:643
Daniell cell
  zinc in, *22*:559
Danish agar, *17*:782
Dapsone, *19*:261; 267; 270
Darbo cold cup test for antifreezes, *2*:555
Darzen's reaction
  in vitamin A synthesis, *21*:496
Darzens synthesis, *10*:905
Dasanit, *15*:914
Data, interpretation and correlation, *6*:**705**
Davidite, *21*:8, 159
Dawsonite, *18*:9; 459; 461
D&C Black No. 1, *5*:867
D&C Blue No 1, *5*:867
D&C Blue No. 2, *5*:868
D&C Blue No. 3. See *Solvent Violet 13.*
D&C Blue No 3, *5*:871
D&C Blue No. 4, *5*:868
D&C Blue No. 6, *5*:868
D&C Blue No. 7, *5*:868
D&C Blue No. 9, *5*:868, 873
D&C Brown No. 1. See *Resorcin Brown.*
D&C Brown No. 1, *5*:868
  used in hair coloring, *10*:793

D&C colors, *5*:857; 858; 864; 867
D&C Green No. 1, *5*:868
D&C Green No. 5, *5*:868, 873
D&C Green No. 6, *5*:868
D&C Green No. 8, *5*:868, 873
DCO. See *Dehydrated castor oil.*
D&C Orange No. 10, *5*:869
D&C Orange No 11, *5*:869, See *Erythrosine Yellowish.*
D&C Orange No. 17, *5*:869
D&C Orange No. 4, *5*:869. See *Orange II.*
  used in hair coloring, *10*:793
D&C Orange No 5, *5*:869
D&C Red No. 2, *5*:869
D&C Red No 3, *5*:869
D&C Red No. 4, *5*:869
D&C Red No. 5, *5*:869
D&C Red No. 6, *5*:869
D&C Red No. 7, *5*:869
D&C Red No. 8, *5*:869
D&C Red No. 9, *5*:869
D&C Red No. 10, *5*:870
D&C Red No. 11, *5*:870
D&C Red No. 12, *5*:870
D&C Red No. 13, *5*:870; *10*:793. ((See also *Lithol Reds.*)
  used in hair coloring, *10*:793
D&C Red No. 27, *5*:870
D&C Red No. 17, *5*:870, 873
D&C Red No. 19, *5*:870, 873. See *Rhodamine B.*
D&C Red No. 21, *5*:870, 873
D&C Red No. 22, *5*:870; 873
  used in hair coloring, *10*:793
D&C Red No. 28, *5*:870
D&C Red No. 30, *5*:870, 874
D&C Red No. 31, *5*:870
D&C Red No. 33, *5*:871
D&C Red No. 34, *5*:871
D&C Red No. 36, *5*:871
D&C Red No. 37, *5*:871
D&C Red No. 39, *5*:871
D&C Violet No. 2, *5*:871. See *Solvent Violet 13.*
D&C Yellow No. 5, *5*:871
D&C Yellow No. 6, *5*:871
D&C Yellow No. 7, *5*:871

adsorption, *18*:522
from chlorobenzene, *5*:256
degradation, *18*:524
corn earworm control, *11*:679
residue in food, *10*:18
in the soil, *18*:523
in water, *18*:523
in water, *21*:701
D&C Yellow No. 8, *5*:871
DDNP. See *Diazodinitrophenol.*
D&C Yellow No. 10, *5*:871
D&C Yellow No. 11, *5*:871, 874
DDNP, *8*:584, 587
DDT, *11*:677, 691
DDT analogs, *11*:691
Deacon-Herter process, *S*:170
Deacon process
for chlorine prod, *11*:335
Dead-stop" titration techniques, *2*:675
Dealkylation, *1*:891
Deammonation, *2*:333
Debaromyces
in alkane utilization, *S*:839
Debye
definition of, *S*:1005
Debye equations, *20*:741
Debye-Hückel theory, *7*:804, 823
DEC. See *2-Chloro-N,N-diethylethylamine.*
Decachlorooctahydro-1,3,4-metheno-2*H*-cyclo-
buta(cd) pentalen-4-one, *5*:251. See *Kepone.*
Decachlorotetracyclodecanone, *5*:251
Decaffeinated coffee, *3*:913; *5*:761
Decafluoroazoxybenzene, *9*:794
Decafluorobiphenyl, *9*:775; 776; 791; 797
Decafluorobutane
as aerosol propellent, *1*:474
Decafluorocyclohexanes, *9*:698
Decafluorodiazoaminobenzene, *9*:794
Decafluorodiphenyl disulfide, *9*:793
Decafluoroethylbenzene, *9*:776
Decafluroro-*o*-xylene, *9*:776
Decahydronaphthalene, *9*:698; *13*:672, 676
from naphthalene, *13*:670
stain solvent, *18*:661
toxicity of, *11*:295
Decahydronaphthalenes, *13*:676
Decalin. See *Decahydronaphthalene.*
Decamethonium, *6*:548; 549
Decamethonium bromide, *2*:409; *6*:550
Decamethylenebis(trimethylammonium bro-
mide). See *Decamethonium bromide.*
Decamethylene dichloride, *16*:256

Decanal
flavor and perfume materials, *9*:358
n-Decane, *12*:160
Decanedioic acid. See *Sebacid acid.*
1,10-Decanediol diacrylate
phys prop, *1*:289
2,5-Decanedione, *12*:158
Decanenitrile, *18*:661
Decanoic acid. See *Capric acid.*
1-Decanol, *1*:567; *9*:722
composition of commercial mixture, *1*:553
phys prop, *1*:543; 564
specifications, *1*:554
2-Decanone, *12*:105
2-Decarboxamido-2-acetyloxytetracycline, *20*:7
2-Decarboxamido-2-acetyltetracycline, *20*:7
4-Decenoic acid, *8*:814
Dechachlorotetracyclodecanone. See *Kepone.*
Decontamination
of chemical warfare agents, *4*:886
Decoquinate
anticoccidial, *21*:248
Decyl acrylate
phys prop, *1*:287
n-Decyl alcohol. See *1-Decanol.*
as synthetic flavor, *9*:354
Decyl alcohols, *1*:567
Decyl aldehyde
as synthetic flavor, *9*:354
Decylamine
prop, *2*:130
Decylene oxide, *8*:247
Decyl ether, *8*:493
Decyl octyl adipate
prod and sales of, *8*:377
Decyl octyl phthalate
of metals, *13*:316
prod and sales of, *8*:377
Deep drawing
Deep Maroon, *5*:871
Deflocculation, *15*:255
Defoliant
sodium acid cyanamide, *6*:559
Deformation
in alloys, *11*:11
Degras, *22*:412
Degree Celsius, *S*:998
Degree Fahrenheit, *S*:999
Degrees of freedom, *15*:133
Degussa process
manuf hydrogen cyanide, *6*:579
7-Dehdrocholesterol, *18*:837

Dehumidification, *1*:493
Dehydrated castor oil
  drying oil, *7*:398; 404; 415
  film prop, *7*:423
Dehydration
  of meats, *13*:178
  to preserve foods, *10*:35
Dehydroabietic acid
Dehydroacetic acid, *1*:157; *10*:923; *12*:94
  17:494; *19*:545
6-Dehydro-17-acetoxy-19-norprogesterone,
  18:853
2,4-Dehydroadamantane, *S*:11
  irradiation of, *21*:553
Dehydroallohyodeoxycholic acid, *3*:482; 483
Dehydroascorbic acid, *10*:789. See *Ascorbic acid.*
Dehydrobufotenine, *1*:759
Dehydrochlorination of hydrocarbons, *5*:91
7-Dehydrocholesterol
  irradiation of, *21*:549
  manuf, *21*:560
  synthesis of Vitamin D $_3$, *21*:564
Dehydrocholic acid, *3*:482; 483; 485
Dehydroconiferyl alcohol, *16*:685
Dehydrodeoxycholic acid, *3*:482; 483
Dehydrodivanillin, *21*:190
  from vanillin, *21*:182
Dehydroemetine, *20*:70
2-Dehydroemetine, *20*:76
2-Dehydroemetine bismuth iodide, *20*:76
2-Dehydroemetine dihydrochloride, *20*:76
2-Dehydroemetine resinate, *20*:76
Dehydroepiandrosterone, *11*:93; *18*:870
  in synthesis of testosterone, *11*:98
2,3-Dehydro-*N*-[3-(ethylthio)butyl]ematine,
  20:77
Dehydrogenation
  of olefins, *14*:331
  olefins from, *14*:326
2,3-Dehydro-*N*-(3-hydroxybutyl)emetine, *20*:77
Dehydrohyodeoxycholic acid, *3*:482; 483
2-Dehydroisoemetine, *20*:77
2-Dehydro-*O*-methylpsychotrine, *20*:77
Dehydronerolidol, *19*:833
9,10-Dehydro-9-oxoanthracene, *19*:164
3-Dehydroretinol, *21*:490
7-Dehydro-*β*-sitosterol
7-Dehydrostigmasterol
  irradiation of, *21*:553
Dehydrothio-*p*-toluidine, *20*:185
  dye synthesis, *20*:187
  prepn of, *20*:186

Dehydrothio-*p*-toluidine, *2*:898; *19*:300; 433
Dehydrothio-*p*-toluidine monosulfonic acid,
  20:187
Dehydrothio-*p*-toluidinesulfonic acid, *19*:4
Dehydroundecylenic acid, *8*:816
3-Dehydrovitamin A, *21*:490
4,4'-Dehydroxydiphenyl, *7*:193
Deicing fluids, *2*:558
Deionization of water, *11*:875, 887
  use of ethylenamines in *7*:36
Delalutin, *6*:78
  in bearing materials, *3*:271
Delrin. See *Acetal resins.*
Delta acid, *13*:714
Demerol hydrochloride, *S*:13
Demethanation
  of olefins, *14*:331
  of olefins, *14*:333
Demethylchlortetracycline, *3*:17, *20*:5,7
  amebicide, *20*:82
6-Demethyl-6-deoxytetracycline, *20*:10,14,16,
  23, 27
6-Demethyl-4-oxotetracycline, *20*:12
6-Demethyltetracycline, *20*:7, 12, 15
Demulsification, *8*:151
Denaturants
  for ethyl alcohol, *8*:450
Denatured ethyl alcohol, *8*:451
Denier, *20*:35
De Nora chlorine cells, *1*:692
Density, *6*:755, 787
  conversion factors, *6*:757
Density and specific gravity, *6*:**755**
Dental alloys. See *Dental materials.*
Dental cements. See *Dental materials.*
Dental investments. See *Dental materials.*
Dental materials, *6*:**777**
  alloys, use of germanium in, *10*:521; 690;
    692
  cements, *6*:778
  investments, *6*:788
  stones, *6*:787; 786
Dental stones. See *Dental materials.*
Dentifrices, *6*:**848**
Deodorants (cosmetic), *6*:354
Deodorization
  of fats, *13*:181
Deoxophylloerythreotioporphyrin, *10*:531
12a-Deoxy-5a,6-anhydrotetracycline, *20*:23
6-Deoxy -L-ascorbic acid, *2*:759
Deoxychloic acid, *3*:481, 483, 484
  in food processing, *10*:8

Deoxycorticosterone
  biosynthesis of, *11*:79
  phys prop, *11*:79
  synthesis of, *11*:86
Deoxycytidilic acid antagonist, *S*:86
Deoxydihydrostreptomycin, *19*:39
25-Deoxy-20-hydroxyecdysone, *18*:849
Deoxyribonucleic acid, *2*:179
  in lithography, *17*:330
  in viral cores, *21*:452
2-Deoxystreptamine, *19*:33; 41
Deoxy sugars, *12*:638
6-Deoxytetracyclines, *20*:15
DEPC. See *3-Chloro-N,N-diethylpropylamine.*
Depilatories, *6*:357
Depo-Provera
  as contraceptive drug, *6*:85
Depolymerase
  uses, *8*:190
Depropanex
  as cardiovascular agent, *4*:523
Depsides
  in tea, *19*:743
Depsipeptides
  antibiotic, *16*:335
Depsipetides
  antibiotic cyclic, *16*:336
N-Desacetylthiocolchicine, *1*:786
Desalination, *22*:1
Desal process, *22*:76
  for desalination, *22*:61
Deserpidine, *1*:803; *4*:519; *16*:649; 666
3-Desethylemetine, *20*:75
  anhydrite, *4*:25
  aquametry by, *2*:681
Desiccants. See *Drying agents.*
Desiccation. See *Drying*
Desmycosin, *12*:647
Desorption
  mass transfer in, *13*:100
D-Desosamine, *12*:639
O-Desosaminyldihydroerythronolide, *12*:642
Desoxybenzoin, *12*:107, 146
Desoxycholic acid
  steroid synthesis, *18*:857
11-Desoxycorticosterone, *18*:851
Desthiobiotin, *3*:522
  in prepn of glutamic acid, *2*:204
Desulfurization, *15*:4
  in hydrorefining, *11*:437
  of petroleum, *S*:358
  of zinc ores, *22*:566

Desulfonation
  hydrolytic, *19*:293
  classes of reactions, *11*:434
Detergency, *6*:**853**
Detergent brighteners, *3*:745; 746
  determination in water, *21*:702
Detergents. (See also *Surfactants.*)
  aliphatic amine oxides, *S*:44
  biodegradable, *19*:288
  in drycleaning, *7*:310; 311
  hydrocarbons in, *11*:280
  as oxidation dye base, *10*:776
  polyvinylpyrrolidone in, *21*:436
  in shampoos, *10*:770
  sodium silicate, *18*:163
Detergents, alkylate, *19*:294.   (See also *Surfactants.*
Determinants
  nomographs, *14*:22
Dethiobiotin, See *Desthiobiotin.*
Detinning, *20*:294, 295
  alkaline electrolytic process, *20*:295
  chlorine process, *20*:295
  economic aspects, *20*:302
  recent developments, *20*:301
  Russian, *20*:302
Detonators, *8*:583-593
  ethyl alcohol for, *8*:447
Deuterium, *6*:895
  electromigration of, *S*:286
  phys prop, *6*:897
  properties, *6*:898
Deuterium and tritium, *6*:**895**
Deuterium-2
  in nuclear reactors, *14*:76
Deuterons
  accelerated, *17*:54
Developers
  for color photography, *5*:818; 826
Developing agents
  Kendall structure, *15*:371
Devulcanization, *17*:581
Dexamethasone
  corticosteroid analog, *11*:90
Dexbrompheniramine maleate
  antihistamine, *11*:38
Dexchlorpheniramine maleate
  antihistamine, *11*:38
Dextran, *3*:601; *19*:226
  as gum, *10*:741
Dextranase
  microorganism producing, *8*:210

Dextrin
    in nuts, *14*:129
    oxalic acid from, *14*:359
Dextrins, *18*:682; *19*:224
Dextrose. See *Glucose.*
Dextrose, *4*:136
    in beet sugar, *19*:21▼
    in confectionery, *6*:38
Dextrose and starch syrups, *6*:**919**
Dextrose caramel, *4*:64
Dextrose equivalent, *4*:67
D Glass
    in ceramic composite armor, *S*:141
β-D-Glucose pentaacetate
    menadione derivatives, *21*:597
D-Homoandrostane-3,17α-dione, *18*:886
dl-D-Homotestosterone, *18*:865
Diabetes
    acetoacetic acid in urine, *1*:154
    acetone in urine, *1*:159
3,5-Diacetamido-2,4,6-triiodobenzoic acid, *17*:132
Diacetatozirconic acid, *22*:630, 636
Diacetin, *10*:630
Diacetone
    stain solvent, *18*:661
Diacetone alcohol, *1*:166; *10*:924; *12*:105, 113, 137
Diacetone-L-sorbose, *2*:752
Diacetone-2-keto-L-gulonic acid, *2*:753
Diacetosuccinic ester, *1*:156
2,3-Diacetoxy-1,3-butadiene, 12:111
1,4-Diacetoxy-2-butyne, *1*:606, 607
6,6-Diacetoxy-1,3-cyclohexadien-5-one, *16*:193
1,4-Diacetoxy-2-methylnaphthalene-3-acetaldehyde, *21*:588
p-Diacetoxybenzene
    polyester from, *16*:168
1,2-Diacetoxyethane
    polyesters from, *16*:168
    pyrolysis of, *22*:482
5,6-(Diacetoxymethyl)1,3-cyclohexadiene
2,4-Diacetoxypiperylene, *12*:111
Diacetyl, *2*:140, 144; *13*:515
    photoinitiator, *16*:518
    as synthetic flavor, *9*:355
s-Diacetylacetone, *10*:924
Diacetylcholine chloride, *2*:409
    prepn from butynediol, *1*:605
1,2-Diacetylhydrazine, *21*:532
    prepn of biacetyl, *12*:144
    nomenclature, *11*:288
Diacetyl ketones

Diacetyl peroxide, *12*:111
    from acetaldehyde, *1*:80
Diacetyltartaric acid, *19*:736
Diacetyltartaric anhydride, *12*:93
    in prepn of L-threitol, *1*:571
Diacid-diamine, *16*:89
Dialkoxy titanium dichloride, *S*:615
O,O'-Dialkyl *O'*-(*p*-nitrophenyl)phosphorothioate, *15*:328
N,N-Dialkylacetamides, *S*:39
Dialkylaluminum hydride
    olefin prod, *S*:653
Dialkylaminoalkyl sulfide, *2*:145
Dialkylaminoalkylthiols
    prepn, *2*:144
p-Dialkylaminobenzylidenes, *16*:286
Dialkylchloroalkylamines, *2*:139, 145-148, 150
Dialkyldimethylammonium halides
    surfactants, *19*:562
N,N-Dialkylhomopiperidinium chloride, *2*:142
Dialkylidenesuccinic acids, *19*:137
N,N-Dialkyl-*p*-phenylenediamines, *15*:222
    oxidation, *15*:220
2,6-Dialkylphenols, *15*:162
N,N-Dialkylpiperidinium chloride, *2*:142
N,N-Dialkylpyrrolidinium chloride, *2*:142
Dialkyl sulfates, *19*:491
Dialkyl sulfites, *19*:488; 493
Dialkyl sulfosuccinates, *19*:519
Dialkyl vinylphosphonates, *11*:530
6,6'-Diallyl-α,α'-bis(diethylamino)-4,4'-bi-- *o*-cresol, *20*:88
Diallylamine, *6*:562
    uses, *2*:124
2-Diallylamino-4,6-diamino-*s*-triazine, *6*:569
5,5-Diallylbarbituric acid, *3*:61
Diallyl carbonate, *4*:390
N,N-Diallyl-2-chloroacetamide
    amide herbicide analysis, *22*:198
Diallylchlorendate
    fire retardant, *S*:483
Diallyl ether. See *Allyl ether.*
Diallyl isocyanurate, *20*:670
Diallyl isophthalate
    manuf, *1*:923
    phys prop, *1*:917
    reactions, *1*:918
    specifications, *1*:924
    uses, *1*:926
Diallylmelamine, *6*:569
N,N-Diallylmelamine, *6*:562
Diallyl phthalate
    in ceramic composite armor, *S*:145

manuf, *1*:923
phys prop, *1*:917
polyester reactive diluent, *20*:822
polyester resins, *15*:461
polyesters, *15*:455
reaction, *1*:918
specifications, *1*:924
uses, *1*:926
Diallyl phthalate polymer
  properties, *1*:926
Diallyl phthalate resins, *12*:190
Diallyl succinate, *19*:145
  copolymers, *19*:146
Diallyl sulfide
  as synthetic flavor, *9*:354
N,N-Diallylurea, *6*:562
Dialuric acid
  from uric acid, *21*:109
Dialysis, *7*:**1**
  PEC hydrogel membranes, *16*:128
Diamine Black BH, *7*:484
Diamine Fast Red F, *2*:877; 898
Diamine Green B, *2*:876
  manuf, *2*:82
  prepn by sulfide reduction, *2*:96
Diamines and higher amines, aliphatic, *7*:**22**
Diamines, aromatic. See *Diarylamines.*
3,5-Diamine-1,2,4-triazole. See *Guanazole.*
Diaminoanisole
  dye from furs, *10*:306
2,4-Diaminoanisole, *15*:219
  in hair coloring, *10*:795
1,4-Diaminoanthraquinone, *2*:458
2,6-Diaminoanthraquinone
  dyes, *19*:434
1,4-Diamino-2,3-anthraquinonedicarbonitrile,
  2:479
1,4-Diamino-2,3-anthraquinonedisulfonic acid, *2*:479
Diaminoanthraquinones. See *Aminoanthraquinones.*
2,2'-Diaminoazobenzene, *15*:220
Diaminobenzenes, *15*:216
2,4-Diaminobenzenesulfonic acid, *15*:217
o,p'-Diaminobiphenyl, *3*:412
p,'-Diaminobiphenyl. See *Benzidine.*
4,4'-Diaminobiphenyl, *9*:797
4,4'-Diamino-3,3'-biphenyldicarboxylic acid,
  3:419
4,4'-Diaminobiphenylene oxide, *3*:418
2,7-Diaminobiphenylene sulfone, *3*:418
4,4'Diamino-3-biphenylsulfonic acid, *3*:417
α,γ-Diaminobutyric acid, *16*:307

in polymyxins, *16*:321; 322
Diaminochlorotriazine, *10*:931
1,3-Diaminocyclitol streptamine, *19*:33
4,4'-Diamino-1,1'-dianthrimide, *2*:463
3,7-Diaminodibenzofuran, *3*:418
3,7-Diaminodibenzothiophene dioxide. See
  Benzidine sulfone.
3,7-Diaminodibenzothiophene-2,8-disulfonic
  acid-5,5-dioxide derivatives as optical bright
  eners, *3*:741
1,4-Diamino-2,3-dichloroanthraquinone, *2*:479
2,2'-Diamino-10,10'-dihydrobianthrone, *2*:451,
  456
3,3'-Diamino-4,4'-dihydroxyarsenobenzene dihydrochloride. See *Arsphenamine.*
3,3'-Diamino-4,4'-dihydroxyarsenobenzne dihydrochloride, *2*:729
4,4'-Diamino-5,5'-dimethyl-2,2'-biphenyldisulfon-
  ic acid, *3*:418
p,p'-Diaminodiphenyl, *7*:535
4,4'-Diaminodiphenyl, *7*:193
p,p'-Diaminodiphenylamine
  in hair coloring, *10*:795
2,4-Diaminodiphenylamine, *10*:306
4,4'-Diaminodiphenylamine, *7*:49; 535; *15*:218
4,4'-Diaminodiphenyl-3,3'-bisglycollic acid,
  2:904
4,4'-Diaminodiphenyl ether, *15*:796
p,p'-Diaminodiphenyl-methane
  antithyroid agent, *20*:270
  in hair coloring, *10*:795
4,4'-Diaminodiphenylmethane, *2*:414
,*o,o*'-Diaminodiphenylsulfide, *2*:414
Diaminodiphenyl sulfone, *8*:311
4,4'-Diaminodiphenyl sulfone, *3*:30, 32;
  19:267
4,4'-Diamino-*sym*-diphenylurea, *7*:535
3,3'-Diaminodipropylamine. See *Iminobispropylamine.*
3,3'-Diaminodipropylamine, *7*:23, 24
Diaminogen B, *7*:485
2,6-Diaminohexanoic acid. See *Lysine.*
4,6-Diamino-*m*-toluenesulfonic acid, *15*:217
Diaminomaleonitrile, *6*:576
1,8-Diaminonaphthalen
  in hair coloring, *10*:795
1,4-Diamino-6-napthalenesulfonic acid,
  13:717
5,8-Diamino-2-napthalenesulfonic acid,
  13:717
Diaminonapthalene sulfonic acids, *13*:717
1,8-Diamino-2,6-octadiene
  from butadiene, *3*:791

7,8-Diaminopelargonic acid, 2:204
2,5-Diaminopentanoic acid. See *Ornithine.*
Diaminophenazine, 15:220
2,4-Diaminophenetole
  in hair coloring, 10:795
2,4-Diaminophenol, 2:221; 15:219
  in hair coloring, 10:795
3,5-Diaminophenol, 16:202
2,4-Diaminophenol dihydrochloride, 15:373
2,5-Diaminophenol-4-sulfonic acid
  in hair coloring, 10:795
4,6-Diamino-1-phenol-2-sulfonic acid, 2:222
2,6-Diamino-1-phenol-4-sulfonic acid, 2:223
1,2-Diaminophenylethane, 14:541
2,4-Diamino-6-phenyl-s-triazine. See *Benzoguanamine.*
2-(Di-p-aminophenyl)methylfuran, 2:414
1,3-Diaminopropane, 7:22, 23
  properties, 7:24
Di(3-aminopropyl)isocyanurate, 20:665
2,6-Diaminopyridine, 19:146
4,6-Diaminopyrogallol hydrochloride, 16:212
2,4-Diamino-s-triazine-6-thiol, 6:568
Diaminostilbene, 16:51
4,4'-Diaminostilbene, 19:15
Diaminostilbenedisulfonic acid, 19:1,6
4,4-Diaminostilbene-2,2-disulfonic acid, 7:478, 535; 19:2,6,15
  derivatives as optical brighteners, 2:897; 3:739, 746
2,6-Diamino-1,3,5,7-tetrachloroanthraquinone, 2:485
2,4-Diaminotoluene-5-sulfonic acid, 15:217
3,5-Diamino-1,2,4-triazole, 2:244
Diaminotriphenylmethane dyes, 20:702
  acid dyes, 20:705
  basic dyes, 20:702
Diammonium imidodisulfonate, 19:244
Diammonium imidosulfonate, 19:600
Diammonium octaborate, 3:647
Diammonium orthophosphate, 15:266; 274
Diammonium phosphate, 2:229; 6:578; 15:290
  fertilizers, S:343
  as flame retardant for textiles, 9:304
  manuf, 2:311
  as water-correcting agent, 10:17
Diammonium tetraborate tetrahydrate, 3:647
  natural, 4:283-294
  synthetic, 4:294-303
Diamond, S:274 (See also *Carbon.*)
  as abrasive, 1:25, 28
  crystal structure of, 4:161, 286
  dielectric prop of, 4:821

  use in grinding wheels, 1:31
  for machining tools, 20:576
  from ocean, 14:159
  synthetic prod, 10:511
Diamond Alizarine Black SN, 2:886
Diamond alkali chlorine cells, 1:684
Diamond Black F, 2:886
Diamond Black PV, 2:873; 885
Diamond Red ECB, 7:527
Diampromid, 2:387, 388
Diamyl adipate
  phys prop, 1:409
Diamylamine
  stain solvent, 18:659
1,1'-Diamyl-4,4'-cyanine iodide, 6:606
Di-tert-amylhydroquinone
Di-n-amyl ketone, 12:104
  as antioxidant, 2:597
Diamyl phthalate
  as cellulosic plasticizer, 4:657
  thermochemical constants as propellant, 8:681
1,4:3,6-Dianhydrosorbitol, 1:578
1,4-Dianilino-5,8-dichloroanthraquinone, 2:472
(4,4'-Dianilinodiphenyl)methane, 7:44
2,2-(4,4'-Dianilinodiphenyl)propane, 7:44
4,4'-Dianilinotriphenylmethyl chloride, 7:44
6,8-Dianilino-1-naphthalenesulfonic acid, 7:49; 13:764
1,3-Dianilino-8-naphthalenesulfonic acid, 13:714
Dianisidine, 7:535
  diazotates derived from, 2:893
  pigments, 15:572
o-Dianisidine, 3:417; 7:565
Dianisidine chlorosulfonate
Dianisidine diisocyanate, 12:46, 47
Dianthraquinonylamines, 2:462
Diaphragm cells
  for chlorine manuf, 1:681
Diaquotetrammine nickel nitrate, 13:759
Diaroylethylenes, 12:822
Diarylamines, 7:**40**
Diaryldithiocarbamates, 7:43
Diarylguanidine
  prepn from guanidine, 10:736
Diarylketonimine, 7:533
Diaspore, 1:931; 2:42, 47; 5:553; 13:18
  in ceramics, 4:763, 773
Diastase
  action and uses, 8:186
Diatomaceous earth, 22:70, 119. See *Diatomite.*

filters in water treatment systems, *22*:70
in matches, *13*:164
use in water filtration systems, *22*:119
Diatomite, *7*:53, 57; *6*:802; *15*:150
  as abrasive, *1*:25
  in ceramics, *4*:769
  in dynamite, *8*:637
  as filter aid, *9*:268
  in pigments, *15*:513
  as thermal insulator, *11*:834
Diatoms. See *Diatomite.*
Diatrizoic acid, *17*:132
3a,9-Diazabenz[*f*]indene, *10*:936
22,25-Diazacholestanol dihydrochloride,
  15:922
1,4-Diazacyclohexane. See *Piperazine.*
4,7-Diazacyclopenta[*jk*]phenanthrene, *10*:933
1,8-Diazacyclopent[*a*]indene, *10*:909
2,9-Diazafluorene, *10*:910
1,3a-Diazaindene, *10*:936
1,3-Diazanaphthalene, *10*:899, 920
1,4-Diazanaphthalene, *10*:899, 920
1,5-Diazanaphthalene, *10*:899
1,2-Diazanaphthalenes, *10*:899; 928
2,3-Diazanaphthalenes, *10*:899; 928
1,10-Diazaphenanthrene, *10*:899
  amebicide, *20*:78
1,4-Diaza(2.2.2)bicyclooctane. See *Triethylenediamine.*
1,4-Diaza[2.2.2]bicyclooctane, *21*:71
Diaza[2.2.2]bicylooctane, *21*:91
Diazinon
  in soil, *18*:528
Diazo Blue Black RS, *7*:535
o-Diazoacetyl-L-serine, *S*:85
Diazoaminobenzene, *2*:415
Diazoamino compounds
  dyes from, *2*:893
Diazodenitrophenol, *8*:587
  characteristics, *8*:584
4-Diazodiphenylamine, *16*:515
Diazo dyes, *2*:877
  development of, *7*:538
Diazomethane, *2*:582; *9*:564; *12*:121, 128;
  13:343; *19*:487, 493
  phenol reaction, *15*:167
  in polyester synthesis, *16*:177
  in prepn of epoxides, *8*:269
  reaction with acetoacetic ester, *1*:156
  reaction with arsenic(III)chloride, *2*:727
  reaction with pentafluorothiophenol, *9*:564,
    793
Diazonium chloride, *12*:144

Diazonium fluoborates, *9*:568; 697
Diazonium fluorosulfate, *9*:677
Diazonium hexafluorotitanates, *9*:684
Diazonium perchlorates, *5*:74
Diazotization, *2*:869
2-Diazo-1-hydroxy-4,6-dinitrobenzene, *8*:584;
  587
2-Diazo-1(2*h*)-naphthalenone, *16*:516
2-Diazo-4,6-dinitrophenol, *13*:891
6-Diazo-5-oxo-L-norleucine, *S*:85
6-Diazo-5-oxonorleucine, *2*:188
6-Diazo-50oxo-L-norleucine, *S*:85
Dibasic ammonium phosphate. See *Dentrif-
  ices.*
Dibasic calcium phosphate. See *Dentrifices.*
Dibasic lead stearate, *7*:280
Dibenamine, *4*:519
4,5'-Dibenzamido-1,1'-dianthrimide, *2*:463
Dibenzanthrone, *2*:494; 496; 499
Dibenzanthrone dyes, *2*:527
Dibenzanthronyl. *7*:469 . See *Benzonyl.*
  antiamebics, *20*:88
Dibenzazepines, *16*:654
Dibenzcycloheptanes, *16*:655
Dibenz[*f,ij*] isoquinoline-2,7(3*H*)-diones
1,4-Dibenzenecarboxylic acid. See *Terephthal-
  ic acid.*
Dibenzenechromium, *5*:475; 480
Dibenzo[*b,h*][1,6] naphthyridines
  antiamebics, *20*:88
Dibenzofuran. *10*:908; *15*:148 See *Diphenyl-
  ene oxide.*
Dibenzopyrenequinone dyes, *2*:529
Dibenzopyrone, *12*:107
8*H*-Dibenzo [*a,g*] quinolizine, *10*:936
Dibenzopyrrole. See *Carbazole.*
Dibenzothiazolyl disulfide
  as accelerator, *17*:510
Dibenzothiopene, *16*:51
3,7-Dibenzothiophenediamine-5,5'-dioxide, *3*:418
1,2-Dibenzoylethane, *12*:161
1,2-Dibenzoyloxyethane, *16*:168
Dibenzoyl peroxide, *8*:244
O,S-Dibenzoylthiamine, *20*:182
1,6-Dibenzoyl-2,5-diketohexane, *19*:138
Di(2-benzoyloxyethyl) terephthalate, *16*:168
Dibenzyl
  from benzyl chloride, *5*:283
  prepn from ethylene oxide, *8*:528
  from toluene, *20*:531
N,N'-Dibenzyladipamide, *1*:409
Dibenzylamine
  from benzonitrile, *S*:595

Dibenzylchlorosilane, *12*:69

Dibenzyldimethylammonium hydroxide
cellulose reaction with, *4*:601

Di(benzyldimethyl(2-phenoxyethyl)ammonium)
4,4'-methylenebis(3-hydroxy-2-naphthoate)
monohydrate, *14*:534
prepn from boron trifluoride, *9*:556
prepn from monofluorophosphoric acid,
9:640

Dibenzyl phosphate, *1*:409

9,13-Di-*cis*-vitamin A, *21*:498

,11,13-Di-*cis*-vitamin A, *21*:498

Diborane, *8*:693; *3*:665, 684. (See also *Boranes.*)
reaction with acetylacetone, *12*:157
reaction with ethylene oxide, *8*:531
reaction with isoprene, *12*:69
properties, *8*:688

Diboron tetrabromide, *3*:683

Diboron tetrachloride, *3*:682

Diboron tetrafluoride, *3*:683

Diboron tetraiodide, *3*:683

2,6-Dibromo *N*-chloro-*p*-benzoquinone imine,
3:775

2,2-Dibromoacetaldehyde diethyl acetal
prepn from ethyl ether, *8*:479

Dibromoacetic acid, *8*:420

1,3-Dibromoadamantane, *S*:6

N,N'-Dibromoadipamide, *1*:409

2,4-Dibromoaniline, *2*:415

2,6-Dibromoaniline, *19*:315

Dibromoanthanthrone, *15*:587

o-Dibromobenzene, *3*:777

p-Dibromobenzene, *3*:777

3,5-Dibromobenzotrifluoride, *9*:778

2,3-Dibromo-1,4-butanediol, *1*:612

1,4-Dibromobutene, *10*:669

3,4-Dibromo-1-butene, *3*:790

1,4-Dibroro-2-butene, *1*:613

Dibromocarbene
steroid synthesis, *18*:886

1,4-Dibromo-2-chloro-2-butene
from chloroprene, *5*:217

1,2-Dibromo-3-chloropropane, *3*:774; *15*:913

3,6-Dibromocoumarin, *6*:427

2,6-Dibromo-2,5-cyclohexadien-1-one, *15*:210

4,4'-Dibromo-1,1'-dianthrimide, *2*:463

Dibromodifluoromethane, *9*:749
fire extinguishing agent, *S*:375

N,N-Dibromo-5,5-dimethylhydantoin, *21*:561

1,3-Dibromo-5,5-dimethylhydantoin, *11*:153

4,4'-Dibromodiphenyl, *3*:777

1,1-Dibromoethane, *3*:777. See *Ethylidene bromide*.

1,2-Dibromoethane
nematocide, *15*:913
in roundworm chemotherapy, *14*:541

(1,2-Dibromoethyl) benzene. See *Styrene bromide*.

Dibromoethylbenzene, *3*:777

Dibromofluorescein, *5*:869

1,4-Dibromo-2,5-fluorobenzene, *9*:784

1,2-Dibromohexafluorocyclobutane, *9*:834

Dibromohydantoin
7-dehydrocholesterol synthesis, *21*:561

α,β-Dibromohydrocinnamic acid. See *Cinnamic acid dibromide*.

α,β-Dibromohydrocinnamic acid, *5*:518

1,2-Dibromo-3-hydroxy-3-methylbutane, *1*:631

Dibromoindigo, *3*:775

5,7-Dibromoisatin, *3*:775

Dibromomethane, *3*:774, 778

1,3-Dibromo-3-methylbutane, *1*:624

5,7-Dibromo-2-methyl-8-quinolinol benzoate
ester, *20*:84

Dibromoneopentylene glycol
flame-retardant polyesters, *20*:831

4,4'-Dibromooctafluorobiphenyl, *9*:791

1,2-Dibromo-1-penten-3-ol,: *1*:620

2,6-Dibromophenol
indophenol, *3*:775

2,3-Dibromo-3-phenyl-1-propanol, *5*:522

1,2-Dibromopropane. See *Propylene bromide*.

1,2-Dibromopropane, *3*:778

1,3-Dibromopropane, *2*:413; *3*:778

2,3-Dibromo-1-propanol
fire retardant, *S*:483

2,3-Dibromopropyl acrylate
phys prop, *1*:289

4,6-Dibromopyrogallol, *16*:192

5,7-Dibromo-8-quinolinol, *20*:83

2,6-Dibromoquinone chlorimide, *6*:681

3,5-Dibromosalicylanilide, *18*:429

4',5-Dibromosalicylanilide, *14*:549

Dibromostyrene
flame-retardant polyester, *20*:831

meso-Dibromosuccinic acid, *19*:136

1,2-Dibromotetrafluorobenzene, *9*:785

1,4-Dibromotetrafluorobenzene, *9*:768; 787; 791

Dibromotetrafluoroethane
fire extinguishing agents, *S*:375

1,3-Dibromotetrofluorobenzene, *9*:786

Dibucaine hydrochloride, *2*:405; 408

Di-2-*tert*-butoxyethyl ester

of dicarboxylic acids, *16*:170
Dibutoxytitanium dichloride, *20*:472
Dibutyl adipate
    as cellulosic plasticizer, *4*:657
    phys prop, *1*:409
Di-*tert*-butyl adipate
    phys prop, *1*:409
Dibutylamine
    uses, *2*:124
2-(Dibutylamino)ethyl acrylate
    phys prop, *1*:288
3-(Dibutylamino)propyl acrylate
    phys prop, *1*:288
γ-Dibutylaminopropyl chloride. See *3-Chloro-N,N-dibutylpropylamine.*
γ-Dibutylaminopropyl chloride, *2*:141
Dibutylaminopropyl *p*-aminobenzoate sulfate.
    See *Butacaine sulfate.*
Dibutylaminopropyl *p*-aminobenzoate sulfate,
    2:405; 407
Dibutyl azelate
    as cellulosic plasticizer, *4*:657
Dibutyl carbonate, *4*:370
Di-*tert*-butyl carbonate, *4*:390
Dibutylchlorogermane, *6*:651
3-(Dibutylchlorogermanyl) propionitrile,
    6:651
4,6-Di-*tert*-butyl-*m*-cresol. See *3-Methyl-4,6-di-tert-butylphenol* .
4,6-Di-*tert*-butyl-*m*-cresol, *1*:910. (See also *3-Methyl-4,6-di-tert-butylphenol.*)
2,6-Di-*tert*-butyl-*p*-cresol. See *4-Methyl-2,6-di-tert-butylphenol ; BHT.*
2,6-Di-*t*-butyl-*p*-cresol, *S*:27
Di(*t*-butyl)-*p*-cresol, *12*:574; *19*:812
4,4'-Di-*tert*-butyldiphenylamine, *7*:42
Dibutylethanolamine
    phys prop, *1*:811
Di-*n*-butyl ether
    plutonium separation, *15*:883
Dibutyl ethyleneborate, *3*:721
Dibutyl fumarate
    prod and sales of, *8*:376
N,N-Di(*n*-butyl)-4-hexyloxy-1-naphthylcar-
    boxamidine hydrochloride, *14*:544
Dibutylisopropanolamine
    phys prop, *1*:811
Dibutyl itaconate, *12*:85
Di-*n*-butyl ketone, *12*:104
Dibutylmagnesium, *12*:711
Dibutyl maleate
    prod and sales of, *8*:377

2,6-Di-*tert*-butyl-4-methylphenol
    from alklyation of *p*-cresol, *1*:895
Dibutylnaphthalenesulfonic acid, *19*:318
Dibutyl oxalate, *10*:924
Di-*t*-butyl peroxide, *14*:792
    plasticizers, *15*:771
    plutonium separation, *15*:883
    in polymerization of ethylene, *8*:501
    polyvinylpyrrolidone catalyst, *21*:434
2,6-Di-*t*-butylphenol, *S*:30
2,6-Di-*tert*-butylphenol, *1*:909; *15*:53
Dibutyl phthalate, *6*:365; 804
Di-*n*-butyl succinate, *19*:146
    as cellulosic plasticizer, *4*:657
    as plasticizer, *15*:756; 762; 764; 722; 774
    plasticizer, *21*:331
    prod and sales of, *8*:376
    as propellants, *8*:659; 660
    thermochemical additive constant, *8*:681
Di-*n*-butyl phthalate, *8*:331; *18*:660
Dibutyl sebacate, *6*:354; *15*:764
    as cellulosic plasticizer, *4*:657
    prod and sales of, *8*:377
    specifications, *8*:373
Di-*n*-butyl sulfate, *19*:488
Dibutyl tartrate, *8*:351
L-Dibutyl tartrate
    in prepn of L-threitol, *1*:571
Di-*n*-butyl β,β'-thiodubutyrate
Dibutyltin-*S,S*'-bis(alkylmercaptoacetates),
    20:323
Dibutyltin diacetate, *6*:806; *20*:322
Dibutyltin dichloride, *20*:312, 320
Dibutyltin dilaurate, *6*:806; *18*:244; *20*:322;
    21:103
    blowing agent, *21*:91
    polyurethan catalyst, *21*:72
Dibutyltin dilaurylmercaptide, *20*:322
Dibutyltin di(2-ethylhexoate), *20*:322
Dibutyltin *S,S*'-(isooctylmercaptoacetate,
    20:215
Dibutyltin lauryl mercaptide
    stabilizer, *21*:390
Dibutyltin maleate, *20*:322
    stabilizer, *21*:390
Dibutyltin oxide, *20*:321
DIC. See *2-Chloro-N,N-diisopropylethylamine.*
Dicalcium ferrite
    in asbestos-cement board, *21*:608
Dicalcium phosphate. See *Calcium phosphate.*
    melt granulation, *S*:341
Dicalcium phosphate dihydrate
    in baking powder, *3*:52

Dicalcium silicate
   in asbestos-cement board, 21:608
   in cement formation, 4:685; 691; 692; 700; 703
Dicamba
   herbicide, 22:200
Dicapryl adipate
   phys prop, 1:409
Di(2-carbmethoxyethyl) isocyanurate, 20:665
Dicarbocyanines, 6:615
Dicarboxylic acids. See Acids, dicarboxylic.
2,3-Dicarboxy-1,4,5,6,7,7-hexachlorobicyclo-[2.2.1]-5-heptene anhydride, 20:831
2,3-Dicarboxy-5,8-endomethylene-5,6,7,8,9,-achloro-1,2,3,4,4a,5,8,8a-octahydro-naphthalene anhydride, S:480
Di(2-carboxyethyl) isocyanurate, 20:665
Dichlobenil
   herbicide, 22:200
Dichlone, 10:223
   as fungicide, 10:223
D_g-2-Dichloroacetamido-1-p-nitrophenyl-1,3-propanediol. See Chloramphenicol.
Dichloramine, 4:915
   formation of, 5:2
Dichloramine B, 4:923
Dichloramine T, 2:616, 617; 4:923, 19:257
2,2-Dichloro-N-cinnamyl-N-methylacetamide-, 20:91
2,2'-Dichloro-N-methyldiethylaminehydro-chloride, S:82
2,2-Dichloro-N-[p-chloro-α-(chlorom-ethyl)phenacyl]acetamide, 20:92
2,2-Dichloro-N-(2-hydroxyethl)-N-[p--(methylsulfonyl)benzyl]a cetamide, 20:90
2,2-Dichloro-N-(2-hydroxyethyl)-N-12p--p-nitrophenoxy)benzyll-3-acetamide, 20:90
2,2-Dichloro-N-(2-hydroxyethyl)-N-(4-pyri-dylmethyl)acetamide, 20:91
2,2-Dichloro-N-(2,4-dichlorobenzyl)-N-(2-- hydroxyethyl)acetamide, 20:90
Dichloroacetaldehyde
   from acetaldehyde, 1:83
N-m-Dichloroacetanilide
   hydrolysis constant of, 4:913
N-o-Dichloroacetanilide
   hydrolysis constant of, 4:913
N-p-Dichloroacetanilide
   hydrolysis constant of, 4:913
Dichloroacetic acid, 8:417
   phys prop, 1:235
1,1-Dichloroacetone, 12:106
1,3-Dichloroacetone, 12:106

Dichloroacetyl chloride
   from pentachloroethane, 5:165
   from trichloroethylene, 5:186
Dichloroacetylene, 5:204
1-(Dichloroacetyl)-4-methylpiperidine, 20:92
β,β-Dichloroacrolein, 21:416
Dichloroactic acid
N-N'-Dichloroadipamide, 1:410
2,4-Dichloro-6-alkylamino-s-triazines, 22:185
2,5-Dichloroaniline
   sulfonation, 19:298
N,N-Dichloroaniline, 2:415
2,5-Dichloroaniline, 2:424; 893
2,6-Dichloroaniline, 19:315
3,4-Dichloroaniline, 2:424; 18:535
2,5-Dichloroaniline, diazotized, 7:539
1,4-Dichloroanthraquinone, 2:479; 480; 483
   prepn 2:485
1,5-Dichloroanthraquinone, 2:485; 488
   melting point curve, 2:443
1,8-Dichloroanthraquinone, 2:481
   melting point curve, 2:443
1,4-Dichloro-6-anthraquinonesulfonic acid, 2:480
1,4-Dichloroanthrone, 2:483
N,N-Dichloroazodicarbonamidine. See Chloroazodin.
N,N-Dichloroazodicarbonamidine, 2:616; 617
2,6-Dichlorobenzal chloride, 5:287; 288
2,6-Dichlorobenzaldehyde, 3:365, 20:700
m-Dichlorobenzene, 5:258
o-Dichlorobenzene, 2:361; 424; 452; 5:258; 7:479
p-Dichlorobenzene, 5:259
p-Dichlorobenzene, 2:424; 7:479
   amine sulfonation, 19:298
   extractive distillation agent, 22:473
   paint remover, 14:487
   phenolsulfonic acids, 15:209
   stain solvent, 18:661
1,2-Dichlorobenzene, 2:483
1,3-Dichlorobenzene, 2:483; 5:258
   in Dowtherm E, 10:851
1,4-Dichlorobenzene, 2:483
5,5'-Dichloro-2,2'-dihydroxy-3,3'-dinitrodiphenyl, 14:549
p-Dichlorobenzene-p-chloroiodobenzene
   binary cooling curves, 13:204
Dichlorobenzenes, 5:258-260
   toxicity of, 5:98; 99; 264
2,2'-Dichlorobenzidine, 3:415
3,3'-Dichlorobenzidine, 3:416; 7:479; 535
   pigments, 15:572

2,4-Dichlorobenzoic acid, *3*:437
2,5-Dichlorobenzoic acid
  synthesis of amiben, *22*:202
Dichlorobenzols. See *Dichlorobenzenes.*
2,6-Dichlorobenzonitrile
  herbicide, *22*:200
  herbicide, *22*:202
2,6-Dichloro-*p*-benzoquinonechloroimide,
  16:912
2,4-Dichlorobenzotrichloride, *5*:287; 288
Dichlorobenzotrifluoride, *9*:778
Dichlorobenzyl chlorides, *5*:287; 288
2,4-Dichlorobenzyltributylphosphonium chloride, *15*:683
4,4'-Dichlorobiphenyl
  benzidine from, *3*:412
Dichlorobisdi-*n*-butylsulfidepalladium(II), *15*:870
1,4-Dichlorobutane, *10*:249; 669
Dichlorobutanes
  from butenes, *3*:840
  chloroprene from, *5*:223
2,3-Dichlorobutane-1,4-diol, *1*:605
1,3-Dichloro-2-butene
  from chloroprene, *5*:217; 222
1,4-Dichloro-2-butene, : *1*:613
  epoxidation, *8*:249
Dichlorobutenes, *3*:789
  chloroprene from, *5*:222; 223
  isochloroprene from, *5*:228
1,4-Dichloro-2-butyne, *1*:605
Dichlorobutyraldehydes
  from butyl alcohols, *3*:825
α,β-Dichlorobutyric acid, *6*:466
Dichlorocarbene, *12*:68
N,N-Dichloro-*p*-carboxybenzenesulfonomide. See
  Halazone.
3,5-Dichlorocatechol, *18*:532
Dichloro-*o*-cresyl carbonate, *6*:429; 430
2,4-Dichloro-6-*o*-chloroaniline-*s*-triazine-
  as fungicide, *10*:226
Dichlorocrotyl alcohol, *1*:605
Dichlorodiamminepalladium, *15*:868
3,3'-Dichloro-4,4'-diaminodiphenylmethane, *12*:62
  curing agent, *21*:79
4,4'-Dichlorodibutyl ether, *10*:249
β,β',-Dichlorodiethyl sulfide, *8*:502
  prepn *8*:502
N,N,-Dichloro-1,8-diformamide-*p*-methane,
  4:924
1,4-Dichloro-2,5-difluorobenzene, *9*:784
1,2-Dichloro-1,1-difluoroethane, *9*:841
1,1-Dichloro-2,2-difluoroethylene
  reaction with isoprene, *12*:68

1,1-Dichloro-2,2-difluoro-3-isopropenylcyclo-
  butane, *12*:68
1,1-Dichloro-2,2-difluoro-3-methyl-3-vinylcy-
  clobutane, *12*:68
Dichlorodifluoromethane, *10*:783
  as aerosol propellent, *1*:472
  from carbon tetrachloride, *5*:131
  fluorochemical surfactants, *S*:373
  in industry, *9*:704
  mixture with ethylene oxide as sterilizing
    agent, *8*:552
  propellant, *9*:747; 871
  refrigerant, *17*:307
1,3-Dichloro-5,5-dimethylhydantoin,   *2*:617;
  4:886, 912, 913, 918, 920; *11*154; *22*:129
1,3-Dichloro-5,5-dimethylhydantoin,   *4*:886;
  912; 913; 918; 920
  as bleaching agent, *3*:557
  in disinfection of swimming pools, *4*:926
1,4-Dichloro-2,3-dihydroxybutane, *8*:249
Dichlorodimethylsilane, *18*:233; 236
4,4'-Dichlorodiphenyl sulfone, *15*:174; *16*:272;
  19:495
Dichlorodiphenyltrichloroethane. See *DDT.*
Dichloroethane
  hydrogen chloride from, *S*:175
1,2-Dichloroethane. See *Ethylene dichloride.*
us-Dichloroethane. See *1,1-Dichloroethane.*
s-Dichloroethane. See *1,2-Dichloroethane,*
1,1-Dichloroethane, *5*:148-149
  photochemical processes, *15*:350
  photochlorination, *15*:337
  prepn from acetaldehyde, *1*:83
  toxicity of, *5*:93
  vinyl chloride from, *5*:176
1,2-Dichloroethane, *5*:149-154; *8*:487; *10*:164
  chlorine reaction with, *5*:200
  by-product of ethylene oxide prod, *8*:531
  ethylene reactions with, *5*:145; 175
  from ethylene, *5*:3
  in manuf of vinyl chloride, *5*:175
  polyester solvent, *16*:169
  poly(vinyl chloride)from, *21*:369
  tetrachloroethylene from, *5*:200
  toxicity of, *5*:93; 153
  1,1,2-trichloroethane from, *5*:158
  trichloroethylene from, *5*:190
2,2-Dichloroethane sulfonyl chloride
N,N-Dichloroethylamine, *4*:923
p-Di(2-chloroethyl)aminophenylalanine, *S*:84
0,0-Di(2-chloroethy)0-(3-choro-4-methylcou-
  marin-7-yl)phosphate, *14*:538

1,1-Dichloroethylene. (See also *Vinylidene chloride.*

s-Dichloroethylene. See *1,2-Dichloroethylene.*

as-Dichloroethylene. See *1,1-Dichloroethylene.*
   stain solvent, *18*:661

1,1-Dichloroethylene, *5*:178-180
   hydrogen chloride reaction with, *5*:155
   in manuf of 1,1,1-trichloroethane, *5*:155
   toxicity of, *5*:93; 180

1,2-Dichloroethylene
   trichloroethylene from, *5*:180-183, 190
   as catalyst, *8*:407
   chlorine reaction with, *5*:162
   monochloroacetylene from, *5*:204
   1,1,2,2-tetrachloroethane from, *5*:162
   toxicity of, *5*:93

Dichloroethyl ether
   paint remover, *14*:487

s-Dichloroethyl ether, *10*:930

sym-Dichloroethyl ether, *2*:361

Di(2-chloroethyl)ether, *10*:648

β,β'-Dichloroethyl ether
   solvent refining, *15*:57

2,2-Dichloroethyl ether
   stain solvent, *18*:660

Dichloroethyl sulfide, *10*:648

Dichloroethyne. See *Dichloroacetylene.*

3',6'-Dichlorofluoran, *22*:434
   violamine synthesis, *22*:434

Dichlorofluorescein
   as indicator, *11*:558

1,1-Dichloro-1-fluoroethane
   from 1,1,1,-trichloroethane, *5*:155

1,2-Dichloro-1-fluoroethane
   from 1,1,2-trichloroethane, *5*:158

Dichlorofluoromethane, *9*:757; :871

1,2-Dichlorohexafluorocyclopentene, *9*:846

2,3-Dichlorohexafluoro-2-butene
   prepn of trifluoroacetic acid, *9*:771

2,2-Dichlorohydracrylic acid
   from tetrachloroethylene, *5*:199

Dichlorohydrins, *5*:314

2,4-Dichloro-1-hydroxyanthraquinone, *2*:479

D$_g$0*threo*-2-Dichloroacetamido-1-*p*-nitrophenyl--1,3-propanediol. See *Chloramphenicol.*

D$_g$-*threo*-2,2-Dichloro-*N*-[*β*-hydroxy-- α-(hydroxymethyl)-*p* -nitrophenethyl]. See *Chloramphenicol.*

Dg-threo-2,2-Dichloro = N-[B-hydroxy-α-(hydroxymethyl *p*-nitrophenethyl] acetamide. See *Chloroamphenicol.*

3,5-Dichloro-4-hydroxybenzenesulfonic acid, *15*:210

2,5-Dichloro-3,6-dihydroxy-*p*-benzoquinone, *1-6*:912

2,2-Dichloro-4'-hydroxy-*N*-methylacetanilide, *20*:92

2,3-Dichloro-4-hydroxyphenoxyacetic acid, *22*:182

2,5-Dichloro-4-hydroxyphenoxyacetic acid, *22*:182

3,5-Dichloro-4-hydroxy phenoxyacetic acid, *22*:183

2,6-Dichloro-8-hydroxypurine, *21*:109

Dichloroisocyanuric acid, *4*:908; 913; 919; *20*:664
   as bleaching agent, *3*:558

Dichloroisodibenzanthrone, *15*:586

1,3-Dichloroisopropyl acrylate
   phys prop, *1*:289

2,2'-Dichloroisopropyl ether, *8*:281

Dichloroisoviolanthrone, *15*:586

4,5-Dichloro-*m*-benzenedisulfonamide, *19*:259

Dichloro-*m*-xylenol, *6*:355

Dichloromaleic anhydride, *6*:694

Dichloromethane. See *Methylene chloride.*

3,6-Dichloro-2-methoxybenzoic acid, *22*:200

3',4'-Dichloro-2-methylacrylanilide, *22*:198

N,N-Dichloromethylamine, *4*:924

2-(Dichloromethyl) benzimidazole, *20*:196

4,4-Dichloro-2-methylbutane
   from vinyl chloride, *5*:173

5,7-Dichloro-2-methyl-8-quinolinol, *20*:83, *21*:246

Di(chloromethyl) tetramethyldisilazine, *18*:262

5-Dichloromethyl-1,3-dioxane, *10*:86

1,1'-Dichloromethyl ether
   phys prop, *8*:471

Dichloromonofluoromethane
   as aerosol propellant, *1*:472

2,3-Dichloro-1,4-naphthoquinone

1,2-Dichloro-4-nitrobenzene, *2*:353

2,5-Dichloro-1-nitrobenzene, *13*:844

3,4-Dichloro-1-nitrobenzene, *13*:844

4-Dichloro-2-nitrobenzene, *2*:353

2,5-Dichloro-3-nitrobenzoic acid
   synthesis of amiben, *22*:202

1,4-Dichloro-2-nitrophenol, *2*:216

4,6-Dichloro-2-nitrophenol, *2*:216

2',5-Dichloro-4'-nitrosalicylanilide, *14*:543

5,2'-Dichloro-4'-nitrosalicylanilide, *15*:911, 921

1,3-Dichloro-1,1,2,3,3-pentafluoro-2-propanol, *9*:760

3,3'-Dichlorooctafluorobiphenyl, *9*:786

1,5-Dichloropentane, *10*:247

Dichlorophene, 2:631; 14:543

2,4-Dichlorophenol, 18:532
   choroacetic acid reaction, 22:180
   phenoxyalkanoic acid metabolism in plants, 22:182

Dichlorophenols, 5:332

2,6-Dichlorophenol-4-sulfonic acid, 16:193

2,4-Dichlorophenoxyacetic acid, 2:204; 5:325, 332; 8:417; 15:678; 19:672
   as an herbicide, 22:179
   phys prop, 1:239
   reaction with alkanolamines, 1:821
   synthesis of, 22:180

2,4-Dichlorophenoxyacetylaspartic acid, 22:182

2,4-Dichlorophenoxyethyl sulfate, 22:183

2,5-Dichlorophenyl isocyanate, 12:46

3,4-Dichlorophenyl isocyanate, 12:46, 47; 22:188

1,3-Di(p-chlorophenyl)urea, 22:190

3,3-Dichlorophthalide, 15:448

1,2-Dichloropropane, 8:281
   allyl chloride from, 5:206
   nematocide, 15:913

1,3-Dichloropropane, 8:488

1,3-Dichloro-2-propanol, 2:352. See Glycerol-dichlorohydrin.

Dichloropropanols, 5:314

1,1-Dichloro-2-propanone, 12:106

1,3-Dichloro-2-propanone(1,3-Dichloroacetone), 12:106

1,3-Dichloropropene
   nematocide, 15:913

2,3-Dichloropropionaldehyde
   phys prop, 1:259
   prepn from acrolein, 1:259

3',4'-Dichloropropionanilide
   analysis of, 22:198

α,α-Dichloropropionic acid, 16:196

2,2-Dichloropropionic acid
   as an herbicide, 22:199

2,3-Dichloropropionitrile
   prepn from acrylonitrile, 1:342

2,3-Dichloropropyl acrylate
   phys prop, 1:289

3,3-Dichloro-n-propyl alcohol
   from vinyl chloride, 5:173

N,N-Dichloropropylamine, 4:924

2,6-Dichloropyridine, 9:798

N,N'-Dichloroquinonediimine, 4:924

Dichloroquinoxaline, 7:632

3,6-Dichlorosalicylic acid, 22:201

Dichlorosilane, 18:182

Dichlorostyrene
   flame-retardant polyesters, 20:832

N-Dichlorosulfamic acid, 4:917

p-(N,N-Dichlorosulfamoyl)benzoic acid, 4:923
   See Halazone.

p-(Dichlorosulfamoyl)benzoic acid, 19:260

3-(Dichlorosulfamoyl)phthalic acid, 4:923

sym-Dichlorotetrafluoroacetone, 9:755

sym-Dichlorotetrafluoroacetonealcoholates, 9:758

1,2-Dichlorotetrafluorobenzene, 9:791

1,3-Dichlorotetrafluorobenzene, 9:786, 791

Dichlorotetrafluoroethane, 9:871

1,2-Dichloro-1,1,2,2-tetrafluoroethane
   as aerosol propellent, 1:473

Dichlorotetrafluoromethane
   refrigerant, 17:308

1,2-Dichlorotetrafluoromethane
   propellant, 9:747

1,1-Dichlorotetrafluoro-2-propanol, 9:763

2,3-Dichlorotetrahydrofuran
   prepn, 10:249

1,3-Dichlorotetrahydroquinazoline-2,4-dione, 4:924

Dichloro-2,4,5,7-tetraiodofluorescein, 22:433

3,4-Dichlorothiophene 1,1-dioxide, 20:221

α,α-Dichlorotoluene. See Benzal chloride.

2,5-Dichlorotoluene
   chlorination and oxidation, 22:200

2,6-Dichlorotoluene, 7:489
   synthesis of dichlobenil, 22:202

N,N-Dichloro-p-toluenesulfonamide. See Dichloramine T.

2,4-Dichlorotriazines
   s-triazine analysis, 22:186

Dichlorotriazinyl, 7:632

2,2'-Dichlorotriethylamine, 1:813
   in chemical warfare, 4:873
   as fungicide, 10:223

2,6-Dichloro-4-nitroaniline
   as fungicide, 10:226

1,4-Dichloro-2-methyl-2-butene, 12:67

3,4-Dichloro-2-methyl-1-butene, 12:67

4,4'-Dichloro-3-trifluoromethyl carbanilide, 18:429

Dichlorphenamide, 19:259
   as diuretic, 7:260

Dichobenil
   herbicide, 22:202

Dichroism of glass, 5:849

Dichromate
   in bleaching fats, 13:180
   colloid systems, photosensitive, 5:512

Dicinnamylidenesuccinic anhydride, *19*:138
N,N'-Dicinnamylidene-1,6-hexanediamine
  cross-linking agent, *21*:272
Dickite, *5*:545
Dicobalt octacarbonyl
Dicoumarol, *15*:919
Dicroton, *6*:445
Dicrotonaldehyde, *6*:445
Dicryl
  amide herbicide, *22*:198
Di-6-chrysenylamine, *7*:40, 46
Dictamine, *1*:771
Dicumarol, *15*:919 See *Bishydroxycoumarin.*
  antivitamin K, *21*:489
Dicumenechromium, *5*:490
Dicumyl peroxide
  curing agent, *18*:229
  in polystyrene foams, *19*:120
  vulcanizer, *17*:569
Dicyanamide
  prepn of guanidine nitrate, *8*:625
Di-n-decyl ketone, *12*:104
Dicyandiamide, *2*:227; *6*:557, 560, 561; *8*:310
  chemical prop, *6*:565
  manuf, *6*:568
  in manuf of guanidine, *10*:738
  melamine from, *20*:667
  in papermaking, *14*:520
  phys prop, *6*:565
  reaction with diethylenetriamine, *7*:29
Dicyandiamide-phosphoric acid
Dicyanoacetylene, *6*:625, 629
1,4-Dicyanobutene, *1*:606, 613
1,2-Dicyanocyclobutane
  prepn from acrylonitrile, *1*:341
Dicyanodiamide. See *Dicyandiamide.*
  in manuf melamine, *2*:366
Dicyanoethyl ether of poly(ethylene glycol),
  6:644
Dicyanoethyl thioether
  prepn from acrylonitrile, *1*:343
2-Dicyanomethylene-1,1,3,3-tetracyanopropane,
  6:630
1,4-Dicyano-2-butene, *1*:613
1,1-Dicyano-2,2-bis(trifluoromethyl)ethylene,
  9:763
Di(2-cyanoethyl) isocyanurate, *20*:665
N,N'-Dicyclocarbodiimide, *14*:664, 666
Dicyclohexyl, *7*:198, 200
Dicyclohexyl adipate
  phys prop, *1*:409
Dicyclohexylamine, *2*:125, 416
Dicyclohexylammonium nitrate, *6*:343

Dicyclohexylcarbodiimide, *19*:322
Dicyclohexyl phthalate, *15*:751, 764
  prod and sales of, *8*:376
Dicyclohexylurea, *19*:323
Dicycloparaffins
  in petroleum, *14*:848
Dicyclopentadiene, *7*:417, 686; *12*:825. (See
  also *Cyclopentadiene.)*
  in oxo process, *14*:380
  toxicity of, *11*:297
Dicyclopentadienedicarboxylic acid, *6*:702
Dicyclopentadiene formate, *6*:703
Dicyclopentadienylbismuth chloride, *3*:546
Dicyclopentadienyltitanium, *S*:616
Dicyclopentadienyltitanium dichloride, *S*:615
Di-sec-butyl adipate
  phys prop, *1*:409
Di-sec-butyl ether, *3*:839
Di-sec-butyl ketone, *12*:104
N,N'-Di-sec-butyl-p-phenylenediamine, *15*:53,
  218
Didecenyl ketone, *8*:371
Di-n-decyl adipate
  phys prop, *1*:409
5,13-Didehydro-8,16-flavanthrinedione, *2*:522
Didodecylamine
  properties, *2*:130
Didodecyl carbonate, *4*:390
Didymium
  as glass colorant, *5*:850
Die casting
  use of abherents in, *1*:3
  zinc-base alloys, *22*:599
Dieldrin, *5*:246; *8*:249; *18*:525
  residue in food, *10*:18
  in water, *21*:701
Dieldrin, *5*:246
  moth-proofing wool, *22*:406
Dielectric absorption, *7*:719
  of electric insulation, *11*:776
Dielectric conductance, *7*:719, 721
Dielectric constant, *7*:721
  of electric insulation, *11*:776
Dielectric drying, *7*:375
Dielectric strength, *7*:723
  of electric insulation, *11*:777
Diels-Alder condensations, *3*:785
Diels-Alder reactions
  of chloroprene, *5*:217
Diene addition reaction
  of cyclopentadiene, *6*:692
Diene polymers, *7*:**64**
Diene rubber, *12*:550

1,3-Dienes
  addition to TCNE, 6:627
Diepiiodohydrin
Diepoxybutane, 10:613
1,2,3,4-Diepoxybutane, D and L
  in prepn of threitol, D and L, 1:571
  phys prop, 8:265
Dies
  graphite, 4:231
  hydrogenolysis results, 10:410
  prop for gas manuf, 10:389
  wire, diamond used for, 4:292
Diesel fuel. See Diesel oil.
Diesel oil, 10:484; 15:2
  synthetic, 4:446
Diet foods, 10:62-68
Diethanolamides
  regular fatty, 19:552
Diethanolamine, 1:810; 10:455
  bis(hydrogen sulfate) of, 19:486
  in carbon dioxide removal from ammonia
    synthesis, 2:279
  condensates, 19:550
  hydrogen sulfide, 19:382
  hydrogen sulfide recovery during manuf,
    15:47
  morpholine synthesis, 13:663
  phys prop, 1:811
  uses, 1:823
Diethanolnitramine dinitrate, 8:681
Diethazine, 16:671
Diethenoid acids, 8:811
Diethoxyborine
  prepn, 8:531
1,1-Diethoxy-2-chloroethane, 1:107. See Diethyl chloroacetal.
Diethoxydimethoxysilane, 18:220
1,1-Diethoxyethane. See Acetal.
Di(2-ethoxyethyl) adipate
  phys prop, 1:409
Diethoxymethane. See Ethylal.
Diethoxymethyl acetate, 6:611
Diethoxymethylene
  from methylenechloride, 5:114
2,5-Diethoxytetrahydrofuran, 19:142
N-N-Diethyl- (See also d-lysergamide. See Ly-
  sergic acid diethylamide.
O,O-Diethyl O-(3-chloro-4-methylcoumarin-
  7-yl) phosphorothionate, 14:538
Diethyl acetal
  from acetaldehyde, 1:82
Diethylacetaldehyde, 3:875. See α-Ethylbutyral-
  dehyde.

Diethylacetylacetone, 12:148
1-Diethylacetyl-5,5-pentamethylenebiuret
  as hypnotic, 11:517
Diethyl adipate, 1:408, 409; 12:130, 144
  prepn, 8:350
Diethylallylacetamide
  as hypnotic, 11:516
Diethylaluminum chloride, 7:75, 686
Diethylaluminum chloride, 6:699
Diethylamine, 2:406, 585; 6:643; 12:69
  from acetaldehyde, 1:81
  as catalyst in prepn of ethylene cyano-
    hydrin, 8:529
  manuf from ethyl alcohol, 8:455
  manuf by hydroammonolysis, 2:347
  prepn of diethylaminoethanol, 8:528
  in prepn of difluoroacetic acid, 9:770; 9:834
  Rhodamine B synthesis, 22:434
  synthesis of nikethamide, 21:534
  uses, 2:123
2-Diethylamine-2',6'-acetoxylidide hydrochloride, 2:405, 406
p-Diethylaminobenzaldehyde, 3:366
  Acid Violet dyes from, 20:715
Diethylaminoethanol, 8:528
2-Diethylaminoethanol, 2:406
2-(Diethylamino)ethyl acrylate
Diethylaminoethylamine, 14:546
2-Diethylaminoethyl benzilate. See Benactyzine.
β-Diethylaminoethyl benzilate
  prepn, 2:153
β-Diethylaminoethyl chloride. See 2-Chloro-
  N,N-diethylethylamine.
β-Diethylaminoethyl chloride, 16:671
2-Diethylaminoethyl chloride, 2:406
2-Diethylaminoethyl p-aminobenzoate. See
  Procaine.
  as cardiovascular agent, 4:516
2-Diethylaminoethyl p-aminobenzoate hydrochloride. See Procaine hydrochloride.
2-Diethylaminoethyl 4-amino-3-butoxybenzo-
  ate hydrochloride. See Benoxinate hydrochloride.
2-Diethylaminoethyl p-nitrobenzoate hydrochloride, 2:406
β-Diethylaminoisopropyl chloride, 2:141. See
  2-Chloro-N,N-diethylpropylamine.
β-Diethylaminoisopropyl chloride
m-Diethylaminophenol, 2:214
  rhodamine B synthesis, 22:434
  Sulfo Rhodamine B synthesis, 22:434
3-Diethylaminopropionitrile, 6:636

3-(Diethylamino)propyl acrylate
  phys prop, *1*:288
Diethylaminopropylamine, *7*:23
β-Diethylaminopropyl chloride, *2*:141. See *1-Chloro-2-(N,N-diethyl)propylamine.*
γ-Diethylaminopropyl chloride. See *3-Chloro-N,N-diethylpropylamine.*
5-Diethylamino-2-aminopentane, *2*:149, 153
3-Diethylamino-1-propylamine, *8*:310
Diethylamino-2-propanol, *8*:280
2-Diethylamino-2',6'-acetoxylidide hydrochloride. See *Lidocaine hydrochloride.*
1-Diethylamino-3-pentanone
  steroid synthesis, *18*:858
Diethylaniline, *6*:621
N,N-Diethylaniline, *2*:412, 420; *7*:479
5,5-Diethylbarbituric acid, *3*:60, 61, 67, 70; *7*:254
Diethylbenzene
  in etching, *16*:523
  in ethylbenzene, *22*:500
Diethylbenzenes, *19*:74
N,N-Diethylbenzenesulfonamide, *21*:534
Diethylbismuth chloride, *3*:544
Diethyl bis(2-cyanoethyl)malonate, *6*:641
Diethylbromoacetamide
  as hypnotic, *11*:516
Diethylbromoacetylurea
  as hypnotic, *11*:517
Diethyl butylmalonate, *12*:857
Diethylcadmium, *3*:906
  as catalyst, *3*:902
Diethylcarbamazine, *14*:535
Diethylcarbamazine citrate, *14*:535
Diethylcarbamyl chloride, *14*:535
1-(*N,N*-Diethylcarbamylmethyl)2,4-dinitropyrrole-, *6*:85
1-Diethylcarbamyl-4-methylpiperazine, *14*:535
Diethylcarbanilide, *8*:696
N,N'-Diethylcarbanilide, *8*:696
Diethyl carbonate, *4*:390, 391
  reaction with ethyl ether, *8*:479
  reaction with ethylenediamine, *7*:27
Diethyl chloroacetal
  phys prop, *1*:107 *14*:538
Diethyl chlorophosphate, *9*:307
Diethylcyclohexane
  properties, *8*:688
O,O-Diethyl *O*-(2,4-dichlorophenyl)phosphorothionate, *15*:914
Diethyldifluoromalonate, *9*:605
Diethyldimethyllead, *10*:476

2,2'-Diethyldiphenylamine, *7*:44, 46
N,N-Diethyl-D-lysergamide. See *Lysergic acid diethylamide.*)
Diethylene diamine, *11*:526; *15*:638. See *Piperazine.*
Diethylene glycol, *8*:479, 553; *10*:646-648; *15*:793
  bisallyl carbonate, *4*:390
  by-product in prepn of ethylene glycol, *8*:527
  as dehydrator for natural gas, *10*:455, 636
  in esterification of fatty acids, *8*:819
  morpholine synthesis, *13*:662
  polyester from, *16*:180
  polyesters, *21*:62
  surfactants, *19*:548
Diethylene glycol bis(allyl carbonate), *1*:918, 923, 924
Diethylene glycol diacrylate
  phys prop, *1*:289
Diethylene glycol dibutyl ether, *10*:647; *12*:292
Diethylene glycol dimethyl ether, *S*:615
Diethylene glycol dinitrate, *8*:582
  characteristics, *8*:632, 633, 634
  manuf (Biazzi process), *8*:595
  in propellants, *8*:660
Diethylene glycol monobutyl ether, *10*:647
Diethylene glycol monobutyl ether acetate, *10*:647
Diethylene glycol monolaurate (soap-free)
  HLB value, *8*:129
Diethylene glycol monomethyl ether, *10*:647; *13*:5
Diethylene glycol monomethyl ether acetate, *10*:647
Diethylene glycol monostearate
  as abherent, *1*:8
Diethylene glycol ricinoleate, *18*:660
Diethylenetriamine, *8*:310
Diethylenetriamine, *7*:22, 23, 24, 27, 29, 31, 32, 33; *8*:310
  piperazine from, *15*:640
  polyurethan catalyst, *21*:72
N,N,N',N,N-Diethylenetriaminepentaacetic acid
  as chelating agent, *6*:8
Diethyleneurea, *2*:228
Diethylethanolamine
  in prepn of novaldiamine, *1*:159
Diethyl ether, *9*:794; *10*:164. (See also *Ether.*)
  ethyl chloride from, *5*:145
  hydrochloric acid reaction with, *5*:145

plutonium separation, *15*:883
Diethyl ethoxymagnesium malonate, *4*:387
Diethyl ethylmalonate, *9*:605
Diethyl formal, *1*:107. See *Ethylal.*
Diethyl fumarate
  styrene copolymer, *20*:807
Diethylglyoxal, *12*:140
Di(2-ethylhexyl) adipate
  as cellulosic plasticizer, *4*:657
  phys prop, *1*:409
Di(2-ethylhexyl) azelate
  as cellulosic plasticizer, *4*:657
Di(2-ethylhexyl) carbonate, *4*:390
  phys prop, *1*:811
Di-2-ethylhexyl fumarate
  vinyl acetate copolymer, *21*:331
Di-2-ethylhexyl maleate
  vinyl acetate copolymer, *21*:331
Di(2-ethylhexyl) phosphate, *19*:529
Di(2-ethylhexyl)phosphoric acid
  uranium extraction, *21*:17
Di(2-ethylhexyl) phthalate
  in diffusion pumps, *21*:135
Di(2-ethylhexyl) sebacate
  as cellulosic plasticizer, *4*:657
O,O-Diethyl N-hydroxynaphthalimidephosphate, *14*:538
Diethylindandione, *10*:150
Diethyl ketone, *12*:104, 644
Diethyllead, *12*:285
Diethylmagnesium, *12*:711
Diethyl maleate, *15*:322
Diethyl malonate, *6*:641; *8*:371, 378, 477; *9*:605; *12*:850
  benzyl chloride reaction with, *5*:283
  cyanoethylation, *6*:638
  prod and sales of, *8*:376
  reaction with acetaldehyde, *1*:84
  reaction with ethylene oxide, *8*:530
Diethylmalonyl chloride, *10*:150
Diethyl mesoxalate, *10*:932
N,N-Diethylmetanilic acid, *2*:214
O,O-Diethyl *O*-(*p*-methylsulfinylphenyl) phosphorothionate, *15*:914
N,N-Diethylnicotinamide, *21*:533
N,N-Diethyl-*p*-phenylenediamine, *15*:218
  developer in color photography, *5*:819
3,3'-Diethyloxacyanine iodide, *6*:611
Diethyl oxalacetate, *10*:911
Diethyl oxalate, *6*:465
  prepn, *8*:351
  prepn of ethyl ether, *8*:479

3,3'-Diethyloxathiacarbocyanine iodide, *16*:287
3,3'-Diethyloxathiacyanine iodide, *6*:610
Diethyl peroxide
  polymerization of vinyl fluoride, *9*:836
O,O'-Diethylphosphorodithioc acid, *6*:643
O,O-Diethyl phosphorothioate, *18*:528
Diethyl phthalate, *6*:804; *7*:543; *15*:751; *19*:324
  as denaturant, *8*:450
  plasticizer, *4*:657; *15*:774
  prepn, *8*:351
  prod and sales of, *8*:376
  thermochemical additive constant, *8*:681
O,O-Diethyl *O*-(2-pyrazinyl)phosphorothionate, *15*:914
Diethyl pyrocarbonate
  as food preservative, *10*:14
  sterilant, *18*:817
4,6-Diethylresorcinol, *2*:629
3,3'-Diethylselenathiacarbocyanine iodide, *16*:288
Diethyl silicone fluids, *18*:250
Diethylstilbestrol, *10*:19; *15*:921
  veterinary uses, *21*:251
Diethyl succinate, *19*:137, 139, 141
Diethyl sulfate, *8*:431, 479, 481; *12*:288; *19*:302, 487
  as alkylating reagent, *1*:898
  manuf, *19*:492
  in manuf of ethyl chloride, *5*:144
Diethyl sulfide, *9*:570
Diethyl sulfite, *19*:495
3,3'-Diethylthiacarbocyanine iodide, *16*:283
3,3'-Diethylthiadicarbocyanine iodide, *6*:615
3,3'-Diethylthiatetra(penta)carbocyanine iodide, *6*:617
3,3'-Diethylthiatricarbocyanine iodide, *6*:616
3,3'-Diethylthiazolinocarbocyanine iodide, *6*:609
  structural formula, *5*:765
1',3-Diethylthia-2'-cyanine iodide, *6*:610
Diethyl thiosulfite, *8*:427
Diethylzinc, *9*:760
3,3'-Diethyl-0,11-neopentylenethiadicarboxyanine iodide, *6*:616
Diethyl[1-acetoxy-2,2,2-trifluoro-1-(trifluoromethyl)ethyl]-malonate, *9*:762
Diethyl (1-hydroxyethyl)malonate
  from acetaldehyde, *1*:84
5,5-Diethyl-1-methylbarbituric acid, *3*:61, 70
N,N-Diethyl-1-piperazine carboxamide, *14*:535

1,1'-Diethyl-11-(4-quinolyl ethrobromide)-
4,4'-dicarbocyanine bromide, 6:618
N,N-Diethyl-1,3-propanediamine. See Diethylaminopropylamine.
N,N-Diethyl-1,3-propanediamine, 20:84
2,2-Diethyl-1,3-propanediol diacrylate
phys prop, 1:289
Diethyl 2-formyl succinate, 12:828; 19:139
Diethyl (2-hydroxyethyl)malonate
prepn, 8:530
1,1-Diethyl-2,2'-carbocyanine iodide, 16:287
1,1'-Diethyl-2,2'-carbocyanine iodine, 6:606,
609, 611
1,1'-Diethyl-2,2'-cyanine iodide, 6:610; 16:287
1,1'-Diethyl-2,4'-cyanine iodide, 6:610
3,3-Diethyl-2,4-pentanedione, 12:148
Diethyl-2,5-diketocyclohexane-1,4-dicarboxylate,
19:139
1,1'-Diethyl-3,8:3',10-diethylenebenzimidazolocar
bocyanine iodide, 6:612
N,N-Diethyl-4-methyl-1-piperazinecarboxamide,
14:535
1,1'-Diethyl-4,4'-carbocyanine iodide, 6:609
1,1'-Diethyl-4,4'-cyanine iodide, 6:610
1,1'-Diethyl-4,4'-tricarbocyanine iodide, 6:616
3,3'-Diethyl-4,5,4',5'-dibenzothiacarbocyanine
bromide structural formula, 5:765
1,3-Diethyl-5-[bis(3-ethyl-2-benzothiazolinyli-
denemethyl)methylene] barbituric acid,
6:619
3,3-Diethyl-5-methyl-2,4-piperidinedione
as hypnotic, 11:514
3,3'-Diethyl-8-azathiacarbocyanine iodide,
6:620
3,3'-Diethyl-8-phenylthiacyanine iodide, 6:611
3,3'-Diethyl-8,10-diphenoxyoxacarbocyanine
iodide, 6:612
3,3'-Diethyl-8,10-ethylenethiacarbocyanine iodide, 6:612 6:619
3,3'-Diethyl-8,10-o-phenylenethiacarbocyanine
iodide, 6:613
3,3'Diethyl-9-methyloxathiacarbocyanine iodide, 6:614
3,3'-Diethyl-9-methylthiacarbocyanine iodide,
6:612
3,3'-Diethyl-9-(1-ethyl-2-ethylthio-3-indolyl)thi-
acarbocyamine iodide, 6:619
3,3'Diethyl-9,11-dimethylthiadicarbocyanine
iodide, 6:615
3,3'-Diethyl-9,13-dimethyloxatricarbocyanine
iodide, 6:617
3,3'-Diethyl-9-(3-ethyl-2- benzothiazolinyli-
denemethyl)thiacarbocyanine iodide, 6:619

Di-α-furylmethane, 10:246
Dietz method for determination of chlorates,
5:46
Differential thermal analysis, 2:695, 4:790
aquametry by, 2:695
Diffusion, 7:**79**
double diffusion process in untreated wood,
22:379
"eddy diffusivity," 20:651
Fick's laws, 20:650
gaseous, 14:86
molecular, 20:650
steady-state from sphere, 13:104
"thermal," 20:650
unimolal unidirectional, 13:103, 143
unsteady, 13:105, 109
Diffusion coefficient
of strong electrolytes, 7:830
Diffusion coefficients
equations for, 6:744
in gases, 6:749
Diffusion separation methods, 7:**91**
Diffusivity, 13:104, 110, 111, 113. (See also
Diffusion.)
Difluoboric acid, 3:615
Difluoroacetic acid, 9:770
2,5-Difluoroaniline
in prepn of 1,2,4-trifluorobenzene, 9:784
2,4-Difluoroaniline, 9:784
3,5-Difluoroaniline, 9:785
1,4-Difluorobenzene, 7:84
m-Difluorobenzene, 9:783
o-Difluorobenzene, 9:783
p-Difluorobenzene, 7:84
1,2-Difluorobenzene, 9:783
1,3-Difluorobenzene, 9:783
1,4-2,5-Difluorobenzene, 9:784
2,5-Difluorobenzotrifluoride, 9:779
2,2'-Difluorobiphenyl, 9:797
3,3'-Difluorobiphenyl, 9:797
4,4'-Difluorobiphenyl, 9:797
α,B-Difluorobutyric acid, 6:466
Difluorodiazine, cis- and trans-, 9:630
vinylidene fluoride elastomer curing agent,
21:270
1,1-Difluoroethane, 9:835
as aerosol propellant, 1:473
1,2-Difluoroethane, 9:632
2,2-Difluoroethanol, 9:751
1,1-Difluoroethylene. See Vinylidene fluoride.
Difluoromethane, 9:509
Difluoromethylene
steroid synthesis, 18:880

2,4-Difluoronitrobenzene
  prepn, 9:783
2,5-Difluoronitrobenzene
  prepn, 9:784
3,4-Difluoronitrobenzene, 9:783
Difluorophosphoric acid, 9:637, 640, 646
Difluorophosphoric acid esters, 9:645
2,3-Difluoropyridine, 9:798
2,6-Difluoropyridine, 9:798
2,4-Difluoroquinoline, 9:800
α,α-Difluorosuccinic acid, 19:137
2,5-Difluoro-1,4-benzoquinone, 9:786
4,4'-Difluoro-2-nitrobiphenyl, 9:797
1,1-Difluoro-2,2-dichloroethyl methyl ether,
  2:401
2,2'-Difluoro-3,3',5,5'-tetranitrobiphenyl, 9:797
3,3'-Difluoro-2,2',4,4'-tetranitrobiphenyl, 9:797
4,4'-Difluoro-2,3'-dinitrobiphenyl, 9:797
1,3-Diformamidoadamantane, S:8
Diformatozirconic acid, 22:647
Di-2-furfuroyl, 10:241; 12:140
Di-2-furylglyoxal, 12:140
Digalen
  as cardiovascular agent, 4:515
Digermane, 10:523
  use of enzymes, 8:195, 196, 203
Digestion
  of sludge from sewage treatment, 22:111
Digilanid
  as cardiovascular agent, 4:515
Digital computers, 6:25, 27
  use of ferrites in, 8:898
Digitalis
  as cardiovascular agent, 4:510, 513, 515
Digitonin, 18:838
  and provitamins D, 21:558
Digitoxigenin, 18:845, 866
Digitoxin, 18:845
  as cardiovascular agent, 4:513, 515
  veterinary uses, 21:252
Diglycerides, 10:619, 630
  in margarine, 13:60
Diglycerol, 8:494; 10:629
Diglyceryl tetranitrate, 8:604
  characteristics, 8:632, 633, 634
  explosive characteristics, 8:633
  stability characteristics, 8:634
Diglycine
  manuf, 2:350
Diglycine manganese chloride dihydrate
  as ferroelectric, 9:5
Diglycolic acid, 2:592; 10:636
  prepn, 8:416

Diglycolide, 10:633
Diglycollic acid. See Diglycolic acid.
Diglyme, 18:178
"Diglyme", S:615
Digoxin
  as cardiovascular agent, 4:513, 516
Diguanylbiuret
  prepn from guanidine, 10:737
1,1-Dihalo-2-methylpropene, 3:841
Di-n-heptadecyl ketone, 12:104
Di-n-heptyl ketone, 12:104
N,N'-Dihexadecylsuccinamide, 9:899
Dihexamethylenetriamine
  condensation, 16:11
Dihexyl adipate
  as cellulosic plasticizer, 4:657
Di-n-hexyl adipate
  phys prop, 1:409
Di-n-hexyl azelate, 15:751, 763
Di-n-hexyl ketone, 12:104
p,p'-Di-n-hexyloxyazoxybenzene
  liquid crystal NMR, 14:64
  prod and sales of, 8:376
Dihexyl phthalate
Dihydro-2-amino-1,2"-dianthraquinonylamine.
  See 2-Amino-0,10-dihydro-1,2"-imidod i anthraquinone.
Dihydro-2-amino-1,2'-dianthraquinonylamine,
  2:454
6,15-Dihydro-5,9,14,18-anthrazinetetrone, 2:453.
  See Indanthrone. dodianthraquinone.
1,3-Dihydro-2H-1,4-benzodiazepin-2-one, 10:934
3,4-Dihydro(2H)-1,2,4-benzothiadiazine-1,1-
  dioxide, 7:263
2,3-Dihydrobenzofuran, 10:908
3,5-Dihydrobenzo[y]quinolizine, 10:936
Dihydrochalcones, 19:601
dl-Dihydrocinerolone, 12:162
Dihydrocivitone, 18:24
Dihydrocumarin
  as synthetic flavor, 9:354
Dihydrodesoxymorphine, 2:381, 382
3,4-Dihydro-2,5-dimethyl-2-formyl-2H-pyran,
  1:257
3,4-Dihydro-2,5-dimethyl-2H-pyran-2-methanol,
  1:257
9,10-Dihydro-0,10-dioxoanthracene, 2:431
9,10-Dihydro-9,10-dioxoanthracene, 2:431.
  See Anthraquinone.
9,10-Dihydro-9,10-dioxo-1-anthracenesulfonic
  acid, 2:440, 441, 443. See 1-Anthraquinone-
  sulfonic acid.
  manuf, 2:444

Dihydrodithiins
  prepn, 6:631
Dihydroendotricyclopentadiene, 6:692
22,23-Dihydroergosterol
  irradiation of, 21:553
Dihydroergotoxine, 1:778, 789, 791
Dihydroerythromycin, 12:642
Dihydroerythronolide, 12:642
3,4-Dihydro-2-ethoxy-2H-pyran, 1:256, 257
2,5-Dihydrofuran, 1:613
3,4-Dihydro-6-methyl-3-p-totylquinazoline
  methiodide, 9:795
  in manuf of Telodrin, 5:250
2,3-Dihydroindole, 16:294
Dihydrojasmone, 12:161; 14:737
Dihydromenadione, 21:592, 595
5,8-Dihydro-2-methyl-1,4-naphthohydroqui-
  none, 21:594
5,8-Dihydro-2-methyl-1,4-naphthoquinone,
  21:594
Dihydronaphthalenes, 13:676
3,4-Dihydro-1(2H)-naphthalenone, 13:676
Dihydronicotinamide
2,3-Dihydro-3-oxobenzisosulfonazole, 19:595
Di(2-hydroxyethyl) terephthalate, 16:170, 174
  nicotinic acid biochemistry, 21:519
4,6-Dihydro-o-tolualdehyde, 6:453
1,1-Dihydroperfluorobutyl acrylate
  phys prop, 1:289
1,1-Dihydroperfluorodecyl acrylate
  phys prop, 1:289
1,1-Dihydroperfluorohexyl acrylate
  phys prop, 1:289
1,1-Dihydroperfluorooctyl acrylate
  phys prop, 1:289
1,1-Dihydroperfluoropentyl acrylate
  phys prop, 1:289
1,1-Dihydroperfluoropropyl acrylate
  phys prop, 1:289
gem-Dihydroperoxides, 14:783
α,α'-Dihydroperoxyperoxides, 14:785
Dihydrophenanthridine, 10:933
5,10-Dihydrophenazine, 2:859
Dihydropropane, 2:165, 166
Dihydro-2,8-purinedione, 21:108
2,4-Dihydropyran, 10:247
2,3-Dihydro-4H-pyran, 10:921
Dihydropyrans, 10:921
  from methyl vinyl ether, 21:416
Dihydroquinacridone, 2:446
6,13-Dihydroquino[2,3-b]acridine, 2:446
Dihydrorescinnamine, 1:802
Dihydroresorcinol, 16:199

Dihydrostreptomycin, 19:38
  in bacterial infections, 3:1
  as veterinary drug, 21:244
Dihydrostreptomycin sulfate, 10:107, 112
2,2-Dihydrosuccinic acid, 19:142
Dihydroterphenyl, 7:196
Dihydro-α-terpineol, 19:819
Dihydrotestosterone, 18:869
Dihydro-2,5-thiophenedione, 19:141
2,5-Dihydrothiophene 1,1-dioxide, 19:250
1,2-Dihydro-2,2,4-trimethyl-6-ethoxyquinoline,
  2:600; 17:579
1,2-Dihydro-2,2,4-trimethylquinoline, 2:415
Dihydrovitamin K₁(₂₀), 21:587
  treatment with phosphorous oxychloride,
  21:591
Dihydrovitamin K₁(₂₀) diphosphate, 21:591
Dihydroxy G acid, 13:721
Dihydroxy R acid, 13:722
Dihydroxy S acid, 13:721
Dihydroxyacetone, 10:620
2,4-Dihydroxyacetophenone, 16:200
3,4-Dihydroxyacetophenone, 16:200
Dihydroxyaluminum sodium carbonate
  as antacid, 2:430
2,4-Dihydroxy-6-amino-s-triazine, 22:187
1,4-Dihydroxyanthraquinone, 2:357, 458,
  15:447
  manuf, 2:474
1,5-Dihydroxyanthraquinone
  manuf, 2:474
Dihydroxyanthraquinones, 2:465
2,4-Dihydroxybenzaldehyde, 16:200
3,4-Dihydroxybenzaldehyde, 16:200; 21:185
m-Dihydroxybenzene. See Resorcinol.
o-Dihydroxybenzene, 11:462
p-Dihydroxybenzene, 11:483
1,3-Dihydroxybenzene, 11:472
1,4-Dihydroxybenzene, 11:483
3,4-Dihydroxybenzenesulfonic acid, 15:208,
  211
2,4-Dihydroxybenzoic acid
  Chromogen Red B synthesis, 22:435
  gelling agent, 21:355
  prepn from butynediol, 1:603
3,5-Dihydroxybenzoic acid, 16:202
p,p'-Dihydroxybenzophenone, 20:675
2,4-Dihydroxybenzophenone
  polypropylene light stabilizer, S:813
2,5-Dihydroxy-1,4-benzoquinone, 16:211
1,4-Dihydroxybutane, 2:415
1,4-Dihydroxy-2-butanone
  in prepn of 1,2,4-butanetriol, 1:595

α,β-Dihydroxybutyric acid, DL-*erythro*- and DL-*threo*-, 6:465

3,4-Dihydroxy-α-chloroacetophenone, 8:234

3α,6α-Dihydroxy-5β-cholanic acid, 3:483

3α,7α-Dihydroxy-5β-cholanic acid, 3:483

3α,12α-Dihydroxy-5β-cholanic acid, 3:483

6,7-Dihydroxycoumarin, 6:432

3,3'-Dihydroxydianthrone, 2:471

Dihydroxydibenzanthrone, 7:479

2,2'-Dihydroxydiethyl sulfide, 18:662

4,6-Dihydroxy-2,5-dimethylbenzaldehyde, 15:160

1,4-Dihydroxy-2,3-dimethoxybenzene, 16:210

4,4'-Dihydroxydiphenylalkanes, 16:109

p,p'-Dihydroxydiphenylamine, 2:40

4,4'-Dihydroxydiphenylamine, 7:46

Dihydroxydiphenylmethane
   phenolic resins, 15:183, 189

5β,20-Dihydroxyecdysone, 18:849

4,5-Dihydroxyfluorescein, 16:198

2,4-Dihydroxyimidazole. See *Hydantoin.*

2,4-Dihydroxy-7-methoxy-1,4-benzoxazin-3-one, 22:187

2,2-Dihydroxy-1,3-indanedione, 19:324

3,4-Dihydroxy-α-methylcinnamic acid, 19:44

3,4-Dihydroxy-α-(methylaminomethyl)benzyl alcohol.
   See *Epinephrine.*
   as cardiovascular agent, 4:517

5,7-Dihydroxy-4-methylcoumarin, 16:209

6,7-Dihydroxy-4-methylcoumarin, 16:200

Dihydroxymaleic acid, 19:736

1,5-Dihydroxynaphthalene, 2:873

1,6-Dihydroxynaphthalene
   steroid synthesis, 18:857

5,7-Dihydroxynapthalene-1,3-disulfonic acid, 13:723

Dihydroxynapthalenemonosulfonic acids, 13:721

   4,6-Dihydroxy-2-napthalenesulfonic acid, 13:721

   6,7-Dihydroxy-2-napthalenesulfonic acid, 13:722

   4,5-Dihydroxy-1-napthalenesulfonic acid, 13:721

   4,5-Dihydroxy-2,7-naphthalenedisulfonic acid, 22:181

   3,12-Dihydroxy-7-oxocholanic acid, 3:484

2,2'-Dihydroxy-3,3',5,5',6-pentachlorobenzanilide, 14:549

3,4-Dihydroxyphenethylamine, 8:233

1,3-Dihydroxyphenol, 15:185

L-Dihydroxyphenylalanine
   vanillin in synthesis of, 21:192

3,4-Dihydroxyphenylalanine, 8:233; 10.305

2,6-Dihydroxypurine, 10:913

2,4-Dihydroxyquinoline, 9:800
   pigments, 15:572

9,10-Dihydroxystearic acid
   phys prop, 1:235

Dihydroxytartaric acid, 19:736

3,4-Dihydroxytetrahydrofuran
   prepn from butenediol, 1:612

1,2-Dihydroxy-3,4,5-trimethoxybenzene, 16:212

1,2-Dihydroxy-3,4,6-trimethoxybenzene, 16:212

16,17-Dihydroxyviolanthrone, 7:479

3α,12α-Dihydroxy-7-oxo-5β-cholanic - acid, 3:483

Dihypnone, 12:113

Diimide, 12:828
   as a reducing agent, 11:175

Diindogen, 11:565

Diiodoacetic acid, 8:420

2,4-Diiodoaniline, 2:415

1,4-Diiodo-2-butene, 3:790

2,3-Diiodo-2-butene-1,4-diol, 1:605

1,2-Diiodochloroethylene
   from monochloroacetylene, 5:204

1,2-Diiododisilane, 18:173

Diiodoethanoic acid, 8:420

Diiodofluorescein, 5:869

Diiodohydroxyquin, 2:632
   amebicide, 20:83

Diiodohydroxyquinoline
   amebicides, 20:88

5,7-Diiodo-8-hydroxyquinoline, 2:632

Diiodomethane, 11:863

2,6-Diiodo-4-nitrophenol, 14:535

3,5-Diiodo-4-oxo-1(4H)-pyridineacetic acid
   diethanolamine salt, 17:132

5,7-Diiodo-8-quinolinol, 20.83

Diiodosilane, 18:173

1,3-Diiodotetrafluorobenzene, 9:786

1,4-Diiodotetrafluorobenzene, 9:786

3,3'-Diiodothyronine, 2:189

Diiodotyrosine
   properties, 20:262
   thyroxine from, 20:263

3,5-Diiodotyrosine, 2:189

Diisoamyl adipate
   phys prop, 1:409

Diisobutyl adipate
   phys prop, 1:409

Diisobutylcarbinol, 1:564, 567. (See also *2,6-Dimethyl-4-heptanol.*)

Diisobutylene
   epoxidation, 8:249

Diisobutyl ketone, *12*:104, 111, 136
Diisoctyl phthalate, *15*:747
  plasticizer, *15*:723
Diisocyanates, *21*:69
Diisodecyl adipate
  plasticizer, *15*:741
  prod and sales of, *8*:377
Diisodecyl phthalate
  plasticizer, *15*:723
  prod and sales, *8*:376
Diisonicotinic hydrazide, *21*:532
Diisonitroacetone
  nerve agents detected by, *4*:876
Diisooctyl adipate
  phys prop, *1*:409
  prod and sales of, *8*:377
Diisooctylamine-borane, *3*:729
Diisooctyl phthalates
  prod and sales of, *8*:376
Diisopropanolamine, *1*:810; *19*:381
  in hair waving lotion, *10*:788
  hydrogen sulfide, *19*:384
  phys prop, *1*:811
Diisopropenylacetylene
  prepn from 2,5-dimethyl-3-hexyne-2,5-diol,
    1:626
Diisopropenyl sebacate, *16*:170
Diisopropoxydiacetoxysilane, *18*:219
Diisopropyl adipate
  phys prop, *1*:409
Diisopropylamine, *16*:576
  uses, *2*:123
β-Diisopropylaminoethyl chloride. See *2-Chloro-N,N-diisopropylethylamine.*
Diisopropylammonium nitrate, *6*:343
Diisopropylbenzene, *6*:544
p-Diisopropylbenzene, *6*:546
Diisopropylbenzyl monohydroperoxide, *7*:680
2,6-Diisopropyl-5,5-dimethyl-4-*m*-dioxanol,
    3:868, 875
2,6-Diisopropyl-5,5-dimethyl-4-ethoxy-*m*-diox-
    ane, *3*:875
N,N'-Diisopropyldithiosuccinamide, *19*:145
Diisopropylethanolamine
  phys prop, *1*:811
Diisopropyl fluorophosphate
  and protease activity in enzymes, *S*:297
Diisopropylglyoxal, *12*:140
Diisopropyl hydrogen phosphite, *9*:645
Diisopropylideneacetone, *12*:105
Diisopropyl ketone, *12*:104, 111
Diisopropyl monochlorophosphate, *9*:645

Diisopropyl peroxydicarbonate
  prepn, *4*:387, 389
Diketene, *12*:91, 129; *19*:144
Diketocyanidine, *12*:51
α-Diketones, *12*:138-147
β-Diketones, *12*:147-158
γ-Diketones, *12*:158-162
1,2-Diketones, *12*:138-147
1,3-Diketones, *12*:147-158
1,4-Diketones, *12*:158-162
Diketopiperazines, *10*:929
  of inks, *11*:613
Dilatometry, *20*:100
  polymerization rates by, *20*:809
Di-*O*-lauroyl-D-glucosyl sulfate, *19*:226
Dilauryl thiodipropionate
  as food preservative, *10*:14
Dilauryl 3,3'-thiodipropionate
  prod and sales of, *8*:376
Dilead trioxide, *12*:276
Dilinoleo-linolenin, *7*:405
1,2-Dilithio-1,2-diphenylethane
1,4-Dilithiotetrafluorobenzene, *9*:787, 793
Dill seed, *9*:371, 352, 363; *14*:201
Dill weed, *14*:201
Diloxanide
  amebicide, *20*:92
Diloxanide furoate, *20*:70
  amebicide, *20*:92
Dilsilazane, *18*:196
Dimandelatozirconic acid, *22*:648
Dimanganese decacarbonyl, *13*:5
Dimazon, *2*:640
Dimedone, *12*:151
Dimenhydrinate, *16*:674
  antihistamine, *11*:38
Dimensional analysis, *7*:**176**
Dimensionless groups, *7*:185; *10*:821
Dimer acid
  prepn, *8*:820, 847
Dimercaprol, *2*:584
  antidote for arsenic, *2*:729
2,3-Dimercaptopropanol. See *Dimercaprol.*
2,3-Dimercapto-1-propanol
  prepn from allyl alcohol, *1*:918
Dimeric acetone peroxide, *8*:488
Dimerization
  of propylene, *16*:581
"Dimeryl" diisocyanate, *21*:103
Dimethallyl carbonate, *4*:390
4,9,5,8-Dimethano-3a,4,4a,5,8,8a,9,9a-octahy-
    dro-1-β-naphthindene, *6*:691

Dimethindene maleate
  antihistamine, *11*:38
2,3-Dihydro-2,3-dioxindole, *11*:591
Dimethisterone, *6*:77, 87
Di(2-methoxyethyl) adipate
  phys prop, *1*:409
2,4-Dimethoxyaniline, *16*:296
Di(2-methoxyethyl) phthalate
  as cellulosic plasticizer, *4*:657
3,4-Dimethoxybenzaldehyde, *6*:672
3,3'-Dimethoxybenzidene, 7479, 535. See *o-Dianisidine.*
4,4'-Dimethoxybenzil, *12*:140
3,4-Dimethoxybenzoic acid. See *Veratric acid.*
3,4-Dimethoxybenzyl alcohol, *19*:301
Dimethoxyborane, *3*:666
4,4'-Dimethoxydiphenylamine, *7*:46
p,p'-Dimethoxydiphenylamine, *7*:40, 47
1,1-Dimethoxyethane, *1*107. (See also *Dimethyl acetal.*)
1,2-Dimethoxyethane, *12*:69
  solvent for silanes, *18*:175, 178
Dimethoxymethane, *1*:107; *10*:96. See *Methylal.*
9,10-Dimethoxymethylanthracene, *16*:170
2,4-Dimethoxypentafluoropyridine, *9*:799
3,4-Dimethoxyphenethylamine, *16*:645
2,3-Dimethoxyphenol, *16*:191
2,6-Dimethoxyphenol, *16*:199
3,5-Dimethoxyphenol, *9*:605
  in prepn of papaverine, *1*:798
2,6-Dimethoxyphenylpenicillin, *3*:15, 16; *14*:698
Dimethoxytoluquinones, *16*:210
Dimethyl acetal
  phys prop, *1*:107
Dimethylacetamide, *1*:145, 146
N,N-Dimethylacetamide, *2*:67; *9*:844
3,4-Dimethylacetophenone
  from camphor, *4*:55
Dimethyl acetylenedicarboxylate, *21*:177
5,5-Dimethylacridan, *2*:600
9,9-Dimethylacridan, *7*:44
1,3-Dimethyladamantane, *S*:2, 4, 13
3,5-Dimethyl-1-adamantanol, *S*:10, 14
5,7-Dimethyl-1,3-adamantanediol, *S*:2, 10
Dimethyl adipate
  phys prop, *1*:409
Dimethylalkyl oxides
  toxicity of, *S*:43
6,7-Dimethylalloxazine, *17*:446
β,β-Dimethylallyl pyrophosphate, *14*:183
6-*N*-Dimethylallyladenine, *15*:681

Dimethylaluminum dichloride, *18*:233
Dimethylamine, *2*:120; *10*:109
Dimethylamine
  manuf by ammonolysis, *2*:345
  in manuf of sodium dimethyl dithiocarbamate, *4*:372
  from monuron, *22*:190
  in prepn of dimethylacetamide, *1*:146
  reaction with isoprene, *12*:69
  synthesis of monuron, *22*:188
Dimethylamine-borane, *3*:728
Dimethylamine hydrochloride, *1*:82
Dimethylamine sulfate
  in prepn of leather, *2*:122
p-Dimethylaminobenzaldehyde, *2*:173; *3*:366; *6*:670; *10*:915
  Acid Violet dyes from, *20*:715
  reaction with phenyl isocyanate, *12*:51
p-Dimethylaminobenzaldehydeazine, *S*:617
m-Dimethylaminobenzoic acid, *20*:712
Dimethylaminoborane, *3*:732
1-Dimethylamino-2-chloropropane, *2*:143
2-Dimethylamino-1-chloropropane, *2*:143
3-Dimethylamino-1-chloropropane, *16*:670
7-Dimethylamino-6-demethyl-6-deoxytetracycline, *20*:16, 27
Dimethylaminodichloromethane, *6*:611
6-Dimethylamino-4,4-diphenylheptan-3-one, *2*:149, 150
6-Dimethylamino-2-(2-(2,5-dimethyl-1-phenylpyrrol-3-yl)-vinyl)-1-methylquinolinium
Dimethylaminoethanol, *2*:145
β-Dimethylaminoethanol, *16*:674
2-(Dimethylamino)ethyl acrylate
  phys prop, *1*:288
β-Dimethylaminoethyl benzhydryl ether, *16*:674
2-(Dimethylamino)ethyl-*p*-(butylamino) benzoate and hydrochloride
  prod and sales, *8*:376
2-Dimethylaminoethyl-*p*-butylaminobenzoate hydrochloride. See *Tetracaine hydrochloride.*
β-Dimethylaminoethyl chloride. See *2-Chloro-N,N-dimethylethylamine.*
3-(2-Dimethylaminoethyl) indol-4-ol. See *Psilocin.*
β-Dimethylaminoisopropyl chloride. See *2-Chloro-N,N-dimethylpropylamine.*
3-Dimethylamino-2-methylacrolein, *6*:615
γ-Dimethylamino-*β*-methylpropyl chloride, *2*:139
2-*N*-(2-Dimethylaminoethyl)-*N*-(*p*-met- hoxybenzyl)aminopyridine, *2*:152

prop, *2*:141

m-Dimethylaminophenol, *2*:214

Dimethylaminopropylamine, *7*:23, 24, 32, 34

β-Dimethylaminopropyl chloride, *2*:141. See *1-Chloro-2-(N,N-dimethyl)propylamine.*

γ-Dimethylaminopropyl chloride. See *3-Chloro-N,N-dimethylpropylamine.*

Dimethyl 4-aminophthalate
  polyimides from, *S*:750

Dimethylaminosilane, *18*:198

N-Dimethylaminosuccinamic acid, *15*:683

4-Dimethylaminotetracyclines, *20*:12

4-Dimethylamino-3,5-xylenyl *N*-methylcarbamate, *15*:910

Dimethylaniline
  amine oxides, *S*:32
  halosulfonates, *19*:490
  sulfur trioxide complexes, *19*:491

N,N-Dimethylaniline, *2*:368, 413, 414, 420; *6*:627; *7*:479; *15*:218
  manuf of Tetryl, *8*:621

3,5-Dimethylaniline, *2*:348

β,β-Dimethylanthionine, *14*:670

Dimethyl anthranilate
  as synthetic flavors, *9*:354

3-Dimethylarsenylpropylarsine, *22*:658

Dimethylarsine
  prepn, *2*:728

3,3'-Dimethyl-8-azathiacyanine iodide, *6*:620

2,2'-Dimethylbenzidine. See *m-Tolidine.*

3,1'-Dimethylbenzidine. See *o-Tolidine.*

N,N-Dimethylbenzylamine, *14*:534

p-*α,α*-Dimethylbenzylphenol, *1*:912. (See also *p-Cumylphenol.*)

6,6-Dimethylbicyclo-[1.1.3]hept-2-enethanol acetate prod and sales, *8*:276

2,2'-Dimethyl-1,1'-bianthraquinone, *2*:488

1,1'-Dimethyl-4,4'-bipyridinium di(methonsulfate), *22*:207

Dimethylbismuthine
  bismuthine from, *3*:537

2,6-Dimethyl-4-bromophenol, *15*:172

Dimethylbutadiene
  in methyl rubber, *7*:677

2,3-Dimethylbutadiene, *7*:73

3,4-Dimethylbutadiene, *12*:136

2,2-Dimethylbutane
  in hexanes, *11*:1
  in hydroisomerization, *11*:446

2,3-Dimethylbutane, *8*:504
  in hexanes, *11*:1
  in hydroisomerization, *11*:446
  prod from isobutene and ethylene, *8*:500

3,3-Dimethyl-2-butanone, *12*:104

1,3-Dimethyl-3-butenyl acrylate
  phys prop, *1*:288

1,3-Dimethylbutyl acrylate
  phys prop, *1*:288

3,3-Dimethyl-1-butyl chloride
  prepn, *8*:500

Dimethylbutylamine, *6*:451

2,4-Dimethyl-6-*tert*-butylphenol, *1*:911; *19*:672

Dimethylcadmium, *3*:906

N,N'-Dimethylcarbanilide, *8*:681

Dimethyl carbonate, *4*:390

3,5-Dimethyl-4-chlorophenol, *2*:627

2,3-Dimethylchromone, *1*:156

4,6-Dimethylcoumalic acid, *10*:923. See *Isodehydroacetic acid.*

4,6-Dimethylcoumalic acid, *10*:923

4,6-Dimethyl-coumalic acid, *1*:157; *10*:922

4,6-Dimethylcoumarin, *1*:156

Dimethylcyclohexane, *9*:698, 705

Dimethyl cyclohexanedicarboxylate, *10*:680

5,5-Dimethyl-1,3-cyclohexanedione, *12*:148
  in acetaldehyde determination, *1*:90

1,1-Dimethylcyclopentane

2,2-Dimethylcyclopropanenitrile, *10*:678
  in commercial hexane, *11*:1

4,7-Dimethyl-5-decyne-4,7-diol, *1*:625

3,3'-Dimethyl-4,4'-diaminobiphenyl. See *o-Tolidine.*

Dimethyl 2,2-dichlorovinyl phosphate, *15*:752

2,2'-Dimethyl-4,4'-diaminobiphenyl. See *m-Tolidine.*

2,5-Dimethyl-2,5-di(*t*-butylperoxy)hexane, *18*:229

O,O-Dimethyl-2,2-dichlorovinyl phosphate, *14*:537

9,10-Dimethyl-9,10-dihydroanthracene
  from vinyl chloride, *5*:173

N,N-Dimethyl-2,2-diphenylacetamide
  photodecomposition, *22*:198 *6*:77, 87

2,5-Dimethyl-3,4-diphenylcyclopentadienone dimer, *S*:756

Dimethyldiethoxysilane, *18*:217

3,6-Dimethyl-*p*-dioxane-2,5-dione, *12*:173, 185

N,N'-Dimethyl-*N,N'*-dinitrosoterephthalamide
  microencapsulation of, *13*:454

2,4-Dimethyl-1,3-dioxane, *10*:663

4,4-Dimethyl-1,3-dioxane
  in manuf of isoprene, *12*:76

2,2-Dimethyl-1,3-dioxolane
  prepn, *8*:529

4,4'-Dimethyldiphenylamine, *7*:46

Dimethyl disulfide, *19*:321

Dimethyl dithiophosphate, *15*:322

Dimethyldodecylamine, *S*:41

Dimethyldodecylamine-borane, *3*:729

Dimethyldodecylamine oxide, *S*:33, 46

Dimethylenedicyclohexyl diisocyanate, *21*:103

Dimethylene ether, *2*:231
in amino resin polymerization

Dimethylethanolamine
phys prop, *1*:811
uses, *1*:823

Dimethyl ether, *2*:413
dimethyl sulfate from, *19*:492
hydration, *13*:391
in polymerization of isoprene, *12*:70
in prepn of acetic acid, *4*:435
sulfation, *19*:305

N,N'-Dimethylethylene sebacamide, *16*:22

2,2-Dimethylethylenimine, *11*:527

3,5-Dimethyl-5-ethyl-2,4-oxazolidinedione,
11:522

6α,21-Dimethyl-17α-ethynyl-17β-hydr- oxyan-
drost-4-en-3-one. See *Dimethisterone*.

6α,21-Dimethyl-17α-ethynyl-17β-hyd- roxyan-
drost-4-en-3-one, 6 :77, 87

Dimethyl ethylphenylmalonate
manuf of, *3*:68

1,7-Dimethyl-7-ethyl-5,6,7,8-tetrahydrophenan-
threne, *17*:483

Dimethylformamide, *2*:68, 238; *6*:451, 696;
8:539; *9*:591, 693, 777, 779 *10*:109, 904;
19:491
alkylation, *19*:495
C-N rotation, *14*:72
paint remover, *14*:487
prepn of pentafluorobenzonitrile, *9*:795
reaction with phenyl isocyanate, *12*:51, 118
reactions with tetrafluorobenzenes, *9*:786,
787
solvent for cellulose triacetate in fiber
manuf, *1*:115

N-Dimethylformamide, *12*:814
spandex solvent, *18*:625

N,N-Dimethylformamide
sucrose solvent, *19*:221

6,7-Dimethyl-9-formylmethylisoalloxazine,
17:446

Dimethylfulvene, *6*:696

N,N-Dimethyl-*N*-fulvenecarboxaldehyde,
6:696

Dimethylfumarate, *12*:826

2,5-Dimethylfuran, *12*:158, 160
in prepn of 2,5-hexanedione, *10*:238, 901

meso-α,α'-Dimethylglutaric acid, *12*:640

Dimethylglycine, *19*:383

Dimethylglyoxal, *12*:140, 144

Dimethylglyoxime
nickel complex, *22*:560

Dimethylguanidine
prepn from guanidine *10*:736

2,6-Dimethyl-2,5 heptadien-4-one, *12*:105

4,4-Dimethylheptamethylenediamine
polypyromellitimide of, *S*:767

2,5-Dimethyl-1,4-hexadiene, *10*:142

2,6-Dimethyl-4-heptanol, *1*:567
phys prop, *1*:564
prepn from methacrolein, *1*:263

2,6-Dimethyl-1,5-hexadien-3,4-diol

2,6-Dimethyl-4-heptanone, *12*:104. See
Diisobutyl ketone.

3;5-Dimethyl-4-heptanone, *12*:104
phys prop, *1*:630

2,5-Dimethylhexane-2,5-diol, *1*:627

2,5-Dimethyl-2,5-hexanediol
benzene reaction, *22*:635

2,5-Dimethyl-3,4-hexanedione, *12*:140

5,5-Dimethyl-2,4-hexanedione, *12*:148

2,5-Dimethyl-3-hexene- 2,5-diol, *1*:633

2,5-Dimethylhexyl-2,5-diperoxyoctoate, *20*:826

2,5-Dimethyl-3-hexyne-2,5-diol, *1*:625
phys prop, *1*:626

3,5-Dimethyl-1-hexyn-3-ol, *1*:621
phys prop, *1*:622

5,5-Dimethylhydantoin, *11*:155; *12*:117

5,5'-Dimethylhydantoin-formaldehyde
in hair sprays, *10*:782

2,2-Dimethylhydrazide, *15*:683

Dimethylhydrazine
in missile system, *12*:609

2,3-Dimethylhydroquinone
dl-γ-tocopherol prod, *21*:580

3,5Dimethyl-1-hexen-3-ol

2,5-Dimethylhydroquinone
dl-β-tocopherol prod, *21*:580

O,O-Dimethyl-1-hydroxy-2,2,2-trichloroethyl
phosphonate, *14*:537

4,4-Dimethyl-4-hydroxy-3-chloro-2-butanone,
5:311

2,2-Dimethyl-3-hydroxyhexanal, *3*:869

4,5-Dimethylimidazole, *12*:116

Dimethylisopropanolamine
phys prop, *1*:811
uses, *1*:823

Dimethyl itaconate, *12*:85

Dimethyl ketene, *6*:694; *12*:88, 91

Dimethyl ketone. See *Acetone*.

S,S-Dimethyl-*N*-(methylsulfonyl)-sulfilimine, *19*:328

2,3-Dimethyl-1,4-naphthoquinone, *21*:593

Dimethyl (3-methyl-2-butenyl)amine, *12*:69

Dimethylmagnesium, *12*:710

Dimethyl malonate, *12*:850

2,6-Dimethylmorpholine
    prepn from diisopropanolamine, *1*:812

1,8-Dimethylnaphthalene, *13*:692

Dimethylnaphthalenes, *13*:678, 691

N,N-Dimethyl-*o*-nitroaniline, *15*:217
    C-N rotation, *14*:72

N,N-Dimethyl-*m*-nitroaniline, *15*:218

2,2-Dimethyl-3,5-nonanedione, *12*:148

Dimethyloctadecylamine-borane, *3*:729

3,7-Dimethyl-2,6-octadienyl acrylate
    phys prop, *1*:288

3,7-Dimethyloctane-1,7-diol, *19*:829

3,6-Dimethyloctane-3,6-diol, *1*:627

3,6-Dimethyl-4-octene-3,6-diol, *1*:625, 633

3,7-Dimethyl-2-octene-1,7-diol, *19*:829

3,7-Dimethyl-6-octenyl acrylate
    phys prop, *1*:288

3,7-Dimethyl-7-octenyl acrylate
    phys prop, *1*:288
    phys prop, *1*:626

Dimethylolaniline
    prepn, *2*:234

o,p-Dimethylolphenol, *15*:181

Dimethylolurea, *2*:229,231, 233
    urea-formaldehyde reaction, *21*:39

Dimethyl oxalate
    tetracycline synthesis, *20*:23

3,5-Dimethyl-1,4-oxathiane
    prepn from allyl ether, *8*:494

5,5-Dimethyl-2-oxazolidinethione, *20*:270

3,3-Dimethyloxetane, *10*:677

Dimethyloxosulfonium methylide, *19*:327
    steroid synthesis, *18*:879

2,2-Dimethylpentane
    in commercial hexane, *11*:1

2,4-Dimethylpentane, *6*:681
    in commercial hexane, *11*:1

2,4-Dimethyl-2,4-pentanediol
    hydrous zirconia reaction, *22*:659

2,2-Dimethylpentanoic acid
    properties, *8*:852

4,4-Dimethyl-1-pentene, *8*:494
    copolymers, *S*:787

2,3-Dimethyl-3-pentanone, *12*:104

2,4-Dimethyl-3-pentanone, *12*:104, 111

2,5-Dimethylperchlorylbenzene, *9*:604

4,13-Dimethylphenanthrene, *17*:483
    flavor and perfume materials, *9*:358

N,N-Dimethyl-*o*-phenylenediamine, *15*:217

α,α-Dimethylphenethyl acetate

2,4-Dimethylphenol
    alkylation of, *S*:27

3,4-Dimethylphenol, *6*:453

Dimethylphenols, *2*:626

3,5-Dimethyl-1-phenylpyrazole, *12*:151

N,N-Dimethyl-*m*-phenylenediamine, *15*:218

Dimethyl-2-phenoxyethyl-2 1'-thenylammonium *p*-chlorobenzene sulfonate, *14*:534

2,5-Dimethyl-1-phenyl-3-pyrrolecarboxaldehyde, *14*:540

N,N-Dimethyl-*p*-phenylenediamine, *15*:218, 222, 374
    in hair coloring, *10*:795
    in prepn of *p*-(4-dimethylaminoanilino) phenol, *2*:220

S,S-Dimethyl-*N*-(phenylsulfonyl)sulfilimine, *19*:328

S,S-Dimethyl-*N*-(phenylsulfonyl)sulfoximine, *19*:329

Dimethyl phosphonate, *6*:651

Dimethyl phthalate, *9*:844
    as cellulosic plasticizer, *4*:657
    plasticizer, *15*:764
    plasticizers for, *15*:763
    prod and sales, *8*:376
    in propellants, *8*:660

Dimethylpolysiloxane
    as antifoaming agent, *10*:4

Dimethylpropane, *14*:707

N,N-Dimethyl-1,3-propanediamine. See *Dimethylaminopropylamine*.

5,7-Dimethyl-3-*n*-propyl-1-adamantanol, *S*:10

2,2-Dimethyl-1,3-propanediol, *10*:676

3,6-Dimethylpyridazine, *12*:159

Dimethylpyridine
    reaction with silver fluoborate, *9*:566

2,4-Dimethylpyridine, *21*:530
    oxidation of, *21*:514

2,6-Dimethylpyridine, *21*:528
    amine oxide from, *S*:43

2,6-Dimethyl-4(1*H*)-pyridone, *10*:922

2,6-Dimethyl-4-pyrone, *10*:922, 924

Dimethylpyroneacetic acid
    binary cooling curve, *13*:203

2,5-Dimethylpyrrole, *10*:901; *12*:158

6,7-Dimethyl-9-(1'ᴅ ribityl)-isoalloxazine, *17*:445

Dimethyl selenide
    reaction with boron trifluoride, *9*:557

2,8-Dimethyl-6-sila-1,5,7,11-tetraoxaspiro [5,5] undecane, *10*:662
Dimethyl silicone, *9*:898; *12*:611
Dimethyl silicone cyclics polymerization, *18*:225
Dimethyl silicones, *18*:226,250 fluids, *18*:237
Dimethylstibine, *2*:581
Dimethylstibine trichloride, *2*:581
Dimethylstibinic acid, *2*:581
N,N-Dimethylsuccinamic acid form, *2*:67
Dimethyl succinate, *19*:138, 139 tetracycline synthesis, *20*:23
Dimethyl sulfate, *6*:436; *9*:794; *19*:486, 487; *S*:552
    as alkylating reagent, *1*:898
    aluminum chloride reaction with, *5*:107
    Benzenetetrol synthesis, *16*:209
    health and safety, *19*:494
    in polycarbonate synthesis, *16*:107
    and pyrogallol, *16*:191
    reaction with ethylenediamines, *7*:27
    solvolysis, *19*:488
    sulfation, *19*:305
    vacuum pyrolysis, *19*:491
Dimethyl sulfide, *19*:321, 332
    from carbon disulfide, *4*:373
    from methyl chloride, *5*:104
    as synthetic flavor, *9*:354
Dimethyl 5-sulfoisophthalate, *19*:317
Dimethyl sulfone, *9*:693, 798; *19*:321
Dimethylsulfonium methylide, *18*:869
Dimethyl sulfoxide, *9*:693, 800; *10*:164; *12*:147, 154; *15*:173; *19*:320
    plasticizer, *15*:773
    polysiloxanes, *18*:225
    sucrose solvent, *19*:221
S,S-Dimethylsulfoximine, *19*:328
Dimethylsulfoxonium methylide, *18*:869
5-Dimethylsulfuranylidene-2,2-dimethyl-1,3-dioxa-4,6–cyclohexanedione, *19*:327
3-Dimethylsulfuranylidene-2,4-pentanedione, *19*:218
5-Dimethylsulfuranylidene-2,4,6-trioxohexahydropyrimidene, *19*:328 phorodithioate. See *Malathion*.
Dimethyl terephthalate, *6*:546; *8*:361; *10*:154; *15*:462-472
    analysis, *15*:470
    chemical prop, *15*:464
    economics, *15*:468
    health and safety, *15*:471
    manuf, *15*:465
    phys prop, *15*:463
    polyesters from, *16*:168, 170, 174
    specifications, *15*:469
    uses, *15*:472
Dimethyl tetrachloroterephthalate, *15*:338
Dimethyl 2,3,5,6-tetrachloroterephthalate, *22*:200
3,5-Dimethyltetrahydro-1,3,5-thiadiazine-2-thione as fungicide, *10*:224
5,5-Dimethyl-2-thiooxazolidone, *20*:269
3,5-Dimethyl-1,3,5(2*H*)-tetrahydrothiadiazine-2-thione, *15*:913 as fungicide, *10*:235
Dimethylthiophene, *12*:158
2,5-Dimethylthiophene, *8*:901
Dimethyl 3,3'-thiodipropionate prepn from methyl acrylate, *1*:292
Dimethyltin dichloride, *20*:306
N,N-Dimethyl-*p*-toluidine
2,4-Dimethyl-2,4,6-triethyltetrahydropyridine, *12*:117 in dental materials, *6*:832
Dimethyltubocurarine chloride, *6*:550
Dimethyltubocurarine iodide, *6*:550
O-Dimethyltubocurarine iodide, *1*:788
1,3-Dimethyluric acid, *21*:109
N,N'-Dimethyluron, *2*:233
Dimethylvinyl carbinole, *12*:552
Dimethylyn, *16*:676
2,2-Dimethyl-1-propanol, *2*:375 trimethylethylene by dehydration of, *4*:395 phys prop, *1*:626
Dimetridazole, *21*:248
Di(4-morpholine) polysulfides, *13*:665 vulcanizer, *17*:569
Dimorphotheca seed oil drying oil *7*:400
Dimsyl. See *Sulfoxides*.
"Dimsyl" ion, *19*:326
Di-*n*-undecyl ketone. See *Laurone*.
Di-*n*-undecyl ketone, *12*:104
Dinaphtho(1,2,3-*cd*:3',2',1'-*lm*)perylene-5,10-dione, *2*:496 basic dye from, *20*:725
1,2'-Dinaphthylamine
N,N'-Di-β-naphthyl-*p*-phenylenediamine, *16*:12; *17*:518
N,N'-Di-2-naphthyl-*m*phenylenediamine, *15*:218
N,N'-Di-2-naphthyl-*p*-phenylenediamine, *2*:601; *7*:40, 41, 46, 47; *15*:910
Dinex, *15*:910

Dinicotinoylornithine, *21*:536
Dinitraniline orange, *15*:573
Dinitroadamantane, *S*:7
gem-Dinitroalkanols
    from nitroalcohols, *13*:828
4,6-Dinitro-2-aminophenol
    in hair coloring, *10*:795
2,4-Dinitroaniline, *2*:425
    pigments, *15*:573
Dinitroanilines
    metabolism of substituted used as herbi-
        cides, :*22*:206
    soil chemistry, *18*:536
    substituted as herbicides, *22*:205
1,5-Dinitroanthraquinone, *2*:49
1,8-Dinitroanthraquinone, *2*:492
2,6-Dinitroanthraquinone, *2*:485
1,3-Dinitrobenzene, See *m-Dinitrobenzene.*
Dinitrobenzene
    from benzene, *3*:374, 398
    sensitivity characteristics, *8*:632
    explosive characteristics, *8*:633
m-Dinitrobenzene, *2*:425; *15*:216
    as explosive, *8*:617
    in prepn of 2, 4-diaminophenol, *2*:221
2,4-Dinitrobenzenesulfonic acid, *15*:217
2,2'-Dinitrobenzidine, *3*:416
3,3'-Dinitrobenzidine, *3*:416
3,3'-Dinitrobenzil, *12*:146
3,5-Dinitrobenzoic acid, *3*:437
    phys prop, *1*:238
3,5-Dinitrobenzotrifluoride, *9*:778
3,5-Dinitrobenzoyl chloride, *3*:437
N,N'-Dinitro-1,4-butanediamine, *10*:667
2,2-Dinitrobutyl acrylate
    phys prop, *1*:289
3,3-Dinitrobutyl acrylate
    phys prop, *1*:289
Dinitrochlorobenzene, *7*:478
2,4-Dinitrochlorobenzene, *2*:425
    manuf of Tetryl, *8*:621
2,4-Dinitro-1-chlorobenzene, *13*:843
Dinitrochloromethane
    from trichloroethylene, *5*:187
Dinitrocresol
    herbicidal, *22*:176
    as an herbicide, *22*:176
2,4-Dinitro-*o*-cresol, *6*:441
Dinitro-*o*-cresol, *19*:672
4,6-Dinitro-*o*-cresol, *15*:910
4,6-Dinitro-*o*-cyclohexylphenol, *15*:910
Dinitrodiamminepalladium(II), *15*:869
Dinitrodiammineplatinum(II), *15*:865

cis-Dinitrodiammineplatinum(II), *15*:865
2,4-Dinitrodiazobenzene, *2*:873
Dinitrodiglycol. See *Diethylene glycol dinitrate.*
4,4'-Dinitrodiphenylamine, *2*:352; *15*:218
N,N'-Dinitroethylenediamine. See *Ethylenedi-*
    nitramine.
2,4-Dinitrofluorobenzene, *9*:781
Dinitrogen pentoxide, *10*:155
    in prepn of nitronium fluorosulfate, *9*:679
Dinitrogen tetroxide, *8*:271, 479; *10*:155
    in prepn of nitronium fluorosulfate, *9*:679
2,6-Dinitro-4-hydroxyperchlorylbenzene, *9*:604
2,4-Dinitro-3-methyl-6-*tert*-butyl anisole,
    1:902
2,4-Dinitro-3-methyl-6-*tert*-butylanisole. See
    Musk ambrette.
4,6-Dinitro-2(1-methylheptyl)phenyl croton-
    ate, *6*:466
4,6-Dinitro-2-methylphenol
    as leather fungicide, *10*:233
1,4-Dinitronaphthalenes, *13*:705
Dinitroparaffins, *12*:108
3,3-Dinitro-1,5-pentamethylene diisocyanate,
    13:832
2,2-Dinitropentyl acrylate
    phys prop, *1*:289
Dinitrophenol
    in fungicide, *10*:230
2,4-Dinitrophenol, *7*:479, 497; *11*:552; *13*:890
    dyes, *19*:432
    as fungicide, *10*:234
    in prepn of 2-amino-4-nitrophenol, *2*:95, 217
    in prepn of 2,4-diaminophenol, *2*:221
    sulfur black, *19*:426
2,6-Dinitrophenol, *13*:891
Dinitrophenols
    styrene inhibitor, *19*:68
    as insecticides, *11*:689
1,4-Di(*o*-nitrophenyl)-diacetylene
2,6-Dinitro-1-phenol-4-sulfonic acid, *2*:223
    in synthesis of indigo, *11*:563
2,4-Dinitrophenylhydrazine
    analysis of chlorinated acids, *22*:200
    menadione assay, *21*:595
2,4-Dinitrophenylhydrazone, *6*:456
2,4-Dinitrophenylhydrazones
    determination of acetaldehyde in paper
        chromatography of, *1*:90
Dinitrophenylmethylamine. See *N-Methyldi-*
    nitroaniline.
    in manuf of Tetryl, *8*:621
2,4-Dinitrophenylsemicarbazone, *6*:456
2,4-Dinitrophenythiocyanate

as leather fungicide, *10*:233
2,2-Dinitropropyl acrylate
  phys prop, *1*:289
Dinitroresorcinol
  prepn of lead dinitroresorcinate, *8*:591
2,4-Dinitroresorcinol, *13*:892
4,4'-Dinitro-2,2'-stilbenesulfonic acid, *19*:2
  explosive characteristics, *8*:633
  sensitivity characteristics, *8*:632
  thermochemical constant, *8*:681
Dinitrosopentamethylenetetramine, *9*:853;
  9:857
Dinitrostilbenedisulfonic acid, *1*:9, 3
4,4'-Dinitro-2,2'-stilbenedisulfonic acid, *7*:479;
  19:4
α-Dinitrotoluene. See *2,4-Dinitroluene.*
m-Dinitrotoluene
  reduction, *15*:217
2,4-Dinitrotoluene, *15*:217
  plasticizer, *15*:774
Dinitrotoluenes, *1*:3, 851
Dinitrotriglycol. See *Triethylene glycol dini-trate.*
Dinonyl adipate
  phys prop, *1*:409
Dinonylphenol, *19*:533
Dinonyl phthalate
  cellulosic plasticizer, *4*:657
Dinoseb
  herbicide, *22*:203
Dioctadecylamine
  properties, *2*:130
  phys prop, *1*:409
Dioctyl adipate, *6*:365
  plasticizer, *15*:801
Dioctylamine
  properties, *2*:130
4,4'-Dioctyldiphenylamine, *7*:46, 47
Di-*n*-octyl ketone, *12*:104
Dioctyl maleate
  prod and sales of, *8*:376
3-(Dioctylphosphoroso)propionitrile, *6*:651
Dioctyl phthalate, *15*:455, 721
  plasticizer, *15*:801
Di-*n*-octylphosphine oxide, *6*:651
Dioctyl sebacate, *6*:354
Dioctyl sodium sulfosuccinate, *12*:813; *19*:141
  as cathartic, *4*:592
Dioctyl succinate, *19*:146, 318
Dioctyl sulfosuccinate
  salts, *19*:318
Di(*n*-octyl)tin   *S,S'-bis(isooctyl)mercapto-acetate), 20*:203

Di(*N*-octyl)tin   *S,S*-bis   (isooctylmercaptoace-tate)
  stabilizer, *21*:390
Di(*n*-octyl)tin-*S,S'*-bis(isooctylmercaptoace-tate), *20*:323
Di(*n*-octyl)tin dichloride, *20*:319, 320
Di(*n*-octyl)tin maleate, *20*:323
Di(*n*-octyl)tin oxide
  toxicology of, *20*:324
Diodes
  electroluminescent, *17*:857, 858
Dioform. See *1,2-Dichloroethylene.*
Diolefin
  nomenclature, *11*:288
N,N'-Dioleoyldiethylenetriamine, *2*:601
Dioscorine, *1*:766
Diosgenin, *18*:838, 840
  as steroid precursor, *11*:83
  in steroid synthesis, *11*:85
Diosgenone, *18*:854
Diosphenol, *12*:139
Dioxalatozirconic acid trihydrate, *22*:650
Dioxane, *2*:144; *9*:794; *12*:814
  in aquametry, *2*:697
  prepn from ethylene glycol, *10*:643
  stain solvent, *18*:660
p-Dioxane, *8*:487; *10*:922
1,3-Dioxane
  in copolymerization of trioxane, *1*:98
Dioxane dibromide, *12*:67
Dioxanes, *10*:922
Dioxane-sulfur trioxide
  complex, *19*:600
Dioxazines, *2*:866
Dioxazine violet, *15*:582
1,3-Dioxepane, *10*:669
p-Dioxin, *10*:898
3,12-Dioxo-5β-cholanic acid, *3*:483
3,6-Dioxo-5β-cholanic acid, *3*:483
1,3-Dioxolane, *10*:96, 646, 896
  phenolic resins, *15*:190
2,5-Dioxopiperazinediacetic dilactam, *12*:842
2,4-Dioxopiperidines
  as hypnotics, *11*:514
  as anticonvulsants, *11*:523
3,5-Dioxotriazolidines, *10*:917
Dioxy S acid, *13*:721
Dioxygen difluoride, *9*:508, 633
Dioxygenyl hexafluoroplatinate(V), *15*:863
α,α'-Dioxyperoxides, *14*:783
Dipenta. See *Dipentaerythritol hexanitrate.*
Dipentaerythritol, *1*:593; *7*:412
Dipentaerythritol hexanitrate, *8*:610

characteristics, *8*:632, 633, 634
Dipentamethylenethiuram hexasulfide
  vulcanizer, *17*:569
Dipentene, *17*:385; *19*:815, 820, 835
Dipentylzirconium, *22*:650
2,6'-Diperchloryldiphenoquinone, *9*:604
gem-Diperoxides, *14*:786
Diphacinone, *15*:920
Diphenadione, *5*:600, 601
Diphenamid
  amide herbicide, *22*:198
  antihistamine, *11*:38
Diphenhydramine.HCL
Diphenic acid, *12*:132
  phys prop, *1*:239
Diphenol carbonate, *8*:361
Diphenoxyacetic acid, *8*:418
Diphenyl, *7*:191
  from benzene, *3*:373, 398
  in Dowtherm A, *10*:850
  as dye carrier, *16*:151
  as food preservative, *10*:14
  in naphthalene feedstocks, *11*:454
  in petroleum, *14*:849
  purification of nicotinic acid, *21*:516
  prepn from acetanilide, *1*:150
N,N'-Diphenylacetamidine
Diphenylacetonitrile, *2*:143, 149
Diphenylacetyl chloride, *12*:92
2-Diphenylacetyl-1,3-indanedione, *15*:920
N,N'-Diphenyladipamide, *1*:409
Diphenylamine, *2*:413, 420, 600; *7*:40, 41, 43,
  44, 46, 47, 49; *8*:696
  antioxidant, *17*:518
  as dye intermediate, *7*:49
  indicator, *22*:560
  perchlorylation reaction, *9*:604
  polysiloxanes, *18*:225
  in propellants, *8*:659, 695
Diphenylamine chloroarsine, *10*:929
Diphenylamine hydrochloride, *7*:43
Diphenyl and terphenyls, *7*:**191**
Diphenylarsenic trichloride, *2*:728
Diphenylarsine
  prepn, *2*:728
Diphenylbenzene. See *Terphenylbenzene.*
N,N'-Diphenylbenzidine, *7*:43
2,2'-Diphenylbenzopyrancarboncyanine perchlorate, *16*:284
2,5-Diphenyl-*p*-benzoquinone
Diphenyl Black, *7*:565
  polyester inhibitor, *20*:823
Diphenylbromomethane, *3*:777

1,4-Diphenyl-1,3-butadiene, *5*:521
1,2-Diphenylbutane
  from butadiene, *3*:792
Diphenylcadmium, *3*:907
Diphenylcarbazide
  dichromate indicator, *11*:558
Diphenyl carbonate, *4*:390
  transesterification, *16*:106
Diphenylchloroarsine
  in chemical warfare, *4*:877
N,N'-Diphenyldiphenoquinone diimine, *7*:43
Diphenyldithiocarbamic acid, *7*:43
Diphenylene oxide, *10*:908; *15*:148
Diphenyl epsilon acid, *13*:714
Diphenylethane
  from 1,2-dichloroethane, :5:151
1,1-Diphenylethane, *10*:139
  from vinyl chloride, *5*:173
1,2-Diphenylethanol, *8*:495
2,2-Diphenylethanol, *10*:730
Diphenyl ether, *2*:413, *15*:148
  phys prop, *8*:471
Diphenylformamidine, *6*:613; *16*:288
Diphenylglyosal, *12*:140. See *Benzil.*
Diphenylguanidine, *2*:414
sym-Diphenylguanidine, *10*:740
1,3-Diphenylguanidine, *17*:511
Diphenylhydantoin
  in epilepsy control, *11*:519
5,5-Diphenylhydantoin, *11*:155
  as anticonvulsant, *11*:520
Diphenylimidazolone
  disulfonic acid of, as optical brightener,
    3:743
Diphenylindolylmethane dyes, *20*:725
Diphenyline, *3*:412
Diphenyl isophthalate, *15*:461
Diphenylketene, *6*:694; *12*:88, 92; *19*:330
Diphenyl ketone, *3*:439
N,N'-Diphenyl-*m*-phenylenedioimine, *15*:222
Diphenylmagnesium, *12*:711
N,N'-Diphenylmalamide, *12*:841
Diphenylmethane
  from benzyl chloride, *5*:283
  in manuf of benzophenone, *3*:440
Diphenylmethane diisocyanate
  polyurethans, *21*:58
Diphenylmethane dyes, *7*:471, 473
Diphenylmethane-4-,4'-diisocyanate, *12*:46,
  47, 48
1,1-Diphenyl-2-methyl-1-propene, *19*:286
Diphenyl methylsulfinylmethylmethanol,
  19:326

Diphenylnaphthylmethane
  derivatives as dyes, 20:672
  dye classification, 20:677
Diphenylnaphthylmethane dyes
  acid, 20:723
  acid mordant, 20:725
  basic, 20:721
Diphenylnitrogen, 7:43
1,8-Diphenyl-1,3,6,8-octanetetraone, 19:138
Diphenyl oxide, 15:154. (See also Phenolic ether.)
  as heat transfer medium, 10:850
2,2-Diphenyloxirane, 19:327
4,7-Diphenyl-1,10-phenanthroline
  α-tocopherol assay, 21:581
N,N-Diphenyl-p-phenylenediamine, 2:413;
  7:40, 43, 46, 47; 15:214, 223; 17:518
Diphenyl phthalate
  plasticizer, 15:728, 751, 764
Diphenylphthalide, 2:435
1,2-Diphenylpropane
N,N'-Diphenyl-1,3-propanediamine, 2:413
  from allyl chloride, 5:208
  prepn from allyl alcohol, 1:920
α-Diphenyl-2-pyridylmethanol, 16:673
  chlorinated as stain solvents, 18:661
Diphenyls. (See also Diphenyl and terphenyls.)
  chlorinated, 5:289
Diphenylsilanediol, 18:242
Diphenyl silicone, 18:227
Diphenylsodamide
  titanium amido compounds, 20:472
Diphenylstibinic acid, 2:581
Diphenyl succinate, 19:141
Diphenyl sulfide, 19:393
Diphenyl sulfite, 19:493
Diphenyl sulfone, 3:402, 403, 406; 19:298
Diphenyl terephthalate
  polyester from, 16:168
Diphenylthiocarbazone
  zinc analysis, 22:561
5,5-Diphenyl-2-thiohydantoin, 11:157
Diphenylthiourea, 2:414
N,N'-Diphenylthiourea, 17:509
Diphenyltitanium, 20:443
Diphenylurea. See Carbanilide.
sym-Diphenylurea. See Carbanilide.
1,3-Diphenylurea. See Carbanilide.
Diphenylurea, 8:681
Diphenyl urethan, 8:681
1,3-Diphenyluretidone, 12:51
Diphetarsone
  amebicide, 20:93

Diphosgene, 4:388
  from methyl formate, 4:388
Diphosphatozirconic acid, 22:633, 657
Diphosphine, 15:300
Diphosphopyridine nucleotide, 21:512
Diphosphorus tetrahydride
  reaction with boron trifluoride, 9:556
Diphosphorus tetraphenyl
  reaction with boron trifluoride, 9:556
Diphosphothiamine, 20:182
Dipicrylamine
  cesium extracted by, 4:857
  from seawater, 14:156
Dipivaloymethane, S:465
Dipole conduction
  of electric insulation, 11:775
Dipolyphosphate, 15:232
Dipotassium dithiosuccinate, 19:141
Dipotassium hydrogen phosphate
  as emulsifier in cheese, 10:8
Dipotassium sodium dichloroisocyanurate, 4:919
Dipple's oil
  as denaturant, 8:450
Di-(2-propionoxyethyl) adipate, 1:410
Dipropionylmethane, 12:156
Dipropyl adipate
  phys prop, 1:409
Dipropylamine, 6:648
  uses, 2:123
Di-n-propylamine
  phosgene reaction, 22:194
  reaction with titanium tetrafluoride, 9:683
4-(Di-n-propylamino)-3,5-dinitrobenzoic acid, 22:207
3-Dipropylaminopropionitrile, 6:648
γ-Dipropylaminopropyl chloride, 2:141. See 3-chloro. N,N-dipropylpropylamine.
1,4-Di-n-propyl-2-butyn-1,4-diol, 3:870
Dipropylene glycol, 10:651
  in manuf of polypropylene glycols, 10:658
  toxicity, 10:641
Dipropylene glycol diacrylate
  phys prop, 1:289
  copolymerization rate, 20:807
Dipropylene glycol fumarate alkyd
Dipropylene glycol monobutyl ether, 10:651
Dipropylene glycol monomethyl ether, 10:651
Di-n-propylgloxal, 12:140
Di-n-propyl ketone, 12:104
Dipropylmagnesium, 12:711
P-(Dipropylsulfonamoyl)benzoic acid, 19:258
1,4-Di(2-pyrazinyl) piperazine, 15:644

4,6-Di-*n*-propylresorcinol, *2*:629
Dipyridinium hexachlorozirconate
  in prepn zirconium alkonides, *1*:834
Dipyridone
  oxidation of paraquat, *22*:208
2,2'-Dipyridyl, *16*:804
Dipyridyls
  soil chemistry, *18*:537
Dipyrrylmethanes, *16*:855
Diquat, *14*:545; *19*:671
  herbicide, *22*:207
  metabolism of, *22*:209
  soil chemistry, *18*:537
Diquat dibromide, *16*:804
Direct azo dye, *7*:470
Direct Black (CI), *7*:565
Direct Black 22(CI), *7*:535
Direct Black 38, *7*:497
Direct Black 78, *2*:902
Direct Black 80, *7*:470
Direct Blue 1, *2*:903
Direct Blue 106, *2*:865
Direct Blue 107, *2*:866
Direct Blue 108, *2*:866
Direct Blue 109, *2*:866
Direct Blue 126, *2*:897
Direct Blue 126, *7*:492
Direct Blue 162, *2*:903
Direct Blue 2, *7*:497, 539
Direct Blue 2 (CI), *7*:539
Direct Blue 30 (CI), *7*:535
Direct Blue 43, *2*:900
Direct Blue 71, *2*:900
Direct colors
  used in textile printing, *7*:576
Direct dyes, *7*:534
  in paper coloring, *7*:583
  uses of, *7*:476
Direct Brown 31 (CI), *7*:535
Direct Green 26, *2*:901
Direct Green 28, *7*:491
Direct Green 6, *2*:898
Direct Red (CI), *7*:534
Direct Red 1, *2*:877
Direct Red 180, *2*:904
Direct Red 28 (CI). See *Congo Red.*
Direct Red 75 (CI), *7*:535
Direct Red 80, *2*:901
Direct Sky Blue FFM, *2*:903
Direct Yellow 6, *7*:497
Direct Yellow 12, *2*:899
Direct Yellow 12 (CI), *7*:535
Direct Yellow 44, *7*:491

Direct Yellow 44 (CI), *7*:535
Direct Yellow 59 (CI), *7*:465, 539; *20*:185
Direct dyes
Dirigibles
  use of helium, *10*:882
Disaccharides, *4*:133, 144
  sucrose, *19*:166
Disalicylaldehyde, *15*:162
Disalycilatozirconic acid, *22*:648
Disazo, *7*:522, 527
Disazo dyes, *7*:471,472
Diselenato(IV) zirconic acid, *22*:659
Dishwashing, *6*:868
Disilane, *18*:172
  prepn, *18*:175
Disilanyl compounds, *18*:173
  phys prop, *18*:207
Disilanyl iodide, *18*:173
Disilicon hexafluoride, *9*:653
Disiloxane, *18*:189
Disilylamine, *18*:196
1,2-Disilyldisiloxane, *18*:189
Disilyl ether, *18*:189
Disilyl selenide, *18*:195
Disilyl sulfide, *18*:194
Disilyl telluride, *18*:195
Disinfectant
  use of formaldehyde as, *10*:95
Disinfectants, *8*:447. (See also *Antiseptics and disinfectants.*)
Disinfection, *18*:806
  of municipal water supplies, *22*:85
Disintegration constant, *17*:13
Disodiooctadiene, *3*:787
Disodium arsenate
  in fungicide, *10*:230
Disodium aurothiomalate, *10*:694
Disodium cyanamide, *6*:559
Disodium dihydrogen pyrophosphate, *15*:244
Disodium ethylenebisdithiocarbamate. See *Nabam.*
  as fungicide, *10*:224
Disodium guanylate, *10*:10
Disodium inosinate, *9*:361; *10*:10
Disodium octaborate, *3*:642
Disodium orthophosphate, *15*:266
Disodium pararosolate, *20*:693
Di-*o*-tolylguanidine, *7*:714
  vulcanization, *16*:207
sym-Di-*o*-tolylguanidine, *10*:740
Disodium phosphate, *6*:802; *7*:259, 532
  as emulsifier in cheese, *10*:8
  as water-correcting agent, *10*:17

Disodium phosphite, *12*:278

Disodium salt
  test for chloramines, *4*:915

Disodium triborate, *3*:645

Disodium 1,4-dihydro-3,5-diiodo-4-oxo-2,6-pyridinedicarboxylate, *17*:132

Disodium 3-hydroxy-1,2-propanedisulfonate, *1*:600

Disodium 3,3'-bis(sulfomethylamino)-4,4'-dihydroxy-arsenobenzene. See *Sulfarsphenamine.*

Disophenol, *14*:535

Disperse anthraquinone dye, *7*:470

Disperse Black 9, *7*:497

Disperse Blue, *7*:470

Disperse Blue 15, *20*:192

Disperse Blue 3, *7*:470, 497

Disperse dyes, *2*:905; *7*:466, 540
  in dyeing of acetate fibers, *1*:134
  in textile printing, *7*:576
  uses of, *7*:476

Disperse Red 1, *2*:905

Disperse Red 58, *20*:192

Disperse Red 64, *20*:192

Dispersion
  anomalous, *17*:213
  relative, *17*:217
  specific, *17*:217

Dispersion-hardened alloys, *11*:16

Dispersoids (in gases), *8*:75
  removal, *8*:76

Dissipation factor
  of electric insulation, *11*:776

Dissociation constant
  of ampholytes, *7*:817
  of weak acids and bases, *7*:815
  of weak electrolytes, *7*:819

Dissociation constants
  of weak electrolytes, *7*:820

Distearamidomethane, *22*:142

Distearyldimethylammonium chloride, *10*:778

Disteroylethylenediamine, *9*:899

Distillate
  petroleum, *14*:836; *15*:2

Distillation, *7*:**204.** (See also *Extraction.*)
  of alcoholic beverages, *1*:525
  aquametry by, *2*:682
  azeotropic, *2*:839
  continuous at constant pressure, *21*:231
  control, *11*:765
  counterdiffusion in, *13*:141
  crude oil, *15*:2

  extractive of nitric acid, *13*:807
  extractive of xylenes, *22*:473
  HTU in, *13*:148
  of liquid wastes, *21*:629
  mass transfer in, *13*:100
  NTU in, *13*:145
  pilot plant, *15*:623
  selective azeotropic of xylenes, *22*:472
  vacuum, *15*:2, 59
  vacuum, *21*:147
  vacuum techniques, *21*:124

Distillation column
  considerations, *7*:92, 207, 223

Distillation equipment
  for alcoholic beverages, *1*:519
  as animal feed, *8*:862

Distillers' dried grains

Distillers' malt, *12*:877

Distribution coefficient
  data interpretation of, *6*:740

Distybnatozirconic acid, *22*:658

α,δ-Distyrylfulgide, *19*:138
  in prepn of norlobelanidine, *1*:795

2,6-Distyrylpyridine tetrabromide

Disuccinic acid peroxide, *9*:812, 813

N,N'-Disuccinimides, *19*:140

Disulfamide
  as diuretic, *7*:261

Disulfatomanganates(III), *13*:9

Disulfatozirconic acid, *22*:630, 654, 662

1,8,4-Disulfo acid, *13*:715

3,3'-Disulfo-4,4'-diamino-*sym*-diphenylurea, *7*:535

Disulfonamides
  as diuretics, *7*:257

3,5-Disulfopyrocatechol, *6*:18
  as sequestering agent, *6*:18

3,5-Disulfosalicylic acid, *17*:723

2,3-Disulfosuccinic acid, *19*:141

Disulfur. (See also *Sulfur compounds.*)

Disulfur decafluoride, *9*:508
  sulfur pentafluoride, *9*:664, 673

Disulfur dibromide, *19*:390

Disulfur dichloride, *19*:391

Disulfur difluoride, *9*:664, 674

Disulfur trioxide, *19*:407

Disulfuryl chloride, *19*:404

Disulfuryl dichloride, *5*:358

Diterpenes. See *Terpenes.*

Ditetradecylamine
  properties, *2*:130

p-Dithane
  prepn from ethylene oxide and hydrogen sulfide, *8*:529

Dithiazanine, 6:605

Dithiazanine iodide, 14:535

Dithioammelide. See 2-Amino-A-triazine 4,6-dithiol.

Dithioammelide. See 2-Amino-s-triazine-4,6-dithiol.

Dithiobenzoic acid
  from carbon disulfide, 4:373

3,3'-Dithiobis(2-aminopropionic) acid. See Cystine.

Dithiobuget, 6:567

Dithiocarbamate
  microencapsulation of, 13:454

Dithiocarbamates, 17:513
  as accelerators, 17:510

Dithiocarbamic acid derivatives
  as fungicides, 10:224

Dithiocarbonic acid
  xanthates, 22:419

2,3-Dithiocyano-1-propanol
  prepn from thiocyanogen and allyl alcohol, 1:918

Dithiodiacetic acid, 20:202

Dithiodiethanoic acid, 20:202

Dithiodiglycollic acid, 20:199, 202. See Dithioglycolic acid.

N,N'-Dithiodimorpholine, 13:665

Dithioethylene glycol, 8:418

Dithioglycolide
  chain-growth polymer, 16:221

2,4-Dithiohydantoins, 10:912

1,3-Dithiolan, 8:418

Dithiolanes, 10:909

Dithiols
  synthesis, 20:213

gem-Dithiols
  prepn, 12:118

Dithionous acid, 19:420

Dithiophosphate, 9:393

Dithiosuccinic anhydrosulfide, 19:141

Dithiovanillic acid, 21:192

Dithizone
  zinc analysis, 22:561
  zinc analysis, 22:559

Dithoammelide, 6:567

Dithymol diiodide, 2:619

2,4-Dititrotoluene, 13:851

2,4-Ditoluene diisocyanate, 7:694

m, p'-Ditolylamine, 7:42

Di-p-tolylamine, 7:42

N,N'-Di-p-tolyl-p-phenylenediamine, 7:46

Ditridecyl adipate
  phys prop, 1:409

Di-n-Tridecyl ketone, 12:104

Ditridecyl phthalate
  prod and sales, 8:376

Di(3,5,5-trimethylhexyl) adipate
  phys prop, 1:409

Di(triphenylmethyl) peroxide, 14:792

4,4'-Ditrityldiphenylamine, 7:44

1,5-Di-O-tritylxylitol, 1:579

Dit(2-ethylhexyl) phthalate, 1:566
  as cellulosic plasticizer, 4:657

Diuretics, 7:**248**
  as cardiovascular agents, 4:521
  mercury compounds, 13:242
  veterinary, 21:252

Diuron
  urea herbicide, 22:190

Divaleroylmethane, 12:156

Divers' liquid, 2:321

Divi-divi, 12:317, 322, 323

Divinylacetylene, 5:221

2,5-Divinyl-1,4-dioxane, 1:613

Divinylbenzene, 19:74; 22:500
  uranium extraction, 21:15
  in water treatment, 13:360

Divinyl carbinol
  in prepn of D,1-arabitol, 1:572

Divinyl carbonate
  vinyl pyrrolidone copolymer, 21:433

Divinyl ether, 8:487

Divinyl glycol, 7:415

3,9-Divinylspirobi(m-dioxane)
  phys prop, 1:262

Divinyl sulfone, 11:530

3,9-Divinyl-2,4,8,10-tetraoxospiro(5.5)undecane, 1:262

Divinyltitanium chloride, 20:443

Diwolfram trisilicide, 22:355

Dixanthogen, 22:423

Dixanthogen, 22:428

Dizirconium pentaacetate, 22:636

DMAC. See Dimethylacetamide.

DMC. See 2-Chloro-N,N-dimethylethylamine.

DMIC. See 2-Chloro-N,N-dimethylpropylamine.

DMMPC. See 3-Chloro-N,N,2-trimethylpropylamine.

DMPC. See 3-Chloro-N,N-dimethylpropylamine.

DNA
  x-ray analysis, 22:454

DNOC. See 2,4-Dinitro-o-cresol.

DNOC, 6:441

DNT. See 2,4-Dinitrotoluene.

Di-(1-methylcyclopentyl) peroxide, 14:792

Di(1,3-dimethylbutyl) adipate
  phys prop, *1*:409
1,4,5,6,7,7-Hexachloro-5-norbornene-2,3-di-
  methanol, cyclic sulfite. See *Endosulfan.*
4,8,12,15,19-Docaspentaenoic acid, See *Clupa-*
  nodonic acid.
Docosahexaenoic acid, *7*:664
Docosanoic acid. See *Behenic acid.*
Docosapentaenoic acid, *7*:664
11-Docosenoic acid, *8*:814
13-Docosynoic acid. See *Behenolic acid.*
  chain-transfer agent, *21*:374
1,2,3,4,7,8,9,10,13,13,14,14-Dodecachloro-1,4,4a,-
  5,6,6a,7,10,10a,11,12,12a-dodecahydro-1,4
  7,10-dimethanodibenzo [a,e] cyclooctene,
  S:485
  urban, *S*:329, 335
Dodecanedioic acid
  prepn, *1*:247
2,5-Dodecanedione, *12*:158, 160
3,6-Dodecanedione, *12*:158
Dodecanethiol, *6*:639; *7*:679
1-Dodecanethiol
  odor, *20*:206
1-Dodecanethiol. See *Lauryl mercaptan.*
1,12-Dodecanedithiol
  phys constants, *20*:207
Dodecanoic acid, *8*:812. (See also *Lauric acid.*)
1-Dodecanol, *1*:543, 564, 568. (See also *Lauryl*
  alcohol.)
1-Dodecene
  epoxidization, *8*:247
  sulfation, *19*:302
Dodecyl. (See also *Lauryl.*)
Dodecyl acrylate
  phys prop, *1*:287
n-Dodecyl alcohol. See *1-Dodecanol.*
Dodecyl alcohol
  amines from, *S*:41
Dodecyl aldehyde, *9*:358
Dodecylamine
  properties, *2*:130
Dodecylamine acetate
  solubility, *2*:133
Dodecylamine salts
  solubilities, *2*:132
Dodecylammonium chloride, *2*:632; *9*:386
Dodecylbenzene
  commercial product, *1*:892
DMAP, *6*:81
  prepn of, *3*:374, 398
Dodecylbenzenesulfonates, *19*:516

Dodecyl chloroformate, *4*:386
Dodecyldiphenyl ether, *19*:318
Dodecyl ether, *8*:493
Dodecyl gallate, *10*:14
Dodecyl glycine, *2*:635
Dodecylguanidine acetate, *6*:561
n-Dodecylguanidine acetate, *10*:225
Dodecylguanidine salts, *10*:740
Dodecyl mercaptan. See *Dodecanethiol.*
Dodecylmethylamine, *2*:635
Dodecylphenol, *1*:909
  reaction with ethylene oxide, *8*:529
Dodecylphenoxypenta(ethyleneoxy)ethanol, *19*:527
Dodecylphenoxypolyoxyethyleneethyl sulfate,
  19:528
Dodecylphosphoric acid
  uranium extraction, *21*:17
N-Dodecyl-1,3-propanediamine, *2*:634
Dodine acetate, *10*:225
Doebner's Violet, *20*:677
  acidification of, *20*:696
  structural formulas, *5*:767, 769, 771
  substituent effects on color, *20*:689
Dolemite
  calcined, *17*:227
  refractory, *17*:236
Dolomite, *12*:415; *22*:563
  as abrasive, *1*:25
  in ceramics, *4*:770
  in ocean, *14*:159
  in prepn of trifluoroboroxine, *9*:556
  sulfur extraction, *19*:354
  in talc, *19*:608
  used as blast-furnace flux, *12*:12
Dolomite 12
  manuf of magnesium carbonate, *12*:713
Dolomitic rock, *12*:415
Doolittle equation
  for viscosity, *21*:465
DOP. See *Di(2-ethylhexyl) phthalate.*
DOP. See *Di(2-ethylhexyl) phthalate.*
"Doping agents"
  in dispersion-hardened alloys, *11*:16
Doricin, *16*:333
Doryl. See *Carbamylcholine chloride.*
Dosimeter glass, *10*:599
Dotriacontanoic acid, *8*:813
n-Dotriacontoic acid, *8*:813
Double-bond migration, *14*:317
Douglas fir bark wax, *22*:161
Dow chlorine cell, *1*:684
Dowtherm A
  heat-transfer medium, *10*:850

Dowtherm E
  heat-transfer medium, *10*:851
Dowtherm 209
  heat-transfer medium, *10*:852
Doxinate. See *Dioctyl sodium.*
Doxinate. See *Dioctyl sodium slufoxuccinate.*
Doxycycline, *20*:5, 9, 18, 27
  epimers, *20*:11
Doxylamine succinate
  antihistamine, *11*:38
DPA.(See *Diphenylamine.*)
DPEHN. See *Dipentaerythritol hexanitrate.*
DPPD, *2*:597
Drag chain conveyors, *6*:109
Dragées, *6*:51
Driers and metallic soaps, *7*:**272.** (See also *Dryers.*)
Drilling fluid additives
  use of lignosulfonates in, *12*:372
Drilling fluids, *7*:**287**
Drilling fluids, *3*:352
  clays in, *5*:573
Dromoran. See *3-Hydroxy-N-methylmorphinan.*
Dromoran, *2*:383
Drop black, *15*:545
Dropping mercury electrode, *7*:758
Drug resistance
  chloramphenicol, *4*:933
Drug resistance of pathogenic microorganisms, *3*:8
Drugs. (See also *Pharmaceuticals.*)
Drugs
  antiamebic, *20*:73
  brand-name, *15*:113
  carbon blacks used in, *4*:277, 280
  colors for, *5*:857
  ethical, *15*:113
  generic-name, *15*:113
  graphite used in, *4*:332
  microbial therapeutics, *20*:74
  natural therapeutic, *20*:74
Dry cells. (See also *Batteries.*)
Dry chemical agents
  for fire extinguishing, *S*:364
Drycleaning, *7*:**307**
  use of emulsions, *8*:147
Drycleaning solvents
  classification of, *7*:352
  carbon in reclamation of, *4*:152
Dry curing
  of meats, *13*:174

Dryers. (See also *Driers.*)
Dry ice. See *Carbon dioxide.*
  cloud seeding, *21*:661
Drying, *7*:**326**
Drying agents, *7*:**378**
  anhydrite, *4*:25
Drying oils, *7*:**398**
Drying oils, *4*:29
Dry lubricants
  use of metallic soaps in, *7*:284
Dry process
  for hardboard, *21*:603-607
Dry solvents. See *Drycleaning.*
DTA. See *Differential thermal analysis.*
  aquametry by, *2*:695
DTPA. See *N,N,N',N'',N''-Diethylenetriamine-pentaacetic acid.*
Ductile-to-brittle transition, *13*:189
Ductility, *13*:187
  measuring, *13*:184
Ductility transition temperature, *13*:191
Dulcin, *19*:595
Dulcitol, *1*:573
  phys prop, *1*:570
  prepn, *1*:573
Dumortierite
  refractory, *17*:231
Dunmore element
  hygrometer used in drycleaning, *7*:312
1-Duodecanol
  phys prop, *1*:543
Duphaston, *6*:84
Duplicating, See *Reprography.*
Duplicating fluids, *8*:447
Duralumin, *13*:326
Durene. See *1,2,4,5-Tetramethylbenzene.*
Durene, *6*:629; *16*:300
  oxidation, *15*:482
Dursban
  ectoparasite control, *21*:250
Dust, *7*:**429.** (See also *Gas cleaning.*)
  composition of, *7*:309
  control of, *10*:331
  engineering aspects, *7*:429
  hygienic aspects, *7*:453
  as fire hazard, *9*:292
Dusting powder, *6*:360
Dyeing. See *Dyes--Application and evaluation.*
  of hair *10*:790
Dyeing industry
  use of ammonia in, *2*:296
Dyeing of furs, *10*:305

Dyeing, package
Dyeing (textiles)
  use of formic acid in, *10*:102
Dye intermediates. See *Dyes and dye intermediates.*
Dyes
  acid monoazo, *2*:880
  alcohol-soluble, *18*:658
  from anthranilic acid, *3*:434
  anthraquinone, *2*:50
  azine, *2*:859
  azo, *2*:868
  azoic, *2*:889
  as bacteriostatic agents, *2*:638
  from benzidine, *3*:413, 414
  from benzotrichloride, *5*:286
  from benzyl chloride, *5*:286
  bromine in, *3*:775
  cationic, *20*:194
  for cellulosics, *4*:660
  chrome dyes for wool, *22*:403
  chromium compounds as, *5*:511
  for colored signalling smokes, *4*:904
  for color photography, *5*:816
  from *o*-dianisidine, *3*:417
  diazo, *2*:877
  direct, *20*:185
  disperse, *20*:191
  disperse azo, *20*:193
  economic aspects, *7*:492
  fluorescein, *15*:456
  glycol ether-soluble, *18*:659
  level-dyeing acid dyes for wool, *22*:403
  manuf of intermediates, *7*:487

metal-complex dyes for wool, *22*:403
metalizable, *2*:882
mordant, *2*:502, 883
nonplanar, *5*:781
for paper, *14*:502
planar, *5*:781
phthalocyanine, *15*:456
premetalized, *2*:882, 887; *5*:511
rhodamine, *15*:456
spectral sensitizing, *15*:356
thiazole. See *Thiazole dyes.*
from *o*-tolidine, *3*:415
vat, *20*:188
xanthene, *15*:456
Dyes and dye intermediates, *7*:**462**
Dyesapplication and evaluation, *7*:505, 506
  machinery for, *7*:508
Dyes, natural, *7*:**614**
Dyes, organic
  color and constitution of, *5*:763
Dyes, reactive, *7*:**630**
  reaction mechanism, *7*:637
  uses, *7*:477
Dynamite, *8*:637
  manuf, *8*:636
  semigelatin, *8*:640
Dynel, *1*:333
  manuf, *1*:334
  properties, *1*:333
  uses, *1*:335
Dypnone, *1*:167; *12*:113
Dysproprosium, *17*:144
  electronic configuration, *1*:360
  ionic radii, *1*:370
  in nuts, *14*:131

E

EAN, *4*:491; *6*:127
EPR. See *Electron paramagnetic resonance.*
EPT. See *Ethylene propylene terpolymer.*
Eberhard effect, *15*:381
Ebonite, *9*:857
  properties, *9*:860
Eburnamonine, *1*:773
Ecdysone, *18*:832, 845, 846, 866
  derivatives, *18*:849
  insect hormone, *11*:50
Ecdysterone, *18*:849
Ecgonine, *1*:785
Echimidine, *1*:767
Echinomycin, *16*:333
Echinopsine, *1*:771
Economic evaluation, 7:**642**
Edeine, *16*:313
Edema, *7*:250
Edestin
  in coconuts, *14*:125
Edible fats
  use of lecithin, *12*:356
Edible oils
  use of lecithin, *12*:356
Edifas A. See *Methylethylcellulose.*
Edifas A, *4*:650
Ediger process
  in acetylene manuf, *1*:183
Edison battery, *12*:546
Edrophonium, *6*:551
EDTA. See *N,N,N',N'-Ethylenediaminetetra-*
  acetic acid.
Effective atomic number, *4*:491; *6*:127
Effective conductivity
  of heat, *11*:825
Effluent polishing
  in water treatment systems, *22*:120
Egg albumin
  in confectionery, *6*:44
Eggs, 7:**661**
E Glass
  in ceramic composite armor, *S*:139
Egyptian blue, *5*:853; *15*:541
EHPG. See *N,N'-Ethylenebis[2-(o-hydroxy-*
  phenyl)]glycine.
Eicosandicarboxylic acid. See *Phellogenic acid.*
  in cork, *6*:284

Eicosane, *5*:233
Eicosanoic acid, *8*:813
5,8,11,14-Eicosatetraenoic acid, *8*:815
10-Eicoscenyl-l-ol
  phys prop, *1*:543
1-Eicosene, *21*:434
Eicosonyl alcohol, *1*:543
Eicosopentaenoic acid, *7*:664
Eicosyl alcohol. See *1-Duodecanol.*
Eicosyl alcohol, *1*:543
Einkorn, *22*:253
Einstein equation of viscosity, *21*:483
Einsteinium. See *Actinides.*
Ekonol, *S*:741
9 α-Fluoro-11 β-hydroxycorticoids, *18*:840
Elaidic acid, *8*:814
  phys prop, *1*:231
  structure, *8*:817
Elaidolinolenic acid, *8*:815
Elastic limit, *13*:185
Elastic waves
  complete, *20*:40
  elastic performance coefficient, *20*:41
  photon-modulated longitudinal, *15*:409
Elastin fibers
  in meats, *13*:168
Elastomeric fibers, *18*:614
Elastomers, *17*:543
  butadiene in manuf of, *3*:808
  cast polyurethan, *21*:94
  chain-growth siloxane, *16*:229
  economics, *7*:699
  as gaskets, *14*:450
  from hydrocarbons, *11*:282
  as impregnants, *14*:457
  cis-polybutadiene, *S*:106
  prop of polyurethan, *21*:96
  poly(vinylidene fluoride), *21*:269
  processing, *7*:688
  properties, *7*:690
  solid polyurethan, *21*:94
  urethan, *21*:77
  use of hexane in manuf of, *11*:4
  uses of polyurethan, *21*:97
Elastomers, synthetic, 7:**676.** ((See also *Rub-*
  ber.)

β-Eleastearic acid, *8*:815

Elecampane, *14*:201

Electrical industry
  acetalre resins in, *1*:105
  aluminum in, *1*:983
  humidity conditions in, *1*:486

Electrical insulation. See *Insulation, electric.*

Electrical testing, *7*:**716**

Electrical units, *S*:1002

Electric furnace. See *Furnaces, electric.*

Electric generators
  brush materials for, *4*:222, 332

Electric power generation
  coal as fuel for, *5*:664

Electroanalytical methods, *7*:**726**

Electrochemical cells. See *Batteries and electric cells.*

Electrochemical machining. See *Electrolytic machining.*

Electrochemistry, *7*:**784**

Electrodecantation, *7*:**841**
  for enzymes, *16*:626

Electrodeposition, *7*:765; *9*:203; *15*:862. ((See also *Electroplating.*)
  of refractory coatings, *17*:276

Electrode potential, *7*:793

Electrodes. ((See also *Anodes.*)
  carbon, *4*:206, 228
  graphite, *4*:207, 227
  semigraphite, *4*:228
  Söderberg, *4*:196

Electrodialysis, *7*:**846**
  desalination of water, *22*:62
  removal of dissolved solids in water by, *22*:75

Electroendosmosis. See *Electroosmosis.*

Electrofine chlorinated paraffins, *5*:233

Electroforming, *8*:40; *13*:316

Electrogasdynamics, *16*:463

Electrogravimetry, *7*:768

Electro-Katadyn process, *2*:621

Electrokinesis, *8*:23

Electrokinetic phenomena, *8*:23

"Electroless" plating, *8*:69

Electroluminescence, *12*:617

Electrolysis
  aquametry by, *2*:699
  methods of analysis, *7*:765
  of water, *6*:902
  zinc prod, *22*:584

Electrolytic machining methods, *7*:**866**

"Electromagnetic" system of equations, *S*:1003

"Electromagnetic unit of current," *S*:1005

Electromigration, *S*:**278**

Electron accelerators
  in microwave frequencies, *S*:581

Electron beam heating, *10*:278

Electron beam welding, *8*:17

Electron diffraction, *22*:452

Electronic equipment
  use of adhesives in, *1*:398

Electron microprobe
  X-ray fluorescence analysis, *22*:465

Electron mobility
  in MHD generator, *S*:226

Electron paramagnetic resonance, *7*:874

Electron tube materials, *8*:**1**

"Electronvolt"
  as a "unit", *S*:997

"Electron wind", *S*:280

Electroosmosis, *8*:23
  in drying, *7*:331

Electrophoresis, *8*:23
  sugars, *19*:165

Electrophoretic deposition, *8*:**23**

Electropickling
  citric acid in, *5*:537

Electroplating, *8*:**36**; *13*:249
  chromium, *5*:506
  use of butynediol, *1*:607
  use of fluoboric acid, *9*:564
  use of metal fluoborates, *9*:566

Electrostatic force
  in electromigration, *S*:284

Electrostatic precipitation, *8*:**75**

"Electrostatic" system of equations, *S*:1003

"Electrostatic unit of charge", *S*:1005

Electrostatography, *17*:346

Electrotype metal, *20*:762, 767

Electrowinning
  of copper, *S*:255

Elementsperiodic system, *8*:**93**

Elemi, *17*:385

α-Eleostearic acid, *8*:815
  phys prop, *1*:231

β-Eleostearic acid, *8*:815
  phys prop, *1*:231

Elevators
  bucket, *6*:115

Ellagic acid, *12*:321, 322

Elm bark, *9*:353

Ematamine, *20*:75

Embalming fluids, *8*:**100**

Embedding, *8*:**102**

Embolite, 3:758
Emerald, 2:1
  synthetic prod, 10:511, 517
  use of beryllium fluoride in synthesis of,
    9:552
Emerald green, 15:538
Emery
  as abrasive, 1:25
  in coated abrasives, 1:29
Emetine, 20:70
Emetine, 20:74
Emetine, 1:728, 772, 778, 793; 20:70, 74
  N-substituted derivatives, 20:75
Emetine bismuth iodide, 20:75
Emetine hydrochloride, 20:85
Emetine N-oxide, 20:75
Emf series
  for lithium chloride-potassium chloride eu-
    tectic, 6:315
Emissions
  automobile exhaust, S:56, 59
Emmer, 22:253
Emmeric-Engel reagent
  for α-tocopherol, 21:580
Emodin, 2:466
  cathartic action of, 4:588
Emulsifying agents.   (See also Emulsions;
    Surfactants.)
  in food products, 10:5
Emulsion cleaners, S:165
Emulsions, 8:117
  asphalt, 2:778
  photographic, 15:355, 358
  solubility parameters and, S:904
Enamel
  heat-resistant wire, 21:308
Enameling
  of sheet steel, 8:155
Enameling technique. See Cloisonné.
Enamels, 14:473
  drying oils in, 7:423
  as electric insulation, 11:790
  for home appliances, 5:693
  polyimides in, S:769
Enamels, porcelain or vitreous, 8:155
Enamines
  from morpholine, 13:661
Enanthic acid. (See also Heptanoic acid.)
Enanthic acid, 8:812
  properties, 1:229
Enantholactam
Enargite, 2:712; 6:134, 135

Enargite, 9:393
Encapsulants
  as electric insulation, 11:794
Encapsulation, 8:103
Encaustic, 14:481
Endellite, 5:545
  layer structure, 5:542
Endocrine glands
  removal of, 11:49
Endomethylenetetrahydrophthalic   acid,
    12:825
Endoplasmic reticulum
  in yeast, 22:512
Endosulfan, 5:249
  in water, 21:701
Endrin, 5:247
Endrin, 8:249; 15:919; 18:526
Energy
  conduction, 20:653
  mechanical in transport processes, 20:633
  thermal in transport processes, 20:639
  total energy in transport processes, 20:629
  transport processes, 20:652
Enfleurage
  essential oils by, 14:182
Engines, internal-combustion. See Internal-
    combustion engine.
Engines, steam. See Steam engines.
English vermilion, 15:525
  internal-combustion. See Internal-combus-
    tion engine.
Enniatins, 16:336
Enovid, 6:65, 66, 87
Enovid E, 6:87
Enstatite, 17:237
Enterobacteriaceae
  chemotherapy against, 3:24
Enthalpy
  of pure fluids, 20:132
Entropy, 20:125
  and probability, 20:127
  of pure fluids, 20:132
  reversible processes, 20:126
Environmental conditioning
  and natural gas, S:436
Enzyme detergents, S:294
Enzyme digestive aids, 8:195, 196, 203
Enzymes
  flavoprotein, 17:447
  in beer brewing, 3:297
  in food processing technology, 10:9
  in human blood, 3:577

phosphorylase, *19*:151
proteolytic, *5*:594, 595
sonochemical changes, *20*:781
tryptic in detergents, *S*:294
ultrasonic prepn, *20*:781
in yeasts, *22*:516
Enzymes, industrial, *8*:**173**
Eosamine G, *7*:484
Eosin, *3*:775
as indicator, *11*:558
Eosin G. See *D&C Red No. 22.*
Eosin YS, *5*:870, 873; See *D&C Red No. 22.*
Eosine
structural formula, *5*:765
xanthene dye, *22*:433
Eosine G. See *Eosin YS.*
Eosine G, *5*:870, 873
Eosins
resorcinol dyes, *11*:477
Ephedrine, *1*:778, 788; *4*:517
DL-Epiallobiotin, *3*:520
Epiandrosterone, *18*:857
DL-Epiandrosterone, *18*:858
DL-Epibiotin, *3*:520
Epibromohydrin, *2*:642; *9*:760
Epicatechin, *19*:745, 748
Epichlorohydrin, *2*:642; *5*:304, 318; *13*:856;
16:517; *19*:147
alkylating agent, *20*:699
curing agents, *20*:215
cyanuric acid reaction, *20*:664
epoxy resins, *15*:797
manuf of epoxy resins, *8*:294
α-monochlorohydrin from, *5*:317
perchlorate ester of, *5*:75
phenolic resins, *15*:186
phenoxy resins, *15*:795
polyhydroxyethers, *15*:173
in prepn of triethyl oxonium fluoborate,
9:569
reaction with ethylenediamines, *7*:29
rubber, *17*:545
as starch modifier, *10*:17, *10*:613
in synthesis of glycerol, *10*:624
sterilant, *18*:817
toxicity of, *5*:320
Epigallocatechin, *19*:745, 748
Epigallocatechin gallate, *19*:748
Epilepsy
control of, *11*:519
Epinephrine, *2*:189; *8*:**231**; *11*:468, 470
as cardiovascular agent, *4*:517
derangement in metabolism, *16*:643

Epirhamnose, *4*:137
Epi-*trans*-ergocalciferol, *21*:564
Episulfides
reaction with aziridines, *11*:531
12a-Epi-4-dedimethylaminooxytetracycline, *20*:20
5a-Epi-6-epitetracycline, *20*:12
5a-Epitetracycline, *20*:12
6-Epitetracycline, *20*:12
Epoxidation, *8*:**238**
Epoxides
in prepn of ketones, *12*:128
reactions with aziridines, *11*:531
Epoxides, *8*:**263**
1,2-Epoxybutane. See *Butylene oxide.*
1,2-Epoxybutane, *8*:288
Epoxycyclohexyl structure, *8*:259
Epoxycyclopentyl structure, *8*:259
Epoxydicyclopentyl structure, *8*:259
Epoxy esters
manuf, *8*:253
Epoxyethane. See *Ethylene oxide.*
1,2-Epoxyethylbenzene. See *Styrene oxide.*
12,13-Epoxy-octadeca-*cis*-9-enoic acid, *8*:787
Epoxy oils
manuf, *8*:253
12,13-Epoxyoleic acid, *8*:787
1,2-Epoxypropane. See *Propylene oxide.*
Epoxy resin esters, *8*:304
use as coatings, *8*:307
Epoxy resins, *8*:**294**; *12*:190
as adhesive, *1*:389
in ceramic composite armor, *S*:145
chlorendic anhydride in, *S*:480
as coatings, *14*:476
titanium alkoxide catalysts, *20*:491
2,3-Epoxy-1-propanol, *5*:313
1,2-Epoxy-3,7-dimethyloctan-8-ol, *19*:829
sterilant, *18*:817
5,8-Epoxy-1,2,3,4-tetrafluoro-5,6,7,8-tetrahydro-
naphthalene, *9*:797
Epsomite
from Great Salt Lake, *S*:455
Epsom salts. See *Magnesium sulfate.*
Equation of state gases, *6*:759
Equations
in data presentation, *6*:706
linear, *6*:713
of reference substances, *6*:710
Equilenin, *18*:853, 854
d-Equilenin, *18*:854
Equivalent theoretical plate, *7*:95, 97
Erbium. ((See also *Actinides.)*
in nuts, *14*:131

Ergobasine. See *Ergonovine.*
Ergocalciferol, *18*:837; *21*:549
 synthesis of, *21*:562
Ergocornine, *1*:792
Ergocristine, *1*:792
Ergokryptine, *1*:792
Ergometrine. See *Ergonovine.*
Ergonovine maleate, *1*:789
Ergosterol, *18*:836, 849, 880
 irradiation of, *21*:553
 manuf of, *21*:559
 photolysis, *15*:338
 vitamin D activity, *21*:550
 in yeast, *22*:516
 yeast product, *22*:531
Ergot, *4*:519
Ergot alkaloids, *1*:497, 512
Ergotamine
 as cardiovascular agent, *4*:518
Ergotamine, *1*:763, 778, 789, 790
Ergotamine tartrate, *1*:790
Ergothioneine, *2*:188
Ergotocin. See *Ergonovine.*
Ergotoxine, *4*:519
Erichrome Azurol B, *20*:697
Erigeron, *14*:201
Eriochrome Azurol B, *20*:720
Eriochrome Black, *2*:873
Eriochrome Black T, *2*:873, 884, 885
 as indicator, *11*:560
Eriochrome Blue Black R, *7*:485
Eriochrome Cyanine R, *20*:720
Eriochrome Red B, *2*:885
Erioglaucine, *5*:868
Erioglaucine A, *20*:707
Erucic acid, *1*:253; *8*:814
 in dental material, *6*:827
 phys prop, *1*:231
Erucic acid oils, *8*:784, 786
Erytheogenic acid, *8*:787, 816
Erythralosamine, *12*:641
Erythrite
 phys prop, *5*:722
Erythritol
 phys prop, *1*:570
 prepn, *1*:571
 prepn from butenediol, *1*:612
i-Erythritol, *2*:579
Erythrityl tetranitrate
 as cardiovascular agent, *4*:521
D-Erythro-D-galacto-octitol, *1*:576
Erythrohydroxyanthraquinone, *2*:466

Erythrolaccin, *18*:21, 26
Erythromycin, *12*:633, 635, 641
 amebicide, *20*:80
 in bacterial infections, *3*:1
 in chemotherapy, *12*:653
 as veterinary drug, *21*:245
Erythromycin, *13*:463
Erythronium. See *Vanadium.*
Erythronolide, *12*:641
Erythroresinotannol, *17*:382
Erythrosin, *5*:866; *11*:479
Erythrosine Bluish, *5*:866
Erythrosine Bluish Lake, *5*:869. See *Erythrosine Lake.*
Erythrosine Lake, *5*:869
Erythrosine Yellowish Na, *5*:869
Erz cement. See *Cement.*
Escobilla
 analysis, *9*:173
Esculatin, *6*:432
Esculin, *15*:681
 as optical brightener, *3*:744
Eserine, *1*:772
Esidron
 as diuretic, *7*:256
Esparto, *9*:181
Esparto grass wax, *22*:164
Essential oils. See *Oils, essential.*
Ester-alcohol interchange, *8*:356
Esterases
 classification, *8*:176
 in milk, *13*:514
Ester hydrolysis, *8*:358
Esterification, *8*:**313**
Ester interchange, *8*:**356**. (See also *Interesterification.*)
 prod of carbonylic, *8*:313-354
Esters
 polymerization of cyclic, *16*:171
 solubility parameters, *S*:893
Esters, organic, *8*:**365**
Estradiol, *6*:63; *18*:864
 veterinary uses, *21*:251
9β-Estradiol, *18*:882
Estra-1,3,5(10)-triene-3,17 β-diol 3-methyl ether, *6*:63, 65
17β-Estradiol
 esters of, *11*:117
 phys and chemical prop, *11*:116
 synthesis of, *11*:117
Estradiol 3,17-dipropionate, *11*:117
Estradiol 3-monobenzoate, *11*:117

Estragole, *19*:819, 821
Estragon, *9*:352, 375; *14*:201375
  essential oils as flavor, *9*:375
  oil of, *14*:201
Estrane
  structure of, *11*:79
Estranol, *6*:87
Estrapentaene, *18*:863
Estra-1,3,5(10)-triene-3,17β-diol, *6*:63. See *Estradiol.*
Estriol, *11*:118
Estrogens, *11*:106-117
  nonsteroidal, *11*:127
Estron, *6*:69, 71; *18*:836, 854, 868; *S*:10. See *Acetate and triacetate fibers.*
  isolation of, *11*:113
  phys and chemical prop, *11*:112
  synthesis of, *11*:114
d-Estrone, *18*:855
9β,10α-Estr-4-en-3-one, *18*:882
Etamycin, *16*:332
Etching
  in photoengraving, *16*:522
  powderless, *16*:523
Ethambutol, *3*:30; *19*:40
Ethanal. See *Acetaldehyde.*
Ethanamide. See *Acetamide.*
Ethane, *8*:**383**
  as component of natural gas, *10*:448
  as refrigerant, *10*:859
  chlorine reactions with, *5*:143
  drying by alumina, *2*:52
  ethylene from, *S*:310
  from catalytic reforming, *15*:33
  from petroleum, *15*:18
  in fuel cells, *3*:151
  in manuf of ethanol, *8*:437
  in manuf of ethyl chloride, *5*:143
  in manuf of ethylene, *8*:503, 516
  in manuf of ethylene oxide, *8*:539
  methyl chloride in formation of, *5*:103
  oxidation of, *11*:226
  refrigerant, *17*:309
  toxicity of, *11*:294
Ethanedioic acid. See *Succinic acid.*
1,2-Ethanedithiol
  phys constants, *20*:207
Ethanesulfonic acid, *10*:161; *19*:314, 316
  as catalyst in prepn of esters, *8*:350
Ethanethiol, *6*:646. ((See also *Thiols, cyanoethylation of.)*
  addition reactions, *20*:209

  prepn, *20*:205
  prepn from ethyl alcohol, *8*:456
  stain solvent, *18*:662
  uses, *20*:214
Ethanetricarboxylic acid
  decarboxylation, *19*:144
Ethanoic acid, *8*:**386.** (See also *Acetic acid.)*
  dehydration of, *2*:850
  phys prop, *8*:812
  separation from formic acid and water, *2*:854
Ethanol, *2*:420; *8*:422; *10*:774
  as an hypnotic, *11*:509
  as propellant, *8*:681, 688, 692
  bactericidal action of, *2*:614, 615
  butadiene from, *3*:802
  cyanoethylation of, *6*:644
  dehydration of, *2*:853
  dehydrogenation, *8*:349
  from lactose, *19*:238
  from sucrose, *19*:224
  graph of vapor pressure, *6*:732
  heat of combustion of, *4*:898
  hydrochloric acid reaction with, *5*:143
  hydrolysis of sulfuric esters, *19*:486
  in alcoholic beverages, *1*:518
  in hair coloring, *10*:796
  in hair spray, *10*:783
  in manuf of ethyl chloride, *5*:143
  in manuf of ethyl ether, *8*:480
  in prepn of acetaldehyde, *1*:84
  in prepn of mercury fulminate, *8*:588
  in prepn of silver fulminate, *8*:590
  in refrigeration, *17*:310
  latent solvent, *18*:577
  oxidation to manuf acetic acid, *8*:394
  parenteral solvent, *15*:128
  prepn from ethylene, *8*:501
  prepn from ethylene oxide, *8*:527
  shrinkproofing wool, *22*:409
  solvent recovery, *18*:562
  stain solvent, *18*:659
  thermometers, *19*:782
Ethanolamine. (See also *Alkanolamines.)*
  in gramicidins, *16*:307
  polyester scission, *16*:178
Ethanolamines
  piperazine from, *15*:640
  stain solvents, *18*:660
  substituted as polyurethan catalysts, *21*:91
Ethanoyl chloride. See *Acetyl chloride.*

Ethchlorvynol
  as hypnotic, *11*:511
Ethene. See *Ethylene.*
Ether. (See also *Diethyl ether and Ethyl ether.*)
Ether peroxides, *8*:479
Ethers, *8*:**470**
  solubility parameters of various, *S*:893
Ethinamate
  as hypnotic, *11*:518
Ethionic acid, *19*:494
  prepn, *8*:502
Ethionine, *15*:686
3-Ethocy-4-hydroxybenzaldehyde. See *Ethyl-vanillin.*
Ethoheptazine, *2*:385
Ethopabate
  as anticoccidial, *21*:247
Ethosuximide
  as anticonvulsant, *11*:523
p-Ethoxyacetanilide, *2*:219
o-Ethoxyacetic acid
  as dental material, *6*:826
Ethoxyacetylacetone, *12*:856
3-Ethoxyanhydrovitamin A, *21*:501
o-Ethoxyaniline, *2*:422. See *o-Phenetidine.*
p-Ethoxyaniline
  condensation with dialkylchloroalkyl-amines, *2*:143
o-Ethoxybenzidine, *3*:417
3-Ethoxybenzidine, *3*:417
o-Ethoxybenzoic acid, *6*:826
6-Ethoxybenzothiazole-2-sulfonamide
5-(Ethoxycarbonylmethyl)pyrimidine, *20*:178
β-Ethoxycrotonic ester, *1*:157
Ethoxyethane. See *Ethyl ether.*
2-Ethoxyethanol
  chlortetracycline prod, *20*:6
2-Ethoxyethyl acetate, *9*:844
2-Ethoxyethyl acrylate
  phys prop, *1*:288
1-Ethoxyethyl hydroperoxide, *8*:479
3-(Ethoxymethoxy)propionitrile, *6*:644
(Ethoxymethylene)malononitrile, *12*:856
2-Ethoxynaphthalene, *15*:169
Ethoxypentadiene
  steroid synthesis, *18*:859
p-Ethoxyphenylurea, *19*:595
3-Ethoxypropionaldehyde
  prepn from acrolein, *1*:259
3-Ethoxypropionitrile, *6*:644
Ethoxyquin, *2*:597; *10*:19
Ethoxyzolamide
  as diuretic, *7*:259

3-Ethoxy-4-azaestrone, *18*:872
6-Ethoxy-1,2-dihydro-2,2,4-trimethylquinoline, *2*:597
4-Ethoxy-2-hydroxynitrosobenzene
  sodium borohydride reaction, *S*:501
Ethyl. (See also *Cellulose derivatives-plastics.*)
α-(*N*-Ethylanilino)-*m*-toluenesulfonic acid, *20*:701
Ethyl *α*Bepoxy-*β*-methylhydrocinnamate
  flavoring, *4*:344
N-Ethyl-*N*-benzyl-*m*-toluidine, *20*:701
Ethyl *N*-chlorobenzimidate, *4*:924
Ethyl-*N*-*t*-butylmalonamate, *20*:23
N-Ethyl-*N*-2-hydroxyethylaniline
  prepn, *8*:528
Ethyl Mauveine, *7*:464
N-Ethyl-*N*-phenylbenzylamine, *2*:420
Ethyl-*N*-phenylcarbamate, *8*:681
  as an herbicide, *22*:191
Ethyl Violet
  manuf of, *20*:712
  synthesis, *20*:680
Ethyl acetamidomalonate, *2*:166
Ethyl acetate, *6*:465; *8*:479
  as synthetic flavor, *9*:354
  condensation with ethyl nicotinate, *21*:527
  in nail enamels, *6*:368, 369
  sodium reaction, *18*:451
  stain solvent, *18*:660
  in prod of peroxyacetic, *8*:253
  prepn from acetaldehyde, *1*:82
  prod, *8*:328
  prop, *8*:378
  prod and sales, *8*:376
  spec, *8*:373
Ethyl acetoacetate. See *Acetoacetic ester.*
Ethyl acetoxy(phenylthio)acetate, *19*:324
Ethyl acetylacetate. See *Acetoacetic ester.*
Ethylacetylacetone, *12*:148
Ethyl acrylate
  ethylene copolymer, *14*:218
  gas chromatographic analysis of, *1*:301
  in dental material, *6*:830
  manuf from ethyl alcohol, *8*:455
  phys prop, *1*:287, 301
  polymerization, *1*:303
  prepn, *1*:297
  prod and sales of, *8*:376
  toxicity data, *1*:302
  vinyl acetate copolymer, *21*:331
1-Ethyladamantane, *S*:2, 4, 10
Ethyl *β*-ethoxypropionate
  thiamine synthesis, *20*:178

α-Ethyl-β-hydroxybutyraldehyde. See *Aceto-butyraldol.*
Ethyl β-hydroxybutyrate, *1*:156
α-Ethyl-β-hydroxycaproaldehyde. See *Butyraldol.*
α-Ethyl-β-hydroxycaproaldehyde
prepn and prop, *3*:874
Ethyl β-ketobutyrate. See *Acetoacetic ester.*
Ethylal, *1*:107
phys prop, *1*:107
Ethyl alcohol. See *Ethanol.*
from wood hydrolysis, *22*:384
Ethyl β-mercaptoethyl carbonate, *10*:906
β-Ethyl-β-methylglutarimide, *21*:535
Ethyl β-methyl-β-phenyl glycidate, *9*:354, 355
Ethylamine, *6*:455; *9*:556
in prepn of *m*-ethylaminophenol, *2*:214
uses, *2*:122
3-Ethylaminobenzenesulfonic acid
in prepn of *m*-ethylaminophenol, *2*:214
Ethyl aminobenzoate
as diuretic, *7*:253
7,7'-[(Ethylamino)bis(trimethyleneamino)]bis-benz[*c*]acridine trihydrochloride, *20*:88
1-Ethylaminodi(2-propanol), *8*:280
3-Ethylamino-*p*-cresol
synthesis of Rhodamine 6G, *22*:434
m-Ethylaminophenol, *2*:214
1-Ethylamino-2-propanol, *8*:280
α-Ethyl-β-*n*-propylacrolein, *3*:875. See *2-Ethyl-2-hexenal.*
Ethylaniline
reaction with ethylene oxide, *8*:528
N-Ethylaniline
prepn from acetanilide, *1*:150
n-Ethylaniline, *2*:412-414, 420
in prepn of *m*-ethylaminophenol, *2*:214
2-Ethylanthraquinone, *2*:438
Ethyl azelate
prepn, *8*:350
Ethylbenzene, *10*:138; *19*:55; *22*:467. (See also *Xylenes and ethylbenzene.*)
commercial prod, *1*:892
in acetophenone manuf, *1*:168
manuf, *19*:57
prepn, *8*:500
reaction with isoprene, *12*:68, 110
Ethyl benzoate, *8*:350, 334
prepn, *8*:351, 370
as synthetic flavor, *9*:354
Ethyl-*o*-benzoate
from benzotrichloride, *5*:283

123-Ethylbenzothiazole(2)13(122-123-ethylben-zothiazole(2)13methyl13-3-ethyl-4-oxot hia-zoline(5))dimethinemerocyanine iodide, *16*:288
Ethylbenzylaniline, *20*:701
sulfonated, *10*:717
Ethylbenzylanilinemonosulfonic acid, *20*:701
Ethylbenzylanilinesulfonic acid, *20*:705
Ethyl biscoumacetate, *5*:600, 601; *6*:432
Ethylbismuth dibromide, *3*:544
Ethylbismuth dinitrate, *3*:544
Ethyl bis(2-nitrophenyl)borate, *3*:657
Ethyl bis-(4-hydroxy-2-oxo-2*H*-1-benzopyr-an-3-yl)acetate. See *Ethyl biscoumacetate.*
Ethyl borate, *3*:652
Ethyl bromide, *8*:428; *9*:570; *3*:774
Ethyl bromoacetate, *3*:777; *8*:419
Ethyl bromoacetone
in chemical warfare, *4*:877
Ethyl-3-bromopropionate
reaction with acetaldehyde, *1*:84
2-Ethylbutanal. See *α-Ethylbutyraldehyde.*
prod and sales, *8*:376
2-Ethyl-1-butanol, *1*:565
phys prop, *1*:564
2-Ethylbutyl acrylate
phys prop, *1*:288
5-Ethyl-5-butylbarbituric acid, *3*:61
5-Ethyl-5-*sec*-butylbarbituric acid, *3*:61, 70
Ethyl butylthionocarbamate
xanthate derivative, *22*:428
α-Ethylbutyraldehyde, *3*:875
Ethyl butyrate
flavor and perfume materials, *9*:358
prepn, *1*:82
prepn, *8*:351
properties, *8*:381
as synthetic flavor, *9*:354, 355
2-Ethylbutyric acid, *8*:850
Ethyl caproate
flavor and perfume material, *9*:358
Ethyl carbamate. See *Urethan.*
Ethyl cellulose, *4*:638; *6*:405; *15*:791
films, *21*:430
in microencapsulation, *13*:440
Ethyl cellulose film
Ethyl centralite, *8*:696
in propellants, *8*:659, 681
Ethyl chloride, *5*:140-147
alkali cellulose reaction with, *4*:638
ethylbenzene manuf, *19*:60
in manuf of tetraethyllead, *1*:899; *12*:291
manuf from ethyl alcohol, *8*:454

prepn of, 2:23
prod, 8:501
toxicity of, 5:93, 146
vinyl chloride reaction with, 5:172
Ethyl chloroacetate
  in prepn of sodium fluoroacetate, 9:768
  prod and sales of, 8:376
  reaction with ethylenimine, 11:529
Ethyl chlorocarbonate, 2:215
Ethyl chloroformate, 4:386, 389
  prepn from ethyl alcohol, 8:456
  properties, 1:224
Ethyl chloroformylacetate, 12:850
Ethylchlorosilanes, 18:233
Ethyl chlorosulfonate, 19:402, 600
  from ethylchloride, 5:143
Ethyl chlorosulfurate. See *Ethyl chlorosulfonate.*
Ethyl cinnamate, 5:519, 520; 6:638; 8:350
  as synthetic flavor, 9:354
Ethyl crotonate, 6:465, 469
Ethyl cyanoacetate, 12:850; 14:534; 19:144
Ethyl 2,2-bis(2-cyanoethyl)-3-ketobutanoate,
  1:158
5-Ethyl-5(1-cyclohepten-1-yl)barbituric acid,
  3:61, 70
5-Ethyl-5-(1-cyclohexen-1-yl)barbituric acid,
  3:61, 70
Ethyl diazoacetate, 8:492; 10:911
Ethyl 3,3-dichlorobutyrate, 1:157
Ethyldichlorosilane, 12:69
Ethyldichlorostibine, 2:582
Ethyldiethanolamine
  phys prop, 1:811
Ethyl difluoroacetic acid, 9:770
1-Ethyl-3,5-dimethyladamantane, S:4
1'-Ethyl-3,9-dimethyl-oxa-2'-carbocyanine iodide,
  6:608
1-Ethyl-2,4-diiodoquinolinium iodide, 6:618
Ethyl-2,2-dimethylparaconate, 19:137
1-Ethyl-2,4-dimethylquinolinium iodide, 6:618
S-Ethyl di-n-propylthiocarbamate, 22:194
Ethyl dioxosuccinate, 12:843
Ethyl enanthate, 8:376
Ethylene, 2:399, 413, 414; 6:695: 8:408,
  409,**499;** 9:401 *S:***310**
  alkyl vinyl ether copolymers, 21:423
  as anesthetic, 2:397
  carbon tetrachloride reactions with, 5:131
  chlorine reactions with, 5:3, 151, 174
  copolymers, 14:233, 238
  copolymers with vinylidene fluoride, 9:842

1,2-dichloroethane reactions with, 5:145,
  175
direct hydration in ethanol manuf, 8:434
dried by alumina, 2:52
ethyl acrylate copolymer, 14:218
ethyl ether as source, 8:486
ethylene chlorohydrin from, 5:308
hydration by dilute acids, 8:437
hydrogen chloride reaction with, 5:144
in alkylation of benzene, 2:22
in esterification of fatty acid, 8:819
in ethane oxidation, 11:226
in ethylene oxide manuf, 8:531
in growth reaction, 2:33
in manuf of 1,2-dichloroethane, 5:151
in manuf of ethyl chloride, 5:144
in manuf of ethyl ether, 8:481
in manuf of vinyl chloride, 5:174
in prepn of acetaldehyde, 1:86
in prepn of acrylonitrile, 1:347
in prepn of elastomers, 7:686
in prepn of ethyl chloride, 2:23
in prepn of organoaluminum compounds,
  2:27
in prepn of polyketones, 12:132
in synthesis of higher alcohols, 1:561
manuf from ethyl alcohol, 8:454
manuf of ethanol, 8:430
oligomerization of, S:648
oxidation for manuf of ethylene oxide, 8:534
oxygen in prod of, 14:405
phys prop, 6:625
polymerization, 7:44, 66, 70, 71, 677; 14:323
polymerization, 20:447
poly(vinyl alcohol) copolymer, 21:358
prepn from acetylene, 1:207
prepn from ethanol, 8:429
radiation polymerization, 17:60
reaction with hydrocyanic acid, 19:144
reaction with oxygen difluoride, 9:632
refrigerant, 10:859; 17:309
refrigerant for xylene manuf, 22:488
source of cyclopentadiene, 6:688
sulfation, 19:302
sulfuric acid reaction with, 5:144
synthetic alcohols from, S:634
theoretical aspects of oxidation, 8:542
toxicity of, 11:296
vinyl acetate copolymer, 14:218; 21:346, 379
vinyl acetate copolymer, 21:331
vinyl acetate manuf, 21:323
Ethyleneamine
  from sulfuric esters, 19:495

Ethyleneamines, 7:22. ((See also *Diamines and higher amines, aliphatic.*)
Ethylene bis(chloroformate)
Ethylene bis(chloroformate), 4:386
    reaction with ethylenimine, 11:528
Ethylene bisdithiocarbamate, 7:30
N,N'-Ethylenebis(salicylideneiminato)cobalt(II), 15:162
Ethylenebis(thiocarbamoyl)monosulfide
    polyester from, 16:170
    in soil, 18:539
N,N'-Ethylenebis[2-(*o*-hydroxy-phenyl)]glycine(E-HPG)
    as chelating agent, 6:17
Ethylene bromide, 3:771
    in tetraethyllead motor fuel, 3:751
Ethylene carbonate, 13:339
    copolymerization, 16:174
    properties, 4:390
    reaction with triethylenetetramine, 7:31
Ethylene chloride, 10:164. See *1,2-Dichloroethane.*
Ethylene chlorobromide, 3:777
Ethylene chlorohydrin, 5:304, 306-309; 16:254
    esters of, 8:353
    in prepn of bis(2-chloroethyl) ether, 8:487
    prepn from ethylene, 8:502
    prepn from ethylene oxide, 8:527
    process for ethylene oxide manuf, 8:531
    sulfation, 19:491
Ethylene chlorohydrin acetate
    prepn, 8:527
Ethylene cyanohydrin, 6:652
Ethylene cyanohydrin, 6:668, 673
Ethylene cyanohydrin
    in acrylamide manuf, 1:280
    in acrylic acid prepn, 1:293
    in acrylic acid esters prepn, 1:299
    in acrylonitrile manuf, 1:344
    in monopropanolamines manuf, 1:815
    radiation effects, 14:79
Ethylenediamine, 2:370; 7:22, 23
    succinic anhydride reaction, 19:140
    titanium tetrachloride reaction, 20:472
    surfactants, 19:560
    chlorine dioxide repressed by, 5:30
    cyanoethylation of, 6:648
    from 1,2-dichloroethane, 5:151
    in prepn of ethylenethiourea, 4:372
    in prepn of ethyleneurea, 2:228
    in soil, 18:539
    in spandex, 18:625

manuf by ammonolysis, 2:349
piperazine from, 15:640
polyamide from, 16:2
prop, 7:24
Ethylenediamine hydrobromide, 9:312
Ethylenediamine hydrochloride, 7:31
Ethylenediamine tartrate, 7:27
    X-ray spectroscopy, 22:461
N,N,N',N'-Ethylenediaminetetraacetic acid. See *Ethylenediaminetetraacetic acid.*
Ethylenediaminetetraacetic acid, 4:7; 7:27
    as chelating agent, 6:7
    as food additive, 10:16
    complexes as indicators, 11:560
    deactivation of metalloenzymes, S:297
    in chemical cleaning, S:164
    in shampoo, 6:735
    in soap, 18:425
    iron chelate, 7:37
    sodium salt, 7:523
    surfactants, 19:513
    tetrasodium salt, 7:29
Ethylene dibromide, 8:540
    as lead scavenger, 10:476
    catalyst in prod of ethylene dichloride, 8:501
    in Grignard reaction, 10:724
    nematocide, 15:913
Ethylene dichloride. See *1,2-Dichloroethane.*
Ethylene dichloride, 2:349; 5:149-154; 7:677, 697; 8:350, 537. (See also *1,2-Dichloroethane.*)
    as lead scavenger, 10:476
    in dehydration of acetic acid, 2:850
    in manuf of ethyleneamines, 7:31
    in polysulfides, 16:253
    isolation of provitamins D, 21:558
    paint remover, 14:487
    polymers, 15:795
Ethylene dichloride-benzene
    solvent refining, 15:95
3,3'-(Ethylenediimino)dipropionitrile, 6:648
Ethylene diisothiocyanate
    in soil, 18:539
Ethylene dimethacrylate
    styrene copolymer, 20:814
4,4'-Ethylenedimorpholine, 13:666
Ethylenedinitramine, 8:626
    characteristics, 8:632, 633, 634
    prepn, 7:25
(Ethylenedinitrilo)tetraacetic acid, 2:592
4,4'-Ethylenedioxybenzoic acid, 16:182
3,3'-(Ethylenedioxy)dipropionitrile, 6:644

Ethylenediurea, 2:228

Ethylene fluorohydrin, 9:751

Ethylene fumarate
  heating of, 20:793

Ethylene glycol, 6:640-648; 7:411; 10:638-660
  alkyd resins, 15:793
  as antifreeze, 2:541
  Duff reaction, 15:162
  from 1,2-dichloroethane, 5:149, 151
  in deicing fluids, 2:559
  in esterification of fatty acids, 8:819
  in paints, 14:469
  in refrigeration, 17:309
  polycarbonates, 16:106
  polyester fibers from, 16:144
  polyesters, 21:62
  polyesters from, 16:174
  prepn from ethylene oxide, 8:527, 552
  prepn from sorbitol, 1:577
  prepn from xyletol, 1:577
  prepn of urethan polyesters, 7:694
  reaction with ammonia, 2:579
  succinic acid from, 19:144

Ethylene glycol acrylate
  phys prop, 1:289

Ethylene glycol diacetate
  from 1,2-dichloroethane, 5:151

Ethylene glycol diacrylate
  phys prop, 1:289
  in embalming fluids, 8:101

Ethylene glycol dimethacrylate, 16:120

Ethylene glycol dimethyl ether, 10:643, 644

Ethylene glycol dinitrate. See Glycol dinitrate.

Ethylene glycol methyl ether
  azeotrope with ethylenediamine, 7:22

Ethylene glycol monoacetate, 10:638

Ethylene glycol monobutyl ether acetate,
  10:644, 645
  in paints, 14:468

Ethylene glycol monoethyl ether, 8:479

Ethylene glycol monoethyl ether acetate,
  10:644, 645

Ethylene glycol monophenyl ether, 10:643,
  644

Ethylene glycol monosulfate, 19:495

Ethylene glycol monovinyl ether
  vinylpyrrolidone copolymer, 21:433

Ethylene glycol phenyl ether acetate, 10:644,
  645

Ethylene glycol terephthalate
  polymer manuf, 20:490

Ethylene glycol terephthalate ester, 10:645

Ethylene glycol 2-ethylbutyl ether, 10:643, 644

Ethylene glycol 2-ethylhexyl ether, 10:643, 644

Ethyleneimine. See Ethylenimine.

Ethylene maleate
  heating of, 20:793

Ethylene-maleic anhydride
  in drilling fluids, 7:301

Ethylene nitrate. See Glycol dinitrate.

Ethylene oxide, 2:641; 9:401
  addition to methacrylic acid, 13:335
  alkali cellulose reaction with, 4:646
  ammonolysis of, 2:351
  bisulfite reaction, 19:288
  in copolymerization of trioxane, 1:98
  ethylene chlorohydrin from, 5:308
  ethylene sulfate from, 19:492
  ethylene sulfite as substitute for, 19:488
  in fumigating nuts, 14:139
  in manuf of alkanolamines, 1:814
  in manuf of phenylethyl alcohol, 3:447
  polyethers, 21:58
  polymers, 15:795
  prepn from acrylic acid, 1:290
  prepn of ethylene glycol, 10:642
  prepn of diethylene glycol, 10:646
  prepn of polyethylene glycol, 10:655
  as propellant, 8:660, 688
  reaction with alcohol, 10:639
  reaction with ethyl alcohol to prepare Cello-
    solve and Carbitol, 8:455
  reaction with ethyleneamines, 7:25
  sterilant, 18:815
  surfactants from, 19:521, 529, 548
  tall oil reaction, 19:625
  trimethylamine reaction with, 5:407

Ethylene polyamines, 2:349, 350

Ethylene-propylene copolymers, 7:686

Ethylene-propylene rubber
  in packing, 14:451

Ethylene-propylene terpolymer, 7:687; 17:462,
  575, 592
  prop, 7:693

Ethylene sulfate, 19:488, 492

Ethylene sulfide, 2:642; 10:906
  prepn, 8:531

Ethylene sulfite, 19:488, 495

Ethylenesulfonic acid, 19:284

Ethylenesulfonimide, 7:633

Ethylene terephthalate
  copolyester from, 16:170

Ethylene tetrasulfide, 16:260, 269

Ethylene(thiocarbamoyl)disulfide
   in soil, *18*:539
Ethylenethiourea
   prepn, *4*:372
Ethyleneurea
   in amino resin manuf, *2*:227
   in prepn of ethylenedinitramine, *7*:25
   prepn from ethylenediamine, *7*:27
N,N'-Ethyleneurea, *2*:370. ((See also *Ethyleneurea.)*
1,3-Ethyleneurea
   in prepn of ethylenedinitramine, *8*:626
Ethylene vinyl acetate copolymers, *14*:218
Ethylene-vinyl acetate copolymer(s)
   as adhesives, *21*:346
Ethylene vinyl pyrrolidone copolymer, *21*:433
1,1-Ethylene-2,2'-bipyridinium salts, *22*:207
5,8-Ethylene-5,8-dihydro-2-methyl-1,4-naphtho-
   hydroquinone, *21*:594
5,8-Ethylene-5,8-dihydro-2-methyl-1,4-naphtho-
   quinone, *21*:594
5,8-Ethylene-5,8,9,10-tetrahydro-2-methyl-1,4--
   naphthoquinone, *21*:594
Ethylenic alcohols, secondary, *1*:629
Ethylenic alcohols, tertiary, *1*:630
Ethylenic glycols
   prepn from tertiary acetylenic glycols, *1*:627
Ethylenimine, *2*:413, 462; *6*:455, 648; *10*:898,
   906; *11*:526-548. (See also *Imines, cyclic.)*
   xanthate reaction, *22*:424
Ethylenimonium ion, *10*:907
Ethyl esters
   prepn, *8*:350
Ethyl ether, *8*:477
   as anesthetic, *2*:396, 399
   as denaturant, *8*:450
   hydration to prep ethyl alcohol, *8*:437
   manuf, *4*:480
   prepn of ethanol, *8*:454
   prepn from ethanol, *8*:429
   as solvent for Grignard reagents, *10*:724
   phys prop, *8*:471, 478
   prepn, *8*:474
   spec, *8*:484
   splitting of, *8*:345
   Vitamin D manuf, *21*:566
3-Ethyl-5-[2-(3-ethyl-2-benzothiazolinylidene)--
   ethylidene]rhodanine, *16*:285
1-Ethyl-2,4-bis(3-ethyl-2-benzothiazolinylidene-
   methyl)quinolinium iodide, *6*:618
3-Ethyl-2-(5-(3-ethyl-2-benzothiazolinylidene)--
   1,3-pentadienyl)benzothiazoliumiodi de, *14*535

3-Ethyl-2-[2-ethyl-3-(3-ethyl-2-benzothiazoliny-
   lidene)propenyl]benzothiazolium iodide),
   16:290
3-Ethyl-2[-3-(3-ethyl-2-benzothiazolinylidene)--
   propenyl] benzothiazoliumiodide, *16*:283
3-Ethyl-2-[3-(3-ethyl-2-benzoxazolinylidene)pro-
   penyl]benzothiazolium iodide, *16*:288
3-Ethyl-2-ethylthiobenzothiazolium iodide,
   6:614, 615
2-Ethylethylenimine, *11*:527
Ethyl fluoroacetate, *9*:768, 769
Ethyl formate, *8*:375; *19*:139
   prod and sales of, *8*:376
   as synthetic flavor, *9*:354
   thiamine synthesis, *20*:178
Ethyl fumarate, *12*:843
   prepn, *8*:351
Ethyl glutarate
   in prepn of *l*-lobeline, *1*:795
Ethylglyoxal
   from butyraldehyde, *3*:867
Ethyl α-hydroxyisobutyrate, *6*:672
Ethyl heptafluorobutyrate, *9*:773
Ethyl heptanoate
   prod and sales of, *8*:376
2-Ethylhexanal, *6*:640; *8*:849
2-Ethylhexan-1,3-diol
   titanium chelates, *20*:464
2-Ethylhexane-1,3-diol
   prepn, *4*:399
Ethyl hexanoate
   flavor and perfume material, *9*:358
2-Ethylhexanoic acid, *7*:272
   prepn, *8*:849
   properties, *8*:850
2-Ethyl-2-hexanal, *3*:875; *8*:849
2-Ethyl-1,3-hexanediol diacrylate
   phys prop, *1*:289
2-Ethylhexanol, *9*:898; *19*:141
2-Ethyl-1-hexanol, *1*:564, 565
2-Ethylhexoic acid. See *2-Ethylhexanoic acid.*
2-Ethylhexyl acetate
   spec, *8*:373
   prop, *8*:380
2-Ethylhexyl acrylate
   phys prop, *1*:288, 301
   polymerization, *1*:303
   vinyl acetate copolymer, *21*:331
2-Ethylhexylamines, *2*:127
2-Ethylhexyl chloroformate, *4*:386
2-Ethylhexyl crotonate, *6*:468, 469

2-Ethylhexyl diphenyl phosphate, *15*:323, 762, 764

2-Ethylhexyl diphenyl phosphate plasticizer, *15*:725, 749

2-Ethylhexyl oleate, *8*:256

2-Ethylhexyl tallate, *8*:256

2-Ethylhexyl vinyl ether, *21*:414

Ethyl hydrogen sulfate, *8*:431, 481; *19*:490, 492, 494

Ethyl hydrol, *20*:701

Ethyl hydrol-2,7-naphthalenedisulfonic acid, *20*:723

Ethylhydroxyethyl cellulose, *4*:641, 650

2-Ethyl-3-hydroxyhexanal. See *Butyraldol.*
  prepn and prop, *3*:874

2-Ethyl-2-hydroxymethyl-1,3-propanediol, *1*:595; *3*:876
  prepn, *3*:869

2-Ethyl-2-hydroxymethyl-1,3-propanediol, *1*:595; *3*:876; *7*:412; *15*:771, *21*:60

Ethylidene acetone, *6*:694

Ethylidene bromide, *3*:777

Ethylidene chloride. See *1,1-Dichloroethane.*
  photochlorination, *15*:332

Ethylidene diacetate
  from acetaldehyde, *1*:83
  from acetylene, *1*:86
  manuf of acetic anhydride, *8*:410
  prepn of, *21*:317
  in prepn of acetaldehyde, *1*:86
  vinyl acetate manuf, *21*:327

Ethylidene dichloride. See *1,1-Dichloroethane.*

4,4'-Ethylidenedi-*o*-cresol, *2*:601

Ethylidene sucrose, *19*:231

Ethylidine diethyl ether. See *Acetal.*

Ethylidine dimethyl ether. See *Dimethyl acetal.*

Ethylidine peroxide polymers, *8*:479

Ethyl iodide, *8*:472; *11*:864
  chloroform reaction with, *5*:122

Ethyl iodophenylundecylate, *17*:139

1-Ethyl-2-iodoquinolinium iodide, *6*:609

Ethyl isobutyrate, *8*:381

4-Ethyl-1-isobutyloctyl acrylate
  phys prop, *1*:288

Ethyl isocyanate, *12*:46

Ethyl isocyanate
  from potassium cyanide, *19*:488

Ethylisoformanilide, *6*:613

Ethylisonicotinate, *21*:528

5-Ethyl-3-methyl-5-phenylhydantoin, *11*:157

Ethylisopropylbromoacetamide
  as hypnotic, *11*:517

Ethylisothioacetanilide, *6*:614

Ethyllactamide, *12*:186

Ethyl lactate, *12*:184, 185
  as bleach activator, *5*:34
  prepn, *8*:351, 352
  stain solvent, *18*:660

Ethyl laurate, *8*:371
  in prepn of 1-dodecanol, *1*:568

1-Ethyllepidinium, *6*:610

Ethyl levulinate, *6*:670
  prepn, *8*:351

Ethyllithium, *12*:156, 552

Ethyl magnesium bromide
  reaction with pentafluorobenzene, *9*:788

Ethyl malate, *12*:843
  prepn, *8*:351

Ethyl malonamate, *12*:850

Ethyl malonate
  prepn, *8*:350

Ethyl *n*-butoxyacetate, *8*:492

Ethyl *n*-butyl ether
  phys prop, *8*:471

Ethyl *n*-butyl ketone, *12*:105, 135

Ethyl *n*-butyl ketone, *12*:105

Ethyl mercaptan. See *Ethanethiol.*

β-Ethylmercaptoacrolein diethyl acetal, *14*:535

Ethylmercuric chloride, *13*:244
  as leather fungicide, *10*:233
  as fungicide, *10*:235

Ethylmercuric phosphate, *21*:192

Ethylmercurithiosalicylate, *20*:215

Ethyl methacrylate
  in dental material, *6*:830

Ethylmethylaniline oxide, *S*:32

3-Ethyl-2-methylbenzothiazolium iodide, *6*:610, 613; *14*:535

Ethyl 11-(2-methylcyclohex-2-en-1-yl)undec-*trans*-9-enoate, *7*:408

5-Ethyl-5-(1-methylbutyl)barbituric acid, *3*:61, 70, 72

5-Ethyl-5-(1-methylbutyl)-2-thiobarbituric acid, *3*:61, 70

1-Ethyl-2-[2-methyl-3-(3-methyl-2-benzoxazolinylidene)propenyl]quinolinium iodide, *6*:608

3-Ethyl-5-methyl-1-hepten-3-ol, *1*:630

Ethyl methyl ketone, *12*:109, 117, 122

3-Ethyl-1'-methyl-8,9-diazathia-2'-carbocyanine perchlorate, *6*:620

4-Ethyl-1-methyloctyl acrylate
  phys prop, *1*:288

5-Ethyl-1-methyl-5-phenylbarbituric acid, *3*:61, 70

Ethyl methylphenyl
  as flavoring, 4:344
Ethyl methylphenyl glycidate, 9:354, 355;
  4:344
3-Ethyl-2-methylthiobenzothiazolium iodide,
  6:614
5-Ethyl-5-isopentylbarbituric acid, 3:61, 70, 72
Ethyl methylxanthate, 22:420
Ethyl n-amyl
  phys prop, 8:471
N-Ethylmorpholine, 13:667
  polyurethan catalyst, 21:86, 91
N-Ethyl-2-naphthamide
  form, 2:67
N-Ethyl-1-naphthylamine, 20:701
4-Ethyl-1-octen-3-ol, 1:629
4-Ethyl-1-octyn-3-ol
  phys prop, 1:619
Ethyl nicotinate, 21:527
Ethyl nitrate, 8:479
Ethyl nitrite, 6:693
2-Ethyl-2-nitro-1,3-propanediol
Ethyl nonanoate
  flavor and perfume material, 9:358
Ethyl p-hydroxybenzoate
  as food preservative, 10:12
Ethyl oleate
  parenteral solvent, 15:128
  prepn, 8:351
Ethylolguanidine
  prepn from guanidine, 10:737
Ethyl p-aminobenzoate, 2:405, 406, 407
Ethyl orthoacetate, 6:612
Ethyl orthoformate, 6:574, 612, 616; 9:795;
  16.294
Ethyl orthosilicates, 18:253. ((See also Ethyl
  silicate.)
  stain solvent, 18:659
Ethyl oxalate, 12:144
Ethyl palmitate
  prepn, 8:351
Ethyl pelargonate, 9:358
  prepn, 8:350
  as synthetic flavor, 9.354
3-Ethyl-2,4-pentanedione, 12:148
Ethyl phenylacetate, 2:368
5-Ethyl-5-phenylbarbituric acid, 3:61, 70, 72
  manuf, 3:67
1-Ethyl-2-phenylcyclopropane, 19:325
2-Ethyl-2-phenylglutarimide
  as hypnotic, 11:515

5-Ethyl-5-phenylhexahydropyrimidine-4,6-dione,
  11:523
3-Ethyl-5-phenylhydantoin, 11:157
5-Ethyl-5-phenylhydantoin, 10:912; 11:158
5-Ethyl-2-picoline, 16:803
Ethyl phenyl ether, 15:148
N-Ethyl-m-phenylenediamine
  in prepn of m-ethylaminophenol, 2:214
Ethyl phenylsulfinylacetate, 19:324
Ethyl phenyl sulfoxide, 19:318
Ethyl phosphite, 8:428
Ethyl phosphorodimethylamidocyanidate
  in chemical warfare, 4:875
Ethyl phthalyl ethyl glycolate, 15:751, 764
N-Ethylpiperidine
  manuf from ethyl alcohol, 8:455
Ethyl potassium malonate, 19:143
Ethyl propionate
  prepn, 8:351
  properties, 8:381
1-Ethylpropyl acrylate
Ethyl-n-propylglyoxal, 12:140
Ethyl-n-propyl ketone, 12:105, 126, 209
Ethyl n-propyl sulfide, 20:209
  phys prop, 1:288
2-Ethylpyridine, 16:784
Ethylpyridinium fluoborate, 9:570
Ethyl pyruvate, 6:670
1-Ethylquinaldinium iodide, 6:609, 611
Ethyl salicylate, 17:730
  prepn, 8:351
Ethyl silicate, 18:218. (See also Ethyl ortho-
  silicates.)
Ethyl sodioacetoacetate, 1:154, 157
Ethyl stearate
  properties, 8:382
Ethylstibamine, 2:585
Ethyl suberate
  prepn, 8:350
Ethyl sulfite, 8:428
3-Ethyl-2-sulfobenzothiazolium betaine, 6:610
Ethyl tartrate
  prepn, 8:350
Ethyl teraconate, 19:137
Ethyl terebate, 19:137
2-Ethyltetrahydrofuran, 10:250
3'-Ethyl-1,3,3,11-tetramethylindo-oxadicarbo-
  cyanine iodide, 6:616
Ethyl thioglycolate, 20:203
Ethylthioisonicotinic acid, 3:30
1-Ethylthiopropene
  addition reactions, 20:209

3-Ethylthiopropionitrile, 6:646
Ethyltoluenesulfonamide, 6:826
N-Ethyl-(o,p)-toluenesulfonamide, 15:759, 771
  plasticizer, 15:764
N-Ethyl-p-toluenesulfonamide
N-Ethyl-o-toluidine, 20:704
Ethyltriethoxysilane, 18:250
Ethyl 2,4,6-trimethylbenzoate
  prepn, 8:354
3'-Ethyl-1,3,3-trimethylindothiacarbocyanine
  iodide, 6:613
3'-Ethyl-1,3,3-trimethyl-11,13-neopentylenein-
  dothiatricarbocyanine iodide, 6:617
Ethyltrimethyllead, 10:476
Ethyl undecenoate, 8:371
Ethyl valerate
  prepn, 8:351
Ethyl vanillate, 21:193
Ethylvanillin, 15:160, 164; 21:189, 191, 192
  as synthetic flavor, 9:354
Ethylvinylbenzene, 19:74
Ethyl vinyl ether, 21:416
  as anesthetic, 2:399
  manuf from ethyl alcohol, 8:455
  prepn, 8:428
  from vinyl chloride, 5:173
Ethyl xanthate
  in flotation, 9:386, 393
  prepn from ethyl alcohol, 8:456
Ethyl xanthic acid, 22:420
3-Ethyl-5,7-dimethyl-1-adamantanol, S:10, 12
Ethyne. See Acetylene.
Ethynodiol diacetate, 6:68
Ethynoid acids
  definition, 8:817
Ethynyl butyl ether
  phys prop, 8:471
1-Ethynylcyclohexan-1-ol, 1:621, 622
1-Ethynylcyclohexyl urethan
  as hypnotic, 11:518
Ethynylenebis(magnesium iodide), 12:160
17α-Ethynylestradiol, 6:87. ((See also Estranol.)
Ethynylestradiol 3-methyl ether. See Mestranol.
17α-Ethynylestra-1,3,5(10)-triene-3,17β-d- iol
  3-methyl ether. See Mestranol.
17α-Ethynylestr-4-ene-3β,17β-diol-3- ,17-diac-
  etate. See Ethynodiol diacetate.
17α-Ethynylestr-4-ene-3β,17β-diol-3- ,17-diac-
  etate, 6:68
17α-Ethynylestra-1,3,5(10)-triene-3,17 β-diol
  3-methyl ether. See Mestranol.

Ethynyl ethyl ether
  phys prop, 8:471
17α-Ethynyl-17β-hydroxyestr-5(10)-2n-3-on- e,
  6:65, 87
17α-Ethynyl-17β-hydroxy-19-norandrost-4--
  en-3-one, 6:65, 87
Ethynylmagnesium bromide, 9:761
17α-Ethynyltestosterone, 11:125
17α-Ethynyl-17β-hydroxyestr-5(10)-en-3-on- e.
  See Norethynodrel.
17α-Ethynyl-17β-hydroxy-19-norandrost-4-e-
  n-3-one. See Norethisterone.
17α-Ethynyl-19-nortestosterone. See Norethisterone, 6:65,87
Eucalyptol, 19:831
  as denaturant, 8:450
Eucalyptus, 12:318; 14:201
Eucalyptus oil
  as denaturant, 8:450
Eugallol, 16:198
Eugenol, 8:450; 9:354; 11:470; 12:154;
  14:729; 15:169, 170
  in dental cement, 6:779
  as dental material, 6:823
  as fungicide, 10:235
  flavor and perfume material, 9:358
  as synthetic flavor, 9:355
  vanillin from, 21:184
Eugenol ether, 9:354
Eugenyl methyl ether. See 4-Allylveratrole.
Eulan CN, 20:733
Eulan SN, 20:733
Euler equation
  ideal fluids, 20:644
Euphol, 18:838
Europium, 12:623; 17:144, 150
  electronic configuration, 1:360
  ionic radii, 1:370
Eutectic inhibitors
  recovery of p-xylene, 22:474
Eutectic point
  binary system, 13:200
Eutrophication, S:315
Euxenite, 17:147
Evans resorption gestation method
  α-tocopherol determination, 21:581
Evaporation, 8:559
  in brine heater of multistage flash, 22:30
  comparison of processes, 22:46
  conventional multiflash, 22:24
  film formation by, 9:186, 208
  multiflash-absorption of vapors, 22:44

multiflash-vapor reheat process, *22*:39
processes for desalination, *22*:16
solar, *22*:16
solar in the ocean, *22*:19
vacuum techniques, *21*:123
Evaporation control
  in reservoirs, *21*:664
Evaporation suppression
  use of fatty alcohols in, *1*:556
Evaporators
  control, *11*:761
  prod of high-purity water, *22*:77
  removal of dissolved solids in water with,
    22:74
  steam, *18*:705
  with surfaces for boiling and condensing,
    22:19
  types, *8*:562
  vapor compression, *22*:21
Evapotranspiration, *21*:653
Excelsin
  in Brazil nuts, *14*:125
Exchange adsorption, *1*:468
Exinite, *5*:611
Exiton, *17*:835
Exoisocamphane-6-exo-yl acetate, *19*:831
Explosive D. See *Ammonium picrate.*
Explosive forming
  of metals, *13*:316
Explosives, *8*:**581**
  ammonium nitrate as, *2*:328
  index of, *8*:646
  seismic, *8*:645
Explosives, high
  general index of, *8*:646
Expression
  essential oils by, *14*:183

Ext D&C Green No. 1, *5*:872
Ext D&C Orange No. 3, *5*:872
  used in hair coloring, *10*:793
Ext D&C Red No. 15, *5*:872
Ext D&C Red No. 15, *5*:872
Ext D&C Red No. 8. See *Fast Red S.*
Ext D&C Red No. 8, *5*:872
  used in hair coloring, *10*:793
Ext D&C Violet No. 2, *5*:872
Ext D&C Yellow No. 1, *5*:872
Ext D&C Yellow No. 3, *5*:872
Ext D&C Yellow No. 7, *5*:872, 874
Ext D&C colors, *5*:857, 858, 864
Extraction, *8*:**719**
  diffusion in liquid-liquid, *13*:143
  essential oils from, *14*:182
  mass transfer, *13*:100
  method for desalination of water, *22*:55
  of oil seeds, *11*:4
  use of hexane in, *11*:4
Extracts
  natural in perfumes, *14*:722
Extrusion
  adhesive application by, *S*:933
  of baked carbon, *4*:173
  of cellulosics, *4*:668
  of ceramic materials, *4*:779
  of graphite, *4*:173
  of plastics *15*:805
  of metals, *13*:316
Eye shadow, *6*:368
Eyring equation
  for viscosity, *21*:463
Eyrthrina alkaloids, *1*:770
Ezcurrite, *3*:627

F

FD&C Blue No, 5:866, 873. 1. See *Brilliant Blue FCF*.
  used in hair coloring, 10:793
FD&C Blue No. 2, 5:866, 874
FD&C Green No, 5:866. 1. See *Guinea Green B*.
FD&C Green No. 2, 5:866
FD&C Green No, 5:866. 3. See *Fast Green FCF*.
FD&C Red No. 2, 5:866, 873
FD&C Red No. 3, 5:866
FD&C Red No. 4, 5:867
  used in hair coloring, 10:793
FD&C Violet No. 1, 5:867
  used in hair coloring 10:793
FD&C Yellow No. 5, 5:867, 873
  used in hair coloring 10:793
FD&C Yellow No. 6, 5:867
  used in hair coloring, 10:793
FD&C colors, 5:857, 858, 864
FD&C lakes, 5:864, 867
FS (military smoke agent), 4:903
Fabrics
  air permeability, 20:52
  bending length, 20:57
  breaking strength of textile, 20:48
  coated. See *Coated fabrics*.
  color, 20:58
  colorfastness, 20:58
  compressional resilience, 20:50
  crease retention, 20:54
  drape, 20:58
  elongation, 20:48
  fireproof, 20:45
  flameproof, 20:45
  flame resistant, 20:45
  flame resistant blend, S:958
  flexural rigidity, 20:57
  geometry, 20:48
  "handle," 20:57
  impermeable, 22:136
  luster, 20:59
  moisture transmission, 20:52
  shrinkage, 20:57
  tear strength, 20:49
  thermal transmission, 20:51
  tongue tear test, 20:49
  trapezoid tear test, 20:50
  water repellency, 20:53
  water resistance, 20:53
  wear resistance, 20:55
Fabric whiteness 6:864
Face powder 6:357
  manuf 6:358
F acid, 13:714, 721
Factice, 19:498
Factor-3 selenium, 2:597
Famphur
  ectoparasite control, 21:250
Fanchon Maroon, 5:871. See *Deep Maroon*.
Fanning equation, 20:649
Farnesol, 19:833
Farnoquinone, 21:591
Fast Acid Fuschsine B. See *Acid Fuchsin D*.
Fast Acid Fuchsine B, 5:871
Fast Acid Violet
  synthesis, 20:680
Fast Acid Violet 10B, 20:717
Fast Black Base K, 2:891
Fast Blue FFR, 2:880, 7:470. (See also *Fast Red S*.)
Fast Blue R Base, 3:414
Fast Garnet salt GBC, 2:891
Fast Light Yellow, 5:872
Fast Light Yellow GGXN Conc, 7:525
Fast Red A, 5:872
Fast Red S, 5:872
  used in hair coloring, 10:793
Fast Red Salt ITR, 2:890
Fast Red Salt KB, 2:892, 893
Fast Scarlet Salt 2G, 2:891
Fast color salts, 2:891
Fast drying
  in textiles, 1:265
Fast green FCF, 5:866
  used in hair coloring, 10:793
Fast printing green S, 7:527
Fatigue, corrosion
  of copper alloys, 6:240
Fatigue testing, 13:191, 193
Fats
  in confectionery, 6:40

ethoxylated natural, *19*:548
factory consumption of animal, *13*:183
frying, *13*:182
improving properties, *8*:362
processing of animal, *13*:179
rendering, *13*:179
specialty, *18*:45
sulfated, *19*:524
uses of animal, *13*:182
US prod of animal, *13*:182
Fats, *8*:**776**
alcoholysis with glycerol, *19*:541
polyunstaurated in margarine, *13*:59
Fats and fatty oils, *8*:**776**
Fat-soluble factor A, *21*:484
Fatty acid amines, *2*:72, 73, 74, 75
Fatty acid chlorides
prepn *8*:823
Fatty acid esters
prepn *8*:818
Fatty acid ethoxylates, *19*:545
Fatty acid metal salts
as abherents, *1*:8
Fatty acid peroxydase
amitrole inhibition, *22*:212
Fatty acids, *8*:**811**
from animal fats, *13*:183
composition in fish oil, *9*:320
in eggs, *7*:664
in fish, *9*:320
glycol esters, *19*:548
monobasic, in alkyd resins, *1*:853
in nuts, *14*:128
reduction to form alcohols, *1*:544
2-sulfoethyl esters, *19*:522
in tall oil, *19*:614
in waxes, *22*:156
in wood, *16*:690
Fatty alcohols. See *Alcohols, higher fatty.*
Fatty alkyl ether sulfates
in shampoos *10*:772
Fatty alkyl sulfates
in shampoos *10*:771
Fatty amines. See *Amines, fatty.*
Fatty foods
antioxidants in, *2*:601
Fatty monoglyceride sulfate
in shampoo, *10*:772
Fatty oils, *8*:776-811
sulfation, *19*:302
Faujasite, *18*:157
Fayalite
in ceramics, *4*:772

Febrifugine, *1*:771
Federal Food, Drug and Cosmetic Acts, *5*:857
Feeders
for swimming pools, *22*:131
Feeders, for conveying systems, *6*:97
Feeds, animal *8*:**857**
cobalt requirements, *5*:746
cottonseed meal as, *6*:422
from fats, *13*:183
iron(II) gluconate as supplement, *12*:35
Feed yeast, *22*:540
Feeling factors
in food, *14*:337
Fehling's solution
in acetaldehyde detn *1*:90
Fehling test, *19*:159
Feldspar
as abrasive, *1*:25, 26
as dental porcelain, *6*:798
in enamel frits, *8*:158
pigments, *15*:511
in prepn of trifluoroboronine, *9*:556
rubidium in, *17*:685
separation by flotation, *9*:395
Feldspars, *4*:2; *18*:49
in ceramics, *4*:769
in soaps, *18*:429
Felting, *22*:399
Fenac
chlorinated aliphatic acid herbicide, *22*:199
α-Fenchol, *19*:819
Fenchone, *12*:106; *19*:819
Fenchyl alcohol, *17*:385
Fenethazine. See *Anergen.*
Fennel, *9*:363, 365, 371; *14*:202
Fenugreek, *9*:371; *14*:202
Fenuron
urea herbicide, *22*:190
Ferbam
soil chemistry, *18*:539
as fungicide, *10*:224
Ferberite, *22*:338
Fermentation, *8*:**871**
acetic, *21*:259
in alcohol beverage manuf, *1*:573
alcoholic, *21*:257
alcohol in wine, *22*:314
in brewing, *3*:314
continuous molasses, *22*:523
degree of in beer, *22*:526
detergent enzyme prod, *S*:298
of malic acid by yeast, *22*:525
microorganisms from hydrocarbons, *S*:846

of molasses, 22:522
in penicillin manuf, 14:688
process for oxalic acid, 14:361
prod of vitamin B_{12}, 21:544
pure-culture yeast, 22:524
single-strain yeast, 22:524
in sterilization, 18:812
submerged, 21:263
and yeasts, 22:515
Fermentation ethyl alcohol, 8:438
Fermentation industry
carbon dioxide as by-product, 4:359
Fermentation test
multiple-tube, 21:704
Fermium. See *Actinides.*
Ferrates 12:40
Ferredoxin
and nitrogen fixation, S:609
Ferric acetate 12:23
Ferric ammonium oxalate, 14:368; 17:357
Ferric bromide, 2:320
Ferric chloride, 10:621
coagulant in water treatment, 22:88
dye oxidant, 20:693
in prepn of ketone, 12:126
reaction with ethyl ether, 8:479
sulfur monochloride, 19:392
Ferric citrate
in blueprinting, 17:357
Ferric compounds. (See also *Iron compounds.*)
Ferric dimethyldithiocarbamate. See *Ferbam.*
Ferric ferrocyanide 6:588
Ferric hydroxide, 2:311
Ferric oxalate, 14:368
in blueprints, 17:357
Ferric oxide. (See also *Rouge.*)
in ceramic pigments, 5:854
from ore roasting, 22:565
pigments, 15:525
scale on steel, S:684
sulfuric acid, 19:442
γ-Ferric oxide, 8:884; 12:804
Ferric perchlorate, 5:72
Ferric potassium oxalate, 14:368; 17:357
Ferric sodium oxalate, 17:357
Ferric sulfate 6:588
coagulant in water treatment, 22:73, 94
in prepn of glutamic acid, 2:203
Ferric titanate, 20:414
Ferricyanic acid, 12:31
Ferricyanide. See *Hexacyanoferrate(III).*
Ferrimagnetism, 12:737
in ferrites, 8:883

Ferrite, 6:271; 18:763
Ferrites, 8:**881;** 12:40
as ceramic materials, 4:775
Ferritic steels
as high-temperature alloys, 11:15
Ferroalloy furnaces
carbon refractories for, 4:240
Ferroanthophyllite. (See *Asbestos.*)
Ferroboron, 3:605
Ferrocene, 6:696
Ferrocenyl isocyanate, 12:55
Ferrocerium, 4:848
Ferrochrome lignosulfonate, 7:301; 12:372
Ferrochromium, 5:454, 463
Ferrocyanic acid, 12:26
Ferrocyanide, 2:311. (See also *Hexacyano-ferrates(II).*)
Ferrocyanide blues
as pigments, 14:472
Ferroelectric ceramics, 9:1
prod, 9:14
properties, 9:16, 20
Ferroelectric hysteresis loop, 9:3
Ferroelectrics, 9:1; 11:776
Ferromagnetism, 12:737
in ferrites, 8:882
Ferromanganese, 12:889; 13:26, 35
deoxidizer, 18:724
deoxidizer in welding, 22:242
in ocean, 14:164
Ferromanganese-silicon alloy, 12:899
Ferromangenes. (See also *Spiegeleisen.*)
Ferron
in water analysis, 21:697
Ferronickel, 13:739
Ferroniobium, 13:775
Ferrophosphorus, 7:292; 15:284, 293
vanadium from, 21:160
Ferroselenium, 17:828
Ferrosilicon, 7:292; 18:120, 127
Ferrosilicon
deoxidizer, 18:723
ferrovanadium prod, 21:161
in manuf of magnesium, 12:675
reduction of sodium chloride, 18:439
Ferrosiliocon
smoke from furnaces, 18:402
Ferrosoferric oxide, S:684. (See *Iron com-pounds.*)
Ferrospinels, 8:882
Ferrotellurium, 19:760,
Ferrous acetate, 7:566
Ferrous bromide, 2:320

Ferrous chloride
  as bactericidal agent, 2:622
  in prepn of diphenylamine, 7:40
Ferrous chromite, 17:237
Ferrous compounds. (See also *Iron compounds.*)
Ferrous ethyl xanthate, 22:421
Ferrous fluoride, 9:625, 626
Ferrous hydroxide, 2:310; 6:588
3,5-Dinitro-0-cresol
Ferrous metatitanate, 20:414
Ferrous orthotitanate, 20:414
Ferrous oxide
  in Bessemer process, 18:723
Ferrous perchlorate, 5:71
Ferrous sulfamate
  plutonium separation, 15:884
  in Purex process, 14:95
Ferrous sulfate, 2:310; 7:680
  analysis of wolfram, 22:343
  in bleaching wool, 22:405
  bottling water treated with, 4:341
  chlorate determined with, 5:58
  coagulant for water treatment, 22:96
  pigments, 15:519
  sulfuric acid, 19:442
  from titanium-pigment waste, 19:475
  used in bleaching furs, 10:305
Ferrous sulfate heptahydrate, 12:40
Ferrous sulfide, 2:310
Ferrous titanate, 20:398
Ferrovanadium
  economics of, 21:165
  prod, 21:161
Ferrowolfram, 22:339
Ferroxcona, 8:893
Fertilizers, 9:**25**; S:**338**
  ammonium nitrate as, 2:328
  ammonium sulfate as, 2:330
  calcium cyanamide as, 6:553, 557, 559
  in cotton growing, 6:382
  multinutrient, 9:114-143
  nitrogen, 9:51-78
  phosphate, 9:78-110
  potash, 9:111-114
  secondary and micronutrient, 9:143-147
  use of ammonia in, 2:293
  use of diatomite in, 7:63
  use of polymethylolmethyleneurea in, 2:255
Ferulic acid, 15:681
  from vanillin, 21:182
Fervenulin
  amebicide, 20:82

Fiber blends
  dyeing of, 7:566
Fiberfrax, S:960
Fiber glass, 10:564
Fiber-reactive dyes
  in textile printing, 7:577
Fibers
  carbon, 4:241
  cellulose, 4:598. (See also *Rayon.*)
  flame-retardant non-cellulosic manmade, S:957
  elastic types, 18:614
  graphite, 4:241
  fabric geometry, 20:48
  felts, 20:33
  properties, 20:34-48
  polypropylene, S:808
  poly (vinyl alcohol), 21:363
  primary creep, 20:41
  stick temperature, S:815
  water repellency, 20:53
  wool, 20:34
  woven fabrics, 20:33
  yarns, 20:33
  yarn crimp, 20:36
  yarn geometry, 20:48
Fibers, elastomeric, See *Elastomeric fibers.*
Fibers, man-made, 9:**151**
Fibers, polymers suitable for, 16:252
Fibers, vegetable 9:**171**
Fibrils, cellulose, 4:599
Fibrin, 5:590
Fibrinogen, 3:576; 5:590
  phys prop, 5:589
Fibroin
  in silk, 18:271
Ficin, 10:9
Ficin
  uses, 8:190
  source of enzymes, 8:205
Fick's law
  in absorption, 1:47
Fick's laws of diffusion, 7:80; 13:**101, 109;** 20:650
Field emission
  photsensitive, 15:**409**
Field emission cathodes, 8:9
Fikentscher constant, 21:419, 428
Filament winding, 12:192, 194
  epoxy resins use in, 8:310
Filament-wound products
  in electric insulation, 11:795
Filariasis, 14:535

Filberts, *14*:145
Filipin, *16*:141
Film
   fibers from polypropylene, *S*:824
   poly(vinyl alcohol), *21*:362
Film balance, *1*:423
Film deposition, *13*:249
Film deposition techniques, *9*:**186**
Film-forming inhibitors, *6*:332
Film materials, *9*:**220**
Film-penetration theory
   of mass transfer, *13*:138
Film, photographic. See *Photography.*
Films
   as electric insulation, *11*:792
   insulating surface, *17*:877
   poly-1-butene, *S*:783
   polyimide, *S*:771
Films, plastic
   coatings for, *5*:706
Filtermasse filter, *9*:280
Filters
   ceramic, *4*:839
   trickling, *22*:108
Filters, optical, *9*:**244**
Filtration, *9*:**264**
   centrifugal, *4*:732
   microporous membrane as cold sterilization
      method in water, *22*:80
   removal of wastes, *S*:716
   sterilization by, *18*:822
   in water-treatment systems, *22*:67, 88, 119
Fire bombs
   in chemical warfare, *4*:892
Firebrick
   strength value of, *4*:818
Fire clay
   thermal conductivity of, *4*:803
Fire clays
   in ceramics, *4*:766, 774
Fired-on coatings, *13*:281
Fire extinguishers
   carbon dioxide liquid in, *4*:368
   carbon tetrachloride in, *5*:137
Fire extinguishing agents, *S*:**364**
Fire prevention and extinction, *9*:**286**
Fireproof textiles. See *Fire-resistant textiles.*
Fire-resistance
   tests for, *S*:945
Fire-resistant textiles, *9*:**300**
Fireworks, *16*:835-840
Firming agents
   in food products, *10*:10

Fir needles. See **Abies siberica.**
Fischer-Tropsch reaction, *4*:434
Fischer-Tropsch synthesis, *4*:459; *13*:390
   of synthetic crude oil, *S*:180
Fischer's aldehyde
   prepn, *16*:297
Fischer's base, *16*:294, 296
Fish and shellfish, *9*:**316**
Fish glue, *10*:604
Fish liver oils. See *Fish and shellfish.*
Fish meal
   as animal feed, *8*:867
Fish oils
   separation of, *7*:402, 405
   uses, *7*:423
Fission
   chain reaction, *14*:75
   energy and particle release, *14*:76
   neutrons per, *14*:77
Fixed-bed column
   ion exchange, *11*:894
D-*Glycero*-D--*ido*-heptitol, *1*:576
Glycero-gulo-heptitol, *1*:576
Flame agents
   in chemical warfare, *4*:893
Flame arcs, *4*:210
Flame field expedients
   in chemical warfare, *4*:892
Flame fougasses
   in chemical warfare, *4*:893
Flame materials
   in carbon-arc electrodes, *4*:211
Flame mines
   in chemical warfare, *4*:893
Flameproof textiles. See *Fire-resistant textiles.*
Flame retardancy
   in textiles, *S*:948
Flame retardants,
   semidurable, *9*:304
Flame retardants, durable, *9*:305
Flame-retardant textiles. See *Fire-resistant textiles.*
Flames
   chemistry of, *S*:473
Flame throwers
   in chemical warfare, *4*:889
Flaming Red, *5*:871
Flammability
   of electric insulation, *11*:784
Flammability test for textile, *9*:310
Flares
   illumination, *16*:829
   signal, *16*:831

175

Flatworm infections, *14*:532
Flavanols
  in tea, *19*:743
Flavan-3-ols
  in tea, *19*:743
Flavanthrone, *2*:454, 522
  pigment, *15*:555
Flavanthrone Yellow, *15*:587
Flavicid
  as germicide, *2*:640
Flavindulins, *15*:222
Flavine-adenine dinucleotide, *2*:176; *17*:445
Flavine mononucleotide
  metabolism of urea herbicides, *22*:191
Flavine mononucleotide, *17*:445
  in synthesis of ethylenimine, *11*:535
Flavone, *10*:925; *12*:107
Flavone glycosides, *19*:601
Flavones, *10*:925
Flavoproteins, *17*:447
Flavor. (See also *Taste.*)
  definition, *9*:347
  evaluation, *9*:351
Flavor characterization, *9*:**336**
Flavoring agents
  for confectionery, *6*:46
  in food products, *10*:10
Flavorings
  isobutyric acid esters for, *3*:882
Flavor potentiators, *9*:360
Flavorpurpurin, *2*:466
Flavors
  for carbonated beverages, *4*:336
  use of gum arabic in, *10*:745
Flavors and spices, *9*:347
Flax. See *Linen.*
Flax straw
  in insulating board, *21*:617
Flax wax, *22*:161
Fleabane, *14*:201
Flexible polyurethan foams
  use of isocyanates, *12*:58, 61
Flexographic and rotogravure inks, *11*:621-625
Flexographic printing
  use of glass, *10*:547
Flexol R-2H
  as cellulosic plasticizer, *4*:657
Flight conveyors, *6*:109
Flint
  as abrasive, *1*:24, 29
Flint glass, *5*:848
Flint, potter's, *4*:769

Float gages, *12*:484
Floating-zone refining, *22*:690
Float method
  of density measurement, *6*:769
Flocculants
  polyacrylamides as, *1*:212
Flocculation
  in water treatment, *22*:71
  in water-treatment systems, *22*:90
Florentine yellow, *5*:852
Florigenic acid, *15*:686
Florigens, *15*:686
Floroquinoline, 2-, 3-, 5-, 6-, 7-, 8-, *9*:800
Flory-Huggins interaction parameters, *21*:381
Flory-Huggins interaction parameters, *S*:908
Flotation, *9*:**380**
  in particulate removal, *S*:715
Flotation of ores, *6*:596
Flour. (See also *Wheat and other cereal grains.*)
  enrichment, *22*:299
  grades of wheat, *22*:289
  in prepn of monosodium glutamate, *2*:20
  wheat, *22*:282
Flours
  special types of, *22*:300
Flow
  Newton's law of, *21*:461
  in polyurethan, *21*:82
  time dependence, *21*:463
Flow behavior theories, *21*:463
Flow measurement, *9*:473-483
Flue dust recovery
  in copper smelting, *6*:152
Flue gases
  carbon dioxide recovered from, *4*:357, 362
Fluid-bed technology. See *Fluidization.*
Fluid-char process, *1*:461
Fluidization, *9*:**398**
Fluidized solids
  in heat transfer, *10*:856
Fluidizing system
  conveyor, *6*:114
Fluid mechanics, *9*:445
Fluids
  compressibility factor, *20*:131
  dilatant, *20*:646
  enthalpy of, *20*:132
  entropy, *20*:132
  equations of state, *20*:130
  heat transfer by convection, *10*:821
  ideal, *20*:644
  ideal gas law, *20*:130

Newtonian, *20*:644
non-Newtonian, *20*:645
rheopectic, *20*:646
thixotropic, *20*:646
viscoelastic, *20*:646
Fluke infections
chemotherapy of, *14*:544
Flukes, *14*:532
Flumethiazide
as diuretic, *7*:263
Fluo-. (See also *Fluoro-.*)
Fluoaluminates, *9*:534
"Fluoaluminic acid", *9*:542
Fluoborates, *9*:562
Fluoboric acid, *3*:615; *9*:562, 655
Fluorapatite, *9*:83
Fluorene, *6*:642
from coal tar, *19*:657
from methylene chloride, *5*:114
Fluorene-acenaphthene oil
coal-tar distillate, *19*:664
Fluorene-9-carboxylic acid, *15*:682
9,9-Fluorenedipropionitrile, *6*:642
9-Fluorenone, *12*:107; *12*:132
Fluorescein, *5*:871; *11*:479
bromo derivatives in lipsticks, *6*:364
as indicator, *11*:558
polyester from, *16*:169
structural formulas, *5*:769
synthesis of, *22*:430
Fluorescein dyes, *15*:456
Fluorescence, *5*:795; *9*:485; *12*:616
factors affecting, *5*:797
Fluorescence spectra, *5*:795
Fluorescent bleaches. See *Brighteners, optical.*
Fluorescent lamps, *12*:620
Fluorescent lights, *10*:881
Fluorescent pigments, *9*:483
Fluorescent pigments (daylight), *9*:**483**
Fluoridation
of municipal water supplies, *22*:103
Fluoridation of water
use of fluosilicic acid, *9*:653
use of, *9*:655
Fluoride
determination in water, *21*:695
Fluorinated ethylene-propylene
Fluorinated hydrocarbons, *9*:739-751. (See also *Refrigeration.*)
Fluorinated plastics
as cable insulators, *11*:810
Fluorinated silicone
in packing, *14*:452

Fluorination, electrochemical, *9*:700
Fluorination of organic compounds
Fluorine, *9*:**506**
electrodeposition of zinc, *22*:588
in cottonseed meal, *6*:414
in direct fluorination of compds, *9*:697
cryogenic properties, *6*:472
ethyl chloride reaction with, *5*:143
graphite compds with, *4*:312
in manuf of sulfuric acid, *19*:463
in phosphoric acid, *15*:262
as propellant, *4*:685
in tin oxide glass coatings, *10*:592
trichloroethylene reaction with, *5*:188
Fluorine compounds, inorganic, *9*:**527**
Fluorine compounds, organic, *9*:**686.** (See also *Fluoro chemicals; Fluorinated hydrocarbons.*)
nomenclature, *9*:687
Fluorine, liquid
as propellant, *8*:688, 691
Fluorine monoxide. See *Oxygen difluoride.*
Fluorine perchlorate, *5*:74
Fluorite, *22*:338. See *Calcium fluoride.*
zero point of charge, *9*:384
Fluoro-. (See also *Fluo-.*)
Fluoroacetamide, *9*:769
rodenticide, *15*:917
Fluoroacetic acid, *15*:917
p-Fluoroacetophenone, *9*:782
Fluoroalkoxyzirconium compounds, *22*:637
Fluoro amines, *9*:753
11a-Fluoro-6-demethyl-6-deoxytetracycline, *20*:4
2-Fluoro-3,5-dinitrobenzotrifluoride, *9*:779
Fluoroaniline, 6-, 7-, and 8-, *9*:800
2-Fluoroanthraquinone, *5*:229
Fluoroantimonates, *9*:550
Fluoroantimonites, *9*:549
Fluorobenzene, *9*:781
from benzene, *3*:373
perchlorylation reaction, *9*:603
p-Fluorobenzenesulfonic acid, *9*:781
3-Fluorobenzotrifluoride, *9*:779
p-Fluorobenzyl chloride, *9*:783
2-Fluorobiphenyl, *9*:797
Fluorobiphenyls, *9*:796
1-Fluoro-4-bromonaphthalene, *9*:797
Fluorobromonium hexafluorobismuthate(V), *3*:541
Fluorocarbon
cementable surfaces, *18*:452
foams, *S*:372
Fluorocarbon polymers
use as abherents, *1*:10

Fluorocarbons. (See also *Fluorochemicals; Fluorinated hydrocarbons; Fluorine compounds, organic.*)
Fluorochemical resins, *12*:191
Fluorochemicals, *9*:686. (See also *Fluorinated hydrocarbons; Fluorine compounds, organic.*)
  water- and oil-repellant agents, *22*:145
  water-resistant properties, *22*:150
  surface chemistry, *9*:686; *9*:707
2-Fluorocyclohexanone, *9*:605
Fluoroelastomers, *21*:270
  in packaging, *14*:452
  rubber, *17*:545
2-Fluoroethanol, *9*:751
Fluoro ethers, *9*:753
β-Fluoroethyl alcohol, *9*:751
Fluoroform, *9*:509, 765
  from chloroform, *5*:122
Fluorohydrins, *5*:305
21-Fluoro-17-hydroxy-6α-methylpregn-4-ene-3,20-dione, *6*:80
Fluoroisobutylenes
  prepn, *9*:763
Fluorometers, *5*:798
Fluorometric analysis. See *Fluorometry.*
Fluorometry, *2*:795; *5*:795
Fluoromicas
  waterswelling, *13*:413
1-Fluoronaphthalene, *9*:797
2-Fluoronaphthalene, *9*:797
1-Fluoronaphthalene-4-sulfonic acid, *9*:797
2-Fluoronaphthalene-4-sulfonic acid, *9*:797
5-Fluoronicotinamide, *21*:526
18-Fluoro-1-octadecanol, *9*:727
2-Fluoro-4-nitroaniline, *9*:783
2-Fluoro-5-nitrobenzotrifluoride, *9*:779
4-Fluoro-5-nitrobenzotrifluoride, *9*:779
1-Fluoro-4-nitronaphthalene, *9*:797
4-Fluoroperchlorylbenzene, *9*:604
2-Fluorophenyllithium, *9*:783
Fluorophlogopite
  isomorphic substitution, *13*:412
Fluorophlogopite, *13*:399
Fluorophosphates, *9*:642
Fluorophosphites, *9*:646
Fluorophosphoric acids, *9*:636
Fluoroplastics
  plasticizers, *15*:763
Fluoroprene, *5*:229
Fluoroprotein
  foams, *S*:372
2-Fluoropyridine, *9*:798; *9*:800

3-Fluoropyridine, *9*:798
4-Fluoropyridine, *9*:798
Fluorosilicates, *9*:657, 658
Fluorosilicic acid
  addition to water, *22*:103
Fluorosulfonic acid, *10*:161. (See also *Fluorosulfuric acid.*)
Fluorosulfuric acid, *19*:494. (See also *Fluorosulfonic acid.*)
  thiophene reaction, *20*:223
2-Fluoro-3,3 1',5,5 1'-tetranitrobiphenyl, *9*:797
5-Fluorouracil, *15*:686
  cancer therapy, *S*:85
Fluorphlogopite mica, *8*:31
Fluosilicates, *9*:657
Fluosilicic acid, *9*:532, 651; *18*:57; *19*:231
  manuf, *9*:656
  prepn, *9*:652
  prop, *9*:654
  reactions, *9*:655
  uses, *9*:657
Fluosolids roasters, *6*:145
Fluosulfonic acid. See *Fluorosulfuric acid.*
Fluoxymesterone, *11*:102
Fluroxene, *2*:400, 401
Fluxing minerals, *4*:772
Fluxstone, *12*:416
Fly ash
  collection, *8*:90
Foam
  separation in water-treatment systems, *22*:120
Foamed plastics, *9*:**847**
  castor oil in prepn of, *4*:533
  as thermal insulators, *11*:829
Foams, *9*:**884**
Foams
  applications of polyurethan, *21*:93
  in drying, *7*:359
  as fire extinguishing agents, *S*:368
  flexible polyurethan, *21*:85
  molding of flexible, *21*:89
  polyurethan, *21*:80, 84
  prepn of flexible, *21*:85
  properties of rigid polyurethan, *21*:92
  rigid polyurethan, *21*:90
Fog. See *Dust.*
Fog dissipation, *21*:663
"Fog marking" of nylon yarn, *2*:653
Fog oil
  as military smoke agent, *4*:902
Folic acid, *8*:175; *10*:932

antagonists, *21*:489
relation to vitamin B₁₂, *21*:547
Folic acid antagonist, *S*:85
Follicle-stimulating hormone, *11*:61
Folpet
    as fungicide, *10*:225
Fondant (sugar cream), *6*:48
Food
    bleaching agents for, *3*:566
    citric acid in, *5*:534
    colors for, *5*:857
    microwave processing methods, *S*:578
    synthetic from natural gas, *S*:435
    use of industrial enzymes, *8*:190
    use of natural gums in, *10*:741
Food Additives Amendment (U.S.), *4*:337
Food Blue 1, *5*:866, 874
Food Blue 1 (CI), *5*:866, 874
Food Blue 2, *5*:866, 873
Food Blue 2 (CI). See *Brilliant Blue FCF.*
Food Green 1, *5*:866
Food Green 1 (CI), *5*:866
Food Green 2, *5*:866
Food Green 3, *5*:866. See *Fast Green FCF.*
Food Red 1, *5*:867
Food Red 14, *5*:866
Food Red 14(CI), *5*:866
Food Red 6, *5*:872
Food Red 6 (CI), *5*:872
Food Red 9, *5*:866, 873
Food Violet 2, *5*:867
Food Violet 2 (CI), *5*:867
Food Yellow 4, *5*:867, 873
Food Yellow 4 Lake, *5*:871
Food Yellow 5, *5*:867
Food Yellow 5 Lake, *5*:871
Food additives, *10*:**1**
    benzoic acid, *3*:428
    sodium benzoate, *3*:428
Food aerosols, *1*:479
Food and food processing, *10*:**23**
Food color
    caramel, *4*:63
Food fats, *18*:**33**
Food industry
    application of dialysis, *7*:18
    use of abherents in, *1*:3
    use of dextrose in, *6*:925
Food preservation methods, *10*:34
Food proteins
    biological values of, *2*:185
Foods, diet, *10*:**62**

Food standards, *10*:**69**
Food supplements
    pharmaceuticals, *15*:117
Food yeast, *22*:540, 546
Forensic toxicology, *11*:595
Forest products
    microwave energy contributions, *S*:580
Forging
    metals, *13*:316
Forging of metal
    use of furnaces, *10*:273
Formadine sulfinic acid, *10*:799
Formal. See *Methylal.*
Formalan dyes, *2*:889
Formaldehyde, *2*:173, 414; *6*:442, 570, 643,
    644; *10*:**77**; *12*:327. (See also *Hexamethyl-
    enetetramine.*
Formaldehyde, *6*:570, 643, 644
    in acetal resin prepn, *1*:96
    in acrolein prepn, *1*:268
    in acrylic acid esters prepn, *1*:300
    in acrylonitrile prepn, *1*:347
    in amino resin manuf, *2*:228
    ammonolysis of, *2*:359
    in analysis for ammonium chloride, *2*:313
    in anhydroenneaheptitol prepn, *1*:596
    as bleach activator, *5*:30
    butadiene from, *3*:803
    in butynediol prepn, *1*:609
    carbon monoxide reaction with, *4*:435
    copolymerization of, *1*:98
    with crotonaldehyde, *6*:452
    as denaturant, *8*:450
    in dentifrices, *6*:850
    from dimethyl sulfoxide, *19*:321
    as disinfectant, *2*:640
    in drilling fluids, *7*:301, 304, 412
    in embalming fluids, *8*:100
    in ethane oxidation, *11*:226
    in fur puressing, *10*:309
    in glue manuf, *10*:607, 613
    in match prod, *13*:163
    manuf of glycolic acid, *10*:633
    in manuf of isoprene, *12*:76, 89, 114
    in methane oxidation, *11*:224
    from methanol, *13*:372
    from methylene chloride, *5*:113
    "methylene sulfate," *19*:492
    penol condensation, *15*:148
    in pentaerythritol prepn, *1*:590
    phenolic resins, *15*:176
    photochemical smog, *S*:52

polymerization of, *1*:97
polymers, *15*:795
prepn of carbocyanine dye, *6*:611
prepn from ether, *8*:479
prepn from ethylene, *8*:503
prepn of trimethylene glycol diacetate, *8*:502
Prins reaction, *3*:845
reaction with acetaldehyde, *1*:82
reaction with alkanolamines, *1*:813
in propargyl alcohol manuf, *1*:609
reaction with dicyandiamide, *6*:567
reaction with ethleneamines, *7*:26-29, 31, 44
reaction with ketones, *12*:117
reaction with TNT, *8*:612
resins, *19*:317
sterilant, *18*:815
in tanning of furs, *10*:302
thermoplastics, *15*:792
tetrachloroethylene reaction with, *5*:199
in trimethylolethane prepn, *1*:594, 595
in treatment of leather, *12*:330
vinyl chloride reaction with, *5*:173
xylene reactions, *22*:483
reaction with amines, *2*:234
Formaldehyde cyanohydrin, *1*:815; *6*:589, 668, 671; *10*:85; *19*:144
Formaldehyde diethyl acetal, *1*:107
Formaldehyde dimethyl acetal. See *Methylal.*
Formaldehyde polymer. See *Acetal resins.*
Formaldehyde resins
   reaction with ethyleneamines to form coatings, *7*:35
Formaline. See *Formaldehyde.*
Formamide, *6*:570, 576, 611; *10*:103-108; *12*:116
   in nitrogen solutions, *9*:75, 693
   NMR spectra, *14*:55
   sulfamic acid solvent, *19*:243
   phys prop, *2*:68
   form, *2*:68
Formamidine, *6*:563
Formamidine disulfine
   in etching, *15*:523
Formanilide, *2*:423
Formic acid, *6*:465, 466, 574, 703, 922; *7*:538; *8*:241, 812; *10*:**99**; *12*:116
   carbon monoxide from, *4*:437
   carbon monoxide in prepn of, *4*:432
   in distilled vinegar, *21*:256
   in dyeing of spandex blends, *7*:571
   in dyeing of wool blends, *7*:566
   in dyeing of wool with acid dyes, *7*:523, 528
   from lactose, *19*:238
   in paint removers, *14*:487
   phenylacetic acid, *15*:214
   phys prop, *1*:229
   in sorbitol manuf, *1*:581
   prepn from ethylene, *8*:503
   prepn from formaldehyde, *10*:83, 635
   in prepn of aluminum diformates, *3*:14
   in prepn of aluminum triformate, *2*:14
   separation from acetic acid and water, *2*:854
   in Wallach reaction, *3*:369
   from wood, *22*:383
Formic acid and derivatives, *10*:**99**
Formic acid anhydride, *8*:353
Formic acid esters, *8*:375
Formic anhydride, *2*:215
Ethylene glycol di-*n*-butyl ether, *10*:643, 644
Formin. See *Hexamethylenetetramine.*
Formininoethyl ether
Formonitrile. See *Hydrogen cyanide.*
Formonitrolic acid, *8*:588
Ethylene glycol diethyl ether, *10*:643, 644
Formoxime, *12*:143
Formthionals, *10*:85
Formydrazide, *10*:917
N-Formyl-*o*-aminophenol, *2*:215
4-Formyl-*m*-benenedisulfonic acid, *3*:366
Formyl Violet S4B, *20*:715
Formyl Violet S4BA, *7*:525, 526
Formyl Violet S4BN, *20*:715
Formyl Violet S4B4 *7*525, 526
N-Formylaniline, *2*:423
o-Formylbenzenesulfonic acid, *3*:366; *20*:700
p-Formylbenzenesulfonic acid, *3*:366
β-Formylbutyric acid, *1*:640
Formyl fluoride, *10*:151
L*N*-Formylkynurenine, *21*:522
Formylmethylaniline
   styrene reaction with, *5*:521
3-Formylpropionic acid
   oxidation, *19*:143
5-Formylvanillin
5-Formylvanilltn
   from lignin, *21*:189
2-Formyl-2-propanesulfonic acid, *3*:870
N-Formyl-4',4'-dinitrodiphenylamine, *15*:218
Formyl, *9*,605
D-Forosamine, *12*:639
Forsterite
   in ceramics, *4*:722
   dielectric props of, *4*:821
   in electron tubes, *8,*5

refractory, *17*:228, 237
Fortran
  computers programmed by, *6*:29
"Fossil" resins, *15*:513
Fouadin, *14*:546
Fourdrinier brass
  composition of, *6*:214
Fourier's law of heat conduction, *20*:653
Fourier synthesis
  x-ray analysis, *22*:451
Foxglove, *4*:513
Fractionation
  of blood, *3*:576
Fracturing
  nuclear in copper prod, *S*:253
Fradicidin
  amebicide, *20*:82
Francolite, *9*:83
Francombite, *9*:12
Frank-Wen cluster
  in water, *21*:680
Frankfort black, *15*:544
Frankincense, *17*:389
Franklin
  definition of, *S*:1005
Franklinite, *8*:881; *12*:40; *22*:563
Frasch process, *19*:339, 346
Free energy
  polymer surface, *S*:902
Free energy function
  Gibbs, *21*:204, 211
  standard, *20*:141
Free radicals, *10*:**114**
  in antioxidant function *2*:589
  in hydrocracking, *11*:431
  microwave generation, *S*:583
  oxidatton, *10*:115
  polymerization, *10*:115
Freeze drying, *7*:368. (See also *Lyophilization,*)
  aquametry by, *2*:681
  of meats, *13*:177
  microwave, *S*:578
  to preserve foods, *10*:38
Freezing
  of confectionery, *6*:53
  as desalination method for water, *22*:48
  of foods, *10*:46
  hydrate process, *22*:55
  of meats, *13*:177
  normal mode of, *22*:680
  pressure process, *22*:54
Freezing point, See *Melting point.*

Freezing temperatures, *13*:**198**
Fremy's salt
  potassium biflouride, *9*:649
Freon fluorochloroalkanes
  dried by alumina, *2*:52
Fresco, *14*:481
"Fresh" water, *22*:2
Freundlich equation, *1*:429
Freundlich isotherms, *6*:727
Freund's acid, *13*:714
Friction, *10*:124
Friction coefficient
  in aquametry, *2*:693
Friction material, *10*:**124**
Friedel-Crafts reactions, *10*:**135**
  use of aluminum chloride as catalyst in, *2*:17
  with aziridines, *11*:535
  carbonium ions in, *4*:396
Friedelin
  in cork, *6*:284
Frings generator
  fermentation, *21*:261
Frit (glass), *4*:788
Froth flotation. See *Flotation.*
Frozen foods, *10*:46
Fructoheptonic acid, *19*:231
Fructose, *4*:137; *19*:230, 593
  assimilation by yeast, *22*:517
  (levulose) in beet sugar, *19*:211
  Nyn's method, *19*:162
  sweeteners, *19*:602
  yeast prod, *22*:534
D-Fructose, *4*:137; *19*:151, 153
  oxidation, *19*:230
Fructose cyanohydrin, *19*:230
Fruit flavor
  use of fumaric acid as, *12*:833
Fruit juices, *10*:**167**
  prepn, *10*:43
  use of enzyme in, *8*:200
Fruits
  in confectionery, *6*:45
  as flavor, *9*:353
Fuchsin, *2*:638
Fuchsine, *20*:673
  manuf of, *20*:709
  dye series, *20*:677
  polyacrylonitrile fiber dye, *20*:729
Fuchsines
  acid, *20*:715
  N-alkylated, *20*:711
  basic, *20*:709

as nuclear dyes. *20;731*
Fuchsone
  quinoid grouping in, *20*:685
Fuchsonimine
  quinoid grouping in, *20*:685
Fucoidan, *17*:783
  in algae, *17*:763
Fucoses, *4*:137
Fudge, *6*:49
Fuel cells, *3*:139, 160; *16*:465; *S*:**383**
  for electric vehicles, *S*:69
  and natural gas, *S*:435
  platinum-palladium, *15*:835
Fuel oil, *10*:279; *15*:8, 77
  hydrogenolysis results, *10*:411
  properties for gas manuf, *10*:389
  specifications, *15*:86
Fuels, *10*:**179**
  butanes as motor fuels, *3*:816
  chemical forms, *S*:406
  coals, *5*:644
  liquid from MHD power plants, *S*:241
  microencapsulation of, *13*:454
  natural gas as a motive fuel, *S*:433
  synthetic, *19*:684
  use of alcohol in, *8*:448
Fugacity, *7*:786; *21*:204, 208
  of pure fluids, *20*:138
Fulgides, *19*:137
Fuller's earth
  as adsorbent, *1*:460
  in ceramics, *4*:767
  decolorizer, *15*:61
  desulfurization, *15*:46
  oils decolorized by, *5*:577
  pigments, *15*:512
Fulminic acid, *8*:588; *10*:153
Fumagillin
  amebicide, *20*:80
Fumarase, *12*:843
Fumarate Polyesters
  styrene copolymers, *20*:804
Fumaric acid, *6*:694; *12*:**819**
  in alkyd resins, *1*:862
  in carbonated beverages, *4*:339
  hydrogenation, *19*:143
  hydrogen chloride reaction, *19*:136
  oxidation, *19*:735
  phys prop, *1*:232
  from sorbic actd, *18*:590
  in synthetic oil prepn, *7*:415
Fumaric acid, *12*:819

Fumarin, *15*:919
Fumaronitrile, *12*:826
Fumaroyl chloride, *12*:821
Fumes, *18*:**400**. See *Smokes; Dust.*
Fumigant
  calcium cyanide as, *6*:604
  ethylene oxide as, *8*:552
Fumigants, *11*:720-723
  carbon tetrachloride as, *5*:137
Fungal infections
  chemotherapy, *3*:34
Fungal proteases, *10*:9
  uses, *8*:190
Fungi, *13*:468-471
  citric acid prod by, *5*:531
  yeasts as, *22*:508
Fungichromin, *16*:137
Fungicides, *10*:**220**
  copper compounds as, *6*:266, 277
  soil chemistry, *18*:538
  tetraethyllead in manuf, *12*:297
  use of metallic soaps as, *7*:283
  veterinary, *21*:248
Fungicidin, *16*:137
Funoran, *17*:763, 782
Funtumidine, *18*:841
Fural. See *Furfural.*
2-Furaldehyde. See *Furfural.*
Furaltadone, *13*:856
  antibacterial activity, *3*:6
  structural formula, *3*:5
Furan, *8*:492; *9*:797; *10*:237, 247, 248, 898
  from butynediol, *1*:603
  from succinic acid, *19*:142
2-Furanacrolein
  prepn of, *1*:82
2-Furanacrylic acid, *7*:411
2-Furancarboxylic acid. See **Furoic acid.**
Furan compounds, *10*:**237**
Furapromidium, *14*:549
Furazolidone, *13*:852, 856; *2*:638
  antibacterial activity, *3*:6
  structural formula, *3*:5
  as veterinary drug, *21*:246
Furcellaran, *17*:763, 782
Fur dye
  м-aminophenol as, *2*:213
Furfural, *2*:414; *7*:401, 411; *12*:379; *10*:**237**
  in butadiene separation, *3*:805
  content in distilled alcoholtc beverages,
    1:510
  in extraction of fatty acids, *11*:4

furan polymers, *15*:794
nitroparaffin reaction, *13*:831
oxidation, *19*:143
from pentose, *4*:140
phenolic resins, *15*:176, 185
reaction with acetaldehyde, *1*:82
solvent refining, *15*:57
from sucrose, *19*:224
from wood, *22*:383
Furfuraldehyde. See *Furfural.*
Furfuran. See *Furan.*
Furfurin, *10*:241
Furfuryl acrylate
phys prop, *1*:288
Furfuryl alcohol, *6*:644; *10*:237, 245-247
furan polymers, *15*:794
stain solvent, *18*:660
Furfuryl alcohol resins, *10*:246
3-Furfuryloxypropionitrile, *6*:644
Furil, *10*:241; *12*:140
Furium, *13*:856
Furnace fuels, *10*:279
Furnaces
blast, carbon linings for, *4*:239
carbon arc image, *4*:214
electric, *10*:275, 279
electric arc, *4*:227
ferroalloy, carbon refractories for, *4*:240
fuel fired, *10*:**279**

glass-melting, *4*:784
solar, *4*:215
Furniture finishes, *5*:708
Furocoumarins, *10*:925
Furoic acid, *10*:237
phys prop, *1*:240
Furoin, *10*:241
Furosemide, *19*:259
veterinary uses, *21*:252
Furs and fur processing, *10*:**294**
Fursemid
3-(2-Furylacryloyl)-glycol-L-leucine amide, *S*:297
Fusain, *5*:608, 609
Fusarinic, *15*:681
Fusarium
hydrocarbon utilization, *S*:839
Fused metals
as heat-transfer medium, *10*:855
Fusel oil
content in distilled alcoholic beverages, *1*:510
Fusel oil
source of amyl alcohols, *2*:378
Fuse powder, *8*:651
Fusidic acid, *18*:838
Fusion power, *16*:463
Fustic, *7*:527
Fuze powders, *8*:651

G

G-Strophanthin. See *Ouabain*.
G acid, *13*:722, 723
Gadoleic acid, *8*:814
Gadolinite, *17*:147
Gadolinium
  electronic configuration, *1*:360
  ionic radii, *1*:370
Gage
  alphatron, *21*:130
  cold-cathode ionization, *21*:130
  hot-filament ionization, *21*:130
  Knudsen, *21*:129
  McLeod, *21*:128
  Penning, *21*:130
  Phillips, *21*:130
  Pirani, *21*:130
  quartz fiber, *21*:130
  radiometer, *21*:129
  thermocouple, *21*:130
G agents. See *Nerve agents*.
Gage quartz membrane, *21*:130
Gages
  strain, *22*:224
Gainister. See *Quartzites*.
Galactan sulfate
Galactaric acid, *19*:238
Galactitol, *1*:570, 573. See *Dulcitol*.
Galactoflavine, *17*:448
Galacto-glucomannan acetate, *16*:688
Galactomannanase
  microorganism producing, *8*:210
Galactomannan gum
  cyanoethylation of, *6*:662
D-Galactonic acid, *19*:238
DGalactonic acid γ lactone
  in dulcitol prepn, *1*:573
Galactonic lactone, *4*:138
6-*O*-β-D-Glucopyranosyl-D-glucose. See *Gentiobiose*.
4-*O*-β-D-Galactopyranosyl-D-gl- ucopyranose, *19*:237
Galactose, *4*:136
  in lactose, *19*:237
  yeast assimilation, *22*:517
β-Galactosidase
  ultrasonic prepn, *20*:782

D-Galacturonic acid, *19*:230
Galangal, *14*:202
Galanthamine, *1*:771
Galbanum, *14*:202, 722
Galbanum oil, *14*:729
Galena, *7*:292; *9*:393; *12*:208, 279; *19*:350; *22*:338
Gallacetophenone, *16*:198
Gallacetophenone-3,4-dimethyl ether, *16*:210
Gallamine, *6*:548
Gallamine Blue, *2*:865
Gallamine triethiodide, *6*:550; *16*:196
Gallamine triethiodide, *2*:410
Gallates. See *Gallic acid; Gallium compounds*.
Gallein, *16*:198
Gallic acid, *12*:321; *15*:681; *16*:192, 196, 210
  in fireworks, *16*:837
  gelling agent, *21*:355
  phys prop, *1*:238
Gallic acid glycosides, *10*:307
Gallium, *10*:**311**
  in copper alloys, *6*:210
  as solid film lubricant, *12*:590
  in sphalerite, *22*:562
Gallium and gallium compounds, *10*:**311**
Gallium arsenide
  electronic prop, *17*:857
  semiconductor, *17*:857
  in semiconductors, *2*:716
Gallium compounds, *10*:**311**
Gallium nitride, *13*:823
Gallium phosphide, *12*:628
  nonlinear optical materials, *S*:625
Gallnuts, *12*:318
Gallocyanine, *2*:865
Galltoannic acid, *12*:321
Galvanized coatings, *13*:254-258
Galvanizing, *22*:599
Gambier, *12*:318
  pyrocatechol in, *11*:469
Gamboge, *15*:538
Gamma Acid,
  sodium salt, *7*:478
Gamma globulin
  separation by electrodecantation, *7*:844
Gamma radiation, *17*:58

Gamma rays, *17*:14, 53, 66

Gammexane, *15*:337. See *Benzene hexachloride.*

Ganglionic blockaders
  as cardiovascular agents, *4*:519

Garantose, *19*:595

Gardner color scale, *5*:809

Garjun balsam
  essential oil in, *14*:181

Garlic, *9*:363, 371; *14*:202

Garnet, *2*:1; *10*:758; *22*:338
  as abrasive, *1*:25
  as coated abrasive, *1*:29
  use in water filtration systems, *22*:119

Garnierite, *5*:547; *13*:737

Gas
  compressibility factor, *21*:203
  equations of state for, *21*:199
  ideal gas law, *21*:199

Gas Natural, *10*:**443**
  in ammonia manuf, *2*:275

Gas black
  pigment, *14*:474

Gas centrifuge, *7*:149

Gas chromatography, *5*:429
  bis(2,3-epoxypropyl) in *1*:250
  and mass spectrometry, *13*:87

Gas cleaning, *10*:**329**

Gases
  centrifugal separation of, *4*:755
  disposal of waste, *21*:631
  ionized by microwave energy, *S*:582
  solubility of, *S*:905
  waste treatment, *21*:628
  in water, *21*:700

Gas exchanger
  photosynthetic, *1*:664

Gas-fading of dyes, *2*:506

Gasification
  of coal, *S*:198
  power from coal by, *S*:217

Gaskets, *14*:443

Gas-liquid chromatography
  of acetaldehyde, *1*:90
  sugars, *19*:165

Gas lubrication, *12*:605

Gas, manufactured, *10*:**353**

Gas masks. See *Masks, protective.*

Gas, natural, *10*:**443**; *S*:**421**
  characteristics as furnace fuel, *10*:280
  as feedstock for nitrogen fertilizers, *9*:49, 54
  helium content, *10*:865

helium from, *10*:879
in manuf of carbon disulfide, *4*:376
reserves, *10*:187
as source of liquefied petroleum gas, *12*:470
used to melt glass, *10*:555

Gas oil
  distillation for gas manuf, *10*:472
  high-temperature pyrolysis, *S*:553
  hydrogasification data, *10*:411
  prop for gas manuf, *10*:389
  proteins from, *S*:851
  synthetic, *4*:446

Gas oil, *14*:836; *15*:2, 17
  pyrolysis products, *S*:311

Gasoline, *10*:**463**; *15*:2, 8, 77
  alkylate, *11*:419
  analysis for lead content, *12*:295
  antioxidants, *2*:598
  burning temperature of, *4*:899
  as denaturant, *8*:450, 453
  gum inhibitors, *15*:53
  heat of combustion of, *4*:898
  hydrocarbons in, *11*:266
  light, *15*:17
  as propellants, *8*:661
  specifications, *15*:86
  synthetic, *4*:446
  toxicity of, *11*:304
  use of lecithin in, *12*:358

Gasoline additive
  dimethylformamide as, *10*:113

Gasoline additives
  economic aspects, *12*:293

Gasoline and other motor fuels, *10*:**463**

Gasoline industry
  use of ethyleneamines, *7*:35

Gasoline recovery
  carbon used in, *4*:154

Gas plating, *13*:280-281

Gas sweetening
  by ethanolamine, *1*:821

Gastrin
  activity of, *11*:50

Gas turbines, *16*:440-446

Gas wells
  corrosion in, *6*:330

Gattermann-Koch reaction
  aldehydes, *1*:648

Gattermann Koch synthesis

Gattermann synthesis
  aldehydes, *1*:648
  use of aluminum chloride in, *2*:23

Gaussian probability function
  use in adsorption theory, *1*:438
"Gaussian" system of equations, *S*:1003
Gearksutite, *9*:534
Geiger counters, *17*:72
Gelatin, *10*:**499**
  capsules, *15*:124
  coagulants, *22*:96
  in confectionery, *6*:43
  cyanoethylation of, *6*:663
  in microencapsulation, *13*:440
  as optical filter, *9*:251
  in photography, *15*:358
  shrinkproofing wool, *22*:409
  suspending agent, *21*:375
  tablet coater, *15*:122
  thiolated, *15*:339
Gelatinase
  action and uses, *8*:184
Gelatin capsules
  use of formamide in, *10*:105
Gelatin dynamite, *8*:640
Gelatin effect
  in photography, *15*:381
Gelling agents
  in food products, *10*:5
Gelsemine, *1*:765, 773
Gems, synthetic *10*:**509**
Generator
  Faraday segmented, *S*:224
  fossil-fuel-fired, *20*:164
  nuclear thermoelectric, *20*:167
  solar thermoelectric, *20*:164
Gentamycins, *19*:33, 43
Gentianine, *1*:765, 772
Gentian root, *9*:353
Gentian violet, *14*:533
  as antiseptic, *2*:639
Gentiobiose, *4*:146; *6*:921
Gentisic acid
  as antipyretic, *2*:390
Geothite, *13*:18; *14*:164
Geraium flavor, *9*:353
Geraniol, *14*:735; *19*:818, 826
  as synthetic flavor, *9*:354
Geraniol hydrate, *19*:829
Geranium, *14*:202
Geranyl acetate
  prod and sales of, *8*:377
  in perfumes, *14*:735
Geranyl acetates
  as synthetic flavor, *9*:354

Geranyl acetone, *19*:833
Geranyl butyrate
  as synthetic flavor, *9*:354
Geranyl chloride, *19*:818, 825
Geranyl formate
  prod and sales of, *8*:377
Geranyl isobutyrate
  in perfumes, *14*:735
Geranyl propionate
  in perfumes, *14*:735
Germane, *10*:523; *11*:206
Germanite, *10*:311; *10*:519
Germanium, *10*:**519**
  diffusion in, *17*:875
  electrodeposition of zinc, *22*:587
  electronic prop, *17*:850
  homochain polymers, *11*:636
  infrared detectors, *15*:430, 435
  manganese compounds, *13*:4
  in photoconductive cells, *15*:412
  photoconductive prop, *15*:420
  semiconductor, *17*:834, 850
  silane compounds, *18*:203
  in sphalerite, *22*:562
  zone refining, *22*:680, 697
Germanium compounds, *10*:519-527
Germanium difluoride, *9*:584
Germanium-silicon alloys
  infrared detectors, *15*:438
Germanium tetrafluoride, *9*:584
German silver. See *Nickel silver; Nickel silvers.*
Gersdorffite, *2*:720
Gersthofen wax, *12*:409
Getters, chemical, *8*:2, 7, 20
Gibberellic acid, *8*:208; *12*:875; *15*:678; *19*:803
Gibberellin, *2*:204; *15*:683, 685
Gibberellins, *12*:875; *15*:678
Gibbs
  phase rule, *21*:196
Gibbs-Duhem equation, *20*:137; *21*:209
Gibbs-Duhem principle, *S*:902
Gibbs-Helmholtz equation, *21*:212
Gibbs-Thomson equation, *1*:427
Gibbs free-energy function, *21*:204
  for binary systems, *21*:211
Gibbsite, *1*:932; *2*:8, 42, 43; *5*:551, 553
  activation of, *2*:5
  in ceramics, *4*:763, 773
  in prepn of sodium aluminate, *2*:7,8
  refractory, *17*:232

Gilding brass
  composition of, 6:214
Gillespie still
  vapor-liquid equilibria determination,
    21:234
Gilsonite, 10:**527**; 18:661
Giltenes, 10:530
Gin, 1:514
Ginger, 9:352, 363, 365, 372; 14:203
Ginorite, 3:650
Girard Reagent T
  chlorophyll reactions with, 5:345
Girbotol amine process
  carbon dioxide recovered by, 4:358, 362
Gitalin
  as cardiovascular agent, 4:515
Gitoxin
  as cardiovascular agent, 4:513
Gladiolic acid, 15:160
Glance pitch, 10:528
Glass, 10:**533**
  addition of α-alumina trihydrate, 2:51
  beads in paints, 15:516
  borax in manuf of, 3:637, 644
  borate glasses, 3:620
  from borate melts, 3:620
  cerium oxides in manuf of, 4:852
  colorants for, 5:845
  corrosives transport, 20:601
  as fiber, 9:159
  fibrous as filler, 15:802
  as filter medium, 9:285
  graphite equipment in manuf of, 4:233
  as lubricant, 12:603
  manuf of, 4:788
  melting furnace for, 4:784
  as optical filter, 9:251
  plasticizers for, 15:774
  in prepn of trifluoroboroxine, 9:556
  piping, 15:647
  silica. See *Silica.*
  use of calcium fluoride in, 9:580
  use of cryolite in manuf, 9:545
  use of soda ash in, 1:708
Glass coating
  titanium lactate as, 12:184
Glass electrode, 11:383
Glasses
  thermal analysis, 20:113
Glass fabrics
  dyeing of, 7:572
Glass fiber, 12:189
  addition to plastics, 16:93
  tire cords, 20:331
  used for reinforced plastics, 12:189
Glass fibers
  heat-and flame-resistant, S:959
Glass, flat, 10:571
Glass frosting
  use of ammonium bifluoride in, 9:548
Glass gages
  used in liquid measurement, 12:452
Glass molding
  use of abherents in, 1:7
Glass sand, 4:769; 10:550
Glass-transition temperature
  of polymers, 6:671; 16:252
  of polysulfides, 16:262
Glassware washing, 6:868
Glauber's salt, S:444, 458. See *Sodium sulfate.*
Glaucarubin, 20:79
Glaucodot, 2:720
Glauconite, 5:551
  from ocean, 14:163
Glazes
  for graphite refractories, 4:329
  copper ruby, 5:853
Gliotoxin
  occurence of hydantoins in, 11:141
Globaline, 2:619. See *Tetraglycine hydroper-
    iodide.*
Globaline, 2:619
Globulins
  in cereal grains, 22:268
  in nuts, 14:125
Globulins, serum, 3:576
Gloss white, 15:516
Gloxal
  in embalming fluids, 8:100
Glucamine, 19:231
Glucamylase
  in grain mashes, 22:522
Glucan
  in yeast, 22:516
D-Glucaric acid, 19:229, 238
Glucinic acid, 19:211
L-Glucitol, 1:573
  phys prop, 1:570
  prepn, 1:573
D-Glucitylamine, 2:359; 19:231
Glucoamylase, 6:923
  microorganism, 8:210
  starch cleavage, 18:676
  uses, 8:193

Glucomannan gum
  cyanoethylation of, 6:662
Gluconic acid, 4:138
  as chelating agent, 6:8
  in paint removers, 14:491
  phys prop, 1:234
  as sequestering agent, 6:18
  from sucrose, 19:224
D-Gluconic acid, 6:922; 19:227
Gluconic lactone, 4:138
Glucono-δ-lactone
  in baking powder, 3:52
  in meat curing, 13:371
Glucoronic acid, 19:273, 602
Glucosamine, 4:142
  in yeast, 22:516
Glucosamine acetylsalicylate, 17:733
l-Glucosan
  decomposition of wood, 22:376
Glucose, 2:585; 4:136; 7:413; 12:323; 19:593.
  (See also Dextrose.)
  Benedict's solution, 19:162
  carbon source for algal cultures, 1:652
  copper acetate reagent, 19:162
  in eggs, 7:670
  in D-arabitol prepn, 1:572
  in lactose, 19:237
  oxidation to tartaric acid, 19:735
  in prepn of glutamic acid, 2:203
  streptomycin from, 19:33
  structure of cotton, 6:405
  yeast assimiltation, 22:517
  yeast prod, 22:534
D-Glucose
  polymer in wood, 16:684
D-Glucose, 19:151, 153, 229
  in biogenesis of macrolide antibiotics, 12:649
  oxidation, 19:230
  in streptomycins, 19:38
D-Glucose aerodehydrogenase, 19:164
Glucose monoaleuritate, 18:24
Glucose oxidase, 7:671
  action and uses, 8:183
Glucose oxidases, 10:9
Glucose pentaacetate, 19:227
D-Glucose phenylhydrazone, 4:140
Glucose-1-phosphate, 19:151
Glucose-6-phosphate
  oxidation of, 21:518
α-Glucosidase, 8:194
  microorganism producing, 8:210

β-Glucosidase
  microorganism producing, 8:210
D-Glucosylamine, 4:141
D-Glucosylaniline, 4:141
Glucovanillin, 21:183
Glucuronic acid
  pathway, 2:755
D-Glucuronic acid, 19:228
Glucuronide, 19:273
β-Glucuroxidase
  microorganism producing, 8:210
Glue, 10:**604**
  as adhesive, 1:387
  process in lithography, 17:330
Glue, animal and fish, 10:**604**
Gluside, 19:595
Glutamic acid, 2:157, 177, 198, 199; 12:857
  cerebral agent, 16:645
  oxidation, 19:143
  prepn of crude, 2:202
  from sucrose, 19:224
  in tobacco, 20:508
  in wine, 22:316
D-Glutamic acid
  tetracyclines, 20:25
Glutamic acid hydrochloride, 2:201
Glutamic acid, monosodium salt
  flavor and perfume materials, 9:358
Glutamine, 2:157, 199
  in beet sugar, 19:211
  cerebral agent, 16:645
Glutamine antagonist, S:85
Glutaraldehyde, 1:258; 2:641; 10:613
  from dihydropyrans, 21:416
  in embalming fluids, 8:100
  sterilant, 18:817
  in water-repellent leather, 22:150
Glutaric acid, 1:258; 10:**618**
  phys prop, 1:232
  prepn, 1:247
Glutaric dialdehyde
  in prepn of lobelanine, 1:795
Glutarimide
  derivatives as hypnotics, 11:514
Glutathione
  as antioxidant, 2:601
  occurence, 20:205
  sulfur in, 19:369
  yeast product, 22:531
Glutelin
  in acorns, 14:125
Glutelins
  in cereal grains, 22:268

Gluten
  cyanoethylation of, *6*:663
  determination in wheat, *22*:283
Gluten feed
  as animal feed, *8*:859
Glutethimide
  as hypnotic, *11*:515
Glyceraldehyde, *4*:133; *10*:620
  configuration of amine acids based on, *2*:168
  prepn from glycidaldehyde, *1*:261
Glyceraldehyde-3-phosphate, *21*:522
Glyceraldehyde phosphate dehydrogenase
  in yeast proteins, *22*:516
Glyceratozirconium chloride, *22*:659
Glyceric acid, *10*:620; *19*:238
Glycerides, *8*:777
  surfactants, *19*:537
Glycerin. See *Glycerol*.
Glycerol, *2*:415, 579, 623; *6*:362; *10*:**619**
  as a drying agent, *7*:385, 392, 413
  alkyd resins, *15*:455
  in alkyd resins, *1*:861
  from allyl chloride, *16*:254
  biosynthesis of quinolinic acid, *21*:521
  in breadmaking, *22*:528
  chlorohydrins from, *5*:316
  in dental material, *6*:827, 849
  Duff reaction, *15*:162
  in embalming fluids, *8*:100
  epoxy resins from, *19*:147
  in esterification of fatty acids, *8*:819
  from isopropyl alcohol, *16*:576
  in lecithin, *12*:343
  and oxalic acid reaction, *14*:357
  phthalic anhydride reaction, *20*:793
  pill formulation, *15*:124
  plasticizer, *15*:764
  polyesters, *15*:793
  polyurethans from, *21*:61
  prepn from acrolein, *1*:262
  prepn from propylene, *1*:161
  in prepn of glyceryl nitrate, *8*:603
  in prepn of 7-hydroxyquinoline, *2*:213
  in prepn of 8-hydroxyquinoline, *2*:215
  as preservative for nuts, *14*:137
  process for nitration of, *8*:594
  prod by yeast, *22*:529
  reaction with propylene oxide, *16*:598
  from sorbitol, *1*:577
  in soaps, *18*:429
  from sucrose, *19*:154
  surfactants, *19*:554
  sweetener, *19*:603
  tall oils, *19*:625
  tartaric acid reaction, *20*:793
  use in denaturing alcohol, *8*:450
  in vinegar, *21*:256
  wood hydrolysis, *22*:383
  from xylitol, *1*:577
Glycerol chlorohydrins, *5*:311
Glycerol dichlorohydrin, *2*:352
Glycerol-1,2-dichlorohydrin
  from vinyl chloride, *5*:173
Glycerol esters
  surfactants, *19*:537
Glycerol ethylidene ether, *10*:629
Glycerol-formal
  in embalming fluids, *8*:101
Glycerol α-monochlorohydrin, *10*:905
Glycerol monoacetate, *10*:630
Glycerol 1-monoformate, *10*:619
Glycerol monoricinoleate
  prod and sales of, *8*:377
Glycerol monostearate
  HLB value, *8*:128, 129
Glycerol phosphates, *15*:270
Glycerol succinates
  plasticized, *19*:145
Glycerol tributyrate, *3*:882; *10*:619
Glycerol trioleate
  prod and sales of, *8*:377
Glycerol tri(12-hydroxy)stearate, *22*:164
Glycerolysis, *8*:819
Glycerophosphoric acid salts, *10*:629
Glycerose. See *Glyceraldehyde*.
Glyceryl monostearate, *6*:363
  use as abherent, *1*:8
Glyceryl nitrate, *7*:57; *8*:581, 582, 602
  as cardiovascular agent, *4*:521
  characteristics, *8*:632, 633, 634
  in dynamite, *8*:637
  as gun propellant, *8*:701
  manuf (Biazzi process), *8*:595
  manuf of blasting gelatin, *8*:636, 640
  in molded tablets, *15*:123
  process of burning, *8*:668
  in propellants, *8*:659
  purification process, *8*:594
  thermochemical constants, *8*:681
  use of glycerin, *10*:627
Glyceryl phthalate
  polyester, *16*:159
Glyceryl phthalate resins, *15*:446

Glyceryl stearate
  in shampoo, *10*:774
Glyceryl triacetate, *15*:751
  as cellulosic plasticizer, *4*:657
Glyceryl tributyrate
  as cellulosic plasticizer, *4*:657
Glyceryl trieleosterate, *8*:363
Glyceryl tri(hydroxystearate)
  use as abherent, *1*:8
Glyceryl trinitrate, *15*:774
  as cardiovascular agent, *4*:521
Glyceryl tripropionate
Glycidaldehyde, *10*:624
  prepn from acrolein, *1*:261
  sterilant, *18*:817
Glycidamide, *8*:244
Glycidol, *5*:313; *10*:624, 905
  sterilant, *18*:817
Glycidyl ether resins
  manuf, *8*:295
Glycine, *2*:157, 188, 430; *6*:650; *8*:781; *15*:374
  manuf, *2*:350
  p-nitrophenyl acetate ester, *S*:297
  in silk, *18*:271
  uric acid from, *21*:107
  in wine, *22*:316
Glycine silver nitrate
  as ferroelectric, *9*:5
Glycine-2-¹⁴C, *20*:20
Glycinin
  in soybeans, *18*:601
Glycoaldehyde, *4*:133
Glycobiarsol, *3*:546, 547
  amebicide, *20*:93
Glycocholic acid
  in food processing, *10*:8
Glycogen
  in fish, *9*:321
  in meats, *13*:167
  in yeasts, *22*:516
Glycol diacetate, *8*:363
Glycol dinitrate, *8*:582, 605; *15*:774
  manuf (Biazzi process) *8*:595
Glycolic acid, *7*:529; *10*:632; *19*:238; *20*:199
  in acid dyeing of wools, *7*:523
  carbon monoxide in prepn of, *4*:435
  phys prop, *1*:234
  from oxalic acid, *14*:357
  from trichloroethylene, *5*:187
Glycolide, *10*:633
  copolymerization, *16*:174
  polymers from, *16*:171
Glycollic acid. See *Glycolic acid.*

Glycolonitrile. See *Formaldehyde cyanohydrin.*
Glycolonitrile, *S*:599
p-N-Glycoloylarsanilic acid, *20*:93
Glycols, *9*:401; *10*:**638**. (See also *Ethylene glycol; Propylene glycol.*)
Glycols, ethylenic, *1*:632
Glycol stearate, *10*:774
Glycols, tertiary, *1*:625
Glycoluril, *4*:920; *22*:129
Glycolyl, *8*:*Hydantoin.*
Glycolylguanidine
  prepn from guanidine, *10*:737
Glycoproteins
  complexity of, *11*:50
Glycose glycoside, *4*:145
Glycose oxidase
  action and uses, *8*:184
Glycosides
  as cardiovascular agents, *4*:513
  in nuts, *14*:132
Glycosides
Glycylguanidine
  prepn from guanidine, *10*:737
Glycyrrhiza, *19*:602
Glycyrrhizic acid, *19*:602
  ammonium salt, *19*:602
Glycyrrhizin, *2*:294
Glyeyrrhetinic acid, *19*:602
Glyoaxilic acid
  tetracycline adducts, *20*:10
Glyodin
  as fungicide, *10*:225
Glyoxal, *10*:613, 646
  from oxidation of acetaldehyde, *1*:80
  in papermaking, *14*:520
  phenylenediamines, *15*:221
Glyoxalate cycle, *19*:135
Glyoxaline, *10*:897, 898, 911
Glyoxal sulfate
  from 1,1,2,2-tetrachloroethane, *5*:161
Glyoxyldiureide, *21*:112
Glyoxylic acid, *1*:235; *10*:633
  and oxalic acid, *14*:357
Glyptal, *15*:793
Glyptal resins
  step-growth, *16*:225
Goethite
  effect on pH of recovery of, *9*:386
  zero point of charge, *9*:384
Goethite, *12*:6; *13*:737
Gold, *10*:**681**
  in ceramic pigments, *5*:854
  colorant for glass, *5*:848; *9*:252

as dental material, 6:814
as dental solder, 6:822
in electroplating, 8:60
electromigration of, S:286
impurity in copper, S:265
manganese compounds, 13:3
as neutron monitor, 17:37
pigments, 15:550
removal from lead ore, 12:231
from seawater, 14:156
as solid film lubricant, 12:590
in type metal, 20:757
in vacuum tubes, 8:3, 4
Gold alloy, 6:814
in dental inlay investment, 6:789
in orthodontic wires, 6:812
with porcelain, 6:797
Gold alloys, 10:688-693
Goldang wax, 22:161
Gold compounds, 10:**681**
Gold extraction
calcium cyanide used for, 6:601
by sodium cyanide, 6:585, 589
Gold ore
fluidization process, 9:435
Goldschmidt process
use of aluminum as reducing agent in, 2:1
Gold-198, 17:127
Golgi bodies
in yeast, 22:512
Gonadotropin
human chorionic, 11:65
commercial sources, 11:67
pregnant-mare serum, 11:64
Goniometry, 6:517
Goodmanite, 9:12
Gophacide, 15:919
Goslarite, 22:563
Gossypol, 6:414
Gouache, 14:480
Goulac
in electrolytic refining of copper, 6:159
Government documents
sources, 12:509
Gowerite, 3:650
Gradient calorimetry, 4:51
Gradient-tube method
of density measurement, 6:769, 773
Grahamite, 10:528
Grain
boundary, 11:12
size, 11:12

Grain mashes
yeasts for, 22:519
Grain neutral spirits, 1:513
Grain sorghums
as animal feeds, 8:860
Gramicidin, 16:311
amebicide, 20:82
chemistry of, 16:312
Gramicidin S, 16:312
Gramine, 1:772; 11:590
Granite
graphic, 4:769
Grapefruit, 14:203
essential oils as flavor, 9:352
Grapefruit seed oil, 8:779, 784
Grape sugar. See Dextrose.
Graphite. (See also Carbon.)
in nuclear reactors, 14:83
Graphite, 4:158-243, 304-335; 4:228; 12:588
additive to polyamides, 16:94
artificial, 15:23
in ceramics, 4:774
as electric furnace resistor, 10:255
electrodes in arc furnaces, 10:258
in fire extinguishers, 9:296
as impregnant, 14:457
in manganese battery ore prod, 13:32
pigments, 15:545
in propellants, 8:660, 681
prop as embedding filler, 8:110
refractory electrode, 17:287
silver alloys, 18:294
thermal conductivity of, 4:803
whiskers, 15:802
Graphite acid, 4:312
Graphite bisulfate, 4:312
Graphite cloth, 4:241
Graphite compounds, 4:311
Graphite crystal
carbon fibers, S:109
Graphite felts, 4:243
Graphite fibers
use in reinforced plastics, 12:189
Graphite, natural, 4:304-355
Graphite, nuclear, 4:189, 236
Graphite oxide, 4:312
Graphitization
in steels, 11:14
Graph papers
types of, 6:705
Graphs
interpretation of, 6:705

of vapor pressure, *6*:717
Gravel
  from ocean, *14*:160
Gravity
  variations, *22*:227
Gravity concentration, *10*:**695**
Gravity diffusion, *7*:92
Gravity separation. See *Gravity concentration.*
Gravure, *16*:495, 526-528, 538, 543
  electrostatic powder, *16*:542
  silver halide system, *16*:520
Grease
  in manuf of soap, *18*:417
Greases, *12*:582
  from animal fats, *13*:183
  as impregnants, *14*:457
  use of metallic soaps in, *7*:283
Great Salt Lake Chemicals, *S*:**438**
Green acids, *19*:296
Greenalite, *12*:6
Green earth, *15*:538
Greenockite, *3*:885, 908; *22*:562
Greensalt
  as fungicide, *10*:230
Green vitriol, *12*:40
Grignard reaction, *10*:**721**
  with aziridines, *11*:534
Grignard reagents, *10*:721
  acetaldehyde, *1*:83
  reaction with carbon disulfide, *4*:373
Grinding. See *Abrasives.*
  comminution, *18*:324
Grinding wheels, *1*:31
  bonded, *1*:34
  grain size used in, *1*:33
  standard markings, *1*:32
Griseofulvin, *3*:8, 35; *10*:908; *16*:209
  veterinary uses, *21*:248
Griseomycin, *12*:635
Griseoviridin
  amebicide, *20*:82
Groth centrifuge, *7*:149, 150-152
Groutite, *13*:18
Grouts, cement, *4*:704, 705
Growth hormone, *11*:53-56
Growth inhibitors, *15*:681
Growth regulators, *15*:675
Grünerite, *12*:6
Guaiac, *17*:389
α-Guaiac acid, *17*:389
Guaiacol
  prepn of papaverine, *1*:798

used in denaturing alcohol, *8*:450
  vanillin from, *15*:163; *21*:184
Guaiacol carbonate, *11*:468
Guaiacol potassium sulfonate, *11*:468
Guaiaconic acid, *17*:389
Guaiac resin, *8*:781
Guaiacum, *17*:389
Guaiacwood, *14*:203
Guaiacylpropane
  in lignin, *12*:363
Guaiaretic acid, *17*:389
Guaiol, *19*:833
Guanamide, *10*:930
Guanamine
  prepn from guanidine, *10*:736
Guanamines, *6*:566
Guanazole, *2*:244
Guanethidine
  as cardiovascular agent, *4*:519
Guanidine, *2*:417; *6*:566, 567; *10*:**734, 930**
  phthalocyanine compds, *15*:492
  prepn of nitroguanidine, *8*:625
  reaction with acetylacetone, *12*:157
  sulfonic acids, *19*:318
Guanidine aluminum sulfate hydrate
  as ferroelectric, *9*:5
Guanidine hydrochloride, *6*:567
Guanidine nitrate
  in prepn of chloroazodin, *4*:921
Guanidine orthophosphate, *15*:239
Guanidines
  as rubber chemicals, *17*:511
  as accelerators, *17*:510
Guanidine salts, *10*:**734**
  prepn, *6*:561
Guanimines
  nitrile reactions, *S*:594
Guanine, *10*:913
Guano-urea orthophosphate, *15*:239
Guanylic acid, *2*:179
5'-Guanylic acid, *2*:198
Guanylmelamine, *6*:566
4-Guanyl-1-(nitrosoaminoguanyl)-1-tetrazene. See
  Tetracene.
Guanylthiourea, *6*:567
Guanylurea, *6*:566
Guanylurea-*N*-sulfonic acid, *2*:893
Guar gum, *10*:750; *17*:391
  as drilling fluid flocculant, *7*:305
Guerbet reaction, *3*:825
Guignet's green. See *Chromic oxide green.*
Guinea Green B, *5*:866

Guinea Green BA, 7:525
Guinea Green B Lake, 5:868
Guluronic acid, 17:769
Gum acacia. See *Gum arabic.*
Gum arabic, 7:299; 10:744; 17:391; 19:228
  in confectionery, 6:52
  in food processing, 10:10; 10:744
  in lithography, 17:330
  in paints, 14:468
  in photolithography, 16:530
  reaction with lime, 19:143
Gumase
  action and uses, 8:188
Gum dammar
  in dental material, 6:840, 841
Gum ghatti, 10:748
Gum guaiac
  as food preservative, 10:14
Gum karaya, 7:299; 10:746; 17:391
  in embalming fluids, 8:100
Gummite, 21:8
Gums
  coagulants, 22:96
  millable, 21:95
  in nuts, 14:129
  used as food additives, 10:6
Gums and jellies, 6:51
Gums, natural, 10:**741**
  in confectionery, 6:45
Gums-resin, 17:387
Gum tragacanth, 7:299; 10:745; 17:391
  in embalming fluids, 8:100
Guncotton, 8:597, 697

Gun fiber
  in hardboard, 21:603
Gunk
  decontaminating agent in chemical warfare,
    4:886
Gun propellants. See *Propellants.*
α-Gurjunene, 17:389
β-Gurjunene, 17:389
Guru nuts. See *Kola nut extract.*
Gutta-percha, 15:790; 17:682
  x-ray analysis, 22:453
Guy's hospital pills, 7:254
Gynesine, 21:535
Gypsite, 4:17
Gypsum, 4:14, 16, 17, 18, 19 15:509; 18:158.
  (See also *Calcium sulfate.*)
  calcined in wallboards, 21:613
  as cement set retarder, 4:690
  in ceramics, 4:770
  derivatives in dental materials, 6:784
  in drilling fluids, 7:301, 302
  in manuf ammonium sulfate, 2:329
  prepn of calcium sulfate, 7:392
  phosphoric acid by-product in fertilizer
    manuf, 9:90, 95
  sulfur, 19:343
  sulfur dioxide from, 19:449
  sulfur from, 19:339, 356
  zero point of charge, 9:384
Gypsum board, 21:601, 621
Gypsum sands, 4:17
Gypsum wallboard
  use of lignosulfonates in, 12:372

H

H Acid
  monosodium salt, 7:478
HC smoke mixture
  in chemical warfare, 4:902
HEDTA. See *N-Hydroxyethyl-N,N',N'-ethyl-enediaminetriacetic acid.*
HEOD. See *Dieldrin; Endrin.*
HET-Diol, 5:241, 250
HHDN. See *Aldrin.*
HLB. See *Hydrophile-lipophile balance.*
HMTD. See *Hexamethylenetriperoxide diamine.*
HMX. See *Cyclotetramethylenetetranitramine.*
HTU (Height of a transfer unit)
H acid, 13:725
Haemoglobin
  x-ray analysis, 22:454
Hafnium, 10:**754**
  corrosion resistance to water, 6:302
  occurence with zirconium, 22:615
  reaction with platinum metals, 22:621
  sodium reduction, 18:450
Hafnium carbide, 4:75, 84, 85
Hafnium compounds, 10:**754**
Hagemannite, 9:534
Hail suppression, 21:663
Hair bleaching, 10:790, 798
Hair coloring preparations, 10:790
Hair hygrometer
  aquametry by, 2:693
Hair lacquers
  use of aerosol products, 1:478
Hair preparations, 10:**768**
Hair spray resins, 10:781
Halane. See *1,3-Dichloro-5,5-dimethylhydantoin.*
Halazone, 4:923; 2:616, 617; 19:260
Half-life
  of an element, 17:13
Halide leak" detector for refrigeration systems, 5:6
Halides, acid. See *Acid, halides.*
Halite. See *Sodium chloride.*
Hall effects
  in MHD generator, S:220
Halloysite, 5:545
  in ceramic products, 5:561

Hallucinogenic agents, 16:645
  in chemical warfare, 4:877
Haloacetamides
  antiamebic agents, 20:89
Haloacetylenes
  prepn, 1:172
Haloalkylation, 10:141
Haloforms, 9:757
Halofuran
  prepn, 10:237
Halogenated fire retardants, S:**467**
Halogenation
  of nitroparaffins, 13:870
Halogen fluorides, 9:585
  reaction with metals, 9:587
Halogens
  as hazardous chemicals, 9:292
Haloisoquinolines, 16:883
Halon
  plasticizers, 15:763
Halons
  fire extinguishing agents, S:376
p-Halophenacyl esters, 8:372
Haloquinolines, 16:875
Halosulfinates, 19:490, 493
Halosulfurates, 19:490, 493
Halothane, 2:401; 3:780. See *2-Bromo-2-chloro-1,1,1-trifluoroethane.*
Halowaxes, 5:301
1-Halo-2-methyl-2-propanol, 8:268
Hamycin, 16:138
Hansa Orange, 15:572
Hansa Red B, 2:908
Hansa Yellows, 15:571
Hansa colors
  prepn from acetoacetic ester, 1:158
Hantzsch synthesis
  for nicotinic acid, 21:513, 516
Hardboard, 21:603
Hardboard defibrator process, 21:603
Harden and Young's coenzyme, 21:512
Hardness, 10:**808**
  of minerals, 1:23
  of swimming pool water, 22:131
  of water, 21:691
  of water resources, 22:73

Hardness testing, *13*:185
Hardwoods, *22*:359
Hargreaves process. See *Hydrochloric acid.*
Harmaline, *1*:772; *16*:645
Harmine, *16*:645
Hastelloy
  corrosion resistance to water, *6*:302
  welding, *22*:248
Hastelloy N
  for nuclear reactors, *14*:83
Hausmannite, *13*:8, 17
H-coal process
  for curde oil, *S*:181
Heat
  Fourier's law of heat conduction, *20*:653
  thermodynamics, *20*:118
  transfer coefficients, *20*:654
Heat energy
  interconversion, *20*:147
Heat exchangers, *10*:839-846
  impervious graphite, *4*:218
  shell-and-tube, *10*:839
Heating. (See also *Furnaces.*)
  induction for floating-zone refining, *22*:691
Heat pipe, *S*:**488**
Heat pumping
  thermoelectric, *20*:152
Heat transfer, *10*:**819**
  in evaporators, *8*:559
  by fluidization, *9*:405, 437
  principles of, *11*:824
Heat transfer coefficient, *10*:820
Heat transfer fluids, *7*:193
Heat transfer media other than water, *10*:**846**
Heat treatment
  of metals, *13*:323
  of steel, *13*:327
Heat-tube, *S*:701
Heavy hydrogen. See *Deuterium.*
Heavy water. See *Deuterium.*
Hecogenin, *18*:838, 840
  in steroid synthesis, *11*:84
Hectography, *17*:340
  spirit, *17*:329
Hectorite, *5*:548
  properties, *3*:342
Hedeoma. See *Pennyroyal.*
Hederogenin, *17*:388
Height of a transfer unit. See *HTU.*
Helindone Pink CN, *5*:870, 874
Helio Bordeaux BL, *7*:484
Heliogen, *2*:618
Heliolyte glass, *5*:849

Heliotrine, *1*:767
Heliotrope, *15*:164
Heliotropin, *15*:164
Helium. (See also *Helium group gases.*)
  as component of natural gas, *10*:448
  cryogenic prop, *6*:472, 474
  as lubricant, *12*:606
  dried by alumina, *2*:52
  in electron devices, *8*:4
  manuf, *6*:475
  as nuclear reactor coolant, *14*:81
  separation of isotopes, *7*:138
  shielding gas in welding, *22*:244
Helium-group gases, *10*:862
  compds, *10*:888-894
Helminthic infections, *14*:532
Helotropin, *14*:729
Helvolic acid, *18*:838
Hemaglutinin
  in soybeans, *18*:602
Hematite, *7*:292; *9*:384; *10*:756; *12*:5; *15*:517.
  (See also *Ferric oxide.*)
  zero point of charge
Hematoxylein, *10*:307
Hematoxylin, *10*:307
  in logwood dye, *7*:617
Hemiacetals, *4*:141
  formation, *1*:108
Hemicellulase, *8*:190
Hemicellulases, *10*:9
Hemicellulose
  in wood, *22*:365
Hemicyanine, *16*:292
Hemicyanines, *16*:286
Hemimellitene, *16*:300
Hemimellitic acid, *15*:473
Hemimellitic anhydride, *15*:473
Hemimorphite, *22*:563
Hemin
  structure, *5*:342
Hemioxonols, *16*:286
Hemisodium orthophosphate, *15*:236
Hemlock. (See also *Spruce.*)
Hemlock bark, *12*:318
Hemocyanin, *10*:789
Hemoglobin, *10*:789
Hemophilus bacteria
  chemotherapy against, *3*:28
Hemp, *9*:171-183
  as packing material, *14*:455
Hempseed oil, *7*:403; *8*:779, 784
Hendecanoic acid. See *n-Undecylic acid.*
10-Hendecenoic acid, *8*:814

11-Heneicosane, *12*:104

Heneicosanoic acid, *8*:813

n-Heneicosoic acid, *8*:813

Henequen, *9*:171, 177

  analysis, *9*:173

  characteristics, *9*:174, 175

Henequen agave wax, *22*:161

Henna

  amebicide, *20*:80

  as hair dye, *10*:798

Hentriacontane

  in candelilla wax, *22*:158

Hentriacontanoic acid, *8*:813

1-Hentriacontanol

  phys prop, *1*:543

n-Hentriacontoic acid, *8*:813

Heparin, *5*:599; *19*:228, 306

  antagonists, *5*:602

  phys prop, *5*:589

Hepatitis virus

  inactivation of, *2*:611, 615

Heptabarbital, *3*:61, 70

Heptachlor, *5*:249; *18*:525

  in water, *21*:701

Heptachlor dioxide

  in water, *21*:701

Heptachlor epoxide, *18*:525

1,1,2,3,4,5,6-Heptachlorocyclohexane

  from chlorobenzene, *5*:255

Heptachlorofluorocyclohexane

  prepn, *9*:783

Heptachloropropane

  from trichloroethylene, *5*:188

as-Heptachloropropane

  from pentachloroethane, *5*:165

1,4,5,6,7,8,8-Heptachloro-3a,4,7,7a-tetrahydro-
4,7-endomethanoindene, *5*:249

Heptacosanoic acid, *8*:813

*n*Heptacosoic acid, *8*:813

1,1,2,3,4,5,5-Heptacyano-1,3-pentadiene, *6*:630

Heptadecafluoro-1-(trifluoromethyl)decahy-
dronaphthalene, *9*:688

Heptadecanoic acid, *8*:812. (See also *Margaric
acid.*)

1-Heptadecanol

  phys prop, *1*:543

9-Heptadecanone, *12*:104

3,5-Heptadien-2-one, *12*:105; *12*:129

Heptafluorobutyric acid, *9*:773

Heptafluorobutyric anhydride, *9*:773

Heptafluorocyclohexa-1,3-diene, *9*:788

Heptafluorocyclohexa-1,4-diene, *9*:788

1*H*,4*H*5*H*-Heptafluorocyclohexene, *9*:788

Heptafluoropropane, *9*:773

Heptafluoropropoxyheptafluoropropane,
  9:689

Heptafluoro-2-propyl nitrite, *9*:760

Heptaldehyde

  from castor oil, *4*:532

  prepn, *8*:822

n-Heptaldehyde, *1*:99

Heptamethine cyanines, *6*:616

Heptamethylcyclotetrasiloxane, *12*:70

2,2,4,4,6,8,8-Heptamethylnonane, *10*:485

Heptane. (See also *Heptanes.*)

  as component of natural gas, *10*:448

n-Heptane

  surface tension, *9*:722

  toluene from, *20*:536

Heptanedioic acid. See *Pimelic acid.*

2,4-Heptanedione, *12*:148, 149

3,4-Heptanedione, *12*:140

Heptane number, heptane miscibility of solvents, *18*:574

Heptanes. (See also *Heptane.*)

  branched in petroleum, *14*:847

Heptanes

  toxicity of, *11*:294

2,4,6-Heptanetrione, *10*:924

Heptanoic acid, *8*:812

1-Heptanol, *9*:722

  phys prop, *1*:543

3-Heptanol

  in prepn of ethyl *n*-butyl ketone, *12*:135

2-Heptanone, *12*:104, 135

3-Heptanone, *12*:105, 135

4-Heptanone, *12*:104

Heptazabenzonaphthene, *10*:899

3-Hepten-2-one, *12*:105

5-Hepten-2-one, *12*:105

Heptose

  in yeast, *22*:516

Heptoylacetone, *12*:149

Heptyl acrylate

  phys prop, *1*:287

n-Heptyl alcohol. See *1-Heptanol.*

Heptyl alcohols, *1*:565

n-Heptylamine

  nylon degradation, *16*:10

γ-Heptylbutyrolacetone

  flavoring for carbonated beverages, *4*:344

n-Heptyl ether, *8*:493

  phys prop, *8*:471

Heptylic acid

  in dental material, *6*:827

Heptylphenol, *1*:907

n-Heptylphenol, *2*:625, 626

Herbicides. (See also *Weed killers.*)
  and eutrophication, *S*:333
  soil chemistry, *18*:529
  organic, *18*:516
  residual, *18*:517
  from shale oil, *18*:18
  use of emulsifiers, *8*:150

Hercynite, *17*:237

Herniaren, *6*:432

Heroin
  as analgesic, *2*:380

Herpes simplex
  chemotherapy, *21*:453

Herring oil, *8*:782

Herschel effect in photography, *15*:387, 393

Herstein process
  of acrolein oxidation, *1*266

Hesperetin glucoside dihydrochalcone, *1*:601

Hessite, *19*:759

Heteroauxin. See *3-Indoleacetic acid.*

o-Heterobiotin, *3*:522

Heterocyclic compds, *10*:**895**

Heterogenite
  phys prop, *5*:722

Heteropoly blues
  in colorimetry, *5*:795

Heteropolywolframates, *22*:353

Hetol, *14*:549

Hex. See *Hexachlorocyclopentadiene.*

Hexa, *15*:189

Hexaacetoxydititanoxane, *1*:842

Hexaallyl D-mannitol, *1*:579

2,2,4,4,6,6-Hexa-2-aziridinyl-2,4,6- triphospha-
  1,3,5-triazine resins, *9*:566

Hexaboride cathode, *8*:16

Hexabromocyclodecane
  in polystyrene foams, *19*:120

Hexabromocyclohexane, *3*:374

Hexabromoethane, *3*:757

Hexabromoplatinic(IV) acid, *15*:865

Hexachloroacetone, *9*:755
  in manuf of chloroform, *5*:124

Hexachlorobenzene, *5*:262
  fluorination of, *9*:693, 700, 776
  as fungicide, *10*:226
  toxic effects of, *5*:98

1,2,3,4,7,7-Hexachlorobicyclo(2.2.1)heptadiene-
  2,5, *5*:247

1,4,5,6,7,7-Hexachlorobicyclo(2,2,1)hept-5-ene-
  2,3-di-carboxylic acid, *S*:477

Hexachlorobutene
  from trichloroethylene, *5*:188

Hexachlorobutylene
  from trichloroethylene, *5*:186

Hexachlorocyclohexane. See *Benzene hexa-chloride.*

1,2,3,4,5,6-Hexachlorocyclohexane. See *Benzene hexachloride.*

Hexachlorocyclohexane, *3*:374
  isomers analyzed by x-ray methods, *22*:453

γ-Hexachlorocyclohexane, *18*:527
  photochemistry, *15*:337

Hexachlorocyclopentadiene, *1*:863; *6*:700;
  5:240
  and fire-retardancy, *S*:477
  fluorination of, *9*:698
  prod of, *S*:175

Hexachloro-2-cyclopentenone, *9*:799

Hexachlorodisilane, *18*:168
  prod, *18*:170
  reaction with zinc fluoride, *9*:653

1,2,3,4,10,10-Hexachloro-6,7-epoxy-hs1,4,4a,5,-
  6,7,8,8a-octahydro-1,4-endo,exo-5,8-dimethan-
  onaphthalene. See *Dieldrin and Endrin.*

1,2,3,4,10,10-Hexachloro-6,7-epoxy-1,4,4a,5,6,-
  7,8,8a-octahydro-1,4-endo,exo-5,8-dimethano-
  naphthalene. See *Dieldrin.*

Hexachloroethane, *5*:166
  acetylene reaction with, *5*:200
  carbon tetrachloride from, *5*:134
  from carbon tetrachloride, *5*:130, 131
  liver fluke treatment, *14*:549
  in military smoke agents, *4*:902
  in roundworm chemotherapy, *14*:536
  from trichloroethylene, *5*:188

Hexachlorofluorocyclohexane
  prepn, *9*:783

Hexachlorohexafluorocyclohexane, *9*:790

1,2,3,4,10,10-Hexachloro-1,4,4a,5,8,8a-hexahy-
  dro-1,4-*endo, exo*-5,8-dimethanonaphthalene.
  See *Aldrin.*

1,4,5,6,7,7-Hexachloro-5-norborene-2,3-dimeth-
  anol, cyclic sulfite of. See *Endosulfan.*

Hexachloroiridic(IV) acid, *15*:870

Hexachloroosmic(IV) acid, *15*:874

Hexachloropentafluorocyclohexane, *9*:788

Hexachlorophene, *6*:355, 360

Hexachlorophene, *2*:630, 637
  in hair preparations, *10*:775, 780
  liver fluke treatment, *14*:549
  in soaps, *18*:429

Hexachloroplatinic(IV) acid, *15*:863; *19*:467

Hexachloro-*p*-xylene, *14*:549

Hexachlorozirconates, *22*:644

Hexacosanoic acid. See *Cerotic acid.*

1-Hexacosanol
  phys prop, *1*:543
17-Hexacosenoic acid, *8*:814
1,1,2,4,5,5-Hexacyano-3-azapentadiene, *6*:630
1*H*,1*H*,9*H*-Hexadecafluorononyl butyrate, *9*:729
Hexacyanobenzene, *6*:629
Hexacyanobutadiene, *6*:628
Hexacyanoethane, *6*:629
Hexacyanoferrate(III). (See also *Ferricyanide.*)
Hexacyanoferrates(II). (See also *Ferrocyanide.*) *12*:25, 26
Hexacyanoferrates(III), *12*:25, 31
Hexacyanoferric(II) acid, *12*:26
Hexacyanoferric(III) acid, *12*:31
1,1,2,6,7,7-Hexacyano-1,3,5-heptatriene, *6*:630
Hexacyanoiridic(III) acid, *15*:870
n-Hexadecane. See *Cetane.*
8,9-Hexadecanediol-1,16-dicarboxylic acid
  in cork, *6*:284
Hexadecanoic acid, *8*:812
  specifications of commercial, *8*:836
1-Hexadecanol. See *Cetyl alcohol.*
6,10,14-Hexadecatrienoic acid, *8*:815
9-Hexadecenoic acid. See *Palmitoleic acid.*
8-Hexadecenyl-1-ol
  phys prop, *1*:543
Hexadecyl acrylate
  phys prop, *1*:287
Hexadecyl alcohol, *14*:373
Hexadecylamine
  prop, *2*:130
Hexadecyl ether, *8*:493
Hexadecylphenol, *19*:533
Hexadecylpyridinium chloride, *2*:633
Hexadecyltrimethylammonium bromide. See *Cetrimide.*
3,3,4,4,5,5-Hexadeuterocyclohexanol
  NMR of, *14*:56
1,5-Hexadiene
  prepn by Grignard reaction, *10*:725
  prepn from allyl ether, *8*:494
1,5-Hexadiene-3-yne, *5*:221
1,5-Hexadiene-3,4-diol
  prepn from acrolein, *1*:263
  in prepn of dulcitol, *1*:573
2-4-Hexadienoic acid. See *Sorbic acid.*
2,4-Hexadienoic acid, *8*:815
Hexadimethrine, *5*:602. See *Polybrene.*
Hexadine-N, *2*:619
Hexadine-S, *2*:619
2,4-Hexadiyne-1,6-diol
  prepn from propargyl alcohol, *1*:600

Hexaethyldilead, *12*:285
Hexafluorenium, *6*:549, 551
Hexafluorenium bromide, *6*:550
Hexafluoroacetone, *9*:755
Hexafluoroacetone alcoholates, *9*:758
Hexafluoroacetone anil, *9*:759
Hexafluorobenzene, *9*:689, 691, 750, 775, 788
  prepn, *9*:776
  prepn of pentafluorophenol, *9*:793
  prepn of pentafluorophenylhydrazine, *9*:796
  prepn of pentafluorothiophenol, *9*:793
  prepn of pentafluorotoluene, *9*:794
1*H*,2*H*-Hexafluorocyclohexa-1,3-diene, *9*:785
1*H*,3*H*-Hexafluorocyclohexa-1,3-diene, *9*:689
1*H*,2*H*-Hexafluorocyclohexa-1,4-diene, *9*:785
1,3,5,5,6,6-Hexafluorocyclohexa-1,3-diene, *9*:689
Hexafluorodiethyl ether, *2*:400
Hexafluorodisilane, *9*:653
Hexafluoroethylene, *9*:842
Hexafluorogermanic acid, *9*:585
Hexafluoroglutaryl chloride, *9*:846
Hexafluoro-1,4-naphthoquinone, *9*:798
2,2,3,3,4,4-Hexafluoro-1,5-pentanediol, *9*:846
Hexafluorophosphoric acid, *9*:637, 641
2*H*,2*H*-Hexafluoropropane, *9*:688
Hexafluoro-2-propanol, *9*:763
1,1,1,3,3,3-Hexafluoro-2-propanol, *9*:846
1,1,1,3,3,3-Hexafluoropropane, *9*:688
Hexafluoropropene, *9*:803; *9*:846
Hexafluoropropylene, *9*:741, 842
  vinylidene fluoride copolymers, *21*:269
  vinylidene fluoride terpolymer, *21*:270
Hexafluoropropylene oxide
  reaction with hexafluoroacetone, *9*:763
Hexahydroazepine, *10*:934
Hexahydrobenzhydrazide, *6*:456
Hexahydrobenzoic acid
  from benzoic acid, *20*:561
  phys prop, *1*:236
Hexahydro-1,4-diazine, *15*:638. See *Piperazine.*
1,2,3,4,6,7-Hexahydronaphthalene, *13*:677
Hexahydro-2-oxo-1-azepinepropionitrile, *6*:650
Hexahydrophthalic acid
  phys prop, *1*:236
Hexahydrophthalic anhydride
  in alkyd resins, *1*:862
Hexahydropyrazine, *15*:638. See *Piperazine.*
Hexahydrotriazine
  in polymerization of amino resins, *2*:231, 234, 239
Hexahydro-1,3,5-triazine, *11*:526
Hexahydro-1,3,5-trinitro-*s*-triazine. See *Cyclonite.*

Hexahydroxoplatinic(IV) acid, *15*:866

1,2,4,5,6,8-Hexahydroxyanthraquinone, *2*:467

Hexahydroxybenzene, *16*:212

Hexahydroxybiphenyl, *16*:191

Hexahydroxycyclohexanes, *11*:673
  of petroleum, *11*:424-432

Hexaiodoplatinic(IV) acid, *15*:865

Hexakis(R-isocyano)manganese(I)   iodide, *13*:6

Hexamethonium
  as ganglionic blockades, *4*:519

Hexamethylbenzene, *6*:439; *16*:300

Hexamethylbiacetyl, *12*:140

Hexamethylborazine, *3*:731
  prop, *7*:24

2,2,4,4,6,6-Hexamethylcyclohexane-1,3,5-trione, *16*:202

Hexamethyldilead, *12*:284

Hexamethyldisilazane, *18*:196; *19*:165

Hexamethyldisiloxane, *18*:224, 237

Hexamethyleneamine, *10*:96

Hexamethylene-1,6-biscarbamoylcholine, *6*:549

Hexamethylenebis(*N*¹-ethyleneurea), *22*:143

Hexamethylenediamine, *2*:296; *6*:685; *7*:24; *15*:794. (See also *Nylon; Polyamides.*)
  polymer from, *16*:2
  polymerization, *16*:89
  in prepn of nylon-6,6, *1*:410
  vinylidene fluoride copolymers, *21*:271

Hexamethylenediamine carbamate, *7*:695

Hexamethylene diaminemonocarbamate
  crosslinking agent, *21*:272

Hexamethylenediammonium adipate, *21*:434

Hexamethylene diisocyanate, *12*:46, 47

Hexamethylene glycol, *1*:408. (See also *Hexylene glycol.*)
  prepn from propargyl alcohol, *1*:600
  polyester from, *16*:170

Hexamethyleneimine
  from nylon degradation, *16*:10

Hexamethylenetetramine, *2*:632, 640; *10*:96; *13*:787; *15*:792, *22*:560. (See also *Formaldehyde.*)
  accelerator, *17*:510
  in analysis for ammonium chloride, *2*:313
  Duff reaction, *15*:162
  as food preservative, *10*:14, 77
  manuf of cyclonite, *8*:623
  from methylene chloride, *5*:114
  phenolic resins, *15*:176
  in separation of cresols, *6*:439
  structure, *S*:1

  treatment of phosgene poisoning, *S*:679
  as veterinary drug, *21*:246

Hexamethylenetriperoxide   diamine,   *8*:592; *10*:97

Hexamethylenimine, *10*:934
  from cyanuric acid, *20*:666

Hexamethylolbenzene, *1*:604

Hexamethylolmelamine, *19*:145

Hexamethylphosphoramide, *18*:178
  polysiloxanes, *18*:225

Hexamethylphosphoric triamide
  UV absorber, *21*:120

Hexamethyl-*p*-rosaniline, *9*:896

Hexamine. See *Hexamethylenetetramine.*

Hexamminenickel bromide, *13*:759

Hexane
  soybean extraction, *18*:607
  steam reforming data, *10*:420

n-Hexane, *8*:504. (See also *Hexanes.*)
  as component of natural gas, *10*:448
  in hydroisomerization, *11*:446

1,6-Hexanediamine.   See   *Hexamethylenediamine.*

Hexanedioic acid. See *Adipic acid.*

1,6-Hexanediol. See *Hexamethylene glycol.*

2,5-Hexanediol diacrylate
  phys prop, *1*:289

2,3-Hexanedione, *12*:144

2,4-Hexanedione, *12*:148, 149

2,5-Hexanedione, *12*:158, 161
  prepn, *10*:238; *10*:901

3,4-Hexanedione, *12*:140

Hexane nitrile, *9*:591

Hexanes, *11*:**1**. (See also *Hexane.*)

1-Hexanethiol, *9*:385

1,2,5-Hexanetriol, *7*:415

1,2,6-Hexanetriol, *1*:256, 595; *11*:5
  polyurethans from, *21*:61

Hexanitrodipentaerythrite.   See   *Dipentaerythritol hexanitrate.*

Hexanoic acid. See *Caproic acid.*

1-Hexanol, *1*:563; *9*:722, *11*:5. (See also *Hexyl alcohols.*)
  phys prop, *1*:543, 564

2-Hexanone, *12*:104
  prepn from acetyl chloride, *1*:139

3-Hexanone, *12* :105. See *Ethyl n-propyl ketone.*

Hexaphenylethane, *12*:824

Hexatetracontanoic acid, *8*:813

n-Hexatetracontoic acid, *8*:813

Hexatriacontane
  wettability, *9*:713

Hexatriacontanoic acid, 8:813
n-Hexatriacontoic acid, 8:813
Hexene
  methylation of, 5:105
1-Hexene
  epoxidation, 8:247
  polymerization of, 7:66
2-Hexenol
  prepn, 3:869
1-Hexen-3-ol
  phys prop, 1:629
5-Hexen-2-one, 12:105
Hexides, 1:577
Hexitans, 1:577
Hexitols, 1:572; 11:5
Hexobarbital, 3:61, 70
Hexobendin, 16:197
Hexokinase
  in yeast protein, 22:516
Hexone, 12:104; 15:883. See Methyl isobutyl ketone.
Hexose
  in breadmaking, 22:528
  in yeast, 22:516
Hexoses, 11:5
Hexuronic acid. See Ascorbic acid.
Hexyl acrylate
  phys prop, 1:287
n-Hexyl alcohol. See 1-Hexanol.
Hexyl alcohols, 1:563; 11:5
Hexylcinnamic aldehyde, 14:737
Hexylene glycol. (See also Hexamethylene glycol.)
Hexylene glycol titanate
  polymeric, 20:484
n-Hexyl ether, 8:493
  phys prop, 8:471
Hexyl lactate
  prepn, 8:353
Hexyl mercaptan. See Hexanethiol.
Hexyl methyl ketone, 1:565; 12:104
Hexyl perfluorobutyrate, 9:729
n-Hexylphenol, 2:626
Hexylresorcinol, 2:622
  in roundworm chemotherapy, 14:536
n-Hexylresorcinol, 2:628, 629
4-Hexylresorcinol, 14:536
Hexynediol
  in acid inhibition, 6:329
1-Hexyn-3-ol
  phys prop, 1:619
  prepn, 3:870
4H, 5H-Octafluorocyclohexene, 9:785

1H, 2H, 4H, 5H-Octafluorocyclohexanes, 9:698; 9:788
Hide glue. See Glue.
Higher alcohols. See Alcohols, higher.
High-maltose syrups, 6:927
High-temperature alloys, 11:6. (See also Refactory metals.)
  electrolytic machining, 7:866
High-temperature bearings. See Bearings.
High-temperature pyrolysis cracking process, 1:181
Hildebrand
  solubility parameter unit, S:889
Hildebrand solubility parameters, 21:381
Himgravine, 1:776
Hinsberg test, 2:104
Hippuric acid, 2:176; 21:536
  reaction with hexafluoroacetone, 9:763
Hiragonic acid, 8:815
Hiroe, 17:385
Histamine, 2:178; 11:35
  as cardiovascular agent, 4:523
Histidine, 2:157; 10:912
  in wine, 22:316
  in fish, 9:319
  in synthesis of histamine, 11:35
Histomoniasis
  veterinary treatments for, 21:248
Hoagland's enzymic biological complex, 2:170
Höchst process, 5:263
Höchst waxes, 22:169
Hofmann degradation
  of acrylic acid polymers, 1:310
  of quaternary ammonium hydroxide, S:38
Hofmann mustard oil test, 2:105
Hofmann reaction, 2:69, 106, 108
  manuf of anthranilic acid, 3:434
Hofmann's Violet, 20:698
Hoists. See Conveying systems.
"Hole-wind"
  in electromigration, S:286
Holmium, 17:144
  electronic configuration, 1:360
  ionic radii, 1:370
  in nuts, 14:131
Holograms, 17:373
Holst process for manuf of chlorine dioxide, 5:42
Homeostasis, 7:248
Hominy grits
  as animal feed, 8:859
Homoadamantane-1-carboxylic acid, S:11
4-Homoadamantanone, S:11

Homocysteine, 2:190; 14:668
Homogenization
  of milk, 8:148; 13:522
Homolytic scission, 14:820
Homophthalic acid
  phys prop, 1:239
Homopolymerization
  of vinyl ethers, 21:416
  of vinylidene chloride, 21:278
Homopyrocatechuic acid, 8:233
Homovanillic acid, 8:233
Homoveratroyl chloride
  in prepn of papaverine, 1:798
Honing, 1:36
Hookean slope
  for wool fibers, 22:398
Hooker chlorine cells, 1:686
Hooker's green, 15:538
Hookworms, 14:533
Hopcalite, 18:305
Hopcalites, 13:32
Hops, 14:203
  in beer brewing, 3:302
Hormonal agents
  in cancer chemotherapy, S:82
Hormones, 11:**45**
  anterior-pituitary, 11:52-64
  anterior-pituitary-like, 11:64-68
  adrenal-cortical, 11:77-92; 18:830
  insect molting, 18:845
  nonsteroidal estrogens, 11:127-141
  parathyroid, 20:266
  posterior-pituitary, 11:68-77
  sex, 11:92-127
  thyrotropic, 20:268
Hornblende, 13:736
Hot-dipped coatings, 13:252
Hot-working
  of metals, 13:323
Houdry Houdresid process
  for catalytic cracking, 11:429
Houdry dehydrogenation
  reactor, 4:579
Houdry process, 3:800
Household bleach, 5:12
Huber method
  oxidation of nicotine, 21:513
Hübnerite, 22:337
Hückel molecular orbital theory, 7:891
Huels arc furnace
  for acetylene manuf, 1:181
Huggins slope constant, 21:385

Hull cell test
  for electroplating, 8:54
Humectant
  use of sorbitol as, 1:582
Humectants
  in food products, 10:10
Humic acid, 9:402
  photography, 15:372
Humic acids, 19:211; 22:79
Humidity
  conditions within various industries, 1:486
  control of 1:481, 492
Hunziker-Müllner technique
Hutchinsonite, 20:65
Hyacinthin, 3:366. see) i)Phenylacetaldehyde.
Hyaluronic acid, 19:228
Hyaluronidase
  action and uses, 8:184, 190, 202, 203
  in parenterals, 15:127
Hycanthone, 14:548
Hydantoin, 10:912; 11:**141**
  from uric acid, 21:108
Hydantoins
  as anticonvulsants, 11:520-522
  chlorinated, 4:920
  as hypnotics, 11:510
Hydnocarpic acid
  prepn, 1:236
Hydracrylic acid β-lactone, 10:86. See
    β-Propiolactone.
Hydracrylonitrile. See Ethylene cyanohydrin.
Hydralazine
  as cardiovascular agent, 4:520
Hydrargillite. See Gibbsite.
Hydrastine, 1:769
Hydrated lime. See Lime, hydrated.
Hydration
  coulombic, 21:678
  hydrophobic, 21:680
Hydratropic aldehyde, 14:729
Hydraulic fluids
  azelate esters in, 1:250
  use of adipate esters in, 1:417
Hydraulic limes. See :Lime.
Hydrazides
  alkali metal, 11:174
Hydrazine, 2:425; 6:332, 672; 8:272, 487;
    11:**164**; 12:121, 123
  from chloramine, 5:2
  in fuel cells, 3:153; S:409
  as liquid propellant, 8:687, 688, 692
  polyester scission, 16:178

reaction with ethylene oxide, 8.529
reaction with pentafluoroaniline, 9:794
Hydrazine derivatives, 11:**164**
Hydrazine fluosilicate, 9:655
Hydrazine hydrate, 6:647, 650; 9:794, 795; 14:534
  cyanuric acid reaction, 20:666
  reaction with acetaldehyde, 1:82
Hydrazine perchlorate, 5:73
Hydrazines
  substituted, 11:167-171
Hydrazinium fluoroantimonites, 9:549
2,2¹-Hydrazinodianthraquinone, 2:455
2-Hydrazinoheptafluoronaphthalene, 9:798
4-Hydrazinomethylpyrogallol, 16:196
Hydrazobenzene, 2:415; 3:408, 410, 411, 412; 9:591; 12:857
  to form benzidine, 7:491
Hydrazodicarbonamide, 22:293
Hydrazoic acid, 6:650; 11:**197**; 12:121, 146, 161
Hydrazoic azides, 11:**197**
3,3'-Hydrazonodipropionitrile, 6:650
3,3¹-Hydrazonodipropionitrile, 6:650
Hydrides, S:**499**
  metal, as hazardous chemicals, 9:292
  metal hydride-air fuel cells, S:406
  metal, in reduction of nitro compds, 2:98
Hydrides, 11:**200**
Hydrindene
  in coal tar, 19:657
α-Hydrindone, 6:670
1Hydrindone, 10:149, 150
Hydriodic acid. See *Iodine compounds.*
Hydroabietyl acrylate
  phys prop, 1:289
Hydroabietyl alcohol, 17.504
  as modifier in alkyd resins, 1:866
Hydroammonolysis, 2:336, 347, 359
m-Hydrobenzoic acid
  phys prop, 1:238
p-Hydrobenzoic acid
  phys prop, 1:238
Hydroboracite, 3:627, 628
Hydroboration, 3:708
  selective. S:507
  and synthetic organic chemistry, S:505
Hydrobromic acid. See *Hydrogen bromide.*
Hydrobromination
  of olefins, 14:333
Hydrocal, 6:784, 786
Hydrocarbon fuels
  gasification, 10:388

Hydrocarbon oxidation, 11:**224**
Hydrocarbon reforming systems, S:390
Hydrocarbon resins, 11:**242**
Hydrocarbons, 11:**262**
  in acetylene manuf, 1:178
  in alkylation of aromatic, 1:888
  automobile exhaust, S:56
  in automobile exhaust, S:50
  chlorinated, 5:85
  halogenated as fire retardants, S:484
  hydrocarbone nomenclature, 11:288, 293
  hydrocarbons, toxicity, 11:293, 307
  microbial utilization of, S:838
  nomenclature in alkylation, 1:883
  in petroleum, 14:846, 849
  pyrolysis of in olefin prod, S:641
  solubility parameters of various, S:892
  synthesis of, 4:446, 485
Hydrocarbon separation
  by azeotropic distillation, 2:855
Hydrocardanol, 1:915. See *Alkylphenols.*
Hydrocardanol, 1:915
Hydrocarpic acid, 1:236
Hydrocerulignon, 15:371
Hydrochloric acid, 2:201; 6:651; 11:**307**. (See also *Hydrogen chloride.*)
  Hargreaves process, 11:312
  in amino resin system, 2:240
  in ammonium chloride manuf, 2:319
  in chemical cleaning, S:159
  chemical cleaning, S:161
  dehydration catalyst in prepn of ethers, 8:474
  in detergents, 6:858
  in dye industry, 7:480
  electrolysis of aqueous, S:174
  fluid-bed technology, 9:402
  in glycerol manuf, 10:621
  as inhibitor in pickling, 6:322-326, 328
  prod of, 11:327
  pickling of stainless steels, S:690
  prepn of diphenylamine, 7:40
  prod of chlorine from, 1:703
  refluxing crotonaldehyde, 6:453
Hydrochlorination
  of hydrocarbons, 5:90
Hydrochlorination of hydrocarbons, 5:90
Hydrochlorothiazide, 7:262; 19:259
  as veterinary drug, 21:252
Hydrocinnamaldehyde, 5:521
Hydrocinnamic acid, 5:517
  bromocinnamic acids from, 5:518
  phys prop, 1:239

Hydroclones, 4:747
Hydrocodone, 2:381
Hydrocolloids. See *Resins, water-soluble.*
Hydrocol process for hydrocarbon synthesis,
   4:477. (See also *Carbon.*)
Hydrocortisone
   as veterinary drug, 21:243
Hydrocortisone acetate
   veterinary drug, 21:251
Hydrocracking, 15:2, 4, 34
   of Athabasca tar, 11:423
   iron oxide as catalyst in, 11:420
   molybdenum as catalyst in, 11:421
   nickel sulfide as catalyst in, 11:421
   tungsten sulfide as catalyst in, 11:421
Hydrocyanations
   dimethylacetamide as reaction medium for,
      1:147
Hydrocyanic acid. See *Hydrogen cyanide.*
Hydrocyanite, 6:276
Hydrodealkylation, 11:453-462
Hydrodesulfurization, 15:33, 35, 45, 60;
   19:381. (See also *Hydrorefining.*)
   tar sand refining, 19:717
Hydrodynamic bearings, 3:291
Hydroflumethiazide, 7:264
Hydrofluoric acid. See *Hydrogen fluoride.*
   pickling of stainless steel, S:690
Hydroformylation reaction. See *Oxo process.*
Hydrofuramide, 10:241
Hydrogasification of coal, 10:387
Hydrogasification process
   for coal, S:204
   steam-iron, S:212
Hydrogels
   polyelectrolyte complexes, 16:117
Hydrogen, 11:**338**
   and air fuel cells, S:386
   aniline manuf, 2:79
   in ammonolysis, 2:336
   as by-product in ethylene prod, 8:510
   carbon dioxide reaction with, 4:356
   carbon monoxide reactions with, 4:433, 446,
      489
   coal-tar gas, 19:656
   in corrosion reactions, 6:335
   cryogenic prop, 6:472
   dried by alumina, 2:52
   in drilling fluids, 7:290
   in electron tubes, 8:4
   as fission product, 14:105
   gasification of coal, 10:387
   in Grignard reaction, 10:727

   as heat transfer agent, 10:856
   in hydrocracking, 15:38, 67
   as lubricant, 12:606, 607
   and manganese, 13:5
   and nitrogen fixation, S:604
   as propellant, 8:661, 689
   reaction with arsenic, 2:719
   reduction of nitro compounds, 2:78
   in synthesis of higher alcohols, 1:561
   in tobacco smoke, 20:516
   waste, 21:626
   in wood, 22:364
Hydrogen antimonide, 2:563, 570. (See also
   Stibine.)
Hydrogen arsenide, 2:716, 718
Hydrogenation
   active sites for, 11:427
   catalytic in hexane prod, 11:2
   catalytic of coal, S:180
   of coal-oil slurries, 11:286
   destructive of coal, S:203
   of fats, 13:180
   of fatty oils, 8:801
Hydrogen bond
   in water, 21:669
Hydrogen bromide, 2:320; 3:767, 772; 5:229;
   10:669
   as catalyst, 12:74
   from methyl bromide, 3:772
   sodium bromide from, 18:485
   in xylene complex, 22:477
Hydrogen chloride, 2:375. (See also *Hydrochloric acid.*)
   as catalyst in esterification, 8:326
   dried by alumina, 2:52
   in manuf of magnesium, 12:667
   oxidation of, S:171
   phenolic aldehydes, 15:161
   in prepn of ethyl chloride, 2:23
   prod of chlorine from, 1:703
   as process gas, 7:128
   separation, 7:138, 139
Hydrogen cyanide, 1:139; 6:554, 557, 574-585,
   589, 592, 598, 599, 604, 641, 670, 671, 672;
   9:402. (See also *Cyanides.*)
   in chemical warfare, 4:870, 872
   manuf, 6:576
   in manuf of acrylonitrile, 1:344
   phenolic aldehydes, 15:161
   platinum-rhodium catalyst, 15:835
   in prepn of lactonitrile, 1:346
   prop, 6:574
   reaction with acetaldehyde, 1:83

reaction with ethylene, *19*:144
reaction with ethylenediamine, *7*:29
reaction with ethylene oxide, *8*:529
reaction with formaldehyde, *10*:85
reaction with ketene, *12*:88, 117
reaction with ketones, *12*:120
removal from nuts, *14*:138
Hydrogen disulfide, *19*:389
Hydrogen fluoride
Hydrogen fluoride, *7*:414; *9*:510, 512, 610-625;
  10:161; *12*:123, 236
  carbon tetrachloride reaction with, *5*:131
  as catalyst in commercial alkylation of par-
    affins, *1*:886
  as catalyst in gasoline alkylation processes,
    10:474
  in manuf of aluminum trifluoride, *9*:531
  in prepn of boron trifluoride, *9*:558
  in prepn of perchloryl fluoride, *9*:598
  in prepn of vinyl fluoride, *9*:835
  prop compared to water, *21*:674
Hydrogen gas
  in ammonia manuf, *2*:272
Hydrogen hexasulfide, *19*:390
Hydrogen iodide. See *Iodine compds.*
Hydrogen-ion concentration, *11*:**380**
Hydrogenolysis
  of hydrocarbon fuels, *10*:404
  in prepn alcohols, *1*:546
Hydrogen-oxygen fuel cells, *3*:141
Hydrogen pentasulfide, *19*:390
Hydrogen peroxide, *11*:**391**; *2*:414, 415, 433,
  438; *8*:62, 239, 242; *9*:509; *11*:**391**
  addition compounds, *14*:761
  (alkaline) in epoxidation, *8*:267
  as antiseptic, *2*:623
  as bleaching agent, *3*:558
  in bleaching of furs, *10*:304
  bleaching wool, *22*:405
  chlorite solutions affected by, *5*:29
  in cold waves for hair, *10*:789
  in determination of formaldehyde, *10*:91
  film developer, *15*:371
  in glue manuf, *10*:607
  in hair bleaching, *10*:798
  for hair dyes, *10*:797
  initiator for poly(vinyl acetate), *21*:332
  from isopropyl alcohol, *16*:576
  in manuf of epoxy oils and esters, *8*:253, 407
  in missile system, *12*:609
  as oxidation dye developer, *10*:797
  as propellant, *8*:660, 687

reaction with formaldehyde, *10*:85
reaction with ketones, *12*:108, 125
Hydrogen polysulfides, *19*:342, 378, 389
Hydrogen selenide, *17*:821
  carbon tetrachloride reaction with, *5*:132
Hydrogen sulfide, *2*:330; *6*:588, 646, 643;
  8:272; *9*:602; *10*:448; *19*:338, 342, 353, 375,
  390, 408
  arsenic removal, *15*:261
  from carbon black manuf, *S*:95
  carbon disulfide from, *4*:380
  from carbon disulfide, *4*:374
  in catalytic reforming, *15*:33
  chamber process, *19*:459
  conversion to sulfur by Claus process, *4*:376
  in drilling fluids, *7*:290
  dual-temperature exchange with water,
    6:903
  fluid-bed technology, *9*:402
  fumes, *18*:406
  methane reaction with, *4*:380
  olefin addition, *14*:331
  organotin sulfides, *20*:311
  in prepn of dimethyl-3,3'-thiodipropionate,
    1:292
  prepn of 1-dodecanethiol, *8*:823
  in prepn of thiodiethylene glycol, *10*:648
  reaction with acetaldehyde, *1*:83
  reaction with allyl ether, *8*:494
  reaction with boron trifluoride, *9*:557
  reaction with ethylene oxide, *8*:529
  reactions with formaldehyde, *10*:85, 670
  reaction with hydroxyacetic acid, *10*:633
  reaction with ketones, *12*:118
  recovery from coke-oven gas, *4*:416
  removal from ethanolamine manuf, *15*:47
  removal from petroleum, *15*:49
  in natural gas, *S*:426
  waste, *21*:626
  in water, *21*:700
Hydrogen sulfide corrosion, *6*:331, 338, 339
Hydrogen telluride, *19*:763
Hydrogen tetrasulfide, *19*:390
Hydrogen trisulfide, *19*:390
Hydrogen-2. See *Deuterium.*
Hydrogesterone, *18*:881
Hydroiodic acid
  methacycline synthesis, *20*:8
Hydroisomerization, *11*:446-453
Hydroisoquinolines, *16*:881
Hydrol, *13*:624
  prepn, *4*:147

Hydrol
in manuf of fermentation ethyl alcohol, 8:438
Hydrologic cycle, 22:116
Hydrolysis
versus ammonolysis, 2:333
Hydrolytic desulfonation, 19:292
Hydrolytic stability
of electric insulation, 11:784
Hydrometer, 6:770
Hydronium ion, 1:215; 21:676. (See also Hydrogen ion.)
Hydroperoxides, 14:769-788
in prepn of ketones, 12:126
oHydrophenylurethan, 2:215
Hydrophile-lipophile balance, 8:131
calculation for nonionic emulsifiers, 8:131
determination of required HLB, 8:136
values of surfactants, 8:128
Hydrophobic hydration, 21:680
Hydrophobic 2-hydroxynaphthoic acid, 7:543
Hydrophones
use of ferroelectric ceramics in, 9:1
Hydroprocesses, 11:418
Hydroquinolines, 16:874
Hydroquinone, 2:413, 417, 629; 11:462, 483-492; 15:216. (See also 1,4-Benzenediol.)
antithyroid agent, 20:270
crosslinking agent, 21:272
in developer, 16:502
developing agent, 15:371
as inhibitor, 16:519
methanol clathrates, 13:372
phenolic ethers, 15:171
polyester from, 16:170
polyester inhibitor, 20:823
as polymerization inhibitor, 1:304
in prepn of p-(4-aminoanilino)phenol, 2:218
in prepn of p-anilinophenol, 2:220
in prepn of p,p-iminodiphenol, 2:220
Hydroquinone monobenzyl ether, 15:171
Hydroquinone monomethyl ether, 15:171
polyester inhibitor, 20:823
Hydroquinones
antiamebics, 20:89
Hydrorefining, 11:432-445. (See also Hydrodesulfurization.)
of coal tar, 19:669
Hydroselenide, 8:272
Hydrostreptomycin
palladium catalyst, 15:835
Hydrosulfurous acid, 19:420
Hydrothiazide, 7:263

Hydrothiazides
as diuretic, 7:261
Hydrovanilloin
creosole from, 21:182
Hydroxamic acid, 11:504
Hydroxo-EDTA-Fe(III) chelate
formation of, 6:9
Hydroxostearatodichromium(III) chloride, 22:140
2'-Hydroxyacetanilide, 2:215
3'-Hydroxyacetanilide, 2:213
Hydroxyacetic acid. See Glycolic acid.
Hydroxyacetic-formic acid mixture
stainless steel cleaner, S:164
Hydroxyacetone, 12:105, 143
1-Hydroxyacetone, 12:145
p-Hydroxyacetophenone
as synthetic flavor, 9:354
Hydroxyacetoxyacetic acid, 10:633
Hydroxy acid oils, 8:784, 787
α-Hydroxy acids
as chelating agents, 6:8
2-Hydroxyadipaldehyde, 1:256
4-Hydroxy-N-alkyl-2-quinolone, 16:892
3-(O-Hydroxyanilino)propionitrile, 6:649
p-Hydroxyanisole
in paint removers, 14:488
3-Hydroxyanthranilic acid, 21:522
Hydroxyanthraquinones
manuf of, 2:473
sulfonic acids, 19:315
3-Hydroxy-b-carotene
as a provitamin A, 21:501
ω-Hydroxybehenic acid
in cork, 6:283
m-Hydroxybenzaldehyde
Patent Blue dyes from, 20:707
o-Hydroxybenzaldehyde, 15:162
p-Hydroxybenzaldehyde
from chloroform, 5:123
from lignin, 21:189
separation from vanillin, 21:191
synthesis of bromoxynil and ioxynil, 22:203
2ˡ-Hydroxybenzanilide, 2:215
3ˡ-Hydroxybenzanilide, 2:213
4-Hydroxybenzenedisulfonyl chloride, 15:210
m-Hydroxybenzenesulfonic acid, 15:211
p-Hydroxybenzenesulfonic acid, 15:211
4-Hydroxybenzenesulfonic acid, 19:311
Hydroxybenzenesulfonic acids, 15:208
m-Hydroxybenzenesulfonyl chloride, 15:210
4-Hydroxy-m-benzenedisulfonic acid, 15:208, 210, 964

4-Hydroxybenzene-1,3-disulfonic acid, *15*:208

2-Hydroxy-1,3,5-benzenetrisulfonic acid, *15*:208, 210

Hydroxybenzimidazoles, *16*:82

p-Hydroxybenzoates
as preservatives in cosmetic manuf, *6*:372

2-Hydroxy-1,3,2-benzodioxastibole, *2*:579, 580. See *Antimonylcatechol.*

3,4,5-Hydroxybenzoic acid. See *Gallic acid.*

m-Hydroxybenzoic acid, *17*:736

o-Hydroxybenzoic acid. See *Salicylic acid.*

p-Hydroxybenzoic acid, *15*:681; *17*:737
polyesterification of, *16*:164
polyester from, *16*:170

2-Hydroxybenzophenone
plastics, *15*:800

2-Hydroxybenzophenones
UV absorbers, *21*:115

3-Hydroxy-o-benzoquinone, *16*:191

2-Hydroxy-p-benzoquinone, *16*:200

Hydroxybenzothiozole, *16*:82

1-Hydroxybenzotriazoles, *10*:918

2-Hydroxybenzotrifluoride, *9*:780

3-Hydroxybenzotrifluoride, *9*:778, 780

4-Hydroxybenzotrifluoride, *9*:780

o-Hydroxybenzyl alcohol
phenolic resins, *15*:188
synthesis, *15*:176
styrene reaction, *15*:182

Hydroxybenzyl alcohols
oxidation, *15*:162

α-Hydroxybenzylpenicillin, *14*:664

12-Hydroxy-cis-9-octadecenoic acid. See *Ricinoleic acid.*

o-Hydroxybiphenyl. See *o-Phenylphenol.*

p-Hydroxybiphenyl. See *p-Phenylphenol.*

4-Hydroxybiphenyl, *19*:315

2-Hydroxy-4,6-bis(alkylamino)-s-triazine, *18*:533

4-Hydroxy-1,2-butadiene
from isochloroprene, *5*:228

Hydroxybutanedioic acid. See *Malic acid.*

3-Hydroxy-2-butanone, *12*:105, 145; *13*:515

4-Hydroxy-2-butanone
prepn of 1,3-butylene glycol, *10*:660

4-Hydroxy-2-butynoic acid
prepn from propargyl alcohol, *1*:600

α-Hydroxybutyraldehyde, *3*:876

γ-Hydroxybutyraldehyde, *1*:613

4-Hydroxybutyraldehyde
prepn from allyl alcohol, *1*:919

β-Hydroxybutyric acid, *6*:466
phys prop, *1*:234

β-Hydroxybutyronitrile, *6*:673

2-Hydroxycamphane. See *Isoborneol.*

8-Hydroxycarvotanacetone, *19*:832

Hydroxychloroquine
amebicide, *20*:87

4-Hydroxy-3-chloro-2-butanone, *5*:311

3α-Hydroxy-5β-cholanic acid, *3*:483

25-Hydroxycholecakiferol
and vitamin D₃, *21*:550

25-Hydroxycholecalciferol, *18*:838; *21*:554

25-Hydroxycholesteryl diacetate, *21*:554

o-Hydroxycinnamic acid. See *Melilotic acid.*

p-Hydroxycinnamic acid
in resins, *17*:382

Hydroxycitronellal, *19*:819, 827, 829

Hydroxycitronellal dimethyl acetal, *1*:108

Hydroxycitronellol, *19*:829

7-Hydroxycoumarin, *6*:426. (See also *Umbelliferone.*)

4-Hydroxycyclohexanone, *21*:564

3-Hydroxy-1,2-cyclopentadione, *19*:241

2-Hydroxycymene, *4*:55. See *Carvacrol.*

17-Hydroxy-11-deoxycorticosterone, *11*:93

(2-Hydroxy-1,2-di(carbomethoxy)ethenyl)dimethylsulfonium hydroxide inner salt, *19*:329

3α-Hydroxy-7,12-dioxo-5β-cholanic acid, *3*:483

7-Hydroxy-2,4-dimethylbenzopyrylium chloride, *16*:196

2-Hydroxydiphenyl, *7*:193. (See also *o-Phenylphenol.*)

p-Hydroxydiphenylamine, *7*:40. See *p-Anilinophenol.*

4-Hydroxydiphenylamine, *2*:413; *7*:46; *19*:426

4-Hydroxy-2,7-dibromofluorescein, *22*:434

(4-Hydroxy-3,5-di-tert-butylbenzoyl)dimethylamine, *2*:601

1-Hydroxy-2,2-diphenylethenyloxydimethylsulfonium chloride, *19*:330

20-Hydroxy-24-ethylecdysone, *18*:849

Hydroxydiphenyl ethers, *2*:629

Hydroxydiphenyl sulfides, *2*:629

2-Hydroxy-3,6-disulfo-1-naphthylazobenzenearsonic acid, *20*:252

3-Hydroxyestra-1,3,5(10)-trien-17-one, *6*:69, 71. See *Estrone.*

1-Hydroxy-1,2-ethanedicarboxylic acid. See *Malic acid.*

Hydroxyethanenitrile. See *Formaldehyde cyanohydrin.*

2-Hydroxyethanesulfonic acid, *19*:284, 311

Hydroxyethanoic acid. See *Glycolic acid.*

Hydroxyethanoic acid. See *Hydroxyacetic acid.*

Hydroxy ethers
 prepn, *8*:428
2-Hydroxyethylaniline
3-[*N*-(2-Hydroxyethyl)anilino]propionitrile, *20*:194
 prepn, *8*:528
Hydroxyethyl cellulose, *4*:646; *17*:330, 401
 films from, *13*:832
N-Hydroxyethyl-*N*,*N'*,*N'*-ethylenediaminetriacetic-
 acid
2-(*N*-(2-Hydroxyethyl)ethylamino)ethyl acrylate
 phys prop, *1*:288
 as chelating agent, *6*:7
2-Hydroxyethylhydrazine, *15*:683
N-Hydroxyethyliminodiacetic acid
 as chelating agent, *6*:7
Hydroxyethylmercuric acetate, *13*:245
2-Hydroxyethyl methacrylate, *13*:339; *16*:120
2-Hydroxyethylpyridine, *10*:86; *19*:671
Hydroxyethylstarch
 as gum, *10*:741
(2-Hydroxyethyl)trimethyl ammonium chloride carbamate. See *Carbamylcholine chloride.*
Hydroxy fluoborates, *9*:563
Hydroxyformate, *8*:248
2-Hydroxyheptafluoronaphthalene, *9*:798
Hydroxyhydrindenes, *19*:660
o-Hydroxyhydrocinnamic acid. See *Melilotic
 acid.*
α-Hydroxyhydroperoxides, *14*:782
Hydroxyhydroquinone, *16*:190, 199
 analysis, *16*:200
 prop, *16*:199
 synthesis, *16*:200
 toxicity, *16*:200
Hydroxyhydroquinone-phthalein, *16*:200
3-Hydroxy-5-(hydroxymethyl)-2-methylisonico-
 tinaldehyde, *16*:806
5-Hydroxy-2-(hydroxymethyl)-4-pyrone, *10*:923;
 12:107
3-Hydroxyiminopyrrolenine, *10*:897
5-Hydroxyindoleacetic acid, *16*:644
3-Hydroxy-2-indolecarboxylic acid. See *Indoxylic acid.*
 phys prop, *1*:240
β-Hydroxyisobutylamine, *6*:672
α-Hydroxyisobutyric acid, *1*:623, 627
 esters, *13*:339
 prepn of ketones, *12*:160
B-Hydroxyisobutyronitrile. See *Propylene cyanohydrin.*

α-Hydroxyisobutyronitrile. See *Acetone cyanohydrin.*
2-Hydroxyisobutyronitrile. See *Acetone cyanohydrin.*
4-Hydroxyisophthalic acid, *17*:723
(4-Hydroxy-5-isopropyl-2-methylphenyl)tri-
 methylammonium chloride, *15*:683
7-Hydroxy-1,5-naphthalenedisulfonic acid,
 13:723
6-Hydroxy-2-isopropyl-4-methylpyrimidine,
 18:528
8-Hydroxy-7-iodo-5-quinolinesulfonic acid,
 21:697
Hydroxyisoquinolines, *16*:882
Hydroxyketone dyes, *7*:471, 475
3-Hydroxykynurenine, *21*:522
Hydroxylamine, *6*:631, 660; *8*:272, 418;
 11:**493**
 amine oxides from, *S*:32
 reaction with acetaldehyde, *1*:82
 reaction with gelatin, *10*:501
 reaction with ketones, *12*:123
 reaction with pyrrole, *19*:143
Hydroxylamine fluosilicate, *9*:655
Hydroxylamine hydrochloride, *10*:156
Hydroxylamine salts, *11*:**493**
Hydroxylamine sulfate, *10*:156
Hydroxylamine-*O*-sulfonic acid, *10*:156;
 19:486
Hydroxylapatite, *15*:239, 267, 273
2-Hydroxyldiphenyl, *7*:193
4-Hydroxyldiphenyl, *7*:193. (See also *p-Phenylphenol.*)
γ-Hydroxyleucine, *12*:857
3-Hydroxylindol, *11*:591
4-Hydroxy-2-mercaptothiazole, *10*:915
5-Hydroxymercurivanillin, *21*:192
4-Hydroxy-3-methoxyacetophenone, *21*:193
4-Hydroxy-3-methoxybenzaldehyde, *15*:163;
 21:180
4-Hydroxy-4-methoxybenzoic acid, *21*:192
5-Hydroxy-2-methoxy-3,5-dichlorobenzoic acid, *22*:201
6-Hydroxy-4-methoxy-2,5-dimethylbenzaldehyde,
 15:160
n-Hydroxymethylacrylamide, *1*:291
1-Hydroxymethyladamantane, *S*:11
7-Hydroxy-4-methyl-1,2-benzopyrone, *11*:480
2-Hydroxymethylcyclohexanone, *9*:605
5-Hydroxymethyl-2-furaldehyde. See *5-Hy-
 droxymethylfurfural.*
7-Hydroxy-4-methylcoumarin, *11*:480

Hydroxymethylfurfural, *22*:383

5-(Hydroxymethyl)furfural, *4*:147; *6*:922; *19*:224

N-Hydroxymethylhaloacetamides
  flame retardant for cellulose derivatives, *S*:950

2-Hydroxymethyl-2-methyl-1,3-propanediol, *7*:412; *16*:551. See *Trimethylolethane*.
  polyesters, *21*:62

3-Hydroxy-*n*-methylmorphinan, *2*:383

3-Hydroxy-2-methyl-1,4-naphthoquinone, *21*:587, 592

4-Hydroxy-4-methyl-2-pentanone, *10*:924. See *Diacetone alcohol*.

2-Hydroxy-2-methylpropaninitrile. See *Acetone cyanohydrin*.

17-Hydroxy-6α-methylpregn-4-ene-3,20-dione acetate, *6*:79

3-Hydroxy-2-methylpropanenitrile. See *Propylene cyanohydrin*.

5-Hydroxy-6-methyl-3,4-pyridinedimethanol, *16*:806

4-Hydroxy-6-methyl-2(1*H*)-pyridone, *12*:857

3-Hydroxy-2-methyl-4-pyrone, *10*:923. See *Maltol*.

2-Hydroxy-4-methylquinoline, *1*:158

5-Hydroxy-4-methylumbelliferone, *16*:209

N-(Hydroxymethyl)stearamide, *22*:142

Hydroxymethyl vinyl ketone
  prepn from butynediol, *1*:603

4-Hydroxy-5-methylisophthalic acid, *1*:156

2-Hydroxy-4-methylmercaptobutyronitrile
  as intermediate in prepn of methionine, *1*:271

3-Hydroxy-2,7-naphthalenedisulfonic acid, *20*:723

5-Hydroxy-2,7-naphthalenedisulfonic acid, *13*:722

4-Hydroxy-1,3-naphthalenedisulfonic acid, *13*:722

7-Hydroxy-1,3-naphthalenedisulfonic acid, *13*:723

4-Hydroxy-1,5-naphthalenedisulfonic acid, *13*:722

8-Hydroxy-1,6-naphthalenedisulfonic acid, *13*:722

8-Hydroxy-2,5-naphthalenedisulfonic acid, *13*:722

Hydroxynaphthalenedisulfonic acids, *13*:722

Hydroxynaphthalenemonosulfonic acids, *13*:719

2-Hydroxy-1-naphthalenepropionitrile, *6*:642

2-Hydroxy-1-naphthalenesulfonic acid, *13*:720

Hydroxynaphthalenetetrasulfonic acids, *13*:723

2-Hydroxy-1,3,6,8-naphthalenetetrasulfonic acid, *13*:724

1-Hydroxy-2,4,7-naphthalenetrisulfonic acid, *13*:723

7-Hydroxy-1,3,6-naphthalenetrisulfonic acid, *13*:724

8-Hydroxy-1,3,6-naphthalenetrisulfonic acid, *13*:723

Hydroxynaphthalenetrisulfonic acids, *13*:723

β-Hydroxynaphthoic acid, *7*:497
  in prepn of p-(2-naphthylamino)phenol, *2*:221

2-Hydroxy-1-naphthoic acid, *19*:299
  as hair dye, *10*:798

3-Hydroxy-2-naphthoic acid, *2*:358; *7*:490
  pigments, *15*:569, 575

3-Hydroxy-2-naphthoic anilide, *2*:871

2-Hydroxy-1,4-naphthoquinone, *10*:798

5-Hydroxy-1,4-naphthoquinone, *14*:133

2-Hydroxy-3-naphthoic acid, *7*:541

Hydroxynapthalenesulfonic acids, *13*:719

4-Hydroxy-1-naphthalenesulfonic acid, *13*:720

4-Hydroxy-2-naphthalenesulfonic acid, *13*:719

5-Hydroxy-1-naphthalenesulfonic acid, *13*:720

6-Hydroxy-2-naphthalenesulfonic acid, *13*:721

7-Hydroxy-1-naphthalenesulfonic acid, *13*:721

7-Hydroxy-2-naphthalenesulfonic acid, *13*:721

8-Hydroxy-1-naphthalenesulfonic acid, *13*:720

8-Hydroxy-2-naphthalenesulfonic acid, *13*:720

6-Hydroxynicotinamide, *21*:523

6-Hydroxynicotinic acid, *21*:523

α-Hydroxynitriles, *6*:669

β-Hydroxynitriles, *6*:673

9-Hydroxy-*trans*, *trans*-10,12-octadecadienvic acid, *7*:400

4'-Hydroxy-3'-nitroazobenzene-4-sulfonic acid

4-Hydroxy-3-nitrobenzenesulfonic acid, *15*:211

3-Hydroxy-6-nitrobenzotrifluoride, *9*:780

8-Hydroxyoctadeca-17-en-9,11-diynoic acid, *8*:787

3α-Hydroxy-12-oxo-5β-cholanic acid, *3*:483

12-Hydroxy-*cis*-9-octadecenoic acid, *8*:816

12-Hydroxy-9-octadecenoic acid
  in castor oil, *4*:525, 529, 531

12-Hydroxy-9,10-octadecenoic acid, *19*:524

5-Hydroxy-4-octanone, *12*:106

ω-Hydroxypalmitic acid, *18*:26

ω-Hydroxy-*cis*-9-palmitolaic acid, *18*:26

4-Hydroxy-2-pentanone, *12*:157
  prepn of acetaldehyde and acetone, *1*:82

5-Hydroxy-2-pentanone, *12*:105
3-Hydroxyperchlorylbenzene, *9*:604
4-Hydroxyperchlorylbenzene, *9*:604
α-Hydroxyperoxides, *14*:785
2-Hydroxy-2-phenylacetophenone, *3*:365;
  12:107
α-Hydroxy-α-phenylacetophenone. See *Benz-oin.*
m-Hydroxyphenyl benzoate
  UV absorbers, *21*:118
p-Hydroxyphenylethylamine, *2*:178
Hydroxyphenylfurylmethanol, *15*:185
o-Hydroxyphenyl mercuric chloride, *2*:620
p-Hydroxyphenylpropane
  in lignin, *12*:363
3-Hydroxy-1-phenylpyrrole, *16*:844
2-(Hydroxyphenyl)triazine, *15*:800
p-Hydroxyphenylurethan, *2*:219
5-(α-Hydroxy-α-phenyl-α-2-pyridyl)m- ethyl-
  7-(phenyl-2-pyridylmeth ylene)-5-norborn-
  ene-2,3-dicarboximide, *15*:918
3-Hydroxypicolinic acid, *16*:333
2-Hydroxy-p-phenetidine
  sodium borohydride in synthesis of, *S*:501
17-Hydroxypregn-4-ene-3,20-dione, *6*:77,78
17-Hydroxypregn-4-ene-3,20-dione acetate.
  See *17α-Acetoxyprogesterone.*
17-Hydroxypregn-4-ene-3,20-dione acetate,
  6:78
21-Hydroxypregn-4-en-3,2--dione, *18*:851
11α-Hydroxyprogesterone, *18*:881
17-Hydroxyprogesterone, *11*:93
Hydroxyproline, *2*:158
2-Hydroxypropanenitrile, *6*:668, 672
3-Hydroxypropanenitrile. See *Ethylene cyano-hydrin.*
2-Hydroxy-1,2,3-propanetricarboxylic acid.
  See *Citric acid.*
α-Hydroxypropanoic acid, *12*:170
1-Hydroxy-2-propanone, *12*:105, 143
3-Hydroxypropionaldehyde
  prepn from acrolein, *1*:259
Hydroxypropionic acid
  in prepn of acrylic acid, *1*:294
α-Hydroxypropionic acid. See *Lactic acid.*
3-Hydroxypropionic acid
  polyesters of, *16*:165
Hydroxypropylstarch
  as gum, *10*:741
2-Hydroxypyridine, *10*:903
N-Hydroxy-2-pyridinethione, *16*:804
Hydroxypyrroles, *16*:843, 848

7-Hydroxyquinoline
  prepn from *m*-aminophenol, *2*:213
8-Hydroxyquinoline, *2*:623; *10*:780, 799;
  19:300, 315
  additive to polyamides, *16*:83
  reaction with alkoxides, *1*:841
  zinc analysis, *22*:561
8-Hydroxyquinoline benzoate, *3*:433
Hydroxyquinolines, *16*:874
2-Hydroxy-1,4-quinone, *15*:372
Hydroxystreptomycin, *19*:40
Hydroxysuccinic acid. See *Malic acid.*
Hydroxyterephthalic acid, *15*:464
(4-Hydroxy-1,2,3,4-tetra(carbomethoxy)-1,3-
  butadienyl)dimethylsulfonium hydroxide
  inner salt, *19*:329
22-Hydroxytetracosanoic acid
  in cork, *6*:283
6-Hydroxytetradecanoic acid, *18*:26
4-Hydroxytetrahydrofurfuryl alcohol, *19*:238
4-Hydroxytetrahydropyran, *10*:660
Hydroxytitanium acylates, *20*:469
Hydroxytitanium stearate
  polymeric, *20*:484
β-Hydroxytricarballyic acid. See *Citric acid.*
2-Hydroxy-3,5,6-tribromobenzoquinone-1,4,
  16:199
3-Hydroxy-6-(3,4,5-trihydroxyphenyl)-*o*-benzoqui-
  none, *16*:191
2-Hydroxy-3,4,5-trimethoxyacetophenone,
  16:212
2-Hydroxy-3,4,6-trimethoxyacetophenone,
  16:212
3-Hydroxy-2,4,6-trinitrobenzotrifluoride, *9*:778
5-Hydroxytryptamine, *2*:178, 188, 189
4-Hydroxyundecanoic acid
  flavor and perfume materials, *9*:358
Hydroxyurea
  in cancer therapy, *S*:89
4-Hydroxyuvitic acid
  in prepn of ethyl sodioacetoacetate, *1*:156
8-Hydroxyvaleraldehyde, *2*:165
Hydroxyxanthenes, *22*:432
Hydroxyzine, *16*:675
Hydroxyzirconyl chloride, *22*:633, 635
Hydrozincite, *22*:563
Hygas process, *S*:243
  synthesis gas from coal, *S*:211
Hygrine, *1*:764
Hygrometry
  aquametry by, *2*:692
  applications in drying, *7*:348

Hygromycin
  amebicide, *20*:82
Hygromycins, *19*:33, 44
Hygroscope"
  aquametry by, *2*:693
Hyocholic acid, *3*:481, 483
Hyodeoxycholic acid, *3*:481, 483, 485
DL-Hyoscyamine, *1*:804
1-Hyoscyamine, *1*:766, 778, 804
Hypalon, *7*:695
  in packing, *14*:452
Hypaphorine, *1*:772
Hyperfiltration
  desalination of water, *22*:58
Hypersorption process, *1*:461
Hyperthyroidism, *20*:269
Hypervitaminosis, *21*:488
Hypnean, *17*:763
Hypnone. See *Acetophenone.*
Hypnotics, *11*:**508**
Hypobromite, *2*:173
Hypobromites, *3*:756, 769
Hypobromous acid, *3.*756
Hypochlorite
  in bleaching fats, *13*:180
  reaction with acetaldehyde, *1*:83
Hypochlorites
  ammonia reactions with, *4*:916
  aniline reaction with, *4*:922
  corrosion problems with, *5*:11
  hypochlorous acid content in solutions of,
    4:910
  organic, *5*:24
  solid, *5*:15
  solutions of, *5*:12
  swimming pools disinfected by, *4*:925

  tests for, *4*:914
  urea reactions with, *4*:916
Hypochlorous acid, *5*:7; *8*:267
  ammonia reaction with, *4*:916
  in chloramine solutions, *4*:910
  chlorohydrins from, *5*:305, 308
  as disinfectant, *2*:616
  in ethylene oxide manuf, *8*:531
  in hypochlorite solutions, *4*:910
  prepn of, *5*:8
  prepn of ethylene chlorohydrin, *8*:502
  uses of, *5*:10
  vinyl chloride reaction with, *5*:172
Hypochlorous anhydride. See *Chlorine monoxide.*
Hypohalous acid, *8*:245
Hypoiodite
  reaction with acetaldehyde, *1*:83
Hypoiodous acid, *11*:859
Hypolemnetic withdrawal, *S*:335
Hypolimnion
  deoxygenation in lakes, *S*:319
Hypophosphites, *15*:316
Hypophosphoric acid, *15*:311
Hypophosphorus acid, *15*:311, 315
Hypothrombineminic agents, *6*:432
Hypovanadates, *21*:172
Hypoxanthine, *9*:321
Hypsochromic shift, *15*:568
Hyssop, *14*:203
  essential oils as flavor, *9*:352
Hystazarin, *2*:466
Hysteresis, *9*:1; *12*:739
  loops, *9*:3
Hysteresis loop (ferromagnetic), *8*:885

INH. See *Isonicotinic acid hydrazide.*
IUPAC nomenclature, *14*:2
Ibogaine, *1*:761, 773
Ice
  structure of, *21*:671
Ice colors, *2*:889; *7*:465, 547. (See also *Azoic dyes.*)
Ice cream, *13*:538-548
  use of lecithin in, *12*:355
Iceland spar, *4*:2; *12*:416
Ideal gas law, *20*:130
L and D-Iditol
LAND D-Iditol
  phys prop, *1*:570
  prepn, *1*:575
Idose, *4*:134
Idoxuridine, *21*:455
Igepon A, *19*:308
Ignition powder, *8*:650
Ilang-Ilang. See **Ylang-Ylang.**
Ilemite, *13*:769
Ilemite-magnetite, *15*:497
Illipé, *8*:779, 784
Illipe butter, *5*:387. See *Borneo Tallow.*
Illite, *5*:550
  in ceramic products, *5*:561
  in ceramics, *4*:763
  layer structure, *5*:542
Ilmenite, *12*:6; *20*:347, 349
Ilsemannite
  rhenium in, *17*:411
Imhoff tanks, *22*:105
Imidazoglycerol phosphate dehydratase, *22*:212
Imidazole, *10*:897, 898, 911
Imidazoleacetol phosphate, *22*:212
Imidazoleglycerol phosphate, *22*:212
Imidazolidine, *11*:526
2-Imidazolidinethione
  prepn, *4*:372
Imidazolidinetrione oxalylurea, *21*:114
2-Imidazolidinone. See *Ethyleneurea.*
Imidazolines, *10*:912
Imidazoline-5-thiones
  prepn, *12*:118
Imidazo[1,2-*a*]pyridine, *10*:936

Imides, *2*:66
  cyanoethylation of, *6*:647
  prepn from amides, *2*:70
Imidyl halide salt, *6*:561
Iminazole, *10*:897, 898, 911
Imines, cyclic, *11*:**526**
α-Imino acids, *16*:612
6,6'-Iminobis(1-naphthol-3-sulfonic acid), *7*:49
Iminobispropylamine, *7*:23
  properties, *7*:24
Iminodianthraquinones, *2*:462
N,N'-(Iminodiethylene)bisoctadecanamide, *2*:601
4,4'-Iminodiphenol. See *p,p'-Dihydroxydiphenylamine.*
p,p'-Iminodiphenol, *2*:220
3,3'-Iminodipropionitrill, *6*:648
Iminosulfur fluorides
  prepn with sulfur tetrafluoride, *9*:673
Iminothiadiazolidines, *10*:917
2-Imino-5-methyl-4-oxazolidone, *12*:175
Immedial Yellow GG, *7*:483
  Immersion plating, *8*:69
Impact tests, *13*:189
Impellers for mixing, *13*:578, 584
Imperial green, *15*:538
Imperial smelting furnace
  in zinc prod, *22*:572
Impreg, *22*:380
Impregnated coatings for metals, *13*:263-265
Impregnation, *8*:105
Imuran
  procytoxid in, *22*:531
Inandione
  anticoagulants, *5*:602
  antagonists of, *5*:603
Incandescent lamps
  use of argon and nitrogen, *10*:881
Incapacitating agents
  in chemical warfare, *4*:876
Incendiary agents
  in chemical warfare, *4*:895
Incense
  use of alcohol in, *8*:448
  use of camphor in, *5*:57

Incineration
  solid waste disposal, S:739
  of waste gases, 21:631
  of waste liquids, 21:636
  of waste solids, 21:645
Inclusion compounds, 3:817. (See also *Clathrate compounds.)*
Incoloy
  welding, 22:248
Inconel, 13:273
  corrosion resistance to water, 6:302
  welding, 22:248
Inconel X
  in bearing materials, 3:287
Indalone, 10:924
Indamine
  sulfur dyes, 19:432
Indamine dyes, 7:471, 474; 15:220
Indan
  in petroleum, 14:849
Indandione, 10:150
Indanols
  in coal tars, 6:437; 19:660
Indans
  in naphthalene feedstocks, 11:454
Indanthrene, 2:453. (See also *Indanthrone.)*
  prod of, 2:458
Indanthrene Blue, 7:465
Indanthrene Blue GCD. (See *Carbanthrene Blue.)*
Indanthrene Blue RS, 7:486
Indanthrene Brown BR, 7:486
Indanthrene Golden Orange G, 7:486
Indanthrene Red Violet RH, 7:483
Indanthrene Yellow GGF, 2:894
Indanthrene vat dyes, 2:522
Indanthrone, 2:453
  pigments, 15:586
  sulfur dyes, 19:431
Indanthrone Blue, 15:586
Indazoles, 10:899, 911; 16:768; 19:14
Indene
  as by-product in ethylene prod, 8:510
  from coal tar, 19:657, 671
  titanium compounds, 20:433
Indenes
  in naphthalene feedstocks, 11:454
Indenyllithium
  titanium halide reactions, 20:434
Inderborite, 3:650
Inderite, 3:648
Index Chemicus, S:516

Indican, 7:623; *11*:563
Indicators, *11*:**548**
Indicators, adsorption
  chloride ion determined by, 5:6
Indigo, 2:453; 5:868; 7:565, 621; *11*:562-580;
  15:556, 562, 565; *20*:188
Indigo
  sodium in synthesis, 18:451
  in stains, 18:657
Indigo carmine
  test for chloramines, 4:915
Indigo copper, 6:279
Indigo dyes
  bromination, 19:494
Indigoid dyes, 7:471, 475; *11*:**562;** *20*:188
Indigotin, *11*:565
Indigotin IA, 5:866, 874. (See *Indigotine.)*
5,5'-Indigotindisulfonic acid
  test for chloramines, 4:915
Indigotine, 5:866, 868, 874
Indirubin-thioindigoids, *11*:577
Indium, *11*:**581**
  in copper alloys, 6:210
  as dental alloy, 6:814
  in electron tubes, 8:4
  in electroplating, 8:60
  in germanium photodiode, 15:421
  manganese compounds, 13:3
  as solid film lubricant, 12:590
  in tin-lead alloys, 18:543
  zinc analysis, 22:561
Indium antimonide, 2:573
  infrared detectors, 15:439
Indium arsenide
  in semiconductors, 2:716
Indium compounds, *11*:**581**
Indium nitride, *13*:823
Indium oxide
  in ceramic pigments, 5:852
Indium sulfamate plating solution, 8:61
Indoaniline dyes, 15:220
Indocyanine, 2:863
Indocyanine Green, 6:605
Indogenic acid. See *Indoxylic acid.*
Indole, *1*:771; *10*:899, 909; *11*:**585,** *14*:737
  from coal tar, 19:657
3H-Indole
  as nucleus in cyanine dye, 6:607
Indoleacetic acid, 2:204; 15:686
3-Indoleacetic acid, 2:188, 204; *11*:589; *15*:676
  phys prop, *1*:240
Indoleacetic acids
  herbicidal, 22:176

Indoleacetic oxidase, *15*:681
Indole alkaloids, *1*:772
3-Indolebutyric acid, *15*:677
Indolecarboxylic acids, *11*:589
2,3-Indoledione, *11*:591
Indolenines, *11*:593
Indole-2,3-dione-3-(thiosemicarbazone)
    antiviral agent, *21*:454
Indole-3-propionic acid, *21*:527
Indolizine, *10*:936
Indolo[4,3-*fg*]quinoline, *10*:933
β-Indolylacetic acid. See *3-Indoleacetic acid.*
Indophenol dyes, *7*:471, 474; *19*:425, 426
Indoxyl, *7*:624; *11*:563, 591
α-Indoxylacetic acid. See *3-Indoleacetic acid.*
Indoxylic acid
    phys prop, *1*:240
Induction furnaces, *10*:261
Indulines, *2*:862
Industrial enzymes
    uses, *8*:180
Industrial toxicology, *11*:**595**
Inert gases. See *Helium-group gases.*
Inert gas rule, *4*:491
Inertinite, *5*:611
Influenza
    amantadine hydrochloride chemotherapy,
    21:456
Influenza virus
    inactivation of, *2*:613
Information retrieval sources, *S*:**510**
Infrared detectors, *15*:422-443
Infrared drying, *7*:374
Infrared furnaces, *10*:271
Infrared quenching
    in phosphors, *12*:628
Infrared spectrometry
    and mass spectrometry, *13*:90
Infrared stimulation
    in phosphors, *12*:628
Infusorial earth. See *Diatomite.*
Ingrain colors, *2*:899; *7*:457. (See also *Azoic dyes.*)
Inhibitors
    antioxidants as, *2*:589
    of corrosion, *6*:300, 317
"Inhibitor squeeze"
    method of corrosion inhibition, *6*:334
Inhoffen synthesis
    β-carotene by, *21*:503
Injection molding of plastics, *15*:805
Ink. (See also *Printing inks.*)
Ink dilatancy, *11*:613

Inks, *11*:**611**
    carbon blacks used in, *4*:267, 275
    dispersed concentrates, *15*:597
    use of fluorescent pigments, *9*:498
    use of gum arabic in, *10*:745
Inoculum buildup
    enzyme prod, *S*:298
Inorganic acids
    chemical cleaning solutions, *S*:161
Inorganic high polymers, *11*:**632**
Inorganic materials
    microwave energy absorption, *S*:574
Inorganic refractory fibers, *11*:**651**
Inosilicates, *18*:47, 49
Inosine, *9*:321
Inosinic acid, *2*:188, 198; *9*:321
Inosite, *11*:673
Inositol, *11*:**673**
    from natural sources, *21*:489
    yeast requirement, *22*:517
meso-Inositol, *11*:673; *19*:38
    in cholesterol metabolism, *11*:673
myo-Inositol, *11*:673
Inositol hexanitrate, *4*:521
    of dimensional apparatus, *7*:184
    use of aerosol products, *1*:479
    veterinary, *21*:250
Inositolphosphoric acid
    salts in nuts, *14*:132
Inositols, *16*:213
Insect control
    economics of, *11*:678
Insecticides, *11*:**677**; *22*:175
    benzene hexachloride, *5*:267
    chlorinated hydrocarbon, *18*:516, 518
    in cotton growing, *6*:383
    cyclodiene, *5*:240
    ethyleneamines used as, *7*:36
    general index of, *11*:735
    methyl bromide in, *3*:773
    naphthalene as, *13*:687
    residual, *18*:518
    in the soil, *18*:523
    use of alcohol in, *8*:448
    use of emulsions, *8*:150
Insect repellents
    use of benzyl ether in, *8*:495
Insoluble azo dyes, *7*:547
Inspectional analysis
Instrumentation and control, *11*:**739**
Instrumentation tape, *12*:812
Insulating board, *21*:615

213

Insulating materials
electrical resistance measurement, 2:670
Insulation conductance, 7:719
Insulation, electric, *11*:**774**
use of diarylamines in, 7:48
sulfur hexafluoride as, 9:667
Insulation, thermal, *11*:**823**
use of diatomite in, 7:62
Insulin, *11*:**838**
Intaglio, *16*:495
Interesterification. (See also *Ester interchange.*)
Interfacial tension, *8*:133. (See also *Surface tension.*)
viral chemotherapy, *21*:458
Interferometry, *17*:223, 226
Interferon
International Bureau of Weights and Measures, *S*:986
International Organization for Standardization, *S*:986
International System of Units, *S*:985
Intersitial-cell-stimulating hormone
assay, *11*:60-61
Interstitial compounds, *6*:537
Intestopan, *20*:70, 83
Intrinsic factor, *21*:547
Intrinsic viscosity, *4*:623
Inulase
microorganism-producing, *8*:210
Inulin
levulose from, *4*:137
yeast assimilation, *22*:518
Inulinase
from yeast, *22*:518
Invar, *19*:780
Invertase, *10*:9
in bread making, *22*:528
uses, *8*:190
Invert sugar. See *Sugar, invert.*
Inyoite, *3*:627, 628, 649
Iobenzamic acid, *17*:137
Iocetamic acid, *17*:137
Iochrome, *5*:459
Iodacetic acid
phys prop, *1*:235
Iodamide, *17*:134
Iodic acid, *11*:859
Iodide
analysis for in water, *21*:699
Iodide ion
antithyroid, *20*:269

Iodine, *11*:**847**
addition to municipal waters, *22*:87
aldose determination, *19*:162
as antiseptic, *2*:605, 617, 622
in cottonseed meal, *6*:414
as denaturant, *8*:450
as fission product, *14*:83, 105
as food additive, *10*:11
in Grignard reaction, *10*:724
methylene chloride reaction with, *5*:114
radiopaques, *17*:130
from sea plants, *14*:166
from seawater, *14*:156
swimming pools disinfected by, *4*:926; *22*:128
thiol oxidation, *20*:208
Iodine bromide, *3*:768
Iodine compounds, *11*:**847**
Iodine heptafluoride, *9*:587
properties, *9*:593
Iodine monochloride, *8*:420
action on acctanilide, *1*:150
Iodine monofluoride, *9*:587
reactions, *9*:590
Iodine pentafluoride, *9*:586
phys prop, *9*:593
reactions, *9*:588, 591, 699, 700
Iodine pentoxide, *9*:586
in carbon monoxide detection, *4*:442
carbon monoxide reaction with, *4*:432
Iodine value of fats, *8*:793
Iodine-131, *17*:4, 7, 125
prop, *9*:748
Iodine-132, *17*:**126**
Iodipamide, *17*:138
Iodoacetic acid, *8*:420
Iodoalphionic acid, *17*:136
p-Iodoaniline, *2*:415
Iodobenzene, *9*:591; *11*:864; *12*:147
Iodobromite, *3*:758
4-Iodo-1,2-butadiene, *5*:229
4-Iodobutyric acid, *3*:879
Iodochlorhydroxyquin, *2*:632
amebicide, *20*:83, 85, 89
5-Iodo-2'-deoxyuridine
antiviral agent, *21*:455
Iodoethane, *11*:864
Iodoethanoic acid, *8*:420. See *Iodoacetic acid.*
phys prop, *1*:235
Iodofluorohydrocarbons
Iodoform
from acetaldehyde, *1*:83
as antiseptic, *2*:619

from carbon tetrachloride, *5*:131
from chloroform, *5*:122
Iodoform, *8*:420; *11*:864
N-Iodoformamide, *10*:107
2-Iodohexane
from sorbitol, *1*:577
3-Iodohexane
from mannitol, *1*:577
Iodohydrins, *5*:305
7-Iodo-8-hydroxyquinoline-5-sulfonate, *2*:632
Iodol, *2*:619
Iodomethane, *11*:862
5-Iodo-1-methyl-2(1*H*)-pyridone, *17*:132
2-Iodo-3-nitrobenzoic acid, *3*:437
Iodonium 235, *2*:619
5-Iodo-2-oxo-1(2*H*)-pyridineacetic acid,
17:132
Iodophors, *2*:618
Iodoplatinic acid, *15*:865
Iodoprene, *5*:229
Iodopsin
prod in the eye, *21*:506
Iodopyracet, *17*:132
7-Iodo-8-quinolinol-5-sulfonic acid, *20*:87
Iodosilane, *18*:173, 186
4-Iodo-1,2,3,6-tetrahydrophthalic acid, *5*:229
5-Iodo-2-thiouracil
antithyroid agent, *20*:271
Ioglycamic acid, *17*:139
Ion-active sites, *11*:872
Ion engines
cesium, *4*:865
Ion exchange, *11*:**871**
desalination of water, *22*:60
mass transfer in, *13*:100
prod of high-purity water, *22*:77
removal of dissolved solids in water, *22*:75
resin-in-pulp, *21*:16
uranium extraction, *21*:15
waste-liquid treatment, *21*:630
for water softening, *22*:97, 101, 122
Ion exchange chromatography, *5*:421
Ion exchange membrane fuel cells, *3*:147, 149
Ion exchangers
bottling water treated with, *4*:341
cellulose as, *4*:598
Ion exchange separation
actinide elements, *1*:354
lanthanide elements, *1*:355
Ionic atmosphere, *7*:804
Ionic conductivity
of electric insulation, *11*:774

Ionic elastomers, *7*:684
Ionic processes
for desalination of water, *22*:60
Ionic solids, *4*:761
Ionizing radiation
in meat preserving, *13*:178
Ionone
as synthetic flavor, *9*:354
α-Ionone, *12*:106, 132. (See also *Perfumes.*)
β-Ionone, *1*:846; *12*:106, 132. (See also *Perfumes.*)
vitamin A synthesis, *21*:496
Ionones, *19*:829, 830
Ion plating, *13*:279
Ion retardation, *11*:890
Ion thrustors. See *Ion engines.*
Iopanoic acid, *17*:136
Iophendylate, *17*:139
Iothalamic acid, *17*:133
Ioxynil
herbicide, *22*:203
Ipecac, *1*:793
Ipecac alkaloids, *20*:74
Ipodic acid, *17*:137
Ipomea, *4*:587
Ipomoea orizabinsis
cathartics from, *4*:587
Irgalan Brown Violet DL, *2*:885, 888
Irgalan dyes, *2*:888
Irganol, *2*:882
Iridicinium salts, *15*:872
Iridium, *15*:832, 855, 856
as dental alloy, *6*:814
Iridium carbonyl
catalyst, *14*:377
Iridium carbonyls, *4*:489
Iridium compounds, *15*:870-872
Iridium-192, *17*:127
Iron, *12*:**1**. (See also *Steel.*)
acid reduction of amines, *2*:83
addition to flour, *22*:297
alkaline reduction of amines, *2*:91
analysis for in water, *21*:697
analysis in sparkling wines, *22*:324
aquatic plant nutrient, *S*:326
as catalyst in hydrocarbon synthesis, *4*:452,
465
cerium alloys with, *4*:847
in chlorate candle, *S*:659
coated, *13*:252
content in soils, *9*:35
in copper alloys, *6*:199, 210, 250

dopant in wolfram, *22*:341
effects on brasses, *6*:231
in drilling fluids, *7*:292
electrodeposition of zinc, *22*:588
electromigration of, *S*:287
enameling of, *8*:155
friction against iron, *12*:599
impact tests, *13*:189
as magnetic material, *12*:748
manuf of xanthates, *22*:425
phosphatizing, *13*:296
in phosphoric acid, *15*:261
pig, *12*:1
pig, grades of, *12*:18
pig, lime in, *12*:455
as plant nutrient, *9*:29
in reduction of amines, *2*:82
removal from swimming pool water, *22*:132
removal from water, *22*:72
removal from water supplies, *22*:102
separation from zinc, *22*:562
from Sterling process, *22*:579
in superalloys, *11*:31
thermocouples, *19*:787
welding cast iron, *22*:251
wrought, *18*:716
in zinc alloys, *22*:600
zinc analysis, *22*:561
zirconium deoxidizer in, *22*:620
zone refining, *22*:697
Iron(II). See *Ferrous* entries.
Iron(II) sulfate heptahydrate. See *Ferrous sulfate heptahydrate.*
Iron(III). See *Ferric* entries.
Iron alum, *2*:64
Iron aluminum phosphates, *15*:262
Iron blues, *15*:536, 538
Iron by direct reduction, *S*:**535**
Iron carbide
   in Bessemer process, *18*:723
   in steel, *18*:763
Iron carbonyls, *4*:489
   carbon monoxide flammability influenced by, *4*:429
   prepn, *4*:433
Iron chloride plating solution, *8*:61
Iron-chrome oxide
   cracking catalyst, *15*:39
Iron-cobalt alloys
   as magnetic materials, *12*:756
Iron-cobalt-molybdenum
   as magnetic material, *12*:766

Iron-cobalt powder, *12*:766
Iron-cobalt-vanadium
   as magnetic material, *12*:768
Iron compounds, *12*:**22**. (See also *Ferrous; Ferric* entries.)
Iron 2-ethylhexoate, *7*:278
Irone
   as synthetic flavor, *9*:354
Iron ferricyanide
   photography, *15*:386
Iron fluoborate plating solution, *8*:61
Iron fluosilicate, *9*:658
Iron-gold alloy, *10*:691
Iron hexacyanoferrates
   pigments, *15*:539
"Iron hydrate" *2*:311
Iron hydroxide
   as pigment, *14*:473
Iron linoleate, *7*:278
Iron monofluorophosphate, *9*:642
Iron naphthenate, *7*:276, 278
Iron-nickel alloys
   as magnetic materials, *12*:752
Iron nitride
   and nitrogen fixation, *S*:618
Iron ore
   gravity concentration of, *10*:708
Iron oxide. (See also *Magnetite.*)
Iron oxide. (See also *Rouge.*)
Iron oxide, *2*:311; *6*:363
   as abrasive, *1*:28
   as adsorbent, *1*:460
   black pigments, *15*:545
   as coated abrasive, *1*:29
   in chemical ware, *4*:835
   collection, *8*:90
   colorant for glass, *9*:252
   domestic natural red, *15*:518
   as hydrocracking catalyst, *11*:420
   hydrogen sulfide, *19*:384
   in manuf of carbon products, *4*:169
   natural yellow pigments, *15*:527
   pigment, *15*:541
   pigments, *15*:545
   as pigment, *14*:473; *15*:517
   prop as imbedding filler, *8*:110
   red pigments, *15*:517
   removal by hydroxyacetic acid, *10*:635
   separation by flotation, *9*:395
   slagceram nucleating agents, *S*:881
   in smoke, *18*:402
   synthetic hydrated yellow, *15*:529

in talc, *19*:608
in welding, *22*:242
Iron pentacarbonyl, *12*:293
Iron perchlorates, *5*:71
Iron phosphate, *12*:589
Iron phosphates, *15*:275
Iron pig, *12*:1
  grades of, *12*:18
  lime in, *12*:455
Iron powder, *12*:766
Iron pyrites
  sulfuric acid from, *19*:328
Iron resinate, *7*:279
Iron sands
  in ocean, *14*:160
Iron selenide, *17*:828
Iron silicate
  in Bessemer process, *18*:723
Iron-silicon alloys
  as magnetic materials, *12*:748
Iron sponge, *S*:536
Iron stearate, *7*:280
Iron sulfate
  source of micronutrient fertilizer, *9*:146
Iron sulfate plating solution, *8*:61
Iron sulfide, *2*:331; *19*:379
Iron tallate, *7*:278
Iron-tin alloys, *20*:292
Iron titanates, *20*:414
Iron-55, *17*:122
Iron-59, *17*:123
Irreverre-Sullivan color reaction
  for vitamin $K_{1(20)}$, *21*:589
Isanic acid, *8*:816
Isanic acid, *8*:787, 816
Isanolic acid, *8*:787
Isano oil, *8*:786
  drying oil, *7*:399
Isariin, *16*:337
Isatin, *7*:623; *11*:591
  in asymmetrical indigos, *11*:568
  from indigo, *11*:563
Isatin-3-semicarbazone
  antiviral agent, *21*:454
Isethionic acid, *19*:284, 311
Isinglass, *10*:611
Isoagglutinins, *3*:576
Isoamyl acetate
  properties, *8*:380
  as synthetic flavor, *9*:354, 355
Isoamyl alcohol, *2*:375
Isoamyl butyrate. (See also *Isopentyl butyrate.*)

Isoamyl ether, *8*:493
  phys prop, *8*:471
Isoamyl formate
  as synthetic flavor, *9*:354
Isoamyl isovalerate
  as synthetic flavor, *9*:354
Isoamyl ketone, *12*:104
Isoamyl nitrite, *6*:610
Isoanthraflavin, *2*:466
Isoascorbic acid, *2*:751, 760
Isobenzanthrone, *2*:529
Isoborneol, *4*:57; *19*:819, 830
Isobornyl acetate, *19*:814, 830
  prod and sales of, *8*:376
Isobornyl esters, *4*:56, 57
Isobornyl thiocyanacetate, *19*:815
Isobutane, *9*:871; *3*:815, 833
  as aerosol propellent, *1*:475
  alkylation, *15*:41
  for alkylation, *15*:4
  as component of natural gas, *10*:448
  as feed gas, *10*:474
  in hydrocracking, *11*:432
  in hydroisomerization, *11*:446
  as motor fuel, *10*:492
  oxidation of, *11*:236
Isobutene. See *Isobutylene.*
  prop, *8*:381
Isobutyl acetate
  properties, *8*:389
  specifications, *8*:373
Isobutyl acrylate
  phys prop, *1*:287
Isobutyl alcohol, *3*:822; *8*:492
  latent solvent, *18*:577
  from oxo process, *14*:373
Isobutylamine, *2*:178; *6*:468
Isobutyl cinnamate, *5*:520
Isobutylene, *3*:830; *6*:696, 699; *7*:677; *8*:268
  alkylation, *15*:42
  in alkylation of phenol, *1*:894, 904
  from butyl alcohol, *3*:823
  in manuf of butyl rubber, *12*:80
  in manuf of isoprene, *12*:76
  oxidation, *13*:333
  photochemical addition, *15*:334
  in polyester reaction, *16*:170
  polymerized, *15*:797
  polymers, *14*:309
  in prepn of butyl rubber, *7*:688
  in prepn of ethers, *8*:475
  prod of pivalic acid, *8*:853

in prepn of organoaluminum compds, 2:27
in prepn of butyl rubber, 2:23
pyrolysis of, S:552
reaction with acetaldehyde, 1:84
reaction with hydrogen cyanide, 6:575
reaction with o-cresol, 6:436
reaction with isoprene, 12:64
Isobutylene oxide
    ammonolysis of, 2:351
Isobutyl ether, 8:492
    phys prop, 8:471
Isobutyl isobutyrate, 3:869
Isobutyl isovalerate
    as synthetic flavor, 9:354
Isobutyl mercaptan. See Butanethiol.
Isobutyl methacrylate
    as fluorescent coating system, 9:499
Isobutyl phenylacetate
    prod and sales of, 8:377
    flavor and perfume materials, 9:358
Isobutyl salicylate
    flavor and perfume materials, 9:358
    prod and sales of, 8:377
Isobutyl salicyl cinnamate, 5:517, 519, 520
Isobutyl-α-toluate. See Isobutyl phenylacetate.
Isobutyltriphenylmethylamine, 14:545
Isobutyraldehyde, 3:865, 871
    from oxo process, 14:373
    in prepn of neopentylene glycol, 10:676
    in prepn of 2,2,4-trimethyl-1,3-pentanediol,
        10:678
Isobutyraldehyde, 2:165; 6:640
Isobutyraldehydes, 1:99
Isobutyraldol
    prepn, 3:868, 875
Isobutyraldoxime, 3:875
Isobutyramide, 3:882
    phys prop, 2:68
    form, 2:68
Isobutyric acid, 12:109
    from isobutyraldehyde, 3:867
    from isobutylene, 3:842
    petroleum sweetening, 15:52
    phys prop, 1:229; 3:882
    prepn, 3:882
Isobutyric acid esters, 8:31
Isobutyric anhydride, 12:91
Isobutyronitrile, 9:591; S:599
    from isobutyraldehyde, 3:868
Isobutyrylacetone, 12:148
Isobutyryl chloride, 3:882
Isochloroprene, 5:228

Isocholesterol
    lanolin alcohols, 19:549
Isocil, 22:213
Isocinchomeronic acid, 21:514
Isocrotonic acid, 6:464
    phys prop, 1:229
Isocrotonic acid dichloride, 6:466
Isocyanate
    in prepn of urethane elastomers, 7:692
Isocyanate polymers, 21:57
Isocyanates, 12:45; 19:253
    catalysts in reactions, 21:66, 69
    as modifiers in alkyd resins, 1:866
    polyurethan, 21:58
    reactions of, 21:63
    reaction with aziridenes, 11:530
Isocyanates, organic, 12:45
Isocyanic acid, 10:613
    from cyanuric acid, 20:662
    in prepn of isocyanates, 12:54
Isocyanic acid esters. See Isocyanates, organic.
Isocyanine, 6:607
Isocyanurates
    chlorinated as swimming pool disinfectants,
        22:126
Isocyanuric acid, 20:662
Isocytosine
    as diuretic, 7:253
Isodecanoic acid
    in dental material, 6:827
    phys prop, 1:564
"Isodecanol," 1:567; 14:373
Isodecyl isooctyl phthalate
    prod and sales of, 8:376
Isodehydroacetic acid, 1:157; 10:922
Isodehydrocholesterol, 21:561
Isodibenzanthrone, 2:494
Isodibenzanthrone Violet, 15:586
Isodrin, 5:247; 18:526
Isodurene, 16:300
Isoemetine, 20:75
Isoequilenin, 18:854
Isoeugenol, 14:729; 15:162, 163, 169, 170
    flavor and perfume materials, 9:358
Isoeugenyl methyl ether. See 4-Propenylvera-
    trole.
Isoforming process
    for p-xylene, 22:492
Isogeraniols, 19:825
Isohexane
    in hexanes, 11:1
"Isohexanol," 14:373

Isohexylmagnesium bromide
  steroid synthesis, *18*:858
Isoiditide, *1*:578
14-Iso-17-iso-19-norprogesterone, *18*:880
Isojasmone, *14*:729, 737
Isolan, *2*:889; *15*:910
Isoleucine, *2*:157; *19*:750
  in wine, *22*:316
Isomaltose, *6*:921
Isomaltose
  hydrolysis in grain mashes, *22*:522
Isomannide, *1*:578
Isomax catalyst, *11*:431
Isomelamines, *6*:570
Isomerases
  classification, *8*:176
Isomerism
  dynamic and melting points, *13*:213
Isomerization
  cis-trans, *14*:317
  skeletal, *14*:318
  of uncracked paraffins, *11*:430
  of xylenes, *22*:479
Isomorphism, *6*:523
Isoniazid, *3*:30; *19*:40, 671; *21*:528
Isonicoteine, *1*:765
Isonicotinic acid, *16*:802; *21*:527
  condensation with hydrazine, *21*:529
  prepn of, *21*:513
Isonicotinic acid hydrazide, *3*:30; *19*:671; *21*:528
Isonicotinoyl esters
  hydrolysis of, *21*:530
Isonicotinoylglycine, *21*:532
Isonicotinylhydrazine, *21*:528
Isonitrosoacetaldehyde, *8*:588
Isonitrosoacetic acid, *8*:588
Isonitrosoacetone, *1*:154; *12*:144
Isonitrosocyclopentadiene, *6*:693
Isooctane, *3*:838
  HF alkylate, *15*:44
  prepn from isobutane and isobutylene, *4*:397
  specific rotation of α-tocopherol in, *21*:575
"Isooctanol," *14*:373
Isooctylcohol, *1*:566; *3*:840
  phys prop, *1*:564
"Isooctyl" thioglycolate, *20*:203
Isoparaffin alkylate
  commercial prod, *1*:892

Isopentane, *14*:707
  as component of natural gas, *10*:448
  in manuf amyl alcohols, *2*:375
  in manuf of isoprene, *12*:75
Isopentenylpyrophosphate, *18*:834
Isopentyl butyrate. (See also *Isoamyl butyrate.*)
Isopentyl salicylate
  flavor and perfume materials, *9*:358
  prod and sales of, *8*:377
Isophorone, *1*:166; *2*:348; *12*:106, 113, 137
  demethanation of *S*:273
Isophthalic acid, *10*:112; *15*:457-462
  in alkyd resins, *1*:862
  phys prop, *1*:239
  esterification, *20*:489
  prod from *m*-xylene, *22*:497
Isophthalonitrile
  from *m*-xylene, *22*:485
Isophthaloyl chloride, *15*:461
Isophytol
  synthetic, *21*:579
Isopolywolframates
  complexions of, *22*:351
Isoprene, *7*:63, 414; *12*:**64,** 120
  expoxidation, *8*:249
  from isobutylene, *3*:845
  lithium catalyzed polymerization, *S*:911
  microorganism-producing, *8*:210
  polymers of, *7*:75
  prepn from aluminum alkyls, *2*:39
  in prepn of butyl rubber, *2*:23; *7*:688
  from propylene, *14*:322
  as synthetic rubber, *7*:677
  terpenes, *19*:803, 818
  terpenes from, *14*:183
  toxicity of, *11*:297
Isoprene tetrabromide, *12*:67
Isopropanol. See *Isopropyl alcohol.*
Isopropanolamines, *1*:810
  stain solvents, *18*:660
Isopropenyl acetate, *12*:89, 111, 155
Isopropenylacetylene, *1*:622
2-Isopropenylbicyclo[2.2.1]hept-5-ene, *12*:66
p-Isopropoxydiphenylamine, *7*:47
2-Isopropoxyethyl acrylate
  phys prop, *1*:288
3-Isopropoxypropionitrile, *6*:644
Isopropoxytitanium stearate, *20*:484
  alcoholysis, *20*:470
Isopropyl acetate
  from isopropyl alcohol, *16*:575
  prod and sales of, *8*:376

prop, 8:379
  specifications, 8:373
Isopropyl acrylate
  phys prop, 1:287
Isopropyl alcohol, 2:430; 6:644; 8:280;
  16:564-578
  as antimicrobial, 2:615
  cyanoethylation of, 6:639
  in dye base, 10:796
  in esterification of fatty acids, 8:818
  in hair spray, 10:784
  hydrolysis of sulfuric esters, 19:486
  latent solvent, 18:577
  in prepn of acetone, 1:16; 12:125
  stain solvent, 18:659
Isopropylamine, 12:116; 16:576
Isopropylamines
  prop, uses, 2:123
N-Isopropylaniline, 2:413
2-Isopropylanthraquinone, 2:438
p-Isopropylbenzaldehyde, 3:366
Isopropylbenzene. See *Cumene*.
Isopropyl bromide, 3:778
Isopropyl chloride, 2:413
  vinyl chloride reaction with, 5:173
Isopropyl chloroformate, 4:386
Isopropyl N-(m-chlorophenyl) carbamate,
  22:192
Isopropyl N-(3-chlorophenyl) carbamate,
  15:682
Isopropyl citrates
  in margarine, 13:59
Isopropyl ether, 8:487
  manuf, 8:489
  in manuf of lactic acid, 12:177
  phys prop, 8:471
Isopropyl ethylthiocarbamate
  xanthate derivatives, 22:428
Isopropylideneacetone. See *Mesityl oxide*.
Isopropylideneacetone, 12:105
4,4'-Isopropylidenebis-2,6-dibromophenol, 3:778
4,4'-Isopropylidenediphenol. See *Bisphenol A*.
1,2-Isopropylidene-D-glucofuranose, 19:229
Isopropyl isocyanate, 12:46
Isopropyl lactate
  prepn, 8:352
N-Isopropyllauramide, S:37
N-Isopropyl-α-(2-methylhydrazino-p-toluam- ide
  monohydrochloride, S:89
Isopropyl methyl ketone, 12:111
Isopropyl methylphosphonofluoridate, 15:329
Isopropyl methylphosphorofluoridate. See
  Sarin.

1-Isopropyl-3-methylpyrazol-5-yl N,N-dimeth-
  ylcarbamate, 15:910
Isopropyl myristate, 6:354, 356; 16:576
  parenteral solvent, 15:128
  prod and sales of, 8:376
Isopropyl N-phenylcarbamate
  as an herbicide, 22:191
Isopropylnaphthalene, 19:318
Isopropylnaphthalenesulfonate, 13:679
Isopropyl nitrate
  from isopropyl ether, 8:488
N-Isopropyl-3-(5-nitro-2-furyl)acrylamide, 14:549
N-Isopropylnorepinephrine
Isopropyl oleate, 16:576
Isopropyl palmitate, 6:354; 16:576
  prod and sales of, 8:376
p-Isopropylphenol
  p-cresol from, S:271
Isopropylphenols
  phenolic resins, 15:185
4-Isopropylpyridine
  quaternization, 19:243
Isopropyl sulfate
  in prepn of acetone, 1:161
Isopropyl titanate
  titanium chelates, 20:464
p-Isopropyltoluene, 4:57
4-Isopropyltoluene
  p-cresol from, S:271
Isopropyl 2,4,6-trimethylbenzoate, 8:354
Isopropyl xanthate, 16:576
Isopropyl zirconate, 1:849
Isopulegol, 19:823
Isopurone
  from uric acid, 21:108
Isopyrocalciferol, 21:557; 18:837
Isoquinoline, 10:899, 920; 16:865, 876-886
  in hydrorefining, 11:437
  sulfoxides, 19:326
  zinc analysis, 22:561
Isoquinolines
  antiviral agents, 21:457
Isoraunescine, 16:666
Isorbide dinitrate, 4:521
Isoreserpine, 16:666
Isosafrole, 15:162, 164, 170
Isosebacic acid, 1:252; 18:452
  from butadiene, 3:808
  prepn, 1:247
Isosorbide, 1:578, 579; 19:545
Isotactic polymers, 4:557
Isoterpinolene, 19:815, 824

Isothermal flash
  in distillation, 7:221
Isothermal-isotope method
  in electromigration, S:281
Isothiazole thiosemicarbazone
  antiviral agent, 21:455
Isothiocyanates
  reaction with aziridines, 11:530
Isothipendyl.HC1
Isotope abundance
  by mass spectrometry, 13:91
Isotopes
  separation, 7:92; 14:85
Isotope separation
  use of electromigration in, S:290
Isotriazine structures, 6:570
Isotriboline, 1:770
Isovaleraldehyde, 2:165
Isovaleric acid
  phys prop, 1:229

2-Isovaleryl-1,3-indanedione, 15:920
Isovanillin
  from safrole, 21:185
Isoxanthopterin, 10:932
Isoxanthopterincarboxylic acid, 10:932
Isoxazoles, 12:151; 18:867
  heterocyclic compounds, 10:898, 901
  prepn from 1,4-diketones, 12:159
Isoxazolones, 10:901
Isozonide, 14:417
Istle, 9:171, 177
  analysis, 9:173
  properties, 9:174, 175
Itaconic acid, 1:232; 5:526; 12:**83**
  in synthetic oil prepn, 7:415
Itaconic anhydride, 5:526; 12:85
Itaconyl chloride, 12:85
Itobarbital, 3:61
Ivory blacks, 15:545

J

J acid, *13*:724. See *6-Amino-1-naphthol-3-sul-fonic acid.*
  dyes from, *2*:897, 899
  sodium salt, *7*:478
Jackson candle unit, *21*:691
Jacobsite, *8*:881
Jacutin. See *Benzene hexachloride.*
Jalap, *4*:586
Jalapin, *4*:587
Jalaric acid, *18*:25
Japan wax, *22*:164
  in candles, *4*:58
  in dental material, *6*:836
Jarlite, *9*:534
Jasmin concrete, *14*:203
Jasmine wax, *22*:161
Jasmone, *12*:161, 162; *14*:729, 736
Jaumave istle, *9*:177
Jellies
  use of enzymes in, *8*:200
Jellies and gums, *6*:51
Jervine, *1*:776; *18*:866, 884, 885
Jet fuel, *8*:661, 692
Jet-fuel additive
  ethylene glycol ethers in, *10*:643

Jewelers solder
Jewelry
  use of gold alloys, *10*:688
Jewelry bronze
  composition of, *6*:214
Jigs
  for concentration, *10*:709
Jojoba wax, *22*:166
Joule-Thomson system
  of cryogenics, *6*:473
Juglansin
  in butternuts and walnuts, *14*:125
Juglone, *15*:681
  in walnut trees, *14*:133
Juglone acetate, *20*:23
Junccine, *1*.767
p-n Junctions, *17*:851
Juniper, *14*:204
Juniper tar. See *Cade.*
Junker calorimeter, *4*:47
Jute, *9*:171, 182; *12*:189
  analysis, *9*:173
  dyeing with acid dyes, *7*:523
  as packing material, *14*:455
  properties, *9*:174, 175

K

K acid, *13*:725
Kaempfe process, *8*:362
Kaempferol
tea, *19*:743
Kainite, *16*:384
from Great Salt Lake, *S*:455
Kalinite
source of potassium alum, *2*:64
Kalle's acid, *13*:714
Kallidin
tissue hormone, *11*:50
Kalvar diazo process, *17*:364
Kamala seed oil
drying oil, *7*:400
α-Kamlolenic acid, *7*:400
β-Kamlolenic acid, *7*:400
Kanamycin
in bacterial infections, *3*:1
effect of procytoxid on, *22*:530
Kanamycins, *19*:33, 41
Kaolin, *21*:15. (See also *China clay; Kaolinite.*)
as abrasive, *1*:26
calcined, *17*:227
cracking catalyst, *15*:28
in dental porcelain, *6*:798
in detergents, *6*:858
fibers from, *17*:286
in papermaking, *14*:501
pigments, *15*:511
in prepn of trifluoroboroxine, *9*:556
Kaolin
Kaolinite, *1*:932, 939; *5*:545
in ceramics, *4*:763
in ceramic products, *5*:561
layer structure, *5*:542
zero point of charge, *9*:384
Kaolin minerals, *5*:545
Kapok, *9*:171, 184
Kapok oil, *8*:779, 784
Kapton, *S*:756
Kapton film
thermal neutron degradation of, *S*:762
Karl Fischer reagent, *21*:431
Kasugamine, *19*:43, 44, 45
Katadyn silver, *2*:621
Kauri-butanol value
of solvents, *18*:574

Kauri copal, *17*:385
Kaurinic acid, *17*:386
α-Kaurolic acid, *17*:386
Keatite, *18*:49, 54
Kekulé structure, *2*:701
Kel-Chlor process
HCL oxidation, *S*:171
Kellogg gasification process
for coal, *S*:208
Kelp
alginic acid from, *6*:802
Keluin
definition of, *S*:990
Kelvin
Kelvin relations, *20*:149
Kemp, *22*:393
Kenaf, *9*:183
analysis, *9*:173
properties, *9*:175
Kendall structure, *15*:371
Kepone, *5*:251
Keratin, *2*:190
stabilization of, *6*:663
sulfur in, *19*:368
α-Keratin
x-ray analysis, *22*:454
Keratins
wool as, *22*:387, 399
Kermasite, *2*:564
Kermes, *7*:625
Kermisic acid, *7*:625
Kernel
structure of cereal, *22*:260
Kernite, *3*:627, 629, 630, 639
Kerogen, *11*:423; *18*:1, 13
Kerosene, *15*:2, 4, 17
color, *15*:53
as denaturant, *8*:450, 453
distillation for gas manuf, *10*:472
extraction, *13*:730
hydrocarbons in, *11*:267
hydrogenolysis results, *10*:410
in missile system, *12*:609
as propellants, *8*:661
prop for gas manuf, *10*:389
steam reforming, *10*:420
as solvent for extraction of water, *22*:57

uranium extraction, *21*:17
  from shale oil, *18*:17
Kerrolic acid, *18*:26
Kesting process for manuf of chlorine dioxide,
  5:42
Ketals, *1*:107
  prepn from ketones, *12*:119
Ketene, *2*:414; *6*:694. (See also *Ketenes.)*
  pyrolysis from acetic anhydride, *8*:408
  reaction with 1-aminoanthraquinone, *2*:450
  reaction with epoxide, *8*:272
  synthesis of acetic anhydride, *8*:409
Ketenes and related substances, *12*:**87**
Ketenimines, *12*:97
Ketimines, *8*:308
β-Ketobutyranilide. See *Acetoacetanilide.*
β-Ketobutyric acid. See *Acetoacetic acid.*
3-Ketocamphor. See *2,3-Camphanedione.*
Ketocarbene, *12*:92
Ketocholanic acids, *3*:482, 483, 485
4-Ketodioxolones, *12*:843
5-Keto-D-gluconic acid, *19*:735
α-Ketoglutarate
  isoniazid antidote, *21*:533
Ketoglutaric acid, *2*:177
α-Ketoglutaric acid
  in tobacco, *20*:508
2-Keto-1-gulonic acid, *2*:753
5-Keto-6-keoxyarabohexose, *19*:44
Ketols, *12*:113
6-Keto-myristic acid, *18*:26
Ketone
  0-carbon from butane, *3*:819
Ketone bisulfites, *12*:120
Ketones, *12*:**101**
  esterification, *8*:346
  as hypnotics, *11*:513
  phenolic resins, *15*:187
  solubility parameters of various, *S*:893
Ketone 250, *21*:554
γ-Ketopimelic acid, *19*:135
γ-Ketopimelic dilactone, *19*:142
α-Ketopyruvate
  isoniazid antidote, *21*:533
1-Ketose, *19*:152
α-Ketosulfonic acids, *12*:119
α-Ketotrahydronaphthalene, *6*:670
2-Keto-1,7,7-trimethylnorcamphane. See *Camphor.*
Ketoximes, *12*:123, 124
Kid leather
  dyeing of, *7*:591

Kidney
  function, *7*:248
Kieleyera coriacea
  cork from, *6*:287
Kieselguhr. See *Diatomite.*
Kieserite, *12*:731, 733
Kiln. See *Furnaces.*
Kilns
  for ceramics, *4*:783
Kilogram
  definition of, *S*:990
Kimberlite, *4*:284
Kinematic viscosity, *9*:446
King's blue, *15*:541
King's yellow, *15*:536
Kiton Fast Red 4BLN, *2*:882
Kjeldahl method
  nitrogen in urea, *21*:53
Kleinschmidt asphalt analysis, *2*:785
Klystrons
  microwave power sources, *S*:565
Knoevenagel reaction
  ethyl acetoacetate and acetaldehyde, *1*:158
Knoop hardness number, *1*:23; *10*:813
Knorr synthesis
  of pyrroles, *1*:157
Knudsen gage, *21*:129
Koch's acid, *13*:716
Kodachrome process, *5*:833
Kojic acid, *10*:923; *12*:107
Koji extract
  in prepn of glutamic acid, *2*:204
Kola nut extract
  in carbonated beverages, *4*:337
Kola nuts, *9*:353
Koppers-Becker coke oven, *4*:410
Kossel cameras
  x-ray fluorescence analysis, *22*:465
Kotinsky effect, *15*:381
Kovar, *8*:3
Kraft pulp, *12*:373
Kraft pulping process, *4*:607, 608
Kraus-Collis prepn of tetraethyllead, *1*:899
Krausening, *3*:321; *22*:527
Krebs cycle, *5*:531, 532
  of amino acid biosynthesis, *2*:160
Kroll's process, *22*:615
Kromycin, *12*:641
Krypton, *10*:862
  as anesthetic, *2*:400
  cryogenic prop, *6*:472, 474
  as fission product, *14*:83

phys prop, *10*:868
  separation from air, *10*:876
  separation of isotopes, *7*:138
Krypton difluoride, *10*:893
Krypton tetrafluoride, *9*:508
Krypton-85, *17*:78, 92, 121
  as fission product, *14*:105
  beta source, *17*:57
Kurnakovite, *3*:627, 630, 648

Kwashiorkor
  amino acid treatment of, *2*:191
Kyanite, *10*:758
  in ceramics, *4*:771
  refractory, *17*:231, 233
Kynol, *S*:962
Kynol fiber, *S*:668
Kynurenine, *12*:857; *21*:522

L

LM Glass
  in ceramic composite armor, *S*:141, 143
LPG. See *Liquefied petroleum gas.*
LSD-25. See *Lysergic acid diethylamide.*
Labdanum, *14*:204, 723; *17*:389
Lac, *18*:21
Laccaic acid, *18*:21, 26
Lacceryl alcohol
  in beeswax, *22*:159
Lac dye, *7*:627
Lachrymators. See *Lacrimators.*
L acid, *13*:720
Lacquer blue, *15*:539
Lacquers. (See also *Coatings.*)
  drying oils in, *7*:424
  hardness testing, *10*:817
  oriental, *17*:387
  use of ketones in, *12*:114
Lacrimators
  in chemical warfare, *4*:877
Lactalbumin, *13*:513
Lactaldehyde phenylhydrazone, *4*:139
Lactaldehyde phenylosazone, *4*:139
Lactamide, *12*:175, 186
Lactams, *2*:66
  cyanoethylation, *6*:647
Lactases, *8*:194; *19*:237
Lactic acid, *12*:**170**
Lactic acid, *2*:578; *7*:527, 529, 671; *13*:569;
  19:238
  in dental materials, *6*:826
  formation in wine, *22*:323
  intermolecular condensation, *20*:793
  kinetics of reaction with methanol, *8*:320
  mash, *22*:520
  in paint removers, *14*:487
  phys prop, *1*:234
  prod in wine, *22*:525
  from sucrose, *19*:154, 224
  from wood hydrolysis, *22*:383
Lactic acid esters, *8*:352
  in prepn of acrylic acid esters, *1*:300
Lactic aldehyde, *4*:139
Lactide, *12*:173, 185
Lactitol, *19*:238
Lactobionic acid, *13*:571; *19*:238

Lactobiose, *19*:237
Lactochrome, *17*:445
Lactoflavine, *17*:445
Lactogenic hormone
  bioassay, *11*:58
  general prop, *11*:58
  method of isolation, *11*:58
Lactoglobulin, *13*:513
Lactone
  "lacolic", *18*:26
γ-Lactone, *9*:358
Lactone dyes, *7*:471
Lactone ketene dimers, *12*:94
Lactones, *8*:365
  cardiac-active, *18*:845
  copolymerization, *16*:174
  no "hetero" component, *16*:334
  of ketene dimers, *12*:94
  polymerization, *16*:171
  from *Streptomyces spp*, *16*:330
β-Lactones, *12*:115
Lactonitrile
  lactic acid from, *12*:171, 178, 184
  in prepn of acrylonitrile, *1*:346
Lactoperoxidase
  in milk, *13*:514
Lactosan, *19*:238
Lactose, *4*:145; *19*:237-242, 593
  in egg drying, *7*:673
  in milk, *13*:514, 565
  structure, *4*:145
  reaction with lime, *19*:143
  tablet filler, *15*:121
  yeast assimilation, *22*:517
Lactosides, *19*:239
Lactulose, *19*:238
Lactyllactic acid, *12*:173
Lac wax, *18*:26
Ladaniol, *17*:389
Ladenburg's prismatic structure, *2*:702
Lagering of beer, *3*:320; *22*:526
Lagoons
  liquid-waste disposal, *21*:643
Lagosin, *16*:137
Lake
  pigments, *15*:565, 577

225

Lake Bordeaux B, 5:871
Lake Red C, 5:869
Lake Red C Ba, 5:869
Lake asphalts, 2:764
Lakes
  eutrophic, S:315
  formation of, 2:472
  mixing of, S:334
  naturally eutrophic, S:329
  oligotrophic, S:315
  prod of, S:317
Laksholic acid, 18:25
Lalande electric cell, 3:122
Lambert's (Bouguer's) law of absorption, 5:788
Laminaran, 17:783
  in algae, 17:763
Laminarinase
  microorganism producing, 8:210
Laminated plastics, 12:**188**
  epoxy resins used in, 8:310
  manuf process, 2:249
  use of melamine resins as, 2:248
Laminates, 12:192
  as electric insulation, 11:795
  phenolic resins, 15:202
Lamination
  adhesive application, S:932
  apparel fabric, S:939
  flame, S:939
Lampblack, 4:244
  as hazardous chemical, 9:291
  manuf of, 4:254
  pigment, 14:474
Lampreycides, 15:921
Lanarkite, 12:279
Lanasyn, 2:889
Lanatosides
  as cardiovascular agents, 4:515
Land-animal fats, 8:786, 778
Land disposal
  of waste liquids, 21:644
Landfill, S:739
  sanitary as solid waste disposal method, 21:648
Land spreading
  sludge disposal method, S:729
Langbeinite, 12:733; 16:381, 384
Langelier Index, 22:81
Langmuir Blodgett layers, 22:461
Langmuir adsorption isotherm, 1:429
Langmuir equation, 21:198

Lankamycin, 12:633, 635, 644
Lankavose, 12:639
Lanolin, 22:167, 411-415
  added to cold creams, 6:349
  ethoxylated, 19:548
  in liquids, 6:364, 368
  in ointments, 15:126
  in shampoos, 6:353; 10:774
Lanolin alcohols, 19:549
Lanosterol, 18:834, 838, 866
Lanthanides. See Rare earth elements.
  electronic configuration, 1:360
  ionic radii, 1:370
  similarity between actinides and, 1:355
Lanthanum, 17:144
  electronic configuration, 1:360
  ionic radii, 1:370
  in nuts, 14:131
Lanthanum boride, 8:17
Lanthanum hydride, 11:207
Lanthanum manganite, 8:894
Lanthanum oxide, 8:361
Lanthanum titanate, 8:361
Lanthanum trimethoxide
  prepn, 1:833
Lanthanum-140, 17:126
Lanthionine
  in wool, 22:401
Lapis lazuli, 15:540
Lapping, 1:41
Larch gum, 10:751; 17:391
Lard, 8:779, 790, 791; 13:182; 18:33, 40
  antioxidants used in, 2:593, 595
  in margarine, 13:59
  in soap, 18:417
  utilization, 13:183
Larmor precession, 14:42
Lasca, 18:78
Laser, 7:877, 894
  microsampling, 17:48
Lasers
  luminescent substances in, 12:622
Lasers and nonlinear optics, S:623
Laterites, 13:737
Latex
Latex
  adhesives, 17:642
  butadiene-acrylonitrile, 17:598
  carboxylated, 17:600
  centrifuged, 17:677
  compounding, 17:592
  creamed, 17:677

fabrication, *17*:634
low-ammonia, *17*:679
natural rubber, *17*:674
natural-rubber thread, *18*:629
neoprene, *17*:599
neoprene fibers, *18*:615
normal, *17*:676
paints, *14*:468
paper, *17*:641
polystyrene, *19*:89
preservation of natural, *17*:674
Saran, *21*:300
styrene, *19*:113
styrene-butadiene, *17*:597
synthetic, *17*:595, 683
synthetic, *7*:683
types of compounds, *17*:601
vinylpyridine, *17*:600
2-vinylpyridine, *20*:328
virgin, *17*:674
vulcanization, *17*:642
wood products, *22*:382
Latex foam rubber, *9*:856, 857
compressive behavior, *9*:869
prop of flexible polymers, *9*:862
Latex paints
water-resistant masonry walls, *22*:152
Laudanosine, *1*:769
Laudexium, *6*:548
Laundering, *12*:**197**
comparison with drycleaning, *7*:310
Lauraldehyde
flavor and perfume materials, *9*:358
Laurel, *9*:365, 369
Laurel bromide, *3*:778
Laurel leaf, *14*:204
Lauric acid. (See also *Dodecanoic acid.*)
Lauric acid, *8*:812; *9*:899; *18*:45. (See also
Dodecanoic acid.)
in coconut oil, *18*:418
in manuf of metallic soap, *7*:274
phys prop, *1*:231
in soap fats, *18*:416
sucrose ester, *19*:154
2-sulfoethyl ester, *19*:522
specifications, *8*:836
Lauric acid oils, *8*:783, 784
Lauric diethanolamide, *10*:773; *19*:551
Lauric ethanolamide
surfactant, *19*:526
Lauric monoethanolamide, *19*:552
Laurite, *15*:842
Lauroleic acid, *8*:814

Laurone, *8*:371; *12*:104
Laurophenone, *12*:107
Laurylacrylamide
vinylpyrrolidone copolymer, *21*:433
Lauryl alcohol, *9*:722; *10*:789. (See also *1-
Dodecanol.*)
composition of commercial mixture, *1*:553
specifications, *1*:554
by sodium reduction, *18*:451
Lauryl bromide, *3*:778
Lauryl chloroacetate, *19*:288
Lauryldimethylamine oxide
surfactant, *S*:37
Lauryl entries. (See also *Dodecyl.*)
Lauryllactam
polymerization, *16*:92
Lauryl lactate, *8*:353
Lauryl mercaptan
chain-transfer agent, *21*:374
prepn from lauric acid, *8*:823
Lauryl pyridinium chloride, *9*:897
Lauryl sulfates
in shampoos, *10*:771
Lauryltri(ethyleneoxy)ethanol, *19*:529
Lavandin, *14*:204
essential oil in, *14*:181
Lavender, *14*:204
essential oil as flavor, *9*:352
Lavender oil
as denaturant, *8*:450
Lavender wax, *22*:161
Lavoisier
law of, *19*:27
Lawrencium. See *Actinides.*
Lawsone, *10*:798
Laxatives. See *Cathartics.*
Lazurite, *15*:540
Leaching, *5*:250
of copper, *6*:168
definition, *8*:761
electrolytic zinc plants, *22*:582
technology, *S*:250
Leaching technology, *S*:250
Lead, *12*:**207**
additives in automobile fuels, *S*:59
bath in furnaces, *10*:289
coatings, *13*:252
copper alloys with, *6*:198, 250
in drilling fluids, *7*:292
as drier in alkyd resins, *1*:865
effects on brasses, *6*:228
electrodeposition of zinc, *22*:587
electromigration of, *S*:287

in electroplating, *8*:61
hot-dipped coatings, *13*:260
impurity in copper, *S*:265
from ocean floor, *14*:165
organometallics in phonograph records, *15*:229
pigments, *15*:535, 549
pollutant in gasoline, *S*:731
as protective covering for cables, *11*:817
red pigments, *15*:521
separation by flotation, *9*:395
as solid film lubricant, *12*:590
in storage batteries, *3*:249
tetraethyllead in motor fuels, *3*:751
tin coatings, *20*:288
in type metal, *16*:497, *20*:756
white opaque pigments, *15*:500
in zinc alloys, *22*:600
Lead acetate, *2*:414
as hair dye, *10*:797
pigments, *15*:536
prepn of lead dinitrosoresorcinate, *8*:591
prepn of lead styphnate, *8*:590
Lead-acid storage batteries, *3*:249
Lead alloys, *12*:**247**
Lead alloys, lead utilization, *12*:**247**
Lead arsenate
in agriculture, *2*:715
as insecticide, *11*:681
soil chemistry, *18*:537
Lead azide
characteristics, *8*:584
as explosive, *8*:581
in explosive priming compositions, *8*:653
Lead babbitts, *3*:272, 274, 275
Lead-bearing brasses
composition of, *6*:227
Lead borate
in glass-ceramics, *10*:548
Lead carbonate
pigments, *15*:531
as pigment, *14*:466
Lead chalcogenides
infrared detectors, *15*:431
Lead chlorite, *5*:31
Lead chlorofluoride, *9*:527
Lead chromate, *5*:481
pigments, *15*:527, 530
as pigments, *14*:472
Lead compounds, *12*:**266**
red lead, *12*:272, 275
red lead in storage batteries, *3*:256

Lead cyanamide, *6*:559
Lead 2,4-dinitroresorcinate
as explosive, *8*:591
Lead dioxide, *7*:697
Leaded zinc oxides, *15*:504
Lead 2-ethylhexanoate, *7*:276, 278; *8*:849
Lead ethyl xanthate, *9*:386
Lead fluoborate, *9*:567
Lead fluoborate plating solution, *8*:61
Lead fluoride, *9*:553, 626, 698
Lead fluosilicate, *3*:530; *9*:658
Lead fluosilicate plating solution, *8*:61
Lead-gold alloy, *10*:691
Lead, hard, *3*:255
Lead hexafluorostannate, *9*:682
Lead hydrogen phosphate
stabilizer for poly(vinyl chloride), *21*:390
Lead linoleate, *7*:278
Lead metaniobate, *9*:7
ferroelectric prop, *9*:11
Lead metaniobate ceramics
prop, *9*:21
Lead metasilicate, *18*:158
Lead molybdate, *13*:650
pigments, *15*:530, 533
Lead monofluorophosphate, *9*:642
Lead monoxide
pigments, *15*:502, 541
Lead neodecanoate, *7*:279
Lead niobate
in glass-ceramics, *10*:548
Lead nitrate
pigments, *15*:536
prepn of lead dinitroresorcinate, *8*:591
relative humidity over solutions, *6*:512
Lead orthosilicate, *18*:158
Lead oxide. (See also *Litharge*.)
accelerator, *17*:509
in electron tube, *8*:4
in glass forming, *10*:542, 544
as lubricant, *12*:603
polyester catalyst, *16*:175
Lead oxides
in storage batteries, *3*:256
Lead pencils
graphite in, *4*:331
Lead perchlorate, *5*:72
Lead peroxide, *6*:803, 805
dye oxidant, *20*:705, 723
pigments, *15*:521
polymer cure, *15*:795
Lead phosphates, *15*:275

Lead phosphite
   pigment, *15*:496
Lead picrate, *8*:593
Lead resinate, *7*:279
Lead salicylate, *17*:729
Lead sesquioxide, *12*:276
Lead silicate, *6*:802; *18*:158
   pigments, *15*:533
Lead silicochromate, *15*:533
Lead stearate, *7*:276
   stabilizer, *21*:390
   use as abherent, *1*:8
   x-ray spectroscopy, *22*:461
Lead styphnate, *8*:590
   in explosive priming compositions, *8*:653
Lead sulfamate, *19*:246
Lead sulfamate plating solution, *8*:61
Lead sulfate
   pigments, *15*:530, 533
   as pigment, *14*:466
   in storage batteries, *3*:257
Lead sulfide. See *Galena.*
   petroleum sweetening, *15*:50
   pigments, *15*:504, 541
Lead sulfite
   pigments, *15*:541
Lead tallate, *7*:276, 278
Lead telluride-tin telluride
   infrared detectors, *15*:434
Lead tetraacetate, *1*:577; *19*:322
   cellulose oxidized by, *4*:596
   steroid synthesis, *18*:853
Lead tetraacetate
Lead thiocyanate
   in explosive priming compositions, *8*:653
Lead-tin alloys
   in electroplating, *8*:61
Lead-tin coatings
   hot-dipped, *13*:260
Lead titanate, *20*:414
Lead 3,5,5-trimethylhexoate, *7*:279
   as packing material, *14*:455
Lead-zinc ores
   separation by flotation, *9*:393
Lead zirconate, *22*:656
   ferroelectric prop, *9*:13
   in glass-ceramics, *10*:548
Lead zirconate-lead titanate ceramics
   properties, *9*:21
Lead zirconate-titanate
   nonlinear optical materials, *S*:628
Lead-210, *17*:22
   pigments, *15*:535

Leaks
   detection in vacuum systems, *21*:137
Leather , *12*:**303**
   fatliquoring, *8*:787
   sodium adipate use in, *1*:414
   water-repellent, *22*:150
Leather dyeing, *7*:588
Leather industry
   use of formaldehyde, *10*:94
   use of formic acid, *10*:103
Leather tannage
   use of aluminum formate in, *2*:16
   use of lactic acid, *12*:182
   chemical, *3*:51
   yeast, *3*:41
Leavening agents, *3*:41
Leblanc soda process
   soda ash manuf, *1*:716
   sodium thiosulfate from, *20*:234
Lechlanché electric cells, *3*:103, 106, 108, 111
Lecithin, *8*:781; *12*:**343**
   added to cold creams, *6*:349
   addition to fats, *18*:45
   from choline, *5*:404
   in confectionery, *6*:45
   in egg lipids, *7*:664
   as food additive, *10*:9
   as food preservative, *10*:14
   in magnetic tape, *12*:813
   in manuf of chocolate, *5*:39, 394
   in margarine, *13*:56, 60
   in nuts, *14*:132
Lecithinase
   in prevention of gelation in egg yolk, *7*:669
Lemon, *14*:205
   essential oils as flavors, *9*:352
Lemongrass, *14*.205
   essential oil in, *14*:181
Lemon peel, *9*:353
Lemon petitgrain, *14*:205
Lemons
   citric acid from, *5*:531
Lenigallol, *16*:198
Lentin. See *Carbamylcholine chloride.*
Leonite
   from Great Salt Lake, *S*:455
Lepidine, *6*:621
   test for detection of propylene oxide, *8*:287
Lepidocrocite, *12*:6
Lepidolite, *12*:530; *13*:399
   in ceramics, *4*:772
   cerium content of, *4*:856, 857
   rubidium from, *17*:686
   rubidium in, *17*:685

Leru's system
  of acids and bases, *1*:218
Lespedeza products
  as animal feed, *8*:864
Letterpress and lithographic inks, *11*:616, 621
Letterpress printing, *16*:495, 537, 543
  silver halide system, *16*:519
Leucine, *2*:157; *19*:750
  in silk, *18*:271
  in wine, *22*:316
Leucite, *16*:370
  in ion exchange, *11*:871
Leuckart reaciton, *2*:107; *S*:41
Leuckart synthesis
  of thiopenols, *22*:424
Leuco Ethyl Violet
  carbon tetrachloride sensitization, *20*:694
Leuco Malachite Green, *20*:703
Leucoanthocyanins, *19*:743
Leucoindigo, *7*:561
Leucomycins
  in chemotherapy, *12*:633, 635
Leuco pararosaniline, *20*:675
Leucopterin, *10*:932
Leucoxene, *10*:758; *20*:349, 398
Leuco 1,4-diaminoanthraquinone, *2*:357; *7*:479
Leuopimaric acid, *17*:482
Leurocristine, *1*:808
Leurosidine, *1*:808
Leurosine, *1*:806
Levamisole
  anthelmintic, *21*:249
Levan polyase
  microorganism producing, *8*:210
Levulinic acid, *6*:922; *19*:224
  phys prop, *1*:235
  prepn from furfuryl alcohol, *10*:239, 246
  from sucrose, *19*:154
  from wood, *22*:383
Levulinonitrile, *6*:640
Levulose, *4*:137
  in beet sugar, *19*:211
Lewis acid
  aziridine complexes, *11*:532
Lewisites, *2*:727
Lewis system
Lexan
  in bearing materials, *3*:283
Licanic acid, *7*:399, 405; *8*:816
Licheniformin, *3*:30
Lichens, *13*:470
Licorice, *19*:602

Licorice root, *9*:353
Lidocaine hydrochloride, *2*:405, 406
Liebermann-Burchard reaction
  for provitamins D, *21*:568
Ligand exchange
  and NMR, *14*:73
Ligands, *6*:1. (See also *Coordination compounds.*)
Light
  penetration as measure of eutrophication, *S*:324
Light Green SF Yellowish, *5*:866; *20*:707
Light absorption
  laws of, *5*:788
  in organic dyes, *5*:763
Lightfastness
  test methods, *7*:597
Light oil
  from coke-oven gas, *3*:382
  composition of, *3*:384, 385
  recovery from coke-oven gas, *4*:415, 419
  toluene from, *20*:533
Light scattering
  molecular weights from, *16*:246
Lignin, *6*:663; *12*:**361**; *16*:681, 684
  in cell walls of wood, *22*:365
  in cork, *6*:284
  in hardboard, *21*:603
  oxalic acid from, *14*:359
  sulfonation, *19*:301
  vanillin from, *15*:163; *21*:186
Lignin sulfonic acid, *12*:813
Lignite
  succinic acid from, *19*:134
Lignite and brown coal, *12*:**381**
Lignite reserves, *10*:185
  characteristics as furnace fuel, *10*:280
Lignites
  composition of, *5*:627, 643
Lignitic materials
  in drilling fluids, *7*:296
Lignoceric acid, *8*:813
  tall oil, *19*:621
  phys prop, *1*:231
Lignoceryl alcohol, *19*:626
Lignosulfonates, *12*:369; *19*:294, 301
  surfactants, *19*:516
Lignosulfonic acid, *12*:369
Lilac, *14*:205
Lily of the valley, *14*:205
  beet sugar purification, *19*:210
  in manuf of magnesium, *12*:667

Lime. (See also *Calcium oxide; Calcium hydroxide.*)
Lime, *12*:**414, 415**
  as a drying agent, *7*:383, 384, 392
  bottling water treated with, *4*:340
  calcium produced from, *3*:920
  cane sugar clarification, *19*:174
  in cements, *4*:697
  in ceramics, *4*:770
  in drilling fluids, *7*:301, 302
  in ethylene oxide manuf, *8*:531
  in flotation process, *9*:393
  hydraulic, *4*:693
  in manuf of sodium hydroxide, *1*:742
  in manuf of thermal insulation, *7*:62
  as paint remover, *14*:490
  phosphate removal in water systems, *22*:119
  phosphates, *15*:263
  in prepn calcium cyanamide, *6*:555, 556
  in prepn of tropical bleach, *5*:20
  reaction with silicon tetrafluoride, *9*:653
  smoke from kilns, *18*:402
Lime feldspar. See *Anorthite.*
Lime glass, *4*:5
Lime, hydrated, *12*:415. (See also *Calcium hydroxide.*)
  as by-product of acetylene manuf, *1*:174
  manuf of, *12*:444
Lime kilns, *1*:724
Lime oil
  distilled, *14*:206
Lime, slaked. See *Calcium hydroxide.*
Lime-soda process
  for water softening, *22*:97
Lime-soda softening process
  sodium aluminate in, *2*:9
Limestone, *4*:3; *12*:**414**
  in blast-furnace process for iron, *12*:12
  calcium carbonate in, *4*:7
  in ceramics, *4*:770
  in extracting lithium, *12*:531
  hydraulic limes from, *4*:693
  in manuf of calcium carbide, *4*:103
  in soda ash prod, *1*:714, 722
Lime sulfur
  as agricultural fungicide, *10*:221
Limit dextrinase
  microorganism producing, *8*:210
Limonene, *19*:815, 835
  citrus, *19*:810
ψ-Limonene, *19*:813
d-Limonene, *19*:832
dl-Limonene, *17*:385

l-Limonene, *19*:813
Limonene oxide, *8*:247
Limonite, *12*:6; *13*:14; *15*:517, 526, 527
  in limestone, *12*:422
Linaloe wood, *14*:206
Linalool, *19*:818, 825, 833
  in soaps, *14*:736
  as synthetic flavor, *9*:354
d-Linalool, *19*:813
Linalyl acetate, *19*:825
  prod and sales, *8*:376
  in soaps, *14*:736
  as synthetic flavor, *9*:354
Linalyl benzoate, *3*:433
Linalyl butyrate
  as synthetic flavor, *9*:354
Linalyl chloride, *19*:818
Lincomycin, *12*:652
Lindane. (See also *Benzene hexachloride.*)
  as an insecticide, *11*:694
  in soil, *18*:527
  in water, *21*:701
Linde cycle
  of ultra-low temperatures, *6*:473
Lindeman glass, *10*:545
Lindlar catalyst, *19*:826
Linen, *12*:**461**
  analysis, *9*:173
  bleaching agents for, *3*:562
  bleaching of, *5*:34
  fiber length of flax, *20*:34
  flax, as packing material, *14*:455
Line-slug, *20*:764
Linnacite
  phys prop, *5*:722
Linoleic acid, *7*:399, 403, 405, 409; *8*:815
  in dental material, *6*:827
  in eggs, *7*:664
  specifications, *8*:832
  phys prop, *1*:231
  tall oil, *19*:621
  prepn of polybasic acid, *8*:847
Linoleic acid glycerides
  in margarine, *13*:59
Linolelaidic acid, *8*:815
Linolenic acid, *8*:815
  in drying oils, *7*:402, 403, 405, 409, 411
  in eggs, *7*:664
α-Linolenic acid
  phys prop, *1*:231
Linolenic acid oil, *8*:786
  classification, *8*:784

Linoleum, 7:408, 424
  use of driers in, 7:277
Linolic acid. See *Linoleic acid.*
Linotype, 16:497
Linotype metal, 20:759, 764
Linoxyn, 7:408, 425
Linseed oil, 1:859; 7:282; 8:779, 784, 786, 790, 791
  in alkyd resins, 1:864
  changes in, 7:406
  composition in fatty acid, 8:831
  drying oil, 7:398, 403, 404, 405
  film prop, 7:423
  properties, 7:411, 413, 417, 419, 420
  phenolic resins, 15:197
  in margarine, 13:59
  uses, 7:423
  for waterproofing bridge decks, 22:153
Linseed oil meal
  as animal feed, 8:863
Linuron
  herbicide, 22:188
Liothyronine, 20:260
  assay, 20:264
  synthesis, 20:264
Lipase, 8:182
  action and uses, 8:185, 186, 190, 196
  in milk, 13:514
Lipases, 10:9
Lipids
  in yeast, 22:516
Lipkin bicapillary pycnometer, 6:766
Lipochromes, 7:615
Lipoic acid, 2:190
α-Lipoic acid
  sulfur in, 19:369
Lipoprotein
  in yeast mitochondria, 22:512
β-Lipotropic hormone
  assay, 11:58
  method of isolation, 11:59
  properties, 11:58
Lipoxidase, 22:293
  in soybeans, 18:602
Lipsticks, 6:363
  manuf, 6:365
Liquefied gas, 3:816; 12:**470**
Liquefied petroleum gas, 3:816; 12:**470**
Liqueurs, 1:517
Liquid crystals
  and melting points, 13:213
Liquid-level measurement, 12:**481**
Liquid oxygen explosives, 8:644

Liquids
  dispersion of liquid, 21:643
  disposal of waste, 21:636
  as electric insulation, 11:788
  electomigration of pure, S:289
  ultrasonic degassing, 20:787
  waste, 21:627
  waste-liquid treatment, 21:628
Liquid smoke
  in meat curing, 13:171
Liquifaction
  of natural gas, S:427
Lirodenine, 1:769
Lissamine Green SF, 20:707
Literature of chemistry and chemical technology, 12:**500**
Litharge, 7:412, 413, 695; 15:501, 521, 531, 535. (See also *Lead oxide.*)
Litharge, 7:412, 413, 695
  as catalyst, 8:361
  in storage batteries, 3:256
  vinylidene fluoride elastomers curing agent, 21:272
  in vulcanization of neoprene, 7:711
2-Lithiobiphenyl, 9:783
Lithium, 7:280; 12:**529**
  as catalyst, 12:70
  as hazardous chemical, 9:291
  in prepn of triethylimine borane, 9:556
  salts from Great Salt Lake, S:464
  styrene-butadiene copolymer catalyst, S:911
Lithium acetylide
  in vitamin A synthesis, 21:496
Lithium acid urate, 21:112
  antarthritic, 21:111
Lithium alkoxide
  effect on styrene-butadiene copolymer, S:916
Lithium alkoxy aluminohydrides, 11:221
Lithium-aluminum alloy
  in high-temperature cells, S:129
Lithium(aluminum)-chlorine(carbon) cell, S:129
Lithium aluminum hydride, 9:690, 691, 798; 11:217-222
  in aquametry, 2:691
  in hydrogenation, 14:420
  polyester reduction, 16:178
  in polymerization of acetaldehyde, 1:99
  reaction with epoxides, 8:273, 416
  reaction with phenyl isocyanate, 12:53
  reduction of acetanilide, 1:150
  reduction of amides by, 2:107
  sulfolane reaction, 19:251

Lithium aluminum tri-*tert*-butoxide, *18*:864
Lithium amide
  prepn from ammonia, *2*:262
Lithium borates, *3*:646
Lithium borohydride, *11*:210
Lithium *sec*-butoxide
  in polymerization of acetaldehyde, *1*:99
Lithium butyl
  tetrachloroethylene reaction with, *5*:198
Lithium carbonate
  in prepn of trifluoroboroxine, *9*:556
  prepn of lithium fluoride, *9*:627
Lithium-chalcogen cells, *S*:130, 135
Lithium chlorate, *5*:61
Lithium chloride, *6*:512
  as a drying agent, *7*:384, 385, 393
Lithium chloride-potassium chloride eutectic
  emf series for, *6*:315
Lithium-chlorine cells, *S*:126, 135
Lithium chromate, *5*:498
Lithium compounds, *12*:**529**
Lithium cyanide, *6*:599
Lithium diethylamide
  titanium amido compds, *20*:472
  zirconium tetrachloride reaction, *22*:638
Lithium dititanate, *20*:410
Lithium fluoride, *9*:626
  dielectric prop of, *4*:821
  and nuclear reactor fuels, *14*:82
  in plasma arcs, *15*:345
  x-ray spectroscopy, *22*:461
Lithium-fluorine cells, *S*:123
Lithium fluorosilicate, *9*:658
Lithium fluorosulfate, *9*:679
Lithium gallium hydride, *10*:323
Lithium hexafluorophosphate, *9*:644
Lithium hexafluorostannate, *9*:682
Lithium hexafluorotitanate, *9*:684
Lithium hydride, *11*:200
  in polycarbonate synthesis, *16*:107
  in reduction of nitro compounds, *2*:98
Lithium hydroxide
  polysiloxanes, *18*:225
  prepn of lithium fluoride, *9*:627
Lithium hypochlorite, *5*:15
  swimming pool water disinfectant, *22*:126
Lithium metatitanate, *20*:410
Lithium minerals
  in ceramics, *4*:772
Lithium monofluorophosphate, *9*:642
Lithium niobate
  nonlinear optical materials, *S*:625, 628

Lithium nitrate
  in heat transfer, *10*:857
Lithium nitrite
  in quartz, *18*:110
Lithium orthotitanate, *20*:410
Lithium oxide, *2*:50; *10*:542
Lithium perchlorate, *5*:69
  as drying agent, *7*:395
Lithium peroxide, *14*:748
Lithium phosphate, *8*:272
Lithium polybutadiene, *7*:690
Lithium salicylate, *17*:729
Lithium-selenium cells, *S*:133
Lithium silicate, *18*:141
Lithium stearate, *7*:276
Lithium-sulfur cells, *S*:132
Lithium superoxide, *14*:763
Lithium tantalate
  nonlinear optical materials, *S*:629
Lithium tetrafluoroberyllate, *9*:553
Lithium titanates, *20*:410
Lithium trihydrogen selenite
  as ferroelectric, *9*:5
Lithium-7
  separation, *14*:87
Lithocholic acid, *3*:481, 483; *18*:831
Lithographic ink
  addition of alumina hydrate to, *2*:51
Lithography, *16*:495, 539-544
  chemistry of photo-, *16*:529
  direct, *17*:335
  history, *16*:528
  offset, *17*:239
  plate materials, *16*:529
  reverse, *16*:541
  silver halide system, *16*:520
Lithol Red, *5*:870
Lithol Red Sr
  used in hair coloring, *10*:793
Lithol Reds, *5*:869, 870
Lithol Reds, *15*:574
Lithol Rubin B, *5*:869
Lithol Rubin B Ca, *5*:869
Lithol Rubin B and B Ca, *5*:869
Lithol Rubine, *15*:575
Lithol Rubine BK, *2*:907
Litho oils, *7*:424
Lithopone, *15*:496, 499, 504, 523; *22*:605
  in papermaking, *14*:501
Lithopones
  cadmium, *3*:909
Lithosol Fast Yellow 3 GD, *2*:908

Litmus (dye), 2:866
Liver fluke infections, 14:549
Liver meal
  as animal feed, 8:866
Livingstonite, 13:223
Loba, 17:386
Lobelanidine, 1:794
Lobelanine, 1:765, 794
Lobelia alkaloids, 1:794
Lobeline, 1:778, 794
  van Ekenstein rearrangement, 1:573
Lobry de Bruyn-Alberta van Ekenstein rear-
  rangement, 1:575
Locust bean gum, 10:748; 17:391
Loellingite, 2:712
Logwood, 7:565, 615; 10:307; 15:556
  in stains, 18:657
Lollingite, 2:720
Longifolene, 19:833
Long-oil alkyds, 1:877
Long-oil varnish, 7:424
Lophenol, 18:838
Lorandite, 20:65
Loss index, 7:722
  of electric insulation, 11:776
Lotions, 6:352
  manuf, 6:353
Lovage, 14:206
Lovibond color system, 5:808
Lozenges
  sugar, 6:51
Lubricant additives
  use of fatty alcohols in, 1:556
Lubricants, 12:557
  esters used as, 8:374
  graphite, 4:228, 330
  use of lycerin as, 10:628
  use of metallic soaps in, 7:283
  use of trimethylpentanediol, 10:679
Lubricants, synthetic
  use of adipate esters as, 1:416
  use of azelete esters, 1:250
Lubricating oil
  antioxidants, 2:598
  antistatic, 2:657
  decolorizing, 15:59
  dewaxing, 15:58
  manuf, 15:54
  Pennsylvania-grade, 15:57
  solvent refining, 15:57
  vacuum fractionation, 15:55

Lubricating oils
  use of furfural in, 10:243
Lubrication and lubricants, 12:557
Lubrication, fluid-film, 12:557
Lucanthrone hydrochloride, 14:546
Lucite, 10:592
Lumbang oil. See Candlenut oil.
Lumber industry
  use of adhesives in, 1:396
Lumequoic acid, 8:814
Lumichrome, 17:446
Lumiflavine, 17:446
Luminescence, 12:617; 15:333, 334
Luminescent materials, 12:616
  deposition, 8:30
Luminol, 10:928; 11:174
Lumista-5,7,22-trien-3β-ol. See Lumisterol.
Lumisterol, 6:83; 18:837; 21:555
Lumisterol 2, 18:880
Lunacrine, 1:771
Lungworms, 14:534
Lupanine, 1:767
Lupin alkaloids, 1:766
Lupinine, 1:767
  carbonization process, 4:422
Lurgi-Spülgas
Luteoic acid, 12:323
Lutetium, 17:143
  electronic configuration, 1:360
  ionic radii, 1:370
  in nuts, 14:131
Lutidine
  oxidation with nitric acid, 21:528
2,4-Lutidine
  oxidation of, 21:514
2,6-Lutidine, 6:508
  amine oxide from, S:43
  in coal tar, 19:670
Lutidines, 16:780
Lutidone, 10:922
Lycoctonine, 1:773
Lycodine, 1:763
Lycopene
  reaction with boron trifluoride, 9:556
Lycorine, 1:771
Lymphangiography, 17:140
Lyndiol, 6:68, 87
  of eggs, 7:672
Lyophilization, 7:368. (See also Freeze drying.)
  of meats, 13:177
Lysergic acid
  in prepn of methylergonovine, 1:791

Lysergic acid amide, *1*:763
Lysergic acid diethylamide, *16*:641, 646
  in chemical warfare, *4*:877
Lysine, *2*:157; *10*:247
  in wine, *22*:316
Lysine diisocyanate, *21*:103
Lysosomes
  in yeast, *22*:512

Lysozyme
  action and uses, *8*:185
  in eggs, *7*:664
  medical use, *8*:204
  uses, *8*:190, 201
Lyxose, *4*:134

M

MCDAP, 6:82

MCPA. See *2-Methyl-4-chlorophenoxyacetic acid.*

MCPA. See *2-Methyl-4-chlorophenoxyacetic acid.*

MDCB. See *1,3-Dichlorobenzene.*

MEK. See *Methyl ethyl ketone.*

MER-25
  as contraceptive drug, 6:85

MER/29, 2:49

MSG. See *Monosodium glutamate.*

Ma Huang, 1:768, 788

MacArthur process
  for extraction of gold and silver, 6:589

Macaroni
  use of lecithin in, 12:356

Mace, 9:372, 363; 14:206

Macerated paper, 11:832

Maceration
  essential oils by, 14:182

Machining industry
  humidity conditions within, 1:487

M acid, 13:724

Maclurin, 7:614

Macrocin, 12:647

Macrolide antibiotics, 12:**632**

Madder, 7:527, 620; 15:556

Madder lake, 15:577

Mafenide, 19:261, 267, 270, 272, 276

Magenta, 7:465; 20:677

Magenta II
  manuf of, 20:709

Magenta couplers
  for color photography, 5:823

Maghemite, 8:881

Magmas, 19:758

Magnamycin, 12:645

Magnesia. (See also *Magnesium oxide.*)
  as abrasive, 1:28
  as adsorbent, 1:460
  matrix in MHD plant, S:237
  methylation of phenol, S:273
  pigments, 15:535
  in prepn of trifluoroboroxine, 9:556
  strength value of, 4:818

Magnesia magma, 2:429; 4:591

Magnesian rock, 12:415

Magnesia spinel
  as refractory material, 4:774

Magnesioferrite, 8:881; 17:237

Magnesite, 12:5, 415; 15:510, 517; 17:230, 237
  as arc furnace lining, 10:257
  calcined, 17:227
  in ceramics, 4:770
  refractory, 17:236
  in wood-inorganic binder board, 21:613

Magnesite brick
  strength value of, 4:818

Magnesium, 7:281; 12:**661**

Magnesium
  alloys with thorium, 20:258
  burning temperature of, 4:899
  cerium alloys with, 4:849
  cesium alloys with, 4:862
  in cesium metal manuf, 4:860
  chromate treatment, 13:298
  content in soils, 9:35
  copper alloys with, 6:210
  corrosion resistance to water, 6:302
  as ester-exchange catalyst, 16:167
  in electric cells, 3:132
  in electron tubes, 8:5
  electroplating, 8:45
  ethyl chloride reaction with, 5:142
  to form Grignard reagent, 2:144
  in Grignard reaction, 10:721
  heat of combustion of, 4:898, 899
  in incendiary agents, 4:897
  in limestone, 12:414
  methyl chloride reaction with, 5:104
  in photoengraving, 16:522
  as plant nutrient, 9:28
  in pyrotechnics, 16:826
  reaction with germanium, 10:521, 635
  removal from water supplies, 22:98
  in rolled zinc, 22:601
  in seawater, 14:154
  in water, 21:694
  in prepn of beryllium metal, 3:461
  and water hardness, 22:73
  in wood, 22:366
  in zinc alloys, 22:600
  zirconium alloy, 22:620

Magnesium alloys, *12*:**684**
  welding of, *22*:251
Magnesium aluminate spinel
  in ceramic armor, *S*:153
  in transparent armor, *S*:153
Magnesium and alloys
  as hazardous chemicals, *9*:290
Magnesium benzoate
  benzoic acid oxidation, *15*:155
Magnesium bismuthide, *3*:537
Magnesium bisulfite
  in pulpmaking, *12*:369
Magnesium bomb, *4*:899
Magnesium borates, *3*:648
Magnesium carbide
  hydration of, *S*:551
Magnesium carbonate, *6*:357, 359, 513; *10*:4
  as catalyst, *8*:361
  as cathartic, *4*:591
  in ceramics, *4*:770
  in limestone, *12*:414
  pigments, *15*:510
  as release agent in food, *10*:16
Magnesium casting
  as brine in refrigerant, *10*:860
  in cement, *4*:696
  from Great Salt Lake brine, *S*:460
  hydrolysis of, *S*:174
  from Kroll process, *22*:615
  in rock salt, *18*:472
  use of fluoborates in, *9*:564
Magnesium chromate, *5*:498
Magnesium chromite, *5*:496
Magnesium citrate
  as cathartic, *4*:591
Magnesium compounds, *12*:**708.** (See also
  Magnesite.)
Magnesium cyanide, *6*:604
Magnesium cyclohexylsulfamate, *10*:10
Magnesium dititanate, *20*:413
Magnesium 2-ethylhexanoate, *8*:849
Magnesium ferrite, *8*:884
Magnesium fluoride, *3*:461; *9*:627
Magnesium fluosilicate, *6*:802; *9*:658, 660
Magnesium germanate, *10*:521
Magnesium germanide, *10*:522, 523
Magnesium hexafluorostannate, *9*:682
Magnesium hydride, *11*:204
Magnesium hydroxide
  cane sugar clarification, *19*:175
  water-treatment systems, *22*:73
Magnesium lauryl sulfate
  surfactant, *9*:523

Magnesium metatitanate, *20*:413
Magnesium methoxide, *1*:848
  prepn, *1*:836
Magnesium montmorillonite, *7*:293
Magnesium orthophosphates, *15*:275
Magnesium orthotitanate, *20*:413
Magnesium oxide. (See also *Magnesia.*)
  in cements, *4*:688, 691, 692, 695, 696, 697,
    700, 868
  in ceramic armor, *S*:153
Magnesium oxide, *6*:357, 802; *12*:160. See
Periclase.
  as antacid, *1*:428
  as catalyst, *8*:362, 826
  as catalyst in catalytic reforming of hydro-
    carbons, *10*:424
  in ceramic armor, *S*:150
  in dental cement, *6*:779, 781, 783
  in dental materials, *6*:796
  dielectric prop of, *4*:821
  expansion coefficient, thermal, *4*:805
  in glass forming, *10*:538, 542
  in opaque armor, *S*:150
  phosphor, *15*:352
  refractory, *17*:244
  in smokes, *18*:402
  thermal conductivity of, *4*:803
  in transparent armor, *S*:153
  in vulcanization of neoprene, *7*:711
  vinylidene fluoride copolymer acid acceptor,
    21:270
  zirconium metal prod, *22*:616
Magnesium oxychloride
  in binding of grinding wheels, *1*:34
  cement, *4*:696
  in wood-inorganic binder board, *21*:613
Magnesium oxysulfate
  in wood-inorganic binder board, *21*:613
Magnesium perchlorate, *5*:69
  as drying agent, *7*:391
Magnesium peroxide, *14*:750
  prod from sodium peroxide, *18*:450
Magnesium phosphate
  in dental cement, *6*:779
Magnesium phosphide, *15*:293
Magnesium phthalocyanine, *15*:490
Magnesium salicylate, *17*:729
Magnesium silicate, *10*:4; *18*:159. (See also
  Magnesium trisilicate.)
  in ceramics, *4*:772
  fluoro elastomer filler, *21*:272
  in food processing, *10*:10

as opacifying agent in shampoo, 6:353
as paint extender, 14:466
pigments, 15:500, 506, 509
as release agent in food, 10:16
slag-former in welding, 22:242
talc, 19:608
Magnesium silicide
silanes from, 18:175
Magnesium silicides, 18:129
Magnesium stearate, 6:357; 7:281
fire extinguishers, 7:276, 285
lubricant, 15:121
Magnesium stearate, 10:4
as release agent in food, 10:16
Magnesium styphnate, 8:590
Magnesium sulfate, 4:591; 6:673; 12:731
contact process, 19:467
in prepn glutamic acid, 2:203
in prepn of magnesium trisilicate, 2:429
sulfuric acid, 19:442
as water correcting agent in brewing, 10:17
Magnesium tetrafluorozincate, 9:684
Magnesium titanates, 20:413
Magnesium trisilicate
as antacid, 2:429
as dental abrasive, 6:849
Magnesium zirconate, 22:656
Magnestostrontite, 8:892
Magnetic attractability of minerals, 12:784
Magnetic domains, 12:737
Magnetic forming of metals, 13:316
Magnetic materials, 12:**737**
ceramic, 4:775, 819, 825
nickel ferrites, 13:761
permanent, 12:750
Magnetic materials
cobalt alloys as, 5:732
Magnetic moment, 7:878; 14:41
pigments. See *Iron oxide.*
Magnetic oxide pigments. See *Iron oxide, black
pigments.*
Magnetic perovskites, 8:894
Magnetic properties, 12:**772**
Magnetic pulley, 12:786
Magnetic recording, 12:801
Magnetic separation, 12:**782**
Magnetic substances
copper as, 6:220
Magnetic tape, 12:**801**
Magnetic units, S:1002
Magnetism
in ferrites, 8:881
theory, 12:737

Magnetite, 7:292; 8:881, 885; 10:758. (See also
Iron oxide.)
in copper ores, 6:146
in emery, 1:25
pigments, 15:545
scale on steel, S:684
zero point of charge, 9:384
Magnetobarite, 8:892
Magnetocrystalline energy, 12:743
Magnetohydrodynamic generator, 16:462
Magnetohydrodynamics, S:**217**
Magnetoplumbite, 8:881, 892
Magnetostriction effect, 20:784
Magnets. See *Magnetic materials.*
Magnox, 14:81
Mahogany acids, 19:296
Maize, 22:253
Majoram, 14:206
Malachite, 6:135, 169, 270
Malachite Green, 11:552; 20:675
absorption spectrum, 20:686
as antiseptic, 20:731
dye series, 20:677
manuf, 20:702
polyacrylonitrile fiber dye, 20:729
substituents effects, 20:687
Malacon, 10:756
Malathion, 15:322, 328
soil chemistry, 18:529
residue in food, 10:18
Maleate-fumarate isomerization, 20:797
Maleic acid, 12:**819**
in alkyd resins, 1:862
from butenes, 3:842
dry distillation, 19:144
grafts, 16:7
oxidation, 19:735
in prepn of acrylic acid, 1:294
phys prop, 1:232
polyesters of, 16:160
in synthetic oil prepn, 7:415
Maleic acid mono-4(3-chloro-4-methylphenyl)-
piperazide, 14:547
Maleic anhydride, 6:692, 700, 702; 8:249;
10:661; 12:**819**
in alkyd resins, 1:862
chloroprene reaction with, 5:217
Diels-Alder addition with cyclopentadiene,
6:694
esterification, 19:141
hexachlorocyclopentadiene reaction with,
5:241
iodoprene reaction with, 5:229

olefin reaction, *14*:331
from oxidation of benzene, *3*:373, 398
phys prop, *1*:212
phthalic anhydride by-product, *15*:450
polyester from, *16*:170
reaction with butadiene, *3*:785
reaction with isoprene, *12*:66
styrene copolymer, *19*:91
styrene copolymer, *20*:803
succinic acid from, *19*:143
terpene resins, *19*:836
Maleic anhydride-methyl vinyl ether
  as drilling fluid flocculant, *7*:305
Maleic anhydride-vinyl acetate copolymer
  as drilling fluid flocculant, *7*:305
  in manuf of oil, *7*:415
Maleic hydrazide, *12*:833; *15*:681
  in cigarette smoke, *20*:523
Maleic oils, *7*:415
Maleimide
  copolymers of, *S*:754
  Diels-Alder addition with cyclopentadiene, *6*:694
Maleoyl chloride, *12*:821
Malic acid, *1*:234; *19*:142; *12*:**837**
  in cotton, *6*:405
  decarboxylation in wine, *22*:323
  reduction, *19*:144
  in vinegar, *21*:256
  in wine, *22*:316
D-Malic acid, *19*:736
Malide, *12*:843
Mallophene, *2*:640
Malnutrition
  caused by lack of amino acids, *2*:182
Malonaldehyde tetramethyl acetal
  synthesis, *21*:415
Malonamide, *12*:850
  form, *2*:67
  phys prop, *1*:232
  prepn from butynediol, *1*:606
  tetracycline biosynthesis, *20*:20
  urea reaction, *21*:39
Malonic acid derivatives, *12*:**849**
Malonitrile
  NMR of, *14*:58
Malononitrile, *6*:628, 641; *9*:605; *12*:850; *14*:58; *S*:600
Malonyl chloride, *12*:850
Malonyl coenzyme A, *20*:20
Malonylurea. See *Barbituric acid.*
Malt, *12*:861
Malt amylase. (See also *Enzymes, industrial.*)

Maltase
  in bread making, *22*:528
  uses, *8*:190, 194
  from yeast, *22*:518
Malthenes
  in asphalt, *2*:785
Malting, *12*:**861**
Maltobiose, *19*:239
Maltol, *19*:38, 923
Maltose, *19*:593
Maltose, *4*:145; *19*:593
  manuf, *19*:240
  properties, *19*:240
  yeast assimilation, *22*:517
  yeast prod, *22*:534
Maltotriose
  hydrolysis in grain mashes, *22*:522
Malts, *12*:**861**
  in beer brewing, *3*:300
Malvalic acid, *6*:414
Mandarin oil, *14*:206
Mandarin petitgrain, *14*:206
Mandelic acid, *2*:578; *22*:633
DL-Mandelic acid
  phys prop, *1*:239
Mandelonitrile, *6*:668, 672
Maneb. See *Manganese ethylenebisdithiocarbamate.*
Manganates, *13*:32
Manganates(VI), *13*:12
Manganese, *12*:**887**
  analysis for in water, *21*:697
  in bearing materials, *3*:271
  in Bessemer process, *18*:721
  content in soils, *9*:35
  in copper alloys, *6*:250
  as drier in alkyd resins, *1*:865
  effects on brasses, *6*:229
  electrodeposition of zinc, *22*:588
  ferroalloy, *18*:751
  neutron activation analysis, *17*:38
  in pig iron, *12*:19; *18*:718
  as plant nutrient, *9*:30
  removal from swimming pool water, *22*:132
  removal from water, *22*:72
  removal from water supplies, *22*:102
  in rolled zinc, *22*:601
  separation by flotation, *9*:395
  in spiegeleisen, *18*:721
Manganese(II) hexacyanoferrate(II), *12*:27
Manganese(II) hexacyanoferrate(III) *12*:31
Manganese acetate, *8*:361
  catalyst, *19*:62

Manganese alloys, *12*:**887**

Manganese ammonium phosphate, *15*:541

Manganese black, *15*:545

Manganese blue, *15*:541

Manganese bronzes
   composition of, *6*:227
   casting of, *6*:254
   prop of, *6*:254, 257

Manganese brown, *15*:526

Manganese carbide, *4*:73

Manganese carbonyl, *4*:489

Manganese chromate, *5*:499

Manganese compounds, *13*:**1**
   in ceramic pigments, *5*:854
   as glass colorant, *5*:849
   sulfur dyes, *19*:436

Manganese dioxide, *2*:415. (See also *Manganese oxide.*)
   as abrasive, *1*:28
   colorant for glass, *9*:252
   in dye industry, *7*:481
   forms of, *3*:112
   in Leclanché electric cells, *3*:111
   in secondary electric cells, *3*:244
   slag-former in welding, *22*:242
   sulfur oxide recovery, *S*:359
   thiol oxidation, *20*:208
   uranium ore leaching, *21*:13

Manganese ethylenebisdithiocarbamate
   as fungicide, *10*:224

Manganese 2-ethylhexanoate, *7*:276, 278; *8*:849

Manganese fluoride, *9*:698, 699

Manganese fluosilicate, *6*:802; *9*:658

Manganese formate
   as catalyst, *8*:361

Manganese hexafluorostannate, *9*:682

Manganese linoleate, *7*:278

Manganese metatitanate, *20*:414

Manganese monofluorophosphate, *9*:642

Manganese naphthenate, *7*:276, 278, 285

Manganese neodecanoate, *7*:279

Manganese orthotitanate, *20*:414

Manganese oxide. (See also *Manganese dioxide.*)
   in talc, *19*:608

Manganese oxides
   ore specifications, *13*:23

Manganese phosphates, *15*:270

Manganese resinate, *7*:279

Manganese silicates
   in Bessemer process, *18*:723

Manganese silicides, *18*:131

Manganese sulfate
   in prepn of glutaric acid, *2*:203

Manganese tallate, *7*:276, 278

Manganese tetrafluorozincate, *9*:684

Manganese titanates, *20*:414

Manganese 3-5-5-trimethylhexoate, *7*:279

Manganese violet, *15*:541

Manganese-52, *17*:122

Manganese-54, *17*:122

Manganite
   as pigment, *15*:526

Mangrove
   pyrocatechol in, *11*:469

Mangrove cutch, *12*:317

Manila copal, *17*:387

Manila maguey, *9*:177

Manjak, *10*:528

Mannan
   to prep D-mannitol, *1*:574
   in wood, *22*:365
   in yeast, *22*:516

Mannanase
   microorganism producing, *8*:210

Mannich reaction
   with morpholine, *13*:661
   for polyacrylamide, *1*:310
   prepn of amines, *2*:107

1,5-Mannitan. See *Styracitol.*

Mannitol, *1*:574; *7*:251
   analysis, *1*:580
   in algae, *17*:763
   as diuretic, *7*:411
   humectant, *19*:225
   manuf, *1*:581
   phys prop, *1*:570
   polyurethans, *21*:62
   prepn, *1*:574
   reaction, *1*:577
   reaction with antimony(III) oxide, *2*:579
   as release agent in food, *10*:16
   sweeteners, *19*:593, 603
   in nuts, *14*:129

D-Mannitol
   from sucrose, *19*:154

Mannitol hexanitrate, *5*:579, 584, 585; *8*:610; *19*:226
   as cardiovascular agent, *4*:521
   as explosive, *8*:632-634

Mannitol pentanitrate, *8*:611

Mannose, *4*:136
   yeast assimilation, *22*:517

D-Mannose
   sweeteners, *19*:593

Mannosidostreptomycin, *19*:39
Mannosidostreptomycinase
  microorganism producing, *8*:210
Manometer. See *Pressure measurement.*
Manufactured gas. See *Gas, manufactured.*
1-Maphthylamine-3-sulfonic acid, *13*:709
Maple syrup, *19*:241
1-Mapthylamine-4,7-disulfonic acid, *13*:715
Marangoni effect, *13*:140
Marble, *4*:3; *12*:416
"Marboran", *21*:455
Marcasite, *12*:6, 43
Marcasite, *22*:563
  selenium in, *17*:812
Marcomycin
  amebicide, *20*:82
Margaric acid, *8*:812
  phys prop, *1*:231
Margarine, *13*:**56**
Margaryl alcohol. (See *1-Heptadecanol.)*
Margule's equation, *21*:210
Marihuana, *16*:646
Marine antifouling agents, *12*:297
Marine beaches
  as raw material source, *14*:163
Marine drying oils, *7*:403
Marine oils, *8*:782, 787
Maringenin, *15*:681
Marjoram, *9*:352, 363
  essential oils as flavor, *9*:372
Mark-Houwink relationship
  molecular weight determination, *21*:292
Marketing and marketing research, *13*:**66**
Marl, *12*:416
Marshes
  aquatic plant nutrient sources, *S*:330
Marshite, *6*:273
Marshmallow, *6*:50
Martensite, *13*:328; *18*:766, 771
  metallurgical stability, *11*:12
Martensitic steel, *12*:766
  corrosion resistance to water, *6*:302
Mascara, *6*:366
Maser, *7*:875
  solid-state, *6*:478
Mashing
  in beer brewing, *3*:306, 307
Masks, protective
  in chemical warfare, *4*:883
Masonite, *21*:603
Masonite process, *21*:605
Masonry
  watertight, *22*:151

Masonry cement, *4*:693
Mass conservation
  in transport processes, *20*:617
Mass diffusion, *7*:92, 119
  column design, *7*:124
  cascade, *7*:134
  in transport processes, *20*:650
Mass flow theory
  of mass transfer, *13*:139
Massicot, *12*:273
Mass spectrometry, *13*:**87**
Mass transfer, *13*:99
  in detergency, *6*:883
  in liquid extraction, *8*:749
Mass transfer coefficients, *20*:651
Mass transport, *20*:650
Mastic, *14*:206
Mastic gum
  in dental material, *6*:842
Mastic(h), *17*:386
Matches, *13*:**160**
  as fire hazard, *9*:286
Maté, *14*:723
Matrine, *1*:767
Mattes (copper), *6*:144, 147, 149
Maturing agents
  in food products, *10*:4
Mauritius fiber, *9*:171, 178
Mauve, *2*:860; *20*:673
Mauveine, *2*:860; *7*:464; *15*:222; *20*:673
Maxwell relations, *20*:129
Mazarine Blue, *5*:853
Mazarine (Sèvres) blue, *5*:853
McFadyen-Stevens reaction
  aldehydes, *1*:648
Mc Kinley's plots
  of desulfurization reactions, *11*:434
McLeod gage, *21*:128
Measurement, *S*:984
Meat, *13*:**167**
Meat products, *13*:**167**
Meat sugar. (See *Inositol.)*
Mecamylamine
  antihistamine, *11*:39
  as ganglionic blockader, *4*:519
Mecamylamine
  as ganglionic blockader, *4*:519
Mechanical plating, *13*:282
Mechanical testing, *13*:**184**
Mechlorethanime hydrochloride
  alkylating agent, *S*:82
Mecholyl, *5*:410. See *Acetyl-β-methylcholine chloride.*

Meclizine, *16*:674
Meclizine HCl
    antihistamine, *11*:39
Median lethal dosage (LD$_{50}$), *11*:598
Medrol
    corticosteroid analog, *11*:89
Medroxyprogesterone acetate
    as veterinary drug, *21*:251
Meerwein-Ponndorf-Oppenauer oxidation reduction
    use of metal alkoxides as catalyst, *1*:843
Meerwein-Ponndorf-Verley reaction
Megacidin, *12*:633, 635
Megesterol acetate, *6*:79, 87
Meiosis
    in yeast, *22*:514
Meisenheimer rearrangement, *S*:38
Melam, *6*:571
Melamine, *6*:569; *15*:793. (See also *Amino resins and plastics.*)
    cyanuric acic from, *20*:667
    from hydrogen cyanide, *6*:583
    manuf, *2*:366
    manuf from dicyandiamide, *6*:566
    phys prop, *6*:569
    prepn from guanidine, *10*:737
    reactions, *6*:570
    reaction with ethyleneamines, *7*:31
    stabilizer, *21*:390
Melamine formaldehyde resin
    in particleboard, *21*:610
Melamine-formaldehyde resins, *15*:903
    in pigment manuf, *9*:497
Melamine orthophosphate, *6*:571; *15*:239
Melamine resin acid colloids, *2*:240
Melamine resins
    as blending agents in alkyd resins, *1*:867
    storage life, *13*:832
Melamines, chlorinated, *4*:921
Melanin, *10*:305
    pigment in wool, *22*:405
Melanoidins, *7*:670
Meldola Blue, *2*:864; *7*:482
Melem, *6*:571
Melengesterol acetate, *6*:82
    as veterinary drug, *21*:251
Melengket, *17*:386
Melilot, *14*:723
Melilotic acid, *6*:425, 427
Melilotyl glucoside, *6*:425
Melissic acid, *8*:813
    phys prop, *1*:231
Melissyl alcohol. See *1-Hentriacontonol.*

Mellitic acid
    phys prop, *1*:239
Melphalan
    alkylating agent, *S*:84
Melting point
    experimental methods, *13*:205
Melting point calorimeter, *4*:42
Melting temperatures, *13*:**198**
Membrane permeability
    data interpretation, *6*:749
Membrane rermeability
    data interpretation, *6*:749
Membranes
    ion-selective for electrodialysis, *7*:847
Menaccanite, *20*:347
Menadiol sodium diphosphate, *5*:603
Menadione, *21*:586
    as anticoagulant antagonist, *5*:603
Menaquinone-35
    vitamin K$_{2(35)}$, *21*:586
Menaquinone-7
    vitamin K$_{2(35)}$, *21*:586
Mendeleev, *8*:93
Mendelevium. See *Actinides.*
Mendozite, *2*:65
Menhaden oil, *7*:399, 403, 404; *8*:782, 790
    in soap, *18*:417
Mental disease
    biochemical basis, *16*:642
Mentha arvensis. See *Mint.*
2,4(8)-*p*-Menthadiene, *19*:815
3,8-*p*-Menthadiene, *19*:815
p-Menthadienes, *19*:823
6,8(9)-*p*-Menthadien-2-one. See *Carvone.*
p-Menthane, *19*:815
Menthanediamine, *7*:23
    properties, *7*:24
    use as epoxy resin curing agent, *7*:37
p-Menthane-1,8-diamine. See *Menthanediamine.*
trans-p-Menthane-1,8-diol, *19*:822
p-Menthane hydroperoxide, *19*:817
3-Menthene, *19*:824
Menthol, *19*:823-825
    as denaturant, *8*:450
    in dentifrices, *6*:848
    as synthetic flavor, *9*:355
l-Menthol, *19*:828
Menthone, *19*:825
    as synthetic flavor, *9*:354
Menthyl cinnamate, *5*:520
Menthyl esters, *19*:825
Mepazine, *16*:670

Mephentermine, *10*:728
Mephenytoin, *11*:521
Mephobarbital, *3*:61, 70
  as anticonvulsant, *11*:519
Meprobamate, *16*:676
Meprobamate diol, *16*:676
Meragidone
  as diuretic, *7*:256
Merbiurelidin
  as diuretic, *7*:256
Merbophen, *7*:254, 255, 257
Merbromin, *2*:620; *3*:780; *13*:236, 241
Mercaptals, *12*:119; *20*:210
Mercaptans. See *Thiols.*
Mercaptides, *20*:210
  formation, *20*:210
2-Mercaptoacetanilide, *20*:203
Mercaptoacetic acid, *7*:257; *20*:198
  uses, *20*:215
o-Mercaptobenzoic acid
  uses, *20*:215
Mercaptobenzoidithiazole
  encapsulation of, *13*:454
2-Mercaptobenzomethylthiazole, *16*:82
Mercaptobenzothiazole, *2*:414, 417; *6*:343;
  19:771
2-Mercaptobenzothiazole, *10*:916; *17*:514
  as fungicide, *10*:232
  uses, *20*:215
Mercaptobenzoylbenzoic acid, *2*:442
2-Mercaptocyclohexanone, *12*:118
Mercaptoethanoic acid, *20*:198. (See also *Thioglycolic acid.*)
2-Mercaptoethanol, *18*:662
  prepn, *8*:529
2-Mercaptoethylamine, *10*:804; *20*:215
2-Mercaptoimidazoline, *2*:30; *7*:711
Mercaptomerin
  as diuretic, *7*:256
Mercaptomerin sodium, *4*:522
2-Mercapto-*N*-2-naphthylacetamide, *20*:203
3-Mercaptopropionaldehyde
  intermediate in prepn of methionine, *1*:271
3-Mercaptopropionic acid, *20*:215
6-Mercaptopurine, *21*:527
  antimetabolite, *S*:84
Mercaptosuccinic acid, *19*:146
2-Mercaptothiazole, *10*:915
Mercaptothiazoles, *17*:514
2-Mercaptothiazolines
  from aziridines, *11*:534
2-Mercapto-4-thiazolol, *10*:915
Mercerization, *6*:405

Mercocresol, *2*:620
Mercumatilin
  as diuretic, *7*:256
Mercurial ointment, *13*:240
Mercuric bromide
  reaction with ethyl ether, *8*:479
Mercuric chloride
  as antibacterial agent, *2*:619
  as diuretic, *7*:257
  as fungicide, *10*:230
  as insecticide, *11*:683
Mercuric chromate, *5*:498
Mercuric cyanide, *2*:619
Mercuric entries. (See also *Mercury com-
  pounds.*)
Mercuric fluoride, *9*:587, 628
  in fluorination of organic compounds,
    9:695, 698
Mercuric fluoride dihydrate, *9*:628
Mercuric fluorosulfate, *9*:680
Mercuric fluosilicate, *9*:658
Mercuric fulminate, *10*:152
Mercuric iodide
  as denaturant, *8*:450
Mercuric-iron salt catalyst
  in hydration of acetylene, *1*:85
Mercuric nitrate
  dye oxidant, *20*:693
Mercuric nitrate
  pigments, *15*:536
Mercuric oxide
  chlorine reaction with, *5*:3
  in electric cells, *3*:119
  pigments, *15*:536
  prepn of allyl ether, *8*:494
Mercuric oxycyanide, *2*:619
Mercuric pentafluorobenzoate, *9*:795
Mercuric peroxide, *14*:751
Mercuric peroxyacetate, *2*:165
Mercuric sulfate, *12*:145
  in manuf of 1-anthraquinonesulfonic acid,
    2:444
Mercurimetric oxidation
  in acetaldehyde determination, *1*:90
Mercurochrome, *22*:434. (See also *Merbro-
  min.*)
Mercurophen, *13*:242
Mercurophylline
  as diuretic, *7*:256
Mercurous azide, *8*:592
Mercurous carbonate
  prepn of mercurous fluoride, *9*:628

Mercurous chloride, 4:588; 13:236, 238
  as diuretic, 7:254
  as insecticide, 11:683
Mercurous chromate, 5:498
Mercurous entries. (See also *Mercury compounds.*)
Mercurous fluoride, 9:628
  in fluorination of organic compounds, 9:695
Mercurous fluosilicate, 9:658
Mercurous oxide
  prepn of mercurous fluoride, 9:628
Mercury, 2:440; 13:**218**
  in bismuth alloys, 18:543
  as catalyst in acetylene hydration, 1:85
  in dental amalgam, 6:807
  in electron tubes, 8:5
  as heat transfer medium, 10:853
  as lubricant, 12:603
  manuf of anthraquinonedisulfonic acids, 2:445
  as nuclear reactor coolant, 14:80
  in prepn of mercury fulminate, 8:588
  from seawater, 14:156
  thallium alloy, 20:64
Mercury arcs
  photochemistry, 15:340
Mercury cathode cell
  for chlorine manuf, 1:688
Mercury cells
  graphite anodes for, 4:206
Mercury compounds, 13:**235**
Mercury fulminate, 8:582
  characteristics, 8:584
  in explosive priming compositions, 8:653
  prepn, 8:588
  prepn from ethyl alcohol, 8:456
Mercury-gold alloy, 10:691
Mercury monofluorophosphate, 9:642
Mercury, recovery by electroxidation, S:**542**
Mercury salicylate, 17:729
Mercury selenide, 12:626
Mercury sulfide, 12:626
  pigments, 15:535
Mercury telluride, 12:626
Mercury telluride-cadmium telluride
Mercury-vapor lamps, 12:623
Merelluride
  as diuretic, 7:256
Merethoxylline
  as diuretic, 7:256
Merocyanines, 16:285
Merox process
  petroleum sweetening, 15:51

Meroxyl 2:620
Merril-Crowe process, S:546
Mersalyl, 7:255, 256
Mertenylcarboxylic acid, 6:630
Merthiolate, 2:620; 13:236, 241. (See also *Thimerosal.*)
Mesaconic acid, 1:253; 5:527
Mescal, 16:646
Mescaline, 1:759, 768; 16:199, 642, 646
  in chemical warfare, 4:878
Mesidine, 2:414
Mesitol, 10:84
Mesitylene, 12:115, 137; 15:861; 16:300. See *1,3,5-Trimethylbenzene.*
  complex with TCNE, 6:626
  NMR of, 14:60
  reaction with antimony trichloride, 2:575
Mesityl oxide, 1:620; 12:105, 113
  in prepn of methyl isobutyl ketone, 12:134, 136
  in prepn of 4-methyl-2-pentanol, 1:565
  thorium separation, 20:252
Mesodiaminopimelic acid, 14:671
Mesophilic baccili
  in canned meats, 13:176
Mesorcinol, 11:481
Mesoxalylurea, 21:113
Mestranol, 6:65, 87; 15:920
Mesyl cellulose, 9:307
Met-L-X Powder, S:367
Metaarsenic acid, 2:722
Metaboric acid
  esters of, 3:663
Metacresol purple, 11:552
Metal alkoxides. See *Alkoxides, metal.*
Metal-bonding adhesives, 1:396
Metal buffers, 6:14
Metal carbides
  as hazardous chemicals, 9:292
Metal-casting processes
  use of polyethylene glycols, 10:657
Metal cleaning. See *Metal surface treatment.*
Metaldehyde, 15:910
  in matches, 13:163
Metal-finishing industry
  treatment of wastes by dialysis, 7:16
Metal-graphite products, 4:159
Metal hydrides. See *Hydrides.*
Metalizable dyes, 2:882
Metallic browns, 15:526
Metallic coatings, 13:**249**
  corrosion-resistant, 6:304

Metallic soaps, 7:272. (See also *Driers and metallic soaps.*)

Metallurgy
  graphite used in, 4:230, 332
  vacuum, 21:150

Metal processing
  use of abherents in, 1:3

Metal products
  coatings for, 5:713, 715

Metal purification
  use of electromigration in, S:290

Metals
  stain solvent, 18:660
  filler for welding, 22:249
  in packing, 14:453
  in water, 21:697
  zone refining of high-melting, 22:696
  zone refining of low-melting, 22:696
  electrochemical corrosion of, 6:290

Metal surface treatments, 6:869; 13:**284, 285**
  case hardening, 13:304-315
  chemical and electrochemical conversion treatments, 13:292-303
  cleaning, pickling, and related processes, 13:284-292

Metal treatments, 13:**315**

Metalworking lubrication, 12:598

Metanephrine, 8:233

Metanil Yellow, 7:482; 5:872

Metanil Yellow Ext Conc, 7:525

Metanilic acid, 2:213, 214, 415, 424; 7:479; 15:211; 19:299

Metaphenylene diisocyanate, 12:46, 47

Metaphosphates, 15:232

Metaphosphoric acid
  prepn of monofluorophosphoric acid, 9:639

Metaplumbic acid. See *Lead compounds.*

Metasilicic acid. See *Silicon compounds.*

Metatitanic acid, 20:391

Metavanadates, 21:172

Metawolframates, 22:352

Metazirconic acid, 22:630

Meteloidine, 1:766

Meteorites
  tritium in, 6:913

Methabarbital, 3:61, 70

Methacholine chloride, 5:410. (See also *Acetyl β-methylcholine chloride.*)

Methacrolein. See *Crotonaldehyde.*

Methacrylamide, 13:342

Methacrylate
  spectral transmittance curve, 9:253

Methacrylic acid, 6:694
  from isobutylene, 3:842
  phys prop, 1:229
  in synthetic oil prepn, 7:415

Methacrylic compounds, 13:**331**

Methacrylic-divinylbenzene linkage, 11:875

Methacrylic-ethylene dimethacrylate linkage, 11:875

Methacrylonitrile, 6:673
  use of triethylenetetramine in vulcanization of polymers of, 7:36

(3-Methacryloxypropyl)trimethoxysilane, 18:232

Methacycline, 20:5, 8, 18
  epimers, 20:11

Methadone, 2:385

DL-Methadone hydrochloride, 2:149, 150

Methaldehyde

Methallylacetone, 12:105

Methallyl alcohol, 1:257

Methallyl chloride, 10:728
  from isobutylene, 3:841, 842

Methallyl chloroformate, 4:386

Methallyl ether, 8:492

Metham, 15:913
  thiocarbamate herbicide, 22:194

Methanal. See *Formaldehyde.*

Methanamide. See *Formamide.*

Methanation
  of synthetic pipeline gas, S:205

Methane, 8:410, 504, 539; 9:402; 13:264
  from acetaldehyde decomposition, 1:80
  carbon monoxide in synthesis of, 4:434
  from catalytic reforming, 15:33
  chlorine reactions with, 5:2, 88, 105, 115, 123, 132
  chloroform from, 5:123
  coal-tar gas, 19:656
  as component of natural gas, 10:448
  cryogenic prop, 6:472, 475
  from dealkylation of toluene, 20:557
  carbon tetrachloride manuf from, 5:132
  from gasification of solid fuels, 10:356
  drying by alumina, 2:52
  hydrogen sulfide reaction with, 4:380
  in makeup gas in ammonia synthesis, 2:273
  in manuf of carbon disulfide, 4:376
  manuf of hydrogen cyanide, 6:577
  methyl chloride from, 5:105
  methylene chloride from, 5:114
  prod in MHD power plants, S:241
  oxychlorination of, S:170
  oxidation of, 11:224

phosgene reaction with, *5*:107

photochlorination, *15*:337

prepn from phenyl isocyanate, *12*:53

as process gas, *7*:128

produced by hydrogasification, *10*:387

reaction with fluorine, *9*:509

reduction of zinc oxide, *22*:571

in removing carbon monoxide from ammonia synthesis, *2*:281

reduction of zinc oxide, *22*:517

as refrigerant, *10*:859; *17*:309

separation of isotopes, *7*:138

from sludge gas, *22*:111

sulfur dioxide reaction with, *4*:381

tetrachloroethylene from, *5*:200

in tobacco smoke, *20*:516

Methanearsonic acid

as a herbicide, *22*:214

Methanedithiol

from carbon disulfide, *4*:373

Methanesulfonic acid, *19*:311, 314, 316

Methanesulfonic anhydride, *10*:154

Methanesulfonyl chloride, *12*:98; *19*:258

from carbon disulfide, *4*:373

Methanethiol, *19*:321, 369

from carbon disulfide, *4*:373

from methyl chloride, *5*:104

in prepn of methionine, *1*:271

in prepn of methylpropionic aldehyde, *2*:165

reaction with allyl alcohol, *1*:919

Methanoic acid. See *Formic acid.*

Methanol, *12*:160; *13*:**370**

Methanol, *2*:412, 420; *6*:439, 456; *13*:370

in amino resin coating manuf, *2*:254

ammonolysis of, *2*:345

as antifreeze, *2*:541

carbon dioxide from, *4*:437

refractory, *17*:231, 233

carbon monoxide in prod of, *4*:343

chloric acid reaction with, *5*:41

in chlorine dioxide manuf, *5*:41

as denaturant, *8* :451, 452

in esterification of fatty acids, *8*:818

fluorination of, *9*:509

in fuel cells, *3*:151

hydrogen chloride reaction with, *5*:105, 106

kinetics of reaction with lactic acid, *8*:320

in manuf of formaldehyde, *10*:87

in manuf of melamine, *2*:366

methyl chloride from, *5*:105, 106, 107

prepn from formaldehyde, *10*:83

in prod of acetic acid, *8*:397

reaction with crotonaldehyde, *6*:450, 451

solvent recovery, *18*:562

stain solvent, *18*:659

use of hydrogen in manuf of, *11*:371

to yield 2-methoxy-2-phenylethanol, *8*:289

Methanol-borontrifluoride

in prepn of aniline, *1*:150

Methylomonas methanica

Methanomonas methanooxidans

Methanthiol

uses, *20*:214

Methapyrilene, *2*:151

Methazonic acid, *13*:864

Methemoglobinemia, *S*:320

Methenamine, *10*:96

treatment of phosgene poisoning, *S*:679

as veterinary drug, *21*:246

Methenamine mandelate, *2*:640

Methenyl chloride. See *Chloroform.*

Methicillin, *3*:15, 16; *14*:682, 698

Methimazole

antithyroid agent, *20*:270

Methine and polymethine dyes, *7*:471, 473

Methine groups

in cyanine dyes, *6*:606

Methionine-1-CH$_3$-^{14}C, *20*:20

Methionine, *1*:159; *2*:157; *5*:404, 405; *8*:233; *14*:668

Methionine

and wool prod, *22*:400

manuf from α-acetobutyrolacetone, *1*:159

prepn from acrolein, *1*:271

prepn from methacrolein, *1*:271

sulfur in, *19*:369

synthesis, *20*:214

in wine, *22*:316

DL-Methionine, *10*:674

DL-Methionine

synthesis of, *2*:165

L-Methionine, *19*:38

Methionine sulfoxide, *19*:331

Methisazone

antiviral agent, *21*:454

Methohexital, *3*:61, 70

Methohexital sodium, *2*:403

Methone, *6*:456

Methopromazine, *2*:150

Methotrelate

folic acid antagonist, *21*:490

Methotrexate

cancer inhibiting effects, *S*:85

Methotrimeprazine, *2*:388

Methoxyacetone, *12*:105

p-Methoxyacetophenone, *1*:170; *15*:163

4'-Methoxyacetophenone
  flavor and perfume materials, 9:358
o-Methoxyaniline, 2:422
1-Methoxyanthraquinone
  prepn, 2:479
3-Methoxy-8-azaestrone, 18:872
p-Methoxybenzaldehyde, 15:163
  flavor and perfume materials, 9:358
Methoxybenzene, 15:148, 171. (See also Anisole.)
3-Methoxy-1,2-benzenediol, 16:191
p-Methoxybenzoic acid. See Anisic acid.
  phys prop, 1:238
Methoxybenzyl acrylate
  phys prop, 1:288
2-Methoxybibenzoylmethane, 12:148
2-Methoxy-4,6-bis(alkylamino)-s-triazines, 22:183
  analysis of, 22:187
2-Methoxy-4,6-bis(amino)-s-triazines, 22:186
2-Methoxy-4,6-bis(isopropylamino)-s-triazine,
  22:185
  from butadiene, 3:790
1-Methoxy-1,3-butadiene, 21:415
  from butadiene, 3:790
3-Methoxybutanol, 6:456
Methoxychlor
  in water, 21:701
1-Methoxy-3-chloro-3-methyl-4-pentene, 12:68
5-Methoxy-1-chloro-2-pentene
5-Methoxy-3-chloro-1-pentene
7-Methoxycoumarim, 6:432
3-Methoxycrotonate, 19:487
β-Methoxycrotonic ester, 1:156
2-Methoxy-6,9-dichloroacridine, 2:149, 153
6-Methoxy-8-(3-diethylaminopropylamino)
  quinoline, 2:151
1-Methoxy-3,5-dimethyladamantane, S:12
Methoxydimethylsufonium iodide, 19:326
3-Methoxyestrone, 18:853
2-Methoxyethyl acrylate
  phys prop, 1:288
Methoxyethylmercuric acetate, 13:245
2-Methoxyethyl palmitate
  as cellulosic plasticizer, 4:657
2-Methoxyethyl silicate, 18:219
Methoxyflurane, 2:401
2-Methoxyhydroquinone, 16:200
3-Methoxy-4-hydroxymandelic acid, 8:233
Methoxy magnesium chloride
  methyl chloride from, 5:107
2-Methoxy-5-methylbenzene, 2:423. See Cresidine.
1-Methoxy-3-methyl-5-chloro-3-pentene, 12:68
Methoxymethyl methyl ketone, 12:105

Methoxymethyl methyl ketone; methoxyacetone, 12:105
2-Methoxymethyl-5-nitrofuran
  as veterinary drug, 21:246
3-Methoxy-6-oxaestrone, 18:872
2-Methoxynaphthalene, 15:169
4-Methoxypentafluoropyridine, 9:799
4-Methoxyperchlorylbenzene, 9:604
o-Methoxyphenol, 21:184
p-Methoxyphenol, 15:171
2-Methoxy-2-phenylethanol, 8:289
p-Methoxyphenyltitanium   triisopropoxide,
  20:433
1-Methoxy-2-propanone, 12:105
1-Methoxy-4-propenylbenzene, 15:170
2-Methoxy-4-propenylphenyl, 15:170
3-Methoxypropionitrile, 6:637
Methoxysilane, 18:190
5-Methoxy-2-tetralone
  steroid synthesis, 18:858
3-Methoxy-6-thioestrone, 18:872
11-Methoxy-2,5-undecanedione, 12:158
p-Methoxytoluene, 15:169
7-Methoxy-1,2,3,4-tetrahydro-N-tosylquinolone,
  18:871
Methsuximide
  as anticonvulsant, 11:523
Methychlothiazide
  as diuretic, 7:263, 264
Methyl N(3-bromo-1-adamantyl)urethan, S:10
O-Methyl   O-(4-(tert-butyl-2-chlorophen-
  yl)methylphosphoramidate, 14:538
Methyl Cotton Blue, 20:718
Methyl Green
  N alkylation, 20:698
  substituent effects on color, 20:690
Methyl Violet, 7:496; 11:552
  substituent effects on color, 20:690
Methyl Violet B, 20:674
  manuf, 20:711
N-Methylacetamide, 9:693
Methyl acetate, 2:368
  from methyl chloride, 5:104
  prepn from methyl ether, 8:476
  prop, 8:378
  prod and sales, 8:376
  saponification rates, 8:368
  solvent for cellulose triacetate in fiber
    manuf, 1:115
p-Methylacetophenone, 1:170
Methyl p-acetoxybenzoate
  polyester from, 16:168
Methyl 2-acetoxypropionate, 8:371

Methyl α-acetoxyvinyl ketone, *12*:111
1-Methyl-2-acetylcyclopentane, *10*:157
Methylacetylene, *S*:**547**
Methyl acetyloleanolate
    prepn, *8*:353
1-Methyl-2-acetylvinyl acetate, *12*:111
Methyl acid sulfate. (See also *Methyl hydrogen sulfate.)*
b-Methylacrolein. See *Crotonaldehyde.*
Methyl acrylate, *8*:371; *12*:184
    alcoholysis, *20*:488
    phys prop, *1*:285, 287, 301
    polyester reactive diluent, *20*:822
    polymerization, *1*:303
    prepn, *1*:297
    in prepn of acrylamide, *1*:281
    toxicity, *1*:303
transβ-Methylacrylic acid. See *Crotonic acid.*
1-Methyladamantane, *S*:2
2-Methyladamantane, *S*:2
α-Methyl-1-adamantanemethylamine hydrochloride, *21*:956
3-Methyl-1-adamantoate, *S*:13
Methyl adipate
    prepn, *8*:350
Methylal, *1*:107; *2*:414; *10*:96
    carbon monoxide reaction with, *4*:435
    hydrolysis, *13*:391
Methyl alcohol. See *Methanol.*
    in refrigeration, *17*:310
Methyl alkyl silicone fluids, *18*:241, 250
1-Methylallyl acrylate
    phys prop, *1*:288
Methylallyl iodide
    steroid synthesis, *18*:859
o-α-Methylallylphenol, *15*:168
3-Methylalpentane, *3*:875. See *αEthylbutyraldehyde.*
3-Methylalpentanol. See *Acetobutyraldol.*
Methylaluminum dichloride, *18*:233
Methyl-*p*-amidinophenylsulfone, *19*:276
Methylamine, *6*:648
    in aminolysis, *2*:369
    from hydrogen cyanide, *6*:574
    manuf of Tetryl, *8*:621
    nylon degradation, *16*:10
    surfactants, *19*:521
    titanium tetrachloride reaction, *20*:473
Methylamine hydrochloride
    in prepn of lobelanine, *1*:795
Methylamines, *2*:120
    manuf, *2*:122
    products from, *2*:121
    uses, *2*:122

N-Methylaminoacetal, *20*:270
p-*N*-Methylaminoacetophenone
    polyene cleavage, *16*:138
(2-Methyl-4-amino-5-(bromomethyl)pyrimidine hydrobromide), *20*:178
2-Methyl-4-amino-5-(formamidomethylpyrimidine), *20*:180
α-(Methylaminoethyl)benzyl alcohol
    as cardiovascular agent, *4*:517
1-Methyl-2-aminoethyl *tert*-butyl peroxide
    from propylenimine, *11*:534
2-Methyl-4-amino-1-naphthol
    menadione synthesis, *21*:593
p-Methylaminophenol, *2*:221; *15*:371
    in hair coloring, *10*:795
p-(Methylamino)phenol sulfate, *15*:372
Methyl *n*-amyl ketone, *12*:104, 135
Methylammonium methyldithio carbamate, *10*:224
3-Methyl-2-amyl-2-cyclopentan-1-, *12*:161
3-Methyl-2-amyl-2-cyclopenten-1-one. See *Dihydrojasmone.*
Methyl amyl sulfoxide, *19*:326
o-Methylaniline. See *Toluidine.*
Methylaniline
    prepn from isocyanates, *12*:53
N-Methylaniline, *2*:368, 412, 420
    condensation with dialkylchloroalkylamines, *2*:143
5-Methyl-*o*-anisidine, *2*:423. See **Cresidine.**
Methyl-*p*-anisyl ketone, See *p-Methoxyacetophenone.*
Methyl anthranilate, *9*:354, 355; *14*:729, 736
    in carbonated beverages, *4*:344
    prod and sales, *8*:377
    as synthetic flavor and perfume materials, *9*:358
2-Methylanthraquinone, *2*:438
    dyes, *19*:434
Methylarsenic tetrachloride, *2*:728
Methylarsine, *6*:651
3,3'-(Methylarsylene) dipropionitrile, *6*:651
4-Methylbenzanthrone, *2*:500
Methylbenzenes
    nuclear chlorination of, *22*:484
5-Methylbenzene-1,3-diol, *19*:164
Methylbenzethonium chloride, *2*:633
α-Methylbenzl acetate
    prod and sales of, *8*:377
Methyl benzoate, *3*:432
    from methyl chloride, *5*:104 *18*:845
    mass spectrometry, *13*:92
    prepn, *8*:350

2-Methylbenzoazole, 2:215
  in prepn of 2-amino-5-nitrophenol, 2:217
α-Methylbenzoin
  photoinitiator, 16:518
Methyl benzoquate
  anticoccidial, 21:248
2-Methyl-1,4-benzoquinone
  menadione synthesis, 21:593
3-Methyl-2-benzothiazolinone hydrazone,
  20:196
o-Methylbenzyl alcohol, 10:731
Methylbenzylaniline, 20:701
Methyl 4,6-benzylidene-α-D-glycoside, 4:144
Methyl benzyl ketone, 15:214
Methylbismuthine
  bismuthine from, 3:537
Methylbismuth oxide, 3:544
Methyl borate. (See also Boric acid esters.)
  from seawater, 14:156
Methyl bromide, 3:772
  in fumigating nuts, 14:139
  nematocide, 15:913
1-Methyl-3-bromomethylpiperidine, 16:670
2-Methyl-1,3-butadiene. See Isoprene.
3-Methyl-1,2-butadiene, 1:624
2-Methylbutanal structure, 1:639
2-Methylbutane
  in hydroisomerization, 11:446
2-Methylbutane
  in commercial hexanes, 11:1
3-Methyl-1,2,3-butanetriol, 1:631
2-Methyl-1-butanol, 2:374
2-Methyl-2-butanol, 2:375
3-Methyl-1-butanol, 2:375
  in prepn of isoamyl ether, 8:493
3-Methyl-2-butanol, 2:375. (See also Isoamyl
  alcohol.)
2-Methylbutanol
  entraining agent, 22:472
3-Methylbutanol
  from isobutylene, 3:840
2-Methyl-3-butanone. See Methyl isopropyl ketone.
3-Methyl-2-butenal, 1:624
2-Methyl-1-butene
  in alkylation of phenol, 1:894
  from neopentyl alcohol, 4:395
2-Methyl-2-butene
  as feed in alkylation of phenol, 1:894
3-Methyl-1-butene, 2:413; 19:286
  copolymers, S:787
2-Methyl-1-buten-3-yne, 1:622
Methylbutenes, S:642

2-Methyl-2-butenoic acid, 6:694
Methyl butenol, 1:633
2-Methyl-3-buten-2-ol
  in manuf of isoprene, 12:78
3-Methyl-1-buten-3-ol
  phys prop, 1:630
2-Methyl-1-buten-3-one, 12:105, 128, 129
5-(Methyl-1-butenyl)-5-ethylbarbituric acid,
  3:61, 70
2-Methylbutyl acrylate
  phys prop, 1:288
3-Methylbutyl acrylate
  phys prop, 1:288
Methyl-t-butyl ether
  phys prop, 8:471
1-Methylbutylidenecyclohexylamine, 6:640
2-Methyl-4-tert-butylphenol, 6:436
2-Methyl-6-tert-butylphenol, 6:436
3-Methyl-6-tert-butylphenol, 1:909; 6:436
4-Methyl-2-tert-butylphenol, 1:910; 6:436, 442
Methyl n-butyl ether
Methyl sec-butyl ketone, 12:104
  phys prop, 8:471
4-Methyl-2,6-di-tert-butylphenol, 1:910;
  6:436, 442; 15:801
Methyl n-butyl ketone, 12:104, 126
Methyl-tert-butyl ketone, 6:670; 12:104
Methylbutynol, 1:208, 621
  phys prop, 1:621
  reactions, 1:622
  uses, 1:623
3-Methylbut-1-yn-3-ol tetrahydropyranyl ether, 18:849
2-Methyl-3-butyn-2-ol, 1:621; 6:644
2Methyl-3-butyn-2-ol, 12:77
  in manuf of isoprene, 12:120
α-Methylbutyraldehyde
  structure, 1:639
Methyl butyrate
  properties, 8:381
p-Methylbutyrophenone, 3:879
Methyl carbamate
  in soil, 18:527
4-Methylcarbostyril, 1:158. See 2-Hydroxy-4-
  methylquinoline.
(N-Methylcarbostyryl) hydrazone, 2:875
3-Methyl-2-(2-carboxethyl)-2-cyclohexenone,
  21:564
Methylcellulose, 4:648. 6:405; 17:401
  as cathartic, 4:592
  as food additive, 10:7
  as gum, 10:741
  suspending agent, 21:375

tablet binder, *15*:121
tablet coating, *15*:122
Methylcellulose film, *9*:232
properties, *9*:222-226
Methyl centralite, *8*:681
β-Methylchalcone, *12*:113
Methyl chloride, *5*:100-111; *9*:400, 871
methanol from, *13*:390
methylene chloride from, *5*:114
prepn of acetyl chloride, *1*:140
as refrigerant, *10*:859
thermometers, *19*:782
3-Methyl-3-chlorobutene-1, *1*:631
4-Methyl-4-chloro-3-chloromethyltetrahydropyran,
12:69
Methyl chloroformate, *4*:386
Methyl chloroformylacetate, *12*:850
Methylchlorosilanes, *18*:236, 253, 261
Methyl 2-chloro-9-hydroxyfluorene-9-carbox-
ylate, *15*:682
2-Methyl-4-chloro-1-naphthol
menadione synthesis, *21*:593
3-Methyl-4-chlorophenol, *6*:434
2-Methyl-4-chlorophenoxyacetic acid, *5*:337;
6:441; *19*:672
methyl silicones from, *18*:232
α-Methylcholest-7-en-3-ol, *18*:838
β-Methylcholine, *5*:411
α-Methylcinnamaldehyde
flavor and perfume materials, *9*:358
Methyl cinnamate, *5*:520
prod and sales of, *8*:377
α-Methylcinnamyl alcohol, *5*:521. See *1-Meth-
yl-3-phenyl-2-propen-l-ol* .
4-Methylcoumarin, *6*:428, 429
Methyl coumarin 3-carboxylate, *8*:353
Methyl crotonate, *6*:469
β-Methylcrotonic acid, *6*:694
3-Methylcrotonic acid, *6*:614
Methyl cyanoacetate, *12*:850
Methylcyclohexane, *9*:722
dehydrogenation to toluene, *20*:535
Methyl cyclohexaneacetate
prepn, *8*:350
2-Methylcyclohexane-1,3-dione, *18*:861, 863
Methylcyclohexanol, *6*:685
2-Methylcyclohexanone, *12*:106
Methyl 2-cyclohexanyl propionate, *16*:556
Methylcyclohexene, *10*:143
2-Methylcyclohexyl acrylate
phys prop, *1*:289
3-Methylcyclohexyl acrylate
phys prop, *1*:289

4-Methylcyclohexyl acrylate
phys prop, *1*:289
Methylcyclohexylene diisocyanate
polyurethans, *21*:103
Methyl cyclohexyl ketone, *12*:108
3-Methylcyclopentadecanone, *14*:727
Methylcyclopentadienes, *6*:688
Methylcyclopentadienylmanganese tricarbonyl, *6*:699; *12*:293; *13*:43
as antiknock agent in gasoline, *10*:476;
13:52
Methylcyclopentadienyl tricarbonyl, *13*:5
Methylcyclopentane, *8*:504
in commercial hexanes, *11*:1
prepn from cyclohexane, *6*:678
in prepn of cyclohexane, *6*:680
in hydrocracking, *11*:431
2-Methylcyclopentanecarboxaldehyde, *1*:640
2-Methylcyclopentane-1,3-dione, *18*:861, 863
3-Methyl-1,2-cyclopentanedione, *12*:140
Methylcyclopentene
in hydrocracking, *11*:431
3'-Methyl-1,2-cyclopentenophenanthrene, *18*:831
3-Methylcyclopentylsulfamate, *19*:600
Methyl cyclopropyl ketone, *12*:105
sulfoxide, *19*:331
4-Methyl-1-decene
prepn from allyl ether, *8*:494
in chemical warfare, *4*:873
Methyl *n*-decyl ketone, *12*:105
Methyl 6-deoxy-6-iodo-α-D-glucopyranoside,
19:231
N-Methyl-2-deoxystreptamine, *19*:44
2-Methyl-4,6,-di-*tert*-butylphenol, *6*:436
N-Methyl-2,2'-dichlorodiethylamine
3-Methyl-4,6-di-*tert*-butylphenol, *1*:910; *6*:436
Methyl dichloromethyl thioether, *15*:162
Methyldichlorosilane, *12*:69
Methyldiethanolamine
phys prop, *1*:811
uses, *1*:823
2-Methyl-3-β,γ-dihydrophytyl-5,6,7,8-tetrahydro-
-1,4-naphthohydroquinone, *21*:588
N-Methyldinitroaniline, *8*:621
2-Methyl-1,4-dimethoxynaphthalene
menadione synthesis, *21*:593
2-Methyl-4,6-dinitrophenol
Methyl 2,4-dinitrophenyl acetate, *8*:350
4-Methyl-1,3-dioxane, *10*:660, 663
2-Methyl-1,5-dioxaspiro[5.5]undecane, *10*:663
2-Methyl-1,3-dioxolane, *1*:107; *10*:646
4-Methyl-1,3-dioxolane, *15*:190

N-Methyl-2,2-diphenylacetamid
  decomposition of diphenamid, *22*:198
2-Methyldiphenylamine, *7*:46
3-Methyldiphenylamine, *7*:46
4-Methyldiphenylamine, *7*:46
Methyldisilane, *18*:179
4-Methyl-2,6-ditertiarybutylphenol
  poly-1-butene antioxidant, *S*:780
Methyl ecgonine sulfate, *1*:785
Methylendomethylenetetrahydrophthalic anhydride
  as liquid epoxy resin hardener, *8*:309
Methylene Blue, *2*:866; *7*:533
  as antiseptic, *2*:638
  as denaturant, *8*:450
  as inhibitor, *16*:519
  perchlorates detected with, *5*:79
  structural formula, *5*:745
  used in hair coloring, *10*:793
Methylene Green, *2*:867
Methyleneaniline
  reaction with phenyl isocyanate, *12*:51
6-Methylenebicyclo(2.2.1)heptene-2, *6*:693
4,4'-Methylenebis(*N*, *N*-dimethylaniline)
  (methane base), *7*:479
N,N'-Methylenebisacrylamide, *1*:275, 291
  prepn from acrylonitrile, *1*:341
N, N'-Methylenebisacrylamide, *16*:519
2,2'-Methylenebis(4-chlorophenol), *14*:543
  as fungicide, *10*:231
Methylenebisamide
  in amino resin polymerization, *2*:231, 239,
    240, 242, 243
2, 2'-Methylenebis(1,2-benzisothiazol)-3-
  (2*H*)-one. *19*:597
2,2'-Methylenebis(6-*tert*-butyl-*p*-cresol),- *2*:600
  as leather fungicide, *10*:233
2,2'-Methylenebis(6-*tert*-butyl-4-ethylphenol) -
  2:600
4,4'-Methylenebis(2,6-di-*tert*-butylphenol), *1*:913
2,2'-Methylenebis(4-ethyl-6-*tert*-butylphenol), -
  1:915
3,3'-Methylenebis(4-hydroxycoumarin), *15*:919
2,2'-Methylenebis[6-(2-methylcyclohexyl)*p*-creso-
  l], *2*:600
2,2'-Methylenebis(4-methyl-6-*tert*-butylphenol)
  - *1*:914; *6*:442
Methylenebis(4-phenyl isocyanate), *4*:533
  castor oil reaction with, *4*:533
  in spandex, *18*:625
4,4'-Methylenebis(phenylisocyanate) *2*:414
N,N'-Methylenebisstearamide, *22*:142

2,2'-Methylenebis(3,4,6-trichlorophenol), *6*:355;
  14:549
1,1'-Methylenebisurea, *2*:229, 231
Methylene bromide, *3*:774, 778
  as aresol propellant, *1*:472
Methylene chloride, *5*:111-119; *8*:350, 407;
  9:871; *10*:142, 164
  in manuf of cellulose acetate flake, *1*:116
  in manuf of cellulose triacetate flake, *1*:119
  paint remover, *14*:486
  as polyester solvent, *16*:169
  as refrigerant, *10*:859
  in refrigeration, *17*:310
  viscosity of polycarbonates, *16*:107
  toxicity of, *5*:93, 117
Methylene chlorobromide, *3*:774
Methylenedianiline, *2*:418; *8*:311
p,p'-Methylenedianiline, *2*:414
Methylene dichloride. See *Methylene chloride.*
Methylene diethyl ether. See *Ethylal.*
2,2'-Methylenedifuran, *10*:246
Methylene dimethyl ether. See *Methylal.*
9,10-Methylenedioxy-1*H*,7*h*-6a-azabenzanthr- ene,
  10:936
  polycondensation, *16*:165
  polyester from, *16*:180
  polyesters, *20*:490
3,4-Methylenedioxybenzaldehyde, *15*:170
Methylenedioxybenzene, *15*:164
3,3'-(Methylenedioxy)dipropionitrile, *6*:644
3,4-Methylenedioxy-1-propenylbenzene, *15*:170
Methylenediphenylene diisocyanate, *21*:58
Methylenedi-*p*--phenylenediisocyanate, *21*:101
2,2'-Methylenedisaccharin, *19*:597
Methylenedisalicylic acid, *17*:721, 734
Methylene diurea, *4*:921
Methylene glycol, *10*:78
Methylene iodide, *9*:714; *11*:863
Methylenenorbornene, *6*:693
6-Methyleneoxytetracycline, *20*:8, 18
o-Methylenequinone
  phenolic resins, *15*:186
Methylenesuccinic acid. See *Itaconic acid.*
6-Methylenetetracyclines, *20*:18
Methyl *trans, trans, cis*-10-epoxy-7-ethyl-3,11-
  dimethyl-2,6-tridecadienoate, *1 8*:845
Methylergonovine, *1*:778, 791
Methylergonovine tartrate, *1*:789
Methyl esters
  prepn, *8*:350
Methylethanolamine
  phys prop, *1*:811

Methyl ether, 8:475
  from dimethyl sulfate, 19:487
  hydrogen chloride reaction with, 5:106
  phys prop, 8:471
  prepn, 8:476
Methyl ether of hydroquinone
  as polymerization inhibitor, 1:304
3-Methyl-2-ethylbenzothiazolium iodide,
  6:612
Methylethylcellulose, 4:650
Methylethylgloxal, 12:140
Methyl ethyl ketone, 6:670, 671, 693; 12:104,
  126, 133
  from butyl alcohol, 3:824, 829
  as paint remover, 14:490
  phenylenediamines, 15:218
  solvent refining, 15:95
  stain solvent, 18:661
Methyl ethyl ketone peroxides
  polyester initiators, 20:825
Methyl ethyl ketoxime, 6:645; 18:245
Methyl 2-ethyl-3-phenylpropyl sulfoxide,
  19:325
2-Methyl-2-ethylsuccinimide, 11:523; 19:146
Methyl ethynyl ketone, 12:105
1-Methyl-5-ethyl-5-phenyl barbituric acid,
  11:519
2-Methyl-5-ethylpyridine, 16:803
  from methyl vinyl ether, 21:416
  synthesis of nicotine, 21:513
-Methyl-3-all-trans-farnesylgeranylgeranyl-1,4--
  naphthoquinone, 21:591
Methyl fluoride, 9:509
Methyl fluoroacetate, 9:769
Methyl 6-fluoropicolinate, 9:798
N-Methylformanilide
  carbon monoxide from, 4:437
  diphosgene from, 4:388
  in manuf of dimethylformamide, 10:109
  Fischer's aldehyde from, 16:297
  in manuf of formamide, 10:105
  in manuf of formic acid, 10:101
N-Methylformanilide, 9:795
2-Methylfuran, 10:86
  in prepn of 4-oxopentanal, 10:238
  properties, 10:248
5-Methylfurfural, 12:160
(5-Methylfurfurylidene)methyl-n-propyl ketone, 12:160
N-Methylglucamine, 2:893; 19:231
N-Methylglucamine, 2:893
4-O-Methyl-α-D-glucoronic acid, 16:688
α-Methyl-D-glucoside, 6:922

Methyl-α-D-glucoside, 4:141, 142, 143, 144;
  7:411; 19:229
Methyl-α-D-glucopyranoside, 19:229
Methyl αD-glucopyranosidurano-6,3-lactone,
  19:229
N-Methyl-L-glucosamine, 19:38
N-Methyl glycine
  surfactants, 19:514
Methylglyoxal, 12:162
Methylguanidine
  prepn from guanidine, 10:736
α-Methyl-9βH-5,7-diene, 18:837
2-Methylheptadecanedioic acid, 16:140
Methyl n-heptadecyl ketone, 12:105
Methyl heptafluorobutyrate, 9:773
2-Methylheptafluoronaphthalene, 9:798
3-Methylheptamethylenediamine
  polypyromellitimide of, S:767
Methylheptenone, 19:825
3-Methyl-2,4-heptanedione, 12:149
6-Methyl-2,5-heptanedione, 12:158
  as synthetic flavor, 9:354
2-Methyl-2-hepten-6-one, 1:631
"Methyl heptine carbonate," 14:729
1-Methylheptyl acrylate
  phys prop, 1:288
  as fungicide, 10:226
Methyl heptyl ketone
  as synthetic flavor, 9:354
5-Methyl-2,3-hexanedione, 12:140
5-Methyl-2,4-hexanedione, 12:148
2-Methyl-3-hexanone, 12:105
  sterilant, 18:816
4-Methyl-1-hexene
Methyl n-heptyl ketone, 12:105
  polymerization of, 7:66
5-Methyl-1-hexene
  polymerization of, 7:66
5-Methyl-5-hexen-2-one, 12:105
2-Methyl-5-hexen-3-yn-2-ol, 7:683
Methylhexylacetaldehyde, 14:729
Methyl n-hexyl ketone, 12:104. See Hexyl
  methyl ketone.
8-Methyl-trans-hydrindanol-one-1, 21:564
Methyl hydrogen sulfate, 19:490, 494
Methylhydroquinone
  DL-δ-tocopherol prod, 21:580
Methyl-p-hydroxybenzoate, 16:170
  as food preservative, 10:12
  in shampoo, 10:774
2-Methyl-3-hydroxy-4,5-bis(hydroxymethyl)-
  pyridine, 16:807

4-Methyl-7-hydroxycoumarin. See *β-Methyl-umbelliferone.*

6-Methyl-7-hydroxycoumarin, *6*:432

4-Methyl-5-hydroxyethyl thiazole, *20*:174

4-Methyl-5-(2-hydroxyethyl)thiazole, *20*:178

O-Methylhydroxylamine
  synthesis of linuron, *22*:188

3-Methyl-5-(1-hydroxy-4-oxo-2,6,6-trimethyl-2-cyclohexen-1-yl)-*cis, trans*-2,4-pentadienoic acid, *15*:685

Methylhydroxypropylcellulose, *4*:650

4,4'-Methylidenebis(1-phenyl-3-methyl-5-pyrazolone), *16*:284

4-Methylimidazole, *19*:225

2-Methylimidazoline
  prepn, *7*:26

1-Methyl-2-imidazolethiol
  antithyroid, *20*:268
  antithyroid agent, *20*:270

3,3'-(Methylimino)dipropionitrile, *6*:648

Methylindanols
  in coal tar, *6*:437

2-Methylindole
  Astrazon dyes, *16*:296
  Fischer's base, *16*:294

3-Methylindole, *14*:726

2-Methyl-3-indoleacetic acid, *19*:225

Methyl iodide, *2*:420; *7*:617; *8*:472; *11*:862
  amine oxides from, *S*:32
  in Grignard reaction, *10*:724
  oxytetracycline synthesis, *20*:24
  in prepn of methyl ether, *8*:476
  and pyrogallol, *16*:191
  from sodium iodide, *19*:488
  steroid synthesis, *18*:859
  tetracycline quaternary salts, *20*:12

Methylionones, *14*:729; *19*:829

1-Methyl isatin-3-thiosemicarbazone
  antiviral agent, *21*:454

Methylisobutylglyoxal, *12*:140

Methyl isobutyl ketone, *8*:379; *10*:758; *12*:104, 134. See *Hexone.*
  as denaturant, *8*:450, 452
  in prepn of 4-methyl-2-pentanol, *1*:565
  stain solvent, *18*:661
  tantalum extraction, *19*:635
  tetracycline solvent, *20*:3
  in water analysis, *21*:697

Methyl isobutyrate, *8*:31, 371

Methyl isocyanate, *12*:46, 47

Methyl isohexyl ketone, *12*:104

Methyl isopropenyl ketone, *12*:105, 128, 129

Methylisopropylbenzene
  cresols from, *S*:273

Methylisopropylglyoxal, *12*:140

N-Methylisopropylideneimine, *10*:728

Methyl isopropyl ketone, *6*:670; *12*:104
  Fischer's base, *16*:294

3-Methyl-6-isopropylphenol. See *Thymol.*

1-Methylisoquinoline, *19*:326

Methyl isothiocyanate, *10*:225; *15*:913

3-Methyl-2-isoxazolin-5-one, *1*:158

Methylisoxazolone, *1*:158

Methylkasugaminide, *19*:45

Methylketene, *12*:88
  prepn, *12*:91

Methyl 2-keto-ʟ-gulonate, *2*:754

3-Methyl-6-ketotetrahydropyridazine, *19*:225

Methyllactamide, *12*:186

Methyl lactate, *12*:177, 184, 185
  prepn, *8*:352

Methyl linoleate, *6*:469

Methyllithium, *9*:794, 798; *12*:156, 552
  titanium trichloride reaction, *20*:442

Methylmagnesium bromide, *7*:48; *10*:730

Methylmagnesium chloride, *12*:292, 710

Methylmagnesium iodide, *16*:671
  isochloroprene reaction with, *5*:228

Methylmaleic anhydride, *12*:826

Methyl mercaptan. See *Methanethiol.*

2-Methylmercaptobenzothiazoles, *10*:917

3-Methylmercapto-1-propanol
  prepn from allyl alcohol, *1*:919

α-Methylmercuric dicyandiamide, *10*:222

Methylmercuric iodide, *13*:244

Methylmercuric 8-hydroxyquinolinate, *10*:222

Methyl mesityl ketone, *6*:640

Methyl methacrylate, *6*:579; *6*:828, 830, 831; *13*:333, 341
  alcoholysis, *20*:488
  as modifier in alkyd resins, *1*:866

Methyl-*m*-methoxybenzoate
  in watertight concrete, *22*:152
  tetracycline synthesis, *20*:23

Methyl methoxyacetate
  carbon monoxide in prepn of, *4*:435

2-Methyl-3-methylamino-1,4-napthoquinone, *21*:593

2-Methyl-2-methylthiopropionaldehyde *O*-methylcarbamoyl oxime, *15*:914

N-Methylmorphinan, *2*:382

N-Methylmorpholine, *13*:667

Methylnaphthalene
  coal-tar distillate, *19*:664
  phthalic anhydride from, *15*:449

1-Methylnaphthalene
  from kerosene, *19*:252
  polyester end-group analysis, *16*:179
2-Methylnaphthalene, *13*:678
  chromic acid oxidation, *21*:593
  menadione derivatives, *21*:597
Methylnaphthalenes, *13*:691
2-Methylnaphthaquinone, *13*:678
Methyl-1-naphthyl ketone, *12*:107
2-Methyl-1,4-naphthohydroquinone, *21*:588, 592
  menadione derivatives, *21*:595
2-Methyl-1,4-naphthohydroquinone diacetate, *21*:592
2-Methyl-1,4-naphthoquinone
  menadione deriv, *21*:586, 592
  phytylation of, *21*:588
  vitamins k_1 and k_2, *21*:585
  vitamin k_3, *21*:585
2-Methyl-1,4-naphthoquinone-3-acetic acid, *21*:588
Methyl β-naphthyl ketone, *14*:737
Methyl nicotinamide, *2*:190
N-Methyl nicotinamide, *21*:536
Methyl nicotinate
  prepn, *8*:353
Methylnitramine, *8*:620
  prepn, *8*:620
Methyl nitrate
  from ethyl ether, *8*:479
  from isopropyl ether, *8*:488
  as propellent, *8*:660
2-Methyl-1-nitroanthraquinone, *2*:438
5-Methyl-5-nitro-1,3-dioxane, *13*:827
2-Methyl-4-nitro-3-hexanol, *3*:870
2-Methyl-5-nitroimidazole-1-ethanol, *20*:86
2-Methyl-1-nitro-2-propanol
  from aliphatic ketones, *13*:831
2-Methyl-2-nitropropyl acrylate
  phys prop, *1*:289
3-Methyl-1-nonen-3-ol
  phys prop, *1*:630
Methylnonylacetaldehyde, *14*:729
Methyl nonyl ketone
  as synthetic flavor, *9*:354
Methyl *n*-nonyl ketone, *12*:105
Methylnorbornanes, *S*:4
Methylococcus capsulatus
Methyl octenyl ketone, *12*:129
Methyl octine carbonate, *14*:729
Methyloctylacetaldehyde, *14*:729
Methyl *n*-octyl ketone, *12*:105
N-Methylolacrylamide, *2*:248

Methyl oleanolate
  prepn, *8*:353
Methyl oleate
  epoxidation of, *8*:254
  prepn, *8*:353
Methylolmelamine, *9*:307
Methylolmethylenebisamide
  in amino resin polymerization *2*:231
p-Methylolphenol, *15*:181, 188
Methylolstearamide, *22*:142
Methylolurea, *15*:793
Methyl orange, *9*:897; *11*:552
Methyl orthoformate, *21*:415
Methyl orthosilicates, *18*:253
5-Methyl-2,4-oxazolidinedione, *12*:175
2-Methyloxetane,
  prepn of 1,3-butylene glycol, *10*:660
  synthesis, *10*:663
dl-Methyl-3-oxoetia-4,9(11),16-trienate, *18*:858
Methyl 3-oxoglutaramate, *20*:24
Methyl 5-oxo-6-heptenoate, *18*:861
Methyl palmitate
  prepn of, *8*:353
Methylparafynol
  as hypnotic, *11*:511
Methyl parathion, *15*:322, 328
4-Methyl-2,3-pentadione, *12*:140
Methyl pentafluorophenyl sulfide, *9*:793
2-Methylpentan-1,3-diol
1-Methyl-3,3-pentamethylenediaziridine, *19*:487
  titanium chelates, *20*:464
3-Methylpentan-1,5-diol
  titanium chelates with, *20*:464
2-Methylpentan-2,4-diol
  titanium chelates with, *20*:464
2-Methylpentane, *8*:504
4-Methyl-2-pentanamide
  form, *2*:67
2-Methyl-2,4-pentanediol, *12*:138
  in hexanes, *11*:1
  in hydroisomerization, *11*:446
3-Methylpentane
  in hexanes, *11*:1
  in hydroisomerization, *11*:446
4-Methyl-2,3-pentanedione, *12*:140
2-Methylpentanoic acid, *8*:850
3-Methylpent-1-en-4-yn-3-ol
  in vitamin A synthesis, *21*:496
4-Methyl-2-pentanol, *1*:563; *8*:369; *9*:385
  as catalyst in prepn of fatty alcohols, *8*:822
  frothing agent, *9*:393
  phys prop, *1*:564

3-Methyl-2-pentanone, *12*:104
4-Methyl-2-pentanone. See *Methyl isobutyl ketone.*
4-Methyl-2-pentenal, *1*:639
2-Methyl-1-pentene
   in manuf of isoprene, *12*:73
2-Methyl-2-pentene, *12*:73
4-Methyl-1-pentene
   copolymers, *S*:787
   homopolymerization, *S*:791
   polymers of, *S*:789
   polymer transparency, *S*:798
   polymerization of, *7*:66
   polymers, *14*:309
3-Methyl-1-penten-3-ol
4-Methyl-2-pentene, *19*:286
4-Methyl-3-penten-2-one, *12*:105. See *Mesityl oxide.*
   phys prop, *1*:630
4-Methyl-2-penten-2-ol acetate, *12*:111
2-Methylpentyl acrylate
   phys prop, *1*:288
Methylpentynol, *1*:208
   as hypnotic, *11*:511
3-Methyl-1-pentyn-3-ol, *1*:621
   NMR of, *14*:68
   phys prop, *1*:622
4-Methyl-1-pentyn-3-ol
   phys prop, *1*:619
Methyl perchlorate, *5*:73
Methyl phenidylacetate, *16*:673
3-Methylphenol. See *m-Cresol.*
2-Methylphenol. See *o-Cresol.*
4-Methylphenol. See *p-Cresol.*
Methylphenols, *2*:626. (See also *Cresols.*)
Methyl phenylacetate, *15*:213
N-Methyl-4-phenyl-4-carbethoxypiperidine hydrochloride, *S*:13
Methylphenylene diisocyanate, *21*:57
Methyl phenyl ether. See *Anisole.*
   from methylation of phenol, *S*:272
Methyl phenylethyl glycidate
   as flavor, *9*:362
3-Methyl-5,5-phenylethylhydantoin
   as anticonvulsant, *11*:521
Methylphenylglyoxal, *12*:140
Methyl phenylglyoxylate, *10*:729
5-Methyl-5-phenyl-2-hexene, *3*:792
   from butadiene, *3*:792
1-Methyl-3-phenyl-2-propen-1-ol, *5*:521
1-Methyl-1-phenylhydrazine, *16*:294
1-Methyl-3-phenylpyrazolone, *7*:539

1-Methyl-2-phenylindole
   dyes from, *20*:726
1-Methyl-2-phenyl-3-indolcarboxaldehyde, *16*:296
Methyl phenyl ketene, *12*:88. (See also *Acetophenone.*)
3-Methyl-1-phenyl-3-phospholine oxide
   as catalyst, *12*:52
Methyl phenyl polysiloxane, *9*:898
3-Methyl-1-phenyl-2-pyrazolin-5-one, *1*:159
3-Methyl-1 phenyl-5-pyrazolone, *2*:871; *10*:911
Methyl phenyl silicone, *18*:227
Methyl phenyl silicone fluids, *18*:241
N-Methyl-2-phenylsuccinimide, *11*:522; *19*:146
Methyl 2-(phenylsulfinyl)ethyl sulfide, *19*:320
Methylphosphoric acid
   durable flame retardants, *S*:949
Methylphosphoric acid
   durable flame retardants, *S*:949
N-Methylphthalimide
   formation of, *5*:104
"Methyl phthalyl ethyl glycolate," *15*:757
2-Methyl-2-phytyl-2,3-dihydro-1,4-naphthoquinone, *21*:589
2-Methyl-3-phytyl-1,4-naphthoquinone
   vitamin $k_{1(20)}$, *21*:586
N-Methylpicramide
   prepn, *8*:620
N-Methylpiperazine, *16*:674
1-Methylpiperazine, *2*:143
2-Methylpiperazine, *19*:225
N-Methyl-n'-piperazino-3-chloropropane, *16*:667
Methyl pivalate, *13*:339
Methylpolysiloxane
   film, *18*:261
   from methyl chloride, *5*:104
Methylpotassium
6-Methylpretetramide, *20*:21
2-Methylpropanal. See *Isobutyraldehyde.*
Methylpropane, *3*:815, 833. See *Isobutane.*
Methylpropane, *3*:815, 833
2-Methyl 1,2-propanediol
   from isobutylene, *3*:842
2-Methyl-2-propanethiol, *20*:205
   uses, *20*:215
2-Methyl-1-propanol. See *Isobutyl alcohol.*
   from oxo process, *14*:373
2-Methyl-2-propanol. See *tert-Butyl alcohol.*
2-Methylpropene. See *Isobutylene.*
Methyl propenyl ketone, *12*:105

2-Methylpropen-2-oic acid, *13*:331
4-Methyl-2-propenyl-1,3-dioxane, *10*:663
Methyl propiolate
  sulfoxides, *19*:329
Methyl propionate
  properties, *8*:381
2-Methylpropionic acid, *3*:882
Methyl *n*-propyl ether
4-Methyl-2-propyl-1,3-dioxane, *10*:663
  phys prop, *8*:471
Methyl *n*-propyl ketone, *12*:104
O-Methylpsychotrine, *20*:75
2-Methylpyridine, *10*:86. (See also *Picolines*.)
6-Methyl-2,4(1*H*,3*H*)-pyridinedione, *12*:857
Methylpyridines
  amine oxides from, *S*:43
  in coal tar, *19*:657
N-Methyl-4-pyridone-5-carboxamide, *21*:522
Methyl-3-pyridyl ketone, *21*:526
β-Methylpyrrolidine, *1*:764
N-Methylpyrrolidone, *9*:693; *9*:777
  in butadiene separation, *3*:806
  coal-tar refining, *19*:669
N-Methyl-2-pyrrolidone, *10*:673; *16*:852
  sucrose solvent, *19*:221
  sulfoxides, *19*:331
  varnishes, *S*:768
Methyl pyruvate
  prepn, *8*:350
N-Methyl-4-quinoline, *1*:771
2-Methylquinoline, *6*:609, 642. See *Quinaldine*.
  prepn of nicotinic acid, *21*:515
Methyl red, *11*:552
Methyl salicylate, *7*:543; *9*:354, 355; *14*:195;
  17:720, 730
  as denaturant, *8*:451
  flavor and perfume materials, *9*:358
  phys prop, *1*:288
  polystyrene stabilizer, *19*:90
  prepn, *8*:350
  prod and sales of, *8*:377
  as synthetic flavor, *9*:362
Methyl silicones, *18*:221
  substitution in, *18*:227
Methylsilylating agents, *18*:260, 261
Methyl stearate
  prop, *8*:382
19-Methylsteroids, *6*:75
Methylsterols, *18*:838
Methylstibene, *2*:581
α-Methylstyrene, *19*:81, 115; *22*:500
  alkyl acrylate latexes, *19*:113
  styrene copolymer, *19*:91
  unsaturated polyester copolymer, *20*:822

Methylstyrenes
  polymers, *19*:115
Methylsulfinyl carbanion, *19*:326
2-(Methylsulfinyl)ethanol, *19*:318
2-(Methylsulfinyl)-1,3-indanedione, *19*:324
2-Methylsulfonyl-4,6-bis(isopropylamino)-*s*-tria-
  zine, *22*:185
Methyl sulfoxide, *19*:320
N-Methyltaurine, *10*:772; *19*:284
  in prepn of azo dyes, *2*:893
Methyl terephthalate
  polyesters, *15*:794
Methyltestosterone, *11*:101; *18*:875
N-Methyl-*N*,2,4,6-tetranitroaniline. See *Tetryl*.
2-Methyl-5,6,7,8-tetrahydro-1,4-naphthohydro-
  quinone, *21*:592
(Methylthio)benzothiazole, *20*:193
Methyl thiol. See *Methanethiol*.
2-Methylthio-4,6-bis(isopropylamino)-*s*-triazine-
  , *22*:185
2-Methylthiophene, *20*:224
3-Methylthiophene, *20*:224
2-(Methylthio)phenol, *19*:329
Methylthiopropionic aldehyde
  in synthesis of methionine, *2*:165
2-Methylthiopyrimidine, *6*:615
Methylthiosulfate, *19*:767
2-Methylthio-4,6-bis(alkylamino)-*s*-triazines,
  22:183
2-Methylthio-4,6-bis(alkylamino)-*s*-triazines,
  22:185
2-Methylthio-4,6-bis(alkylamino)-*s*-triazines,
  22:186
2-Methylthio-4,6-bis(amino)-*s*-triazines, *22*:186
4-Methylthio-3,5-xylenyl *N*-methyl-car-
  bamate, *15*:922
Methyltitanium trichloride, *20*:432
N-Methyltocopheramines, *21*:577
Methyl *p*-tolyl ketone, *1*:170
Methyltrichlorosilane,
  water-repellent agent, *22*:143
Methyltriethoxysilane, *18*:217
Methyl 2,4,6-triethylbenzoate
  prepn, *8*:353, 354
Methyl-3,3,3-trifluoro-*n*-propyl silicone,
  18:227
Methyl trifluoropropyl silicone fluid, *18*:241
Methyl-3,3,3-trifluoro-*n*-propyl siloxane trim-
  er, *18*:225
Methyl 2,4,6-triisopropylbenzoate
Methyl 2,4,6-trimethylbenzoate
  prepn, *8*:353, 354
2-Methyl-2-(4,8,12'-trimethyltrideca-3',7,11'-tri-
  enyl)-6-chromanol, *21*:574

2-Methyl-2-(4',8',12',-trimethyltridecyl)-6-chro-
manol, *21*:574
Methyltriphenylarsonium tetraiodonickelate,
13:759
Methyltris(trimethylsiloxy)-silane, *22*:144
N-Methyltryptamine, *1*:776
α-Methyltryptophan, *21*:522
β-Methylumbelliferone, *6*:432; *10*:925; *11*:480
as optical brightener, *3*:744
4-Methylumbelliferone, *11*:480
5-Methyluracil
as diuretic, *7*:253
6-Methyluracil
formation from acetoacetic ester and urea,
1:156
N-Methyl-δ-valerolactam
4-Methyl-2-vinyl-1,3-dioxane, *10*:663
4-Methyl-2-vinyl-1,3-dioxolane
Methyl vinyl ether
Diels-Alder addition, *21*:416
maleic anhydride, *7*:305
polymers of, *21*:418
prepn from acetylene, *1*:86
in prepn of acetaldehyde, *1*:86
reaction with propargyl alcohol, *1*:600
Methyl vinyl ether, *21*:413
Methyl vinyl ether-maleic anhydride
interpolymers, *21*:422
Methyl vinyl ether-maleic anhydride copolymer, *1*:205
Methyl vinyl ether-maleic anhydride interpolymers, *21*:422
Methyl vinyl ether-maleic anhydride polymer
in hair spray, *10*:782
Methyl vinyl ketone, *2*:413; *8*:408; *12*:105
from acetyl chloride, *1*:139
alkylating agent, *20*:699
in copolymerization of trioxane, *1*:99
prepn of biacetyl, *12*:144, 145
reaction with butadiene, *3*:785
steroid synthesis, *18*:858, 881
Methylvinylpyridine, *7*:301
2-Methyl-5-vinylpyridine, *16*:803
Methyl xanthate
decomposition of, *22*:421
Methyl xanthic acid, *22*:420
Methylxanthogen acetic acid, *22*:419
Methyl yellow, *11*:552
oxidation of, *21*:514
Methymycin, *12*:633, 635, 640
Methynolide, *12*:640
Methyprylon
as hypnotic, *11*:514

Methyridine, *14*:536
Metobromuron
urea herbicide, *22*:190
Metopirone, *7*:267
Metopon, *2*:380
Metrazole, *16*:647
Metre
definition, *S*:989
Metrication, *S*:1006
Metrioltrinitrate. (See also *Trimethylolethane
trinitrate.*)
Metrizoic acid, *17*:133
Metrologia, *S*:986
Metronidazole, *20*:70
amebicide, *20*:94
amebicides, *20*:86, 94
Metrulen, *6*:69
Mevalonic acid, *18*:833
Meyer hardness, *10*:811
Meyer reaction, *2*:725
Miamycin, *12*:633, 635
Mica, *5*:550
in drilling fluids, *7*:298
layer structure, *5*:542
muscovite as pigment, *14*:515; *15*:512
as paint extender, *14*:466
rubidium in, *17*:685
separation by flotation, *9*:395
use as abherent, *1*:11
Mica, barium disilicic, *13*:399
Mica insulation, *11*:794
Micas, *12*:588; *13*:**398**
Mica schist
as abrasive, *1*:25
as embedding filler, *8*:110
Micas, natural and synthetic, *2*:1; *13*:**398**
Micelles
in dry cleaning, *7*:311
surfactants, *19*:573
Michael condensation
of acrolein, *1*:264
Michler's hydrol, *20*:700, 703, 712
Michler's ketone, *7*:478; *12*:107, 132; *20*:699,
721, 725
Microanalysis
in crystallography, *17*:874
Microbacterium, *2*:204
Microbial cells
source of enzymes, *8*:206
Microcalorimeters, *4*:42
Microcapsules. See *Microencapsulation.*
Microchemistry, *13*:**424**
Micrococcus cerificans

Microcrystalline waxes
  as abherent, *1*:7
Microdiffraction
  x-ray fluorescence analysis, *22*:465
Microencapsulation, *13*:**436**
  use in adhesive bonding, *1*:383
Microfilm
  use in literature searching, *12*:523
Microgels, *16*:245
Microhardness, *10*:813
Micronaire test
  for cotton, *6*:402, 410, 411
Micronutrients
  as fertilizer, *9*:143
Microorganisms, *13*:**457**
  confectionery spoilage caused by, *6*:56
  hydrocarbon-grown as protein source, *S*:842
  in prod of amino acids, *2*:160
Microplants. (See also *Pilot plants.*)
Microplants, *S*:**557**
Microscopy
  electron in crystallography, *17*:871
  infrared in crystallography, *17*:871
  x-ray in crystallography, *17*:872
Microscopy, chemical, *13*:**491**
Microstrainers
  in water-treatment systems, *22*:120
Microwaves, *S*:**563**
Mie equations, *20*:741
Mikamycin, *16*:332
Mikohbite, *7*:396
Mildew resistance in textiles, *1*:265
Milk, *13*:**506**
  butter, *13*:548-551
  by-products, *13*:565-572
  cheese, *13*:531-538
  composition, *13*:508-517
  condensed milk, *13*:551-557
  in confectionery, *6*:39
  dried milk products, *13*:557-565
  ice cream and related frozen deserts, *13*:538-548
  market milk industry, *13*:518-531
  sterilization, *8*:202
Milk and milk products
Milk chocolate, *5*:389
Milk fats, *8*:783
Milk of magnesia. See *Magnesia magma.*
Milk products, *13*:**506**
  in confectionery, *6*:39, 44
Milk, soured
  lactic acid in, *12*:170
Millon's base, *13*:248

Millon's reagent, *2*:173
Mills
  attrition, *18*:347
  ball, *15*:595; *18*:347
  bowl, *18*:347
  cage, *18*:356
  compd, *18*:348
  cylindrical roll, *15*:594
  hammer, *18*:341
  high-speed impingement, *15*:595
  impeller, *15*:594
  roller, *18*:347
  rotating disc, *15*:594
  sand grinder, *18*:358
  tube, *18*:347, 349
  two-roll, *15*:593
  vibrating ball, *18*:353
Milo
  starch from, *18*:681
Milori blue, *15*:539
Mimeography, *17*:335, 337
  stencil, *17*:329
Mimosa
  pyrocathechol in, *11*:469
Mimosa wax, *22*:161
Mineral black, *15*:546
Mineral blue, *15*:541
Mineral dyes, *7*:566
Mineral khaki, *7*:566
Mineral oil
  in cosmetics, *6*:363
  as electric insulation, *11*:788
  in food processing, *10*:10
  as release agent in food, *10*:16
  as laxative, *4*:592
  as sealant, *4*:30
Minerals
  thermal analysis, *20*:113
  in yeasts, *22*:516
  from continental shelves, *14*:157
Mineral wool
  lime in, *12*:452
  as thermal insulator, *11*:829, 831
Mineral wool sheets, *11*:833
Minium. See *Lead compounds.*
Minnesotaite, *12*:6
Minocycline, *20*:16
Mint, *14*:206
Mirabilite, *18*:504
Mirasan, *14*:547
Mirex
  prepn of, *5*:251

Mischmetal, *4*:847

Mishima alloys, *12*:761

Mispickel, *2*:712

Missiles
  graphite in design of, *4*:204
  lubricants for, *12*:608

Mist. See *Dust; Sprays.*

Mithramycin
  in cancer therapy, *S*:87

Mitomycin C
  in cancer therapy, *S*:87

Mixers. (See also *Mixing.*)
  flusher, *15*:593

Mixing, *13*:**577**

Mobility
  atomic in alloys, *11*:12

Modacrylic fibers, *1*:313
  flame-resistant prop, *9*:312
  prepn for dyeing, *7*:518

Moderator materials
  in nuclear reactor applications, *4*:234, 235

Moderators
  beryllium, *3*:451

Modocoll. See *Ethylhydroxyethylcellulose.*

Modulation
  electrooptic, *S*:625

Modulus
  tensile of fibers, *20*:39
  textile, *20*:40

Mohr-Westphal balance, *6*:770

Mohr method, *11*:559
  chloride ion determined by, *5*:56

Mohs hardness scale, *1*:23; *10*:809

Molasses, *13*:**613**
  as animal feed, *8*:868
  fermentation of, *22*:522
  in manuf of fermentation ethyl alcohol,
    8:438
  source of itaconic acid, *12*:84
  yeast from, *22*:541
  yeast prod, *22*:539

Molassic acid, *19*:211

Molded plastics
  manuf process, *2*:251
  properties, *2*:252
  use of amino resins, *2*:250

Molding, *8*:105
  fabrication of plastic, *12*:191

Molding sands
  clays used in, *5*:567

Molds
  graphite, *4*:231

Mole
  definition of, *S*:991

Molecular diffusion, *13*:101
  in gases, *13*:103

Molecular diffusivities, *13*:110

Molecular dimensions
  by x-ray methods, *22*:446

Molecular effusion
  gaseous diffusion separation process, *7*:109

Molecular excitation, *15*:333

Molecular heterogencity
  of polymers, *16*:249

Molecular sieve
  xylene separations, *22*:477

Molecular sieves
  in catalysis, *4*:568
  in catalytic cracking, *11*:429
  as drying agents, *7*:393
  as hydrocracking catalyst, *11*:426
  silica in prepn of, *2*:9

Molecular structure of plastics, *15*:790

Molecular symmetry
  and NMR, *14*:61

Molecular weight
  by freezing-point depression, *13*:210
  by mass spectrometry, *13*:90
  of plastics, *15*:812
  of polymers, *16*:245
  x-ray determination, *22*:446

Molecules
  determination of structure of gaseous,
    22:453

Mollier diagram
  for MHD generator, *S*:225

Molluscacides, *14*:544; *15*:909-911

Molofac. See *Dioctyl sodium sulfocuccinate.*

Molozonide, *14*:417, 418

Molybdate orange, *15*:533

Molybdenite, *13*:640, 652; *19*:351; *22*:338
  from copper ores, *6*:142
  rhenium in, *17*:411

Molybdenum, *12*:601, 753; *13*:**634**
  as a refractory metal, *11*:31
  cast alloys, *20*:571
  as catalyst in hydrocracking, *11*:421
  coating, *17*:268
  content in soils, *9*:35
  as electric furnace resistor, *10*:255
  in electron tubes, *8*:5, 6
  ferroalloy, *18*:751
  furnaces for fusion, *10*:289
  in hydroisomerization, *11*:451

from ocean floor, *14*:165
in plain-carbon and low-alloy steels, *11*:17
as plant nutrient, *9*:30
protection, *8*:29
refractory, *17*:246
refractory metal, *17*:287
rhenium alloys, *17*:418
in scheelite, *22*:337
in stainless steels, *11*:19
in superalloys, *11*:20, 24, 26, 30
tantalum alloys, *19*:644
in titanium carbide, *20*:573
Molybdenum alloys, *13*:**634**
Molybdenum carbide, *4*:75, 84. (See also *Cemented carbides.*)
Molybdenum carbonyl, *4*:489
Molybdenum compounds, *13*:**645**
sulfur dyes, *19*:436
Molybdenum disilicide
refractory, *17*:246
as electric furnace resistor, *10*:255
Molybdenum disulfide, *8*:30; *10*:129; *12*:588, 609
additive to polyamides, *16*:94
as impregnant, *14*:457
from wolfram ores, *22*:339
Molybdenum hexafluoride, *9*:587
Molybdenum naphthenate, *8*:285
Molybdenum orange, *13*:655
pigments, *15*:527, 530, 532
Molybdenum oxide,
desulfurization catalyst, *15*:46
reforming catalyst, *15*:32
steel, *18*:751
Molybdenum silicide
coating, *17*:274
strength value of, *4*:818
Molybdenum sulfide
in hydrogenation, *14*:384
sulfolane reactions, *19*:251
Molybdenum trioxide, *9*:402
in oxidation of benzene, *3*:373
Molybdenum-99m, *17*:3
Molybdic oxide, *2*:54
Molybdosilicic acid
Momentum
conservation in transport processes, *20*:624
transport processes, *20*:643
Monazite, *10*:756; *17*:146; *20*:248
Monazo dyes, *7*:471, 472, 522, 527
Monazoles
as optical brighteners, *3*:741
Monazo mordant dyes, *2*:883

Mond process
for recovery of nickel from ores, *4*:443
Monel, *13*:273
corrosion resistance to water, *6*:302
welding, *22*:248
Monensin
as anticoccidial, *21*:246
Monkshood. See *Aconite.*
Monoacetin, *10*:630
Monoacetyl ethylenediamine
reaction to form 2-methylimidazoline, *7*:26
Monoacetylpyrogallol, *16*:198
Monoalkanolamides
condensates, *19*:552
Monoallylamine
uses, *2*:124
Monoallyl succinate, *19*:145
Mono-2-aminoethyl isocyanurate, *20*:664
Mono(3-aminopropyl) isocyanurate, *20*:665
Monoammonium malate, *12*:842
Monoammonium phosphate, *2*:299, 312; *6*:578
fertilizers, *S*:343
in fire extinguishers, *9*:296
fire extinguishing agent, *S*:365
as water-correcting agent, *10*:17
Monoanilinium malate, *12*:842
Monobasic lead sulfate, *12*:279
Monobutyl adipate
phys prop, *1*:409
Mono-*tert*-butyl-*m*-cresol. See *3-Methyl-6-tert-butylphenol.*
Mono-*tert*-butyl-*p*-cresol. See *4-Methyl-2-tert-butylphenol.*
Mono-*tert*-butyl-*p*-cresol, *1*:910. (See also *4-Methyl-2-tert-butylphenol.*)
Monocalcium dihydrogen phosphate
4,7-Phenanthroline-5,6-quinone, *20*:85, 89
Monocalcium phosphates
in balsing powder, *3*:51
Monocesium orthophosphate, *15*:239
Monochloramine, *2*:263, 616; *3*:557; *4*:915
hydrolysis constant of, *4*:913
Monochloroacetic acid, *10*:636
from trichloroethylene, *5*:187
Monochloroacetylene, *5*:204
Monochlorobenzene. (See also *Chlorobenzene.*)
in prepn of 2-chloroanthraquinone, *2*:22
sodium hydroxide hydrolysis, *15*:153
Monochlorobenzotrifluoride, *9*:778
Monochlorodifluoromethane
as aerosol propellant, *1*:472
Monochloroethylene. See *Vinyl chloride.*

Monochlorohydrins, 5:312
Monochloroisocyanuric acid, 4:913
  swimming pools disinfected by, 4:925
Monochloromaleic acid
  in synthetic oil prepn, 7:415
Monochloromethane. See *Methyl chloride.*
Monochlorostyrene
  flame-retardant polyesters, 20:832
Monochlorostyrenes
  unsaturated polyester reactions, 20:812
Monochlorosuccinic acid, 19:136
Monochlortriazinyl, 7:632
Monochromatic filters, 9:255
Monochrome Brown BC, 2:886
Mono entries. (See also *similar entries with Mono*-prefix.)
Monoethanolamine, 1:810; 2:189; 10:455
  from ammonia synthesis, 2:279
  in hair waving lotion, 10:788
  hydrogen sulfide, 19:382
  phys prop, 1:811
  in recovery of carbon dioxide, 4:362
  in soaps, 18:431
  urea synthesis, 21:43
Monoethanolamine carbonate, 4:362
Monoethanolamine plants
  hydrogen sulfide recovery, 15:47
Monoethanolamine sulfite
  setting of wool, 22:407
Monoethonoid acids, 8:811
Monoethyl adipate
  phys prop, 1:409
Monoethylamine
  manuf from ethyl alcohol, 8:455
Monoethylaniline
  benzyl chloride reaction, 20:701
Monoethyl ether
  in shampoo, 10:774
Monoethyl isopropylidenesuccinate, 19:137
Monoethyl malonate, 12:850
Monoethyl mesoxalate, 12:843
Monofluoboric acid, 3:615
Monofluoroacetic acid, 9:767
Monofluoroacetone, 9:509
Monofluorobenzene. See *Fluorobenzene.*
Monofluoroethane, 9:632
Monofluorophosphoric acid, 9:637, 638
Monofluorophosphoric acid diesters, 9:645
Monofluorosuccinic acid, 19:136
Monoformin, 10:619
Monoformylmelamine, 6:570
Monogermane, 10:523

Monoglyceride citrate
  as food preservative, 10:14
Monoglycerides, 10:619,630
  in margarine, 13:57, 60
  prepn, 8:360
  in shortening, 18:38
Monoglyme, 18:178
Monohydroxybenzene, 15:147
Monohydroxyfluoboric acid, 9:564
Monohydroxynapthalenes, 13:717
Monoiodotyrosine, 2:189
  properties, 20:262
Monoisopropanolamides, 19:552
Monoisopropanolamine, 1:810
  phys prop, 1:811
  in soaps, 18:431
Monoisopropyl citrate, 8:781
  as food preservative, 10:14
  as stabilizing agent in food, 10:16
Monomer
  definition, 16:242
Monomethine cyanines
  as optical brighteners, 3:744
Monomethyl adipate
  phys prop, 1:409
Monomethylamine, 2:120
  prepn from acetamide, 1:143
1-Monoglyceride-3-monosodium phosphate, 12:359
Monomethylaniline, 12:293
Monomethyl hydrazine
  properties, 8:688
Monomethylolurea
  urea-formaldehyde reaction, 21:39
Monomethylsuccinate, 19:139
Mononitrochlorobenzenes, 13:841
Mononitronaphthalenes, 13:704
Mononitrotoluenes
  analysis, 13:847
  hazards, 13:847
  manuf, 13:844
Monoolefin
  nomenclature of, 11:288
Monoperoxyphthalic acid, 8:243, 247, 266
Monophenyl phosphate, 9:306
Monopotassium adipate, 1:414
Monopotassium orthophosphate, 15:266
Monopropyl adipate
  phys prop, 1:409
Monorubidium orthophosphate, 15:239
Monosaccharides, 4:133, 134

Mono-*t*-butylhydroquinone
  polyester inhibitor, *20*:823
  condensation with dialhylchloroalkyl-
    amines, *2*:143
Monosilane, *18*:172
Monosilicic acid, *18*:58
Monosilylarsine, *18*:200
Monosodioacetoacetic ester
  isochloroprene reaction with, *5*:228
Monosodium dimethylarsinic acid, *22*:214
Monosodium glutamate, *2*:198; *9*:358
  flavor potentiator, *9*:361
  manuf, *2*:199
  specifications, *2*:210
  pharmaceuticals, *15*:117
Monosodium phosphate
  as diuretic, *7*:259
  as emulsifier in cheese, *10*:8
  as water correcting agent, *10*:17
  in enamel frits, *8*:158
1-Monostearin-3-monosodium sulfoacetate,
  12:359
Monosulfatozirconic acid, *22*:663
Monoterpenes, *19*:810
Monothiosulfuric acid, *9*:677
Monotron hardness test, *10*:814
Monotype, *16*:497
Monotype metal, *20*:766
  repro, *20*:763
  standard, *20*:763
Monovinylacetylene
  bromoprene from, *5*:229
  chloroprene from, *5*:220, 222, 224, 226
  fluoroprene from, *5*:229
  isochloroprene from, *5*:228
  prepn of, *5*:221, 224
Montanic acid, *8*:813
Montan wax, *12*:406; *22*:168
Montecatini cracking process
  in acetylene manuf, *1*:186
Montmorillonite, *3*:339
  in ceramics, *4*:763
  in ceramic products, *5*:561
  layer structure, *5*:542
Montmorillonite minerals, *5*:547
Monuron
  herbicide, *22*:188
Mooney viscosity, *21*:479
Mo permalloy, *12*:753
Morantel
  anthelmintic, *21*:249
Mordant Black 11, *2*:873
Mordant Black 25, *2*:886

Mordant Black 5, *2*:873, 886
Mordant Black 9, *2*:873, 886
Mordant Blue I (CI), *7*:527
Mordant Blue 10, *2*:865
Mordant Blue 45, *2*:865
Mordant Brown 1, *2*:886; *7*:527
Mordant Brown 7 (CI), *7*:527
Mordant Green 4 (CI), *7*:527
Mordant Orange 1, *7*:491
Mordant Red 15 (CI), *7*:527
Mordant Red 3 (CI), *7*:527
Mordant Red 7 (CI), *2*:885; *7*:527
Mordant Violet 35, *2*:865
Mordant Violet 50 (CI), *2*:865; *7*:527
Mordant Violet 54, *2*:865
Mordant Yellow O, *7*:485
Mordant Yellow 1, *2*:883
Mordant Yellow 26 (CI), *7*:535
Mordant acid dyes, *7*:527
Mordant colors
  in textile printing, *7*:576
Mordant dyes, *2*:502, 883; *15*:577
  uses of, *7*:477
Morin, *7*:614
Moroctic acid, *8*:815
Moroxydine, *21*:457
Morphactins, *15*:682
Morphine, *1*:769, 778, 796
Morpholine, *6*:643; *10*:920, 930; *13*:**659**
  as acid-neutralizing inhibitor, *6*:332
  manuf, *2*:361
  prepn from diethanolamine, *1*:812
  s-triazine analysis, *22*:186
3-Morpholine, *2*:361
Morpholine selenide, *2*:601
Morpholinium ethylsulfate
  quaternary salt, *13*:667
Morpholinium hexafluorophosphate, *9*:645
2-(Morpholinodithio)benzothiazole, *13*:662
2-Morpholinoethyl acrylate
  phys prop, *1*:288
3-Morpholino-3',4',5'-trimethoxyacrylophenone,
  16:198
Mortan, (See also *Cement.*)
Mosaic gold, *15*:536
Moser glass, *5*:849
Mothproofing, *11*:729
Motor fuels
  butanes in, *3*:816
Mousse de chêne, *14*:723
Movidyn, *2*:622
Mowrah (Illipé) fat, *8*:779, 784
Mowrah fat, *8*:779, 784

Mucoprotein, *14*:671
Mullite, *17*:245
  in chemical ware, *4*:835
  dielectric prop of, *4*:821
  prepn of synthetic, *4*:771, 773, 774
  thermal expansion coefficient, *4*:805
  thermal conductivity of, *4*:803
Multicomponent systems
  vapor-liquid equilibria of, *21*:226
Multinutrient fertilizers, *9*:114
  granulation, *9*:120
Mungbeans
  as animal feed, *8*:868
Munitions industry
  humidity conditions within, *1*:486
  use of air conditioning in, *1*:497
Munsell color system, *5*:809
Muntz metal
  composition of, *6*:214, 218
  magnetic prop, *6*:220
  stress relief for, *6*:240
Muramic acid
  mucoprotein from, *14*:671
Murexide
  from uric acid, *21*:109
Murexide test, *21*:107, 111
Muriate of potash, *9*:111; *16*:388
Murmuric oil, *8*:779, 784
Muscarine, *1*:763
Muscle activity
  due to lactic acid, *12*:170
Muscle fiber
  of meat, *13*:167
Muscle sugar, *11*:673
Muscone, *14*:727
Muscovite, *13*:398; *15*:512
Musk Tonquin, *14*:726
Musk ambrette, *1*:902
Mustard, *4*:872; *8*:502; *9*:352, 363, 365, 372; *14*:207
  prepn, *8*:502
Mustard seed oil, *8*:779, 784, 786
Mutarotation, *19*:157
D-Mycaminose, *12*:639
L-Mycarose, *12*:639

D-Mycinose, *12*:639
Mycobacillin, *16*:315
Mycobacterium
  hydrocarbon utilization, *S*:839
Myconomycin, *12*:633, 635
Mycosamine, *14*:140
  hydrolysis product, *16*:138
Mycostatin. See *Nystatin*.
  ethylene from, *S*:310
Mycotic infections
  chemotherapy, *3*:34
Myelin
  in detergency, *6*:883
Myelography, *17*:139
Mylar
  spectral transmittance curve, *9*:253
Myofibrils
  in meat, *13*:167
Myoglobin, *10*:13
  x-ray analysis, *22*:454
Myrcene, *7*:417; *19*:813, 817, 825
Myrcenyl chloride, *19*:818
Myrcia oil. See *Bay oil*.
Myricyl alcohol. See *1-Hentriacontanol*.
Myricyl palmitate
  in beeswax, *22*:159
Myristica. See *Nutmeg*.
Myristic acid, *8*:812; *18*:26
Myristic acid, *8*:812
  in coconut oil, *18*:418
  phys prop, *1*:231
  specifications, *8*:836
Myristoleic acid, *8*:814
Myristone, *12*:104
Myristonitrile, *18*:661
Myristyl alcohol. See *1-Tetradecanol*.
Myristyl-cetyl alcohol
  composition of commercial mixture, *1*:553
  specifications, *1*:554
Myrobalans, *12*:317, 323
Myrrh, *14*:207, 723; *17*:389
Myrtle, *14*:207
Myrtle wax, *22*:164
Myxophyceae
  and nitrogen fixation, *S*:609

N

Nabam. See *Disodium ethylenebisdithiocarbamate.*
Nacrite, 5:545
Naégite, 10:756
Nagyagite, 10:759
Nahcolite, 18:9, 459, 461
Nail enamels, 6:368
Nail polish, 6:368
Nail polish removers, 6:371
Nalorphine, 2:382, 387
Napalm, 7:283
  in chemical warfare, 4:892, 893, 894
Naphtha, 14:835; 15:17
  from coal tar, 19:669, 671
  extraction, 13:731
  feedstock, 15:8
  heavy, 15:2
  in hexane prod, 11:2
  high-temperature fibers from, S:961
  high-temperature pyrolysis, S:552
  light, 15:2
  in refrigerants, 22:53
  from shale oil, 18:17
Naphthalene, 13:**670**
  in coal tar, 19:668, 673
  from coke-oven tar, 13:680
  derivatives in petroleum, 14:849
  divinylbenzene by-product, 19:76
  in dye intermediates, 7:469
  dye intermediates from, 7:484
  in dye industry, 7:480, 481
  heat of combustion of, 4:898
  by hydrodealkylation, 11:453
  hydrogenated, 13:676
  phthalic anhydride from, 15:448
  in prepn of anthraquinone, 2:434
  toxicity of, 11:302
1-Naphthaleneacetic acid
  phys prop, 1:239
Naphthaleneβ-naphthol
  binary cooling curve, 13:204
1-Naphthalenecarboxylic acid. See *1-Naphthoic acid.*
2-Naphthalenecarboxylic acid. See *2-Naphthoic acid.*
Naphthalenecarboxylic acids 13:**690**. (See also *Naphthoic acids.*)

Naphthalene derivatives, 13:**697**
1,5-Naphthalenediamine, 13:708
1,8-Naphthalenedicarboxylic acid, 13:690, 692
2,6-Naphthalenedicarboxylic acid, 13:678
1,8-Naphthalenediol, 13:719; 15:372
1,3-Naphthalenedisulfonic acid, 13:701
1,4-Naphthalenedisulfonic acid, 13:701
1,5-Naphthalenedisulfonic acid, 13:701
1,6-Naphthalenedisulfonic acid, 13:702; 19:311
2,6-Naphthalenedisulfonic acid, 13:702
2,7-Naphthalenedisulfonic acid, 13:702
Naphthalenedisulfonic acids, 13:701
1,2-Naphthalenedisulfonic acids, 13:701
Naphthalene Green
  acid dyes, 20:723
  manuf, 20:700
Naphthalene Green V, 20:723, 728
  synthesis of, 20:724
Naphthalenemonosulfonic acids. See *1-and 2-Naphthalenesulfonic acid.*
Naphthalene oil
  coal-tar distillate, 19:664, 668
Naphthalene Red B. See *Acid Fuchsin D.*
Naphthalene Red B, 5:871
Naphthalenes
  chlorinated, 5:297
Naphthalenesodium, 18:436, 452
1-Naphthalenesulfamic acid, 2:873
Naphthalenesulfonates
  as surfactants, 19:519
1-Naphthalenesulfonic acid, 13:700
2-Naphthalenesulfonic acid, 13:687, 700; 15:643; 19:300
Naphthalene-1-sulfonic acid
  prepn of naphthalenic syntans, 12:327
Naphthalene-2-sulfonic acid
  prepn of naphthalenic syntans, 12:327
Naphthalenesulfonic acids, 13:698; 19:284
  from coal tar, 19:675
1-Naphthalenesulfonyl chloride
  phys prop, 1:224
2-Naphthalenesulfonyl chloride
  phys prop, 1:224
1,4,5,8-Naphthalenetetracarboxylic acid, 15:584
1,3,5,7-Naphthalenetetrasulfonic acid, 13:703

1,3,6,8-Naphthalenetetrasulfonic acid, *13*:703
Naphthalenetetrasulfonic acids, *13*:703
1,4,5-Naphthalenetriol, *14*:133
1,3,5-Naphthalenetrisulfonic acid, *13*:703
1,3,6-Naphthalenetrisulfonic acid, *13*:703
1,3,7-Naphthalenetrisulfonic acid, *13*:703
Naphthalenetrisulfonic acids, *13*:703
Naphthalenic syntans, *3*:327
Naphthalic acid, *13*:690
1,4-Naphthaquinone, *3*:786
Naphthas
    as paint removers, *14*:490
Naphthenes, *14*:835
    in hexane prod, *11*:2
Naphthenic acid, *7*:275
Naphthenic acids, *13*:727
    petroleum sweetening, *15*:52
    in soap, *18*:418
Naphthenic oils
    as lubricants, *12*:564
Naphthionic acid, *13*:709
Naphthogene Blue RR, *2*:897
α-Naphthoic acid. See *1-Naphthoic acid.*
β-Naphthoic acid. See *2-Naphthoic acid.*
1-Naphthoic acid
    phys prop, *1*:238
2-Naphthoic acid
    phys prop, *1*:238
α-Naphthol. See *1-Naphthol.*
β-Naphthol. See *2-Naphthol.*
1-Naphthol, *2*:871, 173; *13*:717
    antithyroid agent, *20*:270
    from coal tar, *19*:675
    sugar analysis, *19*:164
    sulfur dyes, *19*:427
2-Naphthol, *2*:357, 413, 871; *6*:642, 645; *7*:41,
    484; *12*:574, 825; *13*:717, 718
    from coal tar, *19*:675
    dye intermediates from, *7*:485
    ethylene sulfite, *19*:488
    from naphthalene, *13*:687
    phenylenediamines, *15*:219
    pigments, *15*:574
    pigments, *15*:569
    sulfonic acids, *19*:315
Naphthol AS, *2*:889, 890; *7*:548
Naphthol AS-BT, *2*:891
Naphthol AS-FGGR. *2*:894
Naphthol AS-G, *2*:890, 891
Naphthol AS-GR, *2*:891
Naphthol AS-ITR, *2*:890
Naphthol AS-OL, *2*:892
Naphthol AS-PH, *2*:893

Naphthol AS-SW, *7*:485
Naphthol Blue Black, *2*:886; *5*:867
Naphthol Blue Black B, *2*:877, 895
1-Naphthol-2,4-disulfonic acid, *13*:722
1-Naphthol-3,6-disulfonic acid, *13*:722
1-Naphthol-3,8-disulfonic acid, *13*:722
1-Naphthol-4,7-disulfonic acid, *13*:722
1-Naphthol-4,8-disulfonic acid, *13*:722
2-Naphthol-3,6-disulfonic acid, *13*:722
2-Naphthol-3,6-disulfonic acid, disodium salt,
    7:479
2-Naphthol-4,8-disulfonic acid, *13*:723
2-Naphthol-6,8-disulfonic acid, *13*:722, 723
Naphthol Green B, *5*:872; *7*:522
2-Naphtholnaphthalene
    binary cooling curve, *13*:204
Naphthols, *7*:547
Naphthols, *13*:717. (See also *Naphthol; 2-
    Naphthol.*)
1-Naphthol-3-sulfonic acid, *13*:719
1-Naphthol-4-sulfonic acid, *13*:720
1-Naphthol-5-sulfonic acid, *13*:720
1-Naphthol-7-sulfonic acid, *13*:720
1-Naphthol-8-sulfonic acid, *13*:720
2-Naphthol-1-sulfonic acid, *13*:720
2-Naphthol-6-sulfonic acid, *13*:721
2-Naphthol-7-sulfonic acid, *13*:721
2-Naphthol-8-sulfonic acid, *13*:721
2-Naphthol-1,3,6,8-tetrasulfonic acid, *13*:724
1-Naphthol-2,4,7-trisulfonic acid, *13*:723
1-Naphthol-3,6,8-trisulfonic acid, *13*:723
2-Naphthol-3,6,8-trisulfonic acid, *13*:724
Naphthol Yellow S, *5*:872, 874
Naphthol Yellow 5XX, *7*:522
1-Naphthonitrile
    from 1-chloronaphthalene, *5*:300
1-Naphthophenone. See *1-Naphthyl phenyl ke-
    tone.*
1-Naphthophenone, *12*:107
1,4-Naphthoquinone, *6*:695
    chloroprene reaction with, *5*:218
    haloprene reactions with, *5*:229
    in prepn of anthraquinone, *2*:434
Naphthoquinones, *16*:908
    reactions, *16*:908
    prepn, *16*:910
    properties, *16*:908
1,4-Naphthoquinones
    vitamin K activity, *21*:585
Naphth[1,2] oxadizol-5-sulfonic acid, *7*:479
Naphth [2,1]- oxazole
Naphthotriazolylnitrostilbenedisulfonic acids,
    19:6

Naphthylacetic acid. See *1-Naphthaleneacetic acid.*

α-Naphthylamine. See *1-Naphthylamine.*

β-Naphthylamine. See *2-Naphthylamine.*

1-Naphthylamine, 2:218, 600; 13:707, 708
  azo coupling, 2:871
  diazotization of, 2:869
  prepn, 2:82
  prepn of 1-fluoronaphthalene, 9:797
  secondary arylamines from, 20:701
  uses, 2:82

2-Naphthylamine, 2:357, 871; 13:708
  in cigarette smoke, 20:523
  in prepn of 2-fluoronaphthalene, 9:797

Naphthylamine Black 10BR, 7:525, 535

1-Naphthylamine-2,4-disulfonic acid, 13:714

1-Naphthylamine-3,6-disulfonic acid, 13:714

1-Naphthylamine-3,7-disulfonic acid, 13:714

1-Naphthylamine-3,8-disulfonic acid, 13:714, 715

2-Naphthylamine-1,5-disulfonic acid, 13:715

2-Naphthylamine-3,6-disulfonic acid, 13:715

2-Naphthylamine-4,8-disulfonic acid, 13:715

2-Naphthylamine-5,7-disulfonic acid, 13:715

2-Naphthylamine-6,8-disulfonic acid, 13:715

Naphthylamines, 13:**707**
  from coal tars, 19:675
  methods of amination, 13:707
  sulfonation, 19:298

1-Naphthylamine-4-sulfonic acid, 13:709

1-Naphthylamine-5-sulfonic acid, 13:710

1-Naphthylamine-6-sulfonic acid, 13:712

1-Naphthylamine-7-sulfonic acid, 13:712

1-Naphthylamine-8-sulfonic acid, 13:713

1-Naphthylamine-2,7-sulfonic acid, 13:714

2-Naphthylamine-1-sulfonic acid, 13:713; 15:574

2-Naphthylamine-5-sulfonic acid, 13:713

2-Naphthylamine-6-sulfonic acid, 13:714

2-Naphthylamine-7-sulfonic acid, 13:714

2-Naphthylamine-8-sulfonic acid, 13:714

1-Naphthylamine-3,4,6,8-tetrasulfonic acid, 13:717

1-Naphthylamine,2,4,7-trisulfonic acid, 13:716

1-Naphthylamine-2,4,8-trisulfonic acid, 13:716

1-Naphthylamine-3,6,8-trisulfonic acid, 13:716

1-Naphthylamine-4,6,8-trisulfonic acid, 13:716

2-Naphthylamine-1,5,7-trisulfonic acid, 13:716

2-Naphthylamine-3,6,8-trisulfonic acid, 13:717

2-Naphthyl butyl ether
  as synthetic flavor, 9:354

1-Naphthyl δ'-methyl carbamate
  insecticidal carbamate, 22:194

1-Naphthyl-*N*-methyl carbamate, 19:335

2-Naphthyl ethyl ether
  as synthetic flavor, 9:354

1-Naphthyl phenyl ketone, 12:107; 2:495

Naphthylsodium, 18:436

α-Naphthylthiourea, 15:917

1,5-Naphthyridine, 10:899

Naphthyridines, 10:932

Naples yellow, 5:852; 15:535

Napthalam
  amide herbicide, 22:198

1-Napthaleneacetamide, 15:677

1,8-Napthalene diisocyanate
  polyurethans, 21:70

1,5-Napthalenediol, 2:600; 13:719

2,3-Napthalenediol, 13:719

2,7-Napthalenediol, 13:719

Napthalenediols, 13:719

4,5-Napthalenediol-2,7-disulfonic acid, 13:723

1,7-Napthalenediol-3-sulfonic acid, 13:721

1,8-Napthalenediol-3,6-disulfonic acid, 13:723

1,8-Napthalenediol-4-sulfonic acid, 13:721

1,3-Napthalenediol-5,7-disulfonic acid, 13:723

2,3-Napthalenediol-6-sulfonic acid, 13:722

Napthalenes
  chlorinated as stain solvents, 18:661

Napthaleneselenate, 22:659

1-Napthol, 13:717

1-Napthylamine-4,6-disulfonic acid, 13:715

1-Napthylamine-4,8-disulfonic acid, 13:715

1-Napthylamine, 5,7-disulfonic acid, 13:715

Narbomycin, 12:633, 635, 644

Naringen, 19:601

Naringenin-7-rhamnoglucoside, 19:601

Naringin, 8:203
  deamidase, 21:521
  form, 2:68
  phys prop, 2:68
  from pyridine, 21:515
  U.S. prod, 21:516

Naringinase, 8:190
  uses, 8:203

Naringin dihydrochalcone, 19:601

National Chromolan Black NWA, 5:511

Natta catalysts
  for poly-1-butene, S:776

Natural Blue 1 (CI), 7:621

Natural Brown 3 (CI), 7:614
  phys prop, 6:586

Natural dyes
  application, 7:565

Natural gas. See *Gas, natural.*

Natural gas
  as engine fuel, *S*:433
Natural Red 1 (CI), 7:614
Natural Red 3 (CI), 7:625
Natural Red 4 (CI), 7:527, 626
Natural Red 8 (CI). (See also *Madder.)*
Natural Red 8 (CI), 7:527, 620
Natural Red 24 (CI), 7:618
Natural Red 25 (CI), 7:627
Natural Red 28, 2:866
Natural rubber. See *Rubber, natural.*
Natural Yellow 6 (CI), 7:615
Natural Yellow 8 and 11 (CI), 7:614
Natural Yellow 10 (CI), 7:614
Natural Yellow 11 (CI), 7:527, 614
Natural Yellow 13 (CI), 7:614
Natural Yellow 19 (CI), 7:615
Naval brass
  magnetic prop, 6:220
  composition of, 6:227
NBR rubbers. See *Acrylonitrile-butadiene co-polymers.*
Neamine, 19:41
Neatsfoot oil, 8:779
Nebramycin, 19:45
Neburon
  urea herbicide, 22:190
Neisseria bacteria
  chemotherapy against, 3:23
Nematocides, 15:911
Nematodes, 14:537
Neo-acids, 9:564
Neo-Antergan, 2:152
Neoarsphenamine, 2:730
  sulfoxylic acid, 19:422
Neocinchophen, 2:391
Neocyanines, 6:618
Neodyarium, 17:144
Neodymium oxide
  colorant for glass, 9:252; 5:849, 850
Neohesperidin dihydrochalcone, 19:601
Neoinosamine-2, 19:44
Neolan Blue 2G, 2:887
Neolan Yellow GR, 2:888
Neomenthol, 19:825
Neomethymycin, 12:635, 641
Neomycin
  amebicide, 20:80
  in bacterial infections, 3:1
  as veterinary drug, 21:244
Neomycins, 19:33, 41

Neon
  cryogenic prop of, 6:472, 474
  in phototubes, 15:398
  phys prop, 10:868
  as process gas in diffusion column, 7:128
  separation from air, 10:875
  separation of isotopes, 7:138
Neon signs, 10:881
Neopalatine Fast dyes, 2:889
Neopentane, 14:707
Neopentyl alcohol. See *2,2-Dimethyl-1-propan-ol.*
Neopentyl chloroformate, 4:386
Neopentylene glycol, 10:676
Neopentylene glycol di-*p*-toluenesulfonate, 10:678
Neophan glass, 5:850
Neoprene, 7:705-715; 17:575. See *Polychloro-prene.*
  curable, 16:346
  hydroxyhydroquinone in, 16:201
  latex, 17:599
  manuf from acetylene, 1:206
  in packing, 14:451
  property of flexible polymers, 9:863
  as protective covering for cables, 11:819
  roofing, 17:461
Neosamine, 19:41
Neostigmine, 6:551
Neotridecanoic acid
Nepetalic acid
  in catnip, 14:198
Nepheline, 1:935, 941
Nepheline-apatite, 10:311
Nepheline syenite
  in ceramics, 4:769
Nephelite, 4:769
Nephelometers, 20:743
Nephelometry, 20:**738-747**
Neptune Blue BRA, 7:526
Neptunium, 1:351; 15:880
Neptunium compounds
  crystal structure, 1:366
  phys prop, 1:366
Neptunium-237, 15:883
Neptunium-239
  in irradiated cores, 21:30
  by neutron absorption, 21:3
Nernst-Einstein relation
  in electromigration, *S*:280
Nerol, 19:826
Neroli, 9:352, 14:207
  essential oil as flavor, 9:353

Nerolidol, *19*:826, 833
Nerve agents (chemical warfare), *4*:875
   antidote against, *4*:885
   automatic detection of, *4*:875
Nerve gases, *15*:328
Nervonic acid, *8*:814
Neryl chloride, *19*:818, 825
Nesosilicates, *18*:47
Nessler reagent, *2*:292
Neutral density optical filters, *9*:259
Neutral red, *11*:552
Neutramycin, *12*:636
Neutrogen dyes, *2*:894
Neutron
   absorption, *14*:76; *17*:69
Neutron diffraction, *22*:452
Neutrons, *17*:67
   accelerated, *17*:54
   deuterium as moderator, *6*:899
   as radiation, *17*:53
Neutron scattering
   aquametry by, *2*:688
Neutron sources, *17*:32
Nevile and Winther's acid, *13*:720
New Fuchsine
   manuf, *20*:709
   phenylation, *20*:699
   synthesis, *20*:682
Newtonian fluids, *20*:644
Newton's law
   of flow resistance, *21*:461
Niacin, *8*:175; *16*:801; *21*:488, 509. (See also *Enzymes, industrial; Nicotinic acid.*)
   addition to flour, *22*:297
   in fish, *9*:322
   U.S. prod, *21*:516
   vitamin B complex, *21*:509
Niacinamide, *16*:780, 801; *21*:525
   vitamin B complex, *21*:509
Niacin diethylamide, *21*:533
Niaouli; *14*:207
Nibren waxes, *5*:301
Nicarbazin, *21*:247
Niccolite, *2*:720
Nickel, *12*:601; *13*:**735**. (See also *Raney nickel.*)
   catalyst for ammonia synthesis, *S*:606
   as catalyst in diamond synthesis, *4*:296
   as catalyst in hydrocarbon synthesis, *4*:452, 453
   as catalyst in steam reforming of hydrocarbons, *10*:420, 421
   cladding with, *13*:273

in cobalt-base superalloys, *11*:24
copper alloys with, *6*:193, 199, 250
corrosion resistance to water, *6*:302
cracking catalyst, *15*:39
as dental alloy, *6*:814
dopant in wolfram, *22*:341
effects on brasses, *6*:231
electrodeposition of zinc, *22*:588
in electron tubes, *8*:3, 4, 6
in electroplating, *8*:54, 62
in hydroisomerization, *11*:451
impurity in copper, *S*:265
mass transport of, *S*:286
Mond process for recovery from ores, *4*:443
from ocean floor, *14*:165
pigments, *15*:550
on silica-alumina (catalyst), *11*:425
in stainless steels, *11*:19
in superalloys, *11*:20
tin coatings, *20*:288
in titanium carbide, *20*:573
welding of, *22*:248
as undercoat in plating, *8*:38
welding of, *22*:250
x-ray fluorescence analysis, *22*:456
in zinc analysis, *22*:560
zone refining, *22*:697
Nickel acetate
   in electroplating bath, *8*:63
Nickel alloys, *13*:**735**
   corrosion resistance to water, *6*:302
   welding of, *22*:248
   welding of, *22*:250
Nickel azo yellow, *15*:572
Nickel brasses and bronzes
   casting of, *6*:255
   properties of, *6*:256, 259
Nickel-cadmium electric cells, *3*:172
Nickel carbonate
   prepn of nickel fluoride, *9*:628
Nickel carbonyl, *12*:147; *13*:757
   in prepn of acrylamide, *1*:296
   in prepn of acrylic acid, *1*:293
   in prepn of acrylic acid esters, *1*:296
   in prepn of acrylic anhydride, *1*:295
Nickel carbonyls, *4*:489
   carbon monoxide in prepn of, *4*:433
Nickel catalyst, *2*:54
Nickel chelates
   polypropylene light stabilizer, *S*:813
   UV absorbers, *21*:117
Nickel chloride, *7*:41
Nickel chloride electroplating solution, *8*:63

Nickel chromate, 5:498
Nickel-chromium alloys
  as resistance furnace electrodes, 10:272
Nickel-cobalt-copper
  as magnetic material, 12:767
Nickel compounds, 13:**753**
  in ceramic pigments, 5:855
Nickel electroplating
  use of coumarin in, 6:431
Nickel ethylhexanoate, 8:849
Nickel ferrate(VI), 12:40
Nickel ferrite
  as ceramic material, 4:775
Nickel fluoborate electroplating solution, 8:63
Nickel fluoride, 9:628
Nickel fluoride tetrahydrate, 9:628
Nickel fluosilicate, 9:658
Nickel formate, 8:805; 10:103
  amine oxide reduction, S:39
  in electroplating bath, 8:63
Nickel-gold alloy, 10:691
Nickel hexafluorostannate, 9:682
Nickel-iron-copper
  as magnetic material, 12:766
Nickel-iron electric cells, 3:227
Nickel isopropyl xanthate
  antiozonant for rubber, 22:427
Nickel Ndi-N-butyldithiocarbamte, 17:579
Nickel monofluorophosphate, 9:642
Nickel oxide, 9:402
  colorant for glass, 9:252
  desulfurization catalyst, 15:46
  steel, 18:751
Nickel oxide glass, 9:255
Nickel oxides
  as glass colorants, 5:849
Nickel perchlorate, 5:72
Nickel-phosphorus alloy, 8:70
Nickel phthalocyanine, 15:488
Nickel potassium chromate, 5:499
Nickel salicylaldoximate, 22:560
Nickel silver, 6:193
  annealing temperature, 6:224
  magnetic prop, 6:220
Nickel silvers, 15:550
Nickel stearate, 7:281
Nickel sulfamate, 19:244
Nickel sulfamate electroplating solution, 8:63
Nickel sulfate
  in electroplating baths, 8:63
Nickel sulfide
  as hydrocracking catalyst, 11:421
  in hydrogenation, 14:384
  hydrogen sulfide, 19:385

Nickel, TD, 11:31
Nickel tetracarbonyl
  reaction with methyl butynol, 1:623
Nickel thiocyanate
  complexes with benzylamines, 22:476
Nickel titanate, 20:414
  pigments, 15:535
Nickel Yellow, 2:908
Nickel 200, 22:248
Niclosamide, 14:543
Nicotinamide, 16:780, 801; 21:523, 525
  in yeast, 22:516
  yeast requirement, 22:517
Nicotinamide adenine dinucleotide, 21:510, 512
  biosynthesis of, 21:519
Nicotinamide adenine dinucleotide phosphate, 12:650; 21:488, 510
  metabolism of urea herbicides, 22:191
Nicotinamide adenine nucleotide, 21:488
Nicotinamide adenine nucleotide diphosphate, 21:512
Nicotinamide mononucleotide, 21:521
Nicotinate, 1:765; 21:523
  and cardiovascular diseases, 20:520
  oxidation of, 21:515
  synthesis of nicotinic acid, 21:513
  from tobacco, 20:514
  in tobacco, 20:503
  in tobacco smoke, 20:516
Nicotine acid, 2:188; 16:801; 19:300
Nicotine sulfate
  from tobacco, 20:514
Nicotinic acid, 21:509
  antagonists for, 22:407
  phys prop, 1:240
  synthesis, 19:671
  in tobacco, 20:514
  yeast requirement, 22:517
Nicotinic acid derivatives
  in hair tonics, 10:802
Nicotinic acid diethylamide, 21:533
Nicotinic acid hydrochloride, 21:514
Nicotinic esters
  ammonolysis of, 21:525
Nicotiniums, 15:682
Nicotinoids
Nicotinonitrile, 19:316
3-Nicotinonitrile, 21:526
Nicotinoylglycine, 21:536
Nicotinuric acid, 21:536
Niddamycin, 12:633, 646

Nifuraldezone
  antibacterial activity, *3*:6
  structural formula, *3*:5
Nigrosine, *2*:862
  used in hair coloring, *10*:793
Nihydrazone, *13*:856
Nikethamide, *21*:533
Nile Blue, *2*:865
Nile Blue A, *7*:533
Ninhydrin, *2*:173
Niobic acid, *13*:781
Niobium, *13*:766; *19*:630
  beryllium compounds, *S*:76
  in electron tubes, *8*:6
  ferroalloy, *18*:751
  as a refractory metal, *11*:31
  solenoids in NMR, *14*:53
  tantalum, *19*:630
  tantalum alloys, *19*:644
  zinc analysis, *22*:561
Niobium carbide, *4*:75, 81, 83. (See also *Cemented carbides.*)
  in machining tools, *20*:572
Niobium carbonyl, *4*:490
Niobium compounds, *13*:**766**
Niobium hydride, *11*:209
Niobium nitrate
  in reactor technology, *14*:96
Niobium pentachloride, *9*:403
Niobium pentoxide
  as glass former, *10*:538
Niobium-05, *17*:125
Nirexon
  as diuretic, *7*:261
Niridazole, *14*:548; *20*:70
  amebicide, *20*:85
Nisin, *16*:330
  in food preservation, *10*:15
Nisinic acid, *8*:815
Nithiazide, *21*:248
Nitralin
  soil chemistry, *18*:536
Nitrate
  analysis for in water, *21*:694
Nitrates
  aquatic plant nutrients, *S*:325
  as hazardous chemicals, *9*:290
Nitration, *13*:**784**
  of xylenes, *22*:484
Nitric acid, *2*:444; *6*:686, 922; *10*:683; *13*:796
  in dye industry, *7*:480
  as fertilizer, *9*:61
  as inhibitor in pickling, *6*:322, 325

as liquid propellant, *8*:688, 689
in manuf of ammonium nitrate, *2*:324; *8*:596
oxygen in synthesis of, *14*:405
photography, *15*:390
pickling of stainless steels, *S*:690
prepn of Cyclonite, *8*:623
in prepn of *m*- dinitrobenzene, *8*:617
prepn of ethylenedinitramine, *8*:626
in prepn of mercury fulminate, *8*:588
in prepn of silver fulminates, *8*:590
in prepn of succinic acid, *10*:669
prepn of tetranitrocarbazole, *8*:622
in prepn of TNT, *8*:613
in prepn of tribromoacetic acid, *8*:420
in prod of chlorine, *1*:703
in prod of nitric phosphate, *9*:133
as propellant, *8*:660
reaction with boron trifluoride, *9*:557
reaction with epoxides, *8*:273
reaction with ethyl ether, *8*:479
reaction with TNT, *8*:612
Nitric oxide, *6*:576; *9*:602; *10*:114
  in automobile exhaust, *S*:50
  chloroprene reaction with, *5*:217
  nitric acid-zinc reaction, *22*:604
  in prepn of acrylonitrile, *1*:346
  prepn from ammonia, *2*:261
Nitric oxide myoglobin, *10*:13
Nitric phosphate
  as fertilizer, *9*:32, 78, 80, 132
  by sulfur recycle method, *S*:355
Nitric phosphates
  fertilizers, *S*:343
Nitrides, *13*:**814**
  as ceramic materials, *4*:774
Nitriding, *13*:308
Nitrification, *9*:70
Nitrile rubber, *17*:545, 574
  formulation, *7*:693
  processing, *7*:689
  properties, *7*:693
  use of acrylonitrile in, *1*:349
Nitriles, *S*:**590**
  plasticizers for, *13*:832
Nitrilimines, *10*:919
Nitrilotriacetic acid
  as chelating agent, *6*:7
  in chemical cleaning, *S*:164
  surfactants, *19*:513
2,2',2"-Nitrilotriethanol, *6*:645
3,3',3"-Nitrilotripropionitrile, *6*:648
3,3',3"-Nitrilotrisporopionamide
  in prepn of acrylamide, *1*:281

Nitrite
    analysis for in water, *21*:694
Nitrites
    aquatic plant nutrients, *S*:325
    as cardiovascular agents, *4*:520
    as hazardous chemicals, *9*:290
p-Nitroacetanilide, *15*:217
    prepn from acetanidide, *1*:150, 152
m-Nitroacetophenone, *1*:167
o-Nitroacetophenone, *12*:852
Nitro alcohols, *13*:**826**
Nitroalkanes, See *Nitroparaffins.*
N-Nitro-1-aminonaphthalene, *2*:873
m-Nitroaniline, *2*:416, 417, 425
    prepn, *2*:95
    in prepn of 1,2-difluorobenzene, *9*:783
o-Nitroaniline, *2*:416, 425
p-Nitroaniline, *2*:416, 425; *6*:649; *7*:479, 490, 491
    diazotates derived from, *2*:893
    diazotized, *7*:539
    manuf by ammonolysis, *2*:352
    pigments, *15*:573
    reduction, *15*:217
    reduction of, *2*:87
4-Nitroaniline-3-sulfonic acid, *15*:217
Nitroanisole, *2*:422
1-Nitroanthraquinone, *2*:491
Nitroanthraquinones, *2*:490
    manuf, *2*:491
    properties, *2*:490
5-Nitro-1-anthraquinonesulfonic acid, *2*:444
8-Nitro-1-anthraquinonesulfonic acid, *2*:444
    reduction, *2*:96
1-Nitro-2-anthroquinonecarboxylic   acid,
    2:438
Nitro azo compounds
o-Nitrobenzaldehyde, *13*:852
m-Nitrobenzaldehyde, *9*:794
Nitrobenzaldehydes, *3*:365
3-Nitrobenzanthrone, *2*:496
Nitrobenzene, *2*:414; *7*:479; *10*:164; *13*:**834**
    as catalyst, *8*:407
    dye oxidant, *20*:693
    in manuf of aniline, *2*:77, 79, 85, 91, 411, 419
    in manuf of benzidine, *3*:410
    in prepn of acetanilide, *1*:151
    in prepn of *m*-dinitrobenzene, *8*:617
    in prepn of iminodianthraquinones, *2*:463
    as propellant, *8*:660
    reduction of, *2*:77, 79, 85, 91
    solvent refining, *15*:57

    stain solvent, *18*:661
    sulfonation, *19*:298
o-Nitrobenzenearsonic acid, *22*:658
    chlorinated, *19*:299
m-Nitrobenzenesulfonic acid, *7*:479; *15*:164
o-Nitrobenzenesulfonyl chloride, *19*:299
p-Nitrobenzenesulfonyl chloride, *19*:269
2-Nitrobenzidine, *3*:416
3-Nitrobenzidine, *3*:416
m-Nitrobenzoic acid, *3*:437
    phys prop, *1*:238
o-Nitrobenzoic acid
    phys prop, *1*:238
p-Nitrobenzoic acid, *3*:437
    phys prop, *1*:238
3-Nitrobenzotrifluoride, *9*:778
p-Nitrobenzoylacetone, *12*:148
N-p-Nitrobenzoyl threonine, *2*:166
p-Nitrobenzyl bromide, *3*:775
    mustard gases detected by, *4*:874
p-Nitrobenzyl pyridine
Nitrobiphenyl
    perchlorylation reaction, *9*:604
1-Nitro-2-butanol
    commercial, *13*:830
2-Nitro-1-butanol, *13*:831
    as fungicide, *10*:234
2-Nitro-2-butene, *6*:695
2-Nitrobutyl acrylate
    phys prop, *1*:289
Nitrocellulose. See *Cellulose nitrate.*
Nitrocellulose lacquer. See *Cellulose nitrate lacquers.*
m-Nitrochlorobenzene, *13*:842
o-Nitrochlorobenzene, *2*:421, 424, 425; *13*:842
p-Nitrochlorobenzene, *2*:425
p-Nitrochlorobenzene, *13*:842
    ammonolysis of, *2*:352
Nitrochlorobenzenes, *13*:840
    sulfonation, *19*:298
2-Nitro-4-chlorotoluene, *13*:851
2-Nitro-6-chlorotoluene, *13*:851
6-Nitrocoumarin, *6*:429
6-Methyl-2-heptanone, *12*:104
Nitrocyclohexane, *6*:674; *13*:882
6-Nitro-1-diazo-2-naphthol-4-sulfonic   acid,
    2:873
4-Nitrodiphenylamine-2-sulfonic acid, *15*:218
Nitro dyes, *7*:471, 472
Nitroethane, *6*:695
Nitroethanol, *13*:826
2-Nitroethyl acrylate
    phys prop, *1*:289

2-Nitroethyl ethyl ether
3-Nitro-4-fluoroperchlorylbenzene, 9:604
  prepn from ethyl ether, 8:479
2-Nitro-2-ethyl-1,3-propanediol, 13:831
2-Nitro-2-ethyl-1,3-propanediol dipropionate
  as fungicide, 10:234
Nitroform, 12:121; 13:864
5-Nitro-2-furaldehyde, 13:855
5-Nitro-2-furaldehyde diacetate, 13:856
5-Nitro-2-furaldioxime, 13:856
Nitrofurans, 13:**853**
  amebicides, 20:86
  in bacterial infections, 3:1
  veterinary uses, 21:246
Nitrofurantoin, 2:638; 13:856
  antibacterial activity, 3:6
  structural formula, 3:5
  as veterinary drug, 21:246
Nitrofurazone, 2:637; 10:19; 13:854
  antibacterial activity, 3:6
  anticoccidial, 21:298
  structural formula, 3:5
  as veterinary drug, 21:246
5-Nitro-2-furfural diacetate, 13:855
Nitrofurfuryl methyl ether
  as veterinary drug, 21:246
Nitrogen, 13:**857**
  as aerosol propellent, 1:475
  alkylation, 1:896
  in ammonia synthesis, 2:274
  aquatic plant nutrients, S:326
  autooxidation of nitrogen compounds,
    11:403
  binary compounds of, 13:815
  as component of natural gas, 10:448
  compounds with boron, 3:728
  content in soils, 9:35
  in cottonseed, 6:414, 415
  in cryobiology, 6:476-478
  cryogenic property, 6:472, 474
  cyanoethylation on, 6:643, 648
  ferroalloy, 18:751
  as fertilizer, 9:32, 47, 51
  in fertilizer mixtures, 9:76
  in fish, 9:319
  use in foam-mat drying, 7:359
  gas-filled thermometers, 19:781
  from gasification of solid fuels, 10:356
  in heterochain inorganic polymers, 11:638
  as lubricant, 20:606, 607
  liquid, as fertilizer, 9:72
  manganese compounds, 13:4
  in modified stainless steels, 11:21

  in manuf of calcium cyanamide, 6:554, 555,
    556
  nutrient in algal cultures, 1:653
  in multinutrient fertilizers, S:339
  as plant nutrient, 9:26
  as refrigerant, 10:859
  separation of isotopes, 7:138
  steel, 18:751
  in tar sand bitumen, 19:695
  in tobacco smoke, 20:516
  in wood, 22:364
Nitrogen bromide, 3:767
Nitrogen chloride
  formation of, 5:2
  in petroleum, 14:851
Nitrogen compounds
Nitrogen dioxide, 10:114; 12:69
  chloroform reaction with, 5:121
  methylene chloride reaction with, 5:114
  in photochemical smog, S:50
  prepn from ammonia, 2:261, 262
  tetrachloroethylene reaction with, 5:198
Nitrogen fixation, S:**604**
  and algae, S:331
Nitrogen gas
  in ammonia manuf, 2:272
Nitrogen iodide, 2:263
Nitrogen mustard
  effect of procytoxid on, 22:530
Nitrogen mustards
  in chemical warfare, 4:873
  clinical use of, 4:875
Nitrogen oxide
  from char-burning power plants, S:232
Nitrogen oxides
  as fission products, 14:105
  fumes, 18:406
Nitrogen pentoxide, 8:407
Nitrogen peroxide, 8:479
  flour bleaching agent, 22:293
Nitrogen solution
  fertilizers, 9:72
Nitrogen sulfide, 2:263; 19:405
  as explosive, 8:593
Nitrogen tetroxide, 2:416; 12:126
  properties, 8:689
  as storable liquid oxidizer for Titan II mis-
    sile, 1:703
Nitrogen trichloride, 4:916
Nitrogen trifluoride, 9:629
  properties, 8:689
Nitroglycerin. See *Glyceryl nitrate; Glyceryl
  trinitrate.*

Nitroglycol. See *Glycol dinitrate.*
Nitroguanidine, *10*:738, 740
  characteristics, *8*:632, 633, 634
  in explosives, *8*:625, 659, 681
3-Nitro-4-heptanol, *3*:870
Nitrohexahydropyrimidenes, *13*:827
4-Nitrohexanenitrile, *6*:641
4-Nitro-3-hydroxymercuri-*o*-cresol anhydride.
  See *Nitromersol.*
Nitroisonitrosoacetic acid, *8*:588
β-Nitroisopropyl isopropyl ether, *8*:488
γ-Nitro-γ-phenylbutyrophenone, *12*:161
Nitromannite. See *Mannitol hexanitrate.*
Nitromersol, *2*:620; *13*:241
6-Nitrometanilic acid, *15*:217
Nitromethane
  as hazardous chemical, *9*:291, 566
  in prepn of papaverine, *1*:798
  as propellant, *8*:660
  reaction with acetaldehyde, *1*:84
  reaction with nitromethane, *12*:142
N-Nitromethylamine, *8*:620
2-Nitro-2-methyl-1-propanol, *13*:827, 831
  as fungicide, *10*:234
2-Nitro-2-methyl-1,3-propanediol, *13*:827, 831
Nitromide, *21*:247
Nitron, *9*:641; *10*:919
1-Nitronaphthalene, *13*:704
Nitronaphthalenes, *13*:703
Nitronaphthalenesulfonic acids, *13*:703, 707
5-Nitro-1-naphthylamine
  prepn, *2*:95
2-Nitro-5-nitrate-2,5-dihydrofurfural di-
  acetate, *13*:855
Nitronium, *9*:645
Nitronium fluoborate, *9*:566, 567
Nitronium hexafluoroantimonate, *9*:630
Nitronium hexafluoroarsenate, *9*:630
Nitronium hexafluorophosphate, *9*:630, 645
Nitronium nitrate, *10*:155
Nitronium perchlorate, *5*:65, 73; *14*:415
Nitronium tetrafluoroborate, *9*:630
Nitronium tetrafluoroboride, *10*:156
5-Nitro-4-octanol
  prepn, *13*:830
Nitroolefins, *6*:695
Nitroparaffins, *13*:**864**
Nitropentaerythrite. See *Pentaerythritol tetra-
  nitrate.*
3-Nitroperchlorylbenzene, *9*:604
p-Nitroperoxybenzoic acid, *8*:244
p-Nitrophenacyl esters, *8*:372

Nitrophenide
  as anticoccidial, *21*:247
m-Nitrophenol, *2*:213; *13*:890
o-Nitrophenol, *2*:422
o-Nitrophenol, *13*:889
  from nitrobenzene, *3*:372
  reduction to *o*-aminophenol, *2*:97
p-Nitrophenol, *7*:479; *11*:552; *13*:890; *18*:528
  as fungicide, *10*:233, 234
  in prepn of *p*-aminophenol, *2*:219
  in roundworm chemotherapy, *14*:535
6-Nitro-1-phenol-2,4-disulfonic acid, *2*:224
Nitrophenols, *13*:**888**
  as herbicides, *22*:203
Nitrophenylacetic acid, *15*:214
2-Nitrophenylamine, *8*:681
4-Nitrophenylarsonic acid, *21*:248
p-Nitrophenylazobenzoyl chloride
  in separation of flavor compds, *9*:339
p-Nitrophenylazoorcinol
  in test for beryllium, *3*:472
2-Nitrophenyl borate, *3*:657
4-Nitro-*m*-phenylenediamine, *15*:217
4-Nitro-*O*-phenylenediamine
  in hair coloring, *10*:795
5-Nitro-*m*-phenylenediamine
  in hair coloring, *10*:795
2-Nitro-*p*-phenylenediamine
  in hair coloring, *10*:795
p-Nitrophenylhydrazine, *6*:455
o-Nitrophenylhydrazines, *10*:918
m-Nitrophenylhydroxylamine
  in prepn of 4-amino-2-nitrophenol, *2*:218
2-Nitro-1-phenyl-1,3-propanediol, *13*:826
(O-Nitrophenyl)propiolic acid
  in synthesis of indigo, *11*:563
p-Nitrophenyl sulfite, *19*:489
Nitrophoska, *9*:43
Nitroplatinous acid, *15*:865
1-Nitropropane, *6*:641
  stain solvent, *18*:661
2-Nitropropane, *19*:819
  paint remover, *14*:487
  stain solvent, *18*:661
1-Nitropropene, *6*:695
2-Nitropropyl acrylate
  phys prop, *1*:289
Nitropyrazoles
  amebicides, *20*:86
Nitropyridines
  amebicides, *20*:86
4-Nitropyrogallol, *16*:210

Nitropyrroles, *16*:850
Nitropyrroles
  amebicides, *20*:86
2-Nitroresorcinol, *13*:892
5-Nitrosalicylic acid, *17*:723
  from indigo, *11*:563
6-Nitrosalicylic acid
  from indigo, *11*:563
Nitrosoacetaldehyde, *8*:588
Nitrosobenzene, *6*:695
  prepn, *2*:77
  reaction with isoprene, *12*:69
  reaction with phenyl isocyanate, *12*:51
Nitrosocylohexane, *6*:677
Nitrosodialkylanilines
  prepn of amines from, *2*:109
p-Nitrosodimethylaniline
  synthesis of menadione, *21*:593
p-Nitroso-*N,N*-dimethylaniline hydrochloride,
  16:197
N-Nitrosodiphenylamine, *7*:42, 46, 47, 48
Nitroso dyes, *7*:471, 472, 522
Nitrosonium fluorosulfate, *9*:679
Nitrosonium hexafluoroantimonate, *9*:630
Nitrosonium hexafluoroarsenate, *9*:630
Nitrosonium hexafluorophosphate, *9*:630
Nitrosonium pyrosulfate, *9*:679
Nitrosonium tetrafluoroborate, *9*:630
p-Nitrosophenol, *2*:219
Nitrosophenylhydroxylamine
  cupferron, *22*:561
Nitrosopyrroles, *16*:850
1-Nitroso-2-naphthol, *15*:588
  electrodeposition of zinc, *22*:588
Nitrostarch, *8*:601
  explosive characteristics, *8*:633
  sensitivity characteristics, *8*:632
  stability characteristics, *8*:634
Nitrostarch dynamites, *8*:641
Nitrosuccinic acid esters, *19*:141
Nitrosyl bromide, *3*:754
Nitrosyl chloride, *6*:677; *9*:795, 796; *10*:5
  photochemistry, *15*:338
Nitrosyl fluoborate, *9*:568
Nitrosyl fluoride, *9*:630, 690
  reaction with hexafluoroacetone, *9*:760
  silica reaction, *18*:57
Nitrosyl hexafluoroantimonate, *9*:630
Nitrosyl hexafluoroarsenate, *9*:630
Nitrosyl hexafluorophosphate, *9*:630, 645
Nitrosyl perchlorate, *5*:73
Nitrosylsulfate, *20*:194

Nitrosyl sulfuric acid
  dye oxidant, *20*:693
  in prepn of TNT, *8*:613
  thiazole dyes, *20*:194
Nitrosyl tetrafluoroborate, *9*:630
Nitrosyl tribromide, *3*:754
2-Nitrothiophene, *20*:223
m-Nitrotoluene, *13*:849
o-Nitrotoluene, *13*:847
p-Nitrotoluene, *13*:849
p-Nitrotoluene-*o*-sulfonic acid, *19*:2
Nitrotoluenes, *13*:**834**
  derivatives, *13*:852
  sulfonation, *19*:298
  from toluene, *20*:531
4-Nitro-*o*-toluenesulfonic acid, *13*:852
3-Nitro-4-trifluoromethylphenol, *9*:706
5-Nitro-o-toluenesulfonic acid, *7*:479, 497;
  19:2,3
m-Nitro-*p*-toluidine
  pigments, *15*:573
2-Nitro-*p*-toluidine, *7*:479
3-Nitro-*p*-toluidine
  diazotates derived from, *2*:893
Nitrous acid, *2*:415; *7*:43; *10*:613
  mutations caused by, *2*:182
Nitrous oxide, *2*:323
  as aerosol propellent, *1*:475
  as anesthetic, *2*:395
  in commercial oxygen, *14*:401
Nitrous vitriol, *19*:442, 456
Nitryl, *9*:645
Nitryl chloride, *10*:155
Nitryl fluoride, *9*:630
  reaction with hexafluoroacetone, *9*:760
Nitryl hexafluoroantimonate, *9*:630
Nitryl hexafluoroarsenate, *9*:630
Nitryl hexafluorophosphate, *9*:630, 645
Nitryl tetrafluoroborate, *9*:630
N^1,N^2-Bis(*p*-ethoxyphenyl)-acetamidine
  hydrochloride. See *Phenacaine hydrochloride*.
NMN-pyrophosphorylase, *21*:523
NMR. See *Nuclear magnetic resonance*.
  and mass spectrometry, *13*:90
Noble gases. See *Helium-group gases*.
Nobleite, *3*:649
Nocardia
  hydrocarbon utilization, *S*:839
Noctiluca consistens, *15*:276
Nomenclature, *14*:1
Nomographs, *14*:**15**
  as flavor compound, *9*:342

Nonacosanoic acid, 8:813
n-Nonacosoic acid, 8:813
Nonactin, 12:648
Nonadecanoic acid, 8:813
1-Nonadecanol
  phys prop, 1:543
2-Nonadecanone, 12:105
n-Nonadecyl alcohol
  phys prop, 1:543
n-Nonadecylic acid, 8:813
1H, 2H, 4H-Nonafluorocyclohexanes, 9:698
Nonafluoro-2-nitrobiphenyl, 9:792
Nonamethylenediamine
  polypyromellitimide of, 5:767
Nonanedioic acid. See Azelaic acid.
Nonanoic acid, 8:812. See Pelargonic acid.
1-Nonanol
  phys prop, 1:543
2-Nonanone, 12:105
5-Nonanone, 12:104
Nonbronze blue, 15:539
Nonlinear optical materials, 5:623
Nonsteroidal estrogens, 11:127-141
Nontronite, 5:548
  properties, 3:342
Nonyl acrylate
  phys prop, 1:287
Nonyl alcohol
  as synthetic flavor, 9:354
n-Nonyl alcohol
  phys property, 1:543
Nonyl alcohols, 1:567
Nonyl aldehyde
  as synthetic flavor, 9:354
Nonylbenzene
  prepn of, 3:374, 398
Nonylphenol, 1:908
  in drilling fluids, 7:302, 303
Nonylphenol polyethylene
  glycol ether, 7:304
Nonylphenoxypoly(ethyleneoxy)ethanols, 19:533
Nonylphenoxypolyoxyethyleneethyl   sulfate,
  19:528
Nonylphenoxytri(ethyleneoxy)ethanols, 19:527
Nootkatone
  as flavor compd 9:342
Nopol, 19:813
Nopyl acetate, 8:376; 19:813
Noradrenaline, See Norepinephrine.
Noranda process, 19:351
Norboramide, 15:918
Norbutin, 6:65
5,7-Norcholestadiene-3β-ol, 21:554

Nor-compounds, 21:554
Nordihydroguaiaretic acid, 2:598; 8:781
  as food preservative, 10:14
Nordstrandite, 2:42, 47
Norepinephrine, 8:231, 233, 235, 237; 16:645
  as cardiovascular agent, 4:517
28-Norergosterol, 21:553
Norethandrolone
  veterinary uses, 21:251
Norethindrone, See Norethisterone.
Norethisterone, 6:65, 87
Norethynodrel, 6:65, 87
Norinyl, 6:65, 67, 87
Norlestrin, 6:67, 68, 87
Normal superphosphate
  fertilizer prod, 9:100
Normorphine, 2:382
19-Norpregn-4-ene-3,20-dione. See 19-Norpro-
  gesterone.
19-Nor-14 β, 17 α-pregn-4-ene-3,20-dione,
  18:880
19-Norprogesterone, 6:62
19-Norsteroids, 6:62; 18:836, 861
19-Nortestosterone
  from estradiol, 6:63
Nortropane, 10:899
Nortropane hydrochloride, 16:784
Notch-strength ratio, 13:187
Nougat, 6:50
Novaldiamine
  prepn from acetoacetic ester, 1:159
Novobiocin, 6:432
  amebicides, 20:82
  in bacterial infections, 3:1
  as veterinary drug, 21:245
Novolak resin, 15:182
Novolak resins, 15:177, 179, 189, 191, 195
NTA. See Nitrilotriacetic acid.
  esterification, 8:340
  reaction with dicyandiamide, 6:566
  yield increase, 1:147
Nuclear ceramic materials, 4:775
Nuclear disintegration, 2:813
Nuclear magnetic resonance, 14:40
Nuclear magnetic resonance spectra
  exo-5-chlorobicyclo [2,1,1] hexane-exo-6-
    tert-butyl carboxamide, 14:59
  3-methyl-1-pentyn-3-ol, 14:68
Nuclear magnetic resonance spectrometry
Nuclear magnetic resonance spectroscopy
  aquametry by, 2:687

Nuclear processes
  use of hafnium in, *10*:762
  for nitric acid, *13*:807
Nuclear radiation
  use in measuring liquid levels, *12*:494
Nuclear reactor
  use of extraction to reprocess spent fuels,
    8:759
Nuclear reactors, *14*:**74**
  "breeder", *21*:27, 29
  gas-cooled, *21*:28
  graphite applications in, *4*:234
  heavy-water, *21*:28
  light-water, *21*:28
  moderator materials for, *4*:234, 235
  power, *21*:28
  radiochemical plants, *14*:115
  reflector materials for, *4*:234, 235
  reprocessing irradiated fuel, *21*:30
  waste management, *14*:102
Nuclear spin, *14*:40
  coupling, *14*:48
Nucleation, *6*:484, 487
  in film deposition, *9*:186, 188
Nuclei
  charged as radiation, *17*:53
  magnetic, *14*:66
  magnetic properties, *14*:42
Nucleic acids
  from yeasts, *22*:531
  in yeasts, *22*:516
Nucleocidin
  amebicide, *20*:82
Nucleolus
  in yeast, *22*:512
Nucleosides
  antiviral agents, *21*:455
Number systems
  for computers, *6*:27
Nurnburg violet, *15*:541
Nutation, *14*:43
Nutmeg, *9*:352, 363, 365, 372; *14*:207
  essential oil in, *14*:181
Nutrients
  as food additives, *10*:11
Nutrilites
  for yeast, *22*:517
Nutrition
  hydrocarbons in, *11*:282
Nuts, *14*:**122**
  in confectionery, *6*:45
NW acid, *13*:720

Nylon, *15*:794. (See also *Polyamides.*)
  acid dyes on, *7*:532
  antiplasticization, *15*:775
  in bearing materials, *3*:272, 275, 283
  as cable insulator, *11*:820
  dyeing with disperse dyes, *7*:541
  dyeing of fiber blends, *7*:568, 570, 571
  electrical behavior of, *1*:129
  with epoxy resins as adhesives, *8*:311
  as filter medium, *9*:285
  flame resistance, *9*:312
  "fog marking" of, *2*:653
  gelation, *16*:11
  hard fiber, *18*:614
  man-made fiber, *9*:155
  membranes for reverse osmosis, *22*:60
  in packing materials, *14*:453
  paper, *14*:496
  plasticizers for, *15*:726, 759
  polymers, *15*:790
  prepn of, *2*:70
  prepn for dyeing, *7*:519
  stretch, *18*:615
  terminology, *16*:3
  use in reinforced plastics, *12*:189
  use of thiodiglycol in printing with dyes,
    10:648
  work of rupture value, *1*:131
Nylon-1
  prepn, *12*:50
Nylon-6, *13*:882; *16*:29
  phenol synthetic route, *15*:148
Nylon-6 film, *9*:231
  properties, *9*:222-226
Nylon-6,6, *6*:879; *16*:24
  phys prop, *1*:417
  polyvinylpyrrolidone graft copolymer,
    21:434
  prepn, *1*:245, 410
  prepn from cyclohexane, *6*:681
  tire cord, *20*:328
  uses, *1*:417
Nylon-6,6 (poly(hexamethylene adipamide)),
    16:2, 12, 23, 24
  condensation polymer, *16*:219
  critical surface tension, *9*:715
  raw materials, *16*:24
Nylon-6,8, *1*:250
Nylon-7 synthesis
  carbon tetrachloride in, *5*:138
Nylon-8,8, *1*:250
Nystatin, *3*:8, 35; *10*:19; *16*:133
  structure, *16*:140
  veterinary fungicide, *21*:249

O

Oakbark
  pyrocatechol in, *11*:469
Oakmoss, *14*:723
Oats, *12*:863; *22*:274-276
  as animal feed, *8*:860
  classes of, *22*:264
  origin, *22*:253
  uses for, *22*:257
Obtusilic acid, *8*:814
Ocean dumping
  sludge disposal, *S*:730
Ocean mining
  methods of, *14*:168
Ocean raw materials, *14*:**150**
Ocean water
  recovery of brine, *7*:860
Ocetene-1 oxide, *8*:249
Ocotea, *14*:207
Octachlor. See *Chlordan.*
Octachlorocyclopentane, *6*:700
2,3,4,5,6,7,8,8-Octachloro-2,3,3a,4,5,5a-hexa-
  hydro-4,7-endomethanoindene. See *Chlordan.*
Octachloronaphthalene, *9*:777
Octachlorotrisilane, *18*:168, 170
Octacosanoic acid, *8*:813
9,12-Octadecadicnoic acid. See *Linoleic acid.*
Octadecanamide. See *Stearamide.*
2,5-Octadecanedione, *12*:158
Octadecanoic acid, *8*:812. See *Stearic acid.*
Octadecanol
  evaporation control, *21*:665
1-Octadecanol
  phys prop, *1*:543
4,8,12,15-Octadecatetraenoic acid, *8*:815
9,11,13,15-Octadecatetraenoic acid, *8*:815
9,11,13-Octadecatrienoic acid, *7*:405; *8*:815
cis-9, *cis*-11, *trans*-13-Octadecatrienoic ( ) *acid,*
  8:815
Octadec-17-enedi-9,11()yonic acid, *8*:787, 816
9,12,15-Octadecatrienoic acid. (See also *Linolenic acid.*)
  in drying oils, *7*:402, 403, 405, 409, 411
  in eggs, *7*:664
1-Octadecene
9-Octadecenyl-1-ol
  phys prop, *1*:543

N-Octadecyl alcohol. See *1-Octadecanol.*
n-Octadecyl alcohol
  phys prop, *1*:543
Octadecylamine
  properties, *2*:130
Octadecyl ether, *8*:493
1-*N*-Octadecyl-3-ethyleneurea
Octadecyl isocyanate, *12*:46, 47
N-Octadecyl isocyanate
  reaction with ethylenimine, *22*:143
  water-repellent properties of, *22*:141
1-(Octadecyloxymethyl)-pyridinium chloride,
  22:142
Octadecyl vinyl ether, *21*:423
6-Octadecynoic acid, *8*:816
9-Octadecynoic acid, *8*:816
Octafining process
  for *p*-xylene, *22*:492
Octafluoroanthraquinone, *9*:798
Octafluorocyclobutane
  as aerosol propellant, *1*:474
Octafluorocyclohexa-1,3-diene, *9*:797
Octafluoronaphthalene, *9*:691, 775, 776, 797
2,2,3,3,4,4,5,5-Octafluoro-1-pentanol, *9*:846
Octafluoropropane, *9*:688
Octafluorotoluene, *9*:691, 776
1,2,3,6,1,2',3,6'-Octahydrobiphenyl, *3*:787
Octahydrocoumarin, *6*:427
1,3,4,5,6,7,8,8,-Octahydro-1,3,3a,4,7,7a-hexa-
  hydro-4,7-methanoisobenzofuran. See *telodrin.*
1,2,3,4,5,6,7,8-Octahydro-2-quinolone, *6*:641
Octahydro-1,3,5,7-tetrazocine. See *Cyclo-
  tetranethylenetetranitramine.*
Octamethylcyclotetrasiloxane, *18*:222, 237
  silicone synthesis, *18*:224
Octamethylenediamine
  polyamide from, *16*:2
1,1,4,4,5,5,8,8-Octamethyl-1,2,3,4,5,6,7,8-octa-
  hydroanthracene, *22*:635
Octamethylpyrophosphorylamide, *15*:329
Octamethylsucrose, *19*:154
n-Octane, *9*:722
n-Octaneboronic acid, *3*:723
Octanedioic acid. See *Suberic acid.*
2,4-Octanedione, *12*:148

2,5-Octanedione, *12*:158
3,5-Octanedione
 prepn, *12*:160
4,5-Octanedione, *12*:140
Octane number of gasoline, *10*:465, 480; *11*:2
Octanes
 branched in petroleum, *14*:847
 toxicity of, *11*:294
Octanoic acid, *8*:812
1-Octanol, *1*:565; *7*:304; *9*:722
 phys prop, *1*:543, 564
2-Octanol, *1*:565
 phys prop, *1*:543, 564
2-Octanone, *1*:565; *12*:104. See *Hexyl methyl ketone.*
Octaphenylcyclotetrasiloxane, *18*:225
Octatriacontanoic acid, *8*:813
n-Octatriacontoic acid, *8*:813
1-Octene-5,7-dione, *12*:150
1-Octen-3-ol
 as flavor compd, *9*:342
Octenyl succinic anhydride
 as starch modifier, *10*:17
Octyl acrylate
 phys prop, *1*:287
n-Octyl alcohol. (See also *1-Octanol.*)
Octyl alcohol
 composition of commercial mixture, *1*:553
 in paints, *14*:468
 specifications, *1*:554
 as synthetic flavor, *9*:354
Octyl aldehyde
 as synthetic flavor, *9*:354
2-Octyl-1-cyclopropeneheptanoic acid, *6*:414. See *Malvalic acid.*
Octyl-decyl alcohol
 composition of commercial mixture, *1*:553
Octylene oxide, *8*:247
n-Octyl ether, *8*:493
 phys prop, *8*:471
Octyl lactate
 prepn, *8*:353
p-Octylphenol, *19*:672
Odda process, *S*:341
 and heat-tubes, *S*:702
Odor
 chemical nature of, *14*:339
 evaluation, *14*:343
 in flavor perception, *9*:350
 of municipal waters, *22*:102
 testing, *14*:337
 of water, *21*:691

Odor control, *14*:**170**
OFHC, *6*:183, 185
Oil
 analysis for in water, *21*:702
 artists', *14*:480
 essential, *14*:**178**
 synthetic crude from coal, *S*:178
Oilcloth, *7*:424
Oil, light
 recovery from coke-oven gas, *4*:415, 419
Oil of cloves, *12*:135
Oil Red, *2*:906
Oil release
 theory of, *S*:966
Oils
 cottonseed, *19*:498
 decolorizing with clays, *5*:577
 ethoxylated, *19*:548
 fish, *19*:498
 flower in perfumes, *14*:722
 as impregnants, *14*:457
 lard, *19*:500
 light petroleum, *15*:45
 in nuts, *14*:129
 sperm whale, *19*:498
 sulfated, *19*:524
 sulfur-chlorination, *19*:498
 sulfurization, *19*:498
Oil sands, *19*:682
Oils, essential, *14*:**178**
 cyclohexane for extraction, *6*:681
Oil shale, *18*:1
Oils, lubricating. See *Lubrication.*
Oil wells
 corrosion in, *6*:330, 332
Ointments
 mercury compd, *13*:240
Oiticica oil, *7*:399, 404, 419, 420
Oiticica oil, *8*:779, 784, 786
 film props, *7*:423
 uses, *7*:423
Oleamide
 phys prop, *2*:73
Oleandomycin, *3*:9; *12*:636, 643
 amebicide, *20*:80, 82
 in chemotherapy, *12*:655
 as veterinary drug, *21*:245
L-Oleandrose, *12*:639
Olefin elastomers, *7*:686
Olefin polymers, *14*:**217**
 high-density polyethylene, *14*:242
 higher olefins, *14*:309

in microencapsulation, *13*:446
polyethylenes by Ziegler process, *14*:259
polypropylene, *14*:282
Olefins, *S*:**632**
  allylic reactions, *14*:318
  in coal tar, *19*:660
  in commercial hexane, *11*:2
  esterification, *8*:341
  higher, *14*:**313**
  nomenclature of, *11*:288; *14*:314
  in oxo process, *14*:379
Olefin sulfonates
  surfactants, *19*:522
Oleic acid, *7*:403, 405; *9*:898, 899; *8*:257, 814
  in dental impression materials, *6*:804, 827
  in eggs, *7*:664
  kinetics of reaction with *n*-butyl alcohol, *8*:320
  in manuf of metallic soap, *7*:274
  ozonolysis, *8*:821
  ozonolysis of, *14*:430
  phys prop, *1*:231
  prepn of dimer acid, *8*:820
  structure, *8*:817
  sucrose ester, *19*:154
  sulfated, *19*:526
  surfactants, *19*:526
  2-sulfoethyl ester, *19*:522
  tall oil, *19*:621
  in tallow, *18*:418
Oleic-linoleic acid oils, *8*:786, 784
  hydrogenation, *8*:803
Oleic monoethanolamide, *19*:552
Oleo, *13*:56
Oleo-dilinolenin, *7*:405
sym-Oleodipalmitin, *8*:790
sym-Oleodistearin, *8*:790
Oleo oil, *13*:182
  in margarine, *13*:59
Oleoresins, *9*:368
Oleostearine
  in margarine, *13*:59
Oleo stock, *13*:182
  in margarine, *13*:59
Oleoyl *N*-methyltauride, *19*:318
N-Oleoylsarcosinate, *19*:514
Oleum, *9*:785, 786, 797
  contact process, *19*:442
  DDT, *19*:315
  1,2-dichloroethane reaction with, *5*:151
  in dye industry, *7*:480
  in nitrification, *19*:473
  prod of, *19*:472

pyrosulfuryl chloride, *19*:404
reaction with cyanine dyes, *6*:621
succinic acid from, *19*:44
sulfamation, *19*:306
sulfonation, *15*:211; *19*:281, 294, 304, 308
sulfuric acid, *19*:444
tantalum, *19*:642
Oleuropein, *10*:50
Oleyl alcohol
  phys prop, *1*:543
  sulfation, *19*:304
Olibanum, *14*:208, 724; *17*:390
Oligo-1,6-α-glucosidase
  definition, *16*:243
  uses, *8*:190, 194
Olin process for manuf of chlorine dioxide, *5*:41
Olivacine, *1*:773
Olive oil, *8*:779, 784, 790, 786
  in alkyd resins, *1*:864
  processing, *8*:797
Olives
  processing of, *10*:50
Olivine, *13*:736
Olive T, *7*:470
  in ceramics, *4*:772
Onion, *9*:363; *14*:208
Onion powder, *9*:373
O,O-Di(*p*-chlorophenyl) *N*-acetamidinophos-phoramidothionate, *15*:919
Oolitic aragonite
  in ocean, *14*:159
Oolitic limestone, *12*:416
Opal glasses, *10*:547
Operating line
  in distillation, *7*:100
  in mass transfer column, *13*:141
Opium, *1*:796
  as a hypnotic, *11*:509
Opium alkaloids, *1*:796
Opopanax, *14*:208, 725; *17*:390
Opsin
  reaction with 11-*cis*-retinal, *21*:506
Optical bleaches, *3*:737
  uses of, *7*:477
Optical brighteners, *3*:737; *19*:13
Optical data, indices of, *6*:528
Optical rotation, *19*:156
Optical whitening agents, *3*:737
Orange
  bitter, *14*:208
  essential oil as flavors, *9*:352
  sweet, *14*:208

Orange I, *2*:880; *5*:872; *7*:482
Orange II, *2*:874; *5*:869; *7*:496
  used in hair coloring, *10*:793
Orange IV, *11*:552
Orange mineral, *15*:522, 536; *12*:275
Orange peel, *9*:353
Orangeseed oil, *8*:779, 784
Orange wax, *22*:161
Orcein, *2*:866
Orcinol, *11*:481
  sugar analysis, *19*:164
β-Orcinol, *11*:481
Oregano, *9*:373, 363
Organic acids
  chemical cleaning solutions, *S*:163
Organic fluorine
  history of industry, *9*:704
Organic glasses, *9*:252
Organic isocyanates. See *Isocyanates, organic.*
Organoaluminum compounds, *2*:26
    manuf, *2*:34
    phys prop, *2*:28
    prepn, *2*:26
    reactions, *2*:30
    uses, *2*:39
Organoboranes
  carbonylation of, *S*:507
Organochlorosilanes, *18*:236
Organolead compounds, *12*:282
    reactions, *12*:285
    synthesis, *12*:288
    types, *12*:283
Organoleptic examination
  of essential oils, *14*:185
Organoleptic testing, *14*:**336**
Organolithium compounds, *12*:547
Organometallics
  and nitrogen fixation, *S*:613
Organosilicon alkoxides, *18*:184
Organosilicon compounds, *18*:133, 201
Organosilicones
  surfactants, *19*:554
Organosodium compounds, *18*:439
Organosols, *15*:740; *21*:403
  in coating systems, *S*:899
  as fabric coatings, *5*:682
Organothiocyanates
  as insecticides, *11*:690
Organotin compounds, *20*:305, 318
Organotitanium(IV) compounds, *20*:425, 430
    sandwich compounds, *20*:433
    stability, *20*:426
    titanocenes, *20*:426

Origanum, *14*:208
Orkla process for sulfur, *19*:350
  electrical behavior of, *1*:129
  as filter medium *9*:285
Orlon, *1*:313. (See also *Acrylic fibers.*)
Ornithine, *2*:157; *12*:857
Orpiment, *2*:712; *15*:536
  in shellac, *18*:30
Orris, *14*:724
Orris root, *14*:208
  essential oil as flavor, *9*:352
Orsat apparatus for gas analysis, *10*:192
Orthanilic acid, *2*:415, 423; *19*:299
Orthoarsenic acid, *2*:722
Orthoborates, *3*:652
Orthoboric acid, *3*:612. (See also *Boric acid.*)
Orthohydrogen, *6*:476
Orthophosphate
  aquatic plant nutrient, *S*:325
Orthophosphoric acid, *9*:637; *10*:669; *15*:234, 313. (See also *Phosphoric acid.*)
  in dental cement, *6*:779, 781, 782
Orthoplumbic acid, *12*:276. (See also *Plumbic acid.*)
Orthoquinones
  o-phenylenediamines, *15*:221
Orthosilicic acid, *18*:135
Orthotelluric acid, *19*:767
Orthotitanic acid, *20*:391
Orthotitanic esters, *8*:359
Orthotolidine arsenite test, *5*:46
Orthotolidine oxalate test, *5*:46
Orthotolidine test
  for chloramines, *4*:915
Orthovanadates, *21*:172
Orthozirconic acid, *22*:630
Ortolan, *2*:889
Oryzenin, *22*:269
Oscillation
  parametric, *S*:624
Osmic acid, *8*:243
Osmium, *15*:832, 858-859
  zirconium reaction, *22*:621
Osmium carbonyls, *4*:489
Osmium compounds, *15*:874, 875
Osmium tetroxide, *12*:823; *15*:835, 875
    styrene-butadiene copolymer degradation, *S*:914
Osmocene, *15*:875
Osmometer
    Fuoss-Mead, *14*:351
    Zimm-Myerson, *14*:352

Osmophor, *14*:183
  accompanying dialysis, *7*:1
Osmosis, *14*:**345**
  reverse, *22*:75
  reverse, *22*:123
  reverse in desalination of water, *22*:58
Osmotic pressure, *14*:**345**
  molecular weights from, *16*:246
Ossein glue, *10*:610
Ostreogrycin, *16*:332
Ostwald color system, *5*:809
Othmer still
  vapor-liquid equilibria determination, *21*:233
Otto cycle
  unconventional powerplants, *S*:68
Ouabain
  as cardiovascular agent, *4*:513, 516
Ouricuri oil, *8*:779, 784
Ouricury wax, *22*:164
Outokumpu process, *19*:351
Ovin, *6*:77, 87
Ovomucoid
  in eggs, *7*:664
Ovulen, *6*:69, 87
9-Oxa-10-azaanthracene, *10*:930
α-Oxabutyraldehyde
  from butyraldehyde, *3*:867
2-Oxacortisol, :18:876
Oxacyanocarbons, *6*:630
Oxacyclopropane, *10*:898
Oxadiazolines, *10*:919
1-Oxaindene, *10*:899, 907, 908
Oxalacetic acid
  from alcoholic fermention, *21*:257
Oxalates
  from formates, *14*:362
Oxalic acid, *2*:578; *6*:922, 465; *10*:633, 640, 661; *14*:**56**
  in detergents, *6*:857
  from phenol, *15*:148
  polyamide from, *16*:2
  reaction with ethylenediamines, *7*:27
  as rust remover, *S*:164
  separation of cresols, *6*:439
  from succinic acid, *19*:142
Oxaloacetic acid
  in tobacco, *20*:508
Oxalohy-roxamic acid, *2*:592
Oxalohyroxamic acid, *2*:592
Oxaloniobic acid, *19*:633
Oxalotantalic acid, *19*:633

Oxalyl bromide, *12*:114
Oxalyl chloride, *1*:223; *10*:153
Oxalyl fluoride, *9*:509
Oxalylguanidine
  prepn from guanidine, *10*:737
Oxamic acid, *14*:367
  in beet sugar, *19*:211
Oxamide, *14*:367
Oxamides
  monosubstituted, *14*:367
Oxamines
  amine oxides, *S*:32
4*H*, 1-Oxanapthalene, *10*:898
Oxanthranol, *2*:433
3-Oxapentamethylene
  detn, *16*:178
3-Oxapentamethylene glycol
  polyester from, *16*:180
p-Oxathiane
  prepn from ethylene oxide and hydrogen sulfide, *8*:529
2*H*,1,2-Oxazine, *10*:897
4*H*,1,2-Oxazine, *10*:898
Oxazine dyes, *2*:864; *7*:471, 474, 533; *19*:432
Oxazole, *10*:896, 898, 914
Oxazoles, *12*:825
Oxazolidine, *10*:896
2,4-Oxazolidinedione, *10*:914
Oxazolidinediones
  as anticonvulsants, *11*:522
2-Oxazolidinone, *7*:31
2-Oxazolidones, *10*:914
2-Oxazoline, *10*:896
Ox-bile extract, *3*:484
Oxetanes, *10*:907; *16*:174
Oxidation colors
  in textile printing, *7*:576
Oxidation dyes
  in hair coloring, *10*:794
Oxidation of hydrocarbons. See *Hydrocarbon oxidation.*
Oxidation ponds, *22*:122
Oxidative coupling, *2*:874
Oxidoreductases
  classification, *8*:176
Oximes
  C-N rotation, *14*:72
  from hydroxylamine, *11*:505
  reduction of, *2*:108
  prepn from acetoacetic acid, *1*:154
Oximodoacetone. (See also *Isonitrosoacetone.*)
Oxindole, *11*:592

Oxine. See *8-Hydroxyquinoline.*
Oxirane. See *Ethylene oxide.*
Oxiranes
  copolymerization, *16*:174
11-Oxitigogenin, *18*:840
2-Oxo-1,2*H*-benzopyran. See *Coumarin.*
2-Oxo-4,6-bis(alkylamino)-1,2-dihydro-*s*-triazin-
  e, *22*:185
α-Oxobutyraldehyde
  from butyraldehyde, *3*:867
4-Oxobutyric acid, *19*:138
Oxochlorozirconates, *22*:644
2-Oxo-1,1,3,3-cyclohexanetetrapropionitrile, *6*:640
2-Oxohexamethylenimine. See ε-*Caprolaotam.*
Oxonium perchlorates, *5*:75
Oxonols, *16*:284
4-Oxo-9,11,13-octadecatrienoic acid, *8*:816.
  See *Licanic acid.*
5-Oxooxazolines, *10*:914
4-Oxopentanal, *10*:238
Oxophenarsine hydrochloride, *2*:730
Oxo process, *3*:840; *14*:**373**
  aldehydes, *1*:646
  butyl alcohols prod by, *3*:826, 827
  manuf of isodecanol," 1:567
  manuf of synthetic higher alcohols, *1*:560
  propionic acid, *16*:555
  in synthetic manuf of monosodium glut-
    amate, *2*:206
4-Oxo-4*H*-pyran-2-carboxylic acid, *10*:923
4-Oxo-1,4-pyran-2,6-dicarboxylic acid, *10*:923
Oxopyrazolines, *16*:762
4-Oxo-1(4*H*)-*pyridinepropionitrile, 6*:649
Oxo reaction, *14*:328
Oxyacetylene
  flame temp, *1*:172
1,3-Oxy acid, *13*:719
1,4-Oxy acid, *13*:720
  hydrogen chloride oxidation, *S*:171
1,5-Oxy acid, *13*:720
2-Oxyanthroic acid, *2*:488
Oxybiotin, *3*:522
4,4'-Oxybis(benzenesulfonhydrazide), *9*:853,
  857
Oxybutyrate
  dehydration, *13*:339
Oxychloride cements, *12*:720
Oxychlorination, *S*:169
Oxyclozanide, *14*:549
Oxydiacetic acid. See *Diglycolic acid.*
Oxydibutyric acid, *10*:673
Oxydiethanoic acid. See *Diglycolic acid.*

2,2'-Oxydiethylenebischloroformate, *4*:386
N,N-(p,p'-Oxydiphenylene) pyromellitimide,
  S:756
3,3'-Oxydipropionic acid, *2*:592
3,3'-Oxydipropionitrile
  acrylonitrile from, *6*:636, 644
  in gas-liquid chromatographic analysis of
    acetaldehyde, *1*:90
Oxygen, *2*:281; *10*:726; *14*:**390**
Oxygen, *14*:**390**
  analysis for in water, *21*:696
  carbon monoxide reaction with, *4*:430
  cellulose reactions with, *4*:596
  as contaminant in cosmetics, *6*:371
  in cryogenics, *6*:471, 472, 473
  cyanoethylation, *6*:639, 640, 644
  effects on copper, *6*:202
  as heat-transfer medium, *10*:856
  in heterochain inorganic polymers, *11*:636,
    643
  hypolimnetic depletion in lakes, *S*:323
  impurity in copper, *S*:265
  in phosphors, *12*:620
  reaction with aluminum alkyls, *2*:31
  in rockets, *6*:479
  in tobacco smoke, *20*:516
Oxygen balance of explosives, *8*:581
Oxygen compounds
  in petroleum, *14*:851
Oxygen demand
  biochemical, *21*:700
  chemical, *21*:701
Oxygen difluoride, *9*:508, 524
  as rocket fuel oxidizer, *9*:554, 631, 677
Oxygen enrichment
  copper prod, *S*:250
Oxygen-free, high-conductivity copper, *6*:183,
  185
  in explosives, *8*:644
  as propellant, *8*:688, 690
Oxygen generation system, *S*:**658**
Oxygen-index test, *S*:471
  method for textiles, *S*:945
Oxygen, liquid
α-Oxy-α'-hydroperoxyperoxides, *14*:784
Oxyhemoglobin
  ultrasonic irradation, *20*:780
Oxyhydrochlorination, *S*:169
  in prod of chlorinated hydrocarbons, *11*:336
α-Oxy-hydroperoxides, *14*:776
Oxy-Koch's acid, *13*:723
Oxyl synthesis, *4*:486

Oxymorphone, *2*:381
β-Oxynaphthoic acid, *7*:543
  pigments, *15*:569, 575
α-Oxyperoxides, *14*:776
  polymeric, *14*:786
Oxyphenylbutazone, *2*:391
Oxyquinoline benzoate, *6*:360
Oxyquinoline sulfate, *6*:360
Oxytetracycline, *20*:2, 5, 6, 20
  amebicides, *20*:80
  in bacterial infections, *3*:1
  epimers, *20*:11
  as food preservative, *10*:14, 19
  total synthesis, *20*:23
  uses, *20*:28
  as veterinary drug, *21*:244
Oxytetracycline hydrochloride, *20*:6
Oxytetracyclinonitrile, *20*:4
Oxy-Tobias acid, *13*:720
Oxytocin
  biological activity, *11*:68
  corticotropin-releasing factors, *11*:74
  hormone analogs, *11*:73
  sulfur in, *19*:369
  structure of, *11*:69
  synthesis of, *11*:72
  uses in medicine, *11*:74
Oyster shell, *12*:416
Oyster shell flour
  in drilling fluids, *7*:291, 292
Oyster shells
  for manuf of lime, *12*:669
Ozocerite, *22*:169
Ozokerite, *6*:356, 368

Ozonation
  synthesis of peroxy compounds, *14*:781
Ozone, *2*:623; *12*:823; *14*:**410**
  addition to municipal waters, *22*:87
  as a bactericide, *22*:129
  in bleaching fats, *13*:180
  chlorine reaction with, *5*:4
  chloroform reaction with, *5*:121
  from photochemical smog, *18*:404
  odor and taste control agent in water, *22*:102
  in odor control, *14*:175, 431
  ozonolysis of oleic acid, *8*:821
  properties, *8*:689
  reaction with crotonaldehyde, *6*:455
  reaction with ethylene, *8*:503
  sterilant, *18*:816
  tetrachloroethylene reaction with, *5*:198
  trichloroethylene reaction with, *5*:186
"Ozone factor," *17*:578
Ozone fluoride. See *Trioxygen difluoride.*
Ozonide
  zwitterion reactions, *14*:417
Ozonides, *14*:764, 786
  cross, *14*:418
  molozonide, *14*:418
  reactions, *14*:419
  structure of, *14*:416
Ozonization
  amine oxide prod, *S*:40
Ozonolysis, 419, *14*:416, 420
Ortho-Novum, *6*:65, 67, 87
  of oleic acid, *8*:821

P

PABA. See *p-Aminobenzoic acid.*
PAS. See *p-Aminosalicylic acid.*
PDCB. See *p-Dichlorobenzene.*
PETN. See *Pentaerythritol tetranitrate.*
PVF. See *Polyvinyl fluoride.*
PVP. See *Polyvinylpyrrolidone.*
PWP. See *Plasticized white phosphorus.*
Pachnolite, 9:534
Pachystermine A, *18*:842
  of vat dyes, 7:557
Packages, *14*:**432**
Packaging, *14*:**432**
  use of acetal resins in, *1*:105
  weighing, *14*:433
Packaging materials
  in contact with food, *10*:18
Packing materials, *14*:443
Paint, *14*:**462**, (See also *Industrial Coatings;*
  Coatings, Industrial.)
  latex for water-resistant masonry walls,
    22:152
  poly(vinyl acetate) dispersions, *21*:346
  use of lecithin in, *12*:358
Paint industry
  consumption of alkyd resins, *1*:879
  use of driers in, 7:277
Paint pigments
  sulfates as, *12*:278
  methylene chloride as, 5:117
  use of dimethylacetamide, *1*:247
  use of dimethylformamide, *10*:111
Paint removers, *14*:**485**
Paladium
  manganese compounds, *13*:4
Palatine Fast, *2*:887
Palatine Fast Orange RNA, 7:522
Palatine Fast Red BEN, *2*:888
Palladium, *15*:832, 852
  amine oxide reduction, *S*:39
  catalyst in hydrogenation, *14*:420
  colloidal catalyst, *19*:143
  cracking catalyst, *15*:37
  as dental alloy, 6:814
  hydrocarbon control in automobile exhaust,
    S:63
  in electroplating, 8:64

  in electron tubes, 8:6
  in oxidation of ethylene, *1*:87
  phenolic aldehydes, *15*:162
  in photoemitters, *15*:403
  tetracycline prod, *20*:7
Palladium chloride, *12*:122; *15*:854
  vinyl acetate catalyst, *21*:328
Palladium compounds, *15*:**867-870**
Palladium-gold alloy, *10*:692
Palladous chloride
Palma istle, 9:173
Palmarosa, *14*:209
Palmitic acid, 8:812; *18*:26
  sucrose ester, *19*:154
  tall oil, *19*:621
  in tallow, *18*:418
unsym-Palmitodiolein, 8:790
sym-Palmitodistearin, 8:790
unsym-Palmitodistearin, 8:790
Palmitoleic acid, 7:405, 790; 8:814
  in tall oil, *19*:621
  in eggs, 7:664
Palmitoleyl alcohol. See *8-Hexadecenyl-1-ol.*
3-Palmito-2-stearo-1-butyrin, *10*:619
β-Palmitoxytricarballylic acid, *2*:592
Palmityl alcohol. See *Cetyl alcohol.*
Palm kernel oil, 8:779, 785, 790
  in soap, *18*:417
Palm kernel oil meal
  as animal feed, 8:864
Palm oil, 8:779, 784, 790, 786; *18*:40
  in soap, *18*:417
Palm wax, *22*:161
Palustrine, *1*:759, 763
Pamco process
  synthetic crude oil, *S*:186
Pancreatin, *S*:294
Pandermite, *3*:627, 628, 649
2-Pantanol, *2*:374. (See also *Amyl alcohols.*)
Panthothenic acid derivatives
  in hair tonics, *10*:802
Pantothenic acid
  antagonists for, *22*:407
  in fish, 9:322
  in yeast, *22*:516, 517
d-Pantothenyl alcohol, *10*:780

Pantothenyl alcohol derivatives
  in hair tonics, 10:802
Papaverine, 1:769, 778, 798; 11:470
  as cardiovascular agent, 4:523
  vanillin in synthesis of, 21:192
Paper, 14:**494**
  clays used in, 5:581
  coatings for, 5:706
  coloring for, 7:581
  as electric insulation, 11:790
  improving strength of, 1:283
  as laminated reinforcement, 12:189
  polyvinyl alcohol adhesives, 21:363
  polyvinyl acetate as binder for, 21:350
  microencapsulation of, 13:454
  synthetic, 19:126
  waterproof, 22:149
  wet strength increased by ethyleneamines, 7:34
  water-resistant, 22:149
Paper adhesives, 1:394
Paper brighteners, 3:746
  in amino acid separation, 2:162
Paper chromatography, 1:90; 5:421
  use of TCNE in, 6:626
  sugars, 19:165
Paper coating
  use of abherents in, 1:6
Paper coloring, 7:581
  dyes used, 7:582
Paper, cyanoethylated, 6:660
Paper industry
  use of aluminum sulfate in, 2:61
  use of ammonium bisulfite in, 2:295
  use of formaldehyde in, 10:94
  use of natural gums in, 10:741
  use of sodium aluminate in, 2:10
Paper pulp. See Pulp.
  use of magnesium sulfite in, 12:735
Paper treating
  use of amino resins in, 2:254
Papin, 5:594, 595, 598; 10:9
  in prevention of gelation in egg yolk, 7:669
Paprika, 9:365, 369
  antagonist of, 21:489
Parabanic acid, 10:912; 21:114
  from uric acid, 21:111
Parabutylaldehyde, 3:875
Parabutyraldehyde, 3:875
Paraffin
  in candles, 4:58
  in dental material, 6:840, 841, 842, 843

in matches, 13:163, 165
  in match prod, 13:161
  synthetic, 4:446
Paraffinic oils
  as lubricants, 12:564
Paraffins
  alkylation of, 1:883
  chlorinated, 5:231-240
  in coal tar, 19:660
  nitration of, 13:794
  nomenclature of, 11:288
  normal in hexane prod, 11:2
  catalytic dehydrogenation, S:653
  chlorination-dehydrochlorination of, S:653
Paraffin wax
  chlorinated paraffins, 15:107
Paraffin waxes, 15:92, 97
  as abherent, 1:7
  synthetic, 15:93, 104
Paraform
  in developer, 16:502
  phenolic resins, 15:176, 190
  reaction with styrene, 10:86
Paraformaldehyde, 2:228; 7:304; 10:77, 81.
  (See also Formaldehyde.)
  formal from, 16:254
  manuf, 10:89
  prepn of bis(chloromethyl) ether, 8:477
  sterilizing agent for sugar, 19:210
Parahydrogen, 6:476
Paraisobutyraldehyde
  from isobutyraldehyde, 3:868
Paralactic acid, 12:171
  from acetaldehyde, 1:81
  from formaldehyde, 1:645
Paraldehyde. (See also Acetaldyde.)
  as hypnotic, 11:512
Paramagnetic resonance, 8:874
Paramagnetism, 12:737
Paramelaconite, 6:275
Paramenthane hydroperoxide, 7:680
Paramethadione
  as anticonvulsant, 11:522
Paranitraniline Red, 2:889
Paraplex plasticizers
  in cellulose plastics, 4:657
Paraquat, 14:545; 19:671
  herbicide, 22:207
  metabolism of, 22:209
  soil chemistry, 18:537
Para reds, 15:573
  chlorinated, 15:574

Pararosaniline, *20*:674; *22*:431
  N-alkylation, *20*:698
  basification, *20*:696
  manuf, *20*:709
  reactions of, *20*:694
  in Schiff's reagent, *20*:730
  substituent effects on color, *20*:689
  via Homolka's method, *20*:711
Pararosolic acid, *20*:693, 720
  basification, *20*:697
Parasitic infections, chemotherapy, *14*:**532**
Parasympathin. See *Acetylcholine.*
Parathion, *15*:322, 328
  soil chemistry, *18*:528
Parathormone, *20*:266
Parathyroid hormone, *20*:266
Parawolframates. See *Wolfram compounds.*
Parbendazole
  anthelmintic, *21*:249
Parchmentizing
  of paper, *22*:150
Parchment paper
  use of aluminum formate in, *2*:15
α-Parinaric acid, *8*:815
Paris" black, *4*:277 acetoarsenite.)
Paris green, *6*:269; *11*:680; *15*:538. (See also
  Cupric acetoarsenite.)
  soil chemistry, *18*:537
Parkerite, *13*:736
Park's nucleotide, *14*:672
Parobiosamine, *19*:42
Paromamine, *19*:42
Paromomycin, *20*:70
  amebicide, *20*:80, 82
Paromomycins, *19*:33, 41, 42
Parsley, *9*:352, 363, 373
Parsley leaf, *14*:209
Parsley seed, *14*:209
Parsonite, *21*:8
Particleboard, *21*:608
Particulates
  in air pollution, *S*:733
  removal from polluted water, *S*:714
Parting agents, *1*:1
Pascal
  as an SI unit, *S*:992
Passivators, *6*:300, 317
Passivity
  of active metals, *6*:300, 317
Pasteurization, *18*:806
  beer, *3*:324
  of butter, *13*:549
  of milk, *13*:522

Pastilles, *6*:51
Patchouli, *14*:209
Patent Blue A, *20*:719
Patent Blue AE, *20*:708
Patent Blue AF, *7*:526, *20*:708
Patent Blue AF Ext Conc, *7*:522, 526
Patent Blue Na, *5*:868
Patent Blue V, *20*:707, 719
Patent Blue VF, *7*:525; *20*:708
Patent Blues, *20*:707
Patenting
  and successful research, *S*:859
Patent leather, *12*:340
  sources, *12*:510
Patents, *14*:**552**
Patronite, *21*:159
Pattinson's lead white, *12*:270
Paving brick
  clays used in, *5*:566
Peacock Blue, *15*:565, 577, *20*:708
Peanut oil, *8*:779, 785, 786, 790; *18*:39
  in alkyd resins, *1*:864
  in margarine, *13*:59
  parenteral solvent, *15*:128
  sulfated, *19*:526
Peanuts
  as animal feeds, *8*:863
Pearl ash, *16*:387
Pearlite, *13*:328; *18*:766, 771
Peat
  oxalic acid from, *14*:359
  composition of, *5*:627
Peat wax, *22*:168
Pecans, *14*:145
Pechblende," *21*:1
Pectase, *8*:199
  action and uses, *8*:187
Pectases, *14*:650
Pectic acid, *8*:199; *14*:636
Pectic enzymes, *8*:186
  action and uses, *8*:188
  uses, *8*:190, 198
Pectic substances, *14*:**636**
Pectin, *8*:198
  in beet sugar, *19*:211
  in cotton, *6*:407
  in confectionery, *6*:43
  used as food additive, *10*:6
  from tobacco, *20*:514
Pectinase, *8*:185
  action and uses, *8*:199
Pectinases, *10*:9; *14*:650
Pectin esterases, *8*:199; *14*:650

Pectinic acid, 8:198; 14:636
Pectin jellies, 6:52
Pectin-methylesterase
 uses, 8:190, 199
Pectin methylesterases, 14:650
Pectin-polygalacturonase
 uses, 8:190
Pectins, 19:230
 clarification of wines, 22:316
Pegmatites, 13:768
 tantalum in, 19:630
Pelargonic acid, 8:812
 in dental material, 6:827
 esters, 15:762
 oxygen in synthesis of, 14:405
 prepn, 8:821
 spec, 8:836
Pellagra preventing factor, 21:509
Pelletization, 9:435
Pellicle, 15:364
Pellotine, 1:769
Peltier effect, 20:148
Pencils, graphite
 manuf of, 4:331
Penetration theory
 of mass transfer, 13:136
Penicillin, 10:107
 discovery, 20:1
Penicillin G. See Benzylpenicillin.
 amebiasis, 20:75
 amebicide, 20:80
Penicillin acylase, 14:697
Penicillinase, 3:21; 8:186; 14:653, 681
 action and uses, 8:190
Penicillins, 14:**652**
 sulfur in cephalosporins, 19:369
 veterinary uses, 21:243
Penicillium
 hydrocarbon utilization, S:839
Pennyroyal, 14:209
Penta. See Pentaerythritol tetranitrate.
Pentaborane, 8:693
 properties, 8:689
Pentabromochlorocyclohexane, 3:778
Pentabromofluorobenzene, 9:781
Pentabromophenol
 fire retardant, S:483
Pentacarbocyanines, 6:617
Pentachloroanisole
 from hexachlorobenzene, 5:262
Pentachlorobenzene, 5:262
Pentachlorobenzotrifluoride, 9:777, 778

Pentachlorobutadiene
 from trichloroethylene, 5:188
2,3,4,5,6-Pentachloro-1-butoxybenzene
γ-Pentachlorocyclohexane, 18:527
Pentachloroethane, 5:164-166
 chloroform from, 5:124
 tetrachloroethylene from, 5:199
 toxicity of, 5:93, 165
 from trichloroethylene, 5:188
Pentachlorofluoroacetone, 9:755
Pentachloronitrobenzene
 as fungicide, 10:226
Pentachlorophenol, 5:336
 from coal tar, 19:672
 as fungicide, 10:229, 235
 from hexachlorobenzene, 5:262
 wood preservative, 22:379
Pentachloropyridine
 fluorination of, 9:693, 799
Pentachlorothiophenol, 15:768
 extractive distillation agent, 22:473
Pentacosanoic acid, 8:813
n-Pentacosoic acid, 8:813
1,1,2,3,3-Pentacyanopropene, 6:629
Pentacyclopentadiene, 6:691
1H,1H-Pentadecafluoro-1-octanol, polymethacrylic ester of
 surface tension, 9:715, 720
1H,1H-Pentadecafluorooctyl alcohol, 9:729,
 730, 731
1H,1H-Pentadecafluorooctyl ethanesulfonate,
 9:729, 730, 731
Pentadecanamide
 form, 2:67
Pentadecanoic acid, 8:812
1-Pentadecanol
 phys prop, 1:543
8-Pentadecanone, 12:104
n-Pentadecyl alcohol
 phys prop, 1:543
n-Pentadecylic acid, 8:812
m-Pentadecylphenol. See Hydrocardanol.
2,3-Pentadione, 12:140
Pentaerythritol, 1:589-593; 6:450; 7:412, 413;
 15:793
 alkyd resins, 15:455
 in alkyd resins, 1:861
 esters, 19:495
 fire-retardant paints, 22:376
 polyurethans, 21:62
 stabilizers, 21:390
 tall oil, 19:625

Pentaerythritol chloral
 as hypnotic, *11*:512
Pentaerythritol disulfite, *19*:489
 in esterification of fatty acids, *8*:819
Pentaerythritol tetranitrate, *4*:521; *8*:608
 explosive characteristics, *8*:633
 manuf Biazzi process, *8*:595
 in Pentolite, *8*:628
 sensitivity characteristics, *8*:632
 stability characteristics, *8*:634
Pentaerythrose
 as intermediate in prepn of pentaerythritol,
  1:590
 prepn, *1*:82
Pentaethylenehexamine, *7*:22, 23
 properties, *7*:24
 specifications, *7*:32
 toxicity, *7*:33
Penta (explosive). See *Pentaerythritol tetranitrate.*
Pentafluoroaniline, *9*:790, 794
Pentafluoroanisole, *9*:793
Pentafluorobenzaldehyde, *9*:795
2,3,4,5,6-Pentafluorobenzamide, *9*:795
Pentafluorobenzene, *9*:691, 775, 787, 795
Pentafluorobenzene derivatives, *9*:788-797
2,3,4,5,6-Pentafluorocinnamic acid, *9*:795
2,3,4,5,6-Pentafluoromandelonitrile, *9*:795
Pentafluorophenylhydrizine, *9*:787, 796
 prepn of 2,3,4,5,6-pentafluorotoluene, *9*:794
 prepn of pentafluorobenzaldehyde, *7*:195
 reactions, *9*:792
Pentafluorophenylmagnesium bromide, *9*:788;
 20:440
Pentafluoropropylene
 vinylidene fluoride copolymers, *21*:269
Pentafluoropyridine, *9*:693, 776, 798, 799
Pentafluorothiophenol, *9*:793
2,3,4,5,6-Pentafluorotoluene, *9*:794
1,2,3,4,5-Pentafluorotoluene, *9*:691
Pentaflurorphenyllithium, *9*:788, 792, 795,
 797; *20*:440
Pentagermane, *10*:523
Pentaglycerol trinitrate. See *Trimethylolethane
 trinitrate.*
Pentahydroxybenzene, *16*:212
2,3',4,5',6-Pentahydroxybiphenyl, *16*:209
2,3,4,5,6-Pentahydroxyhexylamine, *2*:359
Pentalead octoxide, *12*:276
Pentalin. See *Pentachloroethane.*
Pentamethine cyanines, *6*:615
Pentamethylbenzene, *16*:300

3,3-Pentamethylenediaziridine, *19*:487
3,3-Pentamethyleneoxazirane, *19*:487
Pentamethylene oxide, *10*:921
1,2,5,7-Pentamethylene tetrasulfide, *8*:418
Pentamethylenetetrazole, *10*:918
Pentamethylenimine, *11*:526
Pentamethyl-*p*-rosaniline, *9*:896
Pentane. (See also *Pentanes.*)
 chlorinolysis of, *5*:91
 in commercial hexanes, *11*:1
n-Pentane, *9*:722
 as component of natural gas, *10*:448
 in hydroisomerization, *11*:446
 in manuf of amyl alcohols, *2*:375
 in polystyrene foams, *19*:119
 prop compared to water, *21*:675
Pentanedioic acid. See *Glutaric acid.*
1,5-Pentanediol, *1*:258; *10*:247
 manuf, *21*:416
1,5-Pentanediol diacrylate
 phys prop, *1*:289
2,4-Pentanedione. See *Acetyl acetone.*
2,3-Pentanedione, *12*:140
Pentanes, *14*:**707**
 branched in petroleum, *14*:847
 hexachlorocyclopentadiene from, *5*:241
1,3,3,5-Pentanetetracarbonitrile, *6*:641
2,4,6,2',4'-Pentanitro-6'-methyldiphenylamine, *7*:48
Pentanoic acid, *8*:812
1-Pentanol. (See also *Amyl alcohols.*)
 prop, *2*:374
2-Pentanol
 in prepn of *sec*-amyl ether, *8*:493
3-Pentanol, *2*:374. (See also *Amyl alcohols.*)
3-Pentanone. See *Diethyl ketone.*
2-Pentanone, *12*:104
Pentaphenylantimony, *2*:582
Pentaphenylarsenic, *2*:729
Pentaphenylbismuth, *3*:545
Pentaphenylphosphorus, *15*:303 *14*:546
Pentasodium antimony(III) bis(pyrocatechol-
 3,5-disulfonate)heptahydrate, *14*:546
Pentasodium triphosphate
 in detergents, *6*:857
Pentasodium tripolyphosphate, *15*:244
Pentatitanium trisilicide, *20*:417
18-Pentatriacontane, *12*:104
Pentatriacontanoic acid, *8*:813
1-Pentene
 sulfonic acids, *19*:314
 polymerization of, *7*:66
2-Pentene
 sulfonic acids, *19*:314

Pentenes, S:641
2-Pentenoic acid
    from acetaldehyde, 1:84
1-Penten-3-ol
    phys prop, 1:629
3-Penten-2-ol, 12:157
3-Penten-2-one, 6:694; 12:105
2-Penten-4-ynals
    in prepn of tricarbocyanines, 6:616
Pentitols, 1:572
Pentlandite, 13:736; 15:841; 19:350, 758
Pentobarbital, 3:61, 70, 72
Pentobarbital sodium, 2:403
Pentolinium
    as ganglionic blockader, 4:519
Pentolite, 8:628
    explosive characteristics, 8:632-634
Pentosan
    in manuf of furfural, 10:242
Pentosans
    in nuts, 14:129
Pentose
    furfural from, 4:140
    in manuf of furfural, 10:242
Pentrit. See Pentaerythritol tetranitrate.
tert-Pentyl acetate, 8:370
Pentyl acrylate
    phys prop, 1:287
α-Pentylcinnamaldehyde, 5:521
    flavor and perfume materials, 9:358
Pentylenes, S:641
    alkylation, 15:42
Pentylenetetrazole, 10:664; 21:535
n-Pentyl ether. (See also n-Amyl ether.)
    phys prop, 8:471
1-Pentyl-3-ol
    phys prop, 1:619
4-Pentynoic acid, 8:816
Pepper, 14:209
    black, 9:363, 365, 373
    white, 9:363, 365, 373
Peppermint, 14:210
    essential oils as flavor, 9:352, 353
    oil as denaturant, 8:450
    in laundry preparations, S:295
    in margarine prod, 13:56
Pepsin, 8:178; 10:9. (See also Enzymes, indus-
    trial.)
    uses, 8:190
Peptidases
    classification, 8:176

Peptides, 2:71
    cyclic, 16:306
    heterodectic, 16:307
    heteromeric cyclic, 16:307
    homodectic, 16:307
    homomeric cyclic, 16:307
    linear, 16:306
Peptolides, 16:335
Peptone
    in prepn of glutamic acid, 2:204
Peptones, 19:274
Peracetic acid. See Peroxyacetic acid.
Peracids. See Peroxy acids.
Perbenzoic acid. See Peroxybenzoic acid.
Perbromoethylene, 8:420
Perbunan. See Acrylonitrile-butadiene copoly-
    mers.
Perchloramide, 9:603
Perchlorate cells, 5:78
Perchlorate esters, 5:75
Perchlorates, 5:67
    oxygen generation systems, S:658
Perchloric acid, 5:27, 61-84; 10:161
Perchloric acid anhydride
    formation of, 5:4
Perchloroethane. See Hexachloroethane.
Perchloroethylene (See also Tetrachloroethyl-
    ene.), 1:207; 5:195-203; 8:42; 9:403
    carbon tetrachloride from, 5:134
    from carbon tetrachloride, 5:130, 131
    from pentachloroethane, 5:124
    in reactor technology, 14:90
    trichloroethylene from, 5:190
    use in drycleaning, 7:308, 315, 319, 324
    water repellency, 20:484
Perchlorofluoroacetones, 9:754
Perchlorofluoro-2-propanol, 9:763
Perchloromethanethiol, 19:374
Perchloropentadienoic acid, 9:718, 720
Perchloropenta-2,4-dienoic acid, 9:711
Perchlorylation, 10:156
Perchlorylbenzene, 9:604
Perchloryl fluoride, 5:74; 9:598; 10:156
    chemical prop, 9:601, 695
    fluorosuccinic acid from, 19:136
    phys prop, 8°689; 9:598
N-Perchlorylpiperidine, 9:603
Perchloryltoluene, 9:604
Percrotonic acid. See Peroxycrotonic acid.
Perferrates, 12:40
Perferrites, 12:40
Perfluoroacenaphthene, 9:691, 798

Perfluoroacetic acid, 9:689, 773
Perfluoroacetone, 9:509
Perfluoroacyl-1-aziridines, 11:539
Perfluoroadipic acid, 9:774
Perfluoroanthracene, 9:798
Perfluorobenzene, 9:689, 788. (See also *Hexa-fluorobenzene.*)
    prepn of, 3:373
Perfluorobiphenyl. See *Decafluorobiphenyl.*
Perfluoro-*n*-butanoic acid, 9:722
Perfluorobutyric acid, 9:713, 723, 724, 773. (See also *Heptafluorobutyric acid.*)
    water-repellent agent, 22:145
    wettability, 9:713, 723, 724
Perfluorocaproic acid
    wettability, 9:713, 773
Perfluorocaprylic acid
    wettability, 9:713, 773
Perfluoro compounds, 19:686-846
Perfluorocyclobutane, 9:810
Perfluorocyclohexene, 19:286
Perfluorocyclopentadiene, 9:691
Perfluorodecahydronaphthalene, 9:797
Perfluorodecanoic acid, 9:723, 724
11-(Perfluorodecyl) undecanoic acid, 9:716, 720, 727
Perfluoro-*n*-dibutyl ether, 9:722
Perfluorodicyclohexyl, 9:797
Perfluorododecanoic acid, 9:723, 724
Perfluoroenanthic acid, 9:773
Perfluoroethane, 9:509
Perfluoroethylene, 12:582
Perfluoroglutaric acid, 9:774
Perfluoroheptane, 9:705
Perfluoro-*n*-heptane, 9:722
Perfluorohexanoic acid, 9:723
Perfluoroindane, 9:798
Perfluoroisobutylene, 9:741, 810
Perfluoroisopropylsulfur(IV) fluorides, 9:804
Perfluorokerosene, 9:706
Perfluorokerosine, 9:714
Perfluorolauric acid
    surface tension as coating, 9:718
    wettability, 9:713
Perfluoromalonic acid, 9:774
Perfluoro(methylcyclohexane), 9:688, 722
    in bacterial infections, 3:1
Perfluoromethylhexane, 9:705
Perfluoromethyl perfluorovinyl ether, 21:423
Perfluoronaphthalene, 9:691, 775, 776, 797
Perfluoroneopentane, 9:803
Perfluoro-*n*-octane, 9:722
Perfluorooctanoic acid, 9:719, 723, 724

Perfluoro-*n*-pentane, 9:722
Perfluoroperoxyacetic acid, 12:108
Perfluoropinacol, 9:763
Perfluoropolyphenylene sulfide, 9:793
Perfluoropropane, 9:509, 688. (See also *Octa-fluoropropane.*)
    as rocket fuel oxidizer, 9:524
Perfluoropropionic acid, 9:773
Perfluoropropylene, 7:696; 12:582
Perfluoropropyl ether, 9:689
Perfluoropyrene, 9:798
Perfluoro-*p*-quaterphenyl, 9:791
Perfluorosuccinic acid, 9:774
Perfluoro-*p*-terphenyl, 9:791
Perfluorotributylamine, 9:799
Perfluorovaleric acid
    wettability, 9:713, 773
Perfluoro(1-methyl-decahydronaphthalene), 9:688
Performic acid. See *Peroxyformic acid.*
Perfume industry
    use of acetals in, 1:108
Perfumes, 14:**717**
    acetophenone as ingredient, 1:170
    in aftershave lotions, 6:354
    benzyl ether as solvent in, 8:495
    from hydrolysis of esters, 8:365
    in deodorants, 6:357
    in face powder, 6:358
    in nail enamels, 6:370
    use of alcohol in, 8:448
    use of cresols in, 6:442
    use of coumarin in manuf, 6:431
Perhydroacenaphthene, S:4
Perhydroanthracene, S:4
Perhydrofluorene, S:4
Perhydrophenanthrene, S:4
Peri acid, 7:478, 484; 13:713
Periclase, 12:724; 17:227, 236
    as abrasive, 1:25
    in cement, 4:686, 688
Perilla oil, 7:403, 404; 8:779, 785
Perimycin, 16:138
Perinone orange, 15:584
Perinone Red, 15:584
Periodates, 11:860
Periodic acid, 1:577; 2:415; 10:640; 11:860
Periodic table, 8:94
Periodoacetic acid. See *Triiodoacetic acid.*
Periplogenin, 18:845
Perkin Green, 7:464
Perkin reaction, 1:643
Perkin ring closure, 1:233

Perlite
   as abrasive, *1*:25
   in gypsum board, *21*:623
   as thermal insulator, *11*:829
Permanent Bordeaux FRR Extra, *2*:908
Permanent Orange. See *Permatone Orange.*
Permanent Violet, *15*:541
Permanent Yellow NGG, *2*:909
   in bleaching fats, *13*:180
Permanent waves for hair, *10*:785
Permanganate. (See also *Manganese com-
   pounds.)*
   oxidation of perchloryltoluene, *9*:604
Permanganates, *13*:33, 49
   in odor control, *14*:175
Permanganic acid, *13*:32, 34
Permanganic anhydride, *13*:32
Permanganyl ion, *13*:12
Permatone Orange, *5*:869
Permeability
   of air in textiles, *22*:148
   of water vapor in textiles, *22*:148
   of wood, *22*:369
Perosamine, *16*:138
Perovskite, *20*:349
   ferroelectric, *9*:5,6
Peroxidase, *15*:683
   action and uses, *8*:182, 184, 186
Peroxide
   in polymerization of vinyl esters of saturated
      fatty acids, *8*:820
Peroxides, *14*:**746**
   in bleaching fats, *13*:180
   as hazardous chemicals, *9*:292
   from ethers, *8*:472
   oxygen generation systems, *S*:663
Peroxozirconates, *22*:655
Peroxyacetic acid, *8*:239-241, 266, 285
   from acetaldehyde, *1*:80
   aldrin reaction with, *5*:246
   in bleaching of nylon prior to dyeing, *7*:519
   isodrin reaction with, *5*:247
   manuf, *8*:250, 407, 410
   in manuf of butylene oxide, *8*:289
   reaction with allyl ether, *8*:494
   shrinkproofing wool, *22*:409
   from sorbic acid, *18*:590
   sterilant, *18*:817
Peroxy acids, *14*:755, 794
   as epoxidizing agents, *8*:243
   properties, *14*:797
   synthesis, *14*:795

Peroxyacyl nitrates
   in photochemical smog, *S*:50, 52
Peroxybenzimidic acid, *8*:244
Peroxybenzoic acid, *6*:465; *8*:243, 247, 266;
   10:905; *12*:73
Peroxybenzoyl nitrate
   photochemical smog, *S*:52
Peroxyborates, *14*:758
   paramagnetic, *14*:762, 764
Peroxybutyric acid, *3*:879
Peroxycarbonates, *14*:760
Peroxycarboximidic acid, *8*:244
Peroxycarboxylic acids
   hydrolysis of, *11*:403
Peroxy compounds, *14*:**746**
Peroxycrotonic acid, *6*:466
Peroxydisulfuric acid. (See also *Peroxysulfuric
   acid.)*
   platinum, *15*:852
Peroxydisulfuryl difluoride, *9*:679
Peroxyesters, *14*:804
   synthesis, *14*:804
Peroxyformic acid, *8*:241, 248, 266
   reaction with pentafluoroaniline, *9*:794
α-Peroxy-hydroperoxides, *14*:776
Peroxymaleic acid, *8*:243
Peroxymonophosphoric acid, *14*:757
Peroxymonosulfuric acid, *14*:755. (See also
   Peroxysulfuric acid.)
   shrinkproofing wool, *22*:409
Peroxynitric acid, *14*:761
α-Peroxyperoxides, *14*:776
   polymeric, *14*:786
Peroxyphosphates, *14*:757
Peroxypropionic acid, *8*:249
Peroxysilicates, *14*:761
Peroxysuccinic acid, *19*:142
Peroxysulfuric acids. See *Peroxides.*
Peroxysulfurous acid, *14*:760
Peroxytrifluoroacetic acid, *8*:243, 247; *9*:691,
   794
Peroxytungstic acid. See *Peroxywolframic acid.*
Peroxywolframates, *22*:352
Peroxywolframic acid, *12*:823; *22*:352
Perseitol, *1*:576
Persian Gulf oxide, *15*:517
Persson process
   for manuf of chlorine dioxide, *5*:42
Persthene, *13*:736
Persulfuric acids. See *Peroxymonosulfuric acid;
   Peroxydisulfuric acid.*
Perthane
   in water, *21*:701; *11*693, 734

Peru balsam, *10*:802; *14*:724
Pervalohydroxamic acid, *8*:854
Perylene, *10*:144
Perylene Scarlet, *15*:585
Perylene-3,4,9,10-tetracarboxylic acid, *15*:585
Perylene-3,4,9,10-tetracarboxylic dianhydride, *19*:435
Pest control
  economics, *15*:908
Pesticides, *10*:17. (See also *Insecticides; Poisons, economic.*)
  adsorption, *18*:521
  arsenical, *18*:537
  clays used in, *5*:583
  common, *18*:519
  residual in soil, *18*:516
  from shale oil, *18*:18
  in water, *21*:701
  and wildlife, *18*:518
Petalite, *12*:530
  in ceramics, *4*:772
Pethidine, *2*:384
Petitgrain bigarade, *14*:210
Petrichloral
  as hypnotic, *11*:512
Petrolatum, *15*:59, 92, 102
  in food processing, *10*:10
  in ointments, *15*:126
  liquid. (See also *Mineral oil.*)
Petroleum, *14*:**835**
  asphalt, *9*:57; *15*:2. (See also *Asphalt.*)
  composition, *14*:845-855
  cresols manuf from, *6*:436
  manuf of cyclohexane from, *6*:679
  nomenclature, *14*:835-837
  origin, *14*:838-845
  primary yeast from, *22*:543
  products, *15*:77-92
  refinery processes, *15*:1-77
  reserves, *10*:186
  resources, *14*:856-867
  sulfonates, *19*:297
  sulfur removal, *15*:45. (See also *Hydroprocesses.*)
Petroleum catalysts, *2*:9
Petroleum coke, *4*:165
  calcination of, *4*:169
Petroleum ether, *10*:164
  thermal pyrolysis to prep ethylene, *8*:504
Petroleum fractions
  butadiene from, *3*:803

Petroleum fuels
  cracking processes, *10*:415
  gasification, *10*:388
  hydrogenolysis processes, *10*:404
  pyrolysis processes, *10*:391
Petroleum gases
  as motor fuels, *10*:492
Petroleum hydrocarbons
  gasification of, *10*:388
Petroleum industry
  corrosion inhibitors in, *6*:330
  corrosion in refining, *6*:340
  evaluation of corrosion inhibitors, *6*:334
  liquid extraction in, *8*:758
  use of acetal resins in, *1*:105
  use of ethyleneimines in, *7*:35
  use of fluidization in, *9*:434
Petroleum lubricants, *12*:564
Petroleum naphtha
  pressure steam reforming of, *11*:350
Petroleum oils, *12*:611; *19*:296
  as heat-transfer media, *10*:847
  as insecticides, *11*:716
Petroleum products, *15*:**77**
  stabilization of, *2*:598
  use of lecithin in, *12*:358
Petroleum refinery by-products
  used as oils, *7*:414
Petroleum refinery processes, *15*:1
Petroleum refining
  catalysts used in, *4*:538
  use of aluminum chloride in, *2*:21
  use of ammonia in, *2*:294
  use of hydrogen in, *11*:370
Petroleum resins, *11*:248-250
Petroleum sulfonate
  detergent used in drycleaning, *7*:313
Petroleum sulfonates
  in paint removers, *14*:491
  surfactants, *19*:518
Petroleum waxes, *15*:**92**; *22*:169
Petroselaidic acid, *8*:814
Petroselinic acid, *8*:814
Petzite, *19*:759
Pewter
  tin alloys, *20*:291
Peyote, *16*:646
Peyronie's disease, *21*:583
pH
  adjustment of in treated water, *22*:81
  control in swimming water, *22*:130

pH electrodes, *11*:383, 385
pH instrumentation
pH instrumentation
   pH meters, *11*:386
   temperature effects, *11*:387
Phaeophorbide, *19*:751
Phanquone, *20*:70
   amebicide, *20*:89
Pharmaceutical industry
   humidity conditions within, *1*:486
   use of air conditioning in, *1*:497
   use of fatty alcohols in, *1*:556
   use of mannitol in, *1*:583
   use of sorbitol in, *1*:583
Pharmaceutical intermediates
   dialkylaminoalkyl chlorides as, *2*:148, 150
Pharmaceutical products
   use of lecithin, *12*:357
Pharmaceuticals, *15*:**112**
   microencapsulation of, *13*:453
   polyvinylpyrrolidone uses, *21*:436
   use of aniline in, *2*:417
   use of emulsifier, *8*:149
   use of glycerin, *10*:628
   use of gelatin in, *10*:499
   use of gum arabic in, *10*:745
   use of hexane in manuf of, *11*:4
   use of metallic soaps in, *7*:285
   use of natural gums in, *10*:741
   use of polyethylene glycols, *10*:657
Phase rule, *15*:**133**; *21*:196
Phase test for pigments, *5*:350
α-Phellandrene, *19*:815
β-Phellandrene, *19*:815
Phellogenic acid
   in cork, *6*:284
Phellonic acid
   in cork, *6*:283
Phenacaine hydrochloride, *2*:405, 407
Phenacemide
   as anticonvulsant, *11*:524
Phenacetin. (See also *Acetophenetidine.*)
   phys prop, *2*:68
   as stabilizer in bleaches, *10*:799
   as antipyretic, *2*:390
Phenacite, *3*:453
Phenacyl bromide. See *α-Bromoacetophenone.*
Phenacyl bromide, *1*:170
Phenacyl chloride, *1*:170; *12*:107. (See also
   *α-Chloroacetophenone, ω-Chloroacetophen-
   one.*)
Phenacyl esters, *8*:372

Phenamine Black E, *7*:536
Phenampromid, *2*:387
Phenanthraquinone, *9*:403
Phenanthrene, *12*:590
   from coal tar, *19*:669
9,10-Phenanthrenequinone, *15*:221
Phenanthrenes
   in petroleum, *14*:849
Phenanthridine, *10*:899, 934
Phenanthridines, *10*:933
13*H*-Phenanthro[1',2',4,5]cyclopenta[1,2-*b*-]-
   pyrrocoline, *10*:937
Phenanthrofuran, *10*:908
o-Phenanthroline, *7*:279
   in water analysis, *21*:698
Phenanthrolines, *10*:934
Phenarsazines, *10*:929
Phenarsone sulfoxylate, *2*:731
Phenates, *15*:148. (See also *Phenolates; Phen-
   oxides.*)
Phenazathionium, *2*:859
Phenazine, *2*:415; *10*:899, 920
Phenazinium, *2*:859
Phenazo Black BH, *7*:539
Phenazocine, *2*:383
Phenazoxonium, *2*:859
[*v*-Phenenyltris(oxyethylene)]tris[triethylam-
   monium iodide], *2*:410
Phenethecillin, *3*:15
Phenethyl acetate
   prod and sales of, *8*:377
Phenethyl isobutyrate
   flavor and perfume materials, *9*:358
Phenethyl phenylacetate
   flavor and perfume materials, *9*:358
Phenethyl α-toluate
   flavor and perfume materials, *9*:358
o-Phenetidine, *2*:422
p-Phenetidine, *2*:423
   Acid Violet dyes from, *20*:717
Phenetole, *15*:148
Phenindamine tartrate
   antihistamine, *11*:39
Phenindione, *5*:600, 601, 602
Pheniramine, *2*:152
Pheniramine maleate
   as an anticonvulsant, *11*:508
   antihistamine, *11*:39
   control for epileptic seizures, *11*:519
   as a hypnotic, *11*:508
Phenobarbital, *3*:61, 70, 72
   manuf, *3*:67

Phenol, *6*:434, 645; *7*:193, 497
  acid-catalyzed alkylation, *1*:894
  aldehyde polymerization, *15*:178
  alkaline fusion, *19*:315
  alkylation with isobutylene, *1*:894, 904
  as antiseptic, *2*:605, 622, 623
  azo coupling of, *2*:871
  from benzene, *3*:398
  from coal tar, *19*:672
  in coal tar, *6*:437; *19*:657, 668
  condensation with styrene, *8*:289
  as denaturant, *8*:450
  desulfonation product, *15*:209- -
  dimethyl sulfate reaction, *19*:487
  in embalming fluids, *8*:101
  as fungicide, *10*:235
  metal-catalyzed alkylation, *1*:895
  methyl derivatives, *S*:271
  prepn of cyclohexanol, *6*:683
  in prepn of *p*-(4-dimethylaminoanilino)
    phenol, *2*:220
  nitration, *13*:892
  perchlorylation reaction, *9*:604
  polyester fiber solvent, *16*:155
  in prepn of adipic acid, *1*:411
  reaction with acetaldehyde, *1*:83
  reaction with dichloroacetic acid, *8*:418
  separation of cresols, *6*:439
  solvent refining, *15*:57
  sulfonation, *15*:211; *19*:299
  vapor-phase methylation, *S*:271
  in prepn of *p*-(4-aminoanilino)phenol, *2*:218
Phenol Blue
  structural formulas, *5*:770
Phenol Red, *11*:552; *20*:**727**
Phenolates, *15*:148. (See also *Phenates; Phenoxides.*)
Phenol-butyral resin
  as metal-bonding adhesive, *1*:390
Phenol-2,4-disulfonic acid, *15*:208
Phenol-epoxy resin
  as adhesive, *1*:389
Phenol-formaldehyde resin. See *Phenolic resins.*
  in particleboard, *21*:610
Phenolic aldehydes, *15*:**160**
Phenolic condensation
  in ion exchange, *11*:875
Phenolic ether. (See also *Diphenyl oxide.*)
Phenolic ethers, *15*:**165**
Phenolic fibers, *S*:**667**

Phenolic resins, *15*:**176**
  as adhesives, *1*:388
  as modifier in alkyd resins, *1*:866
  phys prop of phenol-formaldehyde resin,
    9:860
  step-growth, *16*:226
  used as coating compds, *8*:106
Phenolmaleins, *12*:822
Phenolphthalein, *11*:552; *15*:148
  alkalimetry, *20*:698
  as cathartic, *4*:588
  as laxative, *20*:731
  from phthalic acid, *15*:456
  from phthalic anhydride, *15*:447
  polyesters from, *16*:169
  synthesis of, *20*:727
  terephthalate polyesters, *16*:182
Phenolphthalimidine, *16*:169
Phenol removal, *11*:893
Phenols
  alkylation, *19*:488
  antioxidants, *17*:524
  chlorination of, *5*:328
  chloroform reaction with, *5*:123
  as herbicides, *22*:203
  metabolism of herbicidal, *22*:205
  as paint removers, *14*:490
  phenylenediamines, *15*:220
  reaction as antioxidants, *2*:590, 591
  sulfated, *19*:305
  sulfonation, *19*:298
  solubility parameters of various, *S*:895
Phenolsulfonic acids, *6*:813; *15*:**208**
1-Phenol-4-sulfonic acid, *2*:223, 224; *7*:479
Phenosafranine, *15*:363
Phenothiazine, *2*:143, 420, 859; *10*:920, 930
Phenothiazines, *16*:650, 667
Phenoxazine, *2*:859; *10*:889, 930
Phenoxides, *15*:148. (See also *Phenates; Phenolates.*)
  phenolic ethers from, *15*:165
Phenoxyacetone, *12*:105
Phenoxyalkanoic acids
  herbicidal, *22*:176
  metabolism as herbicides, *22*:182
  soil chemistry, *18*:530
2-Phenoxyethyl acrylate
  phys prop, *1*:288
2-Phenoxyethyl alcohol
  prepn, *8*:529
2-Phenoxyethyl bromide, *14*:534

2-Phenoxyethyl isobutyrate
  prod and sales of, 8:377
Phenoxyethyl penicillin, 14:698
α-Phenoxyethylpenicillin, 3:15
Phenoxymethylpenicillin, 3:13; 14:653, 666, 687, 700
2-Phenoxy-1-phenylethanol, 8:289
2-Phenoxy-2-phenylethanol, 8:289
1-Phenoxy-2-propanone. See Phenoxyacetone.
β-Phenoxypropionic acid, 10:926
3-Phenoxypropionitrile, 6:645
Phenoxypropylpenicillin, 3:15, 16
Phenoxy resins, 15:173
Phenpiazines, 10:929
Phenprocoumon, 5:600, 601
Phensuximide
  as anticonvulsant, 11:522
Phenylacetaldehyde, 3:366; 14:729
Phenylacetaldehyde ethylene glycol acetal, 1:108
Phenylacetamide, 15:213
N-Phenylacetamide. See Acetanilide.
Phenyl acetate, 15:148
Phenylacetic acid, 14:669; 15:**213**
  amphetamine from, 16:672
  phys prop, 1:239
  as synthetic flavor, 9:354
Phenylacetone, 6:670; 15:214
  amphetamines, 15:215
Phenylacetonitrile, 6:672
  from benzyl chloride, 5:283
Phenylacetyl, 12:140
Phenylacetyl chloride, 15:214
N-Phenylacetylcysteine, 14:670
Phenylacetylene
  thiol addition, 20:209
Phenylacetylpivaloylmethane, 12:148
Phenylacetylurea, 11:524
o-Phenylacridine
  dyes, 20:710
3-Phenylacrolein. See Cinnamaldehyde.
trans-Phenylacrylic acid. See Cinnamic acid.
3-Phenylacrylic acid. See Cinnamic acid.
β-Phenylacrylophenone, 12:107
Phenylalanine, 2:157
  benzylpenicillin synthesis, 14:669
  in wine, 22:316
D-Phenylalanine
  in polymyxins, 16:322
Phenylalanine hydroxylase, 2:189
Phenylalanine mustard, S:84
Phenylaminoguanidine
  prepd from guanidine, 10:737

N-Phenyl-p-aminophenol, 15.218
Phenylammonium chloride. See Aniline hydrochloride.
2-Phenyl-4-(D-arabotetrahydroxybutyl)-osotriazole, 4:140
N-Phenylaniline. See Diphenylamine.
Phenylarsenic tetrafluoride, 2:728
Phenylarsine, 6:651
3,3'-(Phenylarsylene)dipropionitrile, 6:651
p-Phenylazophenacyl bromide
  in separation of flavor compds, 9:339
5-Phenylazosalicylic acid, 17:723
N-Phenylbenzamide. See Benzanilide.
Phenylbenzene, 7:191
2-Phenyl-1,4-benzopyrone, 12:107
2-Phenylbenzoselenazole, 2:601
m-Phenyl benzotrifluoride, 9:568
2-Phenylbenzoxazole, 2:215
1-Phenyl-2-benzoylcyclopropane, 19:327
1-Phenyl-1-benzoylethylene oxide, 12:146
1-Phenyl-2-benzyl-1,3-butanedione, 12:148
Phenylbenzylglyoxal, 12:140
as-Phenyl-o-biphenylethylene, 10:143
Phenylbismuth dichloride, 3:545
Phenyl borates
  ortho-alkylation of phenols, S:30
Phenylboron dichloride, 3:718
Phenylbutadienes
  from isochloroprene, 5:228
1-Phenyl-1,3-butanedione, 12:148
1-Phenyl-2,3-butanedione, 12:143, 148
Phenylbutazone, 2:391; 12:857
2-Phenyl-3-butenoic acid, 10:732
3-Phenyl-1-buten-3-ol
  phys prop, 1:630
4-Phenyl-3-buten-2-one, 12:107
Phenyl butyl ketone, 12:107
Phenyl tert-butyl ketone, 12:107
N-Phenylbutyramide, 3:882
γ-Phenylbutyric acid, 10:150
  phys prop, 1:239
Phenylcarbamates
  soil chemistry, 18:534
Phenyl chloroformate, 4:386
Phenylchlorosilanes, 18:233
2-Phenylchromone, 12:107
2-Phenylchromones, 10:925
Phenyl cinnamate, 5:520
Phenyl cyanide, 2:416
Phenylcyclohexane, 7:198, 200. (See also Cyclohexyl benzene.)
4-Phenylcyclohexane, 3:786
Phenyldiazonium fluoborate, 9:568

7-Phenyl-7*H*-dibenzo[*c,g*]carbazole, *7*:43
Phenyldichloramine. See *N,N-Dichloroaniline.*
Phenyldicresyl phosphate, *6*:442
Phenyldiethanolamine
 phys prop, *1*:811
 uses, *1*:823
Phenyldiethanolamine succinate
 vanillin determination, *21*:192
9-Phenyl-9,10-dihydrophenanthrene, *10*:143
1-Phenyl-5,5-dimethyl-2,4-hexanedione, *12*:148
3-Phenyl-1,1-dimethylurea, *22*:190
Phenyl dineopentyl phosphite
 poly(vinyl alcohol)copolymer, *21*:358
4-Phenyl-1,4-dioxolane, *15*:190
o-Phenylenebis(dimethylarsine), *22*:658
o-Phenylenediacetic acid
 phys prop, *1*:239
 in hair dyes, *10*:794
m-Phenylenediamine, *15*:216, 218. (See also
 Phenylenediamines.)
 diazotization, *15*:219
 dye for furs, *10*:306
 dyes, *19*:435
 in hair coloring, *10*:795
 as liquid epoxy resin hardener, *8*:309, 310,
  311
 manuf, *2*:82, 90, 96
 in prepn of 1,3-difluorobenzene, *9*:783
o-Phenylenediamine, *10*:929; *15*:**216**
 dye for furs, *10*:306
 in hair coloring, *10*:795
 manuf, *2*:361
 menadione assay, *21*:595
p-Phenylenediamine, *7* 565; *15*:**217**
 azo dyes from, *2*:891
 developing agent, *15*:371
 derivs as optical brighteners, *3*:743
 in hair coloring, *10*:795
 epoxy resins, *20*:491
 manuf, *2*:87, 96
 photography, *15*:374
 in prepn of *p*-(4-aminoanilino)phenol, *2*:218
 in prepn of 1,4-difluorobenzene, *9*:784
m-Phenylenediamine dihydrochloride, *6*:456
m-Phenylenediamine-4,6-disulfonic acid,
 15:217
m-Phenylenediamine hydrochloride
 in hair coloring, *10*:795
p-Phenylenediamine hydrochloride
 in hair coloring, *10*:795
Phenylene diamines
 in color photography, *5*:818, 819, 826
 polyimides from, *S*:749

p-Phenylenediamines
 antiozonants, *17*:529
p-Phenylenediamine sulfate
 in hair coloring, *10*:795
m-Phenylenediamine-4-sulfonic acid, *15*:217
p-Phenyleneterephthalamide, *16*:21
Phenylephrine, *4*:517
 as cardiovascular agent, *4*:517
2-Phenylethanol. (See also *β-Phenylethyl alcohol.*)
1-Phenyl-3-(D-*erythro*-trihydroxypropyl)flava-
 zole, *4*:140
Phenylethanolamine
 phys prop, *1*:811
 uses, *1*:823
Phenyl ether, *15*:165, 169, 172
Phenylethyl acetate, *3*:448
Phenylethyl alcohol, *14*:729, 735
 in rose oil, *14*:181
 by sodium reduction, *18*:451
β-Phenylethyl alcohol, *3*:445-449; *8*:528
Phenylethylamine, *2*:179
2-Phenylethylamine, *15*:213; *19*:318
β-Phenylethylamine, *1*:767; *8*:231; *14*:669
Phenyl ethyl ether
 phys prop, *8*:471
5,5-Phenylethylhydantoin
 as anticonvulsant, *11*:520
β-Phenylethyl lactate
 prepn, *8*:353
N-Phenyl-*N*-(ethylmercuri)-*p*-toluenesulfon-
 amide, *10*:223
Phenylformamide, *2*:423. (See also *Formanilide.*)
Phenylformamidine, *6*:574
Phenylformic acid. See *Benzoic acid.*
Phenyl γ-pyridyl ketone, *16*:673
Phenylglycine, *6*:596
 in indigo synthesis, *18*:451
n-Phenylglycine, *2*:453; *7*:624
Phenylglycolic acid, *1*:239; *22*:633
Phenylguanazole, *2*:244
1-Phenyl-1,3-hexadiene, *3*:870
2-Phenylhexafluoro-2-propanol, *9*:846
4-Phenylhexahydroacetophenone, *1*:140
1-Phenyl-1,3-hexanediol monoacetate, *3*:870
5-Phenylhydantoin, *11*:159
Phenylhydrazine, *6*:631, 650; *12*:123; *19*:318
 reaction with acetaldehyde, *1*:82
Phenylhydrazone, *6*:456
Phenylhydroxylamine
 in prepn of *p*-aminophenol, *2*:219

N-Phenylhydroxylamine
  prepn, 2:77
1-Phenyl-3-hydroxypiperidine, 2:415
Phenyl iodide, 11:864
Phenyl isocyanate, 2:414; 9:759; 12:46, 47
  polyurethans from, 21:72
  reactions, 12:48
4-Phenylisocyanate
  castor oil reaction with, 4:533
Phenyl isocyanurate, 20:665
Phenyl isonitrile
  from carbon tetrachloride, 5:132
1-Phenylisopropyl-2-butene, 3:792
  from butadiene, 3:792
Phenyl ketene, 12:88
Phenylketonuria, 2:189
Phenyllithium, 2:582; 12:152, 553
  alkylation, 19:488
  chloroform reaction with, 5:122
  isochloroprene reaction with, 5:228
Phenylmagnesium bromide, 8:370, 494;
  12:120; 16:673
N-Phenylmalanil, 12:842
N-Phenylmaleimide, 12:826
N-Phenylmalimide, 12:841
Phenylmercaptomethylpenicillin, 3:15, 16
Phenylmercuric acetate, 13:236, 243
Phenylmercuric acetate, 10:222, 223
  from benzene, 3:374
  swimming pool algicide, 22:130
  as fungicide, 10:232, 234, 235
Phenylmercuric benzoate
  as denaturant, 8:450
Phenylmercuric borate
  as leather fungicide, 10:233
Phenylmercuric chloride, 10:223
  as denaturant, 8:450
Phenylmercuric nitrate, 2:620
  antimicrobial agent, 15:128
  as denaturant, 8:450
  as leather fungicide, 10:233
Phenylmercuric oleate
  as fungicide, 10:232
Phenylmercuric propionate
  as fungicide, 10:232
Phenylmercuric stearate
  as leather fungicide, 10:233
Phenylmercuric 8-hydroxyquinolinate
  as fungicide, 10:234
Phenylmercury acetate
  soil chemistry, 18:538
Phenylmercury compds, 13:231

Phenylmercury triethanolammonium lactate,
  10:232
  as fungicide, 10:232
1-Phenyl-3-(2-methoxyphenyl)-1,3-propanedione,
  12:148
2-Phenyl-4-methyl-3,6-dihydro-1,2-oxazine, 12:69
Phenyl methyl ether
  phys prop, 8:471
Phenyl methyl ketone, 12:107
1-Phenyl-3-methyl-5-pyrazolone, 15:573
Phenyl methylsulfinylmethyl ketone, 19:326
N-Phenylmorpholine, 8:487
N-Phenyl-N-methyl-1-naphthylamine, 20:721
Phenyl-1-naphthylamine. See N-Phenyl-
  1-naphthylamine.
Phenyl-2-naphthylamine, 10:607. (See also N-
  Phenyl-2-naphthylamine.)
N-Phenyl-1-naphthylamine, 2:399, 600; 7:40,
  46, 47; 12:574; 20:701,721
  as dye intermediate, 7:49
N-Phenyl-2-naphthylamine, 2:413, 417, 600;
  7:40, 41, 43, 46, 47; 10:607; 17:518
  as dye intermediate, 7:49
Phenylnitromethane, 12:161
  from benzyl chloride, 5:283
Phenyl o-nitrophenyl ether, 15:722
Phenyloxirane. See Styrene oxide.
2-Phenyl-5-oxazolone, 9:763
5-Phenyl-2,4-pentadienoic acid. See Cinnamalacetic acid.
1-Phenyl-1,3-pentanedione, 12:148
Phenylperi acid indophenol, 19:436
  as dye carrier, 16:151
o-Phenylphenol, 7:193, 543
  as food preservative, 10:14
  as fungicide, 10:234, 235, 780
  use in commercial dyeing of triacetate, 1:135
p-Phenylphenol, 7:193
p-Phenylphenol, 1:911
  phenolic resins, 15:184, 206
  resins, 15:176
  thermoplastics, 15:792
Phenylphenols, 15:154
o,p-Phenylphenols
  sulfation, 19:305
N-Phenyl-p-phenylenediamine, 7:49, 565;
  15:218, 222. (See also p-Aminodiphenylenedi-
  amine.)
4-Phenyl-4-phenylsulfonylheptanedinitrile, 6:642
n-Phenyl-p-phenylenediamine, 7:42, 565
Phenylphosphine, 6:651
3-Phenylphthalidyl sulfate, 2:436

Phenylpropane
  polymeric in lignin, 22:365
3-Phenyl-1-propanecarboxylic acid. See γ-
  Phenylbutyric acid.
1-Phenyl-1,3-propanediol diacetate, 10:86
2-Phenyl-2-propanol, 12:120
3-Phenylpropene, 8:495
3-Phenyl-2-propen-1-ol. See *Cinnamyl alcohol.*
Phenyl propenyl ketone, 12:107
β-Phenylpropionic acid. See *Hydrocinnamic acid.*
3-Phenylpropionic acid. See *Hydrocinnamic acid.*
Phenylpropyl aldehyde, 14:729
1-Phenyl-3-pyrazolidone, 15:372
α-Phenyl-2-pyridylacetonitrile, 16:673
3-Phenyl-(1*H*, 3*H*)quinazoline-2,4-dione, 2:415
1-Phenylpyrrolidine, 2:415
Phenyl salicylate, 15:148; 17:730
  as denaturant, 8:450
  as dye carrier, 16:151
  UV absorber, 21:120
4-Phenylsemicarbazone, 6:456
Phenylsilane, 18:181
N-Phenylsuccinimide, 12:842
Phenylsulfamates
  cyclamates, 19:600
Phenyl sulfone, 9:677
Phenylsulfonylazide, 19:329
3-Phenylsulfonylpropionitrile, 6:645
Phenyl sulfur trifluoride, 9:696
N-Phenyl-4-thiazolylcarboxamidine hydro-
  chloride, 14:542
3-Phenylthiopropionitrile, 6:646
Phenylthiourea, 15:917
N-Phenylthiourea, 2:414
Phenyltitanium triisopropoxide, 20:425
  prepn, 20:432
Phenyltoloxamine citrate
  antihistamine, 11:40
Phenyltriethoxysilane, 18:217
N-Phenyl-*N'*-*p*-tolyl-*p*-phenylenediamine, 7:46
Phenyl undecyl ketone, 12:107
Phenylureas
  soil chemistry, 18:533
Phenyl vinyl ether, 21:414
Phenyl vinyl ketone, 12:107, 161
Phenyramidol, 2:388
Pheophorbide *a*, 5:341
Pheophytins, 5:343, 348
Pheromones, 11:51

Phillips gage, 21:130
Phillips process
  butadiene from *n*-butane, 3:799
Phlogopite, 13:399; 15:512
Phloionic acid
  in cork, 6:284
Phloramine, 16:202
Phloretin, 16:201
Phloroacylophenones, 16:208
Phlorobenzophenone, 16:203
Phloroglucine, 2:629; 16:190, 192, 201-209
  antithyroid agent, 20:270
Phloroglucinol. See *Phloroglucine.*
  gelling agent, 21:355
Phloxine, 7:522
Phloxine B, 5:870
Phonograph record compositions, 15:**225**
Phonograph records
  prod by electroforming, 8:41
Phonons, 15:425; 17:851
  definition, 4:801
Phorate
  soil chemistry, 18:528
Phorbin
  structure, 5:342
Phormium, 9:171, 178
  analysis, 9:173
Phorone, 12:105, 113
Phosfon, 15:683
Phosgene, 1:233; 2:222, 223, 414; S674
  carbon tetrachloride from, 5:134
  carbon monoxide in prod of, 4:432
  in chemical warfare, 4:870, 871, 872
  from carbonyl sulfide, 19:373
  in manuf coumarin, 6:429
  in manuf of acetic anhydride, 8:408
  in manuf of chloroformic esters, 4:388
  in manuf of isocyanates, 12:55
  methane reaction with, 5:107
  photochemistry, 15:337
  phys prop, 1:224
  polycarbonates, 16:106
  polyester end-group analysis, 16:179
  prepn from acetaldehyde, 1:83
  in prepn of acetyl chloride, 1:140
  in prepn of acrylic anhydride, 1:295
  in prepn of isocyanates, 12:53
  reaction with 1,4-butanediol, 10:669
  reaction with 1,3-butylene glycol, 10:666
  reaction with di-*n*-propylamine, 22:194
  reaction with epoxide, 8:273
  reaction with ethylenediamine, 7:27, 414
  reaction with ethylene oxide, 8:531

Phosnic bronze, *6*:193
Phosphatase, *8*:178. (See also *Enzymes, industrial.*)
   acid in milk, *13*:514
   action and uses, *8*:182, 184, 186
   alkaline in milk, *13*:514
Phosphate
   detn in water, *21*:695
   in multinutrient fertilizers, *S*:339
   separation by flotation, *9*:395
   ultra-high analysis, *S*:362
Phosphate coatings, *13*:293
Phosphate esters, *15*:309
Phosphate fertilizers, *9*:32, 47, 78
Phosphate glasses, *10*:545
Phosphate ore
   separation by flotation, *9*:394
Phosphate peroxyhydrates, *14*:762
Phosphate rock, *15*:261
   as fertilizer, *9*:80, 81
   analysis, *9*:82
Phosphates, *15*:232-276. (See also *entries under the cation, such as Sodium phosphate; Monosodium phosphate, etc.*)
   in meat curing, *13*:171
   removal from water systems, *22*:118
   rock phosphate, *4*:4
Phosphatidylcholines, *12*:343
Phosphatizing, *13*:293
Phosphides, *15*:**276**
Phosphine, *2*:263; *6*:651; *12*:121
   manganese complex, *13*:7
   from phosphides, *15*:292
   from phosphorus, *15*:280
   reaction with boron trifluoride, *9*:556
   zirconium compd adducts, *22*:658
Phosphine oxides, *19*:320
Phosphines, *15*:330
Phosphite esters, *15*:309
Phospho-cozymase, *21*:513
Phosphofructokinase
   ultrasonic prepn, *20*:781
Phospholipids, *8*:777
   in fish, *9*:321
   inositol in, *11*:673
   in nuts, *14*:132
   in yeast mitochondria, *22*:512
Phosphomolybdates, *16*:517
Phosphomolybdic acid, *2*:313; *6*:446; *13*:656; *22*:658
   zirconium detn, *22*:634
Phosphonitrile chloride
   reaction with ethylenimine, *11*:529

Phosphonitrilic chloride, *11*:640; *15*:301
Phosphonitrilic halides, *15*:310
Phosphonitrilic hexaamide
   as a fertilizer, *S*:362
Phosphor bronzes, *6*:194, 206, 241; *15*:294
   annealing temperatures, *6*:224
   magnetic prop, *6*:220
   stress relief for, *6*:240
Phosphor copper, *15*:294
Phosphorescence, *12*:625
   and fluorescence, *9*:489
Phosphoribosyl pyrophosphate, *21*:521
Phosphoric acid, *8*:273, 781. (See also *Orthophosphoric acid.*)
   in carbonated beverages, *4*:338
   as catalyst, *8*:409, 435
   as catalyst in commercial alkylation of benzene with propylene, *1*:889
   as catalyst for esterification, *8*:326
   as catalyst in manuf of amines, *8*:455
   as catalyst in prepn of ethyl alcohol, *8*:502
   cumene synthesis, *15*:150
   in cumene prod, *6*:545
   dehydration catalyst in prepn of ethers, *8*:474
   in detergents, *6*:857
   as drying agent, *7*:395
   in dyeing of wools with acid dyes, *7*:523
   electrolytes in fuel cells, *S*:396
   as fire-retardant for cellulose, *9*:306
   as food preservative, *10*:14
   intermediate in fertilizer industry, *9*:86
   in lecithin, *12*:343
   in prepn of butyl iodide, *8*:492
   in prepn of ethylene, *8*:454
   reaction with boron trifluoride, *9*:557
   as stabilizing agent in food, *10*:16, 161
   sulfuric acid in prod of, *19*:481
   as synergist, *2*:592
   yeast prod, *22*:534
Phosphoric acids, *15*:**232**
Phosphoric anhydride, *9*:637, 638, 643; *15*:311
Phosphorilase
   action and uses, *8*:185
Phosphorite, *4*:4
Phosphorites
   composition of marine, *14*:163
   from ocean, *14*:162
Phosphorodithioic acid
   S-(2-cyanoethyl)*O,O'*-diethyl ester, *6*:646
   O,O'-diethyl ester, *6*:646
Phosphoroxytriamine
   in prepn of flame-retardant finish, *9*:310

Phosphors
  cadmium, 3:902
Phosphor tin, 15:294
Phosphorus, 12:616; 15:**276**
  alloys, 15:292-295
  aquatic plant nutrients, S:325
  black, 15:278, 299
  copper deoxidized with, 6:183, 199, 210, 212
  compounds with boron, 3:736
  content in soils, 9:35
  in copper alloys, 6:250
  effects on brasses, 6:228, 241
  flame-retardant polyesters, 20:832
  ferroalloy, 18:751
  in ferromanganese, 12:889
  in inorganic polymers, 11:636, 645
  as insecticide, 11:683
  in pig iron, 12:19; 18:718, 721
  as plant nutrient, 9:27
  red, 15:277
  red in matches, 13:160, 164
  silane compounds, 18:200
  rodenticide, 15:916
  white, 15:276, 280, 299
  white in matches, 13:160
  yellow, 15:276
Phosphorus acid, 15:311, 316
Phosphorus compounds, 15:**295**
Phosphorus halides, 15:305
Phosphorus heptasulfide
  thiophene synthesis, 20:224
Phosphorus nitride, 9:802
Phosphorus nitrides, 15:310
Phosphorus oxides, 15:310
Phosphorus oxychloride, 2:414; 6:696; 9:306;
  15:304, 307, 313, 321
  ether from, 16:254
  Fischer's aldehyde from, 16:297
  in prepn of acryloyl chloride, 1:295
  prepn of bis(4-chlorobutyl)ether, 8:492
  reaction with ethylene oxide, 8:530
  as starch modifier, 10:17
Phosphorus oxyfluoride, 9:635, 636
Phosphorus pentabromide, 12:122; 15:302
  15:302, 305, 307
Phosphorus pentachloride, 2:444; 6:447; 7:42;
  12:122, 821; 15:302, 305, 307
  ketone reaction with, 5:251
  reaction with ethyl ether, 8:479
  sulfonic acids, 19:313
  thionyl chloride, 19:398
Phosphorus pentafluoride, 9:635; 15:302

Phosphorus pentasulfide, 6:427; 10:107;
  12:119, 158; 15:317
  in prepn of thioacetanilide, 1:150
  succinic acid reaction, 19:141
Phosphorus pentoxide, 6:629; 8:416; 9:306,
  639; 12:158; 15:233, 301, 311
  ACS grade, 15:266
  dehydration catalyst in prepn of ethers,
    8:474
  as drying agent, 7:395
  as glass former, 10:538
  in glass-ceramics, 10:548
  manuf, 15:313
  perchloric acid reaction with, 5:4
  in prepn of butyl iodide, 8:492
  prepn of difluorophosphoric acid, 9:640
  prepn of hexafluorophosphoric acid, 9:641
  prepn of monofluorophosphoric acid, 9:639
  pyrosulfuryl chloride, 19:404
  reaction with isopropyl ether, 8:488
  sulfonic acids, 19:314
  sulfoxides, 19:328
Phosphorus polysulfide, 10:901
Phosphorus sesquisulfide
  in matches, 13:165
Phosphorus sesquisulfide, 15:317
Phosphorus sulfide, 19:392
Phosphorus sulfides, 15:317
Phosphorus sulfochloride, 15:305, 308
Phosphorus sulfofluoride, 15:305
Phosphorus tetroxide, 15:311
Phosphorus thiochloride, 15:305
Phosphorus tribromide, 8:428; 15:310
Phosphorus trichloride, 2:414; 9:306, 602;
  15:305, 306, 321
  amine oxide reduction, S:39
  in prepn of acetyl chloride, 1:140
  prepn of fatty acid chlorides, 8:823
  sulfur monochloride, 19:392
Phosphorus trioxide, 9:636; 15:300, 311
  burning temp of, 4:899
  as glass former, 10:538
  heat of combustion of, 4:898
  in incendiary agents, 4:898
  as military smoke agent, 4:902
  structure, S:1
Phosphorus, white
Phosphorus-32, 17:120
  phys prop, 17:5
  uses, 17:7
Phosphorylation
  of aziridines for pest controls, 11:542

Phosphoryl chloride, 6:615; 10:903, 904; 15:305
Phosphoryl fluoride, 9:635, 636
12-Phosphowolframic acid
  in pigment manuf, 22:357
Phosphowolframicmolybdic acid, 15:579
Phosphowolframicmolybdic acid lakes, 20:673
Phosphowolframicmolybdic lake, 20:722, 728
Phossy jaw," 13:160
Photocathode, 15:397
Photocathodes
  cesium antimonide, 15:404
Photocells, 15:**396, 409-443**
Photochemical chlorination of hydrocarbons, 5:89
Photochemical smog, 2:815
Photochemical technology, 15:**331**
  commercial synthesis, 15:336
Photoconductive cells applications, 15:410-413
Photoconductive effect, 15:409
  oscillatory, 15:409
Photoconductivity
  extrinsic, 15:435
Photoconductors, 15:415-418
Photocopying
  silver, 17:364
  silver halide, 17:350
Photodielectric effect, 15:409
Photodiode, 15:397
Photoelectric emission, 8:10
Photoelectroluminescence, 15:410
Photoelectromagnetic effect, 15:409
Photoemission, 15:397, 409
Photoemitters
  prepn, 15:402-409
Photoengraving, 16:522-526
Photofluorometers, 5:798
Photographic products
  developers, 2:219-221
  use of gelation, 10:499
Photography, 15:**355**
  color, 5:812-845
Photography materials industry
  humidity conditions within, 1:487
  use of air conditioning in, 1:497
Photogravure, 16:526-528
Photoisopyrocalciferol, 21:557
Photolithography, 16:528-530
Photoluminescence, 12:617
Photolysis
  prod of free radicals, 10:121
Photometers
  photoelectric transmission, 20:742

Photometry, 5:791
  filter, 5:792
Photomicrography, 13:494
Photomultipliers, 15:399
  scintillation counting, 15:402
Photosensitive glasses, 10:547
Photosynthesis, 1:649, 654
Photosynthetic gas exchanger
  use of algae in, 1:664
Phototubes, 15:**396**
Photovoltaic cells, applications, 15:413-415
Photovoltaic effect, 15:409
Phthalamic acid, 15:447
Phthalates, 8:313
Phthalazine, 10:899
1,4-Phthalazinediones, 10:928
Phthalazines, 10:899, 920, 928
Phthalein dyes, 20:**727**
Phthalic acid, 13:676
  sonochemical reactions, 20:778
  phys prop, 1:239
m-Phthalic acid, 15:457-462. (See also Iso-phthalic acid.)
p-Phthalic acid, 15:462-472. (See also Tere-phthalic acid.) 15:444-457
Phthalic acids, 15:**444**
  terephthalic acid from, 15:466
Phthalic anhydride, 2:413; 7:415, 479; 8:244; 9:403; 13:678; 15:444-457
  in alkyd resins, 1:854, 862
  coal tar naphthalene, 19:673
  fluorescein synthesis, 22:430
  in manuf of benzoic acid, 3:424
  in manuf of chloroanthraquinone, 2:483
  in manuf of quinizarin, 2:474
  from naphthalene, 13:670, 686
  phenol condensation, 15:148
  phenol reaction, 20:727
  phthalocyanines, 15:490
  phys prop, 1:212
  polyesters, 21:62
  polyester with, 16:170
  in prepn of anthraquinone, 2:434, 435
  prod, 13:684
  Rhodamine B synthesis, 22:434
  reaction with benzene, 10:162, 661
  tall oil, 19:625
  from o-xylene, 22:497
Phthalimide, 2:459; 6:650; 15:447, 456, 488
  from o-xylene, 22:485
α-Phthalimidoglutarimide
  as hypnotic, 11:515

Phthalocyanine blues
   as pigments, 14:473
Phthalocyanine compounds, 15:**488**
Phthalocyanine dyes, 6:631; 7:466, 471, 475;
   15:456; 19:300, 424
Phthalocyanine green, 15:490
   as pigments, 14:473
Phthalocyanines, 15:580; 16:856
   pigments, 15:556, 801
Phthalohydrazides, 10:928
Phthalonic acid, 13:676
   commercial products, 15:490
Phthalonine compounds
Phthalonitrile, 15:447, 456, 489, 490
   from o-xylene, 22:485
o-Phthalonitrile, S:602
Phthalonitriles, S:601
Phthaloyl chloride, 1:144; 12:821; 15:447
Phthalyl glycolates, 10:637
Phthalylsulfacetamide
   structural formula, 3:2
Phthalylsulfathiazole, 19:267, 269, 272
   structural formula, 3:2
Phthiocol, 21:587, 592
Phylloquinone
   vitamin K$_{1(20)}$, 21:586
Phyllosilicates, 18:49
Physeteric acid, 8:814
Physical data sources, 12:500
Physical quantities
   dimensionally independent, S:986
Physostigmine, 1:772, 778, 782
   veterinary drug, 21:252
Phytadiene
   dl-α-tocopheryl acetate prod, 21:579
Phytane, 14:847
Phytic acid, 2:592; 16:517
Phytin
   in nuts, 14:132
Phytoglycogen, 18:675
Phytohormones, 15:675
Phytohormones, 11:51
Phytol, 5:341
   dl-α-tocopheryl acetate prod, 21:579
   vitamin K$_{1(20)}$ from, 21:589
Phytonadione. See vitamin K$_{1(20)}$
Phytosterol, 16:690
   in nuts, 14:132
Phytosterols
   in tobacco leaf, 20:511
Phytyl halide
   dl-α-tocopheryl acetate prod, 21:579

Piassava, 9:181
Pichia
   in alkane utilization, S:839
Pickle curing
   of meats, 13:173
Pickles
   processing of, 10:50
Pickling, 13:289
Pickling, acid
Pickling (of metals), 13:288-290
   inhibitors in, 6:321
Pickling of steel, S:**689**
Picloram, 22:210
α-Picoline, 10:86
3-Picoline
   oxidation of, 21:514
   oxidation of with potassium permanganate,
      21:528
4-Picoline
3-Picoline hydrochloride
   nicotinic acid synthesis, 21:514
Picolines, 16:780, 801; 6:508
   from acetaldehyde, 1:81
   amine oxides from, S:43
   in coal tar, 19:670
   prepn from acrolein, 1:265
   quarternization, 19:253
   reaction with formaldehyde, 10:86
Picolinic acid, 15:681
   phys prop, 1:240
γ-Picolinic acid, 21:527
Picolinium, 15:682
Picramic acid, 2:216
   in prepn of diazodinitrophenol, 8:587
Picratol, 8:628
Picric acid, 15:148
   as explosive, 8:617
   explosive characteristics, 8:633
   from phenol, 15:210
   piperazine reaction, 15:643
   prepn of ammonium picrate, 8:619
   reaction with ethylenediamines, 7:27
   sensitivity characteristics, 8:632
   stability characteristics, 8:634
Picrochromite, 17:237
Picrotoxin, 21:535
Piezoelectric effect
   inverse, 20:785
Piezoelectricity, 9:1
Pig iron. See Iron.
Pig lead. See Lead.
Pigment Black 1, 2:864; 7:565. (See also
   Aniline Black.)

Pigment E, 5:499, 510
Pigment Green B, 15:588
Pigment Green 10, 2:908
Pigment M-50, 5:510
Pigment Red 1, 2:889
Pigment Red 100, 5:871
Pigment Red 12, 2:908
Pigment Red 3, 2:908
Pigment Red 38, 2:909
Pigment Red 4, 5:871
Pigment Red 40, 5:869, 870
Pigment Red 49, 5:870
Pigment Red 53, 5:869
Pigment Red 57, 2:907; 5:869
Pigment Red 63, 5:871
Pigment Red 64, 5:870
Pigment Yellow 16, 2:909
Pigmentation
    of polypropylene fibers, S:828
    in rubber, 17:550
Pigment colors
    measurement of, 5:809
Pigment dyes, 15:570
Pigments, 7:572, 576; 15:**495.** (See also *Color-*
    ants.)
    azo, 2:907
    in cosmetics, 6:357-370
    in face powders, 6:357
    furnaces used in manuf, 10:293
    graphite as, 4:332
    polyvinylpyrrolidone in, 21:437
    titanium dioxide, 20:397
Pikromycin, 12:636, 641
Pilchard oil, 7:399, 403
Pilocarpine, 1:776, 778, 799
    veterinary drug, 21:252
Pilocereine, 1:770
Pilot plants, 15:**605.** (See also *Microplants.*)
Pima. See *Cotton.*
Pimanthrene, 17:483
Pimarcin
    amebicides, 20:82
Pimaricin, 12:698; 16:137, 141
Pimelic Acid, 1:248
    biotin and, 3:526
    in dental material, 6:827
    prepn, 1:247; 3:786
    phys prop, 1:232
Pimelic ketone. See *Cyclohexanone.*
    essential oils as flavor, 9:373
Pimenta berries, 9:352, 365
Pimenta berry, 14:210

Pimenta leaf, 14:210
Pinacol, 12:109
Pinacolone, 12:104
    in chemical warfare, 4:875
Pinacolyl methylphosphorofluoridate
Pinacoxysilanes, 18:242
    from acetone, 1:160
Pinacyanol, 6:606, 609, 611
Pinakryptol yellow, 15:363
Pinane, 19:828
    from α-pinene, 19:813
Pinane hydroperoxide, 19:812
Pinanol, 19:825
Pindone, 15:920
Pineapples
    citric acid from, 5:531
Pinene
    camphor manuf from, 4:55
α-Pinene, 8:248; 16:690; 19:811, 820, 822, 832,
    834; 20:753
β-Pinene, 6:630; 19:803, 813, 824, 825; 20:753
d-αPinene, 17:385
Pinene hydrochloride, 4:55, 57
Pinene oxide, 8:247
α-Pinene oxide, 19:832
Pine oil, 14:210; 19:**819;** 20:753
    as denaturant, 8:450
    frothing agent, 9:385
    in papermaking, 14:520
    α-pinene, 19:813
    resin solvent, 18:660
    terpenes, 19:810
    as disinfectant, 2:627, 635
Pine tar
    as denaturant, 8:450
Pinnoite, 3:648
Pinosylvin, 16:690
Pinus pumilio, 14:211
Pinus sylvestris, 14:211
Pinworms, 14:533
Pipeline heating, S:**694**
Piperazine, 10:906, 929; 11:526; 15:**638**
    in roundworm chemotherapy, 14:535, 539
Piperazinediones, 10:929
Piperidine, 2:369; 6:616, 643; 10:247; 11:526;
    12:69; 16:801
Piperidine alkaloids, 1:765
1-Piperidine carboxylate, 15:683
Piperidine enamine, 20:23
4-Piperidino-2,3,5,6-tetrafluorobenzenesulfona-
    mide, 9:793
Piperine, 1:765

Piperocaine hydrochloride, 2:405, 407
Piperonal, 11:469; 15:162, 164, 170
Piperylene, 6:453
Pipes
    conductance of circular, 21:140
    pressure loss for fluids, 9:468
Piping systems, 15:**646**
Pipradrol, 16:673
Piria's acid, 13:709
Pirinitramide, 2:385
Pisang wax, 22:161
Pitafloja
    analysis, 9:173
    properties, 9:175
Pitch. (See also Tar and pitch.)
    coal tar, 4:167, 181
    coking, 15:23
    creosote blends, 19:679
    as fuel for furnaces, 10:279, 280
    refractory binder, 17:227
    stain solvent, 18:661
Pitchblende, 17:11; 21:7, 31
Pitch coke, 4:166
Pitch-epoxy coatings, 19:678
Pitches
    carbon fiber precursor, S:110
Pitch impregnation, 4:181
Pitch-polyurethan coatings, 19:678
Piteira, 9:178
    analysis, 9:173
Pituitary extracts
    as cardiovascular agents, 4:523
Pival, 15:920
Pivalic acid, 8:851
    prod, 8:853
    properties, 8:852
    uses, 8:855
Pivalophenone, 12:107
Pivaloylacetone, 12:148
Pivaloylvaleroylmethane, 12:148
2-Pivalyl-1,3-indanedione, 15:920
Plantago seed
    as cathartic, 4:592
Plant growth promoter
    dially pimelate as, 1:249
Plant growth substances, 15:**675**
Plant layout, 15:**689**
Plant location, 15:**700**
Plant nutrition
    ion exchange and, 11:893
Plasdone C, 21:436
Plaskon
    plasticizers, 15:764

Plasma
    coagulation factors, 5:588
Plasma arcs
    photochemistry, 15:345
Plasma, blood
    animal, 3:573
    human. See Blood fractionation.
    volume expanders, 3:600; 21:436
Plasma torches
    carbon arcs for, 4:215
Plasmin, 5:594, 595
    phys prop, 5:589
Plasminogen, 5:593
    from blood fractionation, 3:584, 591
    activators of, 5:597
    phys prop, 5:589
Plasmocid, 2:151, 153
Plaster
    lime in, 12:452
Plaster of Paris, 7:299
Plaster of Paris, 4:14, 20
    as dental material, 6:784. (See also Calcium sulfate.)
    from gypsum, 4:770
Plastic
    in ceramic composite armor, S:139
Plastic explosives, 8:631
Plasticization
    of fats, 13:181
Plasticized white phosphorus, 4:902
Plasticizer
    for poly(vinyl chloride) resins, 8:259
Plasticizer industry
    use of fatty alcohol in, 1:556
Plasticizers, 15:**720**
    benzoic acid, 3:430
    for cellulosics, 4:656
    phosphate, 15:315
    tritolyl phosphate, 19:672
Plastics
    acrylic, 13:357
    aziridines in, 11:541
    epoxy resins used in, 8:309
    as bearing materials, 3:275, 283
    Bingham, 20:646
    carbon blacks used in, 4:267, 276, 280
    casting, 15:806
    cellular, 15:806. (See also Foamed plastics.)
    from cellulose derivs, 4:653
    clays used in, 5:583
    coatings for, 5:714
    concentrated pigments, 15:598
    extrusion, 15:805

foam from shale oil, *18*:18
hardness testing, *10*:818
laminates, *15*:807
molding, *15*:802
as optical filters, *9*:252
optical brighteners for, *3*:748
in packing, *14*:452
as packing materials, *14*:456
pigments in thermosetting, *15*:601
in piping, *15*:647
polyimide foam, *S*:770
polyvinylpyrrolidone in, *21*:437
processing, *15*:802
themoset, *15*:791
use for embedding, *8*:105
use of fluorescent pigments, *9*:501
Banbury mixers, *4*:662, 665
from hydrocarbons, *11*:281
welding, *15*:807
Plastics industry
    cyclohexane solvent, *6*:681
    organic isocyanates used in, *10*:45
    use of metallic soaps in, *7*:284
Plastics technology, *15*:**790**
Plastics testing, *15*:**811**
Plastic tooling
Plastigels, *4*:32; *15*:740; *21*:403
Plastisols, *15*:739, 771; *21*:403
    as fabric coatings, *5*:682, 707
Plastoquinones
    Irreverre-Sullivan color test, *21*:589
Plate and frame filter press, *9*:279
Plate glass
    manuf, *10*:562
Platforming, *15*:835. (See also *Petroleum.*)
Plating. See *Electroplating.*
Platinic. See *Platinum compounds.*
Platinous. See *Platinum compounds.*
Platinum, *15*:832, 849-852
Platinum
    as catalyst in diamond synthesis, *4*:296
    catalyst in hydroreforming, *20*:538
    catalytic reforming, *15*:32
    coating, *17*:268
    corrosion resistance to water, *6*:302
    cracking catalyst, *15*:37
    as dental alloy, *6*:814, 820
    electrolytic refining, *6*:598
    electromigration of, *S*:287
    in electron tubes, *8*:6
    in electroplating, *8*:64

in hydroisomerization, *11*:451
in photoemitters, *15*:403
resistance thermometers, *19*:783
in silicone rubber cure, *18*:229
in sodium-air cells, *S*:134
sulfuric acid catalyst, *19*:442, 460, 467
thermocouples, *19*:787
zirconium reaction, *22*:621
Platinum alloys
    catalysts in fuel cells, *S*:387, 389
Platinum anodes
    in perchlorate manuf, *5*:78
Platinum black, *19*:467
Platinum catalyst
    activated alumina used as support for, *2*:54
Platinum-cobalt
    as magnetic material, *12*:767
Platinum compounds, *15*:862-867
    compounds, *15*:**861**
Platinum-gold alloy, *10*:692
Platinum group metals, *15*:**832**
    compounds, *15*:**861**
Platinum group metals, compounds, *15*:**861**
Platinum hexafluoride, *10*:891
Platinum-iridium, *12*:591
Platinum trifluoride, *9*:587
Plattnerite, *12*:276
Plexiglass, *10*:594
Pliofilm, *14*:139
Plumbago. See *Graphite.*
Plumbic and plumbous entries. See *Lead compounds.*
Plutoform Black CAD, *7*:535
Plutonium, *1*:351; *15*:**879**
Plutonium
    as fission product, *14*:106, 107
    prod of, *14*:83
    recovery, *14*:92
Plutonium compounds, *15*:893-896
Plutonium-239, *14*:74
    "breeder" reactors, *21*:29
    from neptunium-239, *21*:3
Plutonium-242
    crystal structure, *1*:366
    irradiation of, *14*:119
    phys prop, *1*:366
Plywood, *15*:**896**
Plywood board, *21*:601
Plywood industry
    use of adhesives in, *1*:397
    use of amino resins in, *2*:245

Pneumatic conveyors, 6:112
Pneumococci
    chemotherapy against, 3:21
Podecdysone A, 18:849
Podocarpane
    derivs of, 17:482
Point defects
    in metals, 13:321
Poise
    unit of viscosity, 21:461
Poiseuille's law, 21:465
Poisons, economic, 15:**908**
    as reactor fuel, 14:76
Poisson's ratio, 17:425
Polacolor process, 5:837
Polarization
    in electric cells, 3:168
Polarization diagrams
    for corrosion processes, 6:295
    detn of arylamide, 1:282
Polarography, 7:748-764
Polectron, 14:80
Polar Yellow 2G, 2:881
Poliomyelitis virus
    inactivation of, 2:613, 616
Polishing, 1:39
Pollucite
    cesium from, 4:857, 858
    rubidium from, 17:688
    rubidium in, 17:685
Pollution, S:**711**
Pollution control
    in pickling of steel, S:691
Polompom
    analysis, 9:173
Polonium, 17:11, 24-26
    from uranium ore, 21:2
Polonium-210, 17:90, 128
    in tobacco, 20:512, 523
Polonium-218, 17:9
Polonovski reaction, S:39
Poloxyethylenated sorbitan monostearate
    as emulsifier in foods, 10:8
Poloxyethylene stearate
    as emulsifier in foods, 10:8
Poloxy-1,2-propylene glycol
    prepn of urethan elastomers, 7:694
Polyacetaldehyde, 1:99
Polyacetals
    chain-growth, 16:229
Polyacetylene, 7:414
Polyacrolein
    adducts, 19:287

Polyacrylamide, 1:276; 2:241; 17:403
    as gum, 10:741
    photochemical processes, 15:340
Polyacrylic acid
    as gum, 10:741
    interferon induction, 21:458
Polyacrylic acid chloride, 2:414
Polyacrylonitrile, 1:343; 6:637, 642; 7:300.
    (See also Acrylic and modacrylic fibers.)
    in pigment manuf, 9:497
    plasticizers for, 15:759
Polyacrylonitrile fibers
    optical brighteners for, 3:747
Polyacrylonitriles
    in poromeric materials, 16:351
Poly-β-alanine
    prepn, 1:278
Polyalkylene amides. See Amino resins.
Polyalkylene oxides
    block copolymers, 19:553
Polyalkylene polysulfides
    step-growth, 16:225
(Polyalkyl)pyridines
    oxidation-ammonification, 21:530
Polyalkyl vinyl ethers, 15:797
Polyallyl phosphonitrilate
    in prepn of flame-retardant finish, 9:308
Polyamic acids, S:747
Polyamides
Polyamides, 1:410; 16:**1.** (See also Nylon.)
    aromatic as fire retardants, S:486
    brighteners for, 19:13
    fibers, 16:46-87
    formation, 2:70
    from isophthalic acid, 15:461
    as packing materials, 14:453
    plasticizers for, 15:759
    plastics, 16:88-105
    prepn from dicarboxylic acids, 1:245
    research on fibers, 9:164
    resins, 12:191
    use as abherent, 1:9
    x-ray analysis, 22:453
Polyamides, fibers
    flame-resistant qualities, 9:312
    optical brighteners for, 3:746
Polyamine-methylene resin
    as antacid, 2:430
Polyamines, 15:796. (See also Amines.)
Polyarylates, 16:160, 167
Poly-L-aspartic acid, 16:306
Polyazo dyes, 7:471, 472

Polybenzarsol
  amebicide, 20:94
Polybenzimidazole, 15:461
  carbon fiber precursor, S:110
  high-temperature fibers, S:960
Polybenzimidazoles
  fire retardants, S:486
Polybenzoxazoles, 16:42
Poly(bisbenzimidazobenzophenanthrolene),
  S:961
Polyblends
  vinyl resins, 21:404
Polybrene, 5:602
Poly-p-bromostyrene, 19:92
Polybromotrifluoroethylene, 9:833
Polybutadiene, 17:573. (See also Elastomers.)
  drying oil, 7:398
  forms of, 7:65
  as gasket, 14:451
  isomeric, 16:244
  liquid polymers, 17:658
  synthetic rubber, 17:543
cis-Polybutadiene
  elastomer, S:106
  latex, 17:592
1,2-Polybutadiene
  prepn of, 7:66
Polybutadienes
  advantages, 7:690
  processing, 7:691
Poly-1-butene, 14:309. (See also Olefin polymers.)
  commercial, 15:797
  crystalline polymers of, S:773
  manuf of isotactic, S:776
  syndiotactic, S:778
  thermoplastics, S:776
Poly1,3-butylene adipate
  elastomers, 21:78
Polybutylene glycol
  epoxy resins from, 19:147
Polybutylenes, 3:855-857; 15:797. (See also
  Olefin polymers.)
Polybutyl ether disulfide, 16:262
Polybutyl formal disulfide, 16:262
Poly-N-n-butyl maleimide, S:756
Poly-2-butyne-1,4-dioxymethylene, 1:606
Poly-p-tert-butylstyrene, 19:92
Poly-2-butyne-1,4-dioxypropylidene, 1:606
Polycaproamide, 16:47
Polycaprolactam, 16:29. (See also Nylon.)
  copolymer, 16:23
  degradation, 16:10

  lateral substituents, 16.19
  monomer synthesis, 16:29
  polymerization, 16:30
  properties, 16:32
Polycapryllactam, 16:37
Polycarbocyanines, 6:617
Polycarbonate film, 9:232
  properties, 9:222-226
  trade names, 9:243
Polycarbonate resins, 12:191
  in phonograph records, 15:230
Polycarbonates, 15:795; 16:**106**
  antiplasticization, 15:775
  photography, 15:364
  plasticizers for, 15:761
  titanium alkoxide catalysts, 20:490
Polycarboxylic acids
  polyesters, 15:793
Polychlorimidometaphosphates, 4:917
Polychlorobenzenes
  chlorophenols from, 5:328
Polychloroprene, 7:705; 9:857, 858
  fire retardant, S:483
  isomeric, 16:244
ω-Polychloroprene, 7:708
Polychloroprenes, 5:218, 220
Polychlorostyrene, 19:92
  foams, 19:121
Polychlorotrifluoroethylene,   9:690,   832;
  12:582, 609
  CASING technique, S:21
Polychlorotrifluoroethylene film, 9:232
  properties, 9:222-226
Polychlorotrisilanes, 18:177
Polychrome metallic blue automotive finish,
  5:703
Polycrystalline
  definition, 6:516
Poly-1,4-cyclohexylenedimethylene tere-
  phthalate, 16:144, 148, 182
Poly-Lα,γ-diaminobutyric acid, 16:306
Polydiethyleneglycol phthalate
  plasticizer, 20:489
Poly-3,5-dimethyl-1-adamantyl methylacryl-
  ate, S:13
Poly-3,5-dimethyl-1-adamantyl acrylate, S:13
cis-Poly 2,3-dimethylbutadiene, 7:73
trans-Poly 2,3-dimethylbutadiene, 7:73
Poly 2,6-dimethyl-1,4-phenylene ether, 15:173
Poly 2,6-dimethylphenylene oxide, 15:795
Poly 4,4-dimethylpropiolactam, 16:34
Polydimethyl siloxane. See Dimethyl silicone.
  water repellency, 20:485

Polydimethylsiloxanes, *18*:244
  water-repellent agents, *22*:144
trans-Polydiolefins
  geometrical bulkiness, *16*:252
Polydipropylene glycol, *co*-fumaric acid chlorostyrene, *20*:815
Polydipropylene glycol-*co*-fumaric acid styrene, *20*:814
Polydodecylbenzene, *19*:295
Polydymite, *13*:736
Polyelectrolyte complexes, *16*:**117**
Polyelectrolytes, *15*:796; *16*:245
  "snake-cage", *11*:877
Poly ω-enanthamide, *16*:36
Polyene antibiotics, *16*:**133**
Polyene macrolides, *12*:632
Polyepichlorohydrin
  as water repellent, *7*:34
Polyester alkyds, *20*:**791**. (See also *Unsaturated Polyesters.*")
Polyester amides, *1*:411; *16*:160
Polyester fabrics
  prepn for dyeing, *7*:520
Polyester fibers, *16*:**143**
  dacron, dyeing rates of, *1*:135
  dacron, electrical behavior of, *1*:129
  optical brighteners for, *3*:747
  tensile-strain recovery, *1*:132
Polyester film, *9*:231
  properties, *9*:222-226
  trade names, *9*:243
Polyester resin
  glass-reinforced in armored vests, *S*:139
Polyester resins
  in hot melt adhesives, *S*:940
Polyesters, *15*:793; *16*:**159**
  brighteners, *19*:14
  in ceramic composite armor, *S*:145
  commercial, *15*:798
  dyeing with disperse dyes, *7*:544
  dyeing of fiber blends, *7*:568
  as film base, *16*:502
  titanium alkoxide catalysts, *20*:489
  in foams, *21*:85
  isophthalic acid, *15*:461
  man-made fibers, *9*:157
  in photography, *15*:364
  phthalic anhydride, *15*:455
  plasticizers, *10*:665
  plasticizers for linear, *15*:761
  polyurethans from, *21*:62
  research as fibers, *9*:167

  in rigid foams, *21*:91
  used in embedment, *8*:107, 111
Polyester-urethans, *21*:77-80
Polyethenoid acids, *8*:811
Polyether esters, *16*:160
  chain-growth, *16*:229
Polyethers. (See also *Glycols; Resins, water-soluble.*)
  in foams, *21*:85
  in rigid foams, *21*:91
Polyether-urethans, *21*:80
Polyethoxystearylmethylammonium chloride, *9*:897
Polyethyl acrylate
  as water-repellent agent, *22*:138
Polyethylene. (See also *Olefin polymers.*)
  antioxidants for, *2*:600
  as cable insulator, *11*:809
  as cable insulators, *11*:820
  carbon blacks as protective agents for, *4*:276
  CASING technique, *S*:22
  chlorinated, *17*:462
  chlorinated as fire retardant, *S*:483
  chlorosulfonated, *17*:462, 545, 573
  chlorosulfonated as fire retardant, *S*:483
  clathrate separations, *22*:476
  commercial, *15*:797
  deposition, *8*:30
  geometric bulkiness, *16*:252
  high density, *14*:242-259
  high pressure, *14*:217-241
  paper waterproofing, *22*:149
  pharmaceutical containers, *15*:131
  photochemical processes, *15*:339
  plasticizers for, *15*:734
  in polyvinylpyrrolidone, *21*:430
  in prepn of Hypalon, *7*:695
  sheets used for seepage control, *21*:663
  styrene graft copolymer, *19*:91
  sulfochlorination, *19*:289
  surface tension, *9*:709, 710, 715, 720
  thermoplastics, *S*:776
  use as abherent, *1*:9
  use of diarylamines in crosslinking of, *7*:48
  used for embedment, *8*:106
  use of fluorescent pigments, *9*:50
  vinyl chloride graft polymer, *21*:379
  vinyl copolymer of chlorinated, *21*:379
  x-ray analysis, *22*:453
  Ziegler process, *14*:259-309
Polyethylene adipate, *12*:58; *16*:161

Polyethylene, cellular
  elastomers, *21*:78
  manuf by decompression expansion, *9*:855
  properties, *9*:860
Polyethylene film, *9*:229
  epoxy resins from, *19*:147
  in esterification of fatty acids, *8*:819
  plasticizer, *15*:773
  properties, *9*:222-226
  sulfonation, *19*:285
  trade names, *9*:242
Polyethylene glycol, *6*:362, 644; *10*:639, 652,
  654-658. (See also *Polyoxyethylene)*
Polyethylene glycol esters, *19*:541
Polyethylene glycol bis(2-ethylhexanoate),
  8:850
Polyethylene glycol-*co*-fumaric acid methyl
  methacrylate copolymer, *20*:807
Polyethylene glycol-*co*-maleic anhydride-*co*-
  phthalic anhydride, *20*:829
Polyethylene glycols, *8*:553
  in suppositories, *15*:126
Polyethylene glycol terephthalate, *15*:761
Polyethyleneimine, *17*:406
Polyethylene-*co*-neopentylene terephthalate,
  16:182
Polyethylene oxide, *17*:406
  coagulant in water treatment, *22*:96
Polyethylene resins, *12*:191
Polyethylene succinate, *19*:139 *15*:465, 472;
  16:143-159
  prepn, *8*:361
  x-ray analysis, *22*:453
Polyethylene terephthalate, *1*:245; *6*:628;
  8:552; *12*:814
Polyethylene tetrasulfide, *16*:265
  from 1,2-dichloroethane, *5*:151
Polyethylene 4,4'-ethylenedioxydibenzoate,
  16:170
Polyethyl formal disulfide, *16*:262
Polyfluorobutadiene, *7*:695
Polyfluorocarbon resins. See *Fluorocarbon.*
Polyformaldehyde, *15*:795
Polyformaldehydes, *1*:96
Polygalacturonases, *8*:199; *14*:650
Polygalacturonic acids, *8*:198; *14*:636
Polygalitol, *1*:578
Polyglycine
  shrinkproofing wool, *22*:409
  x-ray analysis, *22*:454
Polyglycolide, *10*:632; *16*:161
Polyglycol oil, *12*:611
Polyglycolpalmitic amide, *10*:789

Polyglycol succinates
  copolymers, *19*:145
Polyglycol terephthalates, *9*:167
Polyhexafluoropropylene
  surface tension, *9*:715, 720
Poly(hexamethylene adipamide). See *Nylon-
6,6.*
Polyhexamethyleneisophthalamide, *20*:329
Polyhydrazides, *16*:42
Polyhydric alcohols. See *Alcohols, polyhydric.*
Polyhydroxyacetic acid. See *Polyglycolide.*
Polyhydroxybenzenes, *16*:**190**
Poly(hydroxybenzoic acid), *S*:**741**
Polyhydroxyethers, *15*:173
Polyhydroxypivalic acid, *9*:168
Polyimide resins
  in hot melt systems, *S*:940
Polyimides, *15*:796; *S*:**746**
  fibers of aromatic, *S*:763
  high-temperature fibers, *S*:961
  pyromellitic dianhydride, *15*:483
Polyiodotetrasilanes, *18*:178
Polyisobutylene
  thermochemical constants as propellants,
  8:681
Polyisobutylene, *7*:688; *12*:581; *14*:309. (See
  also *Polybutylenes.)*
  radiation degradation, *17*:62
Polyisocyanate, *9*:854
  water-resistant paper, *22*:150
Polyisoprene, *7*:75; *12*:700 (See also *Elastomers.)*
  as gasket, *14*:450
  isomeric, *16*:244
  rubber, *17*:543, 572
cis-Polyisoprene, *7*:689; *9*:858; *12*:79
  latex, *17*:592
trans-Polyisoprene, *12*:80
Poly-*p*-isopropylstyrene
Poly-*cis*-1,4-isoprene, *15*:791
  radiation effects, *19*:109
Polyketones, *12*:132
Polylactam, *16*:89, 92
Polylactic acid, *12*:174, 184
Polylactide, *16*:161
Poly-L-lysine, *16*:306
Polylysine-glutamic acid
Poly-α-methylstyrene, *15*:775
Poly-α-methylstyrene, *19*:115
  radiation degradation, *17*:62
  coagulant in water treatment, *22*:96
Polymer gasoline, *16*:592

Polymerization, *16*:**219**
    of acetaldehyde, *1*:81
    in autoclaves, *11*:761
    bulk of *N*-vinyl-2-pyrrolidone, *21*:432
    bulk of poly(vinyl chloride), *21*:369
    degree of, *15*:790; *16*:251
    emulsion of vinyl acetate, *21*:334
    emulsion of vinyl chloride, *21*:373
    emulsion of vinylidene chloride, *21*:282
    heterogeneous of vinylidene chloride, *21*:280
    of higher aldehydes, *1*:99
    microwave induced, *S*:576
    of olefinic gases, *15*:2
    of olefins, *14*:332
    of petroleum olefins, *15*:41
    phenol-aldehyde, *15*:178
    polyimides via addition, *S*:753
    polyimides via Diels-Alder, *S*:756
    of poly(vinyl chloride), *21*:369
    precipitation of vinyl chloride, *21*:374
    solution of vinyl chloride, *21*:374
    step-reaction, *15*:792
    suspension of vinyl chloride, *21*:375
    suspension of vinylidene chloride, *21*:284
    of vinylidene chloride, *21*:278
Polymerization catalyst
    aluminum chloride as, *2*:22
Polymerization mechanisms, *16*:**219**
Polymerization mechanisms and processes,
    16:**219**
Polymerization of vinyl acetate, *21*:330
Polymerized oils
    separation of, *7*:402
Polymer matrix
    of ion exchange resin, *11*:878
Polymer processing
    use of abherents in, *1*:4
Polymers, *16*:**242**
    of acetaldehyde, *1*:99
    of acrylonitrile, *1*:343
    atactic, *21*:417
    containing sulfur, *16*:**253**
    of formaldehyde, *1*:97
    future fire retardant, *S*:486
    isocyanate, *21*:57
    isotactic, *21*:417
    surface treatment of, *S*:20
    urethan, *21*:**56**
    TPX methylpentene, *S*:800
    of trioxane, *1*:98
    toxicity of gases evolved from fire-retardant,
        S:475
    use of abherents in, *1*:24

Polymers of higher oelfins, *S*:**773**
Polymetatelluric acid, *19*:767
Polymethacrolein, *1*:261
Polymethacrylates
Polymethacrylates, *12*:576. (See also *Meth-
    acrylic compounds.*)
    in printing, *16*:518
    radiation degradation, *17*:62
Polyorthophosphates, *15*:252
Polymethacrylic acid
    interferon induction, *21*:458
Polymethine dyes, *7*:533; *16*:**282**
    cyanine dyes, *6*:605
    thermochemical constants as propellants,
        8:681
Polymethyl acrylate. (See also *Acrylic com-
    pounds.*)
Polymethylbenzenes, *16*:**300**
Polymethylene polyphenyl isocyanate, *12*:46
Polymethylgalacturonases, *14*:650
Polymethyl methacrylate, *6*:879; *13*:331;
    15:797. (See also *Methacrylic compounds.*)
    mass spectrometry, *13*:96
    in microencapsulation, *13*:440
    NMR of, *14*:63
    plasticizers, *15*:762
    prop of commercial plastics, *13*:355
    radiation effects, *19*:109
Poly-4-methyl-1-pentene, *14*:309; *15*:798
    crystalline polymers of, *S*:773
    thermoplastics, *S*:776, 789-805
Polymethylsiloxane
    surface tension as coating, *9*:718
Polymixin B
    amebicide, *20*:80, 82
Polymixins
    in bacterial infections, *3*:1
Polymorphism
    of crystals, *6*:524
    and melting points, *13*:212
    pigments, *15*:568
Poly-4-morpholinomethyl-phenols, *13*:665
Polymyxin B sulfate
Polymyxins, *16*:308, 319-328
    as veterinary drug, *21*:245
Polynosic fibers, *9*:161
Polynucleotide
    action and uses, *8*:185
Polyolefins. See *Olefin polymers.*
    brighteners, *19*:13
    carbon blacks as protective agents for, *4*:276
    hexane use in manuf of, *11*:4
    paper from, *14*:496

plasticizers, 15:762
synthetic paper, 19:126
titanium alkoxide catalysts, 20:488
Polyolefin sulfides, 15:795
Polyoxadiazole
high-temperature fibers, S:960
Polyoxadiazoles, 16:42
Polyoxalates, 16:167
Polyoxy-2,2-dimethylpropionyl, 16:181
Polyoxyacetyl, 16:161
polyester, 16:181
Polyoxy-1,4-butylene glycol
prepn of urethan elastomers, 7:694
Poly-p-oxybenzoyl, 16:168
Poly-1,4-oxybutylene glycols
elastomers, 21:95
Poly-p-oxyethoxybenzoyl, 16:182
Poly-p-oxyethoxybenzoyl, 16:170
Polyoxyethylene. (See also Polyethylene glycol.)
ethoxylation, 19:532
surfactants, 19:531
Polyoxyethylene alicyclic amines, 19:559
Polyoxyethylene alkylamines, 19:556, 559
Polyoxyethylenealkylphenols
sulfated, 19:527
Polyoxyethylene alkyl sulfates, 19:529
Polyoxyethylene cetyl ether
HLB value, 8:128, 130
Polyoxyethylene fatty acid esters, 19:541, 552
Polyoxyethylene glycol monopalmitate
HLB value, 8:130
Polyoxyethylene lauryl ether
HLB value, 8:129, 130
Polyoxyethylene monolaurate
HLB value, 8:129
Polyoxyethylene monooleate
HLB value, 8:129
Polyoxyethylene monostearate
HLB value, 8:129, 130
Polyoxyethylene oleyl ether
HLB value, 8:128, 129, 130
Polyoxyethyleneoxyadipoyl, 16:161
Polyoxyethylene oxypropylene stearate
HLB value, 8:130
Polyoxyethylenes, 15:796
synthetic waxes, 22:171
Polyoxyethylene sorbitan esters
HLB value, 8:129, 130
Polyoxyethylene sorbitan fatty acid esters, 19:548
Polyoxyethylene sorbitol
in drilling fluids, 7:302

Polyoxyethylene sorbitol esters
HLB value, 8:128, 129
Polyoxyethylene stearate
in ceramic composite armor, S:144
Polyoxyethylene stearyl ether
HLB value, 8:128, 129, 130
Polyoxyethylene tallates, 19:545
Polyoxyethylene tridecyl ether
HLB value, 8:129, 130
Polyoxyethylene(20) sorbitan monooleate, 1:583
uses, 1:584
Polyoxyhexanoyl, 16:181
Polyoxy(methylacetyl), 16:161, 181
α-Polyoxymethylene, 7:48
Poly-p-oxymethylenebenzoyl
polyester, 16:182
Polyoxymethylene glycols, 1:96
Polyoxymethylenes, 10:83
Polyoxypentanoyl, 16:161, 180
Polyoxypropylene glycol
polyurethans from, 21:72
Polyoxypropylene mannitol dioleate
HLB value, 8:129
Polyoxypropylene-polyoxyethylene
block copolymers, 19:553
Polyoxypropylene triols
polyurethans from, 21:62
Polyoxytetramethylene glycols
polyurethans from, 21:61
elastomers, 21:78
Poly-1,5-pentylene adipate
Polypeptide antibiotics, 16:**306**
Polyperfluorophenylene ether, 9:793
Polyphenol oxidase, 19:746, 748, 750, 752
Polyphenols
from shale oil, 18:18
in tobacco leaf, 20:511
Poly-m-phenyleneisophthalamide, 16:10
Poly-1,4-phenylene adipate 1:244
Polyphenylene oxide, 15:795; 16:274; 19:672
resins, 15:172
Poly-p-phenylene terephthalate, 16:168
Polyphenyl ethers, 12:581; 15:172
Polyphosphates, 15:232
surfactants, 19:513
Polyphosphoric acid, 10:161
prod of, 9:97
surfactants, 19:530
trans-Polypiperylene, 7:73
Polypodine, 18:849
Poly I-Poly c, 21:458

Polypropiolactams, *16*:34
Polypropylene, *14*:282-309; *15*:797
Polypropylene
  in ceramic composite armor, *S*:143
  elastic hard fibers, *18*:614
  isotactic, *S*:773
  pigments for, *15*:599
  plasticizers for, *15*:726
  stereochemistry, *16*:235
  thermoplastics, *S*:776
  vinyl chloride graft polymer, *21*:379
Poly-1,2-propylene adipate, *16*:180
Poly-1,2-propylene azelate, *16*:180
Polypropylene, cellular
  manuf by decompression expansion, *9*:855
Polypropylene fiber, *9*:158; *S*:**808**
Polypropylene film, *9*:229
  properties, *9*:222-226
  trade names, *9*:242
Poly-1,2-propylene glutarate, *16*:180
Polypropylene glycol, *10*:658; *12*:580. (See
  also *Polyoxypropylene.*)
  phys prop, *10*:656
Polypropylene glycol adipate
  plasticizer, *20*:489
Polypropylene glycol-*co*-fumaric acid-
  co-phthalic anhydride, *20*:812
Polypropylene oxide, *8*:278-280
Polypropylene resins, *12*:191
  in phonograph records, *15*:230
Polypropylenes
  antioxidants used for, *2*:600
Poly-1,2-propylene sebacate, *16*:180
Polypyromellitimides, *S*:757
  from diamines, *S*:753
  high-temperature fibers, *S*:961
Polypyrrolidine, *16*:35
Polyriboinosinic acid-polyribocytidilic acid
  complexes, *21*:458
Polysaccharidases
  action and uses, *8*:184
Polysaccharides, *4*:133, 146
  colloidal, *17*:763
  in soybeans, *18*:601
  used as food additives, *10*:6
Polyselenatopolyzirconic acid, *22*:659
Polysilanes
Polysiloxane
  in packing, *14*:452
Polysiloxanes, *18*:**222**
  characteristics, *18*:226
  halogenated, *18*:167

Polysodium styrenesulfonate, *19*:92
Polystyrene, *16*:243. (See also *Styrene plastics.*)
  antiplasticization, *15*:775
  chain-growth, *16*:229
  commercial, *15*:798
  compatibility with adipate esters, *1*:416
  cyanoethylation of, *6*:663
  in drilling fluids, *7*:301
  in gypsum board, *21*:623
  manuf by decompression expansion, *9*:855
  manuf from expandable formulations, *9*:852
  pharmaceutical packaging, *15*:131
  in phonograph records, *15*:229
  photography, *15*:364
  pigments for, *15*:599
  plasticizers, *15*:763
  as propellant, *8*:681
  in printing, *16*:518
  properties, *9*:860
  spectral transmittance curve, *9*:253
  surface tension, *9*:715
  sulfonation, *19*:297, 300
  thermoplastics, *S*:776
  used for embedment, *8*:106
  use of fluorescent pigments, *9*:501
Polystyrene, cellular
Polystyrene film, *9*:230
  properties, *9*:222-226
  trade names, *9*:242
Polystyrene resins, *12*:190
Polystyrenesulfonic acid
  in epoxidation of soybean oil, *8*:255
  resin, *8*:240, 248, 251
Poly-*p*-styrenethiol, *15*:339
Polysulfatopolyzirconic acids, *22*:663
Polysulfatozirconates, *22*:662
Polysulfide elastomers, *7*:696, 698
  for dental impressions, *6*:803
  as sealants, *4*:31
Polysulfides, *16*:253-272; *19*:499
  adhesives in ceramic armor, *S*:154
  used in embedding, *8*:109, 111
Polysulfites, *19*:484, 495
Polysulfonates, *16*:160
Polysulfone, *15*:174, 795; *16*:272-281
Polysulfones
  from butylenes, *3*:844
  prepn from allyl compounds, *1*:922
Polytellurides, *19*:758
Polyterephthalamides, *16*:10
Polytetrafluoroethylene, *8*:30; *9*:690, 805-831;
  12:608; *15*:763, 799, 801
  CASING technique, *S*:21

in fuel cells, *S*:389
leatherlike materials, *16*:351
manuf, *9*:812
oxygen detn in water, *21*:697
in packing, *14*:452
porous, manuf of, *4*:20
properties, *9*:817
in sodium-air cells, *S*:134
surface tension, *9*:710, 713, 715, 718, 720
uses, *9*:828
zircon filler, *22*:660
Polytetrafluoroethylene (Teflon), *6*:879
on aluminum, *1*:982
in bearing materials, *3*:272, 275, 283
plasticizers, *15*:763
radiation degradation, *17*:62
Polytetramethylene ether glycol, *10*:250
prepn of urethan elastomers, *7*:694
Polytetramethylene isophthalate-*co*-tere-
phthalate, *16*:182
Polytetramethylene sebacate, *1*:244
Polythiazide
as diuretic, *7*:263, 264
Polytrifluorochloroethylene
surface tension, *9*:715, 720
Polytrifluoroethylene
surface tension, *9*:710, 715, 720
Polyurethan
catalyst for rocket fuel, *13*:832
in ceramic composite armor, *S*:145
chlorendic acid in foam, *S*:480
in packing, *14*:452
rubber, *17*:545
thermochemical constants as propellant,
8:681
use of 1,4-butanediol in manuf, *10*:672
zircon filler, *22*:660
Polyurethan elastomers
adipic acid polyesters in, *1*:417, 418
use of isocyanates in prepn, *12*:62
Polyurethan fiber
spandex, *9*:159; *18*:614
spandex dyeing, *7*:520, 547, 571
Polyurethan film, *9*:232
properties, *9*:222-226
Polyurethan foams, *9*:853
classification, *9*:875
prop of flexible polymers, *9*:862
prop of rigid polymers, *9*:860
use of adipic acid in, *1*:418
Polyurethan resins
used in encapsulation, *8*:109, 111

Polyurethans. (See also *Urethan polymers.*)
adhesives for ceramic armor, *S*:154
Polyurethans, *15*:794
prepn from isocyanates, *12*:48
step-growth, *16*:225
Poly-δ-valerolactone, *16*:161
Polyvinyl acetals, *21*:304-317
Polyvinyl acetate, *6*:365; *7*:421; *8*:363;
21:317-353
adhesive, *15*:764
alkaline hydrolysis of, *21*:360
in ceramic composite armor, *S*:144
commercial, *15*:799
in latex paints, *14*:472; *22*:153
in papermaking, *14*:519
phthalate plasticizers, *15*:455
plasticizers for, *15*:732
from polyesters, *16*:180
polyvinyl butyral from, *21*:311
in prepn of polyvinyl alcohol fiber, *9*:158
in printing, *16*:518
in water-resistant paper, *22*:150
Polyvinyl acetate emulsion
as adhesive, *1*:390
Polyvinyl acetate film, *9*:230
Polyvinyl acetate latex
waterproof masonry walls, *22*:152
Polyvinyl alcohol, *6*:365; *21*:353-364
added to waterless hand cream, *6*:350
carbon fiber precursor, *S*:110
in ceramic armor, *S*:138
cinnamic resins, *16*:517
commercial, *15*:799
in Cordela fibers, *21*:442
cyanoethylation of, *6*:662
cyanoethylation on, *6*:643
in drilling fluids, *7*:301
glyoxal-crosslinked, *16*:120
as gum, *10*:741
leatherlike materials, *16*:351
man-made fiber, *9*:158
plastics from, *19*:147
photography, *15*:358
plasticizers, *15*:764
in polyvinylpyrrolidone, *21*:430
prepn, *8*:360
sulfation, *19*:306, 491
surface tension, *9*:715
suspending agent, *21*:375
as water-soluble resin, *17*:401
protective colloid, *21*:332, 341

Polyvinyl alcohol film, 9:230, 232
  properties, 9:222-226
  trade names, 9:243
Polyvinylbenzylammonium chloride, 19:92
Polyvinylbenzyl chloride, 19:115
Polyvinylbenzyltrimethylammonium chloride
  PEC hydrogel, 16:118
Polyvinylbenzyltrimethylammonium polystyrene sulfonate, 16:118
Polyvinyl butyral, 21:305, 311
  from butyraldehyde, 3:873
  commercial, 15:799
  plasticizers for, 15:726, 764
Polyvinylcarbazole
  commercial, 15:800
Polyvinylcarbazoles
  and nuclear reactors, 14:80
Polyvinyl chloride, 21:369
  as cable insulator, 11:808, 820
  commercial, 15:798
  compatability with adipate esters, 1:416
  corrosives transport, 20:601
  crotonaldehyde, 6:457
  deposition, 8:30
  fibers, 21:441
  as fire retardant, S:484
  flame-retardant, 15:507
  isophthalic acid plasticizers, 15:461
  leatherlike materials, 16:351
  manuf from expandable formulations, 9:853
  optical brighteners for, 3:748
  pharmaceutical containers, 15:131
  phthalic anhydride, 15:456
  plasticizers for, 15:722, 737, 739, 782; 19:146
  plastigels, 4:33
  from polyesters, 16:180
  polyvinylpyrrolidone in, 20:430
  in propellants, 8:660, 681
  radiation degradation, 17:62
  roofing, 17:461
  solvents for, 21:442
  styrene graft copolymer, 19:91
  sunlight effects, 15:801
  surface tension, 9:710, 715
  synthetic paper, 19:126
  used in encapsulation, 8:106
  use of diarylamines in crosslinking of, 7:48
  use of recording tape, 12:807
Polyvinyl chloride, cellular
  manuf by decompression expansion, 9:855,
    856
  prop of rigid polymers, 9:860
  prop of flexible polymers, 9:862

Polyvinyl chloride film, 9:230
  properties, 9:222-226
  trade names, 9:242
Polyvinyl chloride plastisols
  use in pressure-sensitive tapes, 1:7
Polyvinyl chloride resins
  as fabric coatings, 5:681
Polyvinyl cyanoethyl ether, 6:636, 663
Polyvinyl esters
  saponification of, 21:357
Poly-N-vinyl-5-ethyl-2-oxazolidone, 10:782
Polyvinyl fluoride, 9:710, 715, 720, 835-840
  commercial, 15:799
  roofing, 17:461
Polyvinyl fluoride film, 9:232
  properties, 9:222-226
Polyvinyl formal, 21:305
  commercial, 15:799
Polyvinyl hydrogen sulfate
  from polyvinyl alcohol, 21:358
Polyvinylidene chloride, 21:275-303
  deposition, 8:30
  leatherlike materials, 16:351
  in papermaking, 14:521
  plasticizers for, 15:764
  surface tension, 9:710, 715
Polyvinylidene chloride fibers
  flame-resistant qualities, 9:312
Polyvinylidene chloride film, 9:230
  properties, 9:222-226
Polyvinylidene fluoride, 9:840; 21:269-274
  plasticizers for, 15:726, 728
  properties, 9:843
  uses, 9:845
Polyvinylimidazole, 10:782
Polyvinyl methyl ether, 17:407
Polyvinyl methyl ethers, 21:417
Poly-2-vinylpyridine
  in polypropylene, S:820
Poly-4-vinylpyridine
  quaternization, 19:253
Polyvinylpyridinium butyl chloride
  coagulant in water treatment, 22:96
Polyvinylpyridinium chloride
  PEC hydrogel, 16:118
Polyvinylpyrrolidone, 2:618; 3:601; 10:673;
    17:405; 21:427
  commercial, 15:800
  in hair spray, 10:782
  plasticizers, 15:764
  as suspending agent, 21:421
Polyvinylpyrrolidone-vinyl acetate
Polywater, 21:682; 22:13
Polywolframates, 22:352
Polyxylylenes, 15:796

Pompeian blue, *15*:541
Ponasterones, *18*:849
Ponceau R. See *Ponceau 2R.*
Ponceau SX, *5*:867; *10*:793
Ponceau SX Lake, *5*:869
Ponceau 2R, *5*:869
Ponceau 3R, *5*:872
Ponchon principle
  in extraction, *8*:729
Ponds
  spray and thermal pollution, *S*:722
Pontamine Black EX
  structural formula, *5*:745
Pontamine Bordeaux B
  gelling agent, *21*:355
Pontamine Brown D 3 GN
  gelling agent, *21*:355
Pontamine Fast Red F
  gelling agent, *21*:355
Pontamine Green 2 GB
  gelling agent, *21*:355
Pontamine Orange R
  gelling agent, *21*:355
Poppy seed, *9*:373
Poppyseed oil, *8*:779, 785
  in margarine, *13*:59
Porcelain
  as dental material, *6*:796
Porcelain enamel. See *Enamel, porcelain and vitreous.*
Porcelain enamel frits
  composition, *8*:158
Porcelains
  for chemical ware, *4*:832
  strength values of, *4*:818
  thermal conductivity of, *4*:803
Porofor N, *6*:672
Poromeric materials, *16*:**345**
Porous printing, *16*:495
Porphin
  structure, *5*:342
Porphyrins, *16*:855
Portamycin
  amebicides, *20*:82
Portland cement. See *Cement.*
Positrons, *17*:67
Postchlorination
  of municipal waters, *22*:87
Potadine, *2*:619
Potash, *9*:34; *16*:370, 387, 388. (See also *Potassium compounds.*)
  from Great Salt Lake, *S*:451
  in multinutrient fertilizers, *S*:339

Potash alum, *2*:64
Potash" blue, *15*:539
Potash feldspars
  in ceramics, *4*:769
Potassium, *16*:**361**
  content in soils, *9*:35
  for corrosion processes, *6*:292
  ferroelectric properties, *9*:13
  graphite compounds with, *4*:311
  as hazardous chemical, *9*:291
  as lubricant, *12*:603
  in MHD generator, *S*:232
  as plant nutrient, *9*:217
  seed in MHD generator, *S*:218
  sodium alloy, *16*:366. (See also *Sodium potassium alloy.*)
  sodium in prod of, *18*:450
  in water, *21*:694
  in wood, *22*:366
Potassium-40
  in water, *21*:703
Potassium-42, *17*:121
Potassium Kurrol's salt, *15*:249
Potassium acetate, *8*:472. (See also *Potassium compounds.*)
  tartaric acid, *19*:736
Potassium acid fluoride, *9*:649. (See also *Potassium hydrogen fluoride.*)
Potassium acid oxalate, *14*:369
Potassium acid tartrate, *2*:578
  precipitation in wines, *22*:320
Potassium acid urate, *21*:112
Potassium alginate
  as dental impression material, *6*:801
Potassium aluminum sulfate, *2*:64
  bottling water treated with, *4*:341
  in food processing, *10*:10
  in tanning fur skins, *10*:301
  as water correcting agent, *10*:17
Potassium amide, *6*:903
  from ammonia, *2*:262
Potassium antimonate(V), *2*:575
Potassium antimonyl tartrate. See *Tartar emetic.*
Potassium ascorbyl palmitate
  as synergist, *2*:592
Potassium bicarbonate
  dry chemical agents, *S*:366
  fire extinguishing agents, *S*:365
  in fire extinguishers, *9*:296
Potassium bichromate
  pigments, *15*:534, 537
  sensitizer, *16*:514

Potassium bifluoride, 9:510, 649. (See also Potassium hydrogen fluoride.)
Potassium bismuthide, 3:537
Potassium bisulfate
  to open ores, 13:770
Potassium bisulfite, 10:13
Potassium bitartrate, 19:676, 739
Potassium bromate, 3:770
  as bleaching agent in flour, 10:6, 789
  in flour, 22:294
Potassium bromide, 6:512
  dielectric property of, 4:821
  photography, 15:372, 386
Potassium tert-butoxide, 19:138, 325
Potassium bromide tetrafluoride, 9:589
Potassium carbonate, 6:512; 7:42; 12:150
  in dental stones, 6:788
  in prepn of trifluoroboroxiné, 9:556
  in recovery of carbon dioxide, 4:361
  in removal of carbon dioxide from ammonia
    synthesis, 2:279
  in synthesis of 1,3-butylene glycol, 10:664
Potassium chlorate
  in dyeing of furs, 10:307
  in explosive priming compositions, 8:653,
    681
  as explosive, 8:589
  in matches, 13:160, 164
Potassium chlorate, 5:52
  in manuf of perchloryl fluoride, 9:606
  uses, 5:60
Potassium chloride
  dielectric property of, 4:821
  in drilling fluids, 7:302
  fire extinguishing agents, S:366
  from Great Salt Lake brine, S:447
  melt granulation, S:341
  phosphor, 15:352
  in potash, 9:111,112
  in sylvinite, 18:473
  as water correcting agent, 10:17
Potassium chlorochromate, 5:481
Potassium 4-chloro-3,5-dinitrobenzenesulfon-
  ate, 2:223
Potassium chlorophyllin, 5:348
Potassium chloroplatinate
  water color standard, 21:690
Potassium chlororuthenate, 15:876
Potassium chromate
  as indicator, 11:559
  manuf of, 5:484
  in matches, 13:164
  phys prop, 5:481

Potassium chrome alum, 5:494
Potassium citrate
  as emulsifier in cheese, 10:8
  tartaric acid, 19:737
Potassium citratozirconate, 22:648
Potassium cobaltic nitrite, 5:745
Potassium cobalticyanide, 5:743
Potassium cobaltinitrite
  pigments, 15:536
Potassium cobaltocyanide, 5:743
Potassium compounds, 16:**369**
Potassium cyanate, 6:597
Potassium cyanide, 6:596, 590, 626; 9:602, 794
Potassium cyanoaurate, 10:693
Potassium cyclohexylsulfamate, 10:10
Potassium dichloroisocyanurate, 4:908, 919;
  20:664, 669
  swimming pool water disinfectant, 22:126
Potassium dichromate
  carbon dioxide purified with, 4:364
  in dyeing of furs, 10:307
  manuf of, 5:484
  phys prop, 5:481
Potassium difluorophosphate, 9:643
Potassium dihydrogen phosphate
  nonlinear optical materials, S:623
Potassium dititanate, 20:411
Potassium ethylate
  as catalyst in manuf of ethyl vinyl ether,
    8:455
  prepn from ethyl ether, 8:479
Potassium ethyl xanthate, 22:419-422, 425
Potassium ferrate(VI), 12:40
Potassium ferricyanide. See Potassium hexa-
  cyanoferrate (III).
Potassium ferriferrocyanides
  pigments, 15:539
Potassium ferrocyanide. See Potassium hexa-
  cyanoferrate (II).
Potassium ferroferrocyanide, 15:539
Potassium fluoborate, 9:564, 659
Potassium fluogermanate, 10:523
Potassium fluoride, 9:589, 649
  complex with alkylaluminums, 2:34
  in fluorination of organic compounds,
    9:693, 797
  in prepn of aromatic fluorocarbons, 9:777
  in prepn of sodium fluoroacetate, 9:768
Potassium fluoroantimonite, 9:550
Potassium fluorogermanite, 9:585
Potassium fluorohafnate, 10:764
Potassium fluorosulfate, 9:679
Potassium fluorosulfinate, 9:800

Potassium fluorotantalate, *19*:633, 649

Potassium fluorozirconate, *16*:517; *22*:614, 647

Potassium fluosilicate, *6*:802; *9*:655
  phys prop, *9*:658
  reaction with potassium chloride, *9*:659

Potassium formate, *6*:597; *16*:**389**

Potassium gallate(III), *10*:325

Potassium hexachlororuthenate(IV), *15*:876

Potassium hexachlorotitanate, *20*:388

Potassium hexacyanocobaltate, *5*:743

Potassium hexacyanoferrate(II), *6*:590; *12*:28
  chlorobenzene reaction with, *5*:256
  phosphoric acid, *15*:262
  in prepn of carbon monoxide, *4*:437

Potassium hexacyanoferrate(III), *12*:31
  oxidation of α-tocopherol, *21*:577
  photography, *15*:386, 390
  sugar analysis, *19*:159

Potassium hexacyanoiridate(III), *15*:870

Potassium hexacyanorhodate(III), *15*:872

Potassium hexafluorobismuthate(IV), *3*:541

Potassium hexafluorogermanate, *9*:585

Potassium hexafluorophosphate, *9*:644

Potassium hexafluorotitanate, *9*:683, 684

Potassium hexafluorozirconate
  electrolysis of molten, *22*:617

Potassium hexametaphosphate, *10*:789

Potassium hexathiocyanatochromate(III), *5*:495

Potassium hexathiocyanoplatinate(IV), *15*:867

Potassium hydride, *11*:202
  to open ores, *13*:770

Potassium hydrogen fluoride. (See also *Potassium fluoride; Potassium acid fluoride.*)

Potassium hydrogen phthalate
  x-ray spectroscopy, *22*:461

Potassium hydrogen tartrate, *7*:529; *19*:736, 741
  in baking powder, *3*:51
  manuf, *19*:739

Potassium hydroxide, *6*:371, 598, 610, 673; *16*:390
  as catalyst in manuf of ethyl vinyl ether, *8*:455
  as drying agent, *7*:396
  mixture with TNT, *8*:612
  as paint remover, *14*:490
  in prepn of carbon monoxide, *4*:437
  saponification of fats, *8*:825
  in self-polymerization of propylene oxide, *8*:279

in soap industry, *18*:418
in synthesis of 1,3-butylene glycol, *10*:664

Potassium *p*-hydroxybenzenesulfonate, *15*:209

Potassium hydroxyfluoborate, *9*:564

Potassium indigosulfonate
  menadione assay, *21*:595

Potassium iodate, *12*:271
  as bleaching agent in flour, *10*:6
  in flour, *22*:294

Potassium iodide, *2*:618; *8*:420, 472, 492; *16*:392
  chlorine dioxide detected with, *5*:45
  sonochemical reactions of aqueous, *20*:779

Potassium iodine hexafluoride, *9*:589

Potassium iodomercurate(II), *13*:240

Potassium isobutyrate
  petroleum sweetening, *15*:52

Potassium isopropyl xanthate, *22*:420

Potassium lactatozirconate, *22*:656

Potassium laurate, *2*:632

Potassium manganate, *13*:12

Potassium manganate(V), *13*:11

Potassium manganate(VI), *13*:33

Potassium manganese(III) diaquodioxalate, *13*:8

Potassium manganese(I)cyanide, *13*:5

Potassium manganic cyanide, *13*:8

Potassium mercuric iodide, *2*:619

Potassium metabisulfite, *10*:13

Potassium metabisulfite
  addition to wines, *22*:318
  in developer, *16*:502
  menadione derivs, *21*:596

Potassium metaborate, *3*:647
  in prepn of trifluoroboroxine, *9*:556

Potassium metaphosphate, *15*:249

Potassium metaplumbate, *12*:276

Potassium metatantalate, *19*:651

Potassium metatitanate, *20*:411

Potassium monofluorophosphate, *9*:642

Potassium niobate, *13*:782

Potassium niobium fluoride, *9*:659

Potassium niobium oxyfluoride, *19*:633

Potassium nitrate, *6*:513; *16*:393
  in black powder, *8*:647
  in embalming fluids, *8*:101
  fertilizer technology, *S*:352
  in fertilizers, *9*:78, 113
  in HTS (heat transfer salt), *10*:854
  perchloric acid reaction with, *5*:65
  as food preservative, *10*:13
  from potassium chloride, *S*:171

in prepn of trifluoroboroxine, 9:556
in propellants, 8:660, 681
Potassium nitrite, *16 394*
  as food preservative, *10*:13
Potassium nitrosopentachloroosmate(III), *15*:874
Potassium orthohexatantalate, *19*:651
Potassium orthophosphates, *15*:239
Potassium osmate(VI), *15*:875
Potassium osmiamate, *15*:874
Potassium oxalate, *14*:368
Potassium oxalatoferrate, *14*:368
Potassium oxide, *16*:370
  in chemical ware, *4*:835
  in electron tubes, *8*:6
  in glass forming, *10*:538, 542
Potassium ozonide, *14*:414
Potassium pentaborate, *3*:646
Potassium pentaborate dihydrate, *3*:646
Potassium pentachloromonoaquothallate(III), *20*:67
Potassium pentachlororhodate(III), *15*:872
Potassium pentafluorophenoxide, *9*:793
Potassium pentafluorostannite, *9*:682
Potassium pentafluorothiophenoxide, *9*:793
Potassium pentafluorozirconate, *9*:685
Potassium pentanitroruthenate(III), *15*:876
Potassium perchlorate, *5*:62, 69
  decomposition when propellant, *8*:669
  in manuf of perchloryl fluoride, *9*:606
  in rocket propellants, *8*:660
Potassium perferrate, *12*:40
Potassium perfluorooctanoate, *9*:723
  water treatment with, *22*:72
Potassium permanganate, *7*:43
Potassium permanganate, *2*:623
Potassium permanganate, *13*:32, 33, 34
  carbon dioxide purified with, *4*:363
  in decomposition of tetraethyllead, *12*:287
  hydrogen sulfide, *19*:385
  in magnesium perchlorate drying agent, *7*:395
  manuf of, *13*:35
  as odor and taste control agent in water, *22*:102
  in oxidation of methylbutynol, *1*:623
  perchlorates detected with, *5*:80
  pigments, *15*:541
  reaction with allyl ether, *8*:494
  reaction with ammonia, *2*:261
  as starch modifier, *10*:17, 799
  uranium ore leaching, *21*:14
Potassium perniobate, *13*:782

Potassium peroxide, *14*:749; *16*:395
Potassium peroxydigermanate, *14*:761
Potassium peroxydisulfate, *7*:706; *14*:757
Potassium peroxysulfate, *7*:679; *10*:5; *15*:796; *19*:143
  catalyst for vinyl polymers, *21*:442
Potassium perrhenate, *17*:412
Potassium perruthenate, *15*:877
Potassium persulfate, *7*:696; *14*:757
Potassium pertantalate, *19*:651
Potassium perxenate, *10*:893
Potassium phenoxide
  xanthates from, *22*:419
Potassium phenylacetate, *15*:213, 215
Potassium phenyl xanthate, *22*:419
Potassium phosphate
  in prepn of glutamic acid, *2*:203
Potassium phosphates, *15*:274; *16*:396
Potassium phthalimide, *2*:142; *9*:794
  methyl chloride reaction with, *5*:104
Potassium polybutadiene, *7*:690
Potassium pyrophosphate, *7*:680; *15*:249
  as detergent builder, *S*:45
Potassium ruthenate(VI), *15*:877
Potassium ruthenate(VII), *15*:877
Potassium ruthenonitrite, *15*:876
Potassium salts
  from sea plants, *14*:166
Potassium silicate
  pigments, *15*:541
Potassium silicates, *18*:135
Potassium sodium chromate, *5*:499
Potassium sodium hexacyanoferrate(III), *12*:32
Potassium sodium tartrate, *4*:591; *19*:736
  in heat transfer salt, *10*:854
  in margarine, *13*:59; *18*:595
Potassium stannate, *20*:318
  reaction with water, *20*:309
Potassium stearate, *6*:350
Potassium sulfate, *8*:660, 681; *16*:396
  calcium tartrate, *19*:738
  in dental materials, *6*:787, 788
  as fertilizer, *9*:113
  as water correcting agent, *10*:17
Potassium sulfatozirconate, *22*:662
Potassium sulfite
  methylene chloride reaction with, *5*:114
Potassium sulfites, *16*:398
Potassium superoxide, *14*:763; *16*:362, 365, 395
  in chlorate candle, *S*:663
Potassium syngenite, *9*:119

Potassium tantalate, *19*:651

Potassium tantalum fluoride, *9*:659, 660; *19*:649

Potassium tetraborate, *3*:646

Potassium tetracarbonatodiphosphatozirconate, *22*:657

Potassium tetrachloroaurate, *10*:693

Potassium tetracyanoplatinate(II), *15*:863

Potassium tetraglycine triiodide, *2*:619

Potassium tetrahydroborate, *16*:387

Potassium tetranitroplatinate(II), *15*:865

Potassium tetraoxalate, *14*:369

Potassium tetraphenylborate, *16*:384

Potassium tetraphosphate, *10*:789

Potassium tetrathiowolframate, *22*:354

Potassium thiocyanate, *9*:602
    as cardiovascular agent, *4*:523
    reaction with ethylene oxide, *8*:531

Potassium thiowolframates, *22*:354

Potassium titanate
    fibrous, *20*:411

Potassium titanates, *20*:411

Potassium tricarbonatozirconate, *22*:640

Potassium trifluoroberyllate, *9*:552

Potassium trifluorostannite, *9*:682

Potassium trifluorozincate, *9*:684

Potassium trioxalate manganate trihydrate, *13*:8

Potassium tripolyphosphate, *15*:249
    thiol manuf, *20*:212

Potassium tungstenate. See *Potassium wolframate.*

Potassium vinyl. See *Vinylpotassium.*

Potassium wolframate

Potassium zirconate, *22*:656

Potassium zirconium fluoride, *9*:660; *16*:517

pH-Potential diagrams

Potentiometry, *7*:726
    as electroanalytical method, *7*:735

Pott-Broche process
    for synthetic crude oil, *S*:184

Potter's flint, *4*:769

Pottery clays, *4*:768

Potting, *8*:102

Poultry grits
    limestone in, *12*:455

Pourbaix diagrams, *6*:292

Povan, *6*:605

Powder coatings
    as electric insulation, *11*:793

Powder diffraction data, *6*:529

Powder filters, *9*:262

Powder metallurgy, *16*:**401**

Powders
    metallic as pigments, *15*:546
    particle characteristics, *16*:405

Power
    steam, *18*:702

Power generation, *16*:**436**
    thermoelectric, *20*:151, 163

Power plants
    char-burning, *S*:229
    coal, *S*:229
    for fuel cells, *S*:387

Power sources
    hybrid, *S*:405, 414
    microwave, *S*:564

Power transmission
    microwave, *S*:565

Pozzolanas
    clays as, *5*:580

Pozzolan cements, *4*:693

Pradtl number, *14*:82

Prandtl's wall-velocity law, *9*:459

Praseodymium, *17*:144
    electronic configuration, *1*:360
    ionic radii, *1*:370

Praseodymium-144, *17*:127

Praseodymium compounds

Praseodymium oxide, *22*:653
    as glass colorant, *5*:849, 850

Precalciferol, *18*:837

Prechlorination
    of municipal waters, *22*:87

Precipitation titrations, *7*:778

Prednisolone, *18*:883
    corticosteroid analog, *11*:88

Prednisolone *t*-butyl acetate
    veterinary drug, *21*:251

Prednisone BMD, *18*:876

Preergocalciferol
    discovery of, *21*:553

Pregnane
    structure of, *11*:77

Pregnane-3,20-dione, *18*:841

Pregnenolone, *6*:77, 84
    in steroid synthesis, *11*:84

Pregn-4-ene-17,21-diol-3,20-dione, *18*:840

Pregn-4-ene-3,20-dione. See *Progesterone.*

Prehnitene, *16*:300

Premetalized acid dyes, *7*:530

Premetalized dyes, *2*:882, 887; *5*:511
    in textile printing, *7*:577

Preobrazhenskite, *3*:648

Presdwood," *21*:603
Preservatives
    benzoic acid, *3*:428
    for carbonated beverages, *4*:344
    sodium benzoate, *3*:428
Preservatives for foods, *10*:12
Preshave lotion, *6*:354
Pressure. (See also *Vacuum.)*
Pressure
    measurement of low, *21*:127
    static method of determination, *21*:230
Pressure measurement, *16*:**470**
    manometers, *21*:127
    prod of low, *21*:131
Pressure-sensitive adhesives, *1*:382
Pressure-sensitive tapes, *1*:395
    use of abherents as, *1*:7
Pressure vessels, *16*:**482**
Priceite, *3*:627, 628, 649
Prilling
    in melt granulation of fertilizers, *S*:341
Primary colors, *5*:801
Primene JM-T, *2*:125, 126
Primene 81-R, *2*:125, 126
Primidone
    as anticonvulsant, *11*:523
Primuline, *7*:465, 539; *20*:185
Primuline base, *20*:186
Prins reaction, *3*:845
Printed circuits
    in electric insulation, *11*:795
Printing
    screen process, *17*:335
    stencil, *17*:335
Printing inks. (See also *Ink.)*
    carbon blacks used in, *4*:267, 275
    drying oils in, *7*:424
    use of aluminum acetates in, *2*:13
    use of driers, *7*:277, 282
    use of fluorescent pigments, *9*:503
    use of formamide in, *10*:106
    use of 2,2,4-trimethylpentanediol, *10*:679
Printing processes, *16*:**494**
Pristane, *14*:847
Pristinamycin, *16*:308, 332
Proactinomycin, *12*:636
Proanthocyanidins
    polymeric, *19*:748
Probability
    and entropy, *20*:127
Probenecid, *19*:258
Probertite, *3*:650

Procaine, *2*:405
    as cardiovascular agent, *4*:516
Procaine amide
    as cardiovascular agent, *4*:516
Procaine hydrochloride, *2*:405, 406
Procaine penicillin G, *3*:15
Procarbazine
    in cancer therapy, *S*:89
Prochlorite, *19*:608
Prochlorperazine, *16*:667
Procion dyes, *2*:882; *7*:466
Procytoxid
    from yeast extract, *22*:530
Prodeconium, *6*:549
Prodigiosine
    amebicides, *20*:80
Prodox, *6*:78
Producer gas. (See also *Gas, manufactured.)*
    characteristics, *10*:280
    manuf results, *10*:374, 378
Profibrinolysin. See *Plasminogen.*
Proflavine, *2*:639
Progesterone, *10*:19; *11*:118-127; *18*:836, 875
    structural formula, *6*:61
    veterinary uses, *21*:251
dl-Progesterone, *18*:865
Progesterone derivatives, *6*:62, 80
Progestin, *18*:879
Progestins, *18*:881
Prolactin. See *Lactogenic hormone.*
Prolamins
    in cereal grains, *22*:268
Proline, *2*:157
    in prepn of azo dyes, *2*:893
    in wine, *22*:316
Proline betaine, *1*:764, See *Stachydrine.*
Proline betaine
Promazine, *2*:150; *16*:670
Promethazine, *2*:152; *16*:670
    antihistamine, *11*:40
Promethazine.HCl
Promethium
    electronic configuration, *1*:360
    ionic radii, *1*:370
Promethium-147
    beta source, *17*:57
Prometone, *22*:185
Prometryne, *22*:185, 188
Prontosil, *3*:4; *7*:258; *19*:261, 266, 277
Proof (alcohol content), *8*:456
Propadiene. See *"Allene".*
Propanamide. See *Propionamide.*
    dehydrogenation by sulfur, *S*:552

Propane, 8:539; 16:546,**547**
  as aerosol propellant, 1:475
  from catalytic reforming, 15:33
  catalytic reforming, 10:423
  characteristics as fuel, 10:280
  chlorinolysis of, 5:91
  as component of natural gas, 10:448
  deasphalting with, 2:770; 15:57
  dried by alumina, 2:52
  economics of ethylene prod, 8:516
  as engine fuel, 10:492
  ethylene from, S:310
  in fats extraction, 13:180
  in manuf of formaldehyde, 10:89
  oxidation of, 11:228
  from petroleum, 15:18
  property for gas manuf, 10:389
  refrigerant, 10:859; 17:308
  separation of oils, 7:402
  solvent refining, 15:57, 96
  steam reforming, 10:420
  toxicity of, 11:294
  vapor-phase oxidation, 8:285
1,2-Propanediamine. See Propylenediamine.
1,3-Propanediamine. See 1,3-Diaminopropane.
1,3-Propanedicarboxylic acid. See Glutaric acid.
Propanedioic acid. See Malonic acid.
1,2-Propanediol. See Propylene glycol.
1,3-Propanediol. See Trimethylene glycol.
1,3-Propanediol diacrylate
  phys prop, 1:289
2,2-Propanedithiol
  stability, 20:207
Propanenitrile
  synthesis of, S:590
2-Propanenitronic acid, 19:819
Propanesultone, 19:284, 288, 290
Propanethiol
  uses, 20:214
1-Propanethiol
  from propylene, 15:339
2-Propanethiol
  spectral studies, 20:207
1,2,3-Propanetriol, 10:619; 20:205. (See also Glycerol.)
Propanil
  amide herbicide, 22:198
  soil chemistry, 18:535
Propanoic acid. See Propionic acid.
1-Propanol. See Propyl alcohol.
  in esterification of fatty acids, 8:818
  from oxo process, 14:373

2-Propanol. See Isopropyl alcohol.
2-Propanone. See Acetone.
Propargyl alcohol, 1:599-602; 18:870
  in corrosion inhibition, 6:329
  prepn from formaldehyde, 10:86
Propargylaldehyde, 1:600
Propargyl bromide. See 3-Bromopropyne.
Propargylic acid. See Propiolic acid.
Propargylic acid, 8:816
Propazine
  herbicide, 22:188
Propellants. (See also Rocket propellants.)
  aerosol, 1:471
  use of hydrogen peroxide as, 11:409
Propene. See Propylene.
2-Propen-1-ol. See Allyl alcohol.
2-Propene-1-thiol
  occurrence, 20:206
p-Propenylanisole, 15:163
4-Propenylguaiacol, 15:163
Propenyl methyl ketone, 6:694
4-Propenylveratrole, 15:170
  flavor and perfume material, 9:358
Propiolactone, 19:141
β-Propiolactone, 2:369; 10:86
  polyesters from, 16:173
  prepn, 12:89
  in prepn of acrylic acid, 1:293
  in prepn of acrylic acid esters, 1:299
  sterilant, 18:816
Propiolaldehyde, 1:600
Propiolic acid, 1:600; 8:816
  chloroprene reaction with, 5:218
  decarboxylation of, 1:173
  phys prop, 1:229
Propionaldehyde, 2:414; 10:640; 16:**548**
  mass spectrometry, 13:92
  from oxo process, 14:373
  in prepn of trimethylolethane, 1:594
  prepn from allyl ether, 8:495
  prod from propylene oxide, 8:278
Propionamide
  form, 2:67, 68
  phys prop, 2:68
Propionanilide, 2:414
Propionic acid, 16:**554;** 19:238
  to form methacrylic acid, 13:339
  phys prop, 1:229; 8:812
  as stabilizer in aluminum formate, 2:15
  as synthetic flavor, 9:354
Propionic acid esters, 8:381
Propionic anhydride, 12:91
  phys prop, 1:212

Propionitrile
  cyanoethylation, 6:634
  prepn from acrylonitrile, 1:341
Propionylacetone, 12:148, 149
β-Propionylactone, 2:642
Propionylbutyryl, 12:140
Propionyl chloride
  phys prop, 1:224
Propionylcyclohexanone, 12:158
Propiophenone, 6:670; 10:147
Proportional limit, 13:187
Proportioning, 22:220, 235
Propoxyphene, 2:385, 386
Propyl acetate
  properties, 8:379
  from n-propyl alcohol, 16:562
  specification, 8:373
n-Propyl acetate
  use in dehydration of acetic acid, 2:850
Propyl acrylate
  phys prop, 1:287
n-Propyl alcohol
  2-butanol azeotrope, 19:252
  solvent recovery, 18:562
Propyl alcohols, 16:**559**. (See also *Isopropyl alcohol.*)
n-Propylamine
  prepn from acrylonitrile, 1:341
Propylamines
  uses, 2:123
o-Propylaniline
  in hydrorefining, 11:437
2-Propylanthroquinone, 2:438
n-Propylbenzene, 6:544
n-Propyl bromide, 3:778
n-Propyl chloride
  alkylation of benzene, 10:137
n-Propyl cinnamate, 5:520
n-Propyl crotonate, 6:469
Propyl-3,5-diiodo-4-oxo-1(4*H*)-pyridineacetate, 17:140
Propylene, 6:439, 695; 8:504, 539; 16:**579**
  alkylation, 15:42
  alkylation, 15:150
  allyl chloride manuf from, 5:209
  ammonolysis of, 2:336
  butyric acid from, 3:880
  as by-product from ethylene prod, 8:510
  chlorohydrins from, 5:316
  copolymers, 14:299
  dimerization, 14:322
  dried by alumina, 2:52

  as feed gas, 10:474
  industrial uses, S:634
  in manuf of acrylonitrile, 1:346
  in manuf of n-butyraldehyde, 3:871
  in manuf of cumene, 6:544
  in manuf of isoprene, 12:73
  oligomerization, 16:281
  in prepn of acetone, 1:161
  in prepn of acrylic acid, 1:294
  in prepn of elastomers, 7:686
  in prod of isopropyl alcohol and ether, 8:489
  pyrolysis of, S:551
  refrigerant, 17:308
  sulfation, 19:302, 490
  sulfonic acids, 19:314
  in synthesis of glycerol, 10:624
  thiol addition, 20:209
  toxicity of, 11:296
Propylene bromide, 3:778
Propylene carbonate, 2:280
1,2-Propylene carbonate, 4:390
Propylene chlorohydrin
  in prod of propylene oxide, 8:280
Propylene chlorohydrins, 5:304, 309, 310
Propylene cyanohydrin, 6:668, 674
Propylenediamine, 7:23
  properties, 7:24
  reclaiming gelled baking enamels, 7:35
  specifications, 7:32
Propylene dibromide
  dehydrohalogenation of, S:550
Propylene dichloride, 8:281
  paint remover, 14:487
Propylene dimers
  isomerization-polymerization of, S:794
Propylene glycol, 10:649-651
  in deicing fluids, 2:559
  in dentifrices, 6:849
  in esterification of fatty acids, 8:819
  as fungicide, 10:235, 638
  in manuf of polypropylene glycols, 10:658
  parenteral solvent, 15:128
  polyesters, 21:62
  prepn of urethan polyesters, 7:694
  in refrigeration, 17:309
  in shampoo, 10:744, 796
  from sucrose, 19:154
  surfactants, 19:548, 553
  sweetener, 19:602
  toxicity, 10:41
Propylene glycol alginate
  as gum, 10:741

Propylene glycol *n*-butyl ether, *10*:650
Propylene glycol *tert*-butyl ether, *10*:650
Propylene glycol diacrylate
  phys prop, *1*:289
Propylene glycol ethers, *10*:650
Propylene glycol ricinoleate, *10*:651
Propylene oligomers, *16*:581, 582
Propylene oxide, *6*:673; *8*:275–288; *16*:**595**; *19*:553
  ammonolysis of, *2*:351
  block copolymers, *19*:553
  in copolymerization of trioxane, *1*:98
  in esterification of fatty acid, *8*:819
  in manuf of alkanolamines, *1*:814
  in manuf of polypropylene glycols, *10*:658
  in manuf of propylene glycol, *10*:649
  NMR of, *14*:54, 56
  in paint removers, *14*:488
  polyethers, *21*:58
  polymerization of, *6*:66; *7*:69
  reaction with ethyleneamines, *7*:25
  sterilant, *18*:816
  styrene by-product, *19*:62
1,2-Propylene oxide
  mass spectrometry, *13*:92
1,2-Propylene sulfite, *19*:485
Propylene tetramer
  prepn, *4*:395
Propylenimine, *11*:526, 527, 536
  commercial importance, *11*:526
n-Propyl ether, *8*:487
  phys prop, *8*:471
  prepn from allyl ether, *8*:494
Propyl gallate, *2*:593, 596, 598; *8*:781; *10*:14
2-Propylheptan-1,3-diol
  titanium chelates, *20*:464
Propyl *p*-hydroxybenzoate
  as food preservative, *10*:12
  in shampoos, *10*:774
Propyliodone, *17*:140
Propyl isocyanate, *12*:46
Propyl isopropyl ketone, *12*:105
n-Propyl ketone, *12*:160
n-Propyllactamide, *12*:186
n-Propyl lactate
  prepn, *8*:352
Propyllithium
  titanium sandwich compds, *20*:440
Propyl nitrate
  as propellant, *8*:660
  properties, *8*:689
  thiol addition, *20*:209

Propyl oleate
  prod and sales of, *8*:377
n-Propylphenol, *2*:626
6-Propyl-2-thiouracil
  antithyroid agent, *20*:270
Propyne. (See also *Methylacetylene.*)
Propyne, *S*:547
2-Propynoic acid. See *Propiolic acid.*
2-Propynoic acid, *8*:816
2-Propyn-1-ol. See *Propargyl alcohol.*
Prosopite, *9*:534
Prostaglandins
  tissue hormones, *11*:50
Protactinium, *1*:351; *17*:26, 27. (See also *Actinides.*)
Protactinium compounds
  crystal structure, *1*:366
  phys prop, *1*:366
Protamine sulfate, *5*:602
Protease, *8*:185
  action and uses, *8*:186, 187, 188, 194
  in milk, *13*:514
Proteases
  classification, *8*:176
  in laundry preparations, *S*:295
  uses, *8*:190
Protective gas atmospheres, *10*:284, 286
Protein. (See also *Proteins.*)
  additives to food, *10*:57
  analysis in sparkling wines, *22*:**324**
  in cotton, *6*:406
  in fish, *9*:318
  purification, *16*:623
Protein biosynthesis
  by tetracyclines, *20*:26
Protein deficiency, *2*:185
Protein efficiency ratio, *2*:184
Protein hormones, *16*:622
Protein hydrolysates
  acylated, *19*:514
Proteins, *16*:**610**
  food, *2*:184
  in nuts, *14*:125
  synthesis of, *2*:179
  x-ray analysis of fibrous, *22*:454
  in yeast, *22*:516
Proteins from petroleum, *S*:**836**
Proteolytic enzymes, *5*:594, 595
Prothrombin, *3*:576; *5*:590
  phys prop, *5*:589
Protoadamantane, *S*:11
Protoberberine, *1*:769

Protocatechualdehyde, *15*:162; *21*:185
Protocatechualdehyde diacetate, *8*:234
Protocatechualdehyde-3-methyl ether, *21*:180
Protocatechuic acid, *7*:617
Protocatechuyl alcohol, *8*:231
Protoeolytic enzymes
   in breadmaking, *22*:528
Protolysis
   and NMR, *14*:73
Protomycin
   amebicide, *20*:82
Proton exchange
   and NMR, *14*:73
Protons
   accelerated, *17*:54
Protopectin, *8*:198; *14*:636
Protopectinase
   uses, *8*:190, 199
Protoveratrine, *1*:778, 800
Protoveratrine A, *1*:775
Protozoa, *13*:473
Protozoal infections, *20*:**70**
   amebiasis, *20*:70
Proustite
   nonlinear optical materials, *S*:630
Provest, *6*:79, 87
Provitamins A, *21*:501
   carotenoids, *21*:505
Provitamins D, *21*:558
   assay of, *21*:567
Prussian blue, *12*:25, 30, 33; *15*:538
Prussic acid. See *Hydrogen cyanide.*
Pseudoazimidobenzene, *10*:917
Pseudobornyl acetate, *19*:831
Pseudobrookite, *20*:414
Pseudocholesterol benzoate, *18*:878
Pseudocumene, *16*:300
   acetyl chloride condensation, *15*:482
   oxidation, *15*:476
Pseudocyanine, *6*:607
Pseudoeleostearic acid, *8*:815
   co-oxidation, *S*:840
   methane utilization, *S*:838
Pseudoionone, *12*:132
   manuf, *1*:846
Pseudoisocyanine, *6*:607
Pseudoitaconanilic acid, *12*:84
Pseudo-ketones, *6*:613
Pseudomorphosis
   alumina hydrates, *2*:48
Pseudopelletierine, *1*:766
Pseudopurpurin, *7*:621
Pseudothioureas, *6*:562

Pseudoureas, *6*:561, 562
Psilocin, *16*:646
   in chemical warfare, *4*:878
Psilocybin, *16*:646
Psilocybine, *1*:763
Psilomelane, *13*:21
Psittacosis-lymphogranuloma
   viral group, *21*:454
Psychochemicals
   in chemical warfare, *4*:877
Psychopharmacological agents, *16*:**640**
Psychotrine, *20*:75
Psychrometry. See *Aquametry.*
Psyllium seed gum, *10*:750
Pteridine, *10*:932
Pterins, *10*:932
Pteroic acid, *10*:932
Pteroylglutamic acid, *10*:932
Puering, (See also *Bating.*)
   in leather manuf, *12*:312
Pulegone
   as synthetic flavor, *9*:354
Pulp, *16*:**680**. (See also *Paper.*)
   cyanoethylation of, *6*:656
   lime in, *12*:458
Pulp bleaching, *3*:563
   chlorine dioxide prepn for, *5*:40
Pulp colors, *15*:603
Pulping
   of wood, *4*:606, 608
Pulpmaking, *12*:369
Pumice
   pigments, *15*:512
Pumice, flour of
   as dental abrasive, *6*:849
Pumicite
   as abrasive, *1*:26
   selection of, *21*:144
Pumps, *16*:**728**
Pumps, vacuum, *21*:132-137
Punched card documentation, *12*:515
Punicic acid, *8*:815
Purex process for spent nuclear fuels, *14*:92
Purgatives. See *Cathartics.*
Purine
   biosynthesis, effect of biotin on, *3*:525
   structural formula, *3*:912
Purine alkaloids, *1*:764
Purine metabolism antagonist, *S*:84
Purines, *10*:913
   as cardiovascular agents, *4*:521
6-Purinethiol, *S*:84
Purinetrione. See *Uric acid.*

Puromycin
  amebicide, 20:80
Purone
  from uric acid, 21:108
γ-Purone, 7:415
Purple of Cassius, 10:686
Purpuramid, 2:458, 460
Purpurin, 2:458, 466; 7:486, 621; 16:201
  oxidation of, 2:469, 473
Purpurogallin, 16:191
Purpuroxanthin. See Xanthopurpurin.
Putty, 7:425
Pycnometer method
  measuring density, 6:766, 773
Pyelography, 17:132
4H-Pyran, 10:897, 898
2-H-Pyran, 10:897, 898
γ-Pyran, 10:920
Pyran copolymer
  interferon induction, 21:458
Pyranine concentrated, 5:868, 873
Pyranines, 8:378; 16:284
Pyrantel
  anthelmintic, 21:249
Pyrantel tartrate, 14:539
8,16-Pyranthrenedione. See Pyranthrone.
Pyranthrone, 2:438, 523
Pyrathiazine, 16:671
Pyrazinamide, 3:30
Pyrazine, 7:30; 10:898, 920, 929
  derivs as optical brighteners, 3:745
  from piperazine, 15:640
1,9-Pyrazolanthrone, 2:449
Pyrazole, 10:898, 910
Pyrazolecarboxamides, 12:151
Pyrazoles, 16:762
  heterocyclic compound, 10:901
Pyrazolines, 10:911; 16:762
Pyrazolone red, 15:576
Pyrazolones, 10:901, 911; 16:762, 772-776
  pigments, 15:570
4-Pyrazone
  as nucleus in cyanine dye, 6:607
Pyrene, S:367
Pyrenes
  in petroleum, 14:849
Pyrethins
  residue in food, 10:18
Pyrethroids
  antimony pentafluoride in, 9:550
  as insecticides, 11:684
  strength value of, 4:818

Pyrex glass. (See also Glass.)
Pyridazine, 10:898, 920
Pyridazine-3-carboxamide, 3:30
Pyridazines, 10:928
9-Pyrid [3,4-b]indole, 10:910
Pyridine, 2:165; 6:611, 630; 7:479; 10:898,
  920; 16:780
Pyridine, 16:780
  ammonolysis of, 2:367
  amine oxide deriv, S:43
  carboxylation, 21:513, 515
  in coal tar, 19:657
  dialkyl sulfites, 19:493
  halosulfonates, 19:490
  Hantzsch synthesis, 1:158
  in purification of mercury fulminate, 8:588
  manuf, 2:346
  as nucleus in cyanine dye, 6:607
  perchlorylation reaction, 9:604
  as solvent, 2:143
  prepn from acrolein-vinyl ether condensa-
    tion product, 1:258
  prepn of ethylpyridinium fluoborate, 9:570
  prepn from paraldehyde, 1:18
  quaternization, 19:252
  reaction with iodine pentafluoride, 9:591
  reaction with silver fluoborate, 9:566
  sucrose solvent, 19:221
  sulfonation, 15:211
  sulfonyl chlorides, 19:402
  sulfur trioxide complexes, 19:491
2-Pyridine aldoxime methiodide, 16:804
Pyridine-2-aldoxime methiodide, 15:323
Pyridine-borane, 3:728
3-Pyridinecarboxamide. See Nicotinamide.
3-Pyridinecarboxamide, 21:525
3-Pyridinecarboxylic acid. See Nicotinic acid.
2-Pyridinecarboxylic acid. See Picolinic acid.
Pyridinecarboxylic acid, 16:801
γ-Pyridinecarboxylic acid, 21:527
Pyridine-β-carboxylic acid, 21:509
3-Pyridinecarboxylic acid, 21:509
4-Pyridinecarboxylic acid, 21:513, 527
Pyridine-4-carboxylic acid, 16:802
Pyridine-3-carboxylic acid amide, 16:801
3-Pyridinecarboxylic acid diethylamide,
  21:533
Pyridine-β-carboxylic acid diethylamide,
  21:533
4-Pyridinecarboxylic acid hydrazide, 21:528
Pyridine-γ-carboxylic acid hydrazide, 21:528
Pyridine derivatives, 16:780

2,3-Pyridinedicarboxylic acid. See *Quinolinic acid.*

2,5-Pyridinedicarboxylic acid, *21*:515

Pyridine-2,3-dicarboxylic acid, *2*:414

Pyridine nucleotide cycle, *21*:519

Pyridines
   antiamebic agents, *20*:88
   as diuretics, *7*:252
   in hydrorefining, *11*:436
   nitrile reactions, *S*:594
   oxidation of substituted, *21*:514

Pyridine sulfate
   "cracking," *19*:670

Pyridine sulfate dibromide
   in analysis of acrylic esters, *1*:302

3-Pyridinesulfonic acid, *19*:300

Pyridine-2-thiol

Pyridine-3-sulfonic acid, *21*:527
   spectral analyses, *20*:208

2-Pyridinethiol-1-oxide
   uses, *20*:215

Pyridinium acetate
   polyester end-group analysis, *16*:179

Pyridinium dicyanomethylide, *6*:630

Pyridinium salts, *16*:**795**

Pyrido[1,2-*a*]benzimidozol-8-yl 3(4*H*)-diazo-4(3*H*)-oxo-1-naphthalene sulfonate, *16*:515

Pyridocoline, *10*:936

4*H*-pyrido[1,2-*a*]pyridine, *10*:936

2-Pyridone, *10*:903

4-Pyridone, *6*:649

2-Pyridone-5-carboxamide, *21*:522

4-Pyridone-5-carboxamide, *21*:522

Pyridones, *10*:922

Pyridoxal. *16*:**806.** (See also *Pyridoxine.*)

Pyridoxal hydrochloride, *16*:808

Pyridoxal 5-phosphate, *16*:806, 809
   biochemical function *16*:818

Pyridoxamine, *16*:**806.** (See also *Pyridoxine.*)
   synthesis, *16*:815

Pyridoxamine dihydrochloride, *16*:809

Pyridoxamine 5-phosphate, *16*:809

4-Pyridoxic acid, *16*:810

Pyridoxine, *2*:189; *8*:175; *16*:780, **806**
   in fish, *9*:322
   in yeast, *22*:516
   yeast requirement, *22*:517

Pyridoxol, *12*:825; *16*:806, 807, 812-815. (See also *Pyridoxine.*)

Pyridoxol 5-phosphate, *16*:810

Pyridylmercuric acetate, *13*:244

Pyrilamine maleate
   antihistamine, *11*:40

Pyrimidazole, *10*:936

Pyrimidine, *10*:989, 920
   in vitamin-enriched yeast, *22*:531

Pyrimidine derivatives
   prepn from dicyanidiamide, *6*:568

Pyrimidine metabolism antagonist, *S*:85

Pyrimidines, *10*:928
   antiamebics, *20*:88
   fused, *18*:868
   nitrile reactions, *S*:594

Pyrimido[4,5-*b*]pyrazine, *10*:932

Pyrite, *12*:6, 43, 208; *15*:517, 519; *19*:350, 758; *22*:563
   burners, *19*:450, 452
   in limestone, *12*:422
   in manuf of carbon disulfide, *4*:381
   selenium in, *17*:812
   separation from chalcocite, *9*:389, 393
   sulfur dioxide from, *19*:447
   sulfur from, *19*:339

Pyrite cinder, *15*:519

Pyrites
   sulfur from, *S*:358

Pyrithiamine, *20*:183; *21*:489

Pyroarsenic acid, *2*:722

Pyrocalciferol, *18*:837; *21*:557

Pyrocatechin, *11*:**462**
   hydrous zirconia and, *22*:654
   reaction with antimony compounds, *2*:579, 580
   in synthesis of epinephrine, *8*:234

Pyrocatechol-3,5-disulfonic acid, *2*:579

Pyrocatechol monomethyl ether, *21*:184

Pyrocatechol-4-sulfonic acid, *15*:208

Pyrocellulose, *8*:597, 697

Pyrochlore, *13*:766, 768
   ferroelectric, *9*:5

Pyrochroite, *13*:17

Pyroelectric effect
   in nonlinear optics, *S*:626

Pyroelectricity, *9*:2

Pyrogallol, *2*:579, 629; *7*:617; *16*:191-201
   developing agent, *15*:371
   tartaric acid, *19*:736
   zirconium complexes, *22*:655

Pyrogallol carbonate, *16*:192

Pyrogallol-4-carboxylic acid, *16*:191, 193

Pyrogallol-5-carboxylic acid, *16*:196

Pyrogallol 1,2-dimethyl ether, *16*:210

Pyrogallol monoacetate, *16*:198

Pyrogallolphthalein, *16*:198

Pyrogallol triacetate, *16*:198

Pyrogallol tri(3,5-dinitrobenzoate), *16*:192

Pyrogallol tri(phenylurethan), *16*:192
Pyrogens
  USP test, *15*:127
Pyroglutamic acid, *2*:199, 207
Pyrographite, *4*:192
Pyrographitic acid, *4*:312
Pyrolidine alkaloids, *1*:764
Pyrolidine enamine, *1*:342
Pyroligneous acid
  in prepn of acetanilide, *1*:151
Pyroligneous acid liquor, *8*:391
Pyrolusite, *13*:2, 9, 19, 29, 41
  as glass decolorizer, *5*:849
Pyrolysis
  ethylene production by high-temperature,
    S:310
  of zirconium halides, *22*:616
Pyromellitic acid, *15*:480-483; *18*:232
  high-temperature fibers
  from polymers of, *S*:961
Pyromellitic dianhydride, *15*:480, 483, 796;
  16:305
  polyamic acid from, *S*:748
Pyrometer, optical, *19*:793-801
Pyromucic acid. See *Furoic acid.*
Pyronate
  as denaturant, *8*:450, 452
α-Pyrone, *7*:415
Pyrones, *10*:920, 922; *12*:150
Pyrophorite, *20*:349
Pyrophosphate, *2*:179; *15*:232; *21*:521
  stabilization of Mn(III), *13*:13
Pyrophosphate copper solutions, *8*:59
Pyrophosphate zinc electroplating solutions,
  8:66
Pyrophosphoric acid, *15*:241
Pyrophosphorus acid, *15*:301, 311
Pyrophyllite, *5*:547; *19*:608, 610, 612
  in ceramics, *4*:771
  pigments, *15*:511
Pyrosulfite
  addn to wines, *22*:318
Pyrosulfuric acid, *19*:421
Pyrosulfurous acid, *19*:421
Pyrosulfuryl chloride, *5*:358; *19*:392, 395, 399,
  404
Pyrosulfuryl fluoride, *9*:632
Pyrotannic acid method for detection of car-
  bon monoxide poisoning, *4*:443
n-Pyrotartaric acid. See *Glutaric acid.*
Pyrotechnic gel"
  as incendiary agent, *4*:899

Pyrotechnics, *16*:**824**
Pyrovanadates, *21*:172
Pyrovatex CP
  flame retardant, *S*:950
Pyrrhotite, *12*:6; *19*:350, 447, 452, 758
Pyrrobutamine phosphate
  antihistamine, *11*:40
Pyrrocoline, *10*:936
Pyrrole, *6*:648; *10*:897, 898; *16*:**841**
  from butadiene, *3*:788
  as nucleus in cyanine dye, *6*:607
  prepn from furan, *10*:248
  reaction with caustic, *19*:143
  reaction with formaldehyde, *10*:86
Pyrrole deriv, *16*:**841**
2*H*-Pyrrolenine, *10*:897
1-Pyrrolepropionitrile, *6*:648
Pyrrolidine, *12*:69
  polymer, *16*:35
  prepn from tetrahydrofuran, *10*:250, 669
  series of cyclic imines, *11*:526
2-Pyrrolidinecarboxylic acid. See *Proline.*
Pyrrolidine hydrochloride, *7*:30
Pyrrolidines, *16*:850
N-(Pyrrolidinomethyl)tetracycline, *3*:17
Pyrrolidone, *16*:844, 850-858
2-Pyrrolidone, *10*:667, 673; *15*:800
  vinylation of, *21*:432
Pyrrolidonecarboxylic acid. See *Pyroglutamic
  acid.*
2-Pyrrolidylethyl chloride, *16*:671
Pyrroline
  prepn from butenediol, *1*:613
2-Pyrroline, *10*:896
Pyrrolizidine, *10*:936
Pyrrolo(2,3-*b*)indole, *10*:909
Pyrrolo[1,2*a*]pyridine, *10*:897, 936
3*H*-pyrrolo[1,2-*a*]pyrrole, *10*:936
Pyrrolo[2,1-*b*]quinazoline, *10*:936
4*H*-Pyrrolo[3,2,2-*ij*]quinoline, *10*:936
Pyruvaldehyde, *10*:640
Pyruvaldehyde bis(phenylhydrazone), *4*:139
  prepn, *4*:139
Pyruvic acid, *2*:159; *12*:640
  phys prop, *1*:235
  in tobacco, *20*:508
Pyruvonitrile, *12*:89
  as abrasive, *1*:25
  in ceramics, *4*:772
  as a pigment, *14*:515
  pigments, *15*:509
  in soaps, *18*:429
Pyrvinium pamoate, *14*:540

Q

Quantum efficiency
  of photoemission, *15*:397
Quantum numbers
  magnetic, *14*:41
Quantum yield, *15*:334
Quartz, *17*:227; *18*:49, 51, 75. (See also *Silica.*)
  as abrasive, *1*:26
  clear fused, *18*:76
  in dental porcelain, *6*:798
  fused, *18*:75
  piezoelectric, *18*:109
  pigments, *15*:513
  raw material, *18*:78
  separation by flotation, *9*:388, 393, 395
  in soaps, *18*:429
  synthetic, *18*:105, 107
  in talc, *19*:608
  x-ray spectroscopy, *22*:461
  zero points of charge, *9*:384
α-Quartz, *18*:105
Quartz, fused
  graphite equipment for prod, *4*:233
Quartz glass, *18*:75
Quartzite
  as abrasive, *1*:26
Quartzite rock, *18*:119
Quartzites
  in ceramics, *4*:768
Quassin, *20*:79
Quaternary ammonium compounds, *16*:**859**
  analysis, *16*:861
  as antiseptics, *2*:632-635
  as antistatic agents, *2*:666
Quaternary ammonium salts
  surfactants, *19*:562
Quaternary ammonium silicates, *18*:135
Quebracho, *12*:316
  in drilling fluids, *2*:10; *7*:296
  pyrocatechol in, *11*:469
Quercetin, *7*:614; *15*:681
  tea, *19*:743
Quercitols, *16*:213
Quercus suber
  cork from, *6*:281, 288
Query language
  information retrieval, *S*:525

Quicklime. See *Calcium oxide; Lime.*
Quinacridone, *2*:446
  pigments, *15*:557, 568, 581
Quinacrine
  amebicide, *20*:87
Quinacrine hydrochloride, *14*:543
Quinaldine, *2*:415; *6*:609, 621
Quinalizarin
  from quinizarin, *2*:470
Quinazoline, *10*:899, 920
Quinazolines, *10*:928
Quince seed gum, *10*:751
Quinhydrone
  as electrode, *7*:739
Quinidine, *1*:772, 773, 783, 784
  as cardiovascular agent, *4*:510, 516
Quinine, *1*:772, 778, 783
Quinine hydrochloride
  in carbonated beverages, *4*:344
  as denaturant, *8*:450
Quinine sulfate
  as denaturant, *8*:450
Quinizarin, *15*:447; *7*:479. (See also *1,4-Dihydroxyanthraquinone.*)
  pigments, *15*:577
Quinizarin Green SS, *5*:868
Quino[2,3-*b*]acridine-7,14-dione. See *Quinacridone.*
Quinoline, *2*:415; *6*:574; *7*:522; *10*:899, 920; *16*:**865**
  in hydrorefining, *11*:438
  oxidation of, *21*:515
  phthalocyanine, *15*:489
  sulfonation, *19*:300
Quinoline alkaloids, *1*:771
Quinoline-8-boronic acid, *3*:722
4-Quinolinecarboxylic acid. See *Cinchoninic acid.*
2,3-Quinolinedicarboxylic acid. See *Acridinic acid.*
2,4-Quinolinediol, *16*:891
Quinoline dyes, *7*:471, 473; *16*:**886**
Quinolines
  in hydrorefining, *11*:436
Quinoline Yellow, *5*:871
Quinoline Yellow A, *7*:522

Quinoline Yellow KT Extra, *7*:526
Quinoline Yellow SS, *5*:871, 874
Quinoline Yellow Spirit Soluble, *5*:871, 874
Quinoline Yellow Spirit Soluble. See *Quinoline Yellow SS.*
Quinoline Yellow WS, *5*:871
Quinolinic acid
  phys prop, *1*:240
  from tryptophan, *21*:521
8-Quinolinol, *16*:890. (See also *8-Hydroxy-quinoline.*)
Quinolinols, *16*:874
8-Quinolinols
  antiamebic agents, *20*:83
Quinolizidine alkaloids, *1*:766. See *Lupin alkaloids.*
9-*a*-Quinolizine, *10*:899
4-*H*-Quinolizine, *10*:936
9-*aH*-Quinolizine, *10*:936
2-Quinolyl-1,3-indanedione, *16*:886
Quinone. (See also *Quinones.*)
  as inhibitor, *16*:519
  photography, *15*:372

Quinone-diazides, *12*:90
Quinone dichlorimide, *4*:924
Quinonediimine, *10*:307, 796
  in color photography, *5*:819, 820, 821
o-Quinonediimine, *15*:220
p-Quinonediimine, *15*:220
Quinone dioxime, *7*:697
p-Quinone dioxime
  vulcanizer, *17*:568
Quinones, *16*:**899**
  antiamebics, *20*:89
p-Quinones
  phenylenediamines, *15*:220
Quinophthalone, *16*:886; *19*:300
Quinoxaline, *10*:899, 920
Quinoxalines, *10*:929
  antiamebics, *20*:88
  phenylenediamines, *15*:221
Quinuclidine, *10*:937
3-Quinuclidone, *10*:899

R

RDX. See *Cyclonite.*
RG acid, *13*:722
RNA
  messenger in yeast, *22*:512
  transfer in yeast, *22*:512
  in yeast mitochondria, *22*:512
m-RNA
  tetracyclines, *20*:26
RNA synthetase
  tetracyclines, *20*:26
RR acid, *13*:726
R Salt, *20*:723
Racemase, *2*:209
R acid, *13*:722
Racumin, *15*:919
Radiation
  absorption theory, *17*:67
  auxiliary detectors, *17*:73
  background, *14*:108
  backscattering, *17*:70
  chain reactions, *17*:59
  chemical effects of ionizing, *17*:58
  crosslinking, *17*:61
  damage from nuclear, *14*:79
  degradation, *17*:62
  detectors, *17*:71
  effects on corrosion, *6*:303
  effects on lubricants, *12*:610
  electromagnetic. See *Electromagnetic radiation.*
  emission, *17*:65
  in food preservation, *10*:15, 50
  gamma absorption, *17*:55
  graft copolymerization, *17*:61
  of heat, *11*:825
  incremental exposure to atomic, *14*:109
  laws of, *19*:792
  low G reactions, *17*:58
  measurement, *17*:71
  microwave as sterilant, *18*:825
  polymerization, *17*:59
  protection standards for atomic, *14*:109
  ratemeters, *17*:74
  safety survey instruments, *17*:75
  scalers, *17*:74
  second harmonic, *S*:624
  solar, simulation of, *4*:215
  sources, *17*:56
  sterilization, *17*:62
  tolerances for nuclear, *14*:80
  types of, *17*:53
  ultraviolet as sterilant, *18*:822
  ultraviolet in water purification, *22*:128
Radiation (heat transfer) *10*:819, 834
Radiation injury
  protection against, *2*:190
Radiationless decay, *15*:334
Radiationless transition, *15*:333
Radioactive carbon
  carbon monoxide containing, *4*:433
Radioactive contaminants
  carbon used in removal of, *4*:155
Radioactive drugs, *17*:**1**
Radioactive elements, natural, *17*:**9**
  radium, *17*:11, 18
  radium, in tobacco, *20*:512
  radium, from uranium ore, *21*:2
  radium, in water, *21*:703
  radon, *10*862; *17*22
  radon, phys prop, *10*:868
  radon, separation from air, *10*:876
Radioactive materials
  in water, *21*:703
Radioactive tracers, *17*:**1**. (See also *Radioisotopes.*)
Radioactive waste disposal
  clays used in, *5*:579
Radioactive wastes
  accumulation by algae, *1*:666
Radiocarbon dating, *4*:357
Radiochemical analysis, *17*:**35**
Radiochemical plants. See *Nuclear reactors.*
Radiochemicals
  in aquametry, *2*:688
Radiochemical technology, *17*:**53**
Radiochemical tracer applications, *17*:**35**
Radioelements
  in tobacco leaf, *20*:512
Radiography. See *Radioisotopes.*
Radioisotopes, *17*:**64**
  sealed sources, *17*:78
Radione, *15*:920

Radiopaques, *17*:**130**
Radium. See *Radioactive elements, natural.*
Radon. See *Radioactive elements, natural.*
Radon fluoride, *10*:894
Radon-222
  waste in radioactive ores, *14*:103
Raffia hemp wax, *22*:161
Raffinose
  in nuts, *14*:129
  in soybeans, *18*:601
  yeast assimilation, *22*:517
Rainfall
  and water resources, *21*:652
Ralstonite, *9*:534
Ramie, *9*:171, 184
  analysis, *9*:173
  as packing material, *14*:455
  properties, *9*:174
Rammelsbergite, *2*:720
Ramsdellite, *13*:19
Random surface renewal theory of mass transfer, *13*:137
Raney nickel, *19*:143. (See also *Nickel.*
  catalyst, *15*:162; *19*:154
  catalyst for hydrogenation of ethylene, *8*:500
  in cortisol synthesis, *11*:86
  desulfurization, *20*:18
  Doxycycline production, *20*:9
  in hydrogenation of aziridines, *11*:535
  as hydrogenation catalyst, *2*:82, 83, 107
  hydrogenation of nitro alcohols, *13*:826
  reduction of amine oxides, *S*:39
  thiophene desulfurization, *20*:220
  vitamin $K_{1(20)}$ assay, *21*:589
Rankine cycle
  unconventional power plants, *S*:67
Raoult's law
  in binary systems, *21*:224
Rapeseed oil, *8*:779, 785, 786, 790, 791
Rapidazol dyes, *2*:892, 894
Rapid Fast dyes, *2*:892
Rapid Fast Scarlet ILH, *2*:892
Rapidogen N, *2*:894
Rapidogen Scarlet IR, *2*:893
Rapidogen dyes, *2*:889, 892, 893
Rare earth elements, *12*:622; *17*:**143**
Rare earth oxides
  as glass colorants, *5*:849
Rare gases. See *Helium-group gases.*
Rasak, *17*:385
Raschig process, *5*:169, 257; *11*:178; *15*:150
Raschig ring packing, *1*:59, 65

Raspite, *12*:281
Rate-meters, *17*:2
Rationalization, *S*:1004
Raunescine, *16*:666
Rauwolfia, *1*:778; *16*:640
  as cardiovascular agent, *4*:519
Rauwolfia alkaloids, *1*:800
Rauwolfia serpentina, *1*:772
Ravison oil, *8*:779, 785, 786
Rayleigh equation, *20*:740
Rayleigh method of dimensional analysis, *7*:179
Rayleigh scattering, *20*:739
Rayon, *17*:**168**
  bleaching agents for, *3*:563
  carbon disulfide in manuf of, *4*:384
  carbon fiber precursor, *S*:110
  dyeing with basic dyes, *7*:534
  dyeing in fiber blends, *7*:566
  fiber, *9*:153
  fire-retardant finishes, *9*:313
  flame-retardant, *S*:957
  as packing material, *14*:455
  use of polyethylene glycols in manuf, *10*:657
  in reinforced plastics, *12*:189
  saponified acetate, *17*:182, 193
  tire cord, *20*:328, 330
  work of rupture value, *1*:131
Reactive dyes. See *Dyes, reactive.*
Reactors
  automobile exhaust manifold, *S*:61
Recrystallization. See *Crystallization.*
Red Ba, *5*:870
  casting of, *6*:251
Red Ca, *5*:870
Red Sr, *5*:870
Red carbon" *12*:93
Red chalk, *15*:525
Red lead. See *Lead compounds.*
Redoxon. See *Ascorbic acid.*
Red pepper, *9*:365, 369
Red squill, *15*:916
Reductive amination, *2*:336
Reductodehydrocholic acid, *3*:483
Redwood, *10*:307
Redwood bark
  shredded as insulator, *11*:832
  as thermal insulator, *11*:829
Red 28 (CI), *7*:535
Red 46 (CI), *7*:535
Reflectance
  measurement in detergent effectiveness, *6*:864

Reflection prints, 5:835
Reflector materials for nuclear reactors, 4:234, 235
Reformate
  catalytic, 11:2
Reformatsky reaction
  prepn of β-hydroxy acids, 1:230
Reforming
  catalytic, 15:31
  reactions in catalytic process, 15:32
  steam, S:390
  thermal, 15:20
Refraction, 17:**210**
Refractories, 17:**227**
  carbon in, 4:238, 329
  in glass industry, 10:554
  glazes for, 4:329
  graphite in, 4:238, 329
  silicon carbide in, 4:128
Refractory
  contaminating materials in water, 22:117
Refractory alloys
  use of hafnium in, 10:762
Refractory coatings, 17:**267**
Refractory fibers, 17:**285**
Refractory fibers, inorganic. See *Inorganic refractory fibers.*
Refractory materials, 4:774
  nonclay, 4:772
Refractory metals, 11:31. (See also *High-temperature alloys.*)
  aluminide coatings of, 11:33
  oxides of, 11:33
  silicide coatings of, 11:33
Refractory minerals
  in ceramics, 4:772
Refractory oxide fibers
  prepn, 11:658
  prop, 11:659
Refrigerant
  n-butane as, 22:51
  water vapor as, 22:49
Refrigerants. (See also *Refrigeration.*)
  direct contact, 22:51
  naphtha in, 22:53
Refrigeration, 17: **295**. (See also *Cryogenics.*)
  thermoelectric, 20:152
  use of ammonia in, 2:296
Regression
  of higher plants in lakes, S:325
Reich process
  carbon dioxide recovered by, 4:359

Reichstein synthesis of ascorbic acid, 2:752
Reimer-Tiemann reaction aldehydes, 1:648
Reinecke salt
  choline assay with, 5:407, 408, 494
Reinforced plastics, 12:**188**
  use of neopentylene glycol, 10:676
  use of 2,2,4-trimethyl-1,3-pentanediol, 10:679
Reissblei. See *Graphite.*
Relative volatility function
  binary system, 21:207
Release agents. See *Abherents.*
Relief printing, 16:495
Relomycin, 12:636, 647
Remalan dyes, 2:882
Remalloy, 12:766
Remijia, 1:783
Rendering
  of fats, 13:179
Renierite, 10:519
Rennet. See *Rennin.*
Rennin, 8:190, 195; 13:532
Repellents
  for insects, 11:724
Replication
  DNA, S:841
Reprography, 17:**328**
Rescinnamine, 1:802; 16:649, 666
Research
  basic, S:857
  defensive, S:858
  growth, S:858
  purposes of, S:857
Research managment, S:**855**
α-Resene, 17:385
β-Resene, 17:385
Reserpine, 1:761, 772, 778, 801; 4:519; 10:19; 16:640, 646, 649
Reserve-type electric cells, 3:100, 127
Resin
  American fossil, 17:383
Resin acids
  in woods, 16:690
Resinat, 2:430
Resin bonding
  of grinding wheels, 1:34, 35
Resin dental cements, 6:779, 783
Resin-in-pulp ion exchange, 21:16
Resin mortars, 10:246
Resinols
  coal-tar distillate, 19:661
Resinotannol, 17:382

Resins. (See also *Plastics; Resins, natural; Resins, watersoluble.*)
accroides, *17*:382
acrolein-pentaerythritol, *19*:495
acrylic, *15*:799
alkyd, *15*:455, 461, 793
aloe, *17*:387
amber, *17*:382
amorphous, *15*:726
A-stage, *15*:792
balm of Gilead, *17*:387
B-stage, *15*:792
in ceramic composite armor, *S*:145
chain-growth alkyd, *16*:229
crystalline, *15*:726
C-stage, *15*:792
diazonovalak, *17*:331
dispersions of vinyl, *21*:403
as electric insulation, *11*:794
epoxy, *15*:793
extrusion of vinylidene chloride copolymers, *21*:298
fossil," *15*:513
hydrocarbon. See *Hydrocarbon resins.*
ion-exchange from shale, *18*:18
ion-exchange in sugar purification, *19*:213
ion-exchange in sugar refining, *19*:190
impregnation in woods, *22*:379
lacquer of vinylidene chloride polymers, *21*:299
melamine-formaldehyde, *15*:903
melamine-formaldehyde in particleboard, *21*:610
molding of vinylidene chloride copolymers, *21*:296
natural, *17*:**379**
novolac, *15*:792
in nuts, *14*:129
phenol-formaldehyde, *15*:903
phenol-formaldehyde in particleboard, *21*:610
phenolic, *15*:158, 176, 792. (See also *Phenolic resins.*)
phenoxy, *15*:173, 795
pigments in thermoplastic, *15*:599
plasticizers for epoxy, *15*:770
plasticizers for phenolic, *15*:770
plasticizers for thermoplastic, *15*:721
plasticizers for thermosetting, *15*:721, 726, 770
polyester, *15*:455, 461
polyfluorocarbon and sodium, *18*:452
polyphenylene oxide, *15*:172
polystyrene, *19*:90
poly(vinyl chloride) solution, *21*:404
prepn by ethyleneamines, *7*:35
shellac, *18*:21. (See also *Shellac.*)
silicone, *15*:794; *18*:230, 246, 251
urea-formaldehyde, *15*:903
urea-formaldehyde in particleboard, *21*:609
use of furfural in, *10*:244
vinyl latexes, *21*:404
vinyl copolymer, *15*:228
water-resistant props, *22*:149
Resins
Resins, XF, *22*:503
Resins, ion-exchange
in blood fractionation, *3*:596
Resins, thermoplastic, *12*:592
Resins, thermosetting, *12*:592
formation from condensation of cresols, *6*:436
Resins, water-soluble, *17*:**391**
Resistance, electrical
measurement, *7*:716
specific, *2*:649, 651
surface, *2*:655, 670
volume, *2*:655
Resistance furnaces, *10*:269
Resistivity. (See also *Conductivity.*)
definition, *7*:716
Resorcin
as denaturant, *8*:451
Resorcin Dark Brown, *2*:877, 895
Resorcin Brown, *5*:868
used in hair coloring, *10*:793
Resorcinol, *6*:455; *11*:472-483; *15*:216. (See also *m-Dihydroxybenzene.*)
antithyroid agent, *20*:270
azo coupling, *2*:871
developing agent, *15*:372
fluorescein synthesis, *22*:430
gelling agent, *21*:355
in hair coloring, *10*:795, 802
phenolic resins, *15*:185
in prepn of *m*-aminophenol, *2*:213
in prepn of *m*-ethylaminophenol, *2*:214
prepn of lead dinitrosoresorcinate, *8*:591
reduction of phloroglucinol to, *16*:202
resins, *15*:176
sugar analysis, *19*:164
sulfonic acids, *19*:315
tartaric acid, *19*:736
in treatment of leather, *12*:330
zirconium compounds, *22*:655

Resorcinol derivatives, *13*:892
  as antiseptics, *2*:628
Resorcinol monoacetate, *10*:780, 802
b-Resorcylic acid, *11*:479; *16*:193
γ-Resorcylic acid, *16*:202
Respiratory coefficient
  in oxygen generation systems, *S*:663
Respiratory stimulatory factor, *22*:530
Retamo wax, *22*:165
Retene, *17*:483
Reticulin fibers
  in meats, *13*:168
Retinal, *21*:490
  as a provitamin A, *21*:502
Retinals
  stereoisomers, *21*:500
Retinene, *21*:490
Retinoic acid, *21*:490
  as growth promoter, *21*:506
Retinoic acids
  stereoisomers of vitamin A, *21*:500
Retinol, *21*:490
Retinylideneacetone, *21*:495
Retroprogesterone, *6*:84; *18*:880
Retrosteroids, *6*:83; *18*:880
Retrovitamin A, *21*:494, 500
Reverberatory smelting of copper, *6*:146
Reverse osmosis, *14*:**345**
Reversible process of separation, *7*:93
Reversion flavor, *9*:343
Reynolds number, *9*:468
Reynolds transport theorem, *20*:619, 622, 628,
  635
1-Rhamnoascorbic acid, *2*:759
1-Rhamnose, *4*:137
Rhamnus bark
  cascara sagrada from, *4*:590
Rhein, *2*:466
Rhenium, *17*:**411**
  in electron tubes, *8*:6
  filament in mass spectrometry, *13*:88
  wolfram alloy, *22*:345
Rhenium carbonyl, *4*:489
Rhenium heptoxide, *17*:411, 414
Rhenium hexafluoride, *9*:587
Rhenium oxide, *17*:411, 416, 421
Rhenium pentachloride, *17*:415
Rheolaveur, *10*:714
Rheology, *17*:**423;** *20*:645; *21*:460. (See also
  Viscometry.)
  theory of adhesion, *S*:20
Rheometers
  extrusion, *21*:472

Rheum roots. See *Rhubarb*.
Rhinoviruses, *21*:452
Rhizobium
  and nitrogen fixation, *S*:608
Rhodacarbocyanines, *16*:288
Rhodacyanines, *6*:620; *16*:288
Rhodamine, *10*:915
  used in hair coloring, *10*:793
Rhodamine B, *5*:870, 873; *7*:533; *22*:430, 434
Rhodamine B Extra
  in pigment manuf, *9*:497
Rhodamine B-Stearate, *5*:871
Rhodamine B 6G DW
  in pigment manuf, *9*:497
Rhodamine dyes, *15*:456
Rhodamine 6G, *22*:434
  as indicator, *11*:558
Rhodamines
  xanthene dyes, *22*:434
Rhodinol, *14*:729, 735
Rhodinol acetate
  in perfumes, *14*:735
Rhodinol isobutyrate
  in perfumes, *14*:735
Rhodinol propionate
  in perfumes, *14*:735
Rhodium, *12*:591; *15*:832, 856-858
  in electroplating, *8*:64
Rhodium carbonyl, *4*:489
  catalyst, *14*:377
Rhodium compounds, *15*:872-875
Rhodium sulfate electroplating solution, *8*:64
Rhodizite
  rubidium in, *17*:685
Rhodocene, *15*:874
Rhodochrosite, *13*:14, 30, 41
Rhodols, *22*:435
Rhodopsin, *21*:488
  prod of in eye, *21*:506
Rhodutorula
  in alkane utilization, *S*:839
Rhubarb
  as cathartic, *4*:590
Rhyncophylline, *1*:773
Ribitol, *1*:572
  phys prop, *1*:570
Riboflavine, *8*:175; *9*:322; *17*:**445**
  addn to flour, *22*:297
  by fermentation, *21*:489
  metabolism of urea herbicides, *22*:191
  in milk, *13*:515, 570
  in yeast, *22*:516

Riboflavine 5'-adenosine diphosphate, *17*:445, 455

Riboflavine 5'-phosphate, *17*:445, 455

Riboflavine tetrabutyrate, *17*:456

D-Ribonic acid, *19*:230

Ribonucleic acid, *2*:179
　in viral cores, *21*:452

Ribonucleoproteins, *2*:180

Ribose, *4*:134
　in yeast, *22*:516

Ribosome
　granules in yeast, *22*:512

Ribosomes
　in yeasts, *22*:516

Rice, *12*:863
　as animal feed, *8*:860
　classes of, *22*:264
　origin, *22*:253
　technology, *22*:277
　U.S. prod, *22*:259

Rice bran oil, *8*:780, 785

Rice bran oil wax, *22*:161

Ricin
　in castor pomace, *4*:528

Ricinine, *1*:765
　in castor pomace, *4*:528 548
　in castor oil, *4*:525, 529, 531
　phys prop, *1*:235
　prepn of heptaldehyde and undecylenic acid, *8*:822
　prepn of sebacid acid, *1*:251; *8*:822

Ricinoleic acid, *7*:405; *8*:787, 816; *9*:899; *19*:524, 526, 548

Ricinolic acid. See *Ricinoleic acid.*
　castor oil from, *4*:524, 587

Ricinus communis

Rickardit, *19*:759

Rickettsial infections
　chemotherapy, *3*:32

Rifamycin, *12*:648

Riffle
　in gravity concentration, *10*:713

Rimantidine, *S*:12

Rimantadine hydrochloride, *21*:456

Ring inversion
　and NMR, *14*:73

Rinmann's green, *5*:745; *15*:538

Riot control agents, *4*:877

Riprap, *12*:451

Ristocetin, *14*:675

Ritter reaction
　and adamantanes, *S*:13

Rivanol
　as antibacterial agent, *2*:639

Rochelle salt, *6*:787; *19*:736, 741. (See also *Potassium sodium tartrate.*)

Rocket propellants, *8*:660, 694, 695. (See also *Propellants.*)
　ammonium perchlorate as oxidizer in, *5*:68
　cellulose nitrate, *8*:696
　characteristics, *8*:678
　use of nitrogen trifluoride, *9*:629

Rockets
　use of helium, *10*:883

Rock phosphate. See *Phosphates.*

Rockwell hardness scale, *10*:812

Rodenticides, *15*:914-921

Rolitetracycline, *20*:3, 5, **9,** 28

Roller bearings, *3*:295

Roller leveling
　of metals, *13*:316

Roofing materials, *17*:**459**

Rosaniline, *20*:677
　pigments, *15*:588

Rosanilines
　sulfonated *N*-alkylated, *20*:715

Roscoelite, *13*:399; *21*:159

Rose bengal, *22*:433

Rosemary, *9*:363, 374; *14*:211

Rosemary oil
　as denaturant, *8*:451

Rosenheim reaction
　for provitamins D, *21*:568

Rosenmund reaction
　aldehydes, *1*:647

Rosenstiehl green, *13*:11

Rose oil, *9*:352; *14*:211
　distillation, *14*:181

Rose wax, *22*:161

Rosewood, *14*:211

Rosin, *2*:15; *6*:455; *17*:**475**
　in alkyd resins, *1*:863, 866
　in dental materials, *6*:827, 836, 842
　in matches, *13*:165
　in polymerization of chloroprene, *7*:706
　sizing agent, *14*:501
　in soap, *18*:418
　tall oil, *19*:614
　from wood, *22*:382

Rosin acids
　in metallic soaps, *7*:272
　polyoxyethylene derivatives, *19*:545

Rosin derivatives, *17*:**475**
　abietic acid, *1*:236; *19*:545

Rosin oil, 6:455
Rosolic acid
  dye series, 20:677
Ross effect, 15:381
Rotenoids
  as insecticides, 11:687
Rotenone, 16:201
Rotoxamine
  antihistamine, 11:41
Rottenstone
  as abrasive, 1:26
Rouge
  as abrasive, 1:28
Rouge (cosmetic), 6:366
Rouleaux printing machine
  for textile printing, 7:578
Roundworm infections, 14:532
  chemotherapy of, 14:534
Roxarsone, 21:247
Royalite glass, 5:849
  as base for adhesive, 1:390
Rubber. (See also Elastomers; Rubber chemicals; Rubber compounding.)
  chlorinated, 17:646-651
  chlorinated as paint, 14:476
  clays used in, 5:584
  cyclized, 17:651-655
  depolymerized, 17:655
  gutta-percha, 17:682
  hardness testing, 10:818
  hydrochlorinated, 17:650-651
  hydrogenated, 17:656
  latex from natural, 17:674
  use of metallic soaps in processing, 7:284
  natural, 17:660
  natural as gasket, 14:450
  as natural latex as adhesive backing, S:934
  natural latex thread, 18:629
  natural in water-resistant leather, 22:150
  plastics, 15:791
  plasticizers for, 15:765
  in relation to elastomers, 7:676
  sheet rubber for corrosives transport, 20:601
  silicone, 18:222, 228, 241
  synthetic. See Elastomers.
  synthetic from shale, 18:18
  vulcanized in phonograph records, 15:225
  x-ray analysis, 22:453
Rubber accelerators
  from butyraldehyde, 3:873
Rubber antioxidants, 2:599
Rubber bonding
  of grinding wheels, 1:35
Rubber carbons
  improvements in, S:99
Rubber, cellular
  manuf, 9:856
Rubber chemicals, 17:509
  general index of, 17:536
  use of diarylamines as, 7:45, 47
  2,2,4trimethyl-1,2,dithydroquinoline as, 741
Rubber compounding, 17:543
Rubber compounds
  as fabric coatings, 5:681
Rubber derivatives, 17:646
Rubber hydrochloride film, 9:231
  properties, 9:222-226
Rubber industry
  use of abherents in, 1:3
  use of air conditioning in, 1:497
  use of ammonia in, 2:296
  use of aniline in, 2:417
  use of ethyleneamines in, 7:36
  humidity conditions within, 1:487
Rubber latex
  use of electrodecantation, 7:843
  water-resistant prop, 22:149
Rubber latexes
  in papermaking, 14:519
Rubber, natural, 12:64, 79; 17:660. (See also
  Elastomers.)
  use of diarylamines in, 7:47
  prop of flexible polymers, 9:863
Rubber products
  carbon blacks used in, 4:265, 273, 280
  coatings for, 5:714
Rubidinium fluorotitanate, 9:684
Rubidinium trifluorozincate, 9:684
Rubidium, 17:684
Rubidium compounds, 17:684
Rubidium cyanide, 6:599
Rubidium fluorosulfate, 9:679
Rubidium fluosilicate, 9:658
Rubidium gallate(III), 10:325
Rubidium hexachlorotitanate, 20:388
Rubidium hexafluorogermanate, 9:585
Rubidium hydride, 11:202
Rubidium perchlorate, 5:69
Rubidium peroxide, 14:749
Rubidium superoxide, 14:764
Rubidium tetraphenylborate, 16:384
Rubidium-87
  in water, 21:703
Ruby, 2:1
  prod synthetically, 10:509
Rufamycin, 16:330

Rum, *1*:517
Rust removal
  citric acid for, *5*:537
Rutaecarpine, *1*:771
Ruthenium, *15*:832, 859
  as catalyst in carbon monoxide hydrogena-
    tion, *4*:453, 454, 455, 484
  prop, *15*:859
  uses, *15*:859
Ruthenium carbonyls, *4*:489
Ruthenium compounds, *15*:875-878
Ruthenium dioxide, *15*:876
Ruthenium disulfate, *15*:877
Ruthenium disulfide, *15*:842, 877
Ruthenium nitrate
  in reactor technology, *14*:96
Ruthenium oxychloride, *15*:859
Ruthenium pentafluoride, *15*:876
Ruthenium tetroxide, *15*:859, 876
Ruthenium tribromide, *15*:876
Ruthenium trichloride, *15*:876
Ruthenium trihydroxide, *15*:876
Ruthenium triiodide, *15*:876
Ruthenium-106
  as fission product, *14*:107
Ruthenocene, *15*:877
  catalyst, *14*:377

Rutile, *15*:496; *20*:347, 349, 390, 392. (See also
  Titanium oxide.)
  in ceramics, *4*:773
  zero point of charge, *9*:384
Rutile yellow, *5*:854
Rutin, *11*:470
  from tobacco, *20*:514
Ryania
  as insecticide, *11*:689
Rydberg constant
  as a 11unit," *S*:997
Rye
  as animal feed, *8*:860
  classes of, *22*:264
  use in malting, *12*:863
  origin, *22*:253
  technology, *22*:276
  U.S. prod, *22*:259
  uses for, *22*:258
Rye grain
  used in manuf of distilled alcoholic bev-
    erages, *1*:521
Rye whiskey, *1*:513
Ryznar Index, *22*:81
22R-2β,3β,14α,-22,26-Pentahydroxy-5- β-
  cholest-7-en-6-one, *18*:846

S

S Glass
   in ceramic composite armor, *S*:141
SI base units, *S*:987
SI derived units, *S*:991
SI prefixes, *S*:991
S-Methylcysteine
SNAP-50 (turboelectric generator)
   cesium in, *4*:865
S-N" diagrams, *13*:192
SPC, *5*:273
SS acid. See *Chicago acid.*
STB. See *Supertropical bleach.*
Sabadilla
   as insecticides, *11*:688
Saccharase. See *Invertase.*
Sacchareins, *19*:597
Saccharic acid, *6*:922, *19*:229
Saccharides, *4*:133
Saccharin, *10*:10; *19*:260, 311, 595-598, 603
   in carbonated beverages, *4*:339
   KMnO$_4$ in manuf, *13*:32
Saccharinol, *19*:595
Saccharinose, *19*:595
Saccharin salts, *10*:10
Saccharol, *19*:595
Saccharose, *19*:166
Saccharummic acid, *19*:211
S acid, *13*:725
Safelight filters, *9*:262
Safety, *17*:**694**
Safety glass, *10*:569
Safflorite, *2*:720
   phys prop, *5*:722
Safflower oil, *8*:780, 785, 786; *18*:40
   in alkyd resins, *1*:864
   drying oil, *7*:398, 399, 403, 404
   film prop, *7*:419, 423
Saffron, *9*:374
Safranine, *7*:533
   used in hair coloring, *10*:793
Safranine T, *2*:861
Safranines, *15*:220, 222
Safrole, *11*:470; *15*:170
   as denaturant, *8*:451
   vanillin from, *21*:185
Sage, *9*:352, 363, 365, 374; *14*:211

Saha equations
   for MHD generator, *S*:228
Sakaguchi's reagent, *2*:173
Salicyl alcohol, *17*:741-742
Salicylaldehyde, *15*:162; *17*:720, 739-741
   condensation with ethylenediamine, *7*:35
   in dental material, *6*:826
   use as dye intermediate, *1*:109
Salicylamide
   as antipyretic, *2*:390
Salicylanilide, *17*:722, 735
   as fungicide, *10*:232
   gelling agent, *21*:355
Salicylanilides
   polybrominated in soaps, *18*:429
Salicylates
   UV absorbers, *21*:117
Salicylazosulfapyridine, *19*:270, 272
Salicylic acid, *2*:629; *7*:479, 491; *15*:148; *17*:**720**
   chloroform reaction with, *5*:123
   chlorosulfonation, *19*:299
   as denaturant, *8*:451
   in embalming fluids, *8*:101
   as fungicide, *10*:235
   growth inhibitor, *15*:681
   plastics, *15*:800
   phys prop, *1*:238
   reduction, *15*:162
Salicylic acid-related compounds, *17*:**720**
Salicyloylsalicylic acid, *17*:731
Salicylsalicylic acid, *17*:731
Saligenin, *15*:176
Saline cathartics, *4*:591
Salkowski reaction
   for provitamins D, *21*:568
Salmon, *9*:326
   in detn of disinfectant effectiveness, *2*:610
Salt. (See also *Sodium chloride.*)
   in drilling fluids, *7*:301, 302
   fishery, *6*:502
   flavor potentiator, *9*:361
   grainer, *6*:502
   in meat curing, *13*:171
Saltpeter. See *Potassium nitrate.*
Saltpeter, chile. See *Sodium nitrate.*

Samandarine, *1*:759; *18*:844, 866
Samarium, *17*:143-168
  electronic configuration, *1*:360
  ionic radii, *1*:370
Samarium-151
  as fission product, *14*:106
Sampling, *17*:**744**
Sand
  filters in water treatment, *22*:67, 88
  filters for water treatment, *22*:88
  molding, *5*:567
  from ocean, *14*:160
  grinders, *15*:596
  raw material, *18*:78
  silicates from, *18*:139
  use in water filtration systems, *22*:119
Sandalwood, *14*:212
  essential oil in, *14*:181
Sandarac, *17*:387
Sandstone
  as abrasive, *1*:26
  in ceramics, *4*:768
Sand test
  for brisance of explosives, *8*:655
Sanitization, *2*:606
Sanitizers, *18*:806
  for swimming pools, *22*:124
Sansevieria, *9*:179
  properties, *9*:175
Santalol, *19*:833
Santonin, *14*:533
Santophen I, *5*:331
Sapogenins, *18*:830, 840
Saponification equivalent, *8*:793. (See also
  Soap.)
Saponification number, *1*:302
Saponification value of fats, *8*:793
Saponins, *18*:838-841
  in soapnuts, *14*:132
Saponite, *5*:548
  properties, *3*:342
Sapphire, *2*:1
  in plasma arcs, *15*:345
  prod synthetically, *10*:509, 516
Saran, *21*:275
  latex, *21*:300
Sarcolactic acid, *12*:171
Sarcolemma
  in meats, *13*:168
Sarcoplasm
  of meats, *13*:167
Sarcosine, *19*:514
  in prepn of azo dyes, *2*:893

Sardine oil, *7*:399, 403, 404; *8*:782, 790
Sarin, *4*:875; *9*:646; *15*:328
  in chemical warfare, *4*:875
Sassafras, *14*:212; *22*:383
Sassafras oil
  as denaturant, *8*:451
Satatier reversal, *15*:388, 391
Satin
  dyeing by disperse dyes, *7*:541
Satin white, *15*:516
  pigment, *14*:516
Sauconite, *5*:548
  properties, *3*:342
Sauerkraut
  processing, *10*:49
Savin, *14*:212
Savory, *9*:352; *14*:212
Saxin, *19*:595
Saxony blue, *15*:541
Scales, *22*:220-239
Scandium, *17*:143-168
  in nuts, *14*:131
Scandium carbide, *4*:70
Scandium-46, *17*:122
Scapoletin, *6*:432
Scapolite, *12*:616
Scarlet Acid, *2*:899
Schaeffer's acid, *13*:721
Scheele's acid, *22*:335
Scheele's green, *15*:538
Scheelite, *12*:616; *22*:337
  zero point of charge, *9*:384
Schering process
  manuf of triacetate, *1*:119
Schiff base, *1*:644
  from acetaldehyde and primary amines,
    1:81
Schiff bases. (See also *Anils.*)
  in prepn of aromatic amines, *2*:115
  dyes in, *20*:730
  in acetaldehyde detn, *1*:90
Schiff's reagent
  dyes in, *20*:730
  in acetaldehyde detn, *1*:90
Schilling effusiometer, *6*:775
Schistosomiasis, *14*:533, 544
Schists, *13*:399
Schizosaccharomyces pombe
  fission yeast, *22*:513
Schmid-Meissner process
  for nitration of glycerol, *8*:594
Schoch process
  in acetylene manuf, *1*:183

Schoellkopf's acid, *13*:722
Schoenite
    from Great Salt Lake, *S*:455
Scholler process
    sugars from wood, *22*:383
Schreibersite, *15*:280
Schumacher chlorate cells, *5*:53, 77
Schweinfurt green, *15*:538
Science Citation Index, *S*:522, 527
Scillaren B
    as cardiovascular agent, *4*:516
Scilliroside, *15*:916
Scintillation counters, *17*:71, 72
    liquid, *17*:2
    crystal, *17*:2
Scintillation counting, *15*:402
Scleron, *12*:536
Scopolamine, *1*:766, 778, 804
Scopolamine hydrobromide
    as hypnotic, *11*:519
Scopoletin, *15*:681
Scorbutamin. See *Ascorbic acid.*
"Scram" devices, *14*:77
Screening, *18*:367
Screening smokes
    in chemical warfare, *4*:902
Screw conveyors, *6*:110
Scroop
    in silk dress goods, *7*:527
Scrum globulins, *3*:576
Seacoke process
    for synthetic crude oil, *S*:189
Sealing compds, *4*:28
    chlorosulfonated polyethylene, *4*:32
Seal oil, *8*:782
Sea plants
    inorganic materials from, *14*:166
Season cracking
    of copper alloys, *6*:238
Seawater, *7*:48
    analysis, *21*:706
    manuf of fresh water, *8*:578
    pretreatment, *12*:728
    p2:16)recovery of mae nesium from, *12*:667
Seaweed colloids, *17*:**763**
Sebacid acid, *1*:251, 252 1:8) 2:16)
    from castor oil, *4*:531
    in dental material, *6*:827 (pf)oxidation,
        19:143
    phys prop, *1*:232
    polyesters, *20*:490
    prepn, *1*:247

Secobarbital, *3*:61, 70, 72
Sedatives. See *Hypnotics.*
Sediment
    suspended in water, *21*:705
Sedimentation, *17*:**785**
    centrifugal, *4*:729
    in sewage treatment, *22*:107
    and water-treatment systems, *22*:88
Sedimentation gravity
    in particulate removal, *S*:715
Sediments
    composition of lake, *S*:325
    lake, *S*:330
Seebeck effect, *20*:148
Seed yeast, *22*:527
Seekay waxes, *5*:301
Seepage control, *21*:663
Seismic explosives. See *Explosives, seismic.*
Sekazin, *12*:636
Selacholic acid, *8*:814
Selenanthrene, *17*:825
Selenazole, *17*:825
Selenic acid, *17*:824
Selenides, *17*:821-822
N,N'-Selenidodimorpholine. See *Morpholine
    selenide.*
Selenious acid, *17*:824
Selenite, *4*:16
Selenium, *17*:**809**
    analysis for in water, *21*:698
    electrodeposition of zinc, *22*:588
    ferroalloy, *18*:751
    colorant for glass, *9*:252
    in copper alloys, *6*:197
    as glass colorant, *5*:850
    in high-temperature cells, *S*:130
    impurity in copper, *S*:265
    photoconductive prop, *15*:419
    in photovoltaic cells, *15*:413, 419
    reaction with aluminum alkyls, *2*:31
    reaction with hydrogen, *11*:344
    separation from tellurium, *19*:758
    silane compounds, *18*:195
    toner, *15*:386
    pigments, *15*:534
Selenium compounds, *17*:**809**
Selenium diethyldithiocarbamate, *17*:830;
    19:771
Selenium dioxide, *2*:582; *12*:109, 142, 147, 153,
    843; *17*:815, 831
Selenium disulfide
    as dandruff shampoo, *10*:775
Selenium oxychloride, *12*:122

Selenium oxyfluoride
  silica reaction, *18*:57
Selenium sulfide
  pigments, *15*:535
Selenium tellurides, *19*:765
Selenomethionine, *2*:598
Selenous anhydride, *12*:153
Semesan
  soil chemistry, *18*:539
Semicarbazide, *10*:902
  reaction with acetaldehyde, *1*:82
Semicarbazone
  melting point, *6*:456
Semiconductors, *17*:**834**
  arsenic as, *2*:716
  ceramics as, *4*:819
  diamonds as, *4*:289, 303
  electromigration of, *S*:289
  silicon carbide, *4*:122
  thermoelectric prop, *20*:156
  zone refining, *22*:697
Semidurable flame retardants. See *Flame retardants, semidurable.*
Semigelatin dynamite. See *Dynamite, semigelatin.*
Semigraphite. See *Graphite.*
Semitropical bleach. See *Bleach, semitropical.*
Semolina, *22*:291
Senarmontite, *2*:564, 573
Senecio alkaloids, *1*:766
Senna, *4*:589
Sensitization in photochemical technology, *15*:333
  veterinary drug, *21*:252
Sepiolite, *5*:552
Septic tanks, *22*:105
Sequens, *6*:87
Sequestering agents. See *Complexing agents.*
Sequestration. See *Complexing agents.*
Sericin
  in silk, *18*:271
Sericulture, *18*:269
Serine, *2*:157; *18*:271
  in wine, *22*:316
Serotonin, *2*:178, 188, 189; *11*:590; *16*:641
Serpentine, *5*:547; *17*:237; *19*:608. (See also *Asbestos.*)
Serratamolide, *16*:337
Serum albumin, *3*:576
Sesame oil, *8*:780, 785, 786, 790; *18*:40
  parenteral oil, *15*:128
  as sunscreen products, *6*:356
Sesame seed, *9*:365, 374

Sesame seed oil meal
  as animal feed, *8*:864
Sesban
  analysis, *9*:173
Sesone, *22*:183
Sesquiterpenes, *19*:803, 833
Setoglaucine O, *20*:702
  manuf, *20*:704
Sèvres Blue, *5*:853
Sewage, *22*:104
  algae treatment of, *1*:662
Sewage disposal
  yeast activities in, *22*:530
Sewage treatment. See *Water.*
Sewer pipe clays, *4*:768; *5*:566
Sex hormones
  androgens, *11*:92
  estrogens, *11*:106
  progesterone, *11*:118
Shale coke, *18*:18
Shale oil, *18*:**1**; *19*:684
  hydrocracking of, *11*:419, 423
  reserves, *10*:186
Shampoos, *6*:353, 354; *10*:769
Shattering power. See *Brisance.*
Shaving cream
  use of aerosol products, *1*:479
Shaving lotions, *6*:354
Shea fat, *8*:780, 785
Shear modulus, *17*:425
Shear rate
  for Newtonian fluids, *21*:462
Shear stress, *13*:318
Shellac, *15*:721, 790; *18*:**21**
  in dental material, *6*:836
  as denaturant, *8*:451
  in explosive priming considerations, *8*:653
  as fluorescent coating system, *9*:499
  in food processing, *10*:10
  on grinding wheels, *1*:34
  in hair spray, *10*:782
  in phonograph records, *15*:225, 230
Shellac wax, *22*:167
Shell flour. See *Calcite.*
Shell gasification process, *11*:356
Shell-molding process
  use of abherents in, *1*:3
Shellolic acid, *18*:24
Sherardizing, *13*:264
Shincomycin A, *12*:633
Ship hoists, *6*:120
Shock synthesis
  of diamond, *S*:274

Shoenite, *16*:384
Shoes
   use of adhesives in, *1*:399
   use of elastomers in, *7*:687, 694, 695
Shortenings, *13*:182; *18*:**33**
Shrimp meal
   as animal feed, *8*:867
Shrinkage control in textiles
   from resins from acrolein and formalde-
      hyde, *1*:265
Siderite, *12*:6; *15*:517, 526
Siemens
   as an SI unit, *S*:993
Sienna
   pigments, *15*:518, 527, 529
Sieves, molecular. (See also *Molecular sieves.*)
   in catalysis, *4*:568
Sieves; sieving. See *Size measurement of particles; Size separation.*
Silanes, *11*:206; *18*:172-215
Silanols, *18*:189-191
   alkoxide reactions, *20*:452
   condensation, *18*:230
Silazines, *18*:262
Silica, *18*:**46**. (See also *Quartz.*)
   activated as flocculation aid, *22*:71
   as arc furnace lining, *10*:257
   amorphous, *4*:61-72
   analysis for in water, *21*:696
   in bauxite, *2*:8
   brick, *17*:227
   brick in steel furnaces, *18*:728
   as catalyst, *10*:161
   in ceramics, *4*:768, 774
   in cements, *4*:597, 691, 692, 695, 700
   in chemical ware, *4*:835
   colloidal in paint removers, *14*:489
   colloidal, use of sodium aluminate in, *2*:10
   conductivity of, *4*:803
   in dental investments, *6*:788, 804, 826
   diatomite, *7*:56
   dust inhalation of, *7*:292, 459
   as glass former, *10*:538, 542
   in limestone, *12*:422
   in manuf of silicon carbide, *4*:127
   as paint extender, *14*:466
   as paper pigment, *14*:517
   phase equilibria, *17*:246
   phenolic ethers, *15*:167, 171
   phosphorus manuf, *15*:281
   pigment manuf, *20*:401
   pigments, *15*:513, 518, 525, 527, 533, 535,
     540

   in prepn of trifluoroboroxine, *9*:556
   refactory, *17*:227, 232
   as semiconductor, *10*:321
   slag-former in welding, *22*:242
   sulfuric acid, *19*:441
   synthetic quartz crystals, *4*:105-111
Silica aerogel
   as thermal insulator, *11*:830
   vitreous, *4*:72-105
Silica-alumina
   in catalytic cracking, *11*:426, 429; *15*:28, 36
   desulfurization, *15*:46
   methylation of phenol, *S*:272
Silica, colloidal
Silica gel, *18*:61
   as adsorbent, *1*:460
   carbon tetrachloride reaction with, *5*:132
   contact process, *19*:467
   as a drying agent, *7*:384, 385, 396
   xylene separations, *22*:476
Silica glass, *10*:540; *18*:73-105
Silica glasses
   thermal conductivity of, *4*:803
   as thixotropic binders, *4*:34
Silicate bonding
   of grinding wheels, *1*:34
Silicate dental cement, *6*:779, 781
Silicate glasses, *10*:540
   composition of, *10*:542
   prop, *10*:572
Silicates
   in ceramics, *4*:771, 772
Silicates, synthetic inorganic, *18*:134-166
Silicic acid
   as insecticide, *11*:683
   water analysis, *21*:696
Silicides, *18*:125-132
   coating of refractory metals, *11*:33
Silicocarnotite, *9*:109
Silicochloroform, *9*:653
Silicofluoride. See *Fluosilicates; Fluorosilicates.*
Silicofluoroform, *9*:653
Silicomanganese, *12*:889
Silicon, *18*:**111**
   compared with phosphorus, *15*:295
   in copper alloys, *6*:193, 200, 201, 210, 212,
     220, 242, 250
   diffusion in, *17*:875
   electronic prop, *17*:850
   in electron tubes, *8*:6
   heat of combustion of, *4*:898
   in heterochain inorganic polymers, *11*:644
   homochain polymers, *11*:635

ignition mixtures, *16*:826
infrared detectors, *15*:430, 435
manganese compounds, *13*:4
neutron activation analysis, *17*:38
in photoconductive cells, *15*:410
photoconductive prop, *15*:420
in photoemitters, *15*:403
in photovoltaic cells, *15*:413
in pig iron, *18*:718
reaction with germanium, *10*:521
from seawater, *14*:156
semiconductor, *17*:834, 850
zone refining, *22*:697
Silicon boride, *4*:74
Silicon bronzes. See *Copper alloys.*
  as abrasive, *1*:27
  in ceramic armor, *S*:138
  as coated abrasive, *1*:29
  expansion coefficient, thermal, *4*:805
  in explosive priming compositions, *8*:653
  in grinding wheels, *1*:31
Silicon carbide, *4*:114-132; *8*:540; *12*:628
  coating, *17*:274
  as electric furnace resistor, *10*:255
  in opaque armor, *S*:151
  refactory, *17*:232, 245
  in refactory ware, *4*:329
  strength value of, *4*:818
  silicon tetrachloride prod, *18*:170
  titanium cutting, *20*:365
  ethers and esters, *18*:216-221
  halides, *18*:166-172
  metasilicic acid, *18*:135
  silanes, *18*:172-215
  silicones, *18*:221-260
  silylating agents, *18*:260-268
  synthetic inorganic silicates, *18*:134-166
Silicon compounds, *18*:**133**
Silicon dioxide. See *Silica.*
Silicone enamels
  use of diarylamines in, *7*:48
Silicone oil, *2*:204
Silicone resins, *12*:190
  use as resins, *1*:10
Silicone rubber, *7*:698; *18*:222, 228, 241
  as dental impression, *6*:805
  roofing, *17*:462
  sealant, *4*:32
Silicones, *15*:793; *18*:221-260
  as abherents, *1*:3, 5, 6, 7, 9, 10
  as electric insulation, *11*:789
  as modifiers in alkyd resins, *1*:866
  in packing, *14*:452

in papermaking, *14*:520
prop of flexible polymers, *9*:863
prop of rigid polymer, *9*:860
surfactants in foams, *21*:88
water-resistant prop, *22*:150
Silicon ethers, *18*:216-221
Silicon-gold alloy, *10*:692
Silicon halides, *18*:166-172
Siliconizing, *13*:314
Silicon monoxide
  zirconium dioxide recovery, *22*:652
Silicon nitride, *4*:74; *13*:823
  refractory, *17*:245
Silicon oxide
  in opaque armor, *S*:150
Silicon tetrachloride, *6*:670
  reaction with 1,3-butylene glycol, *10*:662
  reaction with silicon tetrafluoride, *9*:652
Silicon tetrafluoride, *9*:532, 617, 620, 651-653
  in phosphorus manuf, *15*:285
Silicooxalic acid, *9*:653
Silicosis, *7*:459
Silk, *18*:**269**
  antiplasticization, *15*:775
  dyeing with acid dyes, *7*:526
  dyeing with basic dyes, *7*:533
  premetalized dyeing on, *7*:531
  prepn for dyeing, *7*:520
  vat dyeing of, *7*:560
  work of rupture value, *1*:131
Silk fibroin
  x-ray analysis, *22*:454
Silk white, *15*:516
Sillimanite, *17*:233
  in ceramics, *4*:771
Silnic bronze. See *Bronze, silnic.*
Silomelane, *12*:6
Siloxane, *18*:184
Siloxane elastomers
  chain-growth, *16*:229
Siloxanes, *15*:794; *18*:189
  basicities, *18*:191
  in dental impression materials, *6*:806
  oligomers, *18*:228
  structure, *18*:191
Silver, *18*:**279**
  as antimicrobial, *2*:621
  as a bactericide, *22*:129
  as bearing material, *3*:281
  catalysts, *19*:460
  catalysts containing, in manuf of ethylene
    oxide, *8*:539
  as catalyst in oxidation of ethyl alcohol, *1*:84

corrosion resistance to water, 6:302
in dentistry, 6:807, 814, 822
electromigration of, S:286
in electroplating, 8:64
as glass colorant, 5:850
impurity in copper, S:265
as lubricant, 12:603
in prepn of silver fulminate, 8:590
removed from lead ore, 12:231
in solders, 18:543
as solid film lubricant, 12:590
in type metal, 20:757
thermocouples, 19:788
wolfram alloy, 22:345
in vacuum tubes, 8:3,4, 7
pigments, 15:550
Silver alloys, 18:**279**
Silver azide, 8:586
Silver bromide
photography, 15:357
Silver-cadmium electric cells, 3:240
Silver carbonate, 8:492, 540
Silver catalyst, 2:54
Silver chlorate, 6:466
Silver chloride, 2:621
chlorine estimated by, 5:5
in electric cells, 3:132
photography, 15:357
Silver chlorite, 5:31
Silver chromate, 5:481, 498
Silver compounds, 18:**295**
Silver-copper-gold alloys, 10:688
Silver cyanide, 8:540
Silver difluoride, 9:662, 799
Silver ethyl xanthate, 22:420
Silver extraction
by sodium cyanide, 6:585, 589
Silver ferrate(VI), 12:40
Silver fluoborate, 9:566, 570
Silver fluorides, 9:661, 799, 800
in fluorination of organic compounds,
9:634, 698, 699
Silver fluoroantimonite, 9:550
Silver fluorosulfate, 9:680
Silver fluosilicate, 9:658, 660
Silver fulminate, 8:590
Silver halides
in color photography, 5:816
photography, 15:355
Silver hexacyanoferrate(II), 12:29
Silver hexafluoroantimonate, 9:662
Silver hexafluoroarsenate, 9:662

Silver hexafluorobismuthate(V), 3:541
Silver hexafluorophosphate, 9:662
Silver iodide, 2:621
cloud seeding, 21:662
photography, 15:357
Silver iodomercurate, 13:240
Silver-lead alloys, 12:260
Silver-lithium alloy, 12:540
Silver-magnesium alloy
photoemitters, 15:408
Silver monofluorophosphate, 9:642
Silver nitrate, 2:621; 8:540
benzyl chloride reaction with, 5:283
chlorine detected by, 5:5
urea reaction, 21:39
Silver oxalate, 8:540
in electric cells, 3:128, 137, 208, 240
Silver oxide, 2:621; 10:633; 12:150
Silver pentafluorobenzoate, 9:795
Silver perchlorate, 5:72
Silver peroxide, 9:509
Silver phosphates, 15:275
Silver phosphides, 15:294
Silver salt, 7:469
Silver subfluoride, 9:661
Silver sulfate, 12:588
Silver sulfide
photography, 15:361
Silver tetafluoroborate, 9:662
Silver trifluoroacetate, 9:749
Silver-zinc secondary electric cells, 3:208
Silvex, 22:179
Silylamine, 6:647
Silylamines, 18:198, 262
Silylating agents, 18:260-268
Silyl compounds, 18:173-203
Silylene halides," 18:182
Simarouba
amebicides, 20:78
Simarubidin, 20:78
Simarubin, 20:78
Simazine, 22:186
in soil, 18:533
Siporex
manuf of, 4:704
Sirius Supra Blue BRR, 2:900
Sirius Yellow, 20:187
Sisal, 9:171, 179; 12:189
analysis, 9:173
prop, 9:174, 175
Sitosterol
in soybeans, 18:601

β-Sitosterol, *19*:626
Sitosterols, *18*:836
Size measurement of particles, *18*:**310**
Size reduction, *18*:**324**
Size separation, *18*:**366**
Skatole, *14*:726
Skim milk. See *Milk, skim.*
Skin effect, *S*:697
Skraup reaction, *2*:213, 494
Skutterudite
  phys prop, *5*:722
Sky Blue FF, *2*:903
Slag
  iron blast-furnace, *S*:877
  utilization of, *S*:876
Slag cements, *4*:694
Slagceram, *S*:**876**
Slag sitalls, *S*:888
Slaked lime. See *Calcium hydroxide; Calcium oxide; Lime.*
Slip ring materials, *4*:222
Sludge
  activated, *S*:716
  handling, treatment, and disposal, *S*:725
Slushing compounds, *6*:343
Small's relationship, *21*:290
Smalt, *15*:541
Smaltite
  phys prop, *5*:722
Smelting
  cyclone, *S*:250
Smithsonite, *22*:563
Smog. See *Smokes.*
Smoke
  fire-retardancy aspects of, *S*:475
Smoke agents
  in chemical warfare, *4*:900
Smoke generators, mechanical
  in chemical warfare, *4*:904
Smokeless powder, *4*:631
Smokes, *18*:**400.** (See also *Dust.)*
  photochemical smog, *2*:815; *S*:50
  signalling, in chemical warfare, *4*:903
Smoking
  of meats, *13*:175
Snakeroot, *14*:212
  essential oil as flavor, *9*:352
Snow-melting media, *10*:860
Soap, *18*:**415**
  alkanolamine, *1*:820, 829
  as denaturant, *8*:451
  from fats, *13*:183

as oxidation dye bases, *10*.796
prepn, *8*:368
in shampoos, *10*:771, 775
use of acetals in, *1*:109
Soaps
  cobalt, *5*:740, 744
  green, *19*:518
  mahogany, *19*:518
  metallic, *4*:33; *7*:272-287
  used in greases, *12*:582
Soapstone, *19*:608, 610
  pigments, *15*:509
Sobrerol, *19*:832
Soda ash. See *Sodium carbonate.*
Soda" blues, *15*:539
Soda, caustic. See *Sodium hydroxide.*
Soda cellulose, *18*:512
Soda lignins, *12*:372
Sodamide, *2*:143; *6*:591; *9*:794, 795; *12*:112, 113
  allyl chloride reaction with, *5*:208
  in arc furnaces, *10*:258
  prod, *18*:451
  sulfonic acids, *19*:316
Söderberg electrodes, *4*:105, 107, 196
Sodioacetoacetic ester, *2*:144
Sodionaphthalene, *8*:474
Sodio-2-carbethoxycyclohexanone, *2*:144
Sodio-2-carbethoxycyclopentanone, *2*:144
Sodium, *2*:144, 145; *18*:**432**
  as catalyst, *8*:359
  compared to potassium, *16*:385
  as denaturant, *8*:451
  as heat-transfer medium, *10*:855. (See also *Sodium-potassium alloy.)*
  methyl chloride reactions with, *5*:103
  manuf of sodium cyanamide, *6*:591
  as hazardous chemical, *9*:291
  as plant nutrient, *9*:31
  in prepn of fatty alcohols, *1*:545
  in reduction of esters to alcohols, *8*:369
  in water, *21*:694
Sodium *N*-acylsarcosinate, *19*:514
Sodium *p*-acetamidobenzenestibonate, *2*:585
Sodium *N*-chloro-*p*-toluenesulfonamide, *19*:257. (See also *Chloramine-T.)*
Sodium acetate, *8*:409
  catalyst, *20*:799
  in commercial prepn of acetyl chloride, *1*:140
  as ester-exchange catalyst, *16*:167
  isochloroprene reaction with, *5*:228

methyl chloride reaction with, 5:104
methylene chloride reaction with, 5:114
in prepn of aluminum diacetate, 2:12
in prepn of p-(2,4-dinitroanilino)phenol, 2:20
vinyl acetate manuf, 21:328
Sodium acetylide, 9:761
methylacetylene from, S:552
reaction in dimethylacetamide, 1:147
Sodium acid fluoride. See Sodium hydrogen fluoride.
Sodium acid oxalate. See Sodium oxalate.
in baking powder, 3:51
Sodium acid pyrophosphate. See Sodium pyrophosphate.
Sodium acid urate, 21:112
Sodium acrylate, 1:266
Sodium-air cells, S:133
Sodium alanine, 19:383
Sodium alcoholate, 8:337
titanium alkoxides, 20:452
Sodium alginate
as dental impression material, 6:801
in shampoos, 10:774
Sodium alkoxides, 18:439. (See also Alkoxides.)
Sodium alkylbenzenesulfonate, 6:881
Sodium alkylmercaptides
synthesis of thiocarbamate herbicides, 22:194
Sodium alkylnaphthalenesulfonate, 16:256
Sodium 5-allyl-5(1-methylbutyl)-2-thiobarbiturate. See Thioamylal sodium.
Sodium aluminate, 10:106
coagulant in water treatment, 22:73, 96
prepn in manuf of alumina, 1:937
manuf, 2:8
phys prop, 2:6
in prepn of aluminum diacetate, 2:12
as source of gallium, 10:313
synthetic prod of cryolite, 1:937
Sodium aluminum chloride
aluminum from, 18:432
Sodium aluminum hydride, 11:222
Sodium aluminum phosphate
in baking powder, 3:51
Sodium aluminum silicate, 10:4
Sodium aluminum sulfate
in baking powder, 3:51
in food processing, 10:10
Sodium amalgam, 8:427; 18:437, 453; 19:238
analysis, 18:449
in fuel cells, 3:152

in prepn of ethanol from ethylene oxide, 8:527
prepn from ammonia, 2:262 Neoarshenamine.
in reduction of nitro compounds, 2:98
reduction of zirconium halides, 22:616
sodium from, 18:440
Sodium amide, 12:154; 18:438. (See also Sodamide.)
Sodium p-aminobenzoate, 2:406
Sodium (2-aminoethyl)thiosulfate, 20:245
Sodium 3-amino-3'-sulfinomethylamino-4,4'-dihydroxy arsenobenzene. See Neoarshenamine.
Sodium tert-amyloxide, 12:112
Sodium amyl xanthate, 22:425
Sodium anthracenecarboxylate, 18:437
Sodium anthraquinonedisulfonate
in gas scrubbing, 1:44
Sodium anthraquinone-2-sulfonate, 2:357
Sodium antimonate, 12:229
Sodium antimony(III) bis(pyrocatechol-2,4-sodium disulfonate). See Stibophen.
Sodium antimony α,α'-dimercaptosuccinate, 14:546
Sodium antimonyl tartrate, 2:579
Sodium arsanilate, 3:2
Sodium arsenate, 12:229
in weed control, 2:715
Sodium arsenide, 2:719
Sodium arsenite
as an herbicide, 22:214
hypochlorite detn with, 5:46
soil chemistry, 18:537
Sodium ascorbate
in meat curing, 13:171
Sodium azide, 8:586
in prepn of sodium azidodithiocarbonate, 4:373
prod, 18:451
Sodium azidodithiocarbonate
from carbon disulfide, 4:373
Sodium bentonite
in detergents, 6:858
Sodium benzeneselenate(IV), 22:659
Sodium benzenesulfonate, 3:402; 15:171
Sodium benzoate, 3:428, 429, 433
in carbonated beverages, 4:344
food preservative, 20:560
in margarine, 13:58
methyl chloride reaction with, 5:104
in paint removers, 14:488
photochemical, 15:338

Sodium benzylate, *8*:495
  as antacid, *2*:428
Sodium bicarbonate, *18*:467
  blowing-agent, *17*:532, 591
  by-product of soda ash manuf, *1*:734
  in carbonated beverages, *4*:344
  in chemical leavening agents, *3*:52
  as dental abrasive, *6*:849
  in detergency, *6*:856
  in fire extinguishers, *7*:285; *9*:296, 857, 892, 893
  fire-extinguishing agents, *S*:365
  fire-fighting foams, *S*:368
  in meat curing, *13*:171
Sodium bichromate, *7*:563
  pigments, *15*:531, 537
Sodium bifluoride. See *Sodium hydrogen fluoride.*
Sodium biphthalate, *10*:799
Sodium bismuthide, *3*:537
Sodium bisulfate, *10*:13; *18*:502, 506; *19*:490
  reaction with aqueous formaldehyde, *10*:86
  as water-correcting agent, *10*:17
Sodium bisulfite, *6*:456, 646; *7*:526; *19*:421
  in acetaldehyde detn, *1*:90
  bleaching wool, *22*:405
  photography, *15*:384
  in prepn of nylon for dyeing, *7*:519
  in pulpmaking, *12*:369
  reaction with acetaldehyde, *1*:83
  reaction with aldehydes, *1*:643
  reaction with butynediol, *1*:604
  reaction with isocyanates, *12*:49, 120
  setting of wool, *22*:407
  surfactants, *19*:522
Sodium borate
  in detergents, *6*:856
  natural, *18*:433
  from seawater, *14*:156
Sodium (2:1) borate, *3*:645
Sodium borohydride, *3*:702; *9*:640; *11*:213; *12*:110, 161
  as bleaching agent, *3*:562
  broader use for, *S*:499
  prod, *18*:450
  steroid synthesis, *18*:871
  tin hydride from, *20*:312
Sodium bromate, *3*:770
  in cold waves for hair, *10*:789
Sodium bromide, *6*:512; *18*:484
  chlorate detn with, *5*:46
Sodium 1,3-butylene glycoxide, *10*:662
Sodium 1-(*n*-butyl)octylsulfate, *19*:511

Sodium cacodylate
  as an herbicide, *22*:214
Sodium caproylacetate, *12*:162
Sodium carbide, *4*:71, 72; *18*:437
Sodium carbonate, *1*:707-740; *6*:512, 587, 588, 589; *7*:293, 567; *8*:418; *9*:393; *18*:437, 458-468
  carbon reduction of, *18*:439
  coal-tar distilling, *19*:670
  in drilling fluids, *7*:301
  in enzyme detergents, *S*:306
  glycerol prod, *22*:529
  hydrogen sulfide, *19*:383
  in manuf of soap, *18*:429
  photography, *15*:373
  prepn of lead dinitroresorcinate, *8*:591
  in prepn of sodium hexacyanoferrate(II), *12*:29
  in recovery of carbon dioxide, *4*:360
  silicate prod, *18*:140
  in sodium aluminate prepn, *2*:6
  sodium bromide from, *18*:485
  in synthesis of 1,3-butylene glycol, *10*:664
  tall oil recovery, *19*:614
Sodium carbonate decahydrate, *18*:465
  in dental stones, *6*:788
  in detergents, *6*:856
  in dye industry, *7*:480
Sodium carbonate monohydrate, *18*:465
Sodium carbonate peroxyhydrates, *14*:761
Sodium carbonate (soda ash)
  bitumen processing, *19*:707
  in chemical cleaning, *S*:169
  in drossing of lead ore, *12*:222
  in glycerol manuf, *10*:621
  phosphates, *15*:263
  in phosphorus manuf, *15*:286
  reduction by coke, *18*:439
  silicate manuf, *18*:139
  thiazole dyes, *20*:186
  tellurium recovery, *19*:759
  from trona, *18*:464
  use in pelletization, *9*:435 458-468
Sodium carbonyl, *18*:438
Sodium carboxymethylcellulose, *4*:643-646; *7*:295, 300, 303
  as cathartic, *4*:592
  in detergents, *6*:858, 874
  as food additive, *10*:7
Sodium carboxymethylhydroxyethylcellulose, *4*:645
Sodium cellulose dithiocarbonate, *4*:371

Sodium cellulose glycolate. See *Sodium carboxymethylcellulose.*

Sodium cellulose sulfate, *4*:631; *15*:791

Sodium cellulose xanthate. See *Xanthates.*

Sodium chlorate, *5*:52; *6*:512; *9*:605
  in chlorate candle, *S*:658
  chlorine dioxide from, *5*:37, 40, 41, 42, 59
  in explosives, *8*:644
  perchlorates manuf from, *5*:76
  as a weed killer, *22*:176

Sodium chloride, *6*:512; *18*:437, 468-484. (See also *Salt.*)
  in alfin catalyst, *S*:928
  as assistant in dyeing of wool with acid dyes, *7*:523
  as brine in refrigerant, *10*:860
  in chlorine prod, *1*:678
  in dentifrices, *6*:849
  as detergency promoter, *6*:857
  dielectric prop of, *4*:821
  by evaporation, *8*:559
  in prevention of gelation in egg yolk, *7*:669
  in fire extinguishers, *9*:296
  fusion with calcium cyanamide, *6*:559
  from Great Salt Lake, *S*:446, 462
  in seawater, *14*:153
  in manuf of ammonium chloride, *2*:317, 318
  as mineral, *18*:459, 470; *19*:344
  in prepn of diethyl sulfate, *19*:492
  in sodium carbonate prod, *1*:714

Sodium chloride solutions
  vapor pressure plot of, *6*:721

Sodium chlorite, *5*:31
  as bleaching agent, *3*:559
  chlorine dioxide from, *5*:39, 43
  chlorine reaction with, *5*:4
  as starch modifier, *10*:17
  water purified by, *5*:43

Sodium 3-chloro-4-acetamidobenzenestibonate, *2*:585. See *Stibosan.*

Sodium chloroacetate
  alkali cellulose reaction with, *4*:644
  thioglycolic acid, *20*:201
  thioglycolic acid synthesis, *20*:199

Sodium-*N*-chloro-*N*-cyclohexylsulfamate, *4*:923

Sodium chlorodifluoroacetate
  steroid synthesis, *18*:880

Sodium *N*-chloroimidodisulfonate, *4*:917

Sodium 2-chloroethanesulfonate, *19*:290

Sodium chlorophyllin, *5*:348

Sodium-*N*-chloro-*p*-toluenesulfonamide. See *Chloramine-T.*

Sodium chromate, *7*:304
  azo dyes aftertreated with, *2*:883
  in chlorate manuf, *5*:55
  in fungicide, *10*:230, 860
  manuf of, *5*:484
  in manuf of chlorine dioxide, *5*:42
  in paint removers, *14*:488
  phys prop, *5*:481

Sodium citrate, *5*:539
  as antacid, *2*:430
  in dental stones, *6*:788
  in embalming fluids, *8*:101
  as emulsifier in cheese, *10*:8

Sodium compounds, *18*:**458**

Sodium cresylate
  from tar acids, *19*:672

Sodium crotonate, *6*:468

Sodium cyanamide, *6*:559

Sodium cyanide, *6*:585, 588; *9*:389, 393, 795; *12*:121
  analysis, *6*:594
  1,2-dichloroethane reaction with, *5*:151
  isochloroprene reaction with, *5*:228
  manuf, *6*:589
  prepn, *1*:848
  in manuf of hydrogen cyanide, *6*:576, 583
  in prepn of Bahr's salt, *6*:631
  in prepn of cyanoacetic acid, *8*:416
  in prepn of dichloroacetic acid, *8*:418
  prod, *18*:451
  reactions, *6*:587
  reaction with ethylenediamine, *7*:29
  from sodium, *18*:432

Sodium cyanoacetate, *8*:417

Sodium 2-cyanoethane sulfonate, *6*:643, 646

Sodium cyanoborohydride, *S*:503

Sodium cyclamate, *19*:598

Sodium cyclohexylsulfamate, *10*:10; *19*:598

Sodium cyclopentadienide, *13*:756

Sodium dehydrocholate, *3*:485
  menadione in, *21*:592

Sodium diacetate
  as food preservative, *10*:14

Sodium diborate, *3*:642

Sodium 3,5-dibromo-4-hydroxybenzenesulfonate, *15*:210

Sodium dibutyldithiocarbamate, *17*:600

Sodium *N*-di-*n*-butyldithiocarbamate, *17*:595

Sodium dichloroisocyanurate, *4*:908, 919; *20*:669
  swimming pool water disinfectant, *22*:126

Sodium dichloroisocyanurate," *20*:663

Sodium 2,2-dichloropropionate, *16*:556

Sodium 2-(2,4-dichlorophenoxy)ethyl sulfate, *15*:678

Sodium dichromate, *6*:512; *7*:525, 527, 528, 529, 538; *9*:795
  azo dyes aftertreated with, *2*:882
  chromium trioxide from, *5*:488
  in dye industry, *7*:481
  as fungicide, *10*:230
  hydrogen sulfide, *19*:385
  manuf of, *5*:484
  phys prop, *5*:481

Sodium diethyldithiocarbamate, *15*:848
  allyl chloride reaction, *22*:194
  vitamin $K_{1(20)}$ treatment with, *21*:589

Sodium diethyldithiocarbonate

Sodium di(2-ethylhexyl)sulfosuccinate, *19*:510, 519

Sodium difluoroacetate, *9*:770

Sodium dihydrobis(2-methoxyethoxy)aluminate, *S*:504

Sodium 1,4-dihydroxy-2,3-butanedisulfonate, *1*:604

Sodium *p*-dimethylaminobenzenediazosulfonate
  as fungicide, *10*:226
  manuf of, *4*:372

Sodium dimethyldithiocarbamate
  as an herbicide, *22*:194

Sodium dimethyldithiocarbonate, *7*:681

N-Sodium diphenylamine, *7*:43

Sodium disulfatozirconate, *22*:633, 662

Sodium disulfide, *18*:515

Sodium dithionate. (See also *Sodium hydrosulfite.*)
  as bleaching agent, *3*:561

Sodium dithionite, *10*:13; *19*:420
  and nitrogen fixation, *S*:609

Sodium dodecanesulfonate, *9*:386

Sodium dodecylbenzenesulfonate
  detergency of, *S*:37
  surfactant, *S*:33

Sodium dodecyl entries. (See also *Sodium lauryl entries.*)

Sodium dodecylsulfate, *9*:386

Sodium *n*-dodecylsulfate, *19*:511

Sodium dodecylsulfonate, *9*:722

Sodium ethoxide, *8*:530; *12*:39, 124, 132, 161
  benzotrichloride reaction with, *5*:283
  prepn, *1*:832
  in prepn of 1-butanol, *8*:455
  prepn from ethanol, *8*:427

prepn of 1,1,3,3-tetracarbethoxypropane, *8*:477
  vinyl chloride reaction with, *5*:173

Sodium ethylate
  vitamin $K_{1(20)}$ treatment with, *21*:589

Sodium ethylenesulfonate, *19*:290, 494
  vinyl acetate copolymer, *21*:331

Sodium ethyl mercaptide, *2*:580 *Thimerosal.*)

Sodium 5-ethyl-5-(1-methylbutyl)-2-thiobarbiturate. See *Thiopental sodium.*

Sodium ethylmercurithiosalicylate. See *Thimerosal.*

Sodium ethyl xanthate
  drying, *22*:426
  prepn of, *22*:424

Sodium ferrate(III), *12*:40

Sodium ferricyanide. See *Sodium hexacyanoferrate(III).*

Sodium ferrocyanide. See *Sodium hexacyanoferrate(II).*

Sodium fluoberyllate, *3*:455

Sodium fluoborate, *9*:564, 565, 697

Sodium "fluoborax," *9*:558

Sodium fluoferrate, *3*:455

Sodium fluoride, *9*:545
  addn to water, *22*:103
  complex with alkylaluminums, *2*:34
  as insecticide, *11*:682
  as fungicide, *10*:230, 235
  in prepn of sulfur tetrafluoride, *9*:662, 672

Sodium fluoroacetate, *9*:768
  rodenticide, *15*:917

Sodium fluoroaluminate
  as insecticide, *11*:682

Sodium fluoroantimonite, *9*:549, 550

Sodium fluorohafnate, *10*:763

Sodium fluorosilicate, *6*: 655, 659, 802; *9*:858
  addn to water, *22*:103

Sodium fluorosulfate, *9*:679
  as insecticide, *11*:682
  phys prop, *9*:658

Sodium fluosilicate. See *Sodium fluorosilicate.*

Sodium formaldehyde bisulfite, *10*:86

Sodium formaldehyde sulfoxylate, *7*:680; *10*:86; *19*:422
  emulsion polymerization, *21*:283

Sodium formate, *6*:588, 673; *10*:91
  content in sodium cyanide, *6*:595
  in manuf of formic acid, *10*:101
  in oxalic acid prod, *14*:362

Sodium furoate, *10*:241

Sodium gallate(III) *10*:325

Sodium gluconate
  as emulsifier in cheese, *10*:8
Sodium glycocholate, *3*:482, 484
Sodium glycodeoxycholate, *3*:482, 484
Sodium glycohenodeoxycholate, *3*:484
Sodium glycolate, *10*:636
Sodium glycolithocholate, *3*:484
Sodium-gold alloy, *18*:454
Sodium hexacarbonylvanadate
  vanadium carbonyl from, *4*:502
Sodium hexachloroiridate(IV), *15*:870
Sodium hexachlororhodate(III), *15*:872
Sodium hexacyanoferrate(II), *6*:588; *12*:29,
  176
Sodium hexacyanoferrate(III), *12*:32
Sodium hexafluoroaluminate. See *Cryolite.*
Sodium hexafluorophosphate, *9*:644
Sodium hexafluorotitanate, *9*:684
Sodium hexametaphosphate, *6*:802; *7*:296;
  9:306
  addn to water supplies, *22*:81
Sodium hexanitroiridate(III), *15*:871
Sodium hexanitrorhodate(III), *15*:873
Sodium hexavanadate, *21*:160
Sodium hexyl xanthate, *22*:425
Sodium hydride, *11*:202; *12*:112, 154; *18*:437
  catalyst, *18*:450
  reaction with boron trifluoride, *9*:556
Sodium hydrogen *p*-aminobenzenearsonate,
  2:729; *21*:246
Sodium hydrogen arsanilate, *19*:141
Sodium hydrogen *N*-(carbamoylmethyl)ar-
  sanilate. See *Tryparsamide.*
Sodium hydrogen carbonate. See *Sodium bi-
  carbonate.*
Sodium hydrogen fluoride, *9*:663
Sodium hydrogen pyrophosphate
  as emulsifier in cheese, *10*:8
Sodium hydrogen sulfate. See *Sodium bisulfate.*
Sodium hydrogen sulfite. See *Sodium bisulfite.*
Sodium hydrosulfide, *2*:144, 218; *9*:792;
  16:255
  methyl chloride reaction with, *5*:104
  thioglycolic acid synthesis, *20*:199
Sodium hydrosulfide dihydrate, *16*:258
Sodium hydrosulfite, *7*:526, 567, 636; *10*:86;
  18:438, 510, 513; *19*:420. (See also *Sodium
  dithionate.)*
  bleach for wool, *22*:405
  as bleaching agent, *3*:561
  in bleaching glue, *10*:607, 799

in dye industry, *7*:481
methacycline synthesis, *20*:7
in reduction of nitro compounds, *2*:96
in reducing vat dyes, *7*:555
Sodium hydrotrifluoroborate
  prepn, *9*:556
Sodium hydroxide, *1*:740-758; *6*:588; *8*:427,
  450
  in alumina manuf, *1*:937
  for bottle washing, *4*:347
  as catalyst on silica gel, *1*:82
  coal-tar "cracking," *19*:670
  cellulose reaction with, *4*:601
  in detergents, *6*:856
  as denaturant, *8*:450
  in drilling fluids, *7*:301, 302
  as drying agent, *7*:396
  in dye industry, *7*:480
  electrolysis of fused, *18*:440
  fluorination of, *9*:631
  use in pelletization, *9*:435
  in prepn of *m*-aminophenol, *2*:213
  petroleum sweetening, *15*:51
  phosphates, *15*:263
  polysiloxanes, *18*:225
  in prepn of sodium aluminate, *2*:6, 7
  in prepn of magnesia magma, *2*:429
  rayon-grade quality, *17*:175
  reaction with metal cyanamide, *6*:559
  in recovery of by-product hydrogen sulfide,
    19:385
  in refining of crude fats or oils, *8*:798
  saponification of fats, *8*:825
  in separation of fats and oils, *8*:823
  in soap industry, *18*:418
  in synthesis of 1,3-butylene glycol, *10*:664
  in tall oil recovery, *19*:614
  used to extract sulfur compounds, *8*:759
Sodium hydroxyacetate, *10*:636
Sodium *m*-hydroxybenzenesulfonate, *15*:210
Sodium 2-hydroxyethanesulfonate, *19*:290
Sodium hypobromite, *8*:430
Sodium hypochlorite, *2*:218; *8*:429; *9*:794;
  12:200, 204
  as bleaching agent, *3*:555
  in bleaching nuts, *14*:137
  cyclopentadiene reaction with, *5*:243
  solid, *5*:15
  solutions of, *5*:12
  swimming pool water disinfectant, *22*:125
  uranium ore leaching, *21*:13
Sodium hypohalites, *12*:109

Sodium hypoiodite, 6:922; 8:430
Sodium hyponitrite, 18:438
Sodium hypophosphite, 8:69
Sodium indoxyl, 18:451
Sodium 5'-inosinate, 2:198
  flavor potentiator, 9:361
Sodium iodide, 8:420
  alkylation, 19:488
  as denaturant, 8:451
  isochloroprene reaction with, 5:229
  in urography, 17:132
Sodium iodide-131, 17:1
Sodium iodomethamate, 17:132
Sodium iodomethanesulfonate, 17:132
Sodium 7-iodo-8-quinolinol-5-sulfonate, 20:83
Sodium iridonitrite, 15:871
Sodium isethionate, 19:288, 522
Sodium isopropoxide
  in alfin catalyst, S:928
Sodium lactate, 8:352; 12:183
Sodium lactatozirconate, 22:635, 656
Sodium lauroyl isethionates
  in shampoos, 10:773
Sodium N-lauroylsarcosinate, 19:514
Sodium lauryl entries. (See also Sodium dodecyl entries.)
Sodium lauryl ether sulfate, 10:772
Sodium lauryl sulfate, 1:568; 9:722
  in food processing, 10:8
  HLB value, 8:130
  in liquid egg treatment, 7:668
Sodium lauryl sulfoacetate, 19:288
Sodium-lead alloy, 12:288, 290; 18:453
Sodium lignin sulfonate
  in paint removers, 14:491
Sodium lignosulfonate, 7:297
  disperse dyes, 20:192
Sodium maleate
  hydrogenation, 19:143
Sodium manganate, 13:12
Sodium mercaptide
  thioglycolic acid, 20:201
Sodium mercuric phenoxyacetate, 7:254
Sodium metabismuthate(V), 3:538
Sodium metabisulfite, 10:13; 19:421
  in prepn of polyacrylonitrile, 1:319
Sodium metaborate
  as bleaching agent, 3:559
  as an herbicide, 22:214
Sodium metaborate dihydrate, 3:641
Sodium metaborate tetrahydrate, 3:641
  photography, 15:372

Sodium metagermanate, 10:525
Sodium metal
  chlorine prod in manuf of, 1:702
Sodium metaniobate, 13:782
Sodium metaperiodate
  reaction with allyl ether, 8:494
Sodium metapertantalate, 19:651
Sodium metaphosphate
  as dental abrasive, 6:849
Sodium metaphosphates, 15:243, 273
Sodium metasilicate, 18:135, 138, 140
  in enzyme detergents, S:306
  in paint removers, 14:491
Sodium metatantalate, 19:651
Sodium methiodal, 17:132, 140
Sodium methoxide, 1:845; 7:401, 617; 8:244, 289; 12:154, 160
  in fat esterification, 13:181
  hexachlorobenzene reaction with, 5:262
  in prepn of methyl ether, 8:476
  reaction with bis(chloromethyl) ether, 8:477
  reaction with 1,3-butylene glycol, 10:662
Sodium α-dl-1-methyl-5-allyl-5-(1-methyl-2-pentynyl)barbiturate. See Methohexital sodium.
Sodium methylate. See Sodium methoxide.
  as catalyst, 8:357
  in prepn of methyl ether, 8:476
Sodium methyl carbonate
Sodium 2-methyl-2,3-dihydro-1,4-naphthoquinone-2-sulfonate, 21:593
Sodium methyldithiocarbamate
  as fungicide, 10:224
Sodium N-methyldithiocarbamate, 15:913
Sodium 2-methyl-1,4-naphthohydroquinone-3-sulfonate, 21:593
Sodium methyl siliconate, 18:248
  in drilling fluids, 7:302
Sodium N-methyltaurine, 19:290
Sodium methyl xanthate, 22:419
Sodium methyl xanthogenate, 22:419
Sodium molybdate, 13:656
  source of micronutrient fertilizer, 9:146
Sodium monofluorophosphate, 9:637, 642
Sodium monosulfide. See Sodium sulfide.
Sodium monoxide, 18:437, 438. (See also Sodium oxide.)
Sodium myristate
  phase diagram, 18:427
Sodium naphthalenesulfonates, 10:616
  and nitrogen fixation, S:615

Sodium naphthalide. (See also *Sodionaphthalene.*)
   ziegler-type catalysts, *20*:448
Sodium naphthenate, *13*:685. (See also *Sodionaphthalene.*)
Sodium niobate, *13*:782
   in glass-ceramics, *10*:548
Sodium nitrate, *12*:227; *18*:486-498
   in black powder, *8*:647
   in drilling fluids, *7*:302
   in dye industry, *7*:480
   in dynamite, *8*:639
   in embalming fluids, *8*:101
   as food preservative, *10*:13
   in HTS (heat transfer salt), *10*:854
   in manuf of nitrogen fertilizers, *9*:51, 65
   in meat curing, *13*:171
   sulfur burners, *19*:449
Sodium nitrilotriacetate
   in enzyme detergents, *S*:306
Sodium nitrite, *6*:512; *7*:562; *18*:438, 498, 502
   as cardiovascular agent, *4*:521
   as food preservative, *10*:13
   in HTS (heat transfer salt), *10*:789, 854
   in prepn of diazodinitrophenol, *8*:587
   prepn of lead dinitrosoresorcinate, *8*:591
   prepn of tetracene, *8*:592
Sodium *m*-nitrobenzenesulfonate, *2*:457
Sodium nitroprusside
   analytical identification of amino acids, *2*:173
Sodium oleate, *6*:880; *9*:898; *19*:290
Sodium *N*-oleoyl-*N*-methyltaurate, *19*:521
Sodium orthohexatantalate, *19*:651
Sodium orthophosphates, *15*:236, 263. (See also *Sodium phosphate.*)
Sodium orthosilicate, *18*:135. (See also *Sodium silicates.*)
Sodium osmonitrite, *15*:874
Sodium oxalate, *14*:368; *18*:438
Sodium oxide. (See also *Sodium monoxide.*)
   in chemical ware, *4*:835
   in glass forming, *10*:538, 542
Sodium palmitate, *6*:881; *18*:415
   phase diagram, *18*:426, 428
Sodium pentaborate, *3*:642
Sodium pentachlorophenate, *7*:304; *21*:192
   in paint removers, *14*:491
Sodium pentachlorophenoxide, *14*:544
Sodium pentafluorostannite, *9*:682
Sodium pentanitroosmate(III), *15*:874
Sodium pentatantalate, *19*:651
Sodium perborate. See *Sodium peroxyborate.*

Sodium perchlorate, *5*:69
   in prepn of perchloryl fluoride, *9*:598, 606
Sodium periodate
   cellulose oxidized by, *4*:596
Sodium permanganate
   as catalyst in decomposition of H_2O_2, *8*:687
Sodium peroxide, *10*:905; *14*:747; *18*:437
   in cold-waves for hair, *10*:789
   in dentifrices, *6*:848
   in peroxy compound prod, *18*:450
   prod, *18*:450
   in soaps, *18*:425
   from sodium, *18*:432
Sodium peroxyborate, *8*:622; *14*:759
   in enzyme detergents, *S*:299
   enzyme prewash agents, *S*:295
Sodium perxenate, *10*:893
   in enzyme detergents, *S*:306
Sodium phenanthrenecarboxylate, *18*:437
   in tar acids, *19*:672
Sodium phenate, *15*:148. (See also *Sodium phenoxide.*)
Sodium phenoxide, *8*:487; *19*:383
   in manuf of cresols, *6*:438
Sodium *o*-phenylphenol
   as food preservative, *10*:14
   in manuf of salicylic acid, *17*:724
Sodium phosphate, *18*:437. (See also *Sodium hydrogen phosphate; Sodium orthophosphate.*)
   as cathartic, *4*:591
   in drilling fluids, *7*:301
   as flame-retardant for textiles, *9*:304
Sodium phosphate glasses, *15*:250, 255
Sodium phosphates, *15*:270, 305
   in chemical cleaning, *S*:164
   prod, *15*:263
Sodium phosphide, *18*:437
Sodium phosphides, *15*:293
Sodium phthalate
   titanium chelates, *20*:465
Sodium phthalocyanine, *15*:490
Sodium piperidinodithiocarbamate, *6*:646
Sodium plumbite, *7*:314
   petroleum sweetening, *15*:50
   in shale oil refining, *18*:17
Sodium polyacrylates
   as protective colloid, *14*:468
Sodium polybutadiene, *7*:684
   glass transition temp, *7*:690
Sodium polyethacrylate
   coagulant in water treatment, *22*:96
Sodium polyphosphates, *15*:243, 264
Sodium polyselenide, *17*:815

Sodium poly(styrene sulfonate)
  PEC hydrogels, *16*:118
Sodium polysulfide, *2*:424; *7*:497, 677
  1,2-dichloroethane reaction with, *5*:151
  dyes, *19*:424
Sodium polysulfides, *16*:255; *18*:510, 514
Sodium-potassium
  as lubricant, *12*:603
Sodium-potassium alloy, *16*:366; *18*:453
  as heat-transfer medium, *10*:856
  methyl chloride reactions with, *5*:103
Sodium potassium tartrate
  in dental stones, *6*:788
  as emulsifier in cheese, *10*:8
  ferroelectric prop, *9*:4
Sodium propionate
  as food preservative, *10*:13
Sodium pyroantimonate, *2*:575
Sodium pyrophosphate, *7*:296; *8*:412; *15*:244,
  255, 264, 271; *18*:425
  thorium extraction, *20*:250
Sodium pyrosulfate
  to open ores, *13*:770
Sodium resinate
  in paint removers, *14*:491
Sodium rhodonitrite, *15*:873
Sodium ricinoleate
  in prepn of 2-octanol, *1*:565
  x-ray analysis, *22*:453
Sodium rubidium D-tartrate
Sodium saccharin, *19*:596
Sodium salicylate, *17*:728
  as antipyretic, *2*:390
  as denaturant, *8*:451
Sodium sebacate
  as coproduct in prepn of 2-octanol, *1*:565
  from ricinoleic acid, *1*:53
Sodium selenate, *17*:815, 829, 831
  as insecticide, *11*:683
Sodium selenide, *17*:815
Sodium selenite, *17*:815, 831
Sodium selenosulfate, *17*:831
Sodium sesquicarbonate, *18*:461, 462, 466
Sodium sesquicarbonate
  by-product of soda ash manuf, *1*:734
  in detergents, *6*:856
Sodium sesquicarbonate, *18*:463, 466
Sodium sesquisilicate," *18*:136, 138
Sodium sesquisilicate pentahydrate, *18*:140
Sodium silica
  pigments, *15*:516

Sodium silicate, *9*:390; *7*:299
  addn to water, *22*:81
  in drilling fluids, *7*:302
  electrolytic zinc prod, *22*:586
  in phosphates, *15*:273
  phosphoric acid, *15*:262
  reaction with formamide, *10*:106
  in soaps, *18*:425, 429
Sodium silicates, *18*:135
  in chemical cleaning, *S*:164
Sodium silicide, *18*:437
Sodium sorbate, *10*:13
  prepn, *20*:305
Sodium stannate, *12*:229; *20*:318
  reaction with water, *20*:309
  toxicology, *20*:323
Sodium stearate, *6*:350
  phenol synthesis, *15*:150
  in shampoo, *10*:774
Sodium stearylfurmarate
  flour improver, *22*:294
Sodium styphnate, *8*:590
Sodium succinate, *19*:146
Sodium sucrates, *19*:154
Sodium sulfate, *6*:512, 881; *7*:562, 563; *18*:502,
  510
  in carbonated beverages, *4*:344
  as cathartic, *4*:591
  in chemical cleaning, *S*:159
  dimethyl sulfate prod, *19*:492
  dye industry, *7*:480, 528
  in dyeing on rayon, *7*:567
  in Dynel dyeing, *1*:334
  from Great Salt Lake brine, *S*:444, 457
  pigments, *15*:515, 540
  in prepn of sodium alum, *2*:65
  silicates from, *18*:139
  sodium sulfide from, *18*:511
  spray drying, *7*:335, 336
Sodium sulfenates
  sulfoxides, *19*:331
Sodium sulfhydrate," *2*:224; *18*:513
Sodium sulfide, *2*:441; *8*:612; *16*:255; *18*:438,
  510-515
  benzene hexachloride reaction with, *5*:273
  chamber process, *19*:459
  dyes, *19*:424, 431
  in flotation of copper and lead ores, *9*:388
  methyl chloride reaction with, *5*:104
  phosphoric acid, *15*:262
  pigments, *15*:505, 535
  reaction with Tetryl, *8*:620

sodium thiosulfate from, *20*:235
tall oil, *19*:614
Sodium sulfite, *2*:441; *8*:612; *10*:13
in detn of formaldehyde, *10*:91
in dye industry, *7*:481
from phenol synthesis, *15*:153
photography, *15*:372
Sodium 2-(*p*-sulfophenylazo)-1,8-dihydroxy-
3,6-naphthalenedisulfonate, *21*:695
Sodium sulfosuccinate
in etching, *16*:523
Sodium sulfoxylate, *19*:422
Sodium-sulfur cell, *S*:124, 135
Sodium superoxide, *14*:763; *18*:437
Sodium tartrate
as emulsifier in cheese, *10*:8
Sodium taurochenodeoxycholate, *3*:484
Sodium taurocholate, *3*:482, 484
Sodium taurodeoxycholate, *3*:482, 484
Sodium taurolithocholate, *3*:484
Sodium tellurate, *19*:760
Sodium tellurite, *19*:759
as an herbicide, *22*:214
Sodium tetraborate, *3*:642
Sodium tetraborate decahydrate, *3*:618, 623
Sodium tetraborate pentahydrate, *3*:627, 639
Sodium tetrachloroaurate, *10*:693
Sodium tetrachloropalladate(II), *15*:867
Sodium tetrafluoroberyllate, *9*:552, 553
Sodium tetraiodophenolphthalein, *17*:135
Sodium tetrametaphosphate, *15*:244
Sodium tetramethoxyborate, *3*:671
Sodium tetranitropalladate(II), *15*:869
Sodium tetranitroplatinate(II), *15*:865
Sodium tetraphenylborate
cesium extracted with, *4*:857
Sodium tetraphosphate, *10*:789
Sodium tetrasulfide, *2*:219; *7*:697; *18*:515
Sodium thioantimonate, *2*:566
as cardiovascular agent, *4*:523
Sodium thiocyanate, *6*:588; *7*:523
photography, *15*:384
swelling agent for cellulose, *6*:653
Sodium thioglycolate, *20*:201, **202**
analysis, *20*:237
chlorine estimated by, *5*:5
developer, *16*:502
fixer, *15*:356
manuf, *20*:234
phys prop, *20*:233
prod, *20*:236
specifications, *20*:237

Sodium thiosulfate, *10*:693, 799; *12*:281; *20*:**232**
photography, *15*:382, 384
uses, *20*:238
Sodium titanates, *20*:411
Sodium 2-(2,4,5-trichlorophenoxy)ethyl sulfate, *15*:678
Sodium 3-(2,4,5-trichlorophenoxy)proprionate, *15*:678
Sodium 2,3,6-trichlorophenyl acetate, *15*:678
Sodium trifluoroberyllate, *9*:552
Sodium trifluorostannite, *9*:682
Sodium trifluorozincate, *9*:684
Sodium trihydrogen pyrophosphate, *15*:244
Sodium trimetaphosphate, *15*:244, 273
Sodium trimethoxyborohydride, *3*:671
Sodium tripolyphosphate, *15*:245, 265, 272; *18*:425
addn to water supplies, *22*:81
in presoaked products, *S*:300
Sodium trithiocarbonate, *19*:430
xanthate synthesis, *22*:424
Sodium tungstate. (See also *Sodium wolframate*.)
as flame retardant for textiles, *9*:304
Sodium uranylvanadate
from uranium ores, *21*:14
Sodium urate, *21*:112
uric acid synthesis, *21*:110
Sodium vanadate
from uranium ores, *21*:14, 18
Sodium vinylacetylide
vinylacetylenes from, *5*:229
Sodium warfarin, *6*:432
Sodium wolframate, *22*:351
toxicity of, *22*:356
from wolframite, *22*:338
Sodium wolfram bronze, *22*:350
Sodium xanthate
from coal tar, *19*:670
Sodium xylenesulfonate, *6*:353, 653
Sodium-zinc alloy, *18*:454
Sodium zirconium ethoxide
prepn, *1*:833
Sodium zirconium sulfate, *22*:633
Sodium zirconylosilicate, *22*:660, 663
Sodium 2-hydroxy-3-naphthoate, *14*:534
Sodium-22, *17*:120
Sodium-24, *17*:120
Soft black, *15*:544
Softening
of municipal waters, *22*:97

Softwoods, 22:359
Soil chemistry, of pesticides, 18:**515**
Soil fumigant
    bis(2-chloroethyl)ether, 8:487
Soil fumigants, 15:912
Soil release agents
    "co-crystallizing," S:970
    emulsion polymers containing carboxylic
        acid groups, S:971
    fluorochemical, S:970
    water-soluble polymers containing carboxyl-
        ic acid groups, S:971
Soil-release finishes, S:964
Soils
    and detergency, 6:854
    removal from textile fibers, 7:309, 310
Solacet Fast Scarlet B, 2:905
Solan
    amide herbicide, 22:198
Solanaceous alkaloids, 1:804
Solanidine, 1:775; 18:842, 844
α-Solanine, 1:775
Solantine Red 8 BLN, 2:901
Solar evaporation, 8:562
Solar furnace, 4:215
Solarization, 15:387, 391
Solar-ponds, S:445
Solar radiation
    simulation of, 4:215
Solasodine, 1:775; 18:842
Solder
    tin alloys, 20:289
    tin-lead alloys, 18:543
Solders, 12:263; 18:**541**
Solders and brazing alloys, 18:**541**
Sol-gel technology, 14:88
Solid film lubricants, 12:587
Solid fuels gasification, 10:356
    fixed-bed processes, 10:359
    fluidized-bed processes, 10:372
    performance data, 10:382
    suspension, 16:374
Solid particles
    fluidization, 9:398
Solid propellants. See Propellants.
Solids
    analysis of dissolved solids in water, 21:691
    bulk handling, 14:434
    characteristics of, 18:325
    as electric insulation, 11:789
    removal of dissolved from water supplies,
        22:74

suspended in water, 21:692
    treatment of waste solids, 21:630
    waste, 21:627, 645
Solid solutions
    cooling curves, 13:204
Solid-state theory, 17:835
Solids transport
    and fluidization, 9:427
Solid wastes, S:737
Solmargine, 1:775
Solubility map," 18:571
    data interpretation of, 6:740
Solubility of gases
    equations for, 6:724
Solubility parameter, 18:567, 568
Solubility parameters, S:**889**
Solubilized Vat Yellow 8, 2:894
Soluble blue," 15:540
Solubles
    removal of from polluted water, S:716
Solution adsorption isotherm, 1:453
Solution photochemistry, 10:122
Solutions
    chemical potential, 20:137
    Gibbs-Duhem equation, 20:137
    heats of, 20:136
    partial molar quantities, 20:134
    prop of aqueous, 21:680
    thermodynamics of, 20:134
Solvay chlorine cell, 1:689
Solvay methanol reduction process for manuf
    of chlorine dioxide, 5:41
Solvay process
    ammonium chloride as by-product in, 2:318
    soda ash manuf, 1:712
Solvent
    coalescing, S:899
Solvent Black 5, 2:863
Solvent Black 7, 2:863
Solvent Blue 7, 2:862
Solvent Blue 8, 2:867
Solvent Green 3 (CI), 5:868
Solvent Red 23, 5:870, 873
Solvent Red 24, 2:906
Solvent Red 30, 2:907
Solvent Red 43, 5:870, 873
Solvent Red 48, 5:870
Solvent Red 49 (CI), 5:871
Solvent Violet 1, 2:907
Solvent Violet 13, 2:488; 5:871
Solvent Violet 13 (CI), 5:871
Solvent Yellow 33 (CI), 5:871, 874

Solvent Yellow 44 (CI)
  in pigment manuf, 9:497
Solvent dyeing, S:973
Solvent dyes
  uses of, 7:477
Solvent extraction, 8:761
Solvent recovery, 18:**549**
Solvents
  carbon in reclamation of, 4:152
  methylene chloride as, 5:117
  organic in chemical cleaning, S:164
Solvents, drycleaning
Solvents, industrial, 18:**564**
Solvolysis, 2:332
Soman, 15:328
  in chemical warfare, 4:875
Sommelet reaction
  aldehydes, 1:648
Sonar projectors
  aldehydes, 1:648
  cavitational bond-breaking, 20:780
  use of ferroelectric ceramics in, 9:1
Sonn-Müller reaction
Sonochemical reactions, 20:776-780
Sonoluminescence, 20:777
Sorbic acid, 6:452, 453; 7:411; 8:815; 18:**589**
  food preservative, 10:13
  as fungicide, 10:235
  phys prop, 1:229
  prepn, 12:89
1,5-Sorbital, 1:578
1,5-Sorbitan. See Polygalitol.
Sorbitan
  fatty acid esters, 19:544
1,4-Sorbitan, 1:578; 19:545
Sorbitan monolaurate, 19:545
  HLB value, 8:129
Sorbitan monooleate
  HLB value, 8:128
Sorbitan monopalmitate
  HLB value, 8:129
S   bitan monostearate
  as emulsifier in foods, 10:8
  HLB value, 8:128
Sorbitan oleates, 19:545
Sorbitan palmitates, 19:545
Sorbitan sesquioleate
  HLB values, 8:128
Sorbitan stearates, 19:545
Sorbitan trioleate
  HLB value, 8:128

Sorbitan tristearate
  HLB value, 8:128
Sorbitol, 1:573; 6:362, 922; 7:411
  in alkyd resins, 1:862
  manuf, 1:577
  phys prop, 1:570
  plasticizer, 15:764
  polyurethans from, 21:61
  reactions, 1:577
  as release agent in food, 10:16
  stabilizer, 21:390
  from sucrose, 19:154
  sweeteners, 19:593, 603
Sorbitol hexanitrate, 1:579
Sorbitol pentanitrate, 1:579
Sorbose, 4:137
1-Sorbose, 2:752
  prepn from sorbitol, 1:576
Sorel cement. See Magnesium oxychloride cement.
Sorghum, 12:863; 19:151; 22:253, 260, 265, 280
Sorghum oil, 8:780
Sorosilicates, 18:47
Sound
  intense ultrasonic waves, 20:774
  phys nature of, 20:773
Sound waves
  irradiation of an aerosol, 7:442
Sour cream
  cultured, 13:530
Sour gas
  sulfur from, S:358
Soya. See Soy.
Soybean meal
  streptomycin from, 19:33
Soybean oil, 8:256, 780, 785, 786, 790; 18:33, 39, 607
  in alkyd resins, 1:853, 864
  composition in fatty acid, 8:831
  drying oil, 7:399, 403, 404, 417, 419, 420
  epoxidation, 8:254
  film prop, 7:423
  lecithin from, 12:344, 350
  in margarine, 13:59
  phenolic resins, 15:197
  prod of fatty acids, 8:827
  specifications, 8:832
  uses, 7:423
Soybean protein
  cyanoethylation of, 6:663

Soybeans, *18*:**599**
  as animal feed, *8*:863
  fatty acids, *8*:820
  in prepn of monosodium glutamate, *2*:200
Soy lecithin
  in confectionery, *6*:44
  dispersing agent in paint, *14*:468
Soy oil wax, *22*:161
Soy protein
  in papermaking, *14*:518
Sozolic acid, *15*:212
Spandex, *9*:159; *18*:**614.** (See also *Poly-urethan.*)
  dyeing, *7*:520, 547, 590
Sparteine, *1*:765, 767
Spearmint, *14*:212
  essential oil as flavor, *9*:352, 353
Spearmint oil
  as denaturant, *8*:451
Specific gravity, *6*:755
Spectrofluorometers, *5*:795
Spectrophotometers, *5*:788, 793
  color measurement with, *5*:806
Spectrophotometry
  aquametry by, *2*:684
Speculum, *8*:67
Spermaceti
  in dental material, *6*:842
  oxidation, *19*:143
Spermaceti wax, *22*:167
  as abherent, *1*:7
  in food processing, *10*:10
Sperm oil
  in manuf of fatty alcohols, *1*:551
  wax from, *22*:167
Sperrylite, *2*:720; *15*:841
Sphalerite, *9*:388, 393; *12*:208, 616; *15*:504; *19*:350, 448,, 758
Sphalerite, *22*:562
Sphene, *20*:349
Spherodizer
  fluidized-bed granulation, *S*:343
Spheroidization
  of carbides in steels, *11*:14
Spherosomes
  in yeast, *22*:512
Spherulites
  in nylon, *16*:9
Sphingolipids
  in fish, *9*:321
Spices, *9*:362
  definitions, *9*:362

Spiegeleisen, *12*:889; *13*:26, 35; *18*:721, 724. (See also
Spike lavender, *14*:212
Spinel, *2*:1; *10*:510; *17*:237
  properties, *10*:513, 756
Spiramycin, *3*:9
  amebicides, *20*:82, 93
Spiramycins, *12*:633, 646
  in chemotherapy, *12*:657
Spirit Blue, *20*:674
  substituent effects on color, *20*:691
  synthesis, *20*:683
Spirit Blues, *20*:713
  Alkali Blues, *20*:718
  Ink Blues, *20*:718
  Soluble Blues, *20*:718
  sulfonated, *20*:717
Spirits. See *Alcoholic beverages, distilled.*
Spirochetes
  chemotherapy against, *3*:32
Spirodihydrobenzofuran, *10*:908
Spironolactone, *7*:267; *18*:869
Spodumene, *12*:530
  in ceramics, *4*:772
Sporidesmolides, *16*:336
Spray drying, *7*:335, 338, 341, 343
  equipment for, *7*:362-365, 367
Sprayed-metal coatings, *13*:266-273
Spray ponds
  and thermal pollution, *S*:722
Sprays, *18*:**634**
Spruce, *14*:213
Spruce extract, *12*:329
Sputtering, *9*:200; *13*:279
  of refractory coatings, *17*:281
4,4 14α-Trimethylcholest-8,24-dien-3β-ol, *18*:838
Squalene, *18*:834, 838
  in fish, *9*:320
Squill
  as cardiovascular agent, *4*:513, 516
St. John's bread, *9*:353
Stabilizing agents
  in food products, *10*:5
Stachydrine, *1*:764
Stachyose
  in soybeans, *18*:601
Stained glass, *10*:571
  caustic pickling, *10*:290
  as dental alloy, *6*:812
Stainless steel. (See also *Steel; Steels.*)
  martensitic transformation, *11*:21

Stainless steels, 5:469
  highly alloyed austenitic, 11:21
  modified, 11:20
  precipitation-hardened, 11:21, 23
  standard grades of, 11:18
Stains
  for wood finishing, 5:709
Stains, industrial, 18:**654**
Stamping
  of metals, 13:316
Standard electrode potential, 7:794
Stannane, 11:206; 20:**312**
  tin reactions, 20:306
Stannate tin plating solution, 8:53
Stannic acetate, 20:315
Stannic acid, 9:682
Stannic bromide, 20:315
Stannic chloride, 20:316
  as bactericidal agent, 2:622
  dye oxidant, 20:693
  prepn, 20:305
  tin hydride from, 20:312
Stannic chromate, 20:316
Stannic entries. (See also *Tin compounds.*)
Stannic fluoride, 20:316
  in prepn of trifluorosilane, 9:654, 682
Stannic iodide, 20:317
Stannic nitrate, 20:317
Stannic oxide, 20:317
  as dental abrasive, 6:849
  prepn, 20:305
Stannic sulfide
  pigments, 15:536
Stannite, 20:276
Stannous acetate, 20:312
Stannous bromide, 20:313
Stannous chloride, 2:583; 20:313
  phenolic resins, 15:186
  prepn, 20:305
  reaction with gold, 10:686
  toxicology, 20:323
Stannous chloride hydrate, 20:313
Stannous chlorofluoride, 9:682
Stannous chlorotrifluoride, 9:682
Stannous entries. (See also *Tin compounds.*)
Stannous fluoride, 9:682; 20:313
  in dentifrices, 6:848
Stannous fluoroborate, 20:313
Stannous fluorozirconate, 22:635
Stannous formate, 20:314
  as catalyst, 8:361
Stannous gluconate, 20:315

Stannous hydroxide, 8:362
  prepn, 20:305
Stannous iodide, 20:314
Stannous laurate, 20:315
Stannous naphthenate, 20:315
Stannous nitrate, 20:314
Stannous octoate, 6:806
Stannous octoate, 15:794
Stannous octoate, 20:314
  polyurethan catalyst, 21:91
  catalyst, 20:799
Stannous oxalate, 20:314
Stannous oxide, 9:682; 20:314
  prepn, 20:305
  reaction with HF, 20:313
  toxicology, 20:323
Stannous oxide hydrate, 20:314
Stannous palmitate, 20:315
Stannous perchlorate, 20:315
Stannous phenolsulfate
  electroplating solution, 8:65
Stannous pyrophosphate, 20:315
Stannous stearate, 20:315
Stannous sulfate, 20:315
Stannous sulfide, 20:315
Stannous tartrate, 20:315
Stannous 2-ethylhexoate, 20:314
Stanozolol, 18:866
  veterinary uses, 21:251
  used in disinfectant effectiveness, 2:608, 609
Staphylomycin, 16:332
Staple. See *Cotton, classification.*
Star anise. See *Anise.*
Starch, 2:15; 18:**672**
  in confectionery, 6:42
  cyanoethylation of, 6:662
  in dentifrices, 6:849
  dextrose from, 6:919
  derivatives, 18:685-691
  in face powder, 6:357, 358, 359
  floridean, 17:763
  hydrolysis, 19:240
  as indicator, 11:557
  nitration of, 8:601
  in nuts, 14:129
  in paint removers, 14:489
  in papermaking, 14:518
  reaction with propylene oxide, 16:598
  resin, 17:391
  structure of, 4:146
  sulfation, 19:306
  tablet binder, 15:121
  in matches, 13:164

Starches
  coagulants, *22*:96
Starch glycolic acid
  sodium salt as indicator, *11*:557
Starching, *12*:204
Starch jellies, *6*:51
Starch nitrate. See *Nitrostarch.*
Starch-oil
  in ceramic composite armor, *S*:144
Starch sugar. See *Dextrose.*
Starch syrups, *6*:919
Static electricty. (See also *Antistatic agents.*)
  as fire hazard, *9*:288
Statifuge, *4*:741, 744
Staurolite, *10*:756
Staypak, *22*:381
Steam, *18*:**692**
  hydrolyze fats, *8*:368
  use in distillation, *7*:243
Steam cracking
  process for ethylene, *S*:641
Steam engines, *16*:446-456
  automotive, *S*:68
Stearamide
  form, *2*:68
  phys prop, *2*:68, 73
  prepn from stearic acid, *2*:362
Stearamidomethylpyridiniumchloride,
  19:562; *22*:142
Stearatochromic chloride
  water-repellent agent, *22*:140
Stearic acid, *2*:348, 362; *6*:356, 879; *8*:812
  as abherent, *1*:4, 5, 8
  activator for SBR, *7*:689
  in candles, *4*:58; *15*:105
  composition, *8*:833
  in dental impression materials, *6*:804, 836
  in eggs, *7*:664
  lubricant, *15*:121
  in manuf of metallic soaps, *7*:274
  oxidation, *19*:143
  phys prop, *1*:231
  in prepn of glyodin, *10*:225
  as release agent in food, *10*:16
  in shellac, *18*:26
  in soap fats, *18*:416
  specifications of commercial grades, *8*:834
  sucrose ester, *19*:154
  tall oil, *19*:621
  in tallow, *18*:418
  in vanishing cream, *6*:350
  vulcanization of copolymers, *7*:687
  wettability, *9*:713

Stearic acid amide, *6*:451
Stearic acid esters, *8*:381
Stearic acid metal salts
  as abherent, *1*:8
Stearic monoethanolamide, *19*:552
Stearin, *13*:182
  in candles, *4*:58
  in paint removers, *14*:488
α-Stearin, *1*:8. See *Glyceryl monostearate.*
unsym-Stearodiolein, *8*:790
unysm-Stearodipalmitin, *8*:790
Stearolic acid, *8*:816
Stearone, *12*:104
Stearophenone, *10*:147
β-Stearoxytricarballylic acid, *2*:592
Stearyl alcohol. (See also *1-Octadecanol.*)
  composition of commercial mixture, *1*:553
  specifications, *1*:554
  phys prop, *1*:543
  as opacifying agent in shampoo, *6*:353
  in shampoo, *10*:774
  in rinses, *10*:777
Stearylamine, *2*:348
Stearyl citrate
  as food preservative, *10*:14
  in margarine, *13*:59
  as stabilizing agent in food, *10*:16
Stearyldimethylbenzylammonium chloride,
  10:777
Stearyl lactate
  prepn, *8*:353
Stearyl-lactyl lactate, *12*:182
Stearyl monoglyceridyl citrate
  as food additive, *10*:9
Steatite
  pigments, *15*:509
Steel. *18*:715. (See also *Stainless Steel.*)
  in armor, *S*:147
  austentic cold-worked, *11*:13
  boron in, *3*:606
  carbon-chrome-molybdenum in piping,
    15:647
  carbon-molybdenum in piping, *15*:647
  carbon in piping, *15*:646
  carbon tool, *20*:568
  carburizing, *18*:782
  chrome-nickel in piping, *15*:647
  chromium, 5468; 22247, 250
  clad, *13*:273
  coated, *13*:252
  electroplating on, *8*:64
  furnaces used in treatment, *10*:287, 288
  high-speed for machining, *20*:568

lime in, *12*:455
low-alloy for machining, *20*:568
use of open-arc furnaces in melting, *10*:257
phosphatizing, *13*:296
pickling of stainless, *S*:689
plating for protection, *8*:38
as protective covering for cables, *11*:817
stainless for corrosives transport, *20*:601
stainless as pigments, *15*:550
stainless in urea-synthesis, *21*:52
vacuum degassing of, *21*:152
zirconium, *22*:621
Steel blue, *15*:539
Steel galvanizing
use of sodium aluminate, *2*:10
Steel industry
cryogenics, *6*:471
use of argon and helium, *10*:881
Steels. (See also *Iron; Steel.*)
austenitic, *5*:471
austenitic chromium-nickel, *22*:247
corrosion inhibition of, *6*:328
corrosion resistance to water, *6*:302
filler metals for welding carbon steels,
22:249
filler metals for welding low-alloy steels,
22:250
pickling, See *Pickling.*
silicon deoxidizer, *18*:128
stainless, *5*:469
stainless in sodium-sulfur cells, *S*:126
welding of carbon steels, *22*:246
welding chromium-nickel, *22*:250
welding low-alloy steels, *22*:246
welding stainless steels, *22*:246
Steel wool
as abrasive, *1*:28
Steffen process, *13*:616
Steffen's waste, *2*:200
Stellites, *5*:734
Stencil printing
image carriers, *16*:536
Stephen reaction
aldehydes, *1*:648
Sterculia oil, *8*:787
Stereoregularity
in high polymers, *7*:64
Stereotype metal, *20*:762, 764-766
Sterility testing, *18*:826
Sterilization, *18*:**805**
Sterilizing agents
ethylene oxide as, *8*:552

Sterling cycle
unconventional power plants, *S*:69
Sterling process
prod of zinc, *22*:579
Sternutators
in chemical warfare, *4*:877
Steroidal alkaloids, *1*:773-776; *18*:841-844
Steroids, *11*:49; *18*:**830**; *19*:803
anabolic, *11*:102
norsteroids, *11*:104
pharmaceuticals, *15*:117
Sterol esters, *8*:359
Sterols, *18*:835-838. (See also *Steroids.*)
in fats, *8*:781
in fish, *9*:321
Stibamine glucoside, *2*:585
p-Stibanilic acid
amebicides, *20*:94
Stib- entries. (See also *Antimony compounds.*)
Stibenyl, *2*:585
Stibiconite, *2*:564
Stibine, *2*:563, 572; *13*:4. (See also *Hydrogen antimonide.*)
Stibinic acids, *2*:581
Stibnite, *2*:564, 570, 577; *15*:506
Stibonic acids, *2*:581
in amebicides, *20*:94
Stibophen, *14*:546
as drug, *2*:584
Stibosan, *2*:585
Stigmasta-4,22-dien-3-one, *18*:836
Stigmasterol, *18*:836
in soybeans, *18*:601
Stilbamidine, *3*:35
Stilbene
from benzyl chloride, *5*:283
Stilbene derivatives, *19*:**1**
Stilbene dyes, *7*:471, 473
Still gas, *10*:280
Stillingia oil, *8*:780, 785
Stimulants. See *Psychopharmacological agents*,
16:**640.**
Stishovite, *18*:49, 55, 75
Stoddard solvent, *21*:429; *22*:140
used in drycleaning, *7*:314; *15*:81
Stoichiometry, industrial, *19*:**17**
Stokes' law, *9*:468
Stone
architectural, *4*:704
Stoneware
clays used in, *5*:566
Stoneware clay
in ceramics, *4*:767

Stontianite, 7:292; 19:48
Storage batteries, 3:161, 249. See *Batteries*
Storage battery plates
  use of litharge, 12:274
Storax, 17:390; 19:55
  cinnamyl alcohol from, 5:522
  as denaturant, 8:451
Stovarsol, 2:731. See *Acetarsone.*
Stoyzite, 12:281
Strain gages, 22:224
Straw meal
  as animal feed, 8:864
Strecker amino acid synthesis
  of acetaldehyde, 1:83
Streptidine, 19:33, 38
Streptimidone
  amebicides, 20:80
Streptococci
  chemotherapy against, 3:21
Streptodornase, 5:598
  medical uses, 8:203
Streptokinase, 5:594, 595, 598
  phys props, 5:589
Streptomyces, 12:632
  hydrocarbon utilization, S:839
Streptomycin, 13:463; 19:33
  amebicides, 20:80
  assaying of, 3:12
  in bacterial infections, 3:1
  as veterinary drug, 21:244
Streptomycin and related antibiotics, 19:33
Streptothricin
  amebicide, 20:80
Streptozon, 3:4
Stress corrosion, 6:238, 298. See *Corrosion*
Stress relaxation
  in polyurethans, 21:82
Stretford process
  for gas scrubbing, 1:44
Strontium, 19:48
  in causticization of soda ash, 1:745
  from seawater, 14:156
Strontium carbonates
Strontium chloride
  in dentifrices, 6:851
Strontium chromate, 5:481, 499
  as corrosion inhibitor, 5:510
  pigments, 15:534
Strontium compounds, 19:50
Strontium cyanide, 6:604
Strontium ferrate(VI), 12:40
Strontium fluosilicate, 9:658
Stronium hexafluorostannate, 9:682

Strontium hexafluorogermanate, 9:585
Strontium hydride, 11:203
Strontium manganate, 13:12
Strontium manganate(V), 13:11
Strontium monofluorophosphate, 9:642, 643
  as phosphor, 12:622
Strontium orthophosphate
Strontium perchlorate, 5:69
Strontium peroxide, 14:750
Strontium salicylate, 17:729
Strontium tetrafluorozincate, 9:684
Strontium titanate, 20:412
  ferroelectric prop, 9:13
  in synthetic gem manuf, 10:510
Strontium-90, 17:78, 91, 124
  beta source, 17:57
  in cathodes, 8:2
  as fission product, 14:106
Strontium-90-yttrium-90
Strophanthins
  as cardiovascular agents, 4:513, 516
Structural plastics
  epoxy resins used in, 8:309
Struvite
  occurence in canned fish, 9:327
Strychnine, 1:772, 778, 795
  as cardiovascular agent, 4:523
  from nuts, 14:141
  rodenticide, 15:916
  veterinary uses, 21:252
  x-ray analysis, 22:453
Strychnine sulfate
  perchlorates detected with, 5:80
  rodenticide, 15:916
Stucco
  cement, 4:20, 705
  in gypsum board, 21:622
Styphnic acid, 13:892
Styracin. See *Cinnamyl cinnamate.*
Styracitol, 1:578
Styrax, 14:213, 724. (See also *Storex.*)
Styrene, 2:413; 6:695; 19:55
  use of aluminum chloride in manuf of, 2:22
  chloroprene reaction with, 5:218
  cinnamaldehyde from, 5:521
  from cinnamic acid, 5:517
  in copolymerization of trioxane, 1:98
  copolymers, 16:233
  from ethylbenzene, 22:468
  use of fluorescent pigments, 9:501
  inhibition of polymerization by TCNE, 6:627
  NMR spectra, 14:52

phenolic resins, *15*:182
polyester alkyd copolymer, *20*:791
polyester copolymers, *20*:802
polyester resins, *15*:461
polyesters, *15*:455
polymerization of, *7*:48, 66
polymers. See *Styrene plastics.*
polyvinylpyrrolidone graft copolymer, *21*:434
prepn, *8*:500
prod from ethylbenzene, *22*:494, 480
reaction with paraform, *10*:86
in synthetic oil prepn, *7*:415, 416
toxicity of, *11*:303
unsaturated polyester diluent, *20*:817
Styrene
Styrene acrylic
as fluorescent coating system, *9*:499
Styrene bromide, *3*:778
Styrene-butadiene
as fluorescent coating system, *9*:499
latex, *17*:597
in latex paints, *14*:472
in latex paints, *22*:152
as adhesive backing, *S*:935
Styrene-butadiene (SBR)copolymers, *7*:679
numbering system, *7*:681
recipes, *7*:679, 680
Styrene-butadiene latex
in papermaking, *14*:519
Styrene-butadiene rubber. (See also *Elastomers, synthetic; Styrene plastics.*)
advantages, *7*:689, 691
in packing, *14*:451
properties of flexible polymer, *9*:863
Styrene-butadiene solution copolymers, *S*:**910**
Styrene chlorohydrin, *5*:311
sulfonation of, *11*:874
Styrene-divinylbenzene copolymer
Styrene oxide, *8*:249, 289
block copolymers, *19*:553
Styrene plastics, *19*:**85.** (See also *Polystyrene.*)
Styrene polymers
in synthetic waxes, *22*:170
Styrenesulfonic acid, *19*:284
4-Styrenesulfonic acid, *19*:300
Styrone. See *Cinnamyl alcohol.*
Styrylcarbinol. See *Cinnamyl alcohol.*
Styryl dyes, *16*:286, 292-294
Suberic acid, *1*:249
in dental material, *6*:827
phys prop, *1*:232

polyamide from, *16*:2
prepn, *1*:247
Suberylarginine, *18*:845
Sublimation, *13*:213
mass transfer in, *13*:100
Sublimed blue lead, *12*:279
Substance
amount of, *S*:987
Subtilin, *16*:314
in food preservation, *10*:15
Subtilopeptidase B, *S*:297
Subtilopetidase A, *S*:297
Suburban gas," *8*:385
Succinaldehyde, *10*:667; *19*:142, 147
as disinfectant, *2*:616, 617
Succinchlorimide
as disinfectant, *2*:616, 617
Succinic acid, *10*:661; *14*:356; *19*:**134, 736**
from alcoholic fermentation, *21*:257
nystatin degradation, *16*:140
prepn, *1*:247
prepn from 1,4-butanediol, *10*:669
as stabilizer, *2*:15
in vinegar, *21*:256
Succinic anhydride, *19*:134
phys prop, *1*:212
polyester from, *16*:170
as starch modifier, *10*:17
thiophene synthesis, *20*:224
Succinimides
substituted as anticonvulsants, *11*:522
Succinonitrile, *6*:574, 641
Succinoyl chloride, *12*:160
Succinylcholine, *6*:548, 551
Succinylcholine chloride, *2*:409; *6*:550; *15*:922
Succinylsulfathiazole
structural formula, *3*:2
Sucrose, *4*:146. (See also *Sugars.*)
addn to wine, *22*:315
assimilation by yeast, *22*:517
bond determination in, *22*:453
in nuts, *14*:129
polyurethans, *21*:62
in soybeans, *18*:601
structure, *4*:146
tablet filler, *15*:121
in yeast production, *22*:534
Sucrose esters
as denaturant, *8*:451
prepn, *8*:360
Sucrose octaacetate
Sudan Broion
structural formula, *5*:745

Sudan III. (See *Toncy Red)*
Suede, *12*:340
Suede leather
  dyeing of, *7*:591
Sugar, *19*:**151**
  addn to wines, *22*:315
  analysis in sparkling wines, *22*:324
  cane, *19*:**166**
  decolorizing of, *4*:152
  in meat curing, *13*:171
  pill formulation, *15*:124
  refined, in synthesis of glycerol, *10*:625
Sugar alcohols, *1*:569-588
Sugar beet molasses, *22*:534
Sugar beets
  in manuf of fermentation ethyl alcohol,
    8:438
Sugar cane
  in manuf of fermentation ethyl alcohol,
    8:438
Sugar cane molasses, *22*:534
Sugar cane wax, *22*:165
Sugar cream. See *Fondant.*
Sugar, invert
  in confectionery, *6*:37
  in mannitol manuf, *1*:581
  in sorbitol manuf, *1*:581
Sugars
  in carbonated beverages, *4*:339
  in confectionery, *6*:36
  nitric acid oxidation of, *14*:359
  reactions with propylene oxide, *16*:598
  reducing in nuts, *14*:129
  in soaps, *18*:429
  in vinegar, *21*:256
Sugar syrups
  in manuf of caramel color, *4*:63
Suint, *22*:388, 411
Sul-bi-Sul process, *22*:76
  for desalination of water, *22*:61
Sulfacetamide, *19*:267, 269, 272
Sulfachloropyridazine, *19*:268
Sulfadiazine, *10*:735; *19*:261, 266, 268, 271
  manuf, *1*:846
  structural formula, *3*:3
Sulfadimethoxine, *19*:273
  structural formula, *3*:3
Sulfa drugs, *19*:261-279
  as anticoccidials, *21*:246
  veterinary uses, *21*:245
Sulfaethylthiadiazole
  structural formula, *3*:3

Sulfaguanidine, *10*:739
  anticoccidial, *21*:248
  structural formula, *3*:2
Sulfamates, *19*:242. (See also *Amidosulfates.)*
Sulfamation, *19*:280, 306
Sulfamerazine, *10*:739
Sulfamerizine
  structural formula, *3*:3
Sulfamethazine, *12*:157
  structural formula, *3*:3
  as veterinary drug, *21*:246
Sulfamethizole
  structural formula, *3*:2
Sulfamethoxypyridazine
  structural formula, *3*:2
Sulfamic acid, *19*:242, 282. (See also *Amido-
    sulfuric acid.)*
  as corrosion inhibitor, *6*:329
  in disinfection of swimming pools, *4*:925,
    926
  in dyeing of wool with acid dyes, *7*:523
  as a herbicide, *22*:214
  manuf by amminolysis, *2*:367
  in sulfonation, *15*:211
Sulfamic acid and sulfamates, *19*:**242**
4'-Sulfamyl-2,4-diaminoazobenzene. See *Pron-
    tosil.*
4'-Sulfamyl-2,4-diaminoazobenzene, *7*:258
Sulfanilamide
  antithyroid agent, *20*:270
  p-aminobenzoic acid antagonist, *21*:489
  antagonist for nicotinic acid, *22*:407
  in bacterial infections, *3*:1
  as diuretic, *7*:257
  structural formula, *3*:3
2-Sulfanilamide-4-methylpyrimidine, *6*:457
Sulfanilic acid, *2*:415, 424; *7*:479
  water analysis, *21*:695
Sulfanitran, *21*:247
Sulfaphenazole
  structural formula, *3*:2
Sulfaquinoxaline
  as anticoccidial, *21*:246, 248
Sulfarsphenamine, *2*:730
Sulfate
  analysis for in water, *21*:694
Sulfate white liquor, *20*:752
Sulfathiazole
  prepn from *N*-acetylsulfanilyl chloride,
    1:152
  structural formula, *3*:3

Sulfation, *19*:**279, 301-306,307**
  of olefins, *14*:328, 332
  of zirconium, *22*:661
Sulfatoceric acid, *4*:851
Sulfato compounds
Sulfatozirconates, *22*:663
Sulfene, *12*:98
Sulfenone
  in water, *21*:701
Sulfenyl chloride
  thiol oxidation, *20*:208
Sulfides
  as ceramic materials, *4*:774
  in reduction of amines, *2*:94
Sulfinic acids
  prepn by Friedel-Crafts, *10*:155
Sulfisomidine
  structural formula, *3*:3
Sulfisoxazole
  structural formula, *3*:3
Sulfite lignins. See *Lignin.*
Sulfite pulping process, *4*:607, 608
Sulfo Rhodamine B
  xanthene dyes, *22*:434
Sulfoacetic acid, *12*:115
5-Sulfoaminouracil
  uric acid from, *21*:111
Sulfoanthranilic acid
  in prepn of azo dyes, *2*:893
o-Sulfobenzaldehyde, *20*:700
o-Sulfobenzoic acid anhydride
Sulfobromophthalein sodium, *3*:780
Sulfochlorination, *19*:279, 284, 289, 308. (See
  also *Sulfur-chlorination.*)
Sulfolane, *19*:**250,** 381, 384
3-Sulfolene, *10*:902
  from butadiene, *3*:788
Sulfomethylation, *19*:290, 308
Sulfonal," *20*:215
Sulfonamides, *19*:**255**
  in bacterial infections, *3*:4
  as diuretics, *7*:257
  prepn from ethylenamines, *7*:30
  veterinary uses, *21*:245
Sulfonation
  of xylenes, *22*:484
Sulfonation and sulfation, *19*:**279**
Sulfones, *19*:314
  as hypnotics, *11*:513
Sulfonethylmethane
  as hypnotic, *11*:514
Sulfonic acid amines. See *Sulfonamides.*

Sulfonic acids, *19*:**311**
  as catalyst in alcoholysis, *8*:357
  in prepn of aromatic amines, *2*:114
  from toluene, *20*:562
Sulfonmethane
  as hypnotic, *11*:514
Sulfonyl chloride
  in leather tanning, *12*:334
Sulfonyl chlorides, *1*:224
  thiol synthesis, *20*:214
4,4'-Sulfonyldibenzoic acid, *16*:182
p-Sulfophenylacetic acid, *15*:214
Sulfophthalein dyes, *20*:727
4-Sulfophthalic anhydride, *2*:442
5-Sulfosalicylic acid, *17*:723, 735
Sulfoterephthalic acid, *15*:464
Sulfoxides, *19*:**320**
  "dimsyl" ion, *19*:326
Sulfur, *2*:311; *19*:**337**
  as agricultural fungicide, *10*:221
  alkylation of, *1*:897
  in black powder, *8*:647
  by-product for phosphate prod, *S*:358
  carbon monoxide reaction with, *4*:432
  carbon tetrachloride reaction with, *5*:132
  chlorine reactions with, *5*:2
  chloroprene reaction with, *5*:217
  in coating of fertilizers, *9*:78
  in coke, *15*:23
  colorant for glass, *9*:252
  compds with boron, *3*:736
  content in soils, *9*:35
  in copper alloys, *6*:197
  cyanoethylation on, *6*:643
  in dental impression materials, *6*:804
  in diesel fuel, *10*:488
  in drossing of lead ore, *12*:222
  dyes, *19*:424
  in explosive priming compositions, *8*:653
  fusion with *p*-toluidine, *20*:185
  in gasoline, *10*:471
  as hazardous chemical, *9*:291
  in heterochain inorganic polymers, *11*:641
  in high-temperature cells, *S*:130
  homochain polymers, *11*:635
  from hydrogen sulfide by Claus process,
    4:376
  impurity in copper, *S*:265
  in infrared detectors, *15*:432
  as insecticide, *11*:683
  in manuf of carbon products, *4*:168
  in matches, *13*:160, 164

from ocean, *14*:160
as petroleum feedstocks, *10*:390
from petroleum sweetening, *15*:48
in pig iron, *18*:718
pigments, *15*:540
as plant nutrient, *9*:29, 34
in polymerization of chloroprene, *7*:706
in pyrotechnics, *16*:826
reaction with aluminum alkyls, *2*:31
reaction with ethyleneamines, *7*:30
reaction with ethylamines, *7*:36
reaction with hydrogen, *11*:344
recovery in copper operations, *S*:260
removal from petroleum, *15*:45, 49
from shale oil, *18*:18
soil chemistry, *18*:538
in steel, *18*:751
styrene inhibitor, *19*:68
in tar sand bitumen, *19*:695
vulcanization of copolymers, *7*:687
in vulcanization of neoprene, *7*:711
vulcanizer, *17*:566
Sulfur Black I, *7*:497
Sulfur Black T, *7*:465
Sulfur Chlorination, *19*:498
Sulfuration. See *Sulfurization.*
Sulfur compounds, *19*:**367**. (See also *Disulfur*
  entries.)
Sulfur dichloride, *9*:602
  in prepn of sulfur tetrafluoride, *9*:672
Sulfur difluoride, *9*:664, 674
  as ester-exchange catalyst, *16*:167
Sulfur dioxide, *6*:578; *9*:556; *19*:408-420
  addn to wine, *22*:318
  as bleaching agent, *3*:561
  in bleaching glue, *10*:607
  carbon disulfide from, *4*:381
  in chlorine dioxide manuf, *5*:41, 42
  chloroprene reaction with, *5*:217
  in commercial prepn of acetyl chloride,
    1:140
  control of emission in copper smelting op-
    erations, *S*:257
  cream of tartar processing, *19*:739
  dried by alumina, *2*:52
  as food preservative, *10*:13, 36
  fumes, *18*:403, 406
  industrial pollutant, *21*:685
  liquid as solvent refiner, *15*:57
  in manuf of ammonium chloride, *2*:318
  in prepn of Hypalon, *7*:695
  methane reaction with, *4*:381

in prod of ammonium sulfate, *2*:330
reaction with boron trifluoride, *9*:557
reaction with epoxide, *8*:274
reaction with trialkylaluminum, *2*:32
as refigerant, *10*:859
as starch modifier, *10*:17, 164
in starch production, *18*:679
sugar refining, *19*:213
Sulfur dioxide-benzene
solvent refining, *15*:57, 95
Sulfur dyes, *7*:551; *19*:**424**
uses of, *7*:471, 474, 477
Sulfuric acid, *2*:114, 115; *7*:562, 563; *8*:240;
  10:161; *12*:115
in acrylamide manuf, *1*:278
in alkylation, *15*:42, 44
as catalyst in alkylation of paraffins, *1*:886
as catalyst in alkylation of phenols, *1*:894
as catalyst in gasoline alkylation processes,
  10:474
as catalyst in peroxyacetic acid manuf,
  8:250
as catalyst in prepn of *p*-aminophenol, *2*:219
in cellulose acetate flake manuf, *1*:117
in cellulose triacetate flake manuf, *1*:117
corrosion inhibitors for, *6*:322-325
in dehydration of alcohols, *8*:474
as drying agent, *7*:383, 384, 385, 387, 396
in drying of chlorine gas, *1*:701
in dye industry, *7*:480
in dyeing of wool with acid dyes, *7*:523, 528
in esterification of *sec* butyl and *sec* amyl
  acetates, *8*:343
in epoxidation of soybean oils, *8*:254
as a herbicide, *22*:214
lubricating oil treatment, *15*:17
in manuf aluminum sulfate, *2*:60
in manuf ammonium sulfate, *2*:309
in manuf anthraquinonedisulfonic acids,
  2:445
in manuf of ethyl ether, *8*:480
in manuf of fatty acids, *8*:825
in manuf of hydrogen fluoride, *9*:618
petroleum decolorizer, *15*:61
phosphoric acid, *15*:261
phosphorus, *15*:280
pickling medium for steel, *S*:684
pickling of stainless steels, *S*:690
polyesterification catalyst, *20*:798
in prepn of 4-amino-2-nitrophenol, *2*:218
in prepn anthraquinone, *2*:435, 436
in prepn of boron trifluoride, *9*:558
in prepn of ethylenedinitramine, *8*:626

in prepn of *m*-dinitrobenzene, *8*:617
in prepn of TNT, *8*:613
in prepn of tetranitrocarbazole, *8*:622
in prod of phosphoric acid in fertilizer industry, *9*:86
relative heat content of solutions of, *3*:252
uranium ore leaching, *21*:13
as a weed killer, *22*:176
Sulfuric acid and sulfur trioxide, *19*:**441**
Sulfuric and sulfurous esters, *19*:**483**
Sulfuric anhydride. See *Sulfur trioxide.*
Sulfuric esters, *19*:483
Sulfurization and sulfur chlorination, *19*:**498**
Sulfur monobromide, *3*:754
Sulfur monochloride, *8*:427
from carbon tetrachloride, *5*:132
in prepn of mustard gas, *8*:502
Sulfur monofluoride. See *Disulfur difluoride.*
Sulfur monofluorides, *9*:664, 674
Sulfur mustard
alkylating agents, *S*:82
in chemical warfare, *4*:873
Sulfur nitride. See *Nitrogen sulfide.*
in pulpmaking, *12*:369
Sulfurous acid. (See also *Sulfur dioxide.*)
Sulfurous esters, *19*:483
Sulfur pentafluoride. See *Disulfur decafluoride.*
Sulfur tetrafluoride, *9*:508, 664, 671, 695, 696, 800
ammonolysis of, *2*:367
derivatives, *9*:802, 804
Sulfur trioxide, *2*:441, 577; *12*:119; *19*:392, 395, 398, 441, 472, 473. (See also *Oleums.*)
action on fatty acid, *8*:823
cyclamate synthesis, *19*:600
diarylamines as stabilizer, *7*:48
ethyl chloride reaction with, *5*:143
fumes, *18*:403
hexachloroethane reaction with, *5*:167
in military smoke agents, *4*:903
in portland cement, *4*:691, 692
in prepn of fluorosulfuric acid, *9*:677
reaction with epoxide, *8*:274
reaction with ethyl ether, *8*:479
reaction with 1,2,3,4-tetrafluorobenzene, *9*:785
in sulfonation, *15*:211; *19*:280
tetrachloroethylene reaction with, *5*:198
Sulfuryl chloride, *2*:577; *6*:442; *9*:794; *12*:122, 287
chlordene reaction with, *5*:248
in manuf of benzyl chloride, *5*:284

perchloric acid reaction with, *5*:65
prepn carbon catalyst, *4*:156
thiophene reaction, *20*:222
vinyl chloride reaction with, *5*:172
Sulfuryl fluoride, *9*:632, 677
Sulfur-32, *17*:3
Sulfur-35, *17*:121
Sulpho Rhodamine, *7*:526
Sulphon cyanine 5R Extra, *2*:877, 996
Sulphure de Polychlorocyclane, *5*:273
β-Sultone, *19*:286
Sulvanite, *21*:159
Sumaresinolic acid, *17*:388
Sunflowerseed oil, *8*:780, 785, 786, 790; *18*:40
Sunn, *9*:184
analysis, *9*:173
Sunscreen products, *6*:355
Sunset Yellow FCF, *5*:867; *10*:793
Sunset Yellow Lake, *5*:871
Suntan products, *6*:355
Superalloys, *5*:733; *11*:21-31. (See also *High-temperature alloys.*)
Superchlorination
of municipal waters, *22*:87
of swimming pool water, *22*:129
Superconductivity, *6*:479
Superconductors
beryllides as, *S*:79
Supermalloy, *12*:753
Supermendur," *12*:757
Superoxides, *14*:762-764
oxygen generation systems, *S*:663
Superphosphate, *2*:293; *9*:32. (See also *Phosphates; Fertilizers.*)
ammoniation, *9*:116
application fluid bed technology, *9*:400
as fertilizer, *9*:79
fertilizers, *S*:343
Superphosphoric acid. (See also *Polyphosphoric acid.*)
production, *9*:97
Superpolyamides, *16*:3
Superrefractory materials, *4*:774
Supertropical bleach, *5*:20
in chemical warfare, *4*:886
Supertropical bleach decontaminating agent
in machining tools, *20*:572
Super water, *21*:682
ferroalloy, *18*:751
furnaces for fusion, *16*:289
in electron tubes, *8*:7, 12
in superalloys, *11*:24

Supralan, 2:889
Supranol Red BR, 2:896
Suprasterol₂ I, 21:557
Suprasterol₂ II, 21:557
Surface-active agents, 19:507. (See also *Surfactants.*)
  action of adsorption, 1:427
Surface resistivity
  electrical, 2:655, 670
Surface tension, 8:133
  critical of polymer, S:901
  data interpretation, 6:736
Surfactants, 13:678; 19:**507**. (See also *Surface-active agents.*)
  amine oxide, S:36
  amphoteric, 19:564-566
  anionic, 19:512-531
  biodegradable, 19:581-584
  as antistatic agents, 2:657, 666
  cationic, 19:554-564
  in detergents, 6:856, 872
  ethylene oxide manuf from, 8:553
  use of fatty alcohols in, 1:555
  fluorochemical fire-fighting, S:373
  in foaming formulations, 21:85
  HLB values, 8:128
  nonionic, 19:531-554
Suxethonium, 6:549
Swedish black, 15:544
Sweet basil, 9:369
Sweet birch
  essential oils as flavor, 9:352
Sweet birch oil, 22:383
Sweeteners, nonnutritive 19:**593**
Sweetening agents
  in carbonated beverages, 4:339
  sorbitols, 1:582
Sweet pepper, 9:363
  as animal feed, 8:868
Sweet potatoes, dried
Swimming pools
  disinfection of, 4:925
  purification of water in, 2:621
  treatment of water in, 22:124
Sydnones, 10:919
Syenite, nepheline
  in ceramics, 4:769

Sykose, 19:595
Sylvan. See *2-Methylfuran.*
Sylvan chart for dust control applications, 10:330
Sylvanite, 16:370, 384; 19:759
  processing, 16:372
Sylvestrene, 19:816
Sylvinite, 18:473
Sylvite
  from Great Salt Lake, S:455
Sympatholytic drugs
  as cardiovascular agents, 4:518
Sympathomimetic drugs
  as cardiovascular agents, 4:517
Syncrude oil, S:189
Syndets. See *Surfactants.*
Syndiotactic polymers, 4:557
Synergism
  antimony-halogen, S:474
  in antioxidants, 2:592
Synergistin, 16:332
Synethetases
  classification, 8:176
Synthesis gas. (See also *Ammonia; Gas, manufactured; Methanol.*)
Synthesis gas methanation, S:205
Synthetic fiber industry
  use of air conditioning in, 1:497
Synthetic fibers
  anthraquinone dyes for, 2:503
  bleaching agents for, 3:563
Synthetic gas production, S:203
Synthetic wintergreen oil. See *Methyl salicylate.*
Synthol synthesis, 4:486
Syringaldehyde, 10:97
  separation from vanillin, 21:191
Syringic aldehyde
  from lignin, 21:189
Syringylpropane
  in lignin, 12:363
Syrosingopine, 4:519
Syrups
  in carbonated beverages, 4:345
  in manuf of caramel color, 4:63
System
  thermodynamic, 20:118-123
Szaibelyite, 3:628, 648

T

Tableware
  molded plastic, 2:251
Tabun, 4:875; 15:328
  in white table wines, 22:319
Tachydrite, 3:758
Tachysterol, 18:837; 21:555
Taffeta
  dyeing by disperse dyes, 7:541
Tainton process, 22:585
Talbutal, 3:61, 70
Talc, 5:547; 6:357; 17:237; 19:**608**. (See also
  Talcum.)
Talcum. (See also *Talc.*)
  lubricant, 15:121
  use as adherent, 1:11
Talcum powder, 6:360
  phys prop, 1:570
  prepn, 1:575
Talitols
Tall oil, 19:614
  drying oil, 7:399, 419, 420
  film properties, 7:423
  in soaps, 18:417
  prod of fatty acids, 8:845
  surfactants, 19:545
  wood products, 22:382
Tall-oil fatty acid, 8:257, 820
  in alkyd resins, 1:853, 859, 864
Tallow, 8:780; 18:40
  edible in margarine, 13:59
  in soap, 18:417
  inedible, 13:183
Tallow amine acetate
  solubility, 2:133
Tallow bleaching
  chlorine dioxide for, 5:44
Talose, 4:136
Tammann-Hessee equation
  for viscosity, 21:465
Tangerine, 14:213
Tangerine,
  essential oils as flavor, 9:352
Tank cars, 14:434
Tanner's bate
  action and uses, 8:182

Tannic acid
  petroleum sweetening, 15:52
  zirconium complexes, 22:655
Tannin, 12:314
  in radioactive waste treatment, 14:107
  in red table wines, 22:323
  tantalum complexes, 19:633
  wood products, 22:382
Tanning, 7:589; 12:303-343. (See also *Leather.*)
  chrome, 5:510
  citric acid in, 5:538
  photography, 15:380
  phthalic anhydride, 15:456
Tanning industry
  use of hydroxyacetic acid, 10:635
  use of sodium hydroxyacetate, 10:636
Tannins
  content in distilled alcoholic beverage, 1:510
  in drilling fluids, 7:296, 301
  in nuts, 14:129, 132
  in wood, 16:690
Tansy, 14:213
Tantalasilica
  as catalyst, 1:82
Tantalum, 19:**630**
  as a refractory metal, 11:32
  beryllium compounds, S:76
  extraction, 13:766
  platinum-coated, 15:836, 852
  powder metallurgy of, 16:429
  refractory, 17:246
Tantalum beryllides, 3:475
Tantalum carbide, 4:75, 80, 82. (See also
  Cemented carbides, 4:92.)
Tantalum carbonyl, 4:490
Tantalum compounds, 19:**648-652**
Tantalum hafnium carbide, 22:623
Tantalum hydride, 11:209
Tantalum pentafluoride, 9:681
Tantalum pentoxide
  as glass former, 10:538
Tantalum trifluoride, 9:681
Tantalum zirconium carbide, 22:623
Tapers
  (candles), 4:62

Tapeworms, *14*:532, 543
Tar, *19*:**653**
  as fuel for furnaces, *10*:279, 280
  hydrocracking of, *11*:419
  hydrocracking of athabasca, *11*:423
  in tobacco smoke, *20*:516
  stain solvent, *18*:661
Tar acids, *S*:271
Tar and pitch, *19*:**653**
Taraxein, *16*:644
  hydrocracking of athabasca, *11*:423
Tariric acid, *8*:816
Tarnishing. See *Corrosion.*
Tar oil
  from shale oil, *18*:18
Tarragon, *9*:375; *14*:201. (See also *Estragon.*)
  Tartaric acid.)
Tar sands, *19*:**682**
Tartar emetic, *2*:578, 583, 584; *7*:526; *18*:276;
  19:736; (See also *Tartaric acid.*)
  molluscacide, *15*:910
  schistosomiasis treatment, *14*:545
Tartaric acid, *12*:823; *19*:**733**
  as antioxidant, *2*:592, 594
  as chelating agent, *6*:8
  as food preservative, *10*:14
  determination of absolute configuration,
    22:453
  in carbonated beverages, *4*:339
  in prepn of erythritol, *1*:571
  in wine, *22*:316
  phys prop, *1*:234
  polyesters, *15*:793
  reaction with ethylenediamines, *7*:27, 527
meso-Tartaric acid
Tartrazine, *2*:881; *5*:867, 873; *10*:793
  in carbonated beverages, *4*:342, 343, 344
Tartrazine (FD & C Yellow No. 5)
  used in hair coloring, *10*:793
Tartrazine Lake, *5*:871
Tartrazine Lake (D & C Yellow No. 5), *5*:871
Tartronic acid, *19*:238
  phys prop, *1*:234
Taspine, *1*:769
Taste. (See also *Flavor; Organoleptic testing.*)
  chemical nature, *14*:338
  flavor perception, *9*:348
  flavor profile method, *14*:341
  of municipal waters, *22*:102
  of water, *21*:691
  testing, *14*:332
Taurine, *2*:358; *19*:311, 318, 521
Tazettine, *1*:771

TCNE. See *Tetracyanoethylene.*
Tea, *19*:**743**
TCNQ. See *Tetracyanoquinodimethan.*
  caffeine extracted from, *3*:913
  flavor, *9*:353
Teallite, *20*:276
Teaseed oil, *8*:780, 785
Tea wax, *22*:161
Technetium carbonyl, *4*:489
Technetium-99m, *17*:3
Teclozan
  as amebicide, *20*:91
Tectosilicates, *18*:49
Teflon. See *Polytetrafluoroethylene.*
TEL. See *Tetraethyllead.*
Telemetry
  uses of magnetic tape in, *12*:812
Telepogine, *1*:760, 773
Television tubes
  use of ammonium bifluoride in, *9*:548
Tellurium, *19*:**756**
  copper alloys with, *6*:193, 197
  effects on brasses, *6*:229
  electrodeposition of zinc, *22*:588
  in high-temperature cells, *S*:130
  impurity in copper, *S*:265
  lead, *12*:249, 261
  photography, *15*:386
  reaction with hydrogen, *11*:344
  silane compds, *18*:195
Tellurium compounds, *19*:**763, 774**
Telnic bronze, *6*:193
  magnetic properties, *6*:220
Telodrin, *5*:250
Telomycin, *16*:334
Tempera, *14*:480
Temperature measurement, *19*:**774**
Temperature scale
  Celsius, *S*:998
  Fahrenheit, *S*:999
  International practical, *S*:1000
Tempering, *13*:314, 329
Templin, *14*:213
Tennantite, *6*:135
Tennecetin, *16*:137
Tenorite, *6*:135, 275
  short-time, *11*:11
Tensile tests, *13*:185
Terbacil, *22*:213
Terbium, *17*:44
  electronic configuration, *1*:360
  ionic radii, *1*:370

Terephthalic acid, 6:546; 10:154; 15:462-472
  bistriethylammonium salt, 16:170
  diammonium salt, 16:170
  esterification, 20:489
  phys prop, 1:239
  polyester brighteners, 19:13
  polyester fibers, 16:143, 155
  polymer from, 16:2
  xylene oxidation, 20:560
Terephthalonitrile
  from p-xylene, 22:485
Terephthalonitrite, 2:344
Terephthaloyl chloride, 10:668; 15:472
  thiazole dyes, 20:190
Terne, 13:260
Terpene hydrocarbons
  monocyclic, 20:751
Terpene resins, 19:834-838
  chlorinated as insecticides, 11:695
  copolymers, 11:253
  in essential oils, 14:183
  in wood, 16:690
  polymerization, 11:252
  removal from essential oils, 14:189
Terpenes, 19:**803**
Terpenes and terpenoids, 19:**803**
Terpenoid alkaloids, 1:773
Terpenoids, 19:**803**
Terphenylbenzene, 7:194
α-Terpinene, 15:338
Terpineol
  as synthetic flavor, 9:354
Terpinyl acetate
  prod and sales of, 8:376
  as synthetic flavor, 9:354
Terpinyl butyrate
  as synthetic flavor, 9:354
Terra alba. See Calcium sulfate; Gypsum.
Terracotta
  clays used in, 5:566
Terrazzo, 4:704
Tertiomycin, 12:637
Testololactone, 18:876
  deriv of, 11:101
  isolation of, 11:97
Testosterone, 10:19; 11:93, 101
  phys and chemical prop, 11:97
  related compounds, 11:102
Testosterone propionate
  contraceptive action of, 6:85
1,3,4,6-Tetra-O-acetyl-D-fructofuranose,
  19:151

2,3,4,6-Tetraacetyl-α-D-glucosyl   chloride,
  4:142
Tetraalkoxy-1,4-benzoquinones
  reaction with aziridines, 11:531
Tetraalkyldiaminodiaryl ketones, 18:451
Tetraalkyllead, 12:282-299
  prepn by Grignard reaction, 10:727
1,4,5,8-Tetraaminoanthraquinone, 2:447. (See
  also Aminoanthraquinones.)
Tetraamino copper phthalocyanine dyes,
  19:434
Tetraaryl silicate
  a heat-transfer medium, 10:854
1,3,5,7-Tetraazotricyclo [3.3.1.1^{3,7}] decane. See
  Hexamethylenetetramine.
Tetrabasic lead fumarate, 12:270
Tetrabenzyltitanium, 20:433
Tetraboranes, 3:684
Tetrabromoauric(III) acid, 10:693
Tetrabromobisphenol A, 3:778
  as fire retardant, S:467, 483
1,2,2,4-Tetrabromobutane, 1:605
1,1,2,2-Tetrabromochloroethane
  from monochloroacetylene, 5:204
Tetrabromo-o-cresol, 3:778
1,2,6,6-Tetrabromocyclohexene-3,4,5-trione, 16:192
1,1,2,2-Tetrabromoethane, 3:776
Tetrabromofluorescein, 5:870, 873
Tetrabromofluorescein, 11:479
2,4,5,7-Tetrabromofluorescein, 22:433
Tetrabromofluorescein, Eosin, 3:775; 11:479
Tetrabromoindigo, 3:775
Tetrabromophthalic anhydride
  as fire retardant, S:467, 481
  as flame-retardant polyester, 20:831
Tetrabromoplatinic(II) acid, 15:864
2,3,5,6-Tetrabromo-1,4-difluorobenzene, 9:784
2,4,5,6-Tetrabromo-1,3-difluorobenzene, 9:784
1,1,1,2-Tetrabromo-2-fluoroethane, 9:834
Tetrabutylammonium hydroxide, 6:569
Tetrabutylphosphonium hydroxide, 18:239
Tetrabutyltin, 20:312, 319
Tetrabutyltitanate, 8:361
  in dental materials, 6:827
  polyesterification catalyst, 20:799
Tetracaine, 8:376
Tetracaine hydrochloride, 2:406
Tetracalcium aluminoferrite
  in asbestos-cement board, 21:608
  in cement formation, 4:685, 692
1,1,3,3-Tetracarbethoxypropane
  prepn, 8:477

Tetracarbocyanines, *6*:617
Tetracarbonyl hydride, *10*:158
Tetracene, *8*:592
    reaction with boron trifluoride, *9*:556
1,3,5,7-Tetrachloroanthraquinone, *2*:485
1,4,5,8-Tetrachloroanthraquinone,
    prepn *2*:485
Tetrachloroauric acid, *10*:693
    in soil, *18*:535
3,3',4,4'-Tetrachloroazobenzene
Tetrachlorobenzene
    hydrolysis of, *22*:180
1,2,4,5-Tetrachlorobenzene
    in prepn of 2-amino-3,4,6-trichlorophenol,
    2:217
Tetrachlorobenzenes, *5*:262
2,3,5,6-Tetrachloro-1,4-benzoquinone. See
    Chloranil.
    as catalyst, *14*:375
Tetrachloro-*p*-benzoquinone, *20*:693
Tetrachlorobenzotrifluoride, *9*:778
Tetrachlorobiphenyl, *9*:714
1,2,3,4-Tetrachlorobutadiene, *9*:590
Tetrachlorocoumarin, *6*:427
2,3,7,8-Tetrachlorodibenzo-*p*-dioxin, *22*:180
sym-Tetrachlorodifluoroacetone, *9*:755
Tetrachlorodinitroethane
    from tetrachloroethylene, *5*:198
Tetrachloroethane, *1*:206; *5*:159-164
    polyester fiber solvent, *16*:155
    vinylidene chloride from, *21*:276
1,1,2,2-Tetrachloroethane
    poly(vinylchloride) from, *21*:441
Tetrachloroethylene, *5*:195-203. (See also *Per-
    chloroethylene.)*
    chlorine reactions with, *5*:90, 167
    from pentachloroethane, *5*:166
    hexachloroethane from, *5*:167
    in roundworm chemotherapy, *14*:536
    sulfonation, *19*:286
    toxic effects of, *5*:93
Tetra-2-chloroethyltitanate, *5*:306
Tetrachlorofluorescein, *22*:433
4,5,6,7-Tetrachloroisoindolin-1-one, *15*:583
Tetrachloroisophthalonitrile
    as fungicide, *10*:226
Tetrachloromethane. See *Carbon tetrachloride.*
1,1,2,2-Tetrachloro-3-methyl-butan-3-ol, *1*:623
1,2,3,4-Tetrachloronaphthalene, *15*:448
Tetrachloronaphthalenes, *5*:300-303
2,3,5,6-Tetrachloronitrobenzene
    as fungicide, *10*:226
Tetrachloropalladic(II) acid, *15*:868

Tetrachlorophenols, *5*:336
Tetrachlorophthalic anhydride, *9*:798
    as flame-retardant polyesters, *20*:831
    as fire retardant, *S*:467, 482
    in alkyd resins, *1*:862
Tetrachloroplatinic(II) acid, *15*:863
1,1,1,3-Tetrachloropropane
    prepn, *8*:502
Tetrachloroquinone
    as leather fungicide, *10*:233
Tetrachlorostibonium sulfate, *2*:577
Tetrachlorosuccinic acid anhydride, *19*:137
Tetrachloroterephthalodinitrile, *9*:777
Tetrachlorotetrabromofluorescein, *5*:870
2,4,6,8-Tetrachloro-2,4,6,8-tetrazobicyclo-
    (3.3.0)octa-3,7-dione, *4*:920
Tetrachlorothallic(III) acid, *20*:67
2,3,5,6-Tetrachlorotoluene, *19*:315
1,1,3,3-Tetrachlorourea
    formation of, *4*:916
Tetracobalt dodecacarbonyl
4,8,12,15,18,21-Tetracosahexaenoic acid,
    8:815
Tetracosanoic acid; *8*:813. (See also *Lignoceric
    acid.*
Tetracyanobenzoquinone, *6*:629
Tetracyanoethane
    in color photography, *5*:832
Tetracyanoethylene, *6*:625, 630
Tetracyanoethylene oxide, *6*:630
Tetracyanofuran, *6*:631
1,1,3,3-Tetracyanopropene, *6*:629
Tetracyanoquinodimethan, *6*:628
Tetracyanothiophene, *6*:631
Tetracyano-1,4-dithiin, *6*:631
Tetracyclines, *13*:463; *20*:**1, 7, 12, 14**
    in bacterial infections, *3*:1
Tetracyclohexoxysilane, *18*:219
Tetracyclohexyltin, *20*:319
Tetracyclopentadiene, *6*:691
Tetradecanenitrile, *18*:661
2,5,10,13-Tetradecanetetrone, *12*:160
Tetradecanoic acid, *8*:812
Tetradecanol
1-Tetradecanol. See *n-Tetradecyl alcohol.*
1-Tetradecene
    epoxidation, *8*:248
5-Tetradecenoic acid, *8*:814
Tetradecyl acrylate
    phys prop, *1*:287
n-Tetradecyl alcohol.
2,3,4,5-Tetraflurobenzoic acid, *9*:785
    phys prop, *1*:543

Tetradecylamine, 2:130
Tetradecyl ether, 8:493
2,3,4,6-Tetradeoxy-2,4-diamino-D-mannose, 19:45
Tetradymite, 19:759
Tetraethoxygermane, 10:525
1,1,3,3-Tetraethoxypropane, 16:288
Tetraethoxysilane, 18:220
Tetraethylammonium tetrachloronickelate, 13:759
Tetraethylammonium triphenylphosphine-tribromonickelate, 13:759
Amino I Acid, 7:478; 13:715
Tetra(2-ethylbutoxy)silane, 18:219
Tetraethylene glycol
  toxicity, 10:641
Tetraethylene glycol dibutyl ether, 15:883
Tetraethylenepentamine, 7:22, 23
  properties, 7:24
  specifications, 7:32
  toxicity, 7:33
Tetraethyl furantetracarboxylate, 6:631
Tetraethyl germanate, 10:525
Tetraethyllead, 10:467; 12:282-299; 15:54
  and desulfurization, 15:47
  effect in automobile emissions, S:59
  ethyl chloride in manuf of, 5:145, 146
  hazards, 12:302
  in alkylates, 15:41
  manuf, 12:290, 292
  prepn, 1:899
  prepn from trialkylaluminum, 2:34
  property, 12:284
  synthesis, 12:289
  use of sodium in, 18:450
Tetraethyllead motor fuel
  ethylene bromide in, 3:751
Tetraethyl orthosilicate, 18:216
Tetraethylrhodamine
  in analysis of antimony, 2:571
Tetraethylstannane, 20:318
Tetraethylthiuram disulfide, 17:600
Tetraethyltin, 20:318
Tetraethyl titanate, 5:306
2,3,4,6-Tetrafluoroaniline, 9:785, 786
Tetrafluorobenzene, 9:691
1H, 4H-Tetrafluorobenzene, 9:689. (See also 1, 2, 4, 5, -Tetrafluorobenzene.)
1,2,3,4-Tetrafluorobenzene, 9:785
1,2,3,5-Tetrafluorobenzene, 9:786
1,2,4,5-Tetrafluorobenzene, 9:689, 786
  prepn from pentafluorophenylhydrazine, 9:795

Tetrafluorobenzene-1,4-dithiol, 9:787, 793
2,3,4,5-Tetrafluorobromobenzene, 9:785
2,3,5,6-Tetrafluorobromobenzene, 9:786
1,1,2,2-Tetrafluorodichloroethane
  in precipitation polymerization, 21:374
2,3,4,5-Tetrafluorochlorobenzene, 9:785
2,3,4,6-Tetrafluorobenzenesulfonic acid, 9:786, 787
Tetrafluoroethylene, 9:516, 632, 690, 805, 807-812
  copolymers with vinylidene fluoride, 9:842
  in prepn of difluoroacetic acid, 9:770
  phys prop, 6:625
  produced by sintering, 9:859
  toxicity, 9:741
Tetrafluoroethylene epoxide, 9:810
Tetrafluorohydrazine, 9:403, 629
Tetrafluoro-m-phenylenediamine, 9:790
1,2,3,4-Tetrafluoronaphthalene, 9:797
2,3,4,5-Tetrafluoronitrobenzene, 9:785
2,3,4,6-Tetrafluoronitrobenzene, 9:786
2,3,5,6-Tetrafluorophenol, 9:793
2,3,5,6-Tetrafluorophenylenediamine, 9:795
2,4,5,6-Tetrafluorophenylenediamine, 9:794
  in prepn of 1,2,4-trifluorobenzene
2,3,5,6-Tetrafluorophenylhydrazine, 9:784
2,3,4,5-Tetrafluorophenyllithium, 9:785, 787
2,3,5,6-Tetrafluorophenyllithium, 9:787
Tetrafluorophthalic acid, 9:785, 798
2,3,5,6-Tetrafluoropyridine, 9:798
2,3,5,6-Tetrafluoropyridine-4-carboxylic acid, 9:798
Tetrafluoropyrimidine, 9:799
Tetrafluororesorcinol, 9:790, 793
Tetrafluoroterephthalic acid, 9:787
Tetrafluoroterephthalonitrile, 9:787
2,2,3,3-Tetrafluoro-1-propanol, 9:846
2,3,4,5-Tetrafluorothiophenol, 9:785
2,3,5,6-Tetrafluoroxylene, 9:794
Tetrafluoro-1,4-dihydrazinobenzene, 9:795
2,3,5,6-Tetrafluoro-4-hydrazinopyridine, 9:798
2,4,5,6-Tetrafluoro-3-hydrazinoaniline, 9:794
Tetraflurophthalonitrile, 9:785
Tetraflurosilane. See Silicon tetrafluoride.
Tetragermane, 10:523
Tetraglycine hydroperiodide, 2:619
Tetrahedrite, 6:135; 9:393; 18:284
1,2,3,6-Tetrahydroacetophenone, 3:786
6,7,8,9-Tetrahydro-5H-azepotetrazole, 10:918
Tetrahydrobenzaldehyde
  in prepn of epoxy resins, 8:299

1,2,3,6-Tetrahydrobenzaldehyde
    prepn from butadiene, 3:785
2,3,4,5-Tetrahydrobenzoic acid
    phys prop, 1:236
    prepn from acrylonitrile, 1:341
1,2,3,6-Tetrahydrobenzonitrile, 3:786
Tetrahydrocannabinol, 16:646
Tetrahydrocarbomycin B enol acetate, 12:646
Tetrahydrodicyclopentadiene, 6:702
endo-Tetrahydrodicyclopentadiene, S:3
Tetrahydro-p-dioxin. See p-Dioxane.
1,2,3,4-Tetrahydro-2,4-dioxo-5-pyrimidinesulfamic
    acid, 21:111
Tetrahydroendotricyclopentadiene, 6:692
Tetrahydrofuran, 5:310; 6:628; 9:785, 787,
    792, 794; 12:292, 814
    as solvent for Grignard agents, 10:724, 896
    formal from, 16:254
    from 1,4-butanediol, 10:668, 669
    in polymerization of isoprene, 12:70
    oxidation, 19:143
    paint remover, 14:487
    plasticizer, 15:773
    polysiloxanes, 18:225
    polyurethans from, 21:61
    synergistic mixture with dimethylacetamide,
        1:146
Tetrahydrofuran, 10:249; 12:58
Tetrahydrofurfuryl acrylate
    phys prop, 1:288
Tetrahydrofurfuryl alcohol, 10:237, 247;
    19:143
    ammonolysis of, 2:346
    stain solvent, 18:660
Tetrahydrofurfurylamine, 10:247
Tetrahydrofurfuryl chloroformate, 4:386
Tetrahydrofurfuryl nicotinate, 21:526
Tetrahydrofurylsuccinic anhydride, 10:250
Tetrahydroimidazo[4,5-(d] imidazole-2,5(1H,
    3H)-dione. See Glycoluril.
Tetrahydrolinalool, 19:826
1,2,3,4-Tetrahydro-6-[(6-methoxy-2-benzothia-
    zolyl)-az-o] benzo[h] quinoline-3,7-diol,
    20:192
Tetrahydro-6-methyl-4-ureido-2(1H)-pyrimidone,
    6:451
Tetrahydronaphthalene, 9:698; 13:672, 676,
    678
    from naphthalene, 13:670
    in naphthalene feedstocks, 11:454
    stain solvent, 18:661
    toxicity of, 11:303

1,2,3,4-Tetrahydronaphthalene
    purification of nicotinic acid, 21:516
Tetrahydronaphthalenesulfonates, 19:519
Tetrahydrophthalic anhydride
    in alkyd resins, 1:862
1,2,5,6-Tetrahydrophthalic anhydride, 3:785
Tetrahydrophthalic maleate esters
    brominated as fire retardants, S:482
Tetrahydropyran, 10:247, 921
Tetrahydropyranyl malonate, 12:128
Tetrahydropyridines, 12:117
1,2,3,4-Tetrahydroquinoline, 16:892
Tetrahydrothiophene, 19:250
    prepn from tetrahydrofuran, 10:250, 670
Tetrahydrothiophene 1,1-dioxide, 10:456;
    19:250
Tetrahydrotricyclopentadiene, 6:691
Tetrahydrouric acid,
    from uric acid, 21:108
1,2,3,4-Tetrahydroxybenzene, 16:209
1,2,3,5-Tetrahydroxybenzene, 16:210
1,2,4,5-Tetrahydroxybenzene, 16:211
2,3,4,5-Tetrahydroxybenzoic acid, 16:210
2,3,4,6-Tetrahydroxybenzoic acid, 16:211
Tetrahydroxybenzoquinone
    sodium salt of, 16:213
Tetrahydroxy-p-benzoquinone, 16:212
2',3,4,4'-Tetrahydroxychalcone, 12:325
Tetrahydroxyfluoran, 16:198
1-(Tetrahydro-2-furyl)-1,3-butanedione, 12:148
Tetrahydroxyhexadecanoic acid, 18:26
1,2,5,7-Tetrahydroxy-4-methylanthraquinone, 18:26
5,8,9,10-Tetrahydro-2-methyl-1,4-naphthoquinone,
    21:594
1,2,3,4-Tetrahydro-1-naphthol, 13:676
2,3,4,5-Tetrahydroxynitrobenzene, 16:210
Tetrahydro-1,4-oxazine, 10:930; 13:659. (See
    also Morpholine.)
cis-Tetrahydro-2-oxothieno(3,4d)imidazoline-4-v-
    aleric acid, 3:518 (See also Biotin.)
Tetrahydroxyquinone
    as an indicator, 11:560
Tetrahydro-3,3,5,5-tetrakis(hydroxymethyl)-4-
    pyranol, 1:596
Tetrahydro-1,4-thiazine, 10:930
6,7,8,9-Tetrahydro-5H-azepotetrazole, 10:918
Tetraiodoethylene, 9:700
Tetraiodofluorescein, 11:479
2,3,4,5-Tetraiodopyrrol, 2:619. (See also Iodol.
Tetraiodothyronine
    properties, 20:262
Tetraisopropyl titanate, 1:848; 16:517

Tetrakisdiethylamidotitanium, 20:472

Tetrakisdiphenylamidotitanium, 20:472

Tetrakis(hydroxymethyl)phosphonium chloride, 11:539; S:952

Tetrakis(hydroxymethyl)phosphonium chloride
   in flame-retardant finish for cotton, 9:307

Tetrakis(hydroxymethyl)phosphonium hydroxide, S:953

Tetrakistriethanolamine trititanate, 20:467

Tetrakis(trimethylsilyloxo)zirconium, 22:660
   purification of nicotinic acid, 21:516

Tetrakis(1-aziridinyl)-1,4-benzoquinone, 11:531

Tetrakis(2-ethylhexan-1,3-diol)titanate, 20:490

N,N,N',N'-Tetrakis(2-hydroxyethyl)ethylenediamine dioleate, 19:561

N,N,N',N'-Tetrakis(2-hydroxypropyl)ethylenediamine, 20:472

Tetrakis(2-methoxyetoxy)silane, 18:219

Tetrakis(4-methylpyridine)nickel(II), 13:372

Tetralin. See 1,2,3,4-Tetrahydronaphthalene.

Tetralol, 13:676

Tetralone, 13:676

α-Tetralone, 10:149, 150

Tetramandelatozirconic acid, 22:630, 648

Tetra-n-butylstannane, 20:319

Tetramer acid
   prepn, 8:847

Tetrametaphosphates, 15:250

Tetrametaphosphoric acid, 15:313

2,2',6,6'-Tetramethoxy-p,p'-biphenol, 15:371

Tetramethoxysilane, 18:220

1,2,4,4-Tetramethyl-5-acetylcyclohexene, 12:136

1,3,5,7-Tetramethyladamantane, S:4

Tetramethylammonium hydroxide, 18:239

Tetramethylammonium trichloromercurate
   as ferroelectric, 9:5

1,2,4,5-Tetramethylbenzene, 16:300

Tetramethylbenzenes, 16:300-305

Tetramethylbenzidine, 20:703

N,N,N',N'-Tetramethylbenzidine, 3:409

N,N,N',N'-Tetramethyl-1,3-butanediamine, 7:23
   prop, 7:24

Tetramethyl-1,3-cyclobutandione, 12:91, 149

2,2,4,4-Tetramethyl-1,3-cyclobutanediol, 10:680

2,4,7,9-Tetramethyl-5-decyne-4,7-diol, 1:625
   phys prop, 1:626

N,N'-Tetramethyl-4,4-diaminobenzophenone. See Michler's ketone.

Tetramethyldiaminobenzophenone, 18:451

Tetramethyldimethylenediureide, 2:233

Tetramethylene chlorohydrin, 5:310

Tetramethylenediamine, 7:23
   as curing agent, 7:37
   halosilane complexes, 18:186
   polyamide from, S:767
   prop, 7:24
   reaction, 7:30

Tetramethylenediamine dihydrochloride, 7:30

2,4,4,6-Tetramethyl-m-dioxane
   from acetaldehyde, 1:84

Tetramethylenedisulfotetramine, 15:918

Tetramethylene oxide. See Tetrahydrofuran.

Tetramethylene sulfolane, 9:777

Tetramethylene sulfone, 10:156, 164; 19:250

Tetramethylenimine, 11:526
   properties, 7:24

N,N,N',N'-Tetramethylethylenediamine, 7:23; S:917

Tetramethylgermane, 10:525

2,3,4,6-Tetramethyl-D-glucose, 4:142

Tetramethylguanadine
   ester-exchange catalyst, 16:175

2,6,10,14-Tetramethylhexadecane, 14:847

2,2,5,5-Tetramethyl-3,4-hexanedione, 12:140

Tetramethyllead, 10:476; 12:282-299
   prepn, 1:899
   sodium in prod, 18:450

1,4,5,8-Tetrahyroxyanthraquinone, 2:465, 467; 3:786

Tetramethylmethane. See Neopentane.

2,6,10,14-Tetramethylpentadecane, 14:847

Tetramethyl-p-rosaniline, 9:896

Tetra-(4-methylpyridine)nickel dithiocyanate

2,2,4,4-Tetramethyl-3-pentanone, 12:104

Tetramethylsilane
   comparison with tetramethyltitanium, 20:428

3,3',8,10-Tetramethylthiacarbocyanine iodide, 6:612

N,N,N',N'-Tetramethylthiourea

Tetramethylthiuram disulfide, 2:636; 7:697, 706; 19:771. (See also Bis(dimethylthiocarbonyl) Disulfide.)
   as fungicide, 10:224

Tetramethyltitanium
   iodine reaction, 20:432

N,N,N',N'-Tetramethyl-o-tolidine, 3:415

1,3,3,6-Tetramethyl-1-p-tolylindane, 10:143

3,3,5,5-Tetramethyl-1,2,4-trithiolane, 12:119

5,5,8,8-Tetramethyl-5,6,7,8-tetrahydro-2-naphthol, 10:142

Tetramisole, 14:540

Tetramminepalladium(II) chloride, *15*:846, 868, 869

3,3',5,5'-Tetranitrobenzil, *12*:146

Tetranitrocarbazole
    explosive characteristics, *8*:633
    sensitivity characteristics, *8*:632
    stability characteristics, *8*:634

1,3,6,8-Tetranitrocarbazole
    electronic structure, *20*:427

Tetranitrodiglycerin, *8*:604. (See also *Diglyceryl tetranitrate.*)

Tetranitrogen tetrasulfide, *19*:405

1,3,6,8-Tetranitronaphthalene, *13*:707

Tetranitronaphthalenes, *13*:706, 707

Tetranitroplatinic(II) acid, *15*:865

1,3,5,7-Tetranitro-1,3,5,7-tetrazacyclooctane, *8*:623, 624. (See also *Cyclotetramethylenetetranitramine.*)

Tetraoctyltin, *20*:319

Tetraoctyl titanate, *20*:799

1,5,9,13-Tetraoxacyclohexadecane-4,8,12,16-tetraone, *19*:139

Tetraoxygen difluoride, *9*:508, 634

Tetraphenylarsonium chloride, *9*:641
    perchlorates detected with, *5*:80

Tetraphenylbismuthonium
    bromide, *3*:545

Tetraphenylhydrazine, *7*:43

Tetraphenyllead, *12*:285

N,N,N',N'-Tetraphenylmethylenediamine, *7*:44

Tetraphenylstannane, *20*:319

Tetraphenylstibonium bromide, *2*:582

Tetraphenylthiophene, *20*:219

Tetraphenyltin, *20*:319

Tetraphenyltitanium, *20*:428, 443

Tetraphosphate, *15*:234

Tetraphosphorus pentasulfide, *15*:317

Tetraphosphorus trisulfide
    in matches, *13*:160

Tetrapolymer
    of vinylidene chloride, *21*:281

Tetrapolyphosphate, *15*:234

"Tetrapropylene," *19*:296

Tetrapyridinenickel fluorosilicate, *13*:759

Tetrasilane, *18*:177
    dispersing agent in paint, *14*:468

Tetrasodium pyrophosphate, *6*:881, 882; *7*:296; *15*:243, 264, 271
    as emulsifier in cheese, *10*:8
    in detergents, *6* 857

Tetrasodium tetraoxalatozirconate trihydrate, *22*:650

Tetrasulfur tetranitride, *19*:405

N,N'-Tetrathiodimorpholine, *13*:665

1,2,5,7-Tetrathiononane, *8*:418

Tetra(*n*-octyl)tin, *20*:319

Tetratriacontanoic acid, *8*:813

1,3,4,6-Tetrazaindene, *10*:913

1,3,5,8-Tetrazanaphthalene, *10*:932

Tetrazole, *10*:902, 918
    xylene clathrates, *22*:475

Tetritols, *1*:570

Tetrohydrocarbomycin, *12*:646

Tetrolic acid, *6*:465; *8*:816

Tetryl, *8*:619
    explosive characteristics, *8*:632
    in Tetrytol, *8*:628
    sensitivity characteristics, *8*:633
    stability characteristics, *8*:634

Tetrytol, *8*:628
    explosive characteristics, *8*:633
    sensitivity characteristics, *8*:632
    stability characteristics, *8*:634

Textile brighteners, *3*:745, 746, 747

Textile dyes
    used to dye furs, *10*:308

Textile finishing
    and detergency, *6*:867
    use of amino resins in, *2*:247

Textile industry
    acetate yarn packaging for, *1*:123
    brighteners used in, *3*:746, 747
    humidity conditions within, *1*:487
    use of alkanolamine soaps, *1*:821
    use of ethyleneamines in, *12*:358
    use of lecithin in, *12*:358

Textile printing
    use of natural gums, *10*:741

Textiles, *S*:932
    polyvinyl acetate emulsions in, *21*:351
    polyvinyl alcohol sizing agent, *21*:362

Textile technology, *S*:932

Textile testing, *20*:**33**
    denier, *20*:35

Textile waste treatment, *S*:979
    antistatic agents for, *2*:662
    carbon and graphite fibers, *4*:241
    drycleaning of, *7*:307
    fire-resistant, *S*:944

Textolite 2001
    in bearing materials, *3*:275

Thalidomide
    as hypnotic, *11*:515

Thallic bromide, *20*:66

Thallic chloride, *20*:67

Thallic fluoride, *20*:67

Thallic oxide, *20*:67
Thallic sulfate, *20*:68
　　electrodeposition of zinc, *22*:588
　　as solid film lubricant, *12*:590
　　zinc analysis, *22*:561
Thallium, *20*:**63**
Thallium compounds, *20*:**66-69**
Thallium formate, *19*:318
Thallium halides, *20*:66
Thallium hydroxide, *19*:318; *20*:68
Thallium oxides, *20*:67
Thallium oxysulfate, *20*:68
Thallium(I) pentafluorozirconate, *9*:685
　　as insecticide, *11*:683
　　rodenticide, *15*:916
Thallium sulfate, *20*:66, 68
Thallium(III) sulfate, *20*:68
Thallium(I) tetrachlorothallate(III), *20*:67
Thallium(I) hexachlorothallate(III), *20*:67
Thallium tetrasulfatozirconate, *22*:662
Thallium-204, *17*:128
Thallous acetate, *20*:315
Thallous bromide, *20*:66
Thallous chloride, *20*:66
Thallous chromate, *20*:65
Thallous ethoxide, *20*:67
Thallous fluoride, *20*:66
Thallous hydroxide, *20*:67
Thallous iodide, *20*:66
　　zinc analysis, *22*:561
Thallous oxalate, *20*:66
Thallous sulfide, *20*:65
THAM. See *Tris(hydroxymethyl)amino-methane.*
Theaflavins, *19*:748, 752
Thearubigens, *19*:748, 752
　　use of alcohol in, *8*:449
Theater sprays
Theine. See *Caffeine.*
Thenard's blue, *5*:745, 853; *15*:541
Thenium *p*-chlorobenzenesulfonate, *14*:533
3-Thenylbromide, *20*:222
　　antihistamine, *11*:41
Thenylpyramine hydrochloride
Theobroma cacao, *5*:368
Theobromine
　　as cardiovascular agent, *4*:521
　　structural formula, *3*:912
Theobromine sodium salicylate
　　as diuretic, *7*:253
Theophylline, *21*:109
　　as cardiovascular agent, *4*:521

Theophylline ethanolamine
　　as diurectic, *7*:253
Theophylline ethylenediamine
　　as diuretic, *7*:253
Theophylline piperazine
　　as diuretic, *7*:253
Theophylline sodium acetate
　　as diuretic, *7*:253
Theophylline sodium glycinate
　　as diuretic, *7*:253
Theophylline 1-amino-2-propanol
Therapeutic agents, *20*:**70**
Therapeutic agents, protozoal infections, *20*:**70**
Thermal analysis, *20*:**99**
Thermal black
　　manuf, *S*:98
Thermal blacks. See *Carbon black.*
Thermal carbons
　　cascade design, *7*:146
　　column design, *7*:143
　　in U.S., *S*:98
Thermal diffusion, *7*:92, 135
Thermal diffusion constant, *7*:139
Thermal insulation. See *Insulation, thermal.*
Thermal pollution, *S*:771
Therminol, *5*:295
Therminol FR
　　heat transfer medium, *10*:852
Therminol 77
　　heat transfer medium, *10*:853
Thermionic converter,
　　cesium vapor, *4*:865
Thermionic emission, *8*:8
Thermionic generators, *16*:467
Thermistors, *19*:786
　　semiconducting diamond, *S*:277
Thermit, *4*:897
　　burning temperature of, *4*:899
　　heat of combustion of, *4*:899
Thermite, *2*:1
Thermobalance
　　and photochemistry, *15*:331
　　thermal analysis, *20*:101
Thermochemistry
Thermocline
　　in lakes, *S*:318
Thermocompression evaporator, *8*:571
Thermocopying, *17*:347
Thermocouples, *19*:786, 789
Thermodynamics, *20*:**118**

Thermoelectric devices
    fabrication, *20*:161
    power generation, *20*:163
    theory of, *20*:150
Thermoelectric energy conversion, *20*:**147**
Thermoelectric generators, *16*:465
Thermoelectric materials, *20*:159
    thermoelectric energy conversion, *20*:147
Thermoelectric properties
    of materials, *20*:153
    of semiconductors, *20*:156
Thermoelectric theory, *20*:149
    Kelvin relations, *20*:149
Thermoforming
    polypropylene, *14*:305
    aquametry by, *2*:695
Thermogravimetry, *20*:99, 101
    derivative, *20*:99
Thermogravitational column, *7*:136
    for binary gas mixture, *7*:141
Thermometers. See *Temperature measurement.*
Thermomigration
    in electromigration, *S*:283, 291
Thermopiles
    thermoelectric energy conversion, *20*:147
Thermoplastic coverings
    for electric cables, *11*:820
Thermoplastic molding materials, *11*:798
    in reinforced plastics, *12*:190
    for electric cables, *11*:819
Thermosets. See *Resins, thermosetting.*
Thermosetting-resin adhesives, *1*:383
Thevetin
    as cardiovascular agent, *4*:513
Thiabendazole, *14*:542
9-Thia-10-azaanthracene, *10*:930. (See also
    Phenothiazine.
Thiabutazide
    as diuretic, *7*:263, 264
Thiacyanocarbons, *6*:631
Thiacyclobulane, *10*:898
Thiamin, *20*:173
Thiaminase
    in fish, *9*:322
Thiamine, *8*:175; *20*:**173**. (See also *Enzymes,
    industrial.)*
    addition to flour, *22*:297
    as food additive, *10*:11
    carboxylase precursor, *21*:488
    in fish, *9*:322
Thiamine disulfid," *20*:183
Thiamine hydrochloride, *20*:173, 176

Thiamine mononitrate, *20*:173, 176
    manuf from acetobutyrolacetone, *1*:159
    in prepn of glutamic acid, *2*:204
    sulfur in, *19*:369
Thiamine propyl disulfide, *20*:182
Thiamine tetrahydrofurfuryl disulfide, *20*:182
Thiamine triphosphate, *20*:182
Thiaminetriphosphoric acid, *20*:182
Thiamorpholine, *10*:920, 930
Thiamylal, *3*:61, 70
Thianaphthene, *10*:899
1-Thiandene, *10*:899
Thianthrene, *19*:393
Thiaxanthenes, *16*:654
Thiaxanthenones
    antiamebics, *20*:88
Thiazides
    as diuretics, *7*:261
Thiazine dyes, *2*:866; *7*:471, 474
Thiazole, *10*:897, 898
    as nucleus in cyanine dye, *6*:607
Thiazole dyes, *7*:471, 473; *20*:**185**
    as nucleus in cyanine dyes, *6*:607
Thiazoles, *10*:915
2-Thiazolethiol, *10*:915
Thiazolidines, *11*:534
2-Thiazoline-2-thiol, *6*:643, 649
4-Thiazoline-2-thione, *20*:180
Thiazolone, *20*:24
Thiazolylsulfamic acids, *19*:244
Thickening. (See also *Sedimentation.)*
Thickening agents
    in food products, *10*:5
Thick water," *21*:682
3-Thienyllithium, *20*:222
Thietane, *10*:898
3-Thietanone-1,1-dioxide, *12*:98
Thimerosal, *2*:620; *13*:236, 241. (See also *Mer-
    thiolate.)*
Thin-layer chromatography
    amino acid separation, *2*:162
    sugars, *19*:165
Thioacetaldehyde
    from acetaldehyde, *7*:83
Thioacetanilide
    prepn from acetanilide, *1*:150
    cationic, *20*:191, 194
    direct dyes, *20*:185
    disperse, *20*:191
    vat dyes, *20*:188
    in vitamin-enriched yeasts, *22*:531
Thioacetazone, *3*:30

Thioacetic acid, 6:643, 646
  in prepn of acetanilide, 1:151
  prepn from acetyl chloride, 1:139
  reaction with aziridines, 11:534
Thio alcohols, 12:119
Thioammeline, 6:568. (See also 2,4-Diamino-
  s-triazine-6-thiol.)
Thioamylsodium, 2:403
Thiobarbituric acids, 3:60
Thiobendazole
[2,2'-Thiobis(4-tert-octylphenolato)]-m-butyla-
  mine nickel, S,813
  anthelmintic, 21:249
2-Thiobenzoic acid
  reaction with malic acid, 12:842
2,2'-Thiobis(4,6-dichlorophenol), 6:355;
  14:549
4,4'-Thiobis(3-methyl-6-tert-butylphenol), 1:914
2,2'-Thiobis(4-methyl-6-tert-butylphenol), 1:914
4,4'-Thiobis(6-tert-butyl-m-cresol), 2:600
1,1'-Thiobis[n-phenyl-2-naphthylamine], 2:601
Thiocarbamates
  as herbicides, 22:194
  metabolism as herbicides, 22:198
Thiocarbanilide, 2:414
  accelerator, 17:509
  prepn, 4:372
Thiocarbonyl chloride, 19:373
  prepn, 4:372
Thiochroman-1-oxide, 19:324
2-Thiocoumarin, 6:427
6,8-Thioctic acid, 9:29
Thiocyanate
  as indicator, 11:557
  antithyroid, 20:269
Thiocyanate ion
p-Thiocyanatoaniline, 2:416
Thiocyanic acid, 10:758
  methimazole synthesis, 20:270
  reaction with epoxide, 8:274
Thiocyanogen chloride
  formation of, 5:3
Thiocyanogen value of fats, 8:793
Thiodan. See Endosulfan.
1-Thiodene, 10:899
Thiodiacetic acid, 2:592; 20:202
Thiodibutyric acid, 10:673
Thiodiethanoic acid, 20:202
Thiodiethylene glycol, 10:647, 648; 18:662
  from ethylene chlorohydrin, 5:308, 309
Thiodifluorophosphoric acid esters, 9:645
Thiodiglycol, 8:529

Thiodiglycol, 10:648
  prepn, 8:529
Thiodiglycolic acid, 10:633; 20:199, 202
4,4'-Thiodimorpholine, 13:666
Thiodiphenylamine, 2:859; 10:930. See Phenothiazine.
Thiodipropionic acid
  as food preservative, 10:14
  polypropylene antioxidant, S:812
3,3'-Thiodipropionic acid, 2:592, 601
3,3'-Thiodipropionitrile, 6:643, 646, 664
1,1-Thiodi-2-naphthol, 2:601
Thioethylene glycol, 18:662
Thioflavine
  used in hair coloring, 10:793
Thioflavine T, 7:483; 20:188
Thiofuran, 10:898
  reaction with boron trifluoride, 9:557
Thioglycolic acid, 2:580; 6:451; 10:633;
  20:**198**
Thioglycolic acid anilide, 20:203
Thioglycolic acid oxathiolanone, 9:763
Thioglycolic-β-aminonaphthalide," 20:203
Thioglycollate
  reaction with cystine in wool, 22:401
  setting of wool, 22:407
2-Thiohydantoin, 11:159
Thioindigo, 20:188
  color-construction relation, 11:564
  solubilized, 11:579
Thioindigoid dyes
  asymmetrical, 11:574
  symmetrical, 11:570
Thioketene, 12:97
Thioketene dimers, 12:97
Thioketones
  prepn, 12:118
Thiokol, 7:677, 696
  in packing, 14:452
Thiokol polymers
  rubber, 17:545
Thiol
  photochemical formation, 15:339
Thiolacetic acid, 6:447; 12:67
  steroid synthesis, 18:870
  thiol synthesis, 20:213
Thiolacetic acid, 2-cyanoethyl ester, 6:646
Thiolane, 20:222
3-Thiolene, 20:222
2-Thiolene, 20:222
Thiol esters
  prepn, 8:365

Thiols, 6:643; 7:679; 20:**205**
 carbon-carbon additions, 20:209
Thiolutin
 amebicide, 20:82
Thiomaleic anhydride, 20:221
Thiomethoxymethyl benzoate, 19:323
O-(Thiomethoxymethyl)benzophenone oxime,
 19:323
N-Thiomethoxymethyl-α,α,-diphenylnitrone,
 19:323
O-Thiomethoxymethylphenol, 19:323
Thionalide, 20:203
Thionaphthene, 19:673
Thionessal, 20:219
Thionocarbanilates, 22:423
Thionothiolcarbonic acid
 xanthates, 22:419
3-Thiono-1,2-dithio-4-cyclopentenes, 10:909
Thionyl bromide, 19:396
Thionyl chloride, 6:631; 8:428; 10:661, 669;
 12:287, 821; 19:244, 397, 419
 in manuf of endosulfuran, 5:250
 in prepn of dialkyl aminoalkyl chlorides,
 2:145
 in prepn of fatty acid chlorides, 8:823
 in roundworm chemotherapy, 14:541
 manuf, 19:398
 perchloric acid reaction with, 5:65
 cyclic sulfates, 19:493
 reaction with alkanolamines, 1:813
 reaction with aminoanthraquinones, 2:449
 sulfonic acids, 19:314
 uses, 19:400
Thionyl fluoride
 in prepn of sulfur tetrafluoride, 9:672
 reaction with boron trifluoride, 9:557
Thionyl trifluoroacetone, 15:883
Thiopental, 3:61, 70
Thiopental sodium, 2:402
 Socony-Vacuum process, 20:224
Thiophene, 6:455; 10:86; 20:**219**
 from n-butane, 3:819
 from butadiene, 3:788
 from butenes, 3:845
 from sulfur dioxide, 19:419
 prepn from furan, 10:248, 898, 902
Thiophene
 acylation, 20:222
 alkylation, 20:222
2-Thiophenecarboxylic acid
 phys prop, 1:240
Thiophene ring
 in hydrorefining, 11:436

α-Thiophenic acid. See 2-Thiophenecarboxylic
 acid.
Thiophenol, 2:580; 12:120
 alkylation of, 1:895
Thiophenols
 alkylation, 19:487
 synthesis, 20:214
Thiophosgene. See Thiocarbonyl chloride.
Thiophosphoryl chloride, 15:305; 19:392
Thiosalicylic acid, 2:580; 17:742
Thiosemicarbazide, 10:902
Thiosemicarbazone, 3:30; 6:456
Thiosemicarbazones
 antiviral agent, 21:454
Thiostrepton, 16:308, 329
Thiosulfates, 20:**227**
 Bunte-type salts, 20:243
 chemical properties, 20:227, 229
 complex, 20:243
 esters, 20:243
 occurrence, 20:232
Thiosulfatoacetic acid
 thioglycolic acid synthesis, 20:199
Thiosulfuric acid
 chemical prop, 20:228
 structure, 20:227
Thiotepa triethylenethiophosphoramide
 alkylating agent, S:83
2-Thio-2,4-thiazolidinedione, 10:915
Thiouracil
 antithyroid, 20:268
6-Thiouramil
 from uric acid, 21:108
Thiourea, 2:15, 144, 414
 action with thiodiglycol, 10:648
 antithyroid agent, 20:270
 as raw material in amino resin manuf, 2:227
 in prepn of liquid glue, 10:616
 photography, 15:386
 poly(vinyl alcohol) copolymer, 21:358
 prepn, 4:372
 prepn from cyanamide, 6:561
 reaction with aziridines, 11:534
 reaction with ethylene oxide, 8:531
 reaction with epoxides, 8:416
 thiol synthesis, 20:213
Thiourea-formaldehyde resins, 2:231, 233, 235
 for flame resistance for fibers, 9:312
Thiowolframates, 22:354
2-Thioxo-3-thiazolidinepropionitrile, 6:649
Thiuram disulfides, 17:514
 as accelerators, 17:510

Thixotropic binders
  metallic soaps as, *4*:33
  of inks, *11*:613
Thixotropy, *17*:427
Thomsenolite, *9*:534
Thomson effect, *20*:148
  antihistamine, *11*:41
Thonzylamine.HCl
Thoria, *8*:415, 419; *16*:432; *19*:378
  in ceramics, *4*:773
  in dispersion-hardened alloys, *11*:16
Thorianite, *20*:249
Thorite, *20*:248
Thorium, *20*:**248**. (See also *Actinides.*)
Thorium C, *17*:9
  breeder reactors, *21*:29
Thorium benzenearsonate, *20*:252
Thorium carbides, *4*:70, 86; *20*:254. (See also *Cemented carbides.*)
Thorium carbonates, *20*:254
Thorium chloride, *20*:254
Thorium compounds, *20*:254-259
  amido compounds, *20*:472
Thorium compounds, *20*:**254-259**
Thorium dioxide
  deposition on tungsten, *8*:30
  in uranium-233 prod, *14*:82
Thorium fluoride, *20*:255
  crystal structure, *1*:366
  phys prop, *1*:366
Thorium hydride, *11*:207
Thorium hydrides, *20*:255
Thorium hydroxide, *20*:255
  "hexamine" reaction, *20*:252
Thorium iodate, *20*:255
  thorium separation, *20*:251
Thorium iodide, *20*:255
Thorium nitrate, *9*:527; *20*:248, 256
  commercial, *20*:250
  uses, *20*:257
Thorium nitride, *13*:821
Thorium *m*-nitrobenzoate, *20*:252
Thorium oxalate, *20*:256
Thorium oxide. See *Thoria.*
Thorium oxide, *20*:248, 256
  dopant in wolfram, *22*:341
  expansion coefficient, thermal, *4*:805
  in electron tubes, *8*:7, 11, 16
  phenyl ether, *15*:167
  thermal conductivity of, *4*:803
Thorium oxysulfide, *20*:255
Thorium peroxide, *20*:251
Thorium phosphate, *20*:256

Thorium pyrophosphate, *20*:250, 256
Thorium sebacate, *20*:252
Thorium sulfate, *20*:256
Thorium sulfide
  as ceramic material, *4*:774
Thorium-uranium carbide
  as reactor fuels, *14*:87
Thorium-232, *17*:9
Thoron, *20*:252
L-Threaric acid, *19*:734
L-Threitan, *1*:577
Threitols D, *1*:571
  prepn, *1*:570, 571
Threonine, *2*:157
DL-Threonine, *6*:468
Threo-2-(5-nitro-2-furyl)-5-(*p*-nitrophenyl)-2-oxazoline-4-methanol, *20*:87
Thrombin, *3*:576; *5*:590, 595
  phys prop, *5*:589
Thrombolysis, *5*:592
  inactivators of, *5*:599
Thromboplastin, *5*:589, 595
Thucholite, *21*:8
Thuja, *14*:213
Thujic acid, *16*:690
Thulium
  electronic configuration, *1*:360
  in nuts, *14*:131
  ionic radii, *1*:370
Thulium-170, *17*:89, 127
Thyme, *9*:352; *14*:213
  essential oils as flavors, *9*:365, 375
Thyme oil
  as denaturant, *8*:451
Thymidine, *20*:78, 80
Thymol, *2*:628; *6*:442; *9*:354; *19*:829
  as denaturant, *8*:451
  chlorination, *19*:402
Thymol ether, *9*:354
Thymolphthalein, *11*:552
Thyrocalcitonin, *20*:266
  biological activity, *20*:267
Thyroglobulin, *20*:260
  assay, *20*:264
  prop, *20*:261
  purification, *20*:263
Thyroid and antithyroid preparations, *20*:**260**
Thyroid preparations, *20*:260
Thyrotropic hormone, *20*:268
Thyrotropin, *11*:62
  assay, *11*:62
  general properties, *11*:62
  methods of isolation, *11*:62

Thyroxine, *2*:188, 189; *20*:260
  assay, *20*:264
  properties, *20*:262
  synthesis, *20*:263
D-Thyroxine
  commercial, *20*:265
D-Thyroxine
Tielite, *20*:413
Tiglic aldehyde, *16*:140
Tigogenin, *18*:866
  in steroid synthesis, *11*:84
Tile
  carbon tetrachloride in recovery of, *5*:137
  clays used in, *5*:566
  coatings, *13*:252; *20*:288
  copper alloys with, *6*:194, 205, 210, 250
  effect on brasses, *6*:229, 241
  electrodeposition of zinc, *22*:599
  foil in phonograph records, *15*:225
  in dental amalgam, *6*:807
  in electroplating, *8*:39, 65
  in phosphor, *12*:622
  in seawater, *14*:158
  pewter, *20*:291
  phosphorus alloys, *15*:294
  silane derivs, *18*:203
Tile adhesives, *1*:398
Tin, *20*:**273**
  hot-dipped coatings, *13*:258
  manganese compounds, *13*:4
  organometallics in phonograph records, *15*:229
  pigments, *15*:550
  removal from lead ore, *12*:226
  type metal, *16*:497; *20*:291, 757
Tin(IV) oxide
  prepn, *20*:310
Tin and tin alloys, *20*:**273**
Tin-antimony-lead alloys, *12*:264
Tin babbitts, *3*:272, 274, 275
Tin bronzes, *6*:194, 205, 210
  casting of, *6*:252
  properties of, *6*:253, 256, 258
Tincalconite. See *Sodium tetraborate pentahydrate.*
Tin compounds, *20*:**304**
  in ceramic pigments, *5*:855
Tinctorial power
  of food colors, *4*:65
Tincture of cantharides, *10*:802
Tin fluoborate, *9*:567
Tin fluoborate electroplating solution, *8*:65
Tin hydrates, *20*:310

Tin hydride, *20*:312
  organotin reactions, *20*:306
Tin-lead alloys, *12*:262
Tin-nickel alloy
  electroplating, *8*:68
Tin oxide, *20*:317
  ceramic glaze, *20*:304
  in coatings on glass, *10*:592
  toxicology, *20*:323
Tinplate
  use as aerosol containers, *1*:475
Tin stannate electroplating solution, *8*:65
Tin sulfate electroplating solution, *8*:65
Tin tetrachloride, *20*:316
  as catalyst, *8*:529; *15*:162
  prepn, *20*:304
Tin tetrafluoride, *9*:507
Tin tetrahydride, *20*:312
Tin tetraiodide
  stannic acetate from, *20*:315
Tin-zinc alloys
  electroplating, *8*:68
Tire cords, *20*:**328**
Tires
  advantages of styrene butadiene rubber, *7*:689, 691
Tires, rubber
  acetaldehyde, *1*:82
  acid corrosion rates, *20*:374
  airframes, *20*:377
  aldehydes, *1*:641
  calcium zirconate reaction, *22*:656
  carbon black used in, *4*:265, 273
  in ceramics, *4*:773
  in prepn of trifluoroboronine, *9*:556
  in synthetic gems prod, *10*:509, 510
  metal alkoxides as catalyst in, *1*:843
  methylation of phenol, *S*:273
  non-block copolymers in automotive, *S*:924
  properties, *10*:513
  radial, *20*:330
Tiron. See *3,5-Disulfopyrocatechol.*
Tishchenko reaction. (See also *Tishchenko reaction, crossed.*)
Titanates
  as ceramic materials, *4*:775
Titania. (See also *Titanium oxide.*)
  catalyst for ethylbenzene, *19*:62
  slagceram nucleating agents, *S*:881
Titanium, *7*:48; *20*:**347**
  manganese compounds, *13*:4
  welding of, *20*:361; *22*:250
Titanium, *20*:**347**

Titanium(II) allyloxide, *S*:616
Titanium(II) benzyloxide, *S*:616
Titanium(II) *n*-hexyloxide, *S*:616
  dielectric properties of, *4*:821
  in glass-ceramics, *10*:548
  in papermaking, *14*:501
Titanium(III) phosphate, *20*:420
Titanium(III) sulfate, *20*:419
Titanium(IV) alkoxides, *20*:451
  prop, *20*:453
Titanium(IV) alkoxyhalides, *20*:463
Titanium(IV) alkoxyhalides, *20*:463
Titanium(IV) oxide, *20*:380
Titanium(IV) phosphates, *20*:420
Titanium(IV) sulfate, *20*:418
Titanium acetate chloride
Titanium acetylacetonate, *7*:74; *20*:481
  in producing flame-retardant textiles, *9*:305
Titanium acylates, *20*:468
Titanium alkoxides, *20*:425, 451
  of lower valency, *20*:470
  polymeric, *20*:456
Titanium alkoxyhalides, *20*:460, 462
Titanium alkyls, *20*:431
Titanium alloys, *20*:**347, 367**
Titanium alloys, *20*:**347**
  in prod flame-resistant textiles, *9*:305
Titanium amido compounds, *20*:471
Titanium and titanium alloys, *20*:**347**
Titanium anodes
  in chlorate manuf, *5*:54
Titanium boride, *3*:675, 676; *4*:74; *16*:432
Titanium borides, *20*:418
Titanium bromides, *20*:389
Titanium butoxide
  as ester-exchange catalyst, *16*:167
  paints, *20*:479
  vinyl acetate reaction, *20*:452
  water repellency, *20*:484
Titanium carbide, *4*:74, 75, 80; *16*:432; *20*:415
  hardness of, *S*:74
  in machining tools, *20*:572
  strength value, *4*:818
  thermal conductivity, *4*:803
Titanium carboxylate
  polymeric, *20*:484
Titanium chelates, *20*:464
Titanium chlorides
  equilibria between, *20*:387
Titanium 2-chloroethoxide, *20*:452
Titanium chlorotriacetate, *20*:468
  acylates, *20*:468

alkali metal titanates, *20*:410
alkoxides, *20*:425, 470
Titanium compounds, *20*:**380**
  alkaline earth titanates, *20*:411
  amido compounds, *20*:471
  alkoxyhalides, *20*:460
  inorganic, *20*:380
  organotitanium compounds, *20*:425
Titanium diboride
  in opaque armor, *S*:151
Titanium dibromide, *20*:389
Titanium dichloride, *20*:387
  alcohol reaction, *20*:471
Titanium dichlorodiacetate, *20*:468
Titanium dioxide, *6*:357, 363, 804, 826, 836, 843; *7*:41; *9*:403; *20*:380, 392-410
Titanium diodide, *20*:390
Titanium dioxide, *20*:392410
  in acrylic fibers dyeing, *1*:327
  in chemical ware, *4*:835
  delusterant, *16*:13, 30, 49, 144
  pigment, *14*:466, 470, 516; *15*:496-500, 801; *20*:397, 400, 407-408
Titanium disilicide, *20*:417
Titanium disulfide, *20*:419
  amides from, *20*:472
  commercial alkoxides, *20*:475
Titanium ethoxide
  trivalent oxides, *20*:470
Titanium fluorides, *20*:388
  hexafluotitanates, *20*:388
Titanium hydride, *11*:207; *20*:381
  in electron tubes, *8*:4
Titanium hydroxide
  uranium extraction, *20*:391
Titanium iodides, *20*:389
Titanium isopropoxide
  Aquadrol" reaction, *20*:472
  chelates from, *20*:464
  water repellency, *20*:484
Titanium lactate, *12*:184, 465
Titanium monosulfide, *20*:420
Titanium nitride, *4*:74; *20*:416
  in ceramic pigments
  in enameling, *8*:158
  in makeup lotion, *6*:362
  as opacifier, *8*:167
Titanium oxide. (See also *Rutile; Titania.*)
  chloroform reaction with, *5*:122, 854
Titanium oxides, *20*:391
Titanium oxychlorides, *20*:388
Titanium oxyfluoride, *9*:683

Titanium-oxygen compounds, *20*:390

Titanium pentanedionate, *12*:157

Titanium peroxides, *20*:392

Titanium phosphate
  hydrous, *20*:420

Titanium phosphide, *20*:421

Titanium production, *20*:349

Titanium pyrophosphates, *20*:420

Titanium sesquioxide, *20*:390

Titanium silicides, *18*:131; *20*:417

Titanium subsulfide, *20*:420

Titanium sulfates, *20*:418

Titanium sulfides, *20*:419

Titanium tetrabromide, *20*:389
  acylates from, *20*:469

Titanium tetrachloride, *7*:66, 67, 74, 76, 677,
  686; *16*:517; *20*:350, 382-385
  acylates, *20*:468
  alkoxides from, *20*:425, 451
  catalyst, *15*:162
  ethylenediamine reaction, *20*:472
  in polymerization of ethylene, *8*:501
  in prepn of butyl chloride, *8*:492
  methylamine reaction, *20*:473
  reaction with ethyl ether, *8*:479
  titanium alkoxyhalides from, *20*:460
  titanium chelates, *20*:466
  titanium metal from, *20*:352
  Ziegler catalyst, *20*:385

Titanium tetra-*t*-butoxide, *1*:834

Titanium tetraethoxide, *20*:425
  prepn, *1*:833

Titanium tetrafluoride, *20*:388
  in prepn of trifluorosilane, *9*:507, 654, 683

Titanium tetraiodide, *7*:72, 73, 75; *20*:389

Titanium tetraisopropoxide, *20*:448; *S*:617.
  (See also *Tetraisopropyl titanate.*)

Titanium tetraphenoxide
  alkylation catalyst *S*:30

Titanium tribromide, *20*:389

Titanium tribromide hexahydrate, *20*:389

Titanium trichloride, *7*:66, 67, 73; *20*:385
  chelates, *20*:470
  commercial, *20*:386

Titanium trichloride hexahydrate, *20*:387

Titanium trifluoride, *9*:587; *20*:388

Titanium triiodide, *20*:390

Titanium trisulfide, *20*:419

Titanizing, *20*:478

"Titanocene" *S*:616

Titanocenes, *20*:426

Titanous phosphate, *20*:420

Titanous sulfate, *20*:419

Titanyl sulfate, *20*:380, 418

Titrators
  automatic, *7*:748

TMP. See *Trimethylolpropane."*

TNT, *8*:611
  explosive characteristics, *8*:633
  in binary explosives, *8*:627
  in primary compositions, *8*:653
  manuf (Biazzi process), *8*:595
  sensitivity characteristics, *8*:632
  stability characteristics, *8*:634

Tobacco, *20*:**503**
  maleic acid used in, *12*:847
  use of alcohol in, *8*:449

Tobacco industry
  humidity conditions within, *1*:487

Tobacco processing
  use of glycerin in, *10*:628

Tobacco sprays and flavors

Tobernite, *21*:8

Tobias acid, *7*:478; *13*:713; *15*:574

α-Tocohydroquinone, *21*:577

Tocol, *21*:574

Tocopheramines, *21*:577

α-Tocopherethoxide, *21*:577

Tocopherol
  as antioxidant, *2*:601
  in nuts, *14*:131
  in parenterals, *15*:128

α-Tocopherol, *2*:598; *21*494, 574

β-Tocopherol, *21*:574

γ-Tocopherol, *21*:574

δ-Tocopherol, *21*:574

Tocopherols, *8*:781
  in food preservation, *10*:14
  in soybeans, *18*:601

α-Tocopheronic acid, *21*:576

α-Tocopherothiol, *21*:577

α-Tocopheryl acetate, *21*:575
  pharmaceutical-grade, *21*:580

α-α-Tocopheryl acid succinate, *21*:576

α-Tocopheryl acid succinate, *19*:146
  in vitamin capsules, *21*:583

α-α--Tocopheryl acid succinate, *21*:576

α-Tocopherylhydroquinone, *21*:577

α-Tocopherylquinone, *21*:577

α-Tocoquinone, *21*:576

Tocotrienol, *21*:574

α-Tocotrienol, *21*:574

Tolazamide, *19*:259

Tolazoline, *4*:519

Tolbutamide, *19*:259; *S*:12

Tolidine, 7:535
   pigments, 15:572
m-Tolidine, 3:415
o-Tolidine, 3:414
Tolidine diisocyanate, 12:47
o-Tolidine-6,6'-disulfonic acid, 3:418
Tolidines, 13:852
Tollens reagent
   in acetaldehyde determination, 1:90
Tolonium chloride, 5:603
α-Tolualdedyde, 3:366
o-Tolualdehyde, 6:453
p-Tolualdehyde, 3:366
α-Toluamide, 15:213
p-Toluanilide, 2:67
   as denaturant, 8:451
Tolu balsam, 14:725
Toluene, 6:434, 439, 673; 12:292; 20:**527**
   benzaldehyde manuf from, 3:362
m-Toluenediamine, 7:539
   dye for furs, 10:306
   in hair coloring, 10:795
   manuf, 2:82
p-Toluenediamine
   photochlorination, 15:338
Toluenediamines, 15:216
   ammonolysis of, 2:344
Toluene-2,4-diamine, 15:217, 222
   diazotization, 15:219
   dyes, 19:432
   manuf, 2:90
Toluene-3,4-diamine, 15:217
Toluene diisocyanate, 7:693; 21:57
   in flexible foams, 12:58, 62
   manuf, 12:56
   phenolic resins, 15:206
Toluene-2,4-diisocyanate, 7:413; 12:46, 47;
   15:794
2,6-Toluene diisocyante, 12:46
   in prepn of TNT, 8:613
2,4-Toluenedisulfonic acid
   oxidation, 20:700
Toluenes
   alkylated, 20:562
   as raw material in amino resin manuf, 2:227
o-Toluenesulfonamide
   oxidation with KMnO₄, 13:23
   saccharin, 19:595, 597
(o,p)-Toluenesulfonamide
   plasticizer, 15:764, 770
p-Toluenesulfonamide, 2:459; 6:650; 19:257
   chlorine reaction with, 4:923
o-Toluenesulfonamide resin, 9:497

p-Toluenesulfonamide resin, 9:497
Toluenesulfonamides
p-α-Toluenesulfonamidobenzoic acid, 19:258
p-Toluenesulfonate
   thiol synthesis, 20:213
Toluenesulfonic acid, 19:293
   alkylation of phenols, S:30
   as catalyst, 13:827
   as calalyst, 12:54
   from toluene, 20:531
   menadione deriv, 21:597
   methacycline synthesis, 20:8
   polyesterification catalyst, 20:798
   steroid synthesis, 18:875
p-Toluenesulfonic acid, 7:412, 479; 8:326, 359,
   429; 10:161; 19:312, 317, 318, 489; S:271
o-Toluenesulfonic acid, 7:479
p-Toluenesulfonic acid hydrate
   as catalyst in alkylation of phenol, 1:895
p-Toluenesulfonimidazolide, 8:495
α-Toluenesulfonyl chloride, 19:258
p-Toluenesulfonyl chloride, 2:881
   reaction with ethylenediamine, 7:30
p-Toluenesulfonyl isocyanate, 19:330
   phys prop, 1:224
   prepn of, 4:923
   steroid synthesis, 18:870
p-Toluenethiol
   addition reactions, 20:209
α-Toluenethiol, 5:283
m-Toluic acid
   phys prop, 1:238
o-Toluic acid
   phys prop, 1:238
p-Toluic acid
   phys prop, 1:238
α-Toluic acid, 15:213. (See also *Phenylacetic
   acid.*)
Toluidine
   phenylenediamines, 15:220
m-Toluidine, 2:421
o-Toluidine, 6:439
   in prepn of p-(4-amino-m-toluidino)phenols,
   2:219
p-Toluidine, 2:414
   dyes, 19:433
   sulfonic acids, 19:318
Toluidine reds, 15:573
   pigment, 14:474
Toluidines, 13:852
p-Toluidine-m-sulfonic acid, 15:575
p-Tolunitrile, 2:344
Toluylene Red (CI Red 46), 7:535

Tolyl aldehyde
  as synthetic flavor, 9:354
o-Tolylarsenic tetrachloride, 2:728
Tolyl cinnamates, 5:520
p-Tolylenediamine
Tolylene diisocyanate
  coatings, 21:101
  in polyester foams, 21:87
  in prepn of coatings, 4:532
  from toluene, 20:561
Tolylmercuric acetate, 10:222
m-Tolyl methyl ether, 7:490
p-Tolylphenyltitanium triisopropoxide, 20:433
p-Tolyl(sulfonylamido)diphenylamine, 17:600
Tomatidine, 18:842
Tomorin, 15:919
Toner, 2:907
  pigments, 15:565
Toney Red, 5:870, 783
Toning blue, 15:539
Tonka beans
  cobalt alloys, 5:734
  source of coumarin, 6:425
Tool materials for machining, 20:**566**. (See
  also *Cemented carbides.*)
Toothpaste
  in aerosol form, 1:479
  use of glycerin in, 10:628
Topaz
  x-ray spectroscopy, 22:461
Torpex, 8:630
  explosive characteristics, 8:633
  sensitivity characteristics, 8:632
  stability characteristics, 8:634
Torr
  defined, 21:123
Torulopsis
  in alkane utilization, S:839
Tosyl cellulose, 9:307
p-Tosylhydrazone adamantanes, S:11
Tosyl methionine, 2:167
Totaquine, 1:783
Toughness, 13:187
  measurements of, 1:23
Toxaphene, 19:814
Toxicity, 11:595
Toxicology, industrial. See *Industrial toxicolgy.*
Toxisterol₂, 21:557
TPX Methyl-pentene polymer, S:800
Trace analysis
  in crystallography, 17:873

Trace elements
  cobalt in animal feeds, 5:746
  in fertilizers, 9:143-147
  in wood, 16:681
Tracheids
Tracking, 7:724
  classification, 8:176
  veterinary, 21:252
Trademarks, 20:**578**
Tragacanth
  in dentifrices, 6:849
  tablet binder, 15:121
Tranquilizers, 2:150; 11:508. (See also *Psychopharmacological agents.*)
Transacylases
Transcription process, S:841
Transducers
  barium titanate, 20:786
  ultrasonic, 20:786
Trans effect, 6:125
Transesterification
  classification, 8:176
  siloxanes, 18:250
Transfer coenzyme A
Transformers
  classification, 8:176
  use of sulfur hexafluoride, 9:670
Transglycosylases
Transistor
  junction, 17:856
Transition metals
Transition metals
  carbon monoxide compounds with. See *Car-
  bonyls.*
  olefin complexes, 14:319
  peroxy compounds of, 14:751
  silane compounds, 18:204
Transition-point
  of electric insulation, 11:779
Transmembrane phenomena, 7:3
Transpeptidase, 14:675
Transport
  matter, 22:686
Transportation, 20:**596**
Transport processes, 20:**610**
  rates, 20:598
Transuranium elements, 21:6
Transuraniums
  prodn, 14:83
Traps
  in vacuum technology, 21:137

Traube's rule, *1*:427

Travertine, *12*:416

Traxler-Schweyer
asphalt analysis, *2*:785

Tréparsol

Trehalose
in ceramics, *4*:772
in yeast, *22*:516

Tremolite, *19*:608, 610. (See also *Asbestos.*)

Trevorite, *13*:736

Triacanthine, *1*:764

Triacetate
dyeing with disperse dyes, *7*:542
prepn for dyeing, *7*:518

Triacetate fibers, *1*:109
optical brighteners for, *3*:747
prodn, *1*:112
prop, *1*:123
structure, *1*:123
trademarks, *1*:110

Triacetate yarns
dyeing, *1*:134
in propellants, *8*:660, 681
mechanical properties, *1*:129
uses, *1*:136

Triacetin, *8*:351, 363; *10*:630; *15*:751

Triacetone peroxide, *8*:488

Triacetonyl isocyanurate, *20*:664

Tri(acetoxymethyl) isocyanurate, *20*:665

1,2,5-Triacetoxypentane, *10*:247

Triacetyl cyanurate, *20*:666, 670

Triacetylenes
nomenclature, *11*:288

Tri-*O*-acetyl-D-glucosan, *19*:151

Triacetyl isocyanurate, *20*:666

Triacontaglycerol, *10*:630

Triacontanoic acid. See *Melissic acid.*

Triacontanoic acid, *8*:813

21-Triacontenoic acid, *8*:814

1,3,5-Triacrylylhexahydro-*s*-triazine
prepn from acrylonitrile, *1*:341

Trialkylaluminums
manuf, *2*:34

Trialkylboranes, *3*:707
boron trifluoride in prepn of, *9*:556

Triallylamine derivs
uses, *2*:124

Triallyl cyanurate
health factors, *1*:924
phys prop, *1*:917
polyester reactive diluent, *20*:822
prepn, *1*:923
uses, *1*:926

Triallyl isocyanurate, *20*:664, 670

Triallyl phosphate
in prepn of flame-retardant finish, *9*:308

Triallyl trimesate, *15*:480
corticosteroid analog, *11*:90

Triamcinolone

Triamcinolone acetonide
veterinary drug, *21*:251

1,2,3-Triaminoanthraquinone. See *Amino-anthroquinones.*

1,3,5-Triaminobenzene, *16*:203

1,3,5-Triaminobenzene hydrochloride, *16*:205

Tri(aminocarbonylmethyl) isocyanurate, *20*:665

2,4,5-Triamino-6-hydroxypyrimidine, *10*:932

2,4,6-Triaminophenol, *16*:211

2,4,6-Triaminoresorcinol diethyl ether, *16*:212

2,4,6-Triamino-*s*-triazine. See *Melamine.*

4,4',4"-Triaminotriphenylmethane, *20*:675

Triaminotriphenylmethane dyes, *20*:708
dye manuf, *20*:708

2,5,8-Triaminotri-*s*-triazine. See *Melem.*

Triammino nickel trithiocarbamate, *13*:759

Triamterene, *7*:267

Triamyllead acetate, *12*:297

Triarylboranes, *3*:707
boron trifluoride in prepn of, *9*:556

Triarylmethane, *7*:471, 473, 533

1,3,5-Triazido-2,4,6-trinitrobenzene, *8*:592

1,3,5-Triazine, *6*:574; *10*:898, 930

s-Triazines
analysis of as herbicides, *22*:186
as herbicides, *22*:183
metabolism as herbicides, *22*:187
soil chemistry, *18*:533

s-Triazinethiols, *6*:567

Triazinetriol, *20*:662. (See also *Cyanuric acid."*)

s-Triazine-2,4,6-triol, *20*:662

s-Triazine-2,4,6(1*H*,3*H*,5*H*)-trione, - *20*:662

Triazol Dark Blue B, *2*:900

1,2,4-Triazole, *10*:896

Triazolines, *12*:826

Tribasic calcium phosphate
as dental abrasive, *6*:849

Tribenzene, *7*:191

Tribenzoate, *16*:192

Tribenzyl isocyanurate, *20*:664

Tribenzylstibine, *2*:582

Tribenzylstibine oxide, *2*:582

Triboelectricity, *2*:650

Triboluminescence, *12*:617

Tribromoacetic acid, *8*:420

Tribromoaniline
  from seawater, *14*:156
3,5,6-Tribromobenzotrifluoride, *9*:778
Tribromoethanoic acid. See *Tribromoacetic acid.*
Tribromoethanol
  as a hypnotic, *11*:510
  as anesthetic, *2*:402
  bird control, *15*:922
2,2,2-Tribromoethanol, *3*:780
Tribromofluoromethane, *9*:750
  in prepn of hexafluorobenzene, *9*:776
Tribromomethane, *3*:776, 780
Tribromomethyl phenyl ketone, *12*:122
Tribromophenol, *15*:158
2,4,6-Tribromophenol, *3*:778, 780; *15*:210
1,2,3-Tribromopropane, *3*:778
4,5,6-Tribromopyrogallol, *16*:192
3,4'5-Tribromosalicylanilide, *14*:549; *17*:735; *18*:429
Tribromothymol
  in embalming fluids, *8*:100
1,1,1-Tribromo-2-methyl-2-propanol, *3*:780
1,2,3-Tribromo-3-methylbutane, *1*:624
Tributoxytitanium, *20*:470
Tributyl acetylcitrate, *15*:751, 764
Tributyl antimonate, *2*:578
Tributylborane, *9*:836
Tri-*n* 1-butylborane, *3*:707
2,4,6-Tri-*t*-butylphenol, *5*:27
Tributyl citrate, *6*:804; *15*:751
Tri-*n*-butyl citrate, *5*:539
Tributyllead acetate, *12*:297
Tributyl phosphate, *9*:898; *15*:329
  as fungicide, *10*:232
  in paints, *14*:468
  in papermaking, *14*:520
  uranium extraction, *21*:17, 23, 27
Tri-*n*-butyl phosphate
Tributyltin acetate
Tributyltin benzoate, *20*:321
Tributyltin chloride, *20*:312, 320
Tributyltin fluoride, *20*:320
Tributyltin hydroxide
  as fungicide, *10*:232
  in Purex process, *14*:92
  plutonium separation, *15*:883
Tributyltin neodecanoate, *8*:854
Tributyltin oxide
  toxicology, *20*:324
Tributyltin salicylate, *20*:321
Tributyrin, *3*:882; *10*:619

Tricalcium aluminate
  in asbestos-cement board, *21*:608
  in cement formation, *4*:685, 691, 692
Tricalcium diorthophosphate, *15*:239
Tricalcium phosphate, *4*:4; *6*:513
  as emulsifier in cheese, *10*:8
  in food processing, *10*:8
  sugar refining, *19*:187
Tricalcium silicate
  in asbestos-cement board, *21*:608
  in cement formation, *4*:685, 691, 692, 700
Tricamba
  herbicide, *22*:200
Tricanthine, *15*:681
  from citric acid, *5*:527
  properties, *1*:232
Tricarballylic acid
Tricarbethoxymethyl isocyanurate, *20*:664
Tri(2-carbmethoxyethyl) isocyanurate, *20*:665
Tricarbocyanines, *6*:616
1,3,5-Tricarboxybenzene, *15*:478
Tricarboxylic acid cycle. See *Krebs cycle.*
Tri(2-carboxymethyl) isocyanurate, *20*:664, 665
Trichloroacetanilide
  as a herbicide, *22*:199
  decomposition, *9*:590
  in prepn of trichloroacetanilide, *1*:150
  phys prop, *1*:235
  prepn from acetanilide, *1*:150
Trichloroacetatozirconic acid, *22*:636
Trichloroacetic acid, *6*:692; *8*:416, 418, 419
Trichloroacetonitrile, *10*:153
Trichloroacetyl chloride
  from hexachloroethane, *5*:167
  from pentachloroethane, *5*:165
  from tetrachloroethylene, *5*:198
2,4,6-Trichloroaniline, *2*:415
Trichlorobenzene, *7*:543
Trichlorobenzenes, *5*:260
  from monochloroacetylene, *5*:204
2,3,6-Trichlorobenzoic acid
  as a herbicide, *22*:200
Trichlorobenzotrifluoride, *9*:778
2,3,6-Trichlorobenzyl chloride
  herbicide, *22*:199
Trichlorobenzyl cyanide
  synthesis of fenac, *22*:199
β-Trichloroborazine, *3*:730, 732
1,1,3-Trichlorobutane
  from vinyl chloride, *5*:172
1,2,4-Trichloro-2-butene
  from chloroprene, *5*:217

1,3,4-Trichloro-2-butene, *1*:605

2,3,4-Trichloro-1-butene
  from chloroprene, *5*:217

Trichlorobutanes
  chloroprene from, *5*:223

2,2,3-Trichlorobutyraldehyde
  as hypnotic, *11*:513

4,4,4-Trichlorobutyronitrile, *6*:642

3,4,4'-Trichlorocarbanilide, *2*:637; *18*:429

Trichlorocyanuric acid, *2*:616

1,3,5-Trichloro-2,4-dioxohexahydrotriazine, *4*:921
  hexamethylenetetramine adduct
  as hypnotic, *11*:511

Trichloroethane
  catalytic cracking of, *21*:276
  dehydrochlorination of, *21*:276
  1,1-dichloroethylene from, *5*:179
  toxicity of, *5*:156

1,1,1-Trichloroethane, *5*:154-157; *9*:841;
  10:783; *15*:337

1,1,2-Trichloroethane, *5*:157-159; *8*:487;
  16:169
  from 1,2-dichloroethane, *5*:151
  1,2-dichloroethylene from, *5*:180-182

Trichloroethanoic acid. See *Trichloroacetic acid.*

Trichloroethanol-hexamethylenetetramine adduct, *11*:511

Trichloroethyl acetate
  vinylidene chloride from, *21*:276

β-Trichloroethyl-α-ethyl-*n*-butyrate
  as a hypnotic, *11*:510

Trichloroethylene, *5*:183-195
  action on triacetate fibers, *1*:133
  as a drycleaning solvent, *7*:315
  as anesthetic, *2*:400
  in reactor technology, *14*:90
  in refrigeration, *17*:310
  manuf from acetylene, *1*:206
  pentachloroethane from, *5*:165
  stain solvent, *18*:661
  tetrachloroethylene from, *5*:199
  toxic effects of, *5*:93, 97, 192

1,1'2-Trichloroethyl ether, *8*:487

2,2,2-Trichloroethyl urethan
  as hypnotic, *11*:518
  blowing agents in rigid foam, *21*:91

Trichlorofluromethane, *9*:871; *10*:783
  propellent, *9*:747

Trichlorogermane, *10*:523

Trichloroisocyanurate, *20*:669

Trichloroisocyanuric acid, *4*:908, 912, 913,
  918; *20*:663
  as bleaching agent, *3*:558
  as swimming pool water disinfectant, *22*:126

3,3,3-Trichlorolactic acid
  uric acid from, *21*:107

Trichloromelamine, *4*:908, 921

Trichloromethane. See *Chloroform.*

Trichloromethanesulfenyl chloride, *19*:374
  from carbon disulfide, *4*:372

Trichloromethanesulfonyl chloride, *19*:374

Trichloromethiazide
  as diuretic, *7*:263, 264

3,5,6-Trichloro-2-methoxybenzoic acid, *22*:200

α-(Trichloromethyl)benzyl alcohol, *3*:365

Trichloromethyl chloroformate
  from methyl formate, *4*:388

Trichloromethyl isocyanurate, *20*:665

n-Trichloromethylmercapto-4-cyclohexane-1,2-
  dicarboximide, *10*:780
  as germicide, *2*:617

Trichloromethyl perchlorate, *5*:75
  as fungicide, *10*:225

1,1,1-Trichloro-2-methyl-2-propanol, *15*:128
  as hypnotic, *11*:511

Trichloromethylsilane, *18*:233, 237

n-(Trichloromethylthio)-4-cyclohexene-1,2-di-
  carboximide

n-(Trichloromethylthio)phthalimide

N-Trichloromethylthiotetrahydrophthalimide,
  3:785

2,3,4-Trichloronitrobenzene
  prepn of 1,2,3-trifluorobenzene, *9*:784

Trichloronitromethane. (See also *Chloropicrin.*)
  from trichloroethylene, *5*:187
  from uric acid, *21*:109

2,4,5-Trichlorophenol
  extractive distillation agent, *22*:473
  manuf of, *22*:180

2,4,6-Trichlorophenol, *15*:210

Trichlorophenols, *5*:333-335

Trichlorophenothiazinones, *19*:426

2,4,5-Trichlorophenoxyacetic acid, *8*:417;
  15:678
  as a herbicide, *22*:181
  from 2,4,5-trichlorophenol, *5*:335

2,3,6-Trichlorophenylacetic acid
  as a herbicide, *22*:199

1,2,3-Trichloropropane, *7*:697
  in polysulfides, *16*:253
  polymer from, *16*:260

1,1,1-Trichloro-2-propanol
   from acetaldehyde, *1*:83
2,2,3-Trichloropropionaldehyde
   phys prop, *1*:259
2,2,3-Trichloropropionic acid
   as a herbicide, *22*:199
   cinnamaldehyde from, *5*:521
2,3,3-(Trichloropropyl)benzene
2,6,8-Trichloropurine
2,3,5-Trichloro-4-pyridinol, *16*:804
2,4,6-Trichloropyrimidine, *9*:799
Trichloropyrimidyl, *7*:632
Trichlorosilane, *6*:651; *9*:653; *12*:69
   cyanoethylation of, *6*:647
   prod, *18*:170
   silicochloroform, *9*:653
   silicon from, *18*:120
2-(Trichlorosilyl)propionitrile, *6*:651, 647
3-(Trichlorosilyl)propionitrile, *6*:647, 651
Trichlorosulfonium chlorosulfonate, *19*:405
α,α,α-Trichlorotoluene. See *Benzotrichloride.*
2,3,6-Trichlorotoluene
   oxidation with nitric acid, *22*:200
2,4,6-Trichloro-1,3,5-triazine, *9*:799
1,1,3-Trichlorotrifluoroacetone, *9*:755
   alcoholates, *9*:758
Trichlorotrifluorobenzene, *9*:791
   prepn, *9*:785
Trichlorotrifluoroethane, *9*:871
   as a drycleaning solvent, *7*:315
1,1,2-Trichlorotrifluoroethane
1,2,2-Trichloro-1,1,2-trifluoroethane
   from tetrachloroethylene, *5*:198
   in drycleaning, *7*:324
   manuf of polychlorotrifluoroethylene, *9*:832
1,1,2-Trichlorotrifluoro-1-propene
   prepn of trifluoroacetic acid, *9*:771
Tricholine citrate concentrate, *5*:407
Trichomycin, *16*:133, 138
   amebicides, *20*:82
Tricobalt tetroxide, *5*:741
Tricosanoic acid, *8*:813
12-Tricosanone. See *Laurone.*
n-Tricosoic acid, *8*:813
Tricresyl phospate, *6*:434, 442, 804
   as cellulosic plasticizer, *4*:657
   flame retardant, *15*:801
   plasticizer, *15, 722,744; 21*:331
Tricresyl phosphates, *12*:577; *19*:672
Tricyanoethenol, *6*:627
Tri(2-cyanoethyl) isocyanurate, *20*:665, 670
p-Tricyanovinylarylamines, *6*:628

2,4,6-Tricyano-1,3,5-trimethylbenzene, *6*:629
Tricyclene, *19*:814
Tricyclohexyltin hydroxide, *20*:321
Tricyclopentadiene, *6*:691
Tricyclopentadienylbismuth, *3*:546
Tricyclo(3.3.1.1^{3,7})decane, *S*:1
Tricylohexyltin chloride, *20*:320
1,13-Tridecanedial, *1*:639
Tridecanoic acid, *8*:812
   properties, *8*:852
Tridecanol." See *Tridecyl alcohol.*
1-Tridecanol. See *n-Tridecyl alcohol.*
2-Tridecanone, *12*:105
7-Tridecanone, *12*:104
Tridecyl alcohol, *1*:568; *14*:373
   in papermaking, *14*:520
   phys prop of synthetic, *1*:564
   phys prop, *1*:543
n-Tridecyl alcohol
Tridecylbenzenesulfonate, *19*:516
n-Tridecylic acid, *8*:812
Tridecyl phosphite
   in drilling fluids
   polystyrene stabilizer, *19*:90
n-Tridecylpolyoxyethylene ethanol
   in drilling fluids, *7*:304
Tri(3,5-dinitrobenzoate), *16*:192
Tridodecylamine, *2*:130
Tridymite, *17*:227; *18*:49, 53
Trielaidin, *8*:790
Triethanolamine, *6*:371; *19*:382, 403
   as gum, *10*:741
   HLB value, *8*:129
   phys prop, *1*:811
   in removal of carbon monoxide from am-
      monia synthesis, *2*:279
   in soaps, *18*:431
Triethanolamine stearate
   in makeup lotions, *6*:362
Triethanolamine titanate, *20*:467
   catalyst, *20*:485
   titanium alkoxides from, *20*:429
Triethanolamine trinitrate phosphate. See
   Trolnitrate.
Triethenoid acids, *8*:811
Triethoxymethane. See *Triethyl orthoformate.*
Triethoxysilane, *18*:218, 219
Triethylaluminum, *7*:70
   growth reaction, *2*:33
   manuf, *2*:35
   prepn, *1*:561; *2*:27
   in polymerization of ethylene, *8*:501

Triethylaluminum chloride, 7:75
  coordination catalyst, 7:677, 686
Triethylaluminum-titanium tetrachloride
Triethylamine, 2:123; 6:455, 610, 616; 12:91, 98
  cyclamate synthesis, 19:600
  manuf from ethyl alcohol, 8:455
  polyurethan catalyst, 21:91
Triethylamine-borane, 3:728
Triethylamine hydrofluoride, 9:695
Triethylamine oxide, S:40
Triethylammonium ethanesulfonate, 19:313
2,4,6-Triethylbenzoic acid
  prepn of esters of, 8:353
Triethylborane, 3:707
Triethyl citrate, 5:539; 7:543
  in food processing, 10:8
  in liquid egg treatment, 7:668
Triethylenediamine, 7:23; 12:50
  blowing agent, 21:91
  as curing agent, 7:37
  isocyanate catalyst, 21:71
  properties, 7:24
Triethylene glycol, 8:553
  by-products in prepn of ethylene glycol, 8:527
  for dehydrating, 10:455, 639, 648
Triethylene glycol diacrylate
  phys prop, 1:289
Triethylene glycol dinitrate, 8:607
  as cellulosic plasticizer, 4:657
  explosive characteristics, 8:633
  manuf, 8:595
  in propellants, 8:660
  sensitivity characteristics, 8:632
  stability characteristics, 8:634
Triethylene glycol di(2-ethylbutyrate), 10:648; 15:764
  as cellulosic plasticizer, 4:657
Triethylene glycol di(2-ethylhexanoate), 10:648
  as cellulosic plasticizer, 4:657
Triethylenemelamine, 2:248
Triethylene glycol dipelargonate
  alkylating agent, S:82
  piperazine reaction, 15:644
  succinamide condensation, 19:145
Triethylenetetramine, 2:349; 7:22, 23
  as liquid epoxy resin hardener, 8:309
  properties, 7:24
  reactions, 7:31
Triethylenetriamine, 11:526
Tri(2,3-epoxypropyl) isocyanurate, 20:664

Triethylgallium, 10:323
Tri(2-ethylhexyl)phosphate
  as cellulosic plasticizer, 4:657
Triethylimine borane
  prepn, 9:556
Triethyllead hydroxide, 12:289
Triethylmethyllead, 10:476
Triethyl orthoformate, 6:611; 8:375
Triethyl oxonium fluoborate, 9:569
Triethyl phosphate, 2:413; 12:91, 289
  as catalyst, 8:409, 428
  as cellulosic plasticizer, 4:657
  phenylenediamines, 15:218
Triethyl phosphite, 9:690
Triethyl sulfonium fluoborate, 9:570
3,3',9-Triethylthiacarboxyanine iodide, 16:290
Triethyltin chloride
  toxicology, 20:324
Triethyltin oxide
  toxicology, 20:324
Trifluoboric acid, 3:615
Trifluoroacetaldehyde, 1:98
Trifluoroacetamide, 9:751
Trifluoroacetic acid, 9:689, 691, 771, 779
  from benzotrichloride, 5:283
  siloxane oligomers, 18:250
Trifluoroacetic anhydride, 9:691, 751
Trifluoroacetyl fluoride, 9:509
2,4,5-Trifluoroaniline, 9:784, 785, 786, 787
1,2,3-Trifluorobenzene, 9:784
1,2,4-Trifluorobenzene, 9:784
  in prepn of 1,2,3,4-tetrafluorobenzene, 9:786, 787
1,3,5-Trifluorobenzene, 9:784
Trifluoroboroxine
  prepn, 9:556
α,α,α-Trifluoro-2,6-dinitro-N-methyl-p-to- luidine, 22:205
α,α,α-Trifluoro-2,6-dinitro-N,N-di-n-pro- pyl-p-toluidine, 22:205
α,α,α-Trifluoro-2,6-dinitro-p-toluidine, 22:206
3-Trifluoroemthyl-4-nitrophenol, 15:921
Trifluoroethanoic acid, 9:689
1H,1H Trifluoroethanol. See 2,2,2-Trifluoro-ethanol. 9:689
2,2,2-Trifluoroethanol, 9:689, 751
Trifluoroethanol toluenesulfonate ester
Trifluoroethylamine, 9:752
Trifluoroethyl chloride, 9:751
Trifluoroethyl iodide, 9:752
2,2,2-Trifluoroethyl vinyl ether, 2:400, 401; 9:752; 21:414, 424

Trifluoromethanesulfonic acid, *19*:312

1-(Trifluoromethyl)-heptadecafluorodecahydro-
naphthalene, *9*:688

Trifluoromethylhydrothiazide, *9*:781

4,4,4-Trifluoro-3-hydroxy-3-:trifluoromethyl)but-
yric acid, *9*:762

3,3,3-Trifluoro-2-hydroxy-2-(trifluoromethyl)-
propionic acid, *9*:757

Trifluoromethyl hypofluorite, *9*:509

Trifluoromethyl iodide, *9*:690, 700

(Trifluoromethyl)pentafluoroethylsulfur tetrafluoride, *9*:802

2-Trifluoromethylphenothiazine, *9*:781

Trifluoromethylsulfur pentafluoride, *9*:698,
802

Trifluoromethylsulfur trifluoride, *9*:804

Trifluoromethylthiazide, *9*:781

(Trifluoromethyl)undecafluorocyclohexane, *9*:688.
(See also *Perfluoro(methylcyclohexane.)*

2,4,5-Trifluoronitrobenzene, *9*:784

2,4,6-Trifluoronitrobenzene, *9*:785

α,α,α-Trifluoro-6-nitro-2-nitroso-*p*-toluidine, -
22:205

α,α,α-Trifluoro-5-nitro-*N*-propyltoluene-3,4-dia-
mine, *22*:207

Trifluoroperoxyacetic acid, *8*:266

2,4,5-Trifluorophenol, *9*:786

Trifluoropromazine, *9*:781

1,1,1-Trifluoro-2-propene
prepn of trifluoroacetic acid, *9*:771

3,3,3-Trifluoropropanoic acid, *9*:689

2,2,2-Trifluoropropionic acid. See *3,3,3-Tri-
fluoropropanoic acid.*

2*H*,2*H*-Trifluoropropionic acid. See *3,3,3 Tri-
fluoropropanoic acid.*

2,4,6-Trifluoropyrimidine, *9*:694, 799

Trifluorosilane, *9*:653

ω-Trifluorostearic acid, *9*:718

ω-Trifluorostearylamine, *9*:718

Trifluorosuccinic acid, *19*:137

4,4,4-Trifluoro-1-(2-thienyl)-1,3-butanedione,
20:252

α,α,α-Trifluorotoluene. See *Benzotrifluoride.*

2,4,6-Trifluoro-1,3,5-triazine, *9*:799, 800

Triflupromazine, *2*:150

Trifluralin
herbicide, *22*:205
soil chemistry, *18*:536

Trigermane, *10*:523

Trigermylphosphine, *18*:201

Triglycerides, *8*:776; *10*:619, 630
in fish, *9*:320

to prepare monoglycerides, *8*:360
prepn, *8*:363

Triglycine
as ferroelectric, *9*:5
manuf, *2*:350

Triglycine fluoroberyllate

Triglycine sulfate
as ferroelectric, *9*:5

Trigonelline, *1*:765; *21*:512, 535
in coffee, *5*:751, 752

Tri-*n*-hexyl isocyanurate, *20*:665

Trihydrocarbylboranes, *3*:707

2,3,4-Trihydroxyacetophenone, *16*:198

2,4,6-Trihydroxyacetophenone, *16*:208

1,2,4-Trihydroxyanthraquinone. See *Purpurin.*

2,4,6-Trihydroxybenzaldehyde, *16*:203

1,2,4-Trihydroxybenzene. See *Hydroxyhydroquinone.*

1,3,5-Trihydroxybenzene. See *Phloroglucinol.*

1,2,3-Trihydroxybenzene. See *Pyrogallol.*

Trihydroxybenzene
in soil, *18*:533

sym-Trihydroxybenzene, *16*:201

2,3,4-Trihydroxybenzoic acid, *16*:193

2,4,6-Trihydroxybenzoic acid, *16*:208

3,4,5-Trihydroxybenzoic acid, *15*:681; *16*:196.
(See also *Gallic acid.*)

1,2,4-Trihydroxybutane, *1*:595, 603. (See also
1,2,4-Butanetriol.)

2',4',5'-Trihydroxybutyrophenone, *2*:601

Tri(2-hydroxy-3-chloropropyl) isocyanurate,
20:664

Tri(2-hydroxyethyl) isocyanurate, *20*:664, 670

2,6,7-Trihydroxyfluorones, *16*:199

1,2,6-Trihydroxyhexane. See *1,2,6-Hexanetriol.*

Trihydroxyindole
in analysis of epinephrine, *8*:235

Tri(hydroxymethyl) aminomethane, *18*:660

1,3,8-Trihydroxy-6-methylanthraquinone. See
Emodin.

9,10,16-Trihydroxypalmitic acid, *18*:24

2,6,8-Trihydroxypurine. See *Uric acid.*"

2,5,7-Trihydroxypurine, *10*:913

Trihydroxystearic acid
in cork, *6*:284

Trihydroxystearin
from castor oil, *4*:531

Tri(12-hydroxy)stearin, *22*:164

2,4,6-Trihydroxy-*s*-triazine, *20*:662

2,4,6-Trihydroxy-1,3,5-triazine, *9*:800

Triiodoacetic acid, *8*:421

Triiodoethanoic acid. See *Triiodoacetic acid.*

Triiodomethane, *11*:864
3,3',5-Triiodothyronine, *2*:189
3,5,3'-Triiodothyronine
    properties, *20*:262
    synthesis of, *20*:264
Triiron tetroxide, *12*:37, 38; *S*:689. (See also
    Iron compounds.)
Triisobutylaluminum, *7*:74,75,76
    manuf, *2*:35
    prepn, *2*:27
    thermal decomposition, *2*:30
Triisobutylene oxide, *8*:249
Triisopropanolamine, *1*:810
    phys prop, *1*:811
Triisopropylene, *6*:544
Triisopropyl phosphate, *8*:488
2,4,6-Triisopropyl-1,3,5-trioxane
    from isobutyraldehyde, *3*:868
Trilaurin, *8*:790, 791
Trilaurylamine, *15*:883
Trilead tetroxide, *12*:275; *15*:521, 536
Trilinolein, *8*:790
    in linseed oil, *7*:405
Trilithium phosphate
    as catalyst in glycerol synthesis, *10*:624
Trimellitic acid and anhydrides, *15*:473-478
Trimellitic anhydride, *16*:305
Trimellitic anhydride acid chloride, *S*:751
Trimellitic imide, *15*:475
Trimeprazine tartrate
    antihistamine, *11*:41
Trimer acid
    prepn, *8*:820, 847
Trimesic acid, *15*:473, 478
    derivatives, *15*:480
    economics, *15*:479
    manuf, *15*:479
    properties, *15*:478
    specifications, *15*:479
Trimesoyltrichloride, *15*:478
Trimetaphosphimic acid
    as a fertilizer, *S*:362
Trimethadione
    as anticonvulsant, *11*:522
Trimethaphan
    as ganglionic blockader, *4*:519
Trimethidinium
    as ganglionic blockader, *4*:519
Trimethine cyanines, *6*:611
3,4,5-Trimethoxybenzaldehyde, *16*:197
3,4,5-Trimethoxybenzoate, *16*:197
3,4,5-Trimethoxybenzoic acid, *16*:197
Trimethoxyboroxine, *3*:664, 672

Trimethoxybutane, *21*:415
3,4,5-Trimethoxycinnamic acid, *16*:197
3,4,5-Trimethoxy-*N*-(β-diethylaminoethyl)b-
    enzamide *16*:197
1,1,2-Trimethoxyethane
    paint remover, *14*:487
Trimethoxymethane. See *Trimethyl ortho-
    formate.*
2,3,4-Trimethoxy-10-methylacridone, *1*:771
2,4,6-Trimethoxypentafluoropyridine, *9*:799
3,4,5-Trimethoxyphenethylamine. See *Mesca-
    line.*
Trimethylacetic acid. See *Pivalic acid.*
1,1,1-Trimethylacetone, *12*:104
Trimethylacetonylammonium chloride, *5*:411
1,3,5-Trimethyladamantane, *S*:4
Trimethylallyl isocyanurate, *20*:664
Trimethylaluminum, *2*:120
    manuf, *2*:35, 36, 345
    titanium sandwich compounds, *20*:440
Trimethylamine, *19*:491
    choline salts from, *5*:407
    cyclamate synthesis, *19*:600
Trimethylamine-borane, *3*:728
Trimethylamine gallane, *10*:323
Trimethylamine oxide
    in fish, *9*:319, 570
    prepn of, *S*:32
Trimethylarsine, *2*:728
1,3,5-Trimethylbenzene, *15*:861; *16*:300
Trimethylbenzenes, *16*:300
    acetyl chloride condensation, *15*:482
    oxidation, *15*:476
2,4,6-Trimethylbenzoic acid
    prepn of esters, *8*:353
1,3,3-Trimethyl-2-[2-[(2-methyl-6-benzothia-
    zolyl)amino]vinyl]-3H-indolium hydrogen
    sulfate, *20*:194
Trimethylbismuth, *3*:543
Trimethylborane, *3*:707
Trimethyl borate, *9*:785
Trimethyl buten. See *Tripten.*
Trimethylchlorosilane, *19*:165
Trimethyl cyanurate, *20*:664, 666
1,1,4-Trimethylcycloheptone, *19*:815
3,5,5,-Trimethyl-2-cyclohexen-1-one, *1*:166;
    2:348; *12*:137; *S*:273. (See also *Isophorone.)*
3,3,5-Trimethylcyclohexyl acrylate
    phys prop, *1*:289
Trimethyldihydroquinoline, *12*:116
2,2,4-Trimethyl-1,2,dihydroquinoline, *7*:40, 41,
    46

1,3,7-Trimethyl-2,6-dihydroxypurine. See *Caffeine.*

2,2,4-Trimethyl-1,3-dioxane, *10*:663

Trimethylene bis-4-formylpyridinium bromide dioxime, *15*:323

Trimethylene bromide. See *1,3-Dibromopropane.*

Trimethylene bromide, *3*:778
    in prepn of glutaric acid, *10*:618

Trimethylene chlorobromide. See *1-Bromo-3-chloropropane.*

Trimethylene chlorobromide, *3*:774

Trimethylene chlorohydrin, *5*:304, 310

Trimethylene glycol
    prepn from ethylene, *8*:502

Trimethylene glycol diacetate
    prepn, *8*:502
    trimethylene chlorohydrin from, *5*:310

1,3,3-Trimethyl-2-methylene indoline, *16*:294

Trimethylene oxides, *10*:907

Trimethylene sulfate, *19*:488

Trimethylene sulfide, *10*:898

Trimethylenimine, *11*:526

Trimethylethoxyammonium fluoborate, *9*:570
    dehydrated to, *4*:395

Trimethylethylene. See *2-Methyl-2-butene.*

Trimethylgallane, *10*:322

Trimethylgallium, *10*:323

Trimethylguanidine
    prepared from guanidine, *10*:736

4,6,6,-Trimethyl-3-hepten-2-one, *12*:129

3,5,5-Trimethylhexane-1-ol
    titanium alkoxide from, *20*:483

3,5,5-Trimethyl-2-hexen-1-one, *12*:106, 113. (See also *Isophorone.*)

Trimethylhydroquinone
    phytol condensation, *21*:579

Trimethyl(2-hydroxyethyl)ammonium hydroxide. See *Choline.*

1,3,3-Trimethyl-Δ,α-indolineacetaldehyde,- *20*:195

2,2,4-Trimethyl-3-hydroxypentanal, *3*:868, 875

2,2,4-Trimethyl-3-hydroxypentanoic acid, *3*:875

6,7,9-Trimethylisoalloxazine, *17*:446

Trimethyl isocyanurate, *20*:664

1,1,3-Trimethyl-2-(γ-ketobutenyl)-2-cyclohexene-. See *β-Ionone.*

s-Trimethylmelamine-formaldehyde, *2*:248

Trimethylmethoxyammonium iodide, *S*:39 hydrogen sulfate, *20*:194

Trimethyl-(3-methyl-3-hydroxy-1-butynyl)silane, *1*:623

5,5,6-Trimethyl-2-norbornanol acetate, *19*:831

1,3,3-Trimethylnorcamphor, *12*:106

Trimethylolaminomethane, *7*:412

Trimethylolethane." See *2-Hydroxymethyl-2-methyl-1,3-propanediol.*
    polyesters, *21*:62

"Trimethylolethane"

1,1,1-"Trimethylolethane"

Trimethylolethane trinitrate
    explosive characteristics, *8*:633
    manuf (Biazzi process), *8*:595
    prepn and prop, *8*:607
    sensitivity characteristics, *8*:632
    stability characteristics, *8*:634

Trimethylolmelamine
    flame retardant, *S*:954
    in leather treatment, *12*:330

s-Trimethylolmelamine, *2*:232

Trimethylolpropane. See *2-ethyl-2-hydroxymethyl-1,3-propanediol.*

1,1,1-Trimethylolpropane
    polyurethans from, *21*:61

1,2,3-Trimethylolpropane tris(perfluorobutyrate), *9*:729

Trimethyl orthoformate
    prepn of methyl ether, *8*:476

Trimethyl orthoformate (Trimethoxymethane) *8*:375

3,5,5-Trimethyloxazolidinedione, *11*:522

1,3,3-Trimethyloxindole, *16*:295 Camphor.)

1,7,7,Trimethyl-2-oxobicyclo [2,2,1] heptane, *12*:106. (See also *Camphor.*)

Trimethyloxosulfonium iodide, *19*:326

6,10,14-Trimethyl-2-pentadecanone, *21*:588

2,2,4-Trimethylpentane. See *Isooctane.*

2,2,4-Trimethyl-1,3,-pentanediol, *3*:873, 875; *10*:678
    isobutyrate benzoate, *15*:754

2,2,4-Trimethylpentan-1-ol
    polyesters, *20*:490

2,4,4-Trimethyl-2-pentene, *10*:475

Trimethylphloroglucinol, *15*:372

Trimethylphosphine, *15*:299

Trimethylphosphine sulfide, *15*:299

Trimethyl phosphite
    7-dehydrocholesterol synthesis, *21*:561

Trimethylphosphonium chloride, *15*:299

2,5,6-Trimethyl-3-phytyl-1,4-hydroquinone dinicotinate, *21*:524

1,2,4-Trimethylpiperazine
    blowing agent, *21*:91

Trimethyl-p-rosaniline, *9*:896

Trimethylsilyl azide, *18*:188

Trimethylsilylmethanol, *6*:644

3-[(Trimethylsilyl)methoxy]propionitrile, 6:644
Trimethylstibine, 2:582
Trimethylstibine cyanobromide, 2:582
Trimethylstibine dichloride, 2:582
Trimethylstibine dihydroxide, 2:582
Trimethylstibine hydroxychloride, 2:582
Trimethyltitanium, 20:442
2,3,6-Trimethyl-2,3,6-trihydroxy-4-heptyne, 1:623
1,2,7-Trimethylxanthine, 19:751. (See also Caffeine.)
2,2,4-Trimethyl-1-pentanol, 1:567
  phys prop, 1:564
Trimyristin, 8:790, 791
Trinidad asphalt, 2:764
2,4,6-Trinitroaniline, 2:425
  in hair coloring, 10:795
Trinitrobenzene, 8:632-634
  vulcanizer, 17:568
1,3,5-Trinitrobenzene, 2:425; 11:552
  from benzene, 3:374
  as explosive, 8:616
2,4,6-Trinitrobenzoic acid, 16:203
  prepn, 8:612
2,4,6-Trinitrobenzotrifluoride, 9:778
2,4,6-Trinitrochlorobenzene, 2:425
2,4,6-Trinitrodiphenylamine
  prepn, 8:620
Trinitrophenol, 2:216
  in manuf of 2-amino-4,6-dinitrophenol, 2:216
sym-Trinitrophenol, 13:671
2,4,6-Trinitrophenol, 13:891; 15:148. (See also Picric acid.)
Trinitrophenolate. See Lead picrate.
2,4,6-Trinitrophenylethanol
  prepn, 8:612
2,4,6-Trinitrophenylmethylamine, 8:620
2,4,6-Trinitrophenylmethylnitramine. See Tetryl.
2,4,6-Trinitroresorcinol, 13:892
2,4,6-Trinitrosophloroglucinol, 16:203
2,4,6-Trinitrostilbene
  prepn, 8:612
α-Trinitrotoluene, 8:611
1,3,5-Trinitrotoluene, 8:681
  reaction with alkanolamines, 1:814
2,4,6-Trinitrotoluene, 8:611
  oxidation with dichromate, 16:203
α-Trinitrotoluene (TNT), 8:611
1,1,1-Trinitroxymethylethane. See Trimethylolethane trinitrate.
Trioctadecylamine, 2:130

Trioctylamine
  prop, 2:130
Trioctylphosphine oxide, 22:659
  nomenclature of, 11:288
Tri-n-octylphosphine oxide
  thorium separation, 20:252
Tri(n-octyl)tin oxide
  toxicology of, 20:324
Triolefins
Triolein, 8:377, 790
Triorganosilyl cyanides, 18:188
Trioxane, 1:645
  copolymerization of, 1:96, 98
  in embalming fluids, 8:101
  polymerization of, 1:98
1,3,5-Trioxane
  copolymerization, 16:174
s-Trioxane (ζ-Trioxymethylene) 10:77, 83
  manuf, 10:89
3,7,12-Trioxo-5β-cholanic acid, 3:483
Trioxygen difluoride, 9:508, 633
α-Trioxymethylene. See s-Trioxane.
Trioxymethylene
  as trimer, 16:242
2,6,8-Trioxypurine. See Uric acid."
Tripalmitin, 8:790, 791
Triparanol, 2:149, 150
  (1-(p-Diethylaminoethoxy)phenyl)-1-(p-t-olyl)-2-(p-chlorophenyl)ethanol)
Tripelennamine, 2:152
Tripelennamine, HCl
  antihistamine, 11:41
Tripentaerythritol, 1:593
Triperoxymanganate(IV), 13:10
Triphenylaluminum etherate, 7:70
Triphenylbismuth, 3:543, 544
Triphenylbismuth dibenzoate, 3:545
Triphenylbismuth dinitrate, 3:545
Triphenylborane, 3:707
Triphenylchloromethane
  from carbon tetrachloride, 5:132
Triphenylethyl nicotinate, 21:524
Triphenylgallium, 10:323
Triphenylguanidine, 2:414
2,4,5-Triphenylimidazole, 12:146
Triphenyllead lithium, 12:287
Triphenyllead sodium, 12:287
Triphenylmethane, 7:522
  from chloroform, 5:122
  derivs as dyes, 20:672
  dye classification, 20:677
  synthesis, 20:675

Triphenylmethane And Related Dyes, 20:**672**
  sodium in synthesis, 18:451
Triphenylmethane and related dyes, 20:**672**
Triphenylmethane dyes, 2:638; 20:**672**
Triphenylmethane triisocyanate
  coatings, 21:101
Triphenylmethanol
  freezing-point depression of, 4:394
  prepn, 8:370
Triphenylmethylcesium
  formation of, 4:856
Triphenylmethyl chloride
  cesium amalgam reaction with, 4:857
  in carbonium ion production, 4:394
N,N', N"-Triphenylpararosaniline, 20:699
Triphenyl phosphate, 15:148
  as cellulosic plasticizer, 4:657
  plasticizer, 15:721, 764
Triphenylphosphine
  initiator, 15:795
Triphenylphosphinephenylamine, 9:759
Triphenyl phosphite
  plasticizer, 15:749
  stabilizer, 17:519
Triphenylstannane, 6:651
3-(Triphenylstannyl)propionitrile, 6:651
Triphenylstibine, 2:581
Triphenylstibine oxide, 2:582
Triphenylstibine selenide, 2:582
2,3,5-Triphenyltetrazolium chloride, 10:918
Triphenyltin acetate, 10:223; 14:545
Triphenyltin chloride, 20:319, 320
Triphenyltin hydroxide, 10:223; 20:321
Triphenyltin oxide
  cyanuric acid reaction, 20:665
Triphenylurethanemethyl isocyanurate,
  20:665
Triphosphate, 15:234
Triphosphopyridine nucleotide, 21:513
Tripinone, 1:762
Triple point, 13:198
Triple superphosphate
  fertilizers, S:343
  prod as fertilizer, 9:106
Tripoli
  as abrasive, 1:26
  pigments, 15:513
Tripolite, 7:57
Tripolyphosphate, 15:234
Tripotassium phosphate
  hydrogen sulfide, 19:383
Triprolidine. HCl
  antihistamine, 11:41

Tripropoxytitanium, 20:470
Tri-n-propylaluminum
  as catalyst, 12:73
  prepn, 2:27
Tripropyl phosphate, 7:543
Tri-n-propyltin acetate, 15:911
Tri-n-propyltin hydroxide, 15:911
Tripropyltin oxide
  prepn from acrylamide, 1:292
  toxicology, 20:324
2,4,6-Tri-n-propyl-1,3,5-trioxane. See Parabutyraldehyde.
2,4,6-Tri-n-propyl-1,3,5-trioxane, 3:875. (See
  also Parabutyraldehyde.)
Triptene (trimethyl butene)
  from hexene, 5:105
Tripyrrylmethanes, 16:855
Trisaccharides
  ketoses, 19:152
1,1,3-Tris(allyloxy)propane, 1:262
Tris(amino)boranes, 3:734
Tris(1-aziridinyl) phosphine oxide, 11:539;
  15:682
Tris(p-aminophenyl)carbonium pamoate,
  14:548
Tris(1-aziridinyl)phosphine sulfide
  in tumor therapy, 11:543
2,4,6-Tris(1-aziridinyl)-s-triazine, S:82
  in tumor therapy, 11:542
Trisazo dyes, 7:471, 472
Tris(1,3-butylene glycol) biborate, 10:661
Tris(2-carbamoylethyl)phosphine, S:952
Tris(2-carbamoylethyl)phosphine oxide, S:952
Tris(2-carboxyethyl)phosphine, S:952
Tris(carboxy)methane
  prepn, 4:387
Tris(p-carboxyphenyl)bismuth dichloride,
  3:545
Tris(2-chloroethyl)amine
  in chemical warfare, 4:873
Tris-2-chloroethyl phosphate
  prepn, 8:530
Tris(chloromethyl)stibine, 2:582
Tris(2-cyanoethyl)phosphine, 6:651
Tris(2-cyanoethyl)phosphine oxide, 6:647
Triscyclopentadienyltitanium, 20:442
Trisdehydroemetine, 20:77
Tris(2,3-dibromopropyl) phosphate
  durable flame retardants, S:949
Tris-(2,4-dichlorophenoxyethyl)phosphite, 22:183
2,4,6-Tris(diethoxyphosphinyl)-1,3,5-triazine, S:951
Tris(2-diethyl aminoethyl)phosphate, 15:686

Tris(diethylamino)mono(dimethylamido)titanium, 20:473
2,4,6-Tris(difluoromethyl)-s-triazine, 9:770
2,4,6-Tris(dimethylaminomethyl)phenol, 15:181
B-Tris(ethylamino)-N-triethylborazine, 3:736
Tris(ethylamino)borane, 3:735
Tris(ethylenediamine)chromic bromide, 5:492
Trisglycolatozirconic acid, 22:647
Tris(1H,1H-heptafluorobutyl)tricarballyate, 9:730, 731
Tris(2-hydroxyethyl)amine oxide
  prepn from ethylene oxide, 8:529
Tris(hydroxyethyl)sulfonium hydroxide
  prepn, 8:529
Tris(2,4,6-α-hydroxyisopropyl)benzene, 1:623
  phys prop, 1:826
  uses, 1:830
Tris(hydroxymethyl)aminomethane
1,2,4-Tris(hydroxymethyl)benzene, 1:600
1,3,5-Tris(hydroxymethyl)benzene, 1:600
Tris(hydroxymethyl)nitromethane, 13:831
  as fungicide, 10:234
Tris(hydroxymethyl)phosphine, S:953
  prepn from acrolein, 1:263
1,1,3-Tris(hydroxyphenyl)propane
Trisilane, 18:172
  clathrates, 18:176
Trisiloxanes, 18:239
Trisilylamine, 18:196, 199
Trisilylarsine, 18:200
N, N', N''-Trisilylcyclotrisilazane, 18:196
Trisilylphosphine, 18:200
Trislactatozirconic acid, 22:648
Tris(methoxymethyl)methylenebisurea, 2:233
Tris(N-methylanilino)borane, 3:735
Trisodium cyanurate, 20:663, 665
Trisodium hydrogen pyrophosphate, 15:244
Trisodium orthophosphate, 15:238
Tris(1H,1H-pentadecafluorooctoxy)-t-butoxy-silane, 9:729
Tris(1H,1H-pentadecafluorooctyl)tricarballylate, 9:729, 730, 731
Trisodium phosphate, 6:371, 802; 7:680
Trisodium phosphate dodecahydrate
  chlorinates, 5:15
  as emulsifier in cheese, 10:8
  as water correcting agent, 10:17
Tris(2,4-pentanedione)erbium, 12:157
Tris(2,4-pentanedione)lutetium, 12:157
Tris(2,4-pentanediono)europium, 12:157
Tris(2,4-pentanediono)gadolinium, 12:157
Tris(2,4-pentanediono)terbium, 12:157
Tris(2,4-pentanediono)thulium, 12:157

2,4,6-Tris(phenylazo)phenol, 17:723
Tristearin, 6:878; 8:788, 790
Tristimulus values
Tris-s-triazine, 10:899
Tris(trifluoromethyl)stibine dichloride, 2:582
Trisulfatozirconic acids, 22:662
Trisulfox Brown B, 7:535
Triterpenes. See Terpenes.
Triterpenic tricyclic alcohol, 14:725
Trithiane, 10:85
Trititanium pentoxide, 20:390
  phys prop, 17:5
  synthesis, 17:4
  uses, 17:8
Tritium, 6:910-918; 17:5, 8, 92, 97, 119
Tritolyl phosphate
  as plasticizers, 19:672
Triton A-20, 3:30
Tritonal, 8:630
Triton B
  as catalyst for cyanoethylation, 6:634
Tri(triphenyltin) isocyanurate, 20:665
Trityl chloride, 7:44
4-Trityldiphenylamine, 7:44
N-Tritylmorpholine, 15:911
  as molluscicide, 14:545
Triuranium octoxide, 21:31
Triuret
  from urea, 21:38
2,4,5-Trivinyl-1,3-dioxolane, 7:415
Trizirconium tetranitride, 22:624
Troches, 6:51
Trolnitrate, 4:521
Trömelite, 15:249
Trona, 18:9, 458
  mining, 18:461
  occurrence, 18:458
  purification, 18:462
  soda ash from, 18:464
  Wyoming, 18:139
Tropaeolin O, 11:552
Tropaeolin OO, 11:552
Tropane, 1:766
Tropane alkaloids, 1:765
Tropic acid, 2:188
Tropical bleach, 3:556; 5:20
Tropolenes
  in woods, 16:690
Truxillic acid, 16:517
Tryamine, 2:178
Tryparsamide, 2:731
Trypsin, 5:594, 595, 598
  uses, 8:190

Tryptamine
  synthesis of, *11*:534
Trypthophan, *1*:771; *2*:157; *11*:590; *12*:857; *21*:521
  in wine, *22*:316
  yellowing of wool, *22*:406
Tsetse fly
  controls, *11*:678
d-Tubocurarine, *1*:778, 787. (See also *Curare.*)
Tubocurarine chloride, *6*:550
  as muscle relaxant, *2*:408
d-Tubocurarine chloride, *1*:770, 787
Tubulosine, *20*:77
Tucum oil, *8*:780, 785
Tufa, *12*:416
Tukon microhardness test, *10*:813
Tula istle, *9*:173, 177
Tuna, *9*:326
Tung oil, *8*:780, 785, 786, 790
  in alkyd resins, *1*:864
  cast alloys, *20*:571
  changes in properties, *7*:407, 415, 417, 419, 420
  drying oil, *7*:398, 399, 400, 401, 404
  film properties, *7*:423
  for machining tools, *20*:572
Tungsten. See *see Wolfram.*
Tungsten carbide, *4*:75, 80, 81; *8*:29
  as abrasive, *1*:28
  in bearing materials, *3*:275
  cemented, shaping of, *4*:301
  in diamond synthesis apparatus, *4*:299
Tungsten sulfide
  as hydrocracking catalyst, *11*:425
Tunneling
  photon-modulated, *15*:409
Turbidimetry And Nephelometry, *20*:**738**
Turbidity
  of water, *21*:691
Turbidity units, *22*:72
Turbinados, *19*:184
Turbines, gas. See *Gas turbines."*
Turbines, steam. See *Steam turbines."*
Turbulence
  interfacial, *13*:140

Turkey-red oil (sulfated castor oil), *4*:531
Turmeric, *9*:365, 375
Turnbull's blue, *12*:33
Turpentine, *14*:213; *19*:810; *20*:**748**
  heat of combustion of, *4*:898
Turpentine oil
  as denaturant, *8*:451
Turpentines
  resin solvents, *18*:660
Turquoise, *2*:1
Twaddell scale of specific gravity, *6*:758, 759
Twitchell reagent, *8*:368
  catalyst used in hydrolysis of fat, *8*:825
Two-dimensional ideal gas law, *1*:426
Two-film theory of mass transfer, *13*:135
Tyloses
  in dry wood cells, *22*:359
Tylosin, *10*:19; *12*:633, 637, 647
  in chemotherapy, *12*:656
  in food preservation, *10*:15
  as veterinary drug, *21*:244
Tymol blue, *11*:552
  in acetaldehyde determination, *1*:90
Tyndall effect
  in turbidimetry, *20*:739
Tyndall spectra, *20*:742
  vitalloys, *20*:768
Type metal, *20*:**756**
Tyrian purple, *7*:627; *11*:563
Tyrocidine, *16*:313
  iodination, *20*:263
Tyrosine
  in silk, *18*:271
  in wine, *22*:316
  yellowing of wool, *22*:406
L-Tyrosine
Tyrosine (2-amino-3-*p*-hydroxy-phenylpropi-onic acid), *2*:157, 189
  metabolism of, *2*:758
Tyrothricin, *16*:308, 309
  in bacterial infections, *3*:1
Tyuyamunite, *21*:8
TZM, *13*:644

U

UCON fluids
  as heat transfer medium, *10*:854
UV absorbers, *21*:**115**
Uacima
  analysis, *9*:173
Ubiquinones
  Irreverre-Sullivan test, *21*:589
Ucuhuba butter, *8*:780, 785
Ucuhuba wax, *22*:161
Uhde chlorine cell, *1*:691
  synthetic crude oil, *S*:186
Uhde-Pfirrmann process
Uintaite, *10*:528
Uleine, *1*:773
Ulexite, *3*:627, 628, 629, 630, 651
Ullmann-Minajeff reaction, *2*:480
Ullmannite, *13*:736
Ultracentrifuge, *4*:734
  molecular weights from, *16*:247
Ultrafiltration
  and PEC hydrogels, *16*:128
Ultra-low temperatures. See *Cryogenics.*
Ultramarine, *5*:745
Ultramarine blue
Ultramarine blues, *2*:9; *15*:540
  in stains, *18*:657
Ultramarine green, *15*:538, 540
Ultramarine red, *15*:540
Ultramarines, *18*:49
Ultramarine violet, *15*:540
Ultraphosphates, *15*:232, 248
Ultrasonic cleaning
  use of ferroelectric ceramics in, *9*:1
Ultrasonic drilling
  use of ferroelectric ceramics in, *9*:1
Ultrasonics, *20*:**773**
Ultrasonic velocity
  in aquametry, *2*:693
Ultrasonic vibrations
  use in detergency, *6*:877
Ultraviolet filters, *9*:255
Ultraviolet radiation
  in water purification, *22*:128
Ultraviolet spectrometry
  and mass spectrometry, *13*:90
Umbelliferone, *6*:432; *10*:925

Umber
  pigments, *15*:525
Undecafluoropiperidine, *9*:799
Undecafluoro(trifluoromethyl)cyclohexane. See
  (Trifluoromethyl)undecafluorocyclohexane.
γ-Undecalactone, *9*:358; *14*:736
  as flavor, *9*:354, 355, 362
Undecanedioic acid, *1*:253
  prepn, *1*:247
2,5-Undecanedione, *12*:158, 160, 161
Undecanoic acid, *8*:812, 814
  in dental material, *6*:827
1-Undecanol. See *n-Undecyl alcohol.*
2-Undecanone, *12*:105
6-Undecanone, *12*:104
10-Undecenoic acid, *8*:814
n-Undecyl alcohol. See *1-Undecanol.*
Undecyl alcohols, *1*:568
Undecylenic acid
  from castor oil, *4*:532
  as dental material, *6*:827
  as veterinary fungicide, *21*:249
10-Undecylenic acid, *8*:814
  prepn, *8*:822
n-Undecylic acid, *8*:812
  phys prop, *1*:229
Undecylimidazoline, *10*:912
10-Undecynoic acid, *8*:816
Unistat
  anticoccidial, *21*:247
Uniterm index, *12*:514
Units, *S*:**984**
Units and systems, *7*:176
Unsaturated alcohols. See *Alcohols, unsaturated.*
Unsaturated polyesters, *20*:**791**
Unsymmetrical dimethylhydrazine, *8*:692
  prop, *8*:690
Uracil
  as diuretic, *7*:253
  as herbicides, *22*:213
  metabolism as herbicides, *22*:213
  prepn of, *22*:213
Uralkyd"
  coatings, *21*:101 *22*:432

Uranine (CI Acid Yellow 73), 5:871; 9:497; 22:432
Uraninite, 21:5, 7, 31
Uranium. (See also *Actinides.*)
Uranium, 21:**1**
  application in fluid bed technology, 9:403
  in Canada, 21:10
  colorant for glass, 9:252
  enrichment cycle, 21:22
  as fission product, 14:107
  as glass colorant, 5:851
  breeder reactors, 21:29
  as reactor fuel, 14:76
Uranium-235, 15:879; 17:9
  as breeder reactors, 21:29
  centrifugal isotopic separation, 21:26
  electromagnetic separation, 21:26
  gaseous diffusion process, 21:26
  as reactor fuel, 14:76
  separation, 14:85
Uranium-238, 15:822; 17:9
  breeder reactors, 21:29
  irradiation, 21:3
  transuranium elements from, 21:6
Uranium ammonium pentafluoride, 21:33
Uranium and uranium compounds, 21:**1**
Uranium black, 21:7
  oxidation state changes, 21:5
Uranium carbide, 4:70, 86; 13:859
Uranium compounds, 21:1, **31**
Uranium dioxide, 9:404; 21:31
Uranium ferricyanide
  photography, 15:386
Uranium fluoride, 14:82
Uranium fuel elements
  zirconium metal in, 22:621
Uranium hexafluoride, 9:524, 587, 705; 21:33
  fuel-pellet manuf, 21:27
  as lubricant, 12:606
  uranium refining, 21:22
  volatilization, 14:97
Uranium hydride, 11:207
Uranium monoxide, 21:32
Uranium nitrate
  indicator, 22:560
  uranium purification, 21:1
Uranium nitride, 13:821
Uranium oxide
  as nuclear fuel, 4:775
  from pitchblende, 21:1
  refractory, 17:244
  thermal conductivity of, 4:803

Uranium peroxide, 21:32
Uranium reserves, 10:187
  use of gallium in spectroscopy of, 10:326
Uranium tetrachloride
Uranium tetrafluoride, 21:32
  uranium refining, 21:24
Uranium tetramethoxide
  prepn, 1:833
Uranium tricarbonate, 21:35
Uranium trioxide, 21:32
  prepn, 7:368
Uranium-233
Uranocircite, 21:8
Uranophane, 21:8
Uranyl carbonates, 21:35
Uranyl methoxide, 1:833
  as catalyst, 8:361
  uranium purification, 21:3, 25
  uranium trioxide from, 21:32
Uranyl nitrate, 7:368; 21:32, 33
Uranyl salts
  and oxalic acid decomposition, 14:357
Uranyl sulfate, 21:34
  nuclear reactor fuel, 14:82
Urazoles, 10:917
Urbaite, 20:65
Urea
  as animal feed, 8:868
  cyanuric acid from, 20:667
  in dentifrices, 6:848
  as diuretic, 7:251
  dry chemical agents, S:366
  ethylene chlorohydrin reaction with, 5:307
  as fertilizers, 9:68; S:338
  fertilizer technology, S:343
  in fish, 9:319
  in imparting flame resistance to cellulose, 9:306
  polymers from, 15:793
  prepn of, 4:356
  prepn from cyanamide, 6:561
  in prepn of ethyleneurea, 7:27
  in prepn of glutamic acid, 2:203
  in prepn of liquid glue, 10:616
  prepared from guanidine, 10:736
  prod of, 2:178
  as raw material for amino resin manuf, 2:226
  reactions with acetoacetic ester, 1:156
  in sep of cresol isomers, 6:439
  in shampoo, 10:789
  sulfamation, 19:306

sulfamic acid from, *19*:245
in textile printing, *7*:576
Urea and urea derivatives, *2*:15, 67, 68, 71, 173, 623; *6*:451; *21*:**37**
Urea chloride, *4*:69. (See also *Chloroform-amide.*)
Urea clathrates, *21*:40
Urea derivatives, *21*:37
Ureaform
  as fertilizer, *9*:78
Urea-formaldehyde
  step-growth, *16*:226
Urea-formaldehyde adhesives
  modification by furfuryl alcohol, *10*:246
Urea-formaldehyde resin, *15*:903; *21*:54. (See also *Amino resins and plastics.*)
  prop, *9*:860
Urea herbicides, *22*:188
Urea nitrate
  urea-nitric acid reaction, *21*:38
Urea phosphate
  as fertilizers, *S*:343
Urea resins
  as blending agents in alkyd resins, *1*:867
Urease, *5*:594, 595; *9*:897
  action and uses, *8*:187
  in soybeans, *18*:602
  uses, *8*:190
Ureastibamine, *2*:585
p-Ureidobenzenearsonic acid, *2*:731; *21*:248. (See also *Carbarsone.*)
5-Ureidohydantoin, *21*:112
Urethan, *2*:67, 68. (See also *Urethan polymers.*)
  as hypnotics, *11*:516
  stain solvent, *18*:660
  urea-alcohol reactions, *21*:39
Urethane. See *Urethan.*

Urethan elastomers, *7*:692
Urethan foams
  use of neopentylene glycol, *10*:677
  use of polypropylene glycols in, *10*:658
  use of sorbitol in, *1*:585
Urethan oil"
  coatings, *21*:101
Urethan oils, *7*:413
Urethan polymers, *21*:**56**
  use of glycerin in prepn, *10*:628
Urethan prepolymer
  sealants, *4*:31
Urethans
  castor oil in prepn of, *4*:532
  prepn from isocyanates, *12*:48
"Uretidine diones," *21*:65
6,6'-Ureylenebis
Urginea maritima
  drugs from, *4*:513, 516
Uric acid, *10*:913; *21*:**107**
  pyrolysis, *20*:662
Uridene diphosphoglucose, *19*:151
Uridine diphosphate-muramyl pentapeptide, *14*:672
Uridine triphosphate
  in amino acid synthesis, *2*:159
Uridine 5'-pyrophosphate
  glucose ester, *19*:151
Urography, *17*:132
Urotropin, *2*:359. See *Hexamethylenetetra-mine.*
Urotropin, *2*:359
Uroxin, *21*:113
Urushibara nickel
  amine oxide reduction, *S*:39
Urushiol, *17*:387; *22*:654

V

V Permendur, *12*:756
Vaccenic acid, *8*:814
  use in margarine, *13*:62
Vaccines
  in viral treatments, *21*:452
Vaccinia
  DNA virus, *21*:452
Vacuum
  in crystallization, *6*:502, 510
  net speed formula, *21*:139
  use in cryogenics, *6*:479
Vacuum coating, *13*:278
Vacuum deposition
  of refractory coatings, *17*:281
Vacuum distillation of water, *6*:903
Vacuum furnaces, *10*:275
Vacuum metallization, *13*:449
Vacuum technology, *21*:**123**
V agents. See *Nerve agents.*
Valentenite, *2*:564, 573
n-Valeraldehyde, *1*:99
  from butadiene, *3*:791
Valerian, *14*:214
  essential oils as flavor, *9*:352
Valeric acid, *8*:812
  phys prop, *1*:229
  as synthetic flavor, *9*:354
γ-Valerolactam
  form, *2*:67
δ-Valerolactone
  phys prop, *1*:258
  prepn, *1*:258
Valerophenone, *12*:107
Valerylacetone, *12*:148
  from cephalosporin C, *14*:664
  penicillin synthesis, *14*:668
Valine, *2*:157
DL-Valine
L-Valine
Valinomycin, *16*:336
Valone, *15*:920
Valonia, *12*:318, 322
Valve bronze
  properties of, *6*:258
Van Dyke brown, *15*:526
Van Laar equation, *21*:210

Vanadates
  industrial uses, *21*:173
Vanadates(IV), *21*:172
Vanadates(V), *21*:172
Vanadia
  ceramic colorants, *22*:652
Vanadites, *21*:172
Vanadium, *12*:756; *21*:**157**
  ferroalloy, *18*:751
  as glass colorant, *5*:851
  hydrocarbon control in automobile exhaust, *S*:63
  neutron activation analysis, *17*:38
  in phosphoric acid, *15*:261
  sulfuric acid catalysts, *19*:460, 467
  in superalloys, *11*:20
  wolfram alloys, *22*:344
  zinc analysis, *22*:561
Vanadium(IV) oxysulfate, *21*:175
Vanadium alloys, *21*:157
Vanadium and vanadium alloys, *21*:**157**
Vanadium arsenides, *21*:170
Vanadium borides, *21*:170
Vanadium carbide, *4*:4,75, 84, 85; *21*:170. (See also *Cemented carbides.*)
  in steel making, *21*:161
Vanadium carbonyl, *4*:489; *21*:176
Vanadium chloride
Vanadium hydride, *11*:208; *21*:171
Vanadium nitride, *21*:170
Vanadium oxide, *2*:54
  automobile exhaust control by, *2*:822
  catalyst, *19*:143
  industrial uses, *21*:173
  reduction of, *21*:162
Vanadium oxides, *21*:171
Vanadium oxychloride, *7*:75; *21*:173
Vanadium pentoxide, *12*:549; *21*:160, 162, 165, 171, 514
  automobile emission control, *S*:63
  in carnotite, *21*:9
  as catalyst in oxalic acid process, *14*:360
  as catalyst in oxidation of anthracene, *2*:435
  as glass former, *10*:538
  in phthalic anhydride synthesis, *13*:678
  sulfur prod, *19*:355

Vanadium tetrachloride, 7:75; 21:174
Vanadocene, 21:176
Vanadyl acetylacetonate
  polyester accelerator, 20:826
Vanadyl sulfate, 21:175
Vancomycin, 14:675
  in bacterial infections, 3:1
Vanilla, 9:375
Vanilla beans, 9:353
Vanillal, 21:192
Vanillaldehyde, 21:180
Vanillic acid, 21:192
  from vanillin, 21:182
Vanillin, 1:645; 6:425; 11:470; 12:371;
  14:729; 15:162, 163; 21:**180**
Vanillin acetate, 21:182
  in vanillin prod, 21:186
  from vanillin, 21:181
Vanillin hydrozone, 21:186
Vanillin oxime, 21:181
Vanillyl alcohol, 21:182
Vanillylamine, 21:181
Vanillylideneacetone
  from vanillin, 21:181
Vanillylidene cyanohydrin, 21:182
Vanillylidene nitromethane, 21:182
Vanishing cream, 6:348, 350
Vanners
  for gravity concn, 10:718
Vapam
  soil chemistry, 18:539
Vapor
  equations of state, 21:199
Vapor degreasing
  trichloroethylene in, 1:207
Vapor deposition, 9:202
  prod of continuous fibers, 11:661
  of refractory coatings, 17:275
Vaporization
  equilibrium constants, 20:139
  flash, 15:12
  of liquid natural gas, S:430
Vapor-liquid equilibria, 21:**196**
Vapor phase cleaning, S:165
Vapor plating, 13:280
Vapor pressure
  equations for, 6:714
  as function of temperature, 14:613
  in measurement of activity coefficients,
    7:788
Variables
  in data presentation, 6:706

Variamine Blue Base B, 2:891
Varnish
  drying oils in, 7:423
  fish oil used in, 7:402
  furnaces used to manuf, 10:293
  medium oil, 7:424
  short-oil, 7:424
  use of driers in, 7:277
Varnish production
Varnish removers, 14:485
Vasicine, 1:771, 772
Vasopressin
  biological activity, 11:68
  corticotropin-releasing factors, 11:74
  hormone analogs, 11:73
  structure of, 11:70
  sulfur in, 19:369
  synthesis of, 11:72
  uses in medicine, 11:74
Vat Black 25, 7:470
Vat Blue BCS, 2:487
Vat Blue 1, 5:868
Vat Blue 6, 2:487; 5:868, 873; 7:497
Vat Blue 18, 7:492
Vat Blue 20 (CI 59800), 7:492
Vat Blue 30, 20:191
Vat dyes, 7:470, 554, 584; 21:241
  anthraquinone, 2:517
  azo, 2:894
  in textile printing, 7:575
  uses of, 7:477
Vat Green 1, 7:497
Vat Green 9, 7:492
Vat Red 1, 5:870, 874
Vat Red 20, 20:190
Vat Yellow 1, 2:522
Vat Yellow 2, 20:189
Vat Yellow 10, 2:894
Vat Yellow 11, 20:190
Vat dyes, leuco esters, 7:561
Veatchine, 1:773
Vegetable butter oils, 8:778, 784, 785
Vegetable butters, 8:786
Vegetable tanning materials, 12:319
Vegetable tanning of leather, 12:334
Velocity merit rating
  in ceramic armor technology, S:149
Velocity profile, 7:162, 163
Venetian reds, 15:520
Ventilation
  in odor control, 14:172
Venturi tube
  description, 9:479

Veratraldehyde, *15*:163; *21*:193
  from lignin, *21*:189
  in prepn of papaverine, *1*:798
  from vanillin, *21*:182
Veratramine, *1*:776; *18*:844, 866, 885
Veratric acid
  phys prop, *1*:238
Veratrum
  as cardiovascular agent, *4*:520
Veratrum alkaloids, *1*:805
Veratryl alcohol, *19*:301
Verdigris, *15*:538
Verel, *1*:335
  manuf, *1*:336
  prop, *1*:335
  uses, *1*:336
Veridian, *15*:538
Vermiculite, *5*:551; *13*:399
  in gypsum board, *21*:623
  as insulator, *11*:832
Vermillion, *13*:218; *15*:585
Vermouth, *22*:330
Vernamycin, *16*:332
Vernonia oil, *8*:787
Versatate
  vinyl acetate copolymer, *21*:331
Versatic acids
  prop, *8*:852
Versidyne, *2*:388, 389
Vertical current concentrators
  for gravity separation, *10*:707
Vesicant agents
  in chemical warfare, *4*:872
Vesuvianite, *22*:338
Veterinary drugs, *21*:**241**
Vetiver, *14*:214
Vetiverol, *19*:833
Vetiveryl acetate
  chemotherapy against, *3*:26
  prod and sales of, *8*:376
Vicalloy, *12*:768
Vicat softening point
  of Celcon, *1*:102
Vicker's hardness test, *10*:811
  for beryllides, *S*:74
Victoria Blue
  manuf, *20*:700
Victoria Blue B, *20*:721
Victoria Blue R, *20*:721
Victoria Blue 4R, *20*:721
Victoria Green, *5*:853
Victoria Pure Blue, *20*:673
Victoria Pure Blue BO, *20*:721

Video tape, *12*:808
Villard effect, *15*:388
Vinbarbital, *3*:61, 70
Vinblastine sulfate
  in cancer therapy, *S*:88
Vinca alkaloids, *1*:806
  anticancer agents, *S*:82, 88
Vincaleukoblastine, *1*:773, 778, 806
Vincristine sulfate
  in cancer therapy, *S*:88
Vine black, *15*:544
Vinegar, *21*:**254**. (See also *Acetic acid.*)
  as denaturant, *8*:451
  manuf from ethyl alcohol, *8*:454
  in meat curing, *13*:171
Vinegar eels, *21*:269
Vinegar of lead, *12*:268
Vinyl acetal polymers. See *Polyvinyl acetal.*
Vinyl acetate, *6*:456, 457, 467, 468, 469, 695;
  9:404; *12*:155; *21*:317-353, 377, 434, 441
Vinylacetic acid, *6*:466
Vinylacetylene, *1*:206
  in prepn of biacetyl, *12*:144, 145
α-Vinyl-1-aziridine ethanol
  in tumor therapy, *11*:542
Vinyl alcohol. See *Polyvinyl alcohol.*
Vinyl allyl adipate, *1*:410
Vinylbenzene, *19*:55
Vinyl benzoate, *3*:433
Vinyl bromide, *3*:778, 779
  as fire retardant, *S*:483
Vinyl *n*-butyl ether, *21*:417
  phys prop, *8*:471
Vinyl butyral-phosphoric acid
  metal conditioner, *14*:477
Vinyl butyral resin
  from butyraldehyde, *3*:873
Vinyl butyrate, *12*:155
Vinyl caprate
  vinyl acetate copolymer, *21*:331
n-Vinylcarbazole, *10*:910
  uranium refining, *21*:25
  prepn from acetylene, *1*:205
Vinyl carbinol
  steroid synthesis, *18*:863
Vinyl cetyl ether
  vinyl chloride copolymer, *21*:378
Vinyl chloride, *5*:171-178; *6*:695; *9*:400. (See
  also *Polyvinyl chloride.*)
Vinyl chloride-acetate film
  properties, *9*:222-226
Vinyl chloride-vinyl acetate resin
  as coating, *14*:476

Vinylchlorosilanes, *18*:236
Vinyl copolymer resins
   economics, *15*:231
   in phonograph records, *15*:230
Vinyl crotonate, *6*:469
Vinyl cyanide. See *Acrylonitrile.*
1-Vinylcyclohexan-1-ol
   phys prop, *1*:630
4-Vinylcyclohexene, *3*:786
   epoxidation, *8*:249
Vinylcyclohexene dioxide, *8*:249
   phys prop, *8*:265
Vinylcyclohexene monoxide, *8*:249
Vinylcyclopropane
   from butadiene, *3*:791
1-Vinyl-3,5-dimethyladamantane, *S*:9
4-Vinyl-1,3-dioxane
   from butadiene, *3*:789
n-Vinyldiphenylamine
   prepn from acetylene, *1*:205
Vinyl ether
   as anesthetic, *2*:399
   fluorinated, *21*:423
   phys prop, *8*:471
Vinyl ether polymers, *21*:412
   in copolymerization of trioxane, *1*:98
Vinyl ethers, *1*:205. (See also *Polyvinyl ether.*)
Vinyl ethyl ether
   phys prop, *8*:471
Vinyl 2-ethylhexanoate, *8*:850
Vinyl fibers, *9*:169
Vinyl films, *10*:247
Vinyl fluoride
   toxicity, *9*:741, 835
   polymerization, *9*:836
   prop, *9*:836
Vinyl hydrogen fumarate
   vinyl acetate copolymer, *21*:331
Vinyl hydrogen maleate
   vinyl acetate copolymer, *21*:331
Vinylidene, *14*:324 *1,1-Dichloroethylene.*)
   copolymers, *15*:799
   monomer, *21*:275
   reaction with butadiene, *3*:785
   toxicity, *9*:741
Vinylidene chloride, *5*:148, 149; *6*:695; *9*:602,
   841. (See also *1,1-Dichloroethylene.*)
Vinylidene fluoride, *7*:695, 696; *9*:841
Vinylidene polymers, *21*:**269**
Vinyl isobutyl ether, *21*:417
   copolymers of, *21*:421
   vinyl chloride copolymer, *21*:378
Vinyl isocyanate, *12*:51

Vinyl isopropyl ether, *21*:417
Vinyl laurate
   vinyl acetate copolymer, *21*:331
Vinyl lauryl ether
   vinyl chloride copolymer, *21*:378
Vinyllithium, *12*:552
Vinyl methyl ketone, *6*:468, 694
   as flavor compound, *9*:342
Vinyl methyl sulfone
   reaction with gelatin, *10*:501
Vinyl octadecyl ether, *21*:413
Vinylon, *6*:663
5-Vinyl-2-oxazolidinethione, *20*:269
Vinyl phenyl ketone, *6*:694
Vinyl polybutadiene, *7*:690
Vinyl polymers, *21*:**304**. (See also *Vinyl resins.*)
Vinylpotassium
   in prepn of acrylic acid, *1*:295
Vinyl propenyl ketone, *6*:694
Vinylpyridine
   latex, *17*:600
2-Vinylpyridine, *10*:86; *16*:803
   copolymer with butadiene, *8*:711
   synthesis, *19*:671
2-Vinylpyridine latex
   tire cord adhesion, *20*:328
Vinylpyrrolidone
   manuf from acetylene, *1*:205
N-Vinylpyrrolidone
   in drilling fluids, *7*:301
N-Vinyl-2-pyrrolidone, *15*:800; *16*:853
   copolymers in phonograph records, *15*:228
   as fabric coatings, *5*:682
   manuf of, *21*:432
Vinyl resins. (See also *Vinyl polymers.*)
Vinyl rubber film
   properties, *9*:222-226
   rigid, *21*:405
Vinyl *t*-butyl ether, *21*:418
Vinylsilane, *18*:179, 264
Vinylsilicon compounds, *11*:530
Vinyl stearate
   copolymers, *15*:736
Vinyl succinate, *19*:146
Vinyl sulfone, *7*:633
1-Vinyl-2,2,3,3-tetrafluorocyclobutane, *3*:790
   prepn, *3*:790
Vinyltoluene, *7*:415; *19*:77
   as modifier in alkyd resins, *1*:866
   polyester resins, *15*:461
   unsaturated polyester diluent, *20*:821
(m,p)-Vinyltoluenes, *19*:115
Vinyon and related fibers, *21*:**441**

Violamine
   xanthene dye, *22*:434
Violamine RD, *7*:522
Violanthrone, *2*:496; *7*:479, 492
Violet acid, *13*:722
Viologens, *22*:208
Viomycidine, *16*:328
Viomycin, *3*:30; *16*:308, 328
   some available products, *16*:329
Viral chemotherapy, *21*:452
Viral infections, chemotherapy, *21*:**452**
Viridogrisein
   amebicide, *20*:80
Viruses, *13*:460
   bacteriophages, *13*:460
   description of, *21*:452
   DNA, *13*:460
   lysogeny, *13*:461
   RNA, *13*:460
"Virustat", *21*:457
Visa
   soil-release agent, *S*:969
Viscometer. See *Viscometry.*
   Cannon-Fenske, *21*:467
Viscose process, *17*:171
   of asphalts, *2*:782
   carbon disulfide in manuf of, *4*:384
   coefficient, *17*:426
   data interpretation, *6*:739
   for gases, *6*:747
   of inks, *11*:613
   intrinsic, *4*:623
   molecular weights from, *16*:247
   prepn for dyeing, *7*:518, 520
Viscose rayon. (See also *Rayon.*)
Viscosity. (See also *Viscometry.*)
Viscosity-breaking, *15*:2, 20, 22
Viscosity coefficient, *21*:461
Vitallium, *5*:735
Vitamers, *21*:486
   in margarine, *13*:60
   in milk, *13*:515
   in nuts, *14*:131
   substrate of rhodopsin, *21*:488
   synthesis of, *19*:146; *21*:489
Vitamin A, *19*:803, 825, 827, 830, 833;
   21:490-508
Vitamin A$_2$, *21*:490, 501
Vitamin B, *12*:856; *20*:173
Vitamin B complex, *21*:509
   inositol in, *11*:673

Vitamin B$_1$
   in milk, *13*:515
   in nuts, *14*:131
   sulfur in, *19*:369
Vitamin B$_{12}$, *21*:542-549
Vitamin B$_2$, *17*:445
   added to animal feeds, *8*:869
   by fermentation, *21*:489
   from milk, *13*:570
   pharmaceutical, *15*:117
   phosphorylation, *16*:817
   synthesis, *19*:146
Vitamin B$_3$, *16*:801; *21*:525
Vitamin B$_6$, *12*:825, 856; *16*:780, 806
   biosynthesis, *16*:817, 818
Vitamin C. See *Ascorbic acid.*
Vitamin D, *18*:837; *21*:549-573
s-trans-Vitamin D, *21*:551
   as antioxidant, *2*:597
   in margarine, *13*:62
   in peanut oil, *14*:131
   as a standard, *21*:566
Vitamin D$_m$, *21*:553
Vitamin D$_2$, *15*:338; *21*:549
Vitamin E, *21*:574-585
Vitamin G, *17*:445
   as anticoagulant antagonist, *5*:603
   blood coagulant, *21*:488
   synthesis, *21*:489
Vitamin K, *21*:585-600
Vitamin PP, *16*:801; *21*:509, 525
Vitamins, *21*:**484**
   folic acid-related, *19*:261
   in nuts, *14*:131
   survey, *21*:484
   in yeast, *22*:516
   yeast requirements, *22*:517
Vitrain, *5*:608, 609
Vitreous enamel. See *Enamels, porcelain or vitreous.*
Vitrification, *4*:792
Vitrinite, *5*:611, 640, 641, 642
Vivrio cholera
Vodka, *1*:518
Volatility function
   relative for binary systems, *21*:207
Volemitol, *1*:576
Volhard method
   chloride ion determined by, *5*:5
Volidan, *6*:79, 87
Volkhonskoite, *5*:548
   properties, *3*:342

Voltammetry, 7:730
Volume conductance, 7:720
Volume resistance
  electrical, 2:655
Vorce chlorine cell, 1:687
Vortex tube, 7:165
Voyutskii adsorption theory, S:16
Vulcan Fast Red B, 2:909
Vulcan Violet BN, 2:488
Vulcanite
  as dental material, 6:827

Vulcanizates
  fluoroelastomer, 21:273
Vulcanization, 17:627-645
  of chlorosulfonated polyethylene, 17:573
  use of antimony pentasulfide in, 2:577
Vycor glass
  dielectric props of, 4:821
  strength value of, 4:818

W

Waelz kiln
  electrothermic prod of zinc, *22*:576
  zinc oxide prod, *22*:591
Wagner tarnishing theory, *6*:307, 309
Wallboard, *21*:**601**
Wallpaper
  use of formamide, *10*:105
Walnut oil, *7*:403; *8*:780, 785
Walnuts, *14*:145
Warburg and Christian's coenzyme, *21*:513
Ware, ceramic
  chemical, *4*:832
Warfare, chemical, *4*:869
Warfarin sodium, *5*:600, 601
Wash-and-wear textiles
  use of formaldehyde, *10*:94
  use of zinc fluoborate in, *9*:566
Washfast color dyes
  use of aluminum acetate in, *2*:13
Washing
  definition, *8*:761
Washing and scrubbing
  in odor control, *14*:175
Waste disposal methods
  deep-well disposal, *21*:640
Waste management, *21*:625
Wastes
  agricultural, *S*:739
  animal, *S*:739
Wastes, industrial, *21*:**625**; *S*:739
Waste treatment
  textile, *S*:979
Wastewaters
  diversion of, *S*:332
  domestic, *S*:327
  industrial, *S*:328
  nutrient removal from, *S*:332
Water, *21*:**651**; *22*:1 (continuation). (See also *Aquametry*.)
  analysis, *21*:688
  batch gravity water filter, *9*:276
  as polymer, *21*:669
  props, *21*:668
  sewage, *22*:116
  sources, *21*:652
  thermodynamic properties in salt solutions, *7*:818

Water analysis, *21*:688
Water color, *14*:480
Water-conditioning
  use of ethyleneamines in, *7*:36
Water cooling
  mass transfer in, *13*:100
Water corrosion
  of metals, *6*:301
Watercraft
Waterfree creams (cosmetic), *6*:348, 350, 351
Water (desalination), *22*:1
Water gas, *2*:276
  carbon monoxide produced from, *4*:438
  characteristics, *10*:280
  composition of, *4*:438
Water gas process, *10*:363, 424
Water-gel explosives, *8*:644
"Waterglass," *18*:135
Water hardness
  effect on detergency, *6*:873
  management of, *S*:335
Water level
Water measurement. See *Aquametry*.
Water molecule
  structure of, *21*:671
  ortho, *21*:682
Water of esterification
  removal, *8*:325
Water pollution, *S*:713
  surfactants, *19*:581
Waterproofing agents
  metallic salts as, *7*:285
Waterproofing and water repellency, *22*:**135**
Waterproof paper
  use of aluminum formate in, *2*:15
  use of aluminum triacetate in, *2*:13
Waterproof textiles
  use of aluminum formate in prepn of, *2*:15
  use of aluminum formoacetate in, *2*:16
  use of aluminum triacetate in prepn of, *2*:13
Water purification
  use of ammonia, chlorine in, *2*:297
  chlorine dioxide for, *5*:43
  clays used in, *5*:584
  use of lime in, *12*:457

Water repellent
  use of polyepichlorohydrin and ethylene-
    diamine as, 7:34
Water softening, *11*:871, 887
  use of lime in, *12*:457
Water-soluble factor B, *21*:486
Water-soluble flame retardants, *9*:304
Water supplies
  domestic, *S*:320
  industrial, *S*:320
Water vapor
  as fission products, *14*:105
Water, vicinal, *21*:681
Water wells
  hydroxyacetic acid in cleaning, *10*:635
Wattle bark, *12*:317
Watts bath (nickel-plating), *8*:62
Wax
  in candles, *4*:58
  cracking of paraffin to produce olefins,
    S:646
  lac, *18*:26
  microcrystalline in waterproof paper, *22*:149
  from petroleum, *15*:2
  synthetic, *4*:446
Waxes, *22*:**156**
  as abherents, *1*:7
  chlorinated as fire retardants, *S*:467
  chlorinated naphthalene, *5*:297
  ethoxylated, *19*:548
  as impregnants, *14*:457
  microcrystalline, *15*:92, 99
  in nuts, *14*:129
  paraffin, *15*:92, 97
  petrolatum, *15*:92
  as stain solvent, *18*:661
  synthetic paraffin, *15*:93, 104
  used for embedment, *8*:106
Weathering
  outdoor of electric insulation, *11*:783
Weather modification, *21*:661
Webcell filter-press dialyzer, *7*:15
Weberite, *9*:534
Weed killers, *22*:**174**. (See also *Herbicides.*)
Weichmachers, *5*:233
Weighing and proportioning, *22*:**220**
Weight merit rating
  in ceramic-armor technology, *S*:149
Welding, *22*:**241**
Well-counters, *17*:2
Weloganite, *22*:640
Welsbach gas mantle
  ceric oxide in, *4*:852

Wetcleaning, *7*:322
Wetherill grate
  prod of zinc metal, *22*:576
Wettability
  in adhesion, *S*:18
Whale meal
  as animal feed, *8*:867
Whale oil, *8*:782, 790
  in soaps, *18*:417
Wheat
  as animal feed, *8*:861
Wheat and other cereal grains, *22*:**253**
Wheat germ oil, *8*:780, 785
Whetlerite
  in protective masks, *4*:883
Whey
  as animal feed, *8*:865
Whey vinegar, *13*:571
Whipping agents
  protein, in confectionery, *6*:44
Whipworms, *14*:533
Whiskers, *11*:654
  American, *1*:512
  Canadian, *1*:510
  composite strengths, *11*:656, 658
  Irish, *1*:512
  mechanical properties, *11*:655
  prepn of, *11*:655
  Scotch, *1*:511
  thermal properties, *11*:655
Whisky
  defn, *1*:510
White arsenic. See *Arsenic trioxide.*
Whitening agents, *3*:737
White phosphorus. See *Phosphorus, white.*
Whiteware, *4*:760
  chemical, *4*:832
  clays used in, *5*:566
Whitewash, *14*:473
Whitings, *12*:416; *15*:508. (See also *Calcium carbonate.*)
Wicks
  for candles, *4*:58
  for heat pipe, *S*:495
Wieland-Gumlich aldehyde, *1*:772
Wildi method
  for 7-dehydrocholesterol, *21*:562
Willemite, *12*:616; *22*:563
Willgerodt reaction
  with morpholine, *13*:662
Willow black, *15*:544
Wind force
  in electromigration, *S*:284

Wine, 22:**307**
  citric acid in, 5:536
  use of enzymes in prepn, 8:200
  yeast fermentation, 22:523
Winkler titration
  for oxygen in water, 21:696
Wintergreen, 14:214
  essential oils as flavor, 9:352
Wintergreen oil
  as denaturant, 8:451
  prod and sales, 8:377
Winterizing
  separation of drying oils, 7:401
Wirebar (copper), 6:164, 176
Wire dies, diamond, 4:292
Wisconsin process, 13:807
Withafarin, 18:845
Witherite, 3:77, 82; 15:516; 19:51.   (See also
  Barium carbonate.)
  in ceramics, 4:772
Wittig reaction
  for Vitamin D₃, 21:564
Wohl-Ziegler reaction
  of allylic bromination, 1:923
Wolff-Kishner reaction, 1:641
Wolff process
  in acetylene manuf, 1:184
Wolfram, 22:**334**
  ferroalloy, 18:751
α-Wolfram, 22:335
β-Wolfram, 22:335
Wolfram alloys, 22:**334**
Wolfram blue, 22:350
Wolfram borides, 22:355
Wolfram bronze
  ferroelectric, 9:5
Wolfram bronzes, 22:350
Wolfram carbides, 22:339, 343
Wolfram carbonyl, 4:489
Wolfram compounds, 22:**346**
Wolfram dibromide, 22:348
Wolfram dichloride, 22:348
Wolfram diiodide, 22:337, 349
Wolfram dioxide, 22:350
Wolfram dioxydibromide, 22:349
Wolfram dioxydichloride, 22:348
Wolfram dioxydifluoride, 22:348
Wolfram disilicide, 22:355
Wolfram disulfide, 12:588; 22:354
  films, 22:357
Wolfram halides, 22:347
Wolfram hexabromide, 22:348
Wolfram hexacarbonyl, 22:346

Wolfram hexachloride, 22:336, 348
  reduction with aluminum, 22:346
Wolfram hexafluoride, 22:347
Wolfram hexafluoride, 9:587
  reduction of, 22:339
Wolframic acid, 8:243; 22:350
  analysis of wolfram, 22:343
  salts of, 22:351
  from scheelite, 22:338
  from wolfram hexafluoride, 22:347
  wolfram trioxide from, 22:349
Wolframic oxide
Wolframite, 22:337
12-Wolframophosphoric acid hydrate, 22:354
Wolfram oxide
Wolfram oxyhalides, 22:347
Wolfram pentabromide, 22:348
Wolfram pentachloride, 22:348
Wolfram sulfide
Wolfram tetrabromide, 22:348
Wolfram tetrachloride, 22:348
Wolfram tetrafluoride, 22:348
Wolfram tribromide, 22:348
Wolfram trioxide, 22:337, 339, 349, 356
Wolfram trisulfide, 22:354
Wolfsbane. See *Aconite.*
Wollastonite, 17:227
Wood, 7:621; 16:681; 22:**358**
  barking, 16:692
  in bearing materials, 3:285
  chemical cellulose from, 4:606, 608
  chipping, 16:692
  coatings for, 5:708
  crotonic acid from, 6:466
  cyanoethylation of, 6:662
  fibers, 4:598
Wood adhesives, 1:395
  as denaturant, 8:451, 452
Wood alcohol. See *Methanol.*
Wood cellulose
  in manuf of cellophane, 9:228, 233
Wood charcoal
  use in carbon products, 4:166
Wood, destructive distillation
  manuf of acetic acid, 8:391
Wood fibers
  in drilling fluids, 7:298
Wood-fiber wool
  as thermal insulator, 11:829
Wood fireproofing, 12:720
Wood flour, 15:199
  filler, 15:802
  in resins, 15:793

Wood hydrolyzate
  yeast from, 22:541
Wood-inorganic binder board, 21:612
Wood-plastic, 22:381
Wood preservatives
  chromates as, 5:513
Wood products
  bleaching agents for, 3:563
  in cellulose acetate flake manuf, 1:116
  hardness testing, 10:818
  viscosity, 1:120
Wood pulp. (See also Pulp.)
Woodpulp black, 15:544
Wood rosin
  furfural as decolorizing agent, 10:244
Woodward's reagent K, 10:913
Wood wool
  as thermal insulator, 11:832
Wool, 22:387
  bleaching agents for, 3:562
  cyanoethylation of, 6:663
  dyeing with acid dyes, 7:522
  dyeing with basic dyes, 7:533
  dyeing in fiber blends, 7:566, 568, 570, 571
  dyeing with mordant acid dyes, 7:528
  effect of detergents on, 6:857, 867, 874, 877,
    880, 881
  fiber length, 20:34
  as filter medium, 9:285, 312
  as packing material, 14:455
  premetalized dyes on, 7:531
  prepn for dyeing, 7:521
  work of rupture value, 1:131
  vat dyeing, 7:560

Wool Blue G, 20:725
Wool Blue N Extra, 20:724
Wool Blue R, 20:725
Wool dyes
  anthraquinone, 2:511
Wool Fast Blue BL, 2:863; 7:522, 526
Wool Fast Blue FGL, 20:725
Wool Fast Blue GL, 2:863
Wool Fast Blue GLA, 7:526
Wool Green S, 20:723
  synthesis, 20:681
Wool Violet 5BN, 5:867; 10:793
Wool Violet 5BN or Acid Violet 6B
  used in hair coloring, 10:793
Woolen process, 22:391
Wool grease, 22:411
Wool industry
  use of sodium hydroxyacetate, 10:636
Wool oleine, 22:415
Wool stearine, 22:415
Wool wax, 22:167, 388, 411
Wormseed. See Chenopodium.
Wormwood, 14:214
Worsted process, 22:391
Wort, 3:300. (See also Beer.)
Wrinkle-resistant cloth
  use of amino resins in, 2:248
Wulfenite, 12:272
Wurtz-Fittig reaction, 2:728
Wurtzite, 10:323; 15:505; 22:562
Wurtz paraffin
  synthesis, 5:103

X

Xanthates, *22*:**419**
  thiol synthesis, *20*:214
Xanthation, *17*:178
Xanthene, *7*:522, 533; *10*:899, 920
Xanthene dyes, *7*:471, 473; *15*:456; *22*:**430**
9-Xanthenone. See *Xanthone.*
Xanthic acids, *22*:420, 422
Xanthine, *10*:913
Xanthine oxidase, *21*:107
  in milk, *13*:514
Xanthogen, *22*:419
Xanthogen formates, *22*:423
Xanthogenic acid, *22*:419
Xanthone, *10*:149, 926; *12*:107
Xanthopterin, *10*:932
Xanthopterincarboxylic acid, *10*:932
Xanthopurpurin, *2*:466, 471; *7*:621
  as accelerators, *17*:510
  reactions with aziridines, *11*:534
  sodium cellulose, from carbon disulfide,
    4:371
Xanthoresinotannol, *17*:382
Xenon
  as anesthetic, *2*:400
  cryogenic properties, *6*:472, 474
  as fission product, *14*:83
  phys prop, *10*:868
  plasma arcs, *15*:345
  as process gas, *2*:128
  separation from air, *10*:876
  separation of isotopes, *2*:138
Xenon difluoride, *9*:508; *10*:889
Xenon dioxide difluoride, *10*:892
Xenon hexafluoride, *10*:889
Xenon monoxide, *10*:893
Xenon oxide tetrafluoride, *10*:889, 892
Xenon tetrafluoride, *9*:508; *10*:889
Xenon tetroxide, *8*:581; *10*:892
Xenon trioxide, *8*:581; *10*:892
Xenotime, *10*:756; *17*:147
Xeroduplicating, *17*:339
Xerography, *17*:342
Ximenic acid, *8*:814
X-ray analysis, *22*:**438**
X-ray diffraction, *22*:438
X-ray diffraction, index of, *6*:529

X-ray fluorescence spectra, *22*:457
X-rays, *17*:53, 67; *22*:**438**
  sources of, *17*:57
Xylan
  in wood, *22*:365
Xylene, *2*:421; *7*:481; *8*:418; *10*:661; *22*:**467**
  benzene from, *3*:378
  complex with TCNE, *6*:626
  diluent, *18*:577
  in hexane prodn, *11*:2
  from hydroreforming, *20*:551
  oxidation to terephthalic acid, *20*:560
  photochlorination, *15*:338
  polymers, *15*:796
  from shale oil, *18*:18
o-Xylene
  and naphthalene, *13*:685
  phthalic anhydride from, *15*:449
  separation by distillation, *22*:469
m-Xylene, *6*:508
  isolation, *19*:315
  isophthalic acid, *15*:458
  separation by distillation, *22*:469
p-Xylene, *2*:344
p-Xylene, *6*:508
  perchlorylation reaction, *9*:603
  prod, *22*:468
  terephthalic acid from, *15*:465
  from *p*-xylene, *22*:482
m-Xylene-α,α'-diamine, *1*:250; *16*:21
m-Xylenediamine
  prod of, *22*:503
p-Xylene-α,α'-diol, *16*:22
Xylene-formaldehyde resins
  commercial prod, *22*:503
Xylenes
  in catalytic reformate, *15*:31
  in coal tar, *19*:657, 669
  stain solvents, *18*:661
  toxicity of, *11*:301
Xylenes and ethylbenzene, *22*:**467**
Xylenesulfonamide
  in pigment manuf, *9*:497
m-Xylenesulfonic acid
  hydrolysis of, *22*:493
Xylenethiol, *15*:768

3,5-Xylenol
  phenolic resins, *15*:188
o-Xylenol
  sulfur dyes, *19*:427
2,4-Xylenol
  alkylation of, *S*:27
  from methylation of phenol, *S*:273
2,6-Xylenol, *6*:439; *15*:173; *16*:274
  from *o*-cresol, *S*:273
Xylenol orange
  thorium indicator, *20*:252
Xylenols, *S*:271
  in coal tar, *19*:657, 668, 672
  resins, *15*:176

Xylidines, *2*:421
Xylitol, *1*:572
  phys prop, *1*:570
  prepn, *1*:572
L-Xyloascorbic acid. See *Ascorbic acid.*
Xylol, *22*:467
  in prepn of xylitol, *1*:572
D-Xylonic acid γ-lactone
p-Xyloquinone, *12*:142
m-Xylorcinol, *11*:481
Xylose, *4*:134
L-Xylosone, *2*:751
Xylylenediamine, *16*:21

Y

YM31A Glass
in ceramic composite armor, S:141
Yage, 1:772
Yatanine, 20:79
Yeast
brewers', 3:303
carbonated beverages affected by, 4:351
as leavening agent, 3:41
from milk, 13:569
in prepn of 2,3-butanediol, 12:145
reproduction, 22:513
uses, 8:190
wood hydrolysis, 22:384
Yeast extracts, 22:531
Yeasts, 12:370; 22:**507**
Yeast tablets, 22:530
Yercum
analysis, 9:173
Yield strength, 13:187
of metals and alloys, 11:6

Yield value
of inks, 11:613
Ylang-ylang, 9:353; 14:214, 729, 737
Ylides, 19:327
Yogurt, 13:529
Yohimbine, 1:762, 772
Yorkshire Stone Square
fermentation system, 3:315
Young equation, S:966
Young's modulus, 13:186; 17:425
Ytong, 4:704
Ytterbium, 17:144
electronic configuration, 1:360
ionic radii, configuration, 1:370
in nuts, 14:131
Yttrium, 17:143
Yttrium-90, 17:124
Yttrium iron garnet
nonlinear optical materials, S:625
Yttrium oxide, 12:623
in electron tube, 8:8

Z

Z-Furan, *13*:856
Zambesi Black GB, *7*:470
Zambesi Black V, *2*:902
Zanzibar" units, *S*:994
Zapon Fast Scarlet CG, *2*:906
Zapon Fast Violet BE, *2*:907
Zeatin, *15*:681
Zectran, *15*:910
Zefran, *1*:313
Zein, *8*:30; *9*:499
  cyanoethylation of, *6*:663
  in food processing, *10*:10
  in paint removers, *14*:489
Zelan, *16*:804
Zeolites, *18*:49
  aluminosilicate, *18*:157
  cracking catalysts, *15*:28, 37
Zeta potential, *22*:71, 92
Ziegler allylic bromination
  of cholesteryl esters, *21*:561
Ziegler catalyst, *4*:557, 538; *14*:259, 269;
  20:446
  nature of, *14*:270
  organotitanium compounds, *20*:425
  for poly-1-butene, *S*:776
  polymerization by, *7*:66
  polymers prepared by, *7*:66
  titanium trichloride, *20*:385
  of vinylidene chloride copolymers, *21*:282
  in vinyl polymers, *21*:417, 421
"Ziegler chemistry," *S*:649
Ziegler method
  synthesis of higher alcohols from alkyl-
    aluminuns, *1*:560
Ziegler-Natta catalysts
  and vinyl chloride, *21*:377
Ziegler polythene, *S*:773
Ziegler process
  for aluminum alkyl prod, *2*:34
Zimmerman process
  for sludge disposal, *22*:112
Zinc, *7*:281; *22*:**555**
  and air hybrid fuel cell, *S*:398
  as catalyst, *8*:357
  chromate treatment, *13*:298
  coatings, *13*:252

content in soils, *9*:35
copper alloys with, *6*:208, 213, 249
in dental amalgam, *6*:807, 814
electromigration of, *S*:287
in electroplating, *8*:38, 66, 70
as ester-exchange catalyst, *16*:167
heat of combustion of, *4*:898
hot-dipped coatings, *13*:254
as impregnant, *14*:458
inhibitors for, *6*:342
manganese compounds, *13*:3
from ocean floor, *14*:165
roasting of concentrates, *22*:570
Zinc-65, *17*:123
Zinc acetate, *22*:606
  catalyst, *20*:799
  ester-exchange catalyst, *16*:175
Zinc acetate rosin
  in dental cement, *6*:779
Zinc alloys, *22*:**555**
Zinc amalgam
  dithionous acid, *19*:420
Zinc ammonium chloride, *22*:608
  in galvanizing, *13*:256
Zinc ammonium phosphate
  gravimetric determination of zinc, *22*:560
Zinc-base alloys, *22*:599
Zinc blende, *6*:533; *22*:562
  burners, *19*:453
Zinc borate, *22*:606
  as catalyst, *8*:361
  in glass-ceramics, *10*:548
Zinc bromide, *22*:608
Zinc calcine
  from ore roasting, *22*:569
Zinc carbonate, *2*:606, 623
Zinc chloride, *2*:220; *7*:44; *8*:495; *10*:86;
  12:158; *22*:607
  as catalyst, *8*:408
  in drilling fluids, *7*:302
  fire-retardant on wood, *22*:376
  flux, *18*:542
  as fungicide, *10*:230
  Gattermann synthesis, *15*:161
  in manuf of ethylenediamine, *7*:31
  pigments, *15*:505, 534

in prepn of ethyl ether, 8:479
prod of, 22:605
used in analysis of hexacyanoferrates(II), 12:30
Zinc chromate. (See also Zinc yellow.)
Zinc chromite, 5:496. (See also Zinc yellow.)
  as pigment, 14:472
  primers, 20:361
Zinc compounds, 22:**604**
Zinc cyanide, 22:607
  Gattermann synthesis, 15:161
Zinc dibenzyldithiocarbamate, 17:595
Zinc dibutyldithiocarbamate, 7:714
Zinc dimethyldithiocarbamate, 4:70; 15:910
  as fungicide, 10:224, 232
Zinc dithiocarbamate
  reaction with ethylenediamine, 7:30
Zinc dithiophosphates, 15:320
Zinc ethylenebisdithiocarbamate, 10:224
  as insecticide, 7:36
Zinc 2-ethylhexoate, 7:276, 278
Zinc ethyl xanthate, 9:386
Zinc eugenolate, 6:823
  in dental cement, 6:779
Zinc ferrate(III), 12:40
Zinc ferrate(VI), 12:40
Zinc ferricyanide
  dye oxidant, 20:693
Zinc ferrite
  from ore roasting, 22:565
Zinc ferrocyanide
  in zinc analysis, 22:560
Zinc fluoborate, 9:566
Zinc fluoride
  reaction with hexachlorodisilane, 9:653, 684
Zinc fluosilicate, 6:802; 9:658, 660; 22:611
Zinc formaldehydesulfoxylate, 10:86; 19:422
Zinc formate, 22:607
Zinc germanate, 10:525
Zinc-gold alloy, 10:692
Zinc green, 15:534, 537, 538
Zinc halides, 22:607
Zinc hexacyanoferrate(II), 12:30
Zinc hexafluorosilicate, 9:684
Zinc hydrosulfite, 19:420, 422; 22:608
Zinc hydroxide, 2:623; 22:604
  zinc hydrosulfite, 19:421
Zinc iodide, 22:608
Zincite, 22:563
Zinc metaarsenate
  as fungicide, 10:230
Zinc metasilicate, 22:611
Zinc metatitanate, 20:414

Zinc molybdate, 13:650
Zinc monofluorophosphate, 9:642
Zinc naphthenate, 7:276, 278; 12:813
  accelerator, 21:103
  as fungicide, 10:232
Zinc nitrate, 22:608
Zinc octoate, 6:806
  accelerator, 21:103
Zincon
  in zinc analysis, 22:561
Zinc orthophosphate, 22:611
Zinc orthosilicate, 22:611
Zinc orthotitanate, 20:414
Zinc oxide, 6:357, 363, 802, 804, 849; 7:687; 22:571, 606, 609
  in compounding of oxides, 7:714
  in insect repellents, 8:495
  of vulcanization copolymers, 7:687
  in vulcanization of neoprene, 7:711
Zinc peroxide, 2:623; 14:751
  prod, 18:450
Zinc phosphate, 12:588; 15:270; 22:611
  as dental cement, 6:779, 780
Zinc phosphide, 15:294; 22:611
  as insecticide, 11:683
  as rodenticide, 15:916
Zinc potassium chromate, 5:499
Zinc 2-pyridinethiol-1-oxide
  in shampoo, 10:775
Zinc pyridinethione, 10:775
Zinc resinate, 7:279; 17:500
Zinc-retort plants
  in prod of cadmium, 3:887
Zinc salts
  as antiseptic, 2:622
Zinc selenide, 12:626
Zinc silicate, 18:159; 22:611
  as dental cement, 6:779, 782
Zinc sodium chromate, 5:499
Zinc stearate, 6:357, 359; 7:276, 284, 285
  activator for SBR, 7:689
  as mold-release agent, 15:792
  use as abherent, 1:8-9
Zinc sulfate, 22:611
  as fungicide, 10:235
  hydrogen sulfide, 19:385
  in limestone, 22:563
  in manuf metallic soaps, 7:274
  pigments, 15:505
Zinc sulfide, 6:804; 9:386; 22:612
  in papermaking, 14:501
  pigment, 15:496, 503, 506, 509, 515, 534

sulfur dioxide, *19*:448
zinc analysis, *22*:559
in zinc ores, *22*:555
Zinc sulfide phosphors, *12*:626
Zinc sulfoxylate, *19*:422
Zinc tallate, *7*:278
Zinc telluride, *12*:626
Zinc tetrafluoroborate, *6*:684
Zinc tetroxychromate, *5*:510; *15*:534
Zinc titanates, *20*:414
Zinc 3-5-5-trimethylhexoate, *7*:279
Zinc undecylenate, *7*:284
Zinc white, *15*:504
Zinc yellow, *5*:509
Zinc yellows, *15*:527, 530, 534
Zineb. See *Zinc ethylenebisthiocarbamate.*
Zingerone
    from vanillin, *21*:181
Zinnwaldite, *12*:533
Zippe Centrifuge, *7*:149, 151, 152
    cascade, *7*:165
Ziram, *10*:224; *15*:910
Zircaloy, *14*:82, 87
    dissolution of, *14*:105
Zircaloy-2, *22*:618, 621
Zircaloy-4, *22*:621
Zircon, *10*:756; *15*:855; *17*:228. (See also *Zirconium compounds.*)
Zirconate
    organic, *22*:636
Zirconia, *22*:619.   (See also *Zirconium dioxide.*)
    carbonated hydrons, *22*:633, 639
Zirconium, *17*:26; *22*:614
    beryllium compounds, *S*:76
    cesium chromate reaction with, *4*:860
    for cladding uranium, *21*:28
    corrosion resistance to water, *6*:302
    copper alloys with, *6*:212
    in electron tubes, *8*:8
    ferroalloy, *18*:751
    with hafnium, *10*:756
    heat of combustion of, *4*:898
    ignition mixtures, *16*:826
    in modified stainless steels, *11*:21
    in nickel-base superalloys, *11*:29
    sodium reduction, *18*:450
    urea-synthesis reactor, *21*:52
Zirconium-95, *17*:125
Zirconium alloys, *22*:620
Zirconium and zirconium compounds, *22*:**614**
Zirconium benzenetellurate, *22*:659
Zirconium boride, *3*:675, 676; *16*:432; *22*:629

Zirconium borohydride, *22*:639
Zirconium butyrate, *22*:640
Zirconium carbide, *4*:75, 84, 85; *10*:758; *22*:623
Zirconium carbonitride, *22*:623
Zirconium carbonyls, *22*:640
Zirconium chelate compounds, *22*:641
Zirconium chloride, *22*:642
Zirconium cyanonitride, *22*:623, 632
Zirconium cyanonitrile, *22*:652
Zirconium dioxide, *22*:651
    in ceramic pigments, *5*:851, 852
    crystal structure, *22*:619
    in glass ceramics, *10*:548
    oligomers of, *22*:629
    from ores, *22*:632
    in ores, *22*:614
    from zirconium hydride, *22*:626
    zirconium metal prod, *22*:616
Zirconium 2-ethylcaproate, *22*:633
Zirconium 2-ethylhexoate, *7*:278
Zirconium ferrocyanide, *22*:645
Zirconium fluorogermanates, *22*:647
Zirconium germanides, *22*:661
Zirconium-hafnium oxide, *10*:758
Zirconium-hafnium oxychloride, *10*:758
Zirconium-hafnium tetrachloride, *10*:758
Zirconium halogenates, *22*:647
Zirconium hydride, *11*:208; *22*:619
    in electron tubes, *8*:4
    in nuclear reactors, *14*:84
Zirconium hydroxycarboxylates, *22*:647
Zirconium ketonates, *22*:648
Zirconium lactate, *12*:184
Zirconium mandelate, *22*:633
Zirconium mercaptides, *22*:648
Zirconium minerals, *22*:632
    in ceramics, *4*:772
Zirconium molybdates, *22*:649
Zirconium nitrate
    anhydrous, *22*:649
    in reactor technology, *14*:96
Zirconium nitride, *22*:624, 638
    from zirconium hydride, *22*:626
Zirconium nitride halide, *22*:650
Zirconium oxalates, *22*:650
Zirconium oxide
    as abrasive, *1*:28
    expansion coefficient, thermal, *4*:805
    hydrous, *22*:654
    modifier, *17*:286
    as opacifier in enameling, *8*:167

refractory, *17*:244
thermal conductivity of, *4*:803
Zirconium oxides, *22*:651
Zirconium oxide sulfide, *22*:664
Zirconium oxychloride
  as water repellent agent for textiles, *22*:139
Zirconium phenoxides, *22*:655
Zirconium phosphate, *22*:657
Zirconium polyolates, *22*:659
Zirconium proprionate, *22*:640
Zirconium pyrophosphate, *22*:657
Zirconium sesquioxide, *22*:654
Zirconium silicate, *17*:232; *22*:635, 660
  pigment manuf, *20*:401
Zirconium silicides, *22*:661
Zirconium soaps, *22*:640
Zirconium stearate
  water-repellent agent, *22*:140
Zirconium sulfate
  basic, *22*:663
  commercial, *22*:633
  in leather tanning, *12*:334
Zirconium sulfide, *22*:664
Zirconium tetrafluoride, *9*:507, 685; *22*:645
  electrolysis of solutions of, *22*:617
Zirconium trifluoride, *9*:685; *22*:645
Zirconium vanadates, *22*:664
Zirconium wolframates, *22*:649
Zirconyl acetate, *22*:633, 636

Zirconyl chloride, *22*:629, 631, 633, 642
  ammonium vanadate reaction, *22*:664
  anhydrous, *22*:643
  basic, *22*:643
  hydrous zirconium oxide from, *22*:654
  zirconium dioxide recovery, *22*:652
Zirconyl nitrate, *22*:631
Zirconylosilicates, *22*:660
Zirconyl perchlorate, *22*:631, 647
Zirconyl phosphate, *6*:867
Zirconyl pyrophosphate, *22*:657
Zirconyl thiocyanate, *22*:664
Zirconyl trichloroacetate, *22*:636
Zirkite
  in ceramics, *4*:772
Zirkonerde, *22*:614
Zoalene, *10*:19
  coccidostat, *21*:247
Zone leveling, *22*:684
Zone refining, *17*:866; *22*:**680**
Zoxazolamine, *16*:677
Zwikkers reagent, *5*:80.   (See also *Copper-
pyridine.*)
Zwitterion
  ozonide structure, *14*:417
Zwitterions
  Patent Blue dyes as, *20*:708
Zymase
  in breadmaking, *22*:528